Bacterial Genomes

Physical Structure and Analysis

EDITED BY

Frans J. de Bruijn
Michigan State University

James R. Lupski
Baylor College of Medicine

George M. Weinstock
University of Texas Medical School

CHAPMAN & HALL

ITP® INTERNATIONAL THOMSON PUBLISHING
Thomson Science

New York • Albany • Bonn • Boston • Cincinnati • Detroit • London
Madrid • Melbourne • Mexico City • Pacific Grove • Paris • San Francisco
Singapore • Tokyo • Toronto • Washington

Join Us on the Internet

WWW: http://www.thomson.com
EMAIL: findit@kiosk.thomson.com

thomson.com is the on-line portal for the products, services and resources available from International Thomson Publishing (ITP).

This Internet kiosk gives users immediate access to more than 34 ITP publishers and over 20,000 products. Through *thomson.com* Internet users can search catalogs, examine subject-specific resource centers and subscribe to electronic discussion lists. You can purchase ITP products from your local bookseller, or directly through *thomson.com.*

Visit Chapman & Hall's Internet Resource Center for information on our new publications, links to useful sites on the World Wide Web and an opportunity to join our e-mail mailing list.
Point your browser to: **http://www.chaphall.com** or
http://www.thomson.com/chaphall/lifesce.html for Life Sciences

A service of 

Bacterial Genomes

Physical Structure and Analysis

EDITED BY

Frans J. de Bruijn
Michigan State University

James R. Lupski
Baylor College of Medicine

George M. Weinstock
University of Texas Medical School

CHAPMAN & HALL

INTERNATIONAL THOMSON PUBLISHING
Thomson Science

New York • Albany • Bonn • Boston • Cincinnati • Detroit • London
Madrid • Melbourne • Mexico City • Pacific Grove • Paris • San Francisco
Singapore • Tokyo • Toronto • Washington

Cover design: Curtis Tow Graphics and Marlene Cameron (MSU-DOE Plant Research Laboratory);
Background photograph by Frank Dazzo (MSU)

Printed in the United States of America

Chapman & Hall
115 Fifth Avenue
New York, NY 10003

Chapman & Hall
2-6 Boundary Row
London SE1 8HN
England

Thomas Nelson Australia
102 Dodds Street
South Melbourne, 3205
Victoria, Australia

Chapman & Hall GmbH
Postfach 100 263
D-69442 Weinheim
Germany

International Thomson Editores
Campos Eliseos 385, Piso 7
Col. Polanco
11560 Mexico D.F
Mexico

International Thomson Publishing–Japan
Hirakawacho-cho Kyowa Building, 3F
1-2-1 Hirakawacho-cho
Chiyoda-ku, 102 Tokyo
Japan

International Thomson Publishing Asia
221 Henderson Road #05-10
Henderson Building
Singapore 0315

1 2 3 4 5 6 7 8 9 10 XXX 01 00 99 98

Library of Congress Cataloging-in-Publication Data

Bacterial genomes : physical structure and analysis / edited by Frans
J. de Bruijn, James R. Lupski, George M. Weinstock.
p. cm.
Includes bibliographical references and index.
ISBN 0-412-99141-1 (alk. paper)
1. Bacterial genetics. 2. Genomes. I. Bruijn, F.J. de (Frans
J. de) II. Lupski, James R. III. Weinstock, George M.
QH434.B335 1997
572.8'6293--dc21 96-48163
CIP

British Library Cataloguing in Publication Data available

To order this or any other Chapman & Hall book, please contact **International Thomson Publishing, 7625 Empire Drive, Florence, KY 41042.** Phone: (606) 525-6600 or 1-800-842-3636. Fax: (606) 525-7778. e-mail: order@chaphall.com.

For a complete listing of Chapman & Hall titles, send your request to **Chapman & Hall, Dept. BC, 115 Fifth Avenue, New York, NY 10003.**

Contents

Asterisks refer to 'Corresponding Author'

C. Genomic Sequencing Projects

3. Physical Maps of Bacteria and Their Methods for Construction

Contributors

Dr. Fadi Abdi
Department of Biochemical and Biophysical Sciences
University of Houston
Houston, TX 77204-5934

Dr. A. Ally
Life Technologies, Inc.
Rockville, MD 20855

Dr. Staffan Bergh
Department of Biochemistry
Royal Institute of Technology
S-100 44 Stockholm
Sweden

Dr. P.L. Bergquist
Macquarie University
North Ryde
New South Wales 2109
Australia

Dr. Ulfar Bergthorsson
Department of Biology
University of Rochester
Rochester, NY 14627-0211

Dr. Ashok Bhagwat
Department of Chemistry
Wayne State University
437 Chemistry Bldg.
Detroit, MI 48202-3489

Dr. H. Bingham
Department of Microbiology
University of Toronto
Toronto, Ontario
Canada M5S 1A8

Dr. K. M. Borges
Macquarie University
North Ryde
New South Wales 2109
Australia

Dr. K. F. Bott
Curriculum in Genetics
Department of Microbiology and Immunology
University of North Carolina School of Medicine
Chapel Hill, NC 27599-7290

Dr. B. Bourke
Department of Microbiology
University of Toronto
Toronto, Ontario
Canada M5S 1A8

Dr. Bruce Braaten
Department of Pathology
University of Utah
Salt Lake City, UT 84132

Dr. Robert A. Britton
Department of Biology
Massachusetts Institute of Technology
Building 68-530
77 Massachusetts Avenue
Cambridge, MA 02139-4307

Dr. Joan E. Brooks
New England BioLabs
32 Tozer Road
Beverly, MA 01915-5599

Dr. Allan M. Campbell
Department of Biological Sciences
Stanford University
Stanford, CA 94305

Dr. Bruno Canard
Unité de Génétique Moléculaire
Bactérienne
Institut Pasteur
25, rue du Dr. Roux
75724 Paris Cedex 15
France

Dr. Sherwood R. Casjens
Department of Oncological Sciences
Division of Molecular Biology &
Genetics
University of Utah Health Science Center
Salt Lake City, UT 84132

Dr. V. L. Chan
Department of Microbiology
University of Toronto
Toronto, Ontario
Canada M5S 1A8

Dr. Mark S. Chandler
Department of Microbiology &
Immunology
Wright State University
Dayton, OH 45435-0001

Dr. Robert L. Charlebois
Department of Biology
University of Ottawa
30 Marie Curie
Ottawa, ON K1N 6N5
Canada

Dr. Xiangxing Chen
Department of Biochemical and
Biophysical Sciences
University of Houston
Houston, TX 77204-5934

Dr. Monjula Chidambaram
Department of Biochemistry and
Molecular Biology
University of Texas Medical School
6431 Fannin
Houston, TX 77225

Dr. Madhusudan Choudhary
Department of Microbiology & Molecular
Genetics
University of Texas Medical School
6431 Fannin
Houston, TX 77225

Dr. Al Claiborne
Department of Biochemistry
The Bowman Gray School of Medicine
Winston-Salem, NC

Dr. Don B. Clewell
Biologic and Materials Sciences
School of Dentistry
University of Michigan
Ann Arbor, MI 28109-1078

Dr. Stewart T. Cole
Unité de Génétique Moléculaire
Bactérienne
Institut Pasteur
28, rue du Dr. Roux
75724 Paris Cedex 15
France

Dr. Pascale Cossart
Laboratoire de Genetique Moleculaire des Listeria and Centre National de la Recherche Scientifique URA 1300
Institut Pasteur
28 rue du Docteur Roux
Paris 75724
France

Dr. Patrice Courvalin
Unité des Agents Antibacteriens
Institut Pasteur
28, rue du Docteur Roux
75724 Paris Cedex
France

Dr. Antoine Danchin
Institut Pasteur
28, rue du Docteur Roux
75724 Paris Cedex
France

Dr. Guillermo Davila
Dept. de Genética Molecular
Centro de Investigación
Sobre Fijación de Nitrógeno
Universidad Nacional Autónoma de México
Apartado Postal 565-A
Cuernavaca, Morelos
México

Dr. Frans J. de Bruijn
DOE Plant Research Laboratory
Michigan State University
East Lansing, MI 48824-1312

Dr. Aart de Kok
Wageningen Agricultural University
Wageningen, The Netherlands

Dr. David H. Demezas
Department of Microbiology and Molecular Genetics
Oklahoma State University
307 Life Sciences East
Stillwater, OK 74078-0289

Dr. M. Dolan
Department of Molecular Biology
Massachusetts General Hospital
Boston, MA 02114

Dr. Ford Doolittle
Biochemistry Department
Sir Charles Tupper Medical Building
Dalhousie University
Halifax, Nova Scotia
B3H 4H7 Canada

Drs. Karl Drlica
Public Health Research Institute
455 First Avenue
New York, NY 10016

Dr. Bruno Dupuy
Unité de Génétique Moléculaire Bactérienne
Institut Pasteur
25, rue du Dr. Roux
75724 Paris Cedex 15
France

Karin Eiglmeier
Unité de Génétique Moléculaire Bactérienne
Institut Pasteur
28, rue du Dr. Roux
75724 Paris Cedex 15
France

Dr. Bert Ely
Department of Biology
University of South Carolina
Columbia, SC 29208

Dr. Karen Evans
The Sidney Kimmel Cancer Center
3099 Science Park Road
San Diego, CA 92121

Dr. Michael Fonstein
Department of Molecular Genetics and Cell Biology
The University of Chicago
920 East 58th
Chicago, IL 60637

Dr. Wesley Ford
Department of Biochemical and Biophysical Sciences
University of Houston
Houston, TX 77204-5934

Dr. Claire Fraser
Institute for Genomic Research
9712 Medical Center Drive
Rockville, MD 20850

Hafida Fsihi
Unité de Génétique Moléculaire Bactérienne
Institut Pasteur
28, rue du Dr. Roux
75724 Paris Cedex 15
France

Dr. T. Gaasterland
Mathematics & Computer Science Division
Argonne National Laboratory
9700 South Cass Avenue
Argonne, IL 60439

Dr. Thierry Garnier
Unité de Génétique Moléculaire Bactérienne
Institut Pasteur
25, rue du Dr. Roux
75724 Paris Cedex 15
France

Dr. Anne-Marie Gasc
118 route de Narbonne
31062 Toulouse cedex
France

Dr. C. P. Gibbs
Immuno AG
Biomedizinisches Forschungszentrum
Uferstraβe 15
A-2304 Orth an der Donau
Austria

Dr. W. Gilbert
Department of Cellular & Developmental Biology
Harvard University
Cambridge, MA 02138

Dr. Patrick Gillevet
Institute for Bioscience, Bioinformatics, and Biotechnology
George Mason University
Fairfax, VA 22030

Dr. Philippe Glaser
Unité de Régulation de l'Expression Génétique
Département des Biotechnologies
25, rue du Dr. Roux
Institut Pasteur
75724 Paris Cedex 15
France

Dr. James W. Golden
Department of Biology
Texas A&M University
College Station, TX 77843-3258

Dr. Sol H. Goodgal
Department of Microbiology
School of Medicine
University of Pennsylvania
Philadelphia, PA 19104-6076

Prof. Dr. Michael Göttfert
Institut fuer Genetik und Molekulare Biologie
Technische Universitaet
Zelleschwerweg 22
D-01062 Dresden
Deutschland

Dr. E. Hani
Department of Microbiology
University of Toronto
Toronto, Ontario
Canada M5S 1A8

Dr. Robert Haselkorn
Department of Molecular Genetics and Cell Biology
The University of Chicago
920 East 58th Street
Chicago, IL 60637

Dr. Joe Don Heath
Incyte Pharmaceuticals
3174 Porter Drive
Palo Alto, CA 94304

Dr. H. Hennecke
Mikrobiologisches Institut
Eidgenössische Technische Hochschule
ETH-Zentrum
Schmelzbergstrasse 7
CH-8092 Zurich
Switzerland

Dr. Andrew Hessel
Salmonella Genetic Stock Centre
Department of Biological Sciences
University of Calgary
Calgary, Alberta
Canada T2N 1N4

Dr. Charles W. Hill
Penn State University
Hershey Medical Center
Department of Biological Chemistry
Hershey, PA 17033

Dr. Alex R. Hoffmaster
University of Texas Medical School
Houston, TX 77225

Dr. Rhonda J. Honeycutt
Sidney Kimmel Cancer Center
3099 Science Park Road
San Diego, CA 92121

Dr. Y. Hong
Department of Microbiology
University of Toronto
Toronto, Ontario
Canada M5S 1A8

Nadine Honoré
Unité de Genétique Moléculaire Bactérienne
Institut Pasteur
28, rue du Dr. Roux
75724 Paris Center 15
France

Dr. Jerrilyn K. Howell
Department of Pathology and Laboratory Medicine
University of Texas at Houston Medical School
P.O. Box 20708
Houston, TX 77225

E. Hsu
National Center for Human Genome Research
Bethesda, MD

Dr. Wai Mun Huang
Department of Oncological Sciences
Division of Molecular Biology & Genetics
University of Utah Health Science Center
Salt Lake City, UT 84132

Dr. Hor-Gil Hur
University of Minnesota
Department of Soil, Water, and Climate
1991 Upper Buford Circle, 439 BorH
St. Paul, MN 55108

Dr. John Iandolo
Department of Pathology
VCS Building
Kansas State University
Manhattan, KS 66506

Dr. Mitsuhiro Itaya
Mitsubishi Kasei Institute of Life Sciences
11, Minamiooya
Machida-shi
Tokyo 194, Japan

Dr. Liangxia Jiang
Division of Infectious Diseases
Department of Internal Medicine
University of Texas Medical School
6431 Fannin, 1.728 JFB
Houston, TX 77030

Dr. Samuel Kaplan
Department of Microbiology & Molecular Genetics
University of Texas Medical School
6431 Fannin, JFB 1.765
Houston, TX 77225

Dr. Samuel Karlin
Mathematics Department
Stanford University
Stanford, CA 94305-2125

Dr. Seiichi Katayama
Department of Microbiology
Faculty of Medicine
Kagawa Medical University
1750-1, Ikenobe, Miki-cho
Kita-gun, Kagawa, 761-07
Japan

Dr. N. W. Kim
Department of Microbiology
University of Toronto
Toronto, Ontario
Canada, M5S 1A8

Dr. Hans-Peter Klenk
Department of Biochemistry
Sir Charles Tupper Medical Building
Dalhousie University
Halifax, NS B3H 4H7
Canada

Dr. Tokio Kogoma
Departments of Cell Biology & Microbiology
University of New Mexico
Health Sciences Center
Albuquerque, NM 87131

Dr. Anne-Brit Kolstø
Biotechnology Centre of Oslo and Institute of Pharmacy
University of Oslo
P.B. 1125
0316 Oslo
Norway

Dr. Eugene V. Koonin
National Center for Biotechnology Information
National Library of Medicine
National Institutes of Health
Bethesda, MD 20894

Dr. Elizabeth G. Koshy
Department of Molecular Genetics and Cell Biology
The University of Chicago
920 East 58th
Chicago, IL 60637

Dr. Frank Kullmann
The Sidney Kimmel Cancer Center
3099 Science Park Road
San Diego, CA 92121

Dr. Vivek Kumar
Department of Molecular Genetics and Cell Biology
The University of Chicago
920 East 58th
Chicago, IL 60637

Dr. C. Kündig
Mikrobiologisches Institut
Eidgenössische Technische Hochschule
ETH-Zentrum
Schmelzbergstrasse 7
CH-8092 Zurich
Switzerland

Dr. F. Kunst
Unité de Biochimie Microbienne
Département des Biotechnologies
25, rue du Dr. Roux
Institut Pasteur
75724 Paris Cedex 15
France

Dr. Tanya Kuritz
Life Sciences Division
Health and Environmental Risk Analysis Section
Oak Ridge National Laboratory
1060 Commerce Park
Oak Ridge, TN 37830

Matthew Lawes
Department of Genetics
University of Georgia
Biological Sciences Building
Athens, GA 30602-7223

Dr. Pascal Le Bourgeois
Laboratoire de Microbiologie et Genetique Moleculaire
Centre National de la Recherche Scientifique
118 route de Narbonne
31062 Toulouse
France

Dr. Thomas Leisinger
Mikrobiologisches Institut
Eidgenössische Technische Hochschule
ETH-Zentrum
CH-8092 Zürich
Switzerland

Dr. Margaret Lieb
Department of Molecular Microbiology and Immunology
University of Southern California
School of Medicine
2011 Zonal Avenue
Los Angeles, CA 90033

Dr. Shu-Lin Liu
Salmonella Genetic Stock Centre
Department of Biological Sciences
University of Calgary
Calgary, Alberta
Canada T2N 1N4

Dr. R. Lombardi
Department of Microbiology
University of Toronto
Toronto, Ontario
Canada M5S 1A8

Dr. H. Louie
Department of Microbiology
University of Toronto
Toronto, Ontario
Canada M5S 1A8

Dr. David Low
Department of Pathology
University of Utah
Salt Lake City, UT 84132

Dr. James R. Lupski
Department of Molecular & Human Genetics
Baylor College of Medicine
One Baylor Plaza, Room 609E
Houston, TX 77030-3498

Dr. Christopher Mackenzie
Department of Biochemistry and Molecular Biology
University of Texas Medical School
6431 Fannin
Houston, TX 77225

Dr. Stanley R. Maloy (with Matthew Lawes)
Department of Microbiology
University of Illinois
B103 Chemical and Life Sciences Laboratory MC-110
601 S. Goodwin Avenue
Urbana, IL 61801

Dr. Françoise Mathieu-Daudé
The Sidney Kimmel Cancer Center
3099 Science Park Road
San Diego, CA 92121

Dr. Michael McClelland
Sidney Kimmel Cancer Center
3099 Science Park Road
San Diego, CA 92121

Dr. Claudine Médigue
Institut Curie
11, rue Pierre et Marie Curie
75231 Paris Cedex 05
France

Dr. Sylvie Michaux-Charachon
Unite 65 de l'Institut National de la Sante et de la Recherche Medicale
Faculté de Médecine
Avenue Kennedy
30900 Nimes
France

Dr. Eric Michel
Unité des Interactions Bactéries-Cellules
Institut Pasteur
25 rue du Docteur Roux
Paris 75015 France

Dr. Marie-Line Mingot-Daveran
Laboratoire de Microbiologie et Genetique Moleculaire
Centre National de la Recherche Scientifique
118 route de Narbonne
31062 Toulouse
France

Dr. Marilyn Mitchell
Department of Microbiology
School of Medicine
University of Pennsylvania
Philadelphia, PA 19104-6076

Dr. Ivan Moszer
Unité de Régulation de l'Expression Génétique
Département des Biotechnologies
25, rue du Dr. Roux
Institut Pasteur
75724 Paris Cedex 15
France

Dr. Paul Mourachov
Department of Molecular Genetics and Cell Biology
The University of Chicago
920 East 58th
Chicago, IL 60637

Dr. Jan Mrázek
Mathematics Department
Stanford University
Stanford, CA 94305-2125

Dr. Barbara E. Murray
Division of Infectious Diseases
Department of Internal Medicine
University of Texas Medical School
6431 Fannin, 1.728 JFB
Houston, TX 77030

Dr. Arcady R. Mushegian
National Center for Biotechnology Information
National Library of Medicine
National Institutes of Health
Bethesda, MD 20894

Dr. Kirsten S. Nereng
Department of Microbiology and Molecular Genetics
University of Texas Medical School
6431 Fannin
Houston, TX 77225

Dr. Tatiana Nikolskaya
Department of Molecular Genetics and Cell Biology
The University of Chicago
920 East 58th
Chicago, IL 60637

Dr. D. Ng
Department of Microbiology
University of Toronto
Toronto, Ontario
Canada M5S 1A8

Dr. Steve Norris
Department of Pathology
University of Texas Medical School
P.O. Box 20708
Houston, TX 77225

Dr. Howard Ochman
Department of Biology
University of Rochester
Rochester, NY 14627-0211

Dr. Tairo Oshima
Department of Molecular Biology
Tokyo University of Pharmacy & Life Science
1432-1 Horinouchi, Hachioji
Tokyo 192-03
Japan

Dr. R. Overbeek
Department of Mathematics and Computer Science
Argonne National Laboratory
Argonne, IL

Dr. Rafael Palacios
Dept. de Genética Molecular
Centro de Investigación
Sobre Fijación de Nitrógeno
Universidad Nacional Autónoma de México
Apartado Postal 565-A
Cuernavaca, Morelos, México

M.S. Purzycki
Molecular Biosciences and Technology Institute
George Mason University
Fairfax, VA 22030

Dr. Xiang Qin
Division of Infectious Diseases
Department of Internal Medicine
University of Texas Medical School
6431 Fannin, 1.728 JFB
Houston, TX 77030

Dr. Mark A. Ragan
Institute for Marine Biosciences
National Research Council
1411 Oxford Street
Halifax, NS B3H 3Z1
Canada

Elisabeth Raleigh
New England BioLabs
32 Tozer Road
Beverly, MA 01915-5599

Dr. Michel Ramuz
Unite 65 de l'Institut National de la Sante et de la Recherche Medicale
Faculté de Médecine
Avenue Kennedy
30900 Nimes
France

Dr. François Rechenmann
IMAG/LIFIA
46, avenue Félix Viallet
38031 Grenoble Cedex 1
France

Dr. Monica Riley
Senior Scientist
Marine Biological Lab
Woods Hole, MA 02543

Dr. Paul Ritzenthaler
Laboratoire de Microbiologie et Genetique Moleculaire
Centre National de la Recherche Scientifique
118 route de Narbonne
31062 Toulouse
France

Dr. David Romero
Dept. de Genética Molecular
Centro de Investigación
Sobre Fijación de Nitrógeno
Universidad Nacional Autónoma de México
Apartado Postal 565-A
Cuernavaca, Morelos
México

Dr. Ute Römling
Klinische Forschergruppe
Zentrum Biochemie
OE 4350
Medizinische Hochschule Hannover
D-30623 Hannover 61
Germany

Dr. Michael J. Sadowsky
Department of Soil, Water, and Climate
1991 Upper Buford Circle, 439 BorH
University of Minnesota
St. Paul, MN 55108

Dr. Isabelle Saint Girons
Unite de Bacteriologie Moleculaire et Medicale
Institut Pasteur
25, rue du Dr. Roux
75724 Paris Cedex 15
France

Dr. Brigitte Saint-Joanis
Unité de Génétique Moléculaire Bactérienne
Institut Pasteur
25, rue du Dr. Roux
75724 Paris Cedex 15
France

Dr. Hrissi Samartzidou
Department of Biochemical and Biophysical Sciences
University of Houston
Houston, TX 77204-5934

Dr. Kenneth E. Sanderson
Salmonella Genetic Stock Centre
Department of Biological Sciences
University of Calgary
Calgary, Alberta Canada T2N 1N4

Dr. Karen Schmidt
Klinische Forschergruppe
Zentrum Biochemie
OE 4350
Medizinische Hochschule Hannover
D-30623 Hannover 61
Germany

Dr. Thomas M. Schmidt
Department of Microbiology
Michigan State University
E. Lansing, MI 48824

Prof. Dr. Rüdiger Schmitt
Lehrstuhl für Genetik
Universität Regensburg
D-93040 Regensburg
Germany

Dr. E.E. Selkov
Molecular Biosciences & Technology Institute
George Mason University
Fairfax, VA 22030

Dr. Christoph W. Sensen
Institute for Marine Biosciences
National Research Council
1411 Oxford Street
Halifax, NS B3H 3Z1
Canada

Dr. Brian Sheehan
Unité des Interactions Bactéries-Cellules
Institut Pasteur
25, rue du Docteur Roux
Paris 75015
France

Dr. Lawrence J. Shimkets
Department of Microbiology
University of Georgia
Athens, GA 30602

Dr. Kavindra V. Singh
Division of Infectious Diseases
Department of Internal Medicine
University of Texas Medical School
6431 Fannin, 1.728 JFB
Houston, TX 77030

Dr. Rama K. Singh
Institute for Marine Biosciences
National Research Council
1411 Oxford Street
Halifax, NS B3H 3Z1
Canada

Dr. Gerald R. Smith
Fred Hutchinson Cancer Research Center
1124 Columbia Street
Seattle, WA 98104-2092

Dr. S. Smith
Genetics Computer Group
575 Science Drive
Madison, WI 53711

Dr. Bruno W. S. Sobral
National Center for Genome Resources
1800 Old Pecos Trail
Santa Fe, NM 87505

Dr. Erica Sodergren
Department of Biochemistry & Molecular Biology
University of Texas Medical School
Houston, TX 77030

Dr. Erko Stackebrandt (with Naomi Ward-Rainey, Fred A. Rainey)
Deutsche Sammlung von Mikroorganismen und Zellkulturen GmbH (DSMZ)
Mascheroder Weg 16
D-38124, Braunschweig
Germany

Dr. Thaddeus Stanton
National Animal Diseases Center
Agricultural Research Service, USDA
2300 N. Dayton Road
P.O. Box 70
Ames, IA 50010

Dr. Rolf Stettler
Mikrobiologisches Institut
Eidgenössische Technische Hochschule
ETH-Zentrum
CH-8092 Zürich
Switzerland

Dr. George C. Stewart
Department of Pathology
VCS Building
Kansas State University
Manhattan, KS 66506

Dr. Scott Stibitz
Laboratory of Bacterial Toxins
Center for Biologics Evaluation and Research
Food and Drug Administration
8800 Rockville Pike
Bethesda, MD 20892

Dr. Michael Syvanen
Dept. of Microbiology and Immunology
School of Medicine
University of California at Davis
Davis, CA 95616

Dr. Satoshi Tabata
Kazusa DNA Research Institute
Kazusa Academic Park
1532-3 Yanauchino
Kisarazu, Chiba 292
Japan

Dr. Mitura Takanami
Kazusa DNA Research Institute
Kazusa Academic Park
1532-3 Yanauchino
Kisarazu, Chiba 292
Japan

Dr. Diane E. Taylor
Department of Medical Microbiology and Infectious Diseases
University of Alberta
Edmonton, Alberta T6G 2H7
Canada

Dr. Nancy J. Trun
NCI-NIH
Building 37, room 2D21
37 Convent Drive MSC4255
Bethesda, MD 20892-4255

Dr. Michael Tsifansky
Department of Molecular Genetics and Cell Biology
The University of Chicago
920 East 58th
Chicago, IL 60637

Prof. Dr. Burkhard Tümmler
Klinische Forschergruppe
Zentrum Biochemie
OE 4350
Medizinische Hochschule Hannover
D-30623 Hannover 61
Germany

Dr. Marjan van der Woude
Department of Pathology
University of Utah
Salt Lake City, UT 84132

Dr. James Versalovic
Harvard Medical School
Department of Pathology
Massachusetts General Hospital
55 Fruit Street
Boston, MA 02114

Dr. Thomas Vogt
The Sidney Kimmel Cancer Center
3099 Science Park Road
San Diego, CA 92121

Dr. Eldon M. Walker
The University of Texas Medical School
6431 Fannin
Houston, TX 77225

C. Wang
Institute for Bioscience, Bioinformatics, and Biotechnology
George Mason University
Fairfax, VA 22030

Dr. George Weinstock
Department of Biochemistry and Molecular Biology
The University of Texas Medical School
6431 Fannin
Houston, TX 77030

Dr. John Welsh
The Sidney Kimmel Cancer Center
3099 Science Park Road
San Diego, CA 92121

Dr. William R. Widger
Department of Biochemical and Biophysical Sciences
University of Houston
Houston, TX 77204-5934

Dr. Shane R. Wilkinson
Institute of Biological Sciences
University of Wales, Aberystwyth
Dyfed SY23 3DA
United Kingdom

Dr. Conrad L. Woldringh
Section of Molecular Cytology
Institute for Molecular Cell Biology
Biocentrum
University of Amsterdam
Amsterdam, The Netherlands

Dr. C. Peter Wolk
MSU-DOE Plant Research Laboratory
Michigan State University
East Lansing, MI 48824

Dr. K. K. Wong
The Sidney Kimmel Cancer Center
11099 N. Torrey Pines Rd.
La Jolla, CA 92037

Dr. T. Wong
Department of Microbiology
University of Toronto
Toronto, Ontario
Canada M5S 1A8

Dr. Jianguo Xiao
Division of Infectious Diseases
Department of Internal Medicine
University of Texas Medical School
6431 Fannin, 1.728 JFB
Houston, TX 77030

Dr. Akihiko Yamagishi
Department of Molecular Biology
Tokyo University of Pharmacy & Life Science
1432-1 Horinouchi, Hachioji
Tokyo 192-03 Japan

Dr. Yun You
Gladstone Institute of Virology and Immunology
P.O. Box 419100
San Francisco, CA 94141-9100

Dr. Michael Young
Institute of Biological Sciences
University of Wales, Aberystwyth
Dyfed SY23 3DA, United Kingdom

Dr. Su Zheng
Department of Molecular Genetics and Cell Biology
The University of Chicago
920 East 58th
Chicago, IL 60637

Dr. Richard Zuerner
National Animal Diseases Center
Agricultural Research Service, USDA
2300 N. Dayton Road
P.O. Box 70
Ames, IA 50010

Introduction

The concept for this book grew out of a program, entitled "Analysis of Prokaryotic Genomes," that was organized by the editors for the 1992 General Meeting of the American Society of Microbiology. At that time it was certainly not anticipated that entire DNA sequences for "bacterial genomes" could be determined at the present incredible rate. Nevertheless, it was clearly appreciated that bacterial genomes and their physical structure and analysis would provide a wealth of information of biological relevance and a more comprehensive understanding of the diversity of life.

Prokaryotic genomes are of intrinsic interest and, because they are much smaller than eukaryotic genomes, make ideal systems in which to pilot methods for genome analysis. Furthermore, because of a wealth of genetic information and years of physiological, metabolic, cellular, molecular biological, gene regulation, and pathogenicity studies, bacteria represent ideal organisms with which to attempt to relate the biological significance of or determine the biological relevance of genome sequences. The recent determination of the complete DNA sequence for six bacterial genomes: the 1.8-Mb *Haemophilus influenzae,* the 580-kb *Mycoplasma genitalium,* the 1.66-Mb *Methanococcus jannaschii,* the 1.7 Mb *Helicobacter pylori,* the 2.2 Mb *Archaeoglobus fulgidus* and the 4.6 Mb *Escherichia coli* genomes, lends strong support to these contentions. All six were determined by a strategy of whole-genome sequencing and assembly. Comparative genomics between these three microorganisms has enabled valuable insight into the minimal gene complement required of any free-living organism, and has suggested that differences in genome content are reflected as profound differences in physiology and metabolic capacity. Moreover, these genome studies have garnered substantial additional evidence to indicate that the *Archaea* and Eukaryotic kingdoms share a common evolutionary trajectory that is independent of the lineage of prokaryotes. This explosion in genome information is guaranteed to continue unabated for several years. It has been predicted, for instance, that by the year 2000 there will be whole genome sequences from fifty prokaryotes.

This book is organized into three major sections: I. Genome structure, stability and evolution; II. Strategies for genome analysis; and III. Physical maps of bacteria and their methods for construction. In Section I the structural features of genomes, including repetitive sequences, chi sites, cryptic phages, IS sequences and transposons, as well as higher-order structural genome features such as chromosomal organization, are described. The stability and evolution of bacterial genomes is discussed from the standpoint of chromosome ploidy, partitioning, gene order, and rearrangements, as well as bacterial phylogeny. Section II describes strategies for genome analysis, including long-range physical mapping, ordered cosmid and phage library construction, genomic fingerprinting and DNA sequence determination, and informatics analysis. Section III details physical maps and other genome features of ~40 selected bacteria.

The book is likely to be of utility to anyone interested in how living organisms organize, protect, and evolve their genetic information as genomes. It will also be of interest to those investigating and analyzing bacterial genomes and the genomes of the "higher organisms" that are less highly evolved, such as mice and humans.

Acknowledgment

We want to acknowledge Karen Bird of the MSU-DOE Plant Research Laboratory for invaluable assistance with the preparation of this book.

Bacterial Genomes

SECTION 1

Genome Structure, Stability, and Evolution

PART A

Genome Structure

1

Structure and Sizes of Genomes of the Archaea and Bacteria

Lawrence J. Shimkets

Introduction

The vast majority of biochemical and genetic diversity on this planet is found in the prokaryotic domains Archaea and Bacteria, whose members have adapted to most environmental niches on Earth through the use of diverse energy sources and electron acceptors (Woese, 1987; Woese et al., 1990). Prokaryotes are responsible for most components of the major biogeochemical cycles without which human life would not be possible. The rapid replication rates of prokaryotes and interspecies transfer of genetic information has prompted interest in the organization, stability, and replication of genomic DNA. This article examines genome size and genome structure of representatives of the prokaryotic domains Archaea and Bacteria, and compares these with the eukaryotic domain Eucarya.

Genome size

Many approaches have been used to assess genome size, including colorimetric techniques, DNA renaturation kinetics, and two-dimensional gel electrophoresis of kilobase-sized restriction fragments. These techniques have largely been replaced with pulsed field gel electrophoresis (PFGE) of large restriction fragments produced in bacterial genome mapping because of the small size of bacterial genomes (reviewed by Smith and Condemine, 1990; see also Chapters 24–26). The PFGE approach is conceptually simple; one digests the genome with rare cutting restriction enzymes, separates the restriction fragments by size in an agarose gel using PFGE, and calculates the size of each band from known size standards. The genome size is calculated from the sum of the sizes of individual restriction fragments. If several enzymes are used in the analysis, one can also order the restriction fragments to create a large-scale restriction map, thereby determining the topology of the genome. Genome sizes for over 150 bacterial species have now been determined by this method. Although a much larger database exists for genome size estimates based on renaturation kinetics (Herdman, 1985), renaturation data can misrepresent genome size by as much as 10-fold (Walker et al., 1991). Therefore, the data on prokaryotic genome sizes presented in this article is limited to those determined by PFGE. Unfortunately, the large size of Eucarya genomes prevents the use of this technique in most cases. Because PFGE genome sizes are available for only a few eukaryotic organisms, the Eucarya genome sizes in this article are not restricted to those determined with PFGE.

The genome sizes used in this analysis are the sum of the sizes of all the replicons within the cell where the sizes of the smaller replicons have been published. Genome sizes range from 6×10^5 bp for *Mycoplasma genitalium* (Colman *et al.*, 1990; Su and Baseman, 1990; *see also* chapters 40 and 69) to over 10^{11} bp for certain flowering plants (Figure 1-1). The largest prokaryotic genomes belong to the myxobacteria and are 9.4–10 Mbp (Chen *et al.*, 1990; Neumann *et al.*, 1992; *see also* Chapter 70). Genomes of the Archaea and Bacteria are usually smaller than those of the Eucarya. The smallest known haploid Eucarya genome is that of the microsporidium protozoan *Spraguea lophii,* which is 6.2 Mbp (Biderre *et al.*, 1994). The *Myxococcus xanthus* gene density is roughly similar to that of *Escherichi coli* based on sequenced regions (Shimkets, 1993; Chapter 70), allowing one to estimate that this myxobacterium contains approximately 7,800 genes (9,400 kb divided by 1.2 kb). The human genome is estimated to encode only 12 times as many genes.

Paradoxically, in the Eucarya the apparent biological, developmental, and intellectual complexity of an organism is not directly related to the haploid genome size, or C-value. Some amphibian species have C-values 100 times that of others, and flowering plants exhibit an even wider range of C-values.

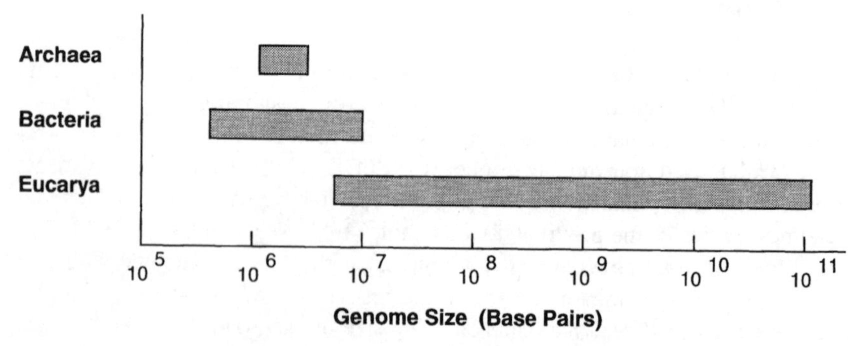

Figure 1-1. Genome size range for members of the three domains. The Bacteria and Archaea data are from PFGE data only, while Eucarya data include both PFGE data and renaturation kinetics.

Furthermore, some amphibians, flowering plants, insects, mollusks, and cartilaginous fishes have larger genomes than mammals. The precise reasons for the C-value paradox in Eucarya remain a mystery.

The C-value paradox was examined in prokaryotes by looking at the relationship between genome size and the morphological and metabolic complexity of the species. One might reasonably expect that genome size should increase with the number of anabolic and catabolic pathways, the number of respiratory systems, and the morphological complexity of the organism. Each species was classified as either a specialist or generalist based on traditional physiological growth tests. Generalists were defined as those species that used multiple carbon and energy sources, several terminal electron acceptors, and had few organic growth requirements in culture. Since the capacity to differentiate spores also requires a significant number of genes, spore forming bacteria were lumped with generalists even if they were nutritional specialists, as in the case of the myxobacteria (reviewed by Shimkets, 1990).

The genome sizes of 141 bacterial species are represented in Figure 1-2 along with their classification as specialists (white boxes) or generalists and spore-forming bacteria (black boxes). The mean genome size for specialists was about 1.9 Mbp with 72 representatives in this category. A theoretical distribution centered around this mean is also shown in Figure 1-2, and correlates nicely with the genome size distribution except for the absence of genomes below 0.6 Mbp. These results suggest that the minimum genome size for living organisms is about 0.6 Mbp. Since all of the organisms with genome sizes of <1.5 Mbp are either obligate intracellular parasites or exhibit fastidious growth in culture (ie. *Mycoplasma, Borrellia, Ureaplasma, Treponema, Rickettsia; see* Chapters 52, 69, 85) it appears that small genome sizes are possible only if the bacterium derives essential nutrients from a host. Free living bacteria have genome sizes of at least 1.6 Mbp. The mean genome size for generalists and spore-forming bacteria was 4.9 Mbp, with 69 representatives in this category.

C-values have an enormous range in narrow groups of eukaryotic organisms. Amphibian species with similar morphologies may have 10-fold differences in their genome sizes. C-value stability was examined in prokaryotic species by looking at the range of genome sizes within a single genus or species. The genome size of 14 natural isolates of *Escherichia coli* ranged from 4,600 to 5,300 kb (Berthorsson and Ochman, 1995). Although *E. coli* and *Salmonella typhimurium* are estimated to have diverged 120–160 million years ago (Ochman and Wilson, 1987), the genome size and gene order are highly conserved (Riley and Krawiec, 1987; *see also* Chapters 17 and 22). Over 50 isolates of *Helicobacter pylori* have been examined, with an overall range in genome size of about 2-fold (Karita *et al.* 1993; *see* Chapter 63). Genome size was relatively constant within a genus. For example, seven species of *Streptomyces* have genomes of nearly identical size (Lin *et al.*, 1993). No more than 2-fold variation was observed within members of a single genus or species. Over 25 *Mycoplasma* isolates

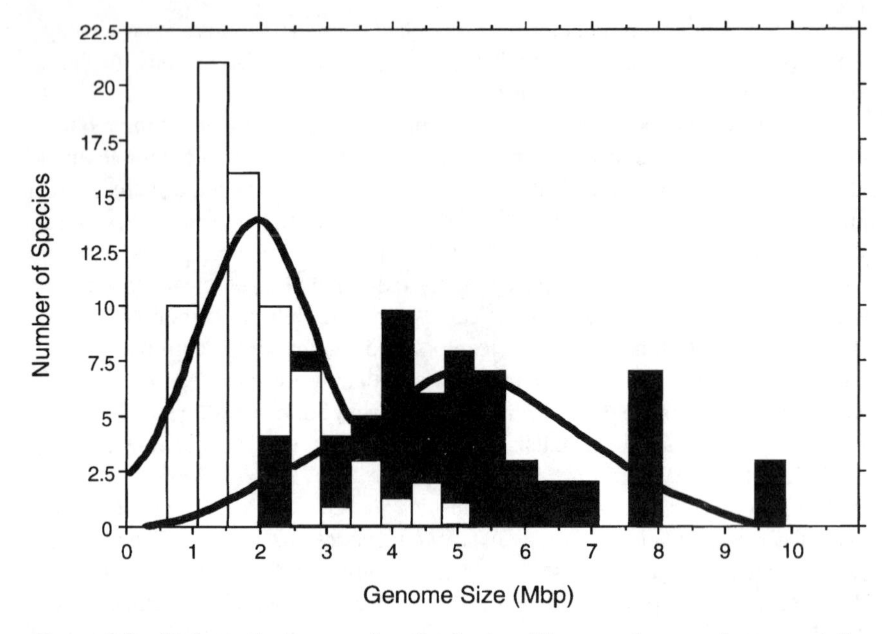

Figure 1-2. Prokaryotic genome size distribution. The open boxes refer to specialist species. The dark boxes refer to generalist species or specialist species-forming spores. The curves for a normal distribution of specialists (left) and generalists (right) are also shown.

belonging to 12 distinct species have about a 2-fold range in genome size; *see* Chapters 40–42, 69). Therefore, it appears that C-value within a defined prokaryotic genus or species is relatively constant compared with the variation observed in certain members of the Eucarya.

Together these results argue that prokaryotic genome size is proportional to the metabolic and morphological complexity of the organism; the C-value paradox does not apply to prokaryotes. One might infer from these data that there are natural constraints on genome size expansion in Archaea and Bacteria. It is possible that these constraints are centered around the need for rapid replication in habitats where nutrient supply is variable. If the cost of replicating a large genome decreases the fitness of the organism by reducing the overall rate of reproduction, there will be a constant selective pressure for maintaining lean genomes.

Genome structure

The genome topology is now known for over 50 bacterial species for which long-range restriction maps have been determined. The most frequent topology is that of one or more circular molecules. However, linear replicons have been reported for *Agrobacterium tumefaciens* C58 (Allardet-Servent *et al.*, 1993), *Borrelia burgdorferi* B31 (Ferdows and Barbour, 1989; *see* Chapter 52), *Rhodococcus fasciens* (Crepsi et al. 1992), and seven different *Streptomyces* species (Lin *et al.*, 1993). Lin *et al.* (1993) asked whether the *Streptomyces lividans* linear chromosome could be circularized without adversely affecting cellular reproduction. They transformed *S. lividans* with a suicide vector containing the kanamycin resistance gene, flanked by sequences homologous with each of the chromosome telomers. Viable Km^R progeny were observed which had circular chromosomes. Thus, the replicative machinery of *Streptomyces* can adjust to chromosomes with different topologies.

While most bacteria have a single, circular replicon, organisms with multiple large replicons have been described, most notably *Agrobacterium tumefaciens* C58 (Allardet-Servent et al. 1993), *Burkholderia cepacia* 17616 (Cheng and Lessie, 1994; Rodley *et al.*, 1995; *see* Chapter 73), *Brucella melitensis* 16M (Michaux *et al.*, 1993), *Leptospira interrogans* (Zuerner *et al.*, 1993) and *Rhodobacter sphaeroides* 2.4.1 (Suwanto and Kaplan, 1989; *see* Chapters 43 and 75). The partitioning of genetic information into multiple replicons does not appear to be essential for organisms with large genomes since *Myxococcus xanthus* contains a single 9.4 Mbp circular chromosome (Chen *et al*, 1991; *see* Chapter 70). Rather, it may reflect the fluid nature of bacterial genomes; homologous recombination can create a large chromosome from two smaller replicons or dissolve a large replicon. The overall genome size in *Bacillus cereus* appears to be conserved at about 5 Mbp. When the physical maps of six *Bacillus cereus* strains were compared, it was discovered that a large portion of the genome was subject to frequent relocation between replicons (Carlson and Kolsto, 1994; *see also* Chapter 49). In one strain, the entire genome was contained in a single circular chromosome. In another strain, the largest replicon was 2.4 Mbp but the remaining 2.6 Mbp DNA was distributed among multiple replicons. Thus, distinctions between chromosomal genes and episomal genes may be arbitrary. In the absence of evidence to the contrary, we should assume that genes on all the replicons within a cell play an important role in long-term survival in the natural environment.

Acknowledgments

This work was supported by NSF grant MCB9304083. I would like to thank William B. Whitman for his helpful suggestions in data analysis.

References

Allardet-Servant, A., S. Michaux-Charachon, E. Jumas-Bilak, L. Karayan and M. Ramuz. 1993. Presence of one linear and one circular chromosome in the *Agrobacterium tumefaciens* C58 genome. *J. Bacteriol.* 175:7869–7874.

Biderre, C., M. Pages, G. Metenier, D. David, J. Bata, G. Prensier, and C. P. Vivares. 1994. On small genomes in eukaryotic organisms: molecular karyotypes of two microsporidian species (Protozoa) parasites of vertebrates. *C.R. Acad. Sci. Paris. Sciences de la vie/ Life Sciences* 317:399–404.

Berthorsson, U., and H. Ochman. 1995. Heterogeneity of genome sizes among natural isolates of *Escherichia coli. J. Bacteriol.* 177:5784–5789.

Carlson, C. R., and A.-B. Kolsto. 1994. A small (2.4 Mb) *Bacillus cereus* chromosome corresponds to a conserved region of a larger (5.3 Mb) *Bacillus cereus* chromosome. *Mol. Microbiol.* 13:161–169.

Chen, H., I. M. Keseler, and L. J. Shimkets. 1990. Genome size of *Myxococcus xanthus* determined by pulsed field gel electrophoresis. *J. Bacteriol.* 172:4206–4213.

Chen, H., A. Kuspa, I. M. Keseler, and L. J. Shimkets. 1991. Physical map of the *Myxococcus xanthus* chromosome. *J. Bacteriol.* 173:2109–2115.

Cheng, H.-P., and T. G. Lessie. Multiple replicons constituting the genome of *Pseudomonas cepacia* 17616. *J. Bacteriol.* 176:4034–4042.

Colman, S. D., P. C. Hu, W. Litaker, and K. F. Bott. 1990. A physical map of the *Mycoplasma genitalium* genome. *Mol. Microbiol.* 4:683–687.

Crespi, M. E., A. B. Messens, M. V. Caplan, M. van Montague, and J. Desomer. 1992. Fasciation induction by the phytopathogen *Rhodococcus fascians* depends on a linear plasmid encoding a cytokinin synthetase gene. *EMBO J.* 11:795–804.

Ferdows, M. S., and A. Barbour. 1989. Megabased size-linear DNA in the bacterium *Borrelia burgdorferi,* the Lyme disease agent. *Proc. Natl. Acad. Sci. USA.* 86:5969–5973.

Herdman, M. 1985. The evolution of bacterial genomes, pp. 37–68. In *The evolution of genome size.* T. Cavalier-Smith, ed. John Wiley and Sons, Inc., New York.

Karita, M., T. Yoshimatsu, K. Okita, and K. Ouchi. 1993. Restriction endonuclease (*Not*I) analysis of chromosomal DNA of *Helicobacter pylori* by PFGE. *Nippon Rinsho.* 51:3102–3108.

Lin, Y.-S., H. M. Keiser, D. A. Hopwood, and C. W. Chen. 1993. The chromosomal DNA of *Streptomyces lividans* 66 is linear. *Mol. Microbiol.* 10:923–933.

Michaux, S., J. Paillisson, M.-J. Carles-Nurit, G. Bourg, A. Allardet-Servant and M. Ramuz. 1993. Presence of two independent chromosomes in the *Brucella melitensis* 16M genome. *J. Bacteriol.* 175:701–705.

Neumann, B., A. Pospiech, and H. U. Schairer. 1992. Size and stability of the genomes of the myxobacteria *Stigmatella aurantiaca* and *Stigmatella erecta. J. Bacteriol.* 174:6307–6310.

Ochman, H., and A. Wilson. 1987. Evolution in bacteria: evidence for a universal substitution rate in cellular genomes. *J. Mol. Evol.* 26:74–86.

Riley, M., and S. Krawiec. 1987. Genome organization. *In* Escherichia coli *and* Salmonella typhimurium: *cellular and molecular biology.* F. C. Neidhardt, J. L. Ingraham, K. B. Low, B. Magasanik, M. Schaechter, and H. E. Umbarger, eds., pp. 967–981. American Society for Microbiology, Washington, DC.

Rodley, P. D., U. Romling, and B. Tummler. 1995. A physical genome map of the *Burkholderia cepacia* type strain. *Mol. Microbiol.* 17:57–67.

Shimkets, L. J. 1990. Social and developmental biology of the myxobacteria. *Microbiol. Rev.* 54:473–501.

Shimkets, L. J. 1993. The myxobacterial genome. In *Myxobacteria II.* M. Dworkin and D. Kaiser, eds., pp. 85–107. American Society for Microbiology, Washington DC.

Smith, C. L., and G. Condemine. 1990. New approaches for physical mapping of small genomes. *J. Bacteriol.* 172:1167–1172.

Su, C. J. and J. B. Baseman. 1990. Genome size of *Mycoplasma genitalium. J. Bacteriol.* 172:4705–4707.

Suwanto, A., and S. Kaplan. 1989. Physical and genetic mapping of the *Rhodobacter sphaeroides* 2.4.1 genome: presence of two unique circular chromosomes. *J. Bacteriol.* 171:5850–5859.

Walker, E. M., J. K. Arnett, J. D. Heath, and S. J. Norris. 1991. *Treponema pallidum* subsp. *pallidum* has a single, circular chromosome with a size of approximately 900 kilobase pairs. *Infect. Immun.* 59:2476–2479.

Woese, C. R. 1987. Bacterial evolution. *Microbiol. Rev.* 51:221–271.

Woese, C. R., O. Kandler, and M. L. Wheelis. 1990. Towards a natural system of organisms: proposal for the domains Archaea, Bacteria, and Eucarya. *Proc. Natl. Acad. Sci. USA* 87:4576–4579.

Zuerner, R. L., J. L. Herrmann, and I. Saint Girons. 1993. Comparison of genetic maps for two *Leptospira interrogans* serovars provides evidence for two chromosomes and intraspecies heterogeneity. *J. Bacteriol.* 175:5445–5451.

2

Chromosomal Organization: Nucleoids, Chromosomal Folding, and DNA Topology

Karl Drlica and Conrad L. Woldringh

SUMMARY The bacterial chromosome (nucleoid), unlike the eukaryotic nucleus, is not bounded by a membrane. Nevertheless, it is a distinct structure that exhibits lobules when multiple replication forks occur and an elongated shape that may arise from coupled transcription-translation taking place at the nucleoid periphery. Since the nucleoid can be isolated with its DNA intact, a variety of topological experiments have been possible. Most have focused on DNA supercoiling, which is controlled largely by the enzymatic activity of DNA gyrase. A second enzyme, DNA topoisomerase I, acts as a safety valve to prevent the accumulation of excess supercoiling. Since negative supercoiling levels respond to changes in the extracellular environment, and since many activities of DNA are affected by supercoiling, supercoiling is potentially important as a component of global regulatory circuits. The nucleoid contains a variety of small proteins, some of which appear to increase the flexibility of DNA. The protein HU facilitates both gyrase activity and site-specific protein-DNA interactions. As a consequence of replication, interlinking occurs between daughter chromosomes, and a decatenase, topoisomerase IV, appears to be responsible for separating them so chromosome segregation can occur.

Introduction

The bacterial chromosome, or nucleoid as it is commonly called, contains a large circular DNA molecule with a length about 1000 times that of the cell in which it resides (Cairns 1963). One of the current challenges is to understand how a huge DNA molecule carries out complex processes such as transcription and replication. For example, we still have no answer to questions such as how genes located in the interior of the nucleoid are accessed and brought to the periphery for coupled transcription-translation. Nor do we know whether replication forks travel through the nucleoid rather than the nucleoid threading through stationary forks. Nevertheless, insights are gradually emerging from cytological and topological studies utilizing information and tools derived from the characterization of topoisomerases and other proteins that participate in shaping the nucleoid.

The topoisomerases alter DNA structure by breaking either one or both DNA strands and then passing nicked or duplex DNA through the break. If strand passage occurs within the same DNA molecule, supercoiling is altered. The major activities involved in this process are DNA gyrase (Gellert *et al.*, 1976b) and DNA topoisomerase I (Wang, 1971). If one DNA molecule is broken and a different DNA is passed through the break, catenation or decatenation occurs, depending on whether two circles are interlinked or unlinked. Topoisomerase IV is probably the primary decatenase. Among the other important DNA-binding proteins is HU, a small, abundant molecule that bends and wraps DNA. To provide a framework for understanding the activities of DNA, we outline below the central features of bacterial chromosome structure as derived from studies with whole cells and with isolated nucleoids.

Nucleoids in Whole Cells

Living Cells

The bacterial nucleoid can be observed by phase contrast microscopy in live cells after matching the refractive index of the surrounding medium to that of the cytoplasm. By applying this technique it has been possible to show through time-lapse microscopy that nucleoids increase in size and segregate in step with cell growth and division (Mason and Powelson, 1956). In slowly growing *E. coli* cells nucleoids look like simple rods, while in rapidly growing and wide cells they exhibit a complex, lobular structure that may reflect multifork replication (Schaechter *et al.*, 1962). Similar nucleoid shapes can be observed by fluorescence microscopy of cells cultured in the presence of DAPI, a DNA-staining agent that does not alter the growth rate of glucose-grown cells (time-lapse microscopy with DAPI is not feasible because the strong illumination necessary for excitation kills the cells [Van Helvoort and Woldringh, 1994]). Collectively these observations show that the bacterial nucleoid is confined to a central structure rather than being dispersed homogeneously throughout the cell. What confines the nucleoid is unknown.

Fixed and Dehydrated Cells

Although the characteristic shapes of nucleoids seen in live bacteria are maintained after fixation with OsO_4 and glutaraldehyde, these fixatives usually contract the nucleoid, as observed both by phase contrast microscopy (Valkenburg *et al.*, 1985) and by fluorescence microscopy of DAPI-stained cells (C.L. Woldringh, unpublished observations). However, permeability changes occur, so interpretation of microscopic images is not always straightforward. For example, contraction due to OsO_4 is lost when fixation is carried out at low salt concentration (Woldringh, 1973; Kellenberger, 1990), indicating that nucleoid size and shape reflect the salt concentration of the medium rather than intracellular structure. The dehydration required for most types of sample preparation also causes shrinkage problems (Woldringh *et al.*, 1977). Thus, the appearance of individual DNA strands seen in thin sections by electron microscopy depends on the sample fixation and dehydration procedure. Three general forms are observed: 1) after OsO_4-fixation, followed by conventional embedding in plastic, the nucleoid appears well separated from the cytoplasm and is filled with fibrillar DNA aggregates. OsO_4 may destroy DNA-protein complexes, thereby allowing naked DNA to aggregate during dehydration (Woldringh 1973; reviewed by Kellenberger, 1990); 2) After glutaraldehyde fixation and conventional embedding, the nucleoid consists of small, dispersed regions in which DNA strands or aggregates are difficult to detect. This appearance could be explained by a better preservation of protein-DNA complexes and the cross-linking of cytoplasmic and nucleoplasmic proteins into a continuous network; 3) After cryofixation and freeze-substitution (Hobot *et al.*, 1985), a much greater dispersal of nucleoid regions is observed than after glutaraldehyde fixation. The "corraline" nucleoid (Kellenberger, 1991) is filled with granular rather than fibrillar structures. This may be due to the preservation of mineral and organic salts that would otherwise be extracted during chemical fixation and conventional dehydration (Kellenberger, 1991). Until we determine which of these methods, if any, accurately reflects the intracellular condition, we will not know to what extent the nucleoid should be regarded as a distinct region separated from the cytoplasmic phase.

Localization of Chromosomal Activities and Proteins

Electron microscopic studies have revealed that transcription occurs at the periphery of the nucleoid, a point first made by autoradiographic detection of nascent transcripts in cells fixed with OsO_4 (Ryter and Chang, 1975). As expected, transcript localization was abolished by puromycin, which detaches ribosomes from mRNA. Subsequently, RNA polymerase was seen concentrated at the nucleoid border by immunocytochemical localization of cryofixed cells (Durrenberger *et al.*, 1988). We might also expect topoisomerase I and gyrase to be localized in transcribing regions, since they are required to relieve supercoils associated with transcription (Liu and Wang, 1987; Pruss and Drlica, 1989). This is the case for topoisomerase I when immunocytochemical methods are used with cryofixed cells (Durrenberger *et al.*, 1988). Gyrase, however, is largely (90%) cytoplasmic in glutaraldehyde-fixed cells (Thornton *et al.*, 1994). Whether this difference between the enzymes is due to biological or methodological factors is unknown. Methodology can be quite important, as emphasized by studies with the small DNA-binding protein called HU. In cryofixed cells, this protein is found at the periphery of the nucleoid (Durrenberger *et al.*, 1988), but with living cells it is seen over the entire nucleoid (Shellman and Pettijohn, 1991).

Isolated Nucleoids

Nucleoid Structure

In the early 1970s, Pettijohn developed a way to gently lyse bacterial cells and extract nucleoids without breaking the DNA (Stonington and Pettijohn 1971). Hydrodynamic analyses revealed that the chromosome is organized into about 50 topologically independent domains of negatively supercoiled DNA (Worcel and Burgi 1972; Pettijohn and Hecht 1973). Intracellular studies later showed that supercoils are present *in vivo* (Sinden *et al.* 1980), as are the topologically independent domains (Sinden and Pettijohn 1981). Isolated nucleoids were suitable for electron microscopy, and the resulting pictures show supercoiled loops (Kavenoff and Bowen 1976). It was initially thought that RNA might be involved in establishing the topological domains, since ribonuclease caused an unfolding of the nucleoid (Pettijohn and Hecht 1973; Drlica and Worcel 1975). However, the major species of RNA associated with the nucleoid are nascent mRNAs, which appear to entangle during cell lysis. Since inhibition of RNA synthesis has no effect on the number of topological domains observed *in vivo* (Sinden and Pettijohn 1981), it has been suggested that nascent RNA does not establish the domains. What does is unknown.

Control of DNA Supercoiling

The importance of negative supercoiling emerges from two types of observation: 1) supercoiling facilitates the DNA bending, looping, and strand separation crucial for many activities of DNA; 2) supercoiling changes in response to environmental alterations. Consequently, the nucleoid can respond globally to a changing environment through supercoiling.

Control occurs at several levels. One is DNA substrate preference: gyrase is much more active on a relaxed substrate and topoisomerase I on a highly supercoiled one. Consequently, the two enzymes tend to balance each other. The physiological importance of this balance became clear when *topA* mutants exhibited high levels of supercoiling and poor growth unless compensated by gyrase mutations that lowered supercoiling (DiNardo *et al.*, 1982; Pruss *et al.*, 1982). Control is also exerted at the level of gene expression: artificial lowering of supercoiling causes an increase in expression of gyrase (Menzel and Gellert, 1983) and a decrease in *top*A expression (Tse-Dinh, 1985). Raising supercoiling above normal levels increases *topA* expression (Tse-Dinh and Beran, 1988). Thus, supercoiling is homeostatically regulated.

Since topoisomerase I is such a potent relaxing activity, the early observation that gyrase also relaxed DNA (Gellert *et al.*, 1976a) was often overlooked. However, it became apparent that topoisomerase I could not be the only intracellular relaxing activity, since relaxation induced by an anti-gyrase agent proceeded at the same rate in the presence or absence of topoisomerase I (Pruss *et al.*, 1986). If gyrase controlled supercoiling by introducing and removing supercoils, then topoisomerase I might act as a safety valve to prevent accumulation of excess supercoiling (Drlica, 1990).

In the late 1980s it became clear that the level of supercoiling established by purified gyrase is determined by the ratio of [ATP] to [ADP] (Westerhoff *et al.*, 1988). High ratios produce high levels of supercoiling and *vice versa,* regardless of initial levels. At about the same time, environmental alterations were found that changed supercoiling. For example, entry into stationary phase lowered supercoiling (Balke and Gralla, 1987), growth at high osmolarity raised supercoiling (Higgins *et al.*, 1988), and growth under anaerobic conditions reduced the relaxing effect of stationary phase (Dorman *et al.*, 1988) (anaerobic conditions were later shown to increase both chromosomal and plasmid supercoiling during exponential growth [Hsieh *et al.* 1991]). A connection between environmental effects and cellular energetics emerged from the observation that a shift to anaerobic conditions caused both supercoiling and the [ATP] : [ADP] ratio to change in concert (Hseih *et al.*, 1991). The connection was extended to treatment of cultures with NaCl or high temperature: both supercoiling and [ATP] : [ADP] transiently increased (Hsieh *et al.* 1991; Camacho-Carranza *et al.*, 1995).

The energetics-supercoiling hypothesis was solidified by manipulating the cellular concentration of H^+-ATPase to vary the ratio of [ATP] to [ADP] over a broad range (Jensen *et al.* 1995). Plasmid supercoiling was quite sensitive to low values of [ATP]:[ADP], but in the range seen with normal growth (ratios of 1 to 10), sensitivity was not as great as expected from *in vitro* measurements (Westerhoff *et al.*, 1988). Thus intracellular factors, probably including topoisomerase I, appear to prevent excessive supercoils from accumulating at high ratios of [ATP] to [ADP]. These considerations strongly support the contention that cellular energetics and chromosome structure are connected (Drlica, 1990; Westerhoff *et al.*, 1990).

Decatenation

During chromosome replication, daughter DNA molecules fail to fully unlink unless the decatenating activity of a topoisomerase is present. The major decatenase appears to be topoisomerase IV rather than gyrase. *In vitro,* this enzyme is much more effective than gyrase, especially at intracellular salt concentrations (Peng and Marians, 1993). Moreover, topoisomerase IV resolves catenanes created both by unidirectional and bidirectional replication, while gyrase acts only on those generated by unidirectional replication (Peng and Marians, 1993). *In vivo,* plasmid catenanes accumulate in a temperature-sensitive topoisomerase IV mutant but not in a gyrase mutant (Adams *et al.*, 1992). However, gyrase does seem to be capable of replacing topoisomerase IV if overexpressed. Gyrase may supplement topoisomerase IV, since shifting a temperature-sensitive gyrase mutant to restrictive conditions causes cell filamentation (Mulder *et al.*, 1990) and the accumulation of dumbbell-shaped nucleoids (Steck and Drlica, 1984).

The 4-quinolones provide a way to monitor chromosome topoisomerase interactions. These antibacterial agents trap gyrase (Snyder and Drlica, 1979) on chromosomal DNA in complexes that contain broken DNA. Quinolones such as oxolinic acid seem to only trap gyrase, and they indicate that the average spacing of gyrase is about 100 kbp (Snyder, 1979). The potent fluoroquinolones trap both gyrase and topoisomerase IV (Khodursky *et al.*, 1995; Chen *et al.*, 1996), and in an oxolinic acid-resistant *gyrA* mutant the size of the fragmented DNA is about 200 kbp (Chen *et al.*, 1996). Thus, by this assay topoisomerase IV appears to be dispersed over the chromosome at about half the density of gyrase.

Our current hypothesis is that topoisomerase IV anchors the chromosome directly to the membrane (Drlica *et al.*, 1978; Kato *et al.*, 1992) and resolves catenanes as interlinking arises during replication. We have argued that gyrase is associated with the replication fork, as well as being distributed around the chromosome (Drlica *et al.*, 1980). This would allow it to relieve the torsional strain arising from replication fork movement and also maintain negative superhelical tension (Drlica *et al.*, 1978).

HU protein

Among the proteins attached to isolated nucleoids is a small, basic, abundant, DNA-binding protein called HU. This protein wraps DNA into nucleosome-like particles *in vitro* (Rouviere-Yaniv *et al.*, 1979; Broyles and Pettijohn, 1986) and bends short DNA into circles (Hodges-Garcia *et al.*, 1989). While there is little direct evidence that HU generates nucleosome-like particles *in vivo,* the bending activity could help HU serve as an accessory protein in processes such as bacteriophage Mu replicative transposition (Craigie *et al.*, 1985; Huisman *et al.*, 1989), gene inversion (Johnson *et al.*, 1986; Wada *et al.*, 1989), and possibly initiation of DNA replication (Hwang and Kornberg, 1992). The bending function of HU may also give the protein a global role in chromosome topology. We recently observed that a deficiency of HU causes poor growth that is readily suppressed by mutations that map in *gyrB* or by expression of *gyrB* from a plasmid (Malik *et al.*, 1996). The absence of HU partially relaxes DNA (Hillyard *et al.*, 1990; Hsieh *et al.* 1991; Rouviere-Yaniv *et al.*, 1992), and a spontaneous *gyrB* suppressor raises it back toward normal levels (Malik *et al.*, 1996). Thus, HU appears to facilitate gyrase action. It also seems to lower the effectiveness of topoisomerase I (Bensaid *et al.*, 1996).

The supercoiling changes associated with an HU deficiency are small; consequently, it is not obvious that the primary growth defect corrected by gyrase is due to perturbation of supercoiling. Several observations make decatenation a candidate activity, even though topoisomerase IV is currently thought to be the major decatenase: 1) HU stimulates decatenation by gyrase in vitro (Marians, 1987); 2) a deficiency of HU leads to production of cells that are filamentous (Imamoto and Kano, 1990) and DNA-less (Huisman *et al.*, 1989); and 3) doublet nucleoids accumulate in gyrase mutants (Steck and Drlica, 1984).

Concluding Remarks

Even though it has been possible to isolate intact nucleoids for more than 25 years, many important structure-function questions remain unanswered. Particularly interesting are DNA-membrane relationships, since they are thought to be important for segregation of replicated daughter chromosomes. Cell fractionation studies have revealed relationships between the origin of replication and the cell envelope (Ogden *et al.* 1988), and it is likely that topoisomerase IV-chromosome interactions will be another source of membrane binding (Kato *et al.* 1992). The observation that inhibition of protein synthesis causes the nucleoid to round up into a more central location (Woldringh *et al.* 1994) raises another intriguing possibility: perhaps the coupled transcription-translation of membrane-bound proteins contributes to indirect membrane attachment of the chromosome (Woldringh *et al.* 1995). Combinations of cytological, biochemical, and genetic approaches should provide insight into this important aspect of chromosome biology.

Acknowledgments

We thank J.-Y. Wang and C.-R. Chen for helpful comments on the manuscript. This work was supported by NIH grant 35257 to K.D. and the Netherlands Organization for Scientific Research (NWO-SLW) to C.L.W.

References

Adams, D., E. Shekhtman, E. Zechiedrich, M. Schmid, and N. Cozzarelli. 1992. The role of topoisomerase IV in partitioning bacterial replicons and the structure of catenated intermediates in DNA replication. *Cell* 71:277–288.

Balke, V. and J.D. Gralla. 1987. Changes in the linking number of supercoiled DNA accompany growth transitions in *Escherichia coli. J. Bacteriol.* 169:4499–4506.

Bensaid, A., A. Almeida, K. Drlica, and J. Rouviere-Yaniv. 1996. Cross-talk between topoisomerase I and HU in *Escherichia coli. J. Mol. Biol.* 256:292–300.

Broyles, S. and D. E. Pettijohn. 1986. Interaction of the *E. coli* HU protein with DNA: evidence for formation of nucleosome-like structures with altered DNA helical pitch. *J. Mol. Biol.* 187:47–60.

Cairns, J. 1963. The chromosome of *Escherichia coli. Cold Spring Harbor Symp. Quant. Biol.* 28:43–46.

Camacho-Carranza, R., J. Membrillo-Hernandez, J. Ramirez-Santos, J. Castro-Dorantes, V. C. Sanchez, and M. C. Gomez-Eichelmann. 1995. Topoisomerase activity during the heat shock response in *Escherichia coli* K-12. *J. Bacteriol.* 177:3619–3622.

Chen, C.-R., M. Malik, M. Snyder, and K. Drlica. 1996. DNA gyrase and topoisomerase IV on the bacterial chromosome: quinolone-induced DNA cleavage. *J. Mol. Biol.* 258:627–637.

Craigie, R., D. Arndt-Jovin, and K. Mizuuchi. 1985. A defined system for the DNA strand-transfer reaction at the initiation of bacteriophage Mu transposition: protein and DNA substrate requirements. *Proc. Natl. Acad. Sci. U.S.A.* 82:7570–7574.

DiNardo, S., K. Voelkel, R. Sternglanz, A. Reynolds, and A. Wright. 1982. *Escherichia coli* DNA topoisomerase I mutants have compensatory mutations in DNA gyrase genes. *Cell* 31:43–51.

Dorman, C., G. Barr, N. NiBhriain, and C. Higgins. 1988. DNA supercoiling and the anaerobic growth phase regulation of *tonB* gene expression. *J. Bacteriol.* 170:2816–2826.

Drlica, K. 1990. Bacterial topoisomerases and the control of DNA supercoiling. *Trends in Genetics* 6:433–437.

Drlica, K., E. Burgi, and A. Worcel. 1978. Association of the folded chromosome with the cell envelope of *Escherichia coli:* characterization of membrane-associated DNA. *J. Bacteriol.* 134:1108–1116.

Drlica, K., S. H. Manes, and E. C. Engle. 1980. DNA gyrase on the bacterial chromosome: possibility of two levels of action. *Proc. Natl. Acad. Sci. U.S.A.* 77:6879–6883.

Drlica, K. and A. Worcel. 1975. Conformational transitions in the *Escherichia coli* chromosome: analysis by viscometry and sedimentation. *J. Mol. Biol.* 98:393–411.

Durrenberger, M., M.-A. Bjornsti, T. Uetz, J. Hobot, and E. Kellenberger. 1988. Intracellular location of the histone-like protein in *Escherichia coli. J. Bacteriol.* 170:4757–4768.

Gellert, M., M. H. O'Dea, T. Itoh, and J.-I. Tomizawa. 1976a. Novobiocin and coumermycin inhibit DNA supercoiling catalyzed by DNA gyrase. *Proc. Natl. Acad. Sci. U.S.A.* 73:4474–4478.

Gellert, M., M. H. O'Dea, K. Mizuuchi, and H. Nash. 1976b. DNA gyrase: an enzyme that introduces superhelical turns into DNA. *Proc. Natl. Acad. Sci. U.S.A.* 73:3872–3876.

Higgins, C. F., C. J. Dorman, D. A. Stirling, L. Waddell, I. R. Booth, G. May, and E. Bremer. 1988. A physiological role for DNA supercoiling in the osmotic regulation of gene expression in *S. typhimurium* and *E. coli. Cell* 52:569–584.

Hillyard, D. R., M. Edlund, K. Hughes, M. Marsh, and N. P. Higgins. 1990. Subunit-specific phenotypes of *Salmonella typhimurium* HU mutants. *J. Bacteriol.* 172:5402–5407.

Hobot, J., W. Villiger, J. Escaig, M. Maeder, A. Ryter, and E. Kellenberger. 1985. Shape and fine structure of nucleoids observed on sections of ultrarapidly frozen and cryosubstituted bacteria. *J. Bacteriol.* 162:960–971.

Hodges-Garcia, Y., P. Hagerman, and D. Pettijohn. 1989. DNA ring closure mediated by protein HU. *J. Biol. Chem.* 264:14621–14623.

Hsieh, L.-S., R. M. Burger, and K. Drlica. 1991. Bacterial DNA supercoiling and [ATP]/[ADP]: changes associated with a transition to anaerobic growth. *J. Mol. Biol.* 219:443–450.

Hsieh, L.-S., J. Rouviere-Yaniv, and K. Drlica. 1991. Bacterial DNA supercoiling and [ATP]/[ADP]: changes associated with salt shock. *J. Bacteriol.* 173:3914–3917.

Huisman, O., M. Faelen, D. Girard, A. Jaffe, A. Toussaint, and J. Rouviere-Yaniv. 1989. Multiple defects in *Escherichia coli* mutants lacking HU protein. *J. Bacteriol.* 171:3704–3712.

Hwang, D. and A. Kornberg. 1992. Opening of the replication origin of *Escherichia coli* DNA by DnaA protein with protein HU or IHF. *J. Biol. Chem.* 267:23083–23086.

Imamoto, F. and Y. Kano. 1990. Physiological characterization of deletion mutations of the *hupA* and *hupB* genes. In *The Bacterial Chromosome* K. Drlica and M. Riley, eds. pp. 259–268. American Soc. for Microbiol. Washington, D.C.

Jensen, P., L. Loman, B. Petra, C. van der Weijden, and H. Westerhoff. 1995. Energy buffering of DNA structure fails when *Escherichia coli* runs out of substrate. *J. Bacteriol.* 177:3420–3426.

Johnson, R., M. Bruist, and M. Simon. 1986. Host protein requirements for *in vitro* site-specific DNA inversion. *Cell* 46:531–539.

Kato, J.-I., H. Suzuki, and H. Ikeda. 1992. Purification and characterization of DNA topoisomerase IV in *Escherichia coli. J. Biol. Chem.* 267:25676–25684.

Kavenoff, R. and B. Bowen. 1976. Electron microscopy of membrane-free folded chromosomes from *Escherichia coli. Chromosoma* 59:89–101.

Kellenberger, E. 1990. Intracellular organization of the bacterial genome. In *The Bacterial Chromosome* K. Drlica and M. Riley, eds. pp. 173–186. ASM, Washington, DC.

Kellenberger, E. 1991. Functional consequences of improved structural information on bacterial nucleoids. *Res. Microbiol.* 142:229–238.

Khodursky, A. B., E. L. Zechiedrich, and N. R. Cozzarelli. 1995. Topoisomerase IV is a target of quinolones in *Escherichia coli. Proc. Nat. Acad. Sci. U.S.A.* 92:11801–11805.

Liu, L. and J. Wang. 1987. Supercoiling of the DNA template during transcription. *Proc. Natl. Acad. Sci. U.S.A.* 84:7024–7027.

Malik, M., A. Bensaid, J. Rouviere-Yaniv, and K. Drlica. 1996. Histone-like protein HU and bacterial DNA topology: suppression of an HU deficiency by gyrase mutations. *J. Mol. Biol.* 256:66–76.

Marians, K. 1987. DNA gyrase-catalyzed decatenation of multiply linked DNA dimers. *J. Biol. Chem.* 262:10362–10368.

Mason, D. and D. Powelson. 1956. Nuclear division as observed in live bacteria by a new technique. *J. Bacteriol.* 71:474–479.

Menzel, R. and M. Gellert. 1983. Regulation of the genes for *E. coli* DNA gyrase: homeostatic control of DNA supercoiling. *Cell* 34:105–113.

Mulder, E., M. El'Bouhali, E. Pas, and C. L. Woldringh. 1990. The *Escherichia coli minB* mutation resembles *gyrB* in defective nucleoid segregation and decreased negative supercoiling in plasmids. *Mol. Gen. Genet.* 221:87–93.

Ogden, G., M. Pratt, and M. Schaechter. 1988. The replication origin of the *Escherichia coli* chromosome binds to cell membranes only when hemimethylated. *Cell* 54:127–135.

Peng, H. and K. Marians. 1993. Decatenation activity of topoisomerase IV during *oriC* and pBR322 DNA replication *in vitro. Proc. Natl. Acad. Sci. U.S.A.* 90:8571–8575.

Pettijohn, D. and R. Hecht. 1973. RNA molecules bound to the folded bacterial genome stabilize DNA folds and segregate domains of supercoiling. *Cold Spring Harbor Symposium of Quantitative Biology* 38:31–41.

Pruss, G., R. Franco, S. Chevalier, S. Manes, and K. Drlica. 1986. Effects of DNA gyrase inhibitors in *Escherichia coli* topoisomerase I mutants. *J. Bacteriol.* 168:276–282.

Pruss, G. J. and K. Drlica. 1989. DNA supercoiling and prokaryotic transcription. *Cell* 56:521–523.

Pruss, G. J., S. H. Manes, and K. Drlica. 1982. *Escherichia coli* DNA topoisomerase I mutants: increased supercoiling is corrected by mutations near gyrase genes. *Cell* 31:35–42.

Rouviere-Yaniv, J., J.-E. Germond, and M. Yaniv. 1979. *E. coli* DNA binding protein HU forms nucleosome-like structure with circular double-stranded DNA. *Cell* 17:265–274.

Rouviere-Yaniv, J., E. Kiseleva, A. Bensaid, A. Almeida, and K. Drlica. 1992. Protein HU and bacterial DNA supercoiling. *Prokaryotic Structure and Function.* pp. 17–43. S. Mohan, C. Dow and J. Cole, eds. Cambridge University Press.

Ryter, A. and A. Chang. 1975. Localization of transcribing genes in the bacterial cell by means of high resolution autoradiography. *J. Mol. Biol.* 98:797–810.

Schaechter, M., J. Williamson, J. Hood, and A. Koch. 1962. Growth, cell, and nuclear divisions in some bacteria. *J. Gen. Microbiol.* 29:421–434.

Shellman, V. and D. Pettijohn. 1991. Introduction of proteins into living bacterial cells: distribution of labeled HU protein in *Escherichia coli. J. Bacteriol.* 173:3047–3059.

Sinden, R. R., J. O. Carlson, and D. E. Pettijohn. 1980. Torsional tension in the DNA double helix measured with trimethylpsoralen in living *E. coli* cells. *Cell* 21:773–783.

Sinden, R. R. and D. E. Pettijohn. 1981. Chromosomes in living *Escherichia coli* cells are segregated into domains of supercoiling. *Proc. Natl. Acad. Sci. USA* 78:224–228.

Snyder, M. and K. Drlica. 1979. DNA gyrase on the bacterial chromosome: DNA cleavage induced by oxolinic acid. *J. Mol. Biol.* 131:287–302.

Steck, T. R. and K. Drlica. 1984. Bacterial chromosome segregation: evidence for DNA gyrase involvement in decatenation. *Cell* 36:1081–1088.

Stonington, O. G. and D. E. Pettijohn. 1971. The folded genome of *Escherichia coli* isolated in a protein-DNA-RNA complex. *Proc. Natl. Acad. Sci. U.S.A.* 68:6–9.

Thornton, M., M. Armitage, A. Maxwell, B. Dosanjh, A. Howells, V. Norris, and D. Sigee. 1994. Immunogold localization of GyrA and GyrB proteins in *Escherichia coli. Microbiology* 140:2371–2382.

Tse-Dinh, Y.-C. 1985. Regulation of the *Escherichia coli* DNA topoisomerase I gene by DNA supercoiling. *Nucl. Acids Res.* 13:4751–4763.

Tse-Dinh, Y.-C. and R. Beran. 1988. Multiple promoters for transcription of the *E. coli* DNA topoisomerase I gene and their regulation by DNA supercoiling. *J. Mol. Biol.* 202:735–742.

Valkenburg, J.A.C., C. L. Woldringh, G. J. Brakenhoff, T. M. Van der Voort, and N. Nanninga. 1985. Confocal scanning light microscopy of the *Escherichia coli* nucleoid: comparison with phase-contrast and electron microscope images. *J. Bacteriol.* 161:478–483.

Van Helvoort, J.M.L.M. and C. L. Woldringh. 1994. Nucleoid partitioning in *Escherichia coli* during steady-state growth and upon recovery from chloramphenicol treatment. *Mol. Microbiol.* 13:577–583.

Wada, M., K. Kutsukake, T. Komano, F. Imamoto, and Y. Kano. 1989. Participation of the *hup* gene product in specific DNA inversion in *Escherichia coli. Gene* 76:345–352.

Wang, J. C. 1971. Interaction between DNA and an *Escherichia coli* protein. *J. Mol. Biol.* 55:523–533.

Westerhoff, H., M. Aon, K. van Dam, S. Cortassa, D. Kahn, and M. van Workum. 1990. Dynamical and hierarchical coupling. *Biochim. Biophy. Acta* 1018:142–146.

Westerhoff, H., M. O'Dea, A. Maxwell, and M. Gellert. 1988. DNA supercoiling by DNA gyrase. A static head analysis. *Cell Biophysics* 12:157–181.

Woldringh, C. L. 1973. Effects of cations on the organisation of the nucleoplasm in *Escherichia coli* prefixed with osmium tetroxide or glutaraldehyde. *Cytobiologie* 8:97–111.

Woldringh, C. L., P. R. Jensen, and H. V. Westerhoff. 1995. Structure and partitioning of bacterial DNA: determined by a balance of compaction and expansion forces? *FEMS Microbiol. Letters* 131:235–242.

Woldringh, C. L., M. A. de Jong, W. van den Berg, and L. Koppes. 1977. Morphological analysis of the division cycle of two *Escherichia coli* substrains during slow growth. *J. Bacteriol.* 131:270–279.

Woldringh, C. L., A. Zaritsky, and N. B. Grover. 1994. Nucleoid partitioning and the division plane in *Escherichia coli. J. Bacteriol.* 176:6030–6038.

Worcel, A. and E. Burgi. 1972. On the structure of the folded chromosome of *Escherichia coli. J. Mol. Biol.* 71:127–147.

3

Prophages and Cryptic Prophages

Allan M. Campbell

A large group of natural bacteriophages (called temperate) can establish a permanent relationship with their hosts (lysogeny), where most viral functions are repressed and the phage genome (prophage) is transmitted vertically from mother to daughters at cell division. Some prophages are inserted into the bacterial chromosome (either by site-specific recombination, like coliphage λ, or by transposition, like coliphage Mu-1), whereas others, like coliphage P1, establish themselves as plasmids. Established lysogens frequently suffer mutations or partial prophage deletions that destroy genes needed for lytic development. The prophage in such cases is called *defective* or *cryptic*.

Natural bacteria can harbor both complete prophages and cryptic prophages. These behave as normal genome components, and their viral origin is not always readily recognized. For example, the well-studied *E. coli* K-12 strain harbors bacteriophage λ (49 kb) and four defective lambda-related (lambdoid) phages (DLP12, 27 kb; Rac 25 kb; Qin, > 15 kb; e14, 14 kb), as well as an unmapped segment that includes the integrase gene of lambdoid phage HK022 (Yagil *et al.*, 1995) and a small fragment derived from bacteriophage P2 (Barreiro and Haggård-Ljungquist, 1992). All of these presumably resulted from lysogenization by phages comparable in size to λ or P2, followed by extensive deletion.

This chapter first reviews the K-12 elements, because they illustrate many of the properties of prophages and the manner in which they are studied. Occasional studies with other prophages are mentioned. Later, ideas about evolutionary relationships between phage and host are discussed.

Elements Endogenous to *E. coli* K-12

Although the K-12 strain was isolated in the 1920s and widely used in laboratory experiments, the presence of λ prophage was not discovered until Lederberg

(1951) found, among the survivors of heavily irradiated cultures, a few colonies that were sensitive to the phage liberated by some of the non-survivors. Wild type K-12 is insensitive to λ infection, because of the repressor made by the prophage. Genetic studies showed that the λ prophage can be mapped to a specific chromosomal location, and that the prophage is linearly inserted into the chromosome.

Most of the defective lambdoid prophages (DLP12, Rac, Qin) were discovered as sources of phage genes that can recombine into λ. In the prototypical experiment, a phage mutant stock (such as the double mutant λ *Qam Qam*) was found to contain a few members that behaved as wild type (i.e., would plate on a host lacking an amber suppressor). These proved to be recombinants, where determinants of products equivalent to gp Q and to the products of the adjacent *S* and *R* genes had recombined into the mutant λ from a defective prophage (Strathern and Herskowitz, 1975). For *Q–R,* the great majority of such recombinants are from DLP12, whose initial characterization was aided by the existence of some K-12 derivatives that have deleted the relevant portion of the DLP12 prophage; but rare rescue from Qin has also been observed (Espion *et al.*, 1983). The Rac prophage was discovered by a different selection, from λ mutants lacking their normal recombination functions. Mapping of all these prophages to specific positions on the *E. coli* chromosome was later accomplished by a combination of genetic and physical methods.

The e14 element was discovered as a 14 kb segment of the *E. coli* chromosome that can be excised following UV treatment (Greener and Hill, 1980). It was later mapped to a site within the isocitrate dehydrogenase (*icd*) gene, the same site used by another λ relative, phage 21. Like all other known phages that insert within rather than between bacterial genes, both e14 and 21 carry a duplication of the 3′ end of that gene, starting from the insertion site. Thus, phage insertion does not disrupt *icd* function, because a phage-derived sequence is substituted for the bacterial 3′ end.

None of these four elements has been completely sequenced, but a body of information (including some sequence) has accumulated about each. At least the larger elements (DLP12, Rac, Qin) have the same gene order as in prophage λ, as expected if they are remnants of lambdoid prophages which have experienced extensive deletion. Two types of components have been ordered: genes of equivalent function (such as *Q, S,* and *R*) and DNA segments sufficiently similar to parts of λ to show up either by Southern hybridization or as sites for homologous recombination. Natural lambdoid phages share a common genetic map based on the same kind of ordering; some functional analogs have diverged beyond recognition but remain in place, interspersed with some segments of very close similarity between any two members of the family.

DLP12 lies at 12 min on the K12 chromosome, λ is at 17 min, e14 at 26 min, Rac at 30, and Qin at 35. This fits with the fact that each natural lambdoid phage preferentially inserts at a unique site, but that various sites are used by different members of the family. Only e14 lies at the insertion site of a known lambdoid phage (21), and its integrase gene is more similar in sequence to that of 21 than to other members of the lambda family. Two of the elements (Rac and e14) have intact integration/excision functions and can excise cleanly from the bacterial chromosome. This operation clearly demarcates the boundary between prophage and bacterial DNA. DLP12 has not been shown to excise and appears to have deleted part of the *xis* gene, but good candidate *att* sites can be identified that bracket the element. No such sites have been identified for Qin, so the prophage/host boundaries are undefined.

Prophages in Other Bacteria

The K-12 story is instructive as to how prophages and cryptic prophages are detected. Lysogeny by some phages (like λ) is maintained by phage-coded repressors that are inactivable by the coprotease activity of RecA protein; therefore, lysogenic cells treated with agents that activate RecA (like ultraviolet light) frequently lyse and liberate phage-like particles, visible by electron microscopy. The demonstration that such particles are complete phages capable of reproducing in other cells depends on the availability of a sensitive host. In the case of λ, the appropriate host arose through loss of the prophage in occasional cells that had survived high doses of irradiation. Some cryptic prophages can also cause lysis after RecA activation, (although none of the cryptic elements in K-12 has this property), with release of phage-like particles or of phage components, such as free heads or tails. Even where the particles appear complete, they may be noninfectious: For example, the elements PBSX, PBSY, PBSZ, widely distributed among *Bacillus subtilis* strains, cause liberation of complete phage-like particles, but the DNA packaged in these particles is an almost random assortment of sequences from throughout the *Bacillus* chromosome. At any rate, liberation of phage-like particles is good evidence for a prophage (perhaps cryptic), and the determinants for particle production can be mapped on the bacterial chromosome [as they have been in *Bacillus* by Thurm and Garro (1975)].

Where no such particles are detected, cryptic prophages can be identified by homology with known phages. Such homologies are expected of cryptic prophages derived from the tester phage, but in principle could also arise by cooption of host genes into phage genomes. Indeed, most theories of viral origins suppose that viral genes arose from host genes. Two conspicuous examples of phages with host-derived sequences are phage 21 (more generally, all phages that insert into structural genes) and P22 (one of whose replication genes is similar to host *dnaB*.) The classification of DLP12, Rac, and Qin as cryptic prophages rests on extensive, scattered homology with λ and the presence of genes that function in the phage life cycle. For e14, the only known sequence similarity with other lambdoid phages is in the integrase gene; an invertible sequence of e14 (to be discussed later) is present in certain phages of other groups.

Southern hybridization with λ probes shows that λ-related sequences are present in many natural isolates of *E. coli* and other Enterobacteriaceae (Anilionis and Riley, 1980). The result is reasonably interpreted to mean that most *E. coli* strains contain either complete or cryptic prophages related to λ and probably underestimates the extent to which souvenirs of past infections remain in bacterial genomes. Another informative study concerns a lambdoid phage (Atlas), which inserts at 28 min. (Stoltzfus, 1991). This part of the chromosome was compared in 72 strains from the ECOR collection. At least two of these contained complete prophages. Most of the others were shorter and probably cryptic. The molecular phylogeny based on chromosomal genes implies that these were the products of several independent lysogenizations by the ancestral phage. The prevalence of shortened versions indicates that prophages often deteriorate rapidly by deletion.

Influence of Phage Genes on Host

For the bacterial geneticist, the frequent occurrence of prophages and the limited criteria for identifying them as such means that some determinants first classified as bacterial genes eventually prove to be parts of prophages, and that in other cases such a proof may be difficult or impossible. Although lysogeny is frequently inconspicuous, there are some classical examples of phage-determined host traits, such as production of diphtheria toxin or of surface antigens in *Salmonella.* The kinds of genes affecting host phenotype can be related to ideas about natural selection on both phage and host.

The ability to lysogenize is not of obvious selective value for a phage. Good arguments can be made that this ability is most advantageous when the prophage confers a selective advantage on its bearer, at least in some environment. In this context, phage genes may be classified into three groups (which may overlap):

(1) Genes important in the phage life cycle: determinants for virion formation, replication, lysis, inhibition of host functions, and regulation. Some of these genes have interested bacterial evolutionists because of their possible utility to the host. The lambdoid prophages provide several possible examples.
   (a) *E. coli recBC* mutants are hypersensitive to radiomimetic agents such as mitomycin C. Selection for mitomycin-resistant mutants yields strains in which the *recE* gene of the Rac prophage has been derepressed, either because of inactivation of the Rac repressor or because of deletions that place *recE* under different control (Kaiser and Murray, 1980).
   (b) DLP12 (as well as lambdoid phage PA-2) contains a gene for a new outer membrane protein (*nmpC*), which can replace the host protein OmpC and repress its synthesis. The presumed function

of *nmpC* in the phage cycle is to reduce the amount of phage loss that results from attachment of progeny phage to fragments of cell envelope containing OmpC (the surface receptor for many lambdoid phages). The DLP12 *nmpC* gene is inactive in the prophage. In *ompC* mutants of *E. coli,* there is selection for expression of *nmpC,* whose product takes over some of the cellular functions of OmpC (Highton *et al.*, 1985; Lindsey *et al.*, 1989). The ability of *Salmonella* phages ε15 and ε34 to alter the biosynthetic pathways for surface polysaccharide antigens used in phage attachment may have a similar selective basis.

(c) Most lambdoid phages encode functions such as *kil*(λ), *sfiC* (e14), *dicB*, and *dicF* (Qin) that inhibit host cell division when derepressed. Their presumed role in phage biology is to direct resources away from host development. It has been suggested that some of these genes may be coopted by the host. Their effect on the host is negative; however, cell division control in *E. coli* (as in higher cells) requires a balanced system of positive and negative elements; and *E. coli* contains a gene, *SfiA* (not known to be phage-related), whose effect is similar to that of *sfiC* or *dicB* and *dicF* in interfering with either the function or the synthesis of the cell septation protein FtsZ.

(2) Genes important in interphage competition: The natural world is full of phages competing for available hosts. Strategies that limit access of other phages to those hosts may be favored. Thus, many prophages express genes whose products restrict development of other phages. These include *rexAB*(λ) and *mcrA* (e14). The *rexAB* gene excludes various other phages, including *rII* mutants of phage T4, by triggering a collapse of the membrane potential, with consequent cell death, soon after infection by the foreign phage. This effect has been dubbed altruistic, because it confers no benefit on the dying cell but might protect its sibs from infection (Parma et al., 1995). The *mcrA* gene of e14 specifies a restriction enzyme that protects the cell (and its carried prophage) from phage attack.

The possible benefit of the *nmpC* gene in preventing neutralization of newly liberated phage particles by cell envelope fragments was discussed above. The fact that expressed *nmpC* renders the cell resistant to other phages that use OmpC as receptor would allow it to function in interphage competition as well; but the direct effect on phage development should constitute a stronger selective force. Indeed, the *nmpC* gene of DLP12 is inactivated by a transposon insertion, so there should be no current selective force for its retention.

(3) Genes important for the host: In a world filled with phages, the genes important in interphage competition should benefit the host as well. We may ask whether phages harbor any genes whose sole selective value is to benefit the host, for reasons unrelated to phage infection. Two possible examples are the *trgG* gene of Rac and the *bor* gene of λ.

The TrgG protein promotes K^+ transport, with higher affinity than its bacterial counterpart (Dosch et al., 1991). The *bor* product increases serum survival of *E. coli,* and hence should enhance pathogenicity (Barondes and Beckwith, 1990). Such functions could provide a selective advantage to lysogenic bacteria. The advantage should apply only under conditions encountered intermittently in nature; or the host should by now have appropriated the genes and deleted the rest of the prophage. The PBS prophages of *Bacillus subtilis* may be retained because the phage-like particles they release can kill bacteria competitors. The major phenotypic effect of diphtheria toxin genes appears to be on the bacterial host, although the nature of any possible selective advantage is not obvious.

Cooption of Phage by Host

It is usually assumed that phage genes ultimately were appropriated from host counterparts, although only rarely was the event recent enough that significant homology remains (an example is gene 12 of phage P22, a variant of host *dnaB.*) The reverse scenario (in which host genes have been derived from invading phages) has no obvious examples, although the *nmpC* gene of DLP12 and *recE* gene of Rac provide laboratory models (discussed earlier). The invertible segment of e14 and other phages is related to the *Salmonella* phase variation system. In phages, this system provides variation in tail fiber proteins functional in attachment to the bacterial surface; in bacteria, it generates variation in antigenic proteins on the bacterial surface. The bacteria may have borrowed this system from a phage, or vice-versa.

Conclusion

Prophages and cryptic prophages are abundant, diverse, and widely distributed. They comprise several percent of the DNA of some natural strains. Whereas some prophage genes affect host phenotype and may have been selected for that purpose, the bulk of the evidence suggests that the major reason for their prevalence is frequent infection, that residence times are short, and that deterioration from active to cryptic prophages is rapid.

Acknowledgments

The author's research is supported by grant GM51117, National Institute of General Medicine.

References

Anilionis, A. and M. Riley. 1980. Conservation and variation of nucleotide sequences within related genomes: *Escherichia coli* strains. *J. Bacteriol.* 143:355–65.

Barondes, J. J. and J. Beckwith. 1990. A bacterial virulence determinant encoded by lysogenic coliphage lambda. *Nature* 346:871–4.

Barreiro, V. and E. Haggård-Ljungquist. 1992. Attachment sites for bacteriophage P2 on the *Escherichia coli* chromosome. DNA sequences, localization on the physical map, and detection of a P2-like remnant in *E. coli* K-12 derivatives. *J. Bacteriol.* 174:4086–93.

Dosch, D. C., G. K. Helmer, S. H. Sutton, *et al.* 1991. Genetic analysis of potassium transport loci in *Escherichia coli:* evidence for three constitutive systems moderating uptake of potassium. *J. Bacteriol.* 173:687–96.

Espion, D., K. Kaiser, and C. Dambly-Chaudiere. 1983. A third defective prophage of *Escherichia coli* K12 defined by the λ derivative, λ-qin III. *J. Mol. Biol.* 170:611–33.

Greener, A. and C. W. Hill. 1980. Identification of a novel genetic element in *Escherichia coli* K-12. *J. Bacteriol.* 144:312–21.

Highton, P. J., Y. Chang, W. R. Marcotte, Jr. and C. A. Schnaitman. 1985. Evidence that the outer membrane porin protein gene *nmpC* of *Escherichia coli* K-12 lies within the defective qsr′ prophage. *J. Bacteriol.* 162:256–62.

Kaiser, K. and N. Murray. 1980. On the nature of *SbcA* mutations in *E. coli* K-12. *Mol. Gen. Genet.* 179:555–63.

Lederberg, E. M. 1951. Lysogenicity in *E. coli* K12. *Genetics* 36:560.

Lindsey, D. R., D. A. Mullins, and J. P. Walker. 1989. Characterization of the cryptic lambdoid prophage DLP12 of *Escherichia coli* and overlap of the DLP12 integrase gene with the tRNA gene *argU*. *J. Bacteriol.* 171:6197–205.

Parma, D. H., M. Snyder, S. Sobolevski, *et al.* 1972. The Rex system of bacteriophage λ: tolerance and altruistic cell death. *Genes Dev.* 6:497–510.

Stoltzfus, A. B. 1991. A survey of natural variation in the *trp-ton* region of the *E. coli* chromosome. Ph.D. thesis, University of Iowa, 234 pp.

Strathern, A. and I. Herskowitz. 1975. Defective prophage in *Escherichia coli* K12 strains. *Virology* 67:136–48.

Thurm, P. and A. J. Garro. 1975. Isolation and characterization of prophage mutants of the defective *Bacillus subtilis* bacteriophage PBSX. *J. Virol.* 16:184–91.

Yagil, E., L. Dorgai, and R. Weisberg. 1995. Identifying determinants of recombination specificity: construction and characterization of chimeric bacteriophage integrases. *J. Mol. Biol.* 252:163–177.

4

Insertion Sequences and Transposons

Mark S. Chandler

With the sequence of two bacterial genomes complete (Fleischmann *et al.*, 1995; Fraser *et al.*, 1995) and several more near completion, it is now possible to view the entire genetic structure of an organism with a new perspective. Yet this perspective is not a static one. As we learn more about various bacterial genomes, the number of characterized bacterial insertion sequences (ISs) continues to increase. Examples are known from a wide range of Gram-negative and Gram-positive bacterial species as well as from the archaebacteria (reviewed by Galas and Chandler, 1989; Murphy, 1989; Charlebois and Doolittle, 1989). ISs are normal constituents not only of many bacterial chromosomes but also of some plasmids and bacteriophages. The prevalence of transposable elements (ISs and transposons) with their capacity for moving from one site in the genome to another, for modifying gene expression, and for promoting genome rearrangements, contributes significantly to a genome in a state of continuous change.

Insertion Sequences

An insertion sequence is a discrete segment of DNA capable of translocating directly to new sites in the genome. ISs carry only the gene(s), in addition to host provided ones, required for their own transposition. They were first identified during investigations of the molecular genetics of gene expression of the galactose operon of *Escherichia coli* (Jordan *et al.*, 1968). Unstable, highly polar mutations were isolated and shown by analysis of heteroduplex molecules and hybridizations to be insertions of the same few types of elements (Fiandt *et al.*, 1972; Hirsch *et al.*, 1972). These elements were named insertion sequences.

Properties of insertion sequences

ISs differ in size and other details but some generalized statements can be made about their structure. They range in size from 768 bp (IS*1*) to 7.1 kb (IS22) but are most commonly between 1 and 1.5 kb. An IS usually ends in short inverted repeats (IRs) (Fig. 4-1). The two copies of the repeat are generally closely related rather than identical. The IRs which vary in length from 9 to 41 bp are characteristic for the particular IS and are important for transposition. During transposition of an IS, a duplication of a small sequence of target DNA is created at the site of insertion (Fig. 4-1). Presumably, this is due to a staggered cut of the host target site that is later filled. The size of these flanking direct repeats are characteristic of each IS although there is some variation. The length of these duplications ranges from 2–13 bp. The lack of homology among the various target sites confirms the apparent randomness of insertion of these elements that was observed in the early genetic studies.

Insertion sequences apparently encode only the protein(s) needed for transposition (Fig. 4-1). Several ISs contain a single long coding region extending almost the full length of the element and on the opposite strand one or more short overlapping open reading frames (ORFs). The long ORF encodes the single protein required for transposition, the transposase. Other elements have ORFs of assorted sizes and positions that make assigning functions difficult from sequence information alone. Several experiments indicate that the transposase protein preferably acts in *cis* on its inverted repeats to catalyze transposition (Derbyshire *et al.*, 1987; Jakowec *et al.*, 1988; Huisman *et al.*, 1989).

ISs have been found on the chromosomes of many different bacteria, with

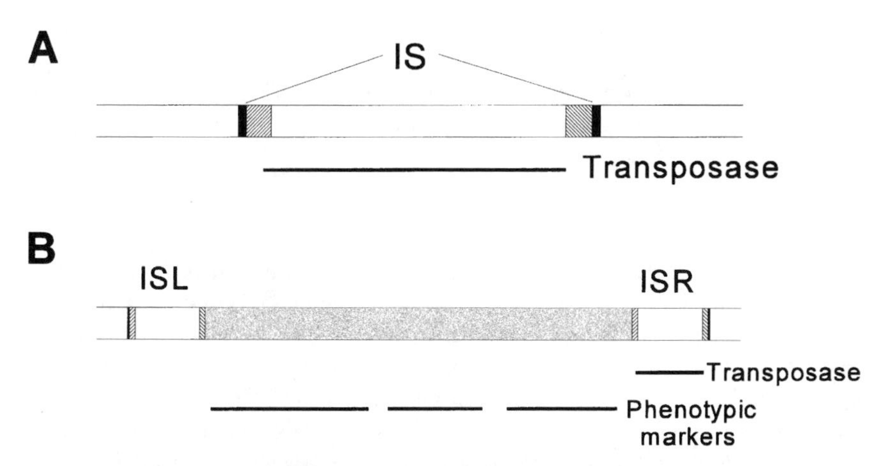

Figure 4-1. DNA organization of a hypothetical bacterial insertion sequence (IS) and composite transposon. **A.** IS showing inverted repeats (striped bars) and duplicated target sequences producing direct repeats (solid bars). The coding region for a transposase is shown below the element. **B.** Composite transposon with ISs on the left (ISL) and right (ISR) of a central region (stippled area) encoding several phenotypic markers (e.g., resistance to antibiotics). In this example only ISR has an active transposase. See text for a description of the size and other properties of various ISs and transposons.

most strains having 5 to 8 copies of a particular sequence. The number of copies can vary considerably depending on the bacterial host. IS*1* can be present as a single copy in some species of bacteria and in several hundred copies in a different species. The number of copies of a particular IS is commonly determined by hybridization (*see also* Chapter 32), which does not show whether individual copies of a particular IS are identical at the sequence level. This type of information will soon be available for a number of organisms whose genomes are currently being sequenced.

The continued evaluation of the genetic structure of bacterial chromosomes has identified new copies of known ISs as well as new prototypes. ISs are also widely distributed on other types of replicons. For example, IS*1* is found on the coliphage P1 and on several conjugal antibiotic-resistance plasmids (R factors). IS*2* and IS*3* are present on the F plasmid (Hu *et al.*, 1975a; Hu *et al.*, 1975b; Iida *et al.*, 1978). The presence of ISs on a plasmid (or other replicon) could cause instability since recombination between ISs could lead to deletion or inversion of the intervening region. Homologous recombination between ISs present on different DNA molecules can lead to insertion of one molecule into the other or recombination between the two molecules. For a discussion of ISs and their role in evolution, *see also* Chapter 20.

Transposons

A transposon is a more complex type of mobile genetic element. Transposons encode, in addition to transposition functions, one or more functions that generally confer a distinctive phenotype (e.g., drug resistance) on the host. A useful way of describing different transposons is to group them based on similar structure and genetic organization.

Composite transposons

One group of these elements called composite or compound transposons is composed of ISs flanking otherwise nontransposable sections of DNA, thus allowing that region to become transposable. Characteristically, two identical or closely related ISs flank a region encoding antibiotic resistance or catabolic functions (Fig. 4-1). The ISs in these composite transposons can be in an inverted (most common) or direct orientation. Since both ends of an IS generally end in inverted repeats, a composite transposon also ends with inverted repeats regardless of the relative orientation (direct or indirect) of the flanking ISs. Analysis of ISs and transposons suggests that several ISs have not been detected as part of composite transposons, while other ISs occur only as part of composite transposons, even though they are able to transpose independently. Still other ISs are found both alone or as transposon-associated units.

The 9.3 kb tetracycline resistance transposon Tn*10* is composed of a central region containing the resistance gene, flanked on either side by inverted copies of the IS*10* insertion sequence. These IS*10* modules are about 1300 bp in length, with 22 bp inverted repeat termini, and carry a transposase gene. However, the two IS*10* elements in Tn*10* are not identical, one of them has a defective transposase (Kleckner *et al.*, 1984). In such cases the functional transposase can act on the ends of an IS*10* element to promote IS*10* transposition, or it can act on the ends of the entire Tn*10* to catalyze Tn*10* transposition. The capacity of a single IS to transpose an entire composite transposon suggests that there may be a lack of selective pressure for the transposase of both terminal ISs to remain functional. In still other cases the modules of the composite transposon are identical.

Tn*903* is 3100 bp in length, with a central region encoding kanamycin resistance, and is flanked by identical inverted copies of IS*903*, both of which carry functional transposases capable of transposing the entire transposon or transposing themselves independently (Derbyshire *et al.*, 1987). In another well-studied example, Tn*9* (about 2500bp), the central region encodes chloramphenicol resistance and is flanked by direct repeats of IS*1*.

Transposition events involving complex transposons can be quite varied. The ISs may transpose independently or transposition of the entire region between the ISs may occur. If a composite transposon resides on a small circular replicon it could behave as two different transposons. This is because the segment of DNA that the ISs flank and the DNA surrounding the ISs become interchangeable. In addition, recombination between ISs can lead to deletion or inversion of the intervening DNA.

*Tn*3 *family of transposons*

The Tn*3* family of transposons does not have the modular structure described above for composite transposons, as none are flanked by IS-like structures. A common evolutionary link with ISs is not obvious, however, the Tn*3* family of transposons as a group shares similarities in structure and function that suggest an evolutionary link between them. For these transposons, the genes that confer the phenotype and thus identify the element are located adjacent to the transposition genes within the body of the element and not between two copies of an IS. These elements can be large (5 kb to 20 kb or more) due to the presence of accessory genes (antibiotic resistance, metal resistance, etc.) in addition to transposition functions. The inverted repeats of this diverse family of elements range from 35 to 48 bp in length and show some homology with those of Tn*3*. These inverted repeats seem to be based on a structure of 38–40 bp with some inverted repeats having an insertion or deletion. A few base pair differences may occur in the inverted repeats of a given element. Homology also exists within the base sequences of the genes involved in the transposition process (*tnpA*, *tnpR*) of this group transposons, although the order of genes within an element may differ. The *tnpA* gene encodes the transposase that catalyzes insertion of the transposon into new sites. The *tnpR* gene encodes resolvase, which has two functions, one involves recombinational events associated with transposition, and the other is repression of synthesis of TnpA. Genes *tnpA* and *tnpR* and their corresponding *cis*-acting elements, the inverted repeats, and the *res* (resolution) site are required for transposition. As a group, these transposons appear to transpose by a similar mechanism. Transposition occurs mostly by a replicative mechanism, giving a cointegrate intermediate (see below) and generating a 5-bp duplication of the target DNA sequence upon insertion. The type transposon, Tn*3,* is 5000 bp and has 38 bp inverted repeats sequences at each end. It carries three genes, one encoding β-lactamase (that confers resistance to ampicillin) as well as *tnpA* and *tnpR*. Between *tnpA* and *tnpR,* which are transcribed divergently from adjacent promoters, is a site called *res.* This 120-bp region overlaps the two promoters and is required in *cis* for transposition.

Transposable bacteriophages

Mu and D108 are two related temperate bacteriophage that use transposition as a normal mechanism to promote replication. The Mu genome is 37 kb with unequal lengths of host DNA at the two ends. Two adjacent Mu genes, *A* and *B*, are required for Mu-mediated transposition. Gene *A* encodes a large transposase protein and *B* encodes a smaller protein involved in interactions with the host DNA. During lysogeny the Mu genome inserts into the bacterial chromosome by non-replicate transposition to produce the integrated prophage DNA flanked by a 5-bp direct repeat of the host target sequence. During the lytic phase, replication of the Mu genome occurs by multiple rounds of replicative transposition yielding mainly cointegrate molecules (see below).

Other types of transposons

The transposable elements described thus far generally display little target site-specificity and move at a low frequency. Tn*7*, however, is unusual in both regards. It is a large (14 kb) genetically complex transposon that inserts with high frequency into a single specific site in the chromosome called the *att* site. However, when the *att* site is not present, Tn*7* transposons to many different sites at low frequency (reviewed by Craig, 1991).

Tn*554* is also not easily categorized. Like Tn*7,* it is genetically complex and inserts into a specific *att* site. However, it is found in gram-positive organisms and is apparently evolutionarily unrelated to Tn*7*. Another group of elements is the conjugative transposons. These gram-positive elements are large, and carry determinants for antibiotic resistance and for the conjugal transfer of the element and its donor molecule (*see* Chapter 13).

Transposition Mechanisms

In considering the mechanisms used for transposition, three types have been proposed: replicative, non-replicative, and conservative. For the replicative and non-replicative mechanisms, a staggered break is made in the target DNA, the transposon is ligated to the projecting single-stranded ends, and repair synthesis fills in the gaps. This explains the duplication of the target DNA, the direct repeats.

Replicative Transposition

With replicative transposition, the transposon sequences are copied as part of its movement (Ljungquist and Bukhari, 1977; Shapiro, 1979; Arthur and Sherratt, 1979). One copy remains at the original site, while the other inserts at the new site. Generally, this process involves a transposase and resolvase as for Tn*3*. For example, the transposase mediates a form of recombination between the donor replicon carrying the element, and the target replicon. This forms a molecule called a cointegrate, in which the donor and target DNA molecules are joined with a copy of the transposon at each junction of the original replicons. The cointegrate may then be resolved by a site-specific recombination at a region known as the *res* site. The resolvase acts at the *res* site on the transposon to efficiently bring about recombination and resolution of the cointegrate. The result of this recombination is the formation of two replicons, each of which contains the transposon. Not all replicative transposons have a resolvase; thus, host recombination enzymes may play a role in cointegrate resolution.

Non-replicative Transposition

A second mode of transposition involves a non-replicative mechanism. The transposon is physically excised from the donor DNA replicon, is inserted into the recipient molecule, and requires DNA synthesis only to fill in the gaps at the ends between the transposon and the recipient molecule (Benjamin and Kleckner, 1989). The donor replicon, now linearized by the excision of the transposon, is usually degraded. Although many ISs and transposons transpose by a single mechanism, others use both replicative and non-replicative pathways.

Conservative Transposition

Conservative transposition is another non-replicative mechanism. In this case the transposon is cut from the donor replicon and inserted into the target molecule in such a way as to preserve every nucleotide of the donor and target molecule. No additional DNA synthesis is required. This process resembles the mechanism of bacteriophage λ integration. The transposases of this type of transposon have amino acid similarity to the λ integrase family.

Regulation of Transposition

The rate of transposition of transposable elements varies considerably, although for most elements it is typically low. Excessive transposition would be detrimental to the cell, causing possible insertion mutations, genetic rearrangements, and accumulation of the elements in the DNA. Regulating the amount of transposase controls the level of transposition of some elements. In certain cases, the level of transcription and translation of the transposase is very inefficient. For IS*10,* the level of transposase is affected by both DNA adenine methylation and by an anti-sense RNA complementary to part of the transposase message (reviewed by Kleckner, 1990).

References

Arthur, A. and D. Sherratt. 1979. Dissection of the transposition process: a transposon-encoded site-specific recombination system. *Mol. Gen. Genet.* 175:267–274.

Benjamin, H. W. and N. Kleckner. 1989. Intramolecular transposition by Tn*10. Cell* 59:373–383.

Charlebois, R. L. and W. F. Doolittle. 1989. Transposable elements and genome structure in Halobacteria. In *Mobile DNA.* D. E. Berg and M. M. Howe, eds. pp. 297–307. American Society for Microbiology, Washington, D.C.

Craig, N. L. 1991. Tn7: a target site-specific transposon. *Mol. Microbiol.* 5:2569–2573.

Derbyshire, K. M., L. Hwang, and N. D. F. Grindley. 1987. Genetic analysis of the insertion sequence IS*903* transposase with its terminal inverted repeats. *Proc. Natl. Acad. Sci. USA.* 84:8049–8053.

Fiandt, M., W. Szybalski, and M. H. Malamy. 1972. Polar mutations in *lac, gal,* and phage lambda consist of a few IS-DNA sequences inserted in either orientation. *Mol. Gen. Genet.* 119:223–231.

Fleischmann, R. D., M. D. Adams, O. White, et al. (1995). Whole-genome random sequencing and assembly of *Haemophilus influenzae* Rd. *Science* 269:496–512.

Fraser, C. M., J. D. Gocayne, O. White, *et al.* 1995. The minimal gene complement of *Mycoplasma genitalium. Science* 270:397–403.

Galas, D. J. and M. Chandler. 1989. Bacterial insertion sequences. In *Mobile DNA.* D. E. Berg and M. M. Howe, eds. pp. 109–162. American Society for Microbiology, Washington, D.C.

Hirsch, H.-J., P. Starlinger, and P. Brachet. 1972. Two kinds of insertions in bacterial genes. *Mol. Gen. Genet.* 119:191–206.

Hu, S., E. Ohtsubo, N. Davidson, and H. Saedler. 1975a. Electron microscopy heteroduplex studies of sequence relations among bacterial plasmids: identification and mapping of insertion sequences IS1 and IS2 in F and R plasmids. *J. Bacteriol.* 122:764–775.

Hu, S., K. Ptashne, S. N. Cohen, and N. Davidson. 1975b. αβ sequence of F is *IS3. J. Bacteriol.* 123:687–692.

Huisman, O., P. R. Errada, L. Signou, and N. Kleckner. 1989. Mutational analysis of IS10's outside end. *EMBO J.* 8:2101–2109.

Iida, S., J. Meyer, and W. Arber. 1978. The insertion element IS1 is a natural constituent of coliphage P1 DNA. *Plasmid* 1:357–365.

Jakowec, M., P. Prentki, M. Chandler, and D. J. Galas. 1988. Mutational analysis of the open reading frames in the transposable element IS*1*. *Genetics* 120:47–55.

Jordan, E. H., H. Saedler, and P. Starlinger. 1968. 0° and strong polar mutations in the *gal* operon are insertions. *Mol. Gen. Genet.* 102:353–363.

Kleckner, N., D. Morisato, D. Roberts, and J. Bender. 1984. Mechanism and regulation of Tn*10* transposition. *Cold Spring Harbor Symp. Quant. Biol.* 49:235–244.

Kleckner, N. 1990. Regulation of transposition in bacteria. *Annu. Rev. Cell Biol.* 6:297–327.

Ljungquist, E. and A. I. Bukhari. 1977. State of prophage Mu DNA upon induction. *Proc. Natl. Acad. Sci. USA* 74:3143–3147.

Murphy, E. 1989. Transposable elements in gram-positive bacteria. In *Mobile DNA.* D. E. Berg and M. M. Howe, eds. pp. 269–288. American Society for Microbiology, Washington, D.C.

Shapiro, J. A. 1979. Molecular model for the transposition and replication of bacteriophage Mu and other transposable elements. *Proc. Natl. Acad. Sci. USA* 76:1933–1937.

5

Interspersed Repetitive Sequences in Bacterial Genomes

James Versalovic and James R. Lupski

Introduction

Prokaryotic and eukaryotic genomes contain dispersed repetitive sequences separating longer single-copy DNA sequences (Lupski and Weinstock, 1992). Various classes of repeated DNA sequences are present in diverse prokaryotic genomes. Coding sequences, such as ribosomal RNA genes (*see* Chapter 21) and insertion sequences (*see* Chapters 4 and 20), may be repeated multiple times per genome but are usually present in relatively low copy numbers. Conserved motifs or subsequences within related genes, such as those encoding transfer RNA (tRNA), may also be repeated multiple times within individual genomes.

During the past 15 years interspersed repetitive DNA sequences have been described in different microbial genomes. Interspersed repetitive sequences are characterized as relatively short (usually <500 bp), non-coding, intercistronic, and dispersed elements in bacterial genomes (Lupski and Weinstock, 1992; *see* Table 5-1). Although such compact genomes were previously thought to contain little repetitive DNA (Lewin, 1994), current data support the fact that repetitive DNA comprises a substantial portion of microbial genomes. Non-random distribution of specific repetitive sequences may reflect regulatory or structural requirements within the bacterial chromosome.

Polynucleotide Sequences and Tandem Repeats

Nucleotide sequence patterns are non-randomly distributed and presumably reflect a higher order of chromosome structure (Burge *et al.*, 1992; Cardon *et al.*, 1993; Blaisdell *et al.*, 1993; Karlin *et al.*, 1994). Relatively short polynucleotide sequence patterns may represent highly repetitive sequences in prokaryotic genomes. The most common trinucleotide, TGG (Burge *et al.*, 1992), found in

Table 5-1. Interspersed repetitive sequences in prokaryotic genomes

Sequence	Size (bp)	Organism in which originally identified	Reference(s)
A motif (BIME)	20	*E. coli* *S. typhimurium*	Gilson *et al.*, 1991
B motif (BIME)	20	*E. coli* *S. typhimurium*	Gilson *et al.*, 1991
boxA	59	*S. pneumoniae*	Martin *et al.*, 1992
boxB	45	*S. pneumoniae*	Martin *et al.*, 1992
boxC	50	*S. pneumoniae*	Martin *et al.*, 1992
Bru-RS1	103	*B. abortus*	Halling and Bricker, 1994
Bru-RS2	105	*B. abortus*	Halling and Bricker, 1994
DR	24	*M. bovis*	Doran *et al.*, 1993
Dr-rep	150–192	*D. radiodurans*	Lennon *et al.*, 1991
ERIC (IRU)	126	*E. coli* *S. typhimurium*	Hulton *et al.*, 1991 Sharples and Lloyd, 1990
$(GTG)_5$	15	*Salmonella* spp. *Shigella* spp.	Doll *et al.*, 1993
L motif (BIME)	32	*E. coli* *S. typhimurium*	Gilson *et al.*, 1991
Mx-rep	87	*M. xanthus*	Fujitani *et al.*, 1991
Ng-rep	26	*N. gonorrhoeae* *N. meningitidis*	Correia *et al.*, 1986
REP (PU)	38	*E. coli* *S. typhimurium*	Gilson *et al.*, 1984 Higgins *et al.*, 1982
RepMP1	300	*M. pneumoniae*	Wenzel and Herrmann, 1988
RepMP2	150	*M. pneumoniae*	Wenzel and Herrmann, 1988
RLEP	545–1063	*M. leprae*	Woods and Cole, 1990
SDC1	400	*M. pneumoniae*	Colman *et al.*, 1990

Escherichia coli, phage, viral, and eukaryotic genomes (Burge *et al.*, 1992) is present in the most common pentanucleotide GCTGG (Blaisdell *et al.*, 1993). GCTGG is present in the seven most commonly observed octanucleotide sequences in *Escherichia coli.* The ten most highly repetitive octanucleotide sequences have been found between 378 and 543 times in 1.6 Mb of nonredundant sequence information from *E. coli* K12 (Blaisdell *et al.*, 1993). Short sequences with known biological function may be maintained by natural selection. The nonamer, AAGTGCGGT, represents the DNA uptake signal sequence (USS) in *Haemophilus influenzae* involved in transformation and lateral DNA exchange in this species. In the 1.83 Mb *H. influenzae* genome (Fleischmann *et al.*, 1995), 1465 copies of the USS (Smith *et al.*, 1995) are distributed randomly with 734 copies in the plus strand and 731 copies in the minus strand.

Tandemly repeated polynucleotide sequences represent larger elements which are also highly repeated in bacterial genomes. The poly-trinucleotides, $(GTG)_5$ or $(GCC)_5$, are highly repetitive in the genomes of *E. coli* and its relatives *Salmonella typhimurium* and *Shigella* sp. (Doll *et al.*, 1993). Such poly-trinucleotide sequences represent interspersed repetitive sequences. Heptanucleotide repeats have also been found in bacterial genomes. Three different heptanucleotide repeats were found in the cyanobacterium *Calothrix* genome and named short tandemly repeated repetitive (STRR) sequences (Mazel *et al.*, 1990). STRR sequences were found by hybridization of genomic DNA with a restriction fragment containing DNA upstream of the *cpeBA* operon. Multiple DNA fragments were highlighted by hybridization and STRR sequences were found in different cyanobacteria. A polymorphic 10-bp direct repeat has been described in *Mycobacterium tuberculosis* and other mycobacterial genomes (Hermans *et al.*, 1992). This major polymorphic tandem repeat (MPTR) is present in clusters with each repeat separated by a 5-bp spacer. MPTR probes hybridized with 10% of clones representing a *M. tuberculosis* genomic library, indicating their presence at multiple locations on the mycobacterial chromosome. Interestingly, the MPTR sequence shares similarity with the recombination signal sequence Chi 5′-GCTGGTGG (Stahl, 1979 and *see* Chapter 6) from *E. coli* and the interspersed repetitive REP elements (Higgins *et al.*, 1982; Stern *et al.*, 1984; Gilson *et al.*, 1984).

Short Interspersed Repetitive Sequences (less than 50 bp)

The prokaryotic interspersed repetitive sequence that was first described and most intensively studied is the repetitive extragenic palindromic (REP) (Stern et al., 1984), or palindromic unit (PU) sequence (Gilson *et al.*, 1984), initially found in the Gram-negative enteric bacteria *Salmonella typhimurium* and *E. coli*. The REP sequence was identified by DNA sequence comparisons of intergenic regions of different operons (Higgins et al., 1982). The REP element consists of a 38-nucleotide consensus palindromic sequence and can form a stable stem-loop structure with a 5-bp variable loop in the center of the consensus sequence (Stern *et al.*, 1984). REP sequences are frequently present in clusters in the *E. coli* genome. By extrapolation from available genomic sequence data (Dimri *et al.*, 1992), greater than 500 individual REP sequences in 295 REP elements have been estimated to occur in the *E. coli* chromosome. This frequency of repetition corresponds to one REP element for every ten genes. Locations and distances between REP sequences vary between *E. coli* strains, leading to diverse DNA fingerprint patterns following PCR amplification with primers complementary to REP (Versalovic *et al.*, 1991) and Southern hybridization of each *E. coli* genome with a REP probe (Dimri *et al.*, 1992). Possible roles for REP sequences include regulation of gene expression by controlling mRNA stability (Newbury *et al.*, 1987; Higgins *et al.*, 1988) and stabilization of chromosome structure (Gilson *et al.*, 1990; Yang and Ames, 1988; Yang and Ames, 1990; Oppenheim *et al.*, 1993).

A high copy number 26-mer repeated DNA element is present in the genomes of *Neisseria gonorrhoeae* and *Neisseria meningitidis*. Two-dimensional S1 nuclease heteroduplex mapping identified a 26-bp repetitive sequence (*N*grep) in the genome of *N. gonorrhoeae* (Correia *et al.*, 1986). Hybridization experiments with a 21-nucleotide oligomer complementary to the consensus demonstrated the presence of this repetitive element in several strains of *N. gonorrhoeae* and *N. meningitidis*. Oligonucleotide primers complementary to *N*grep yielded complex DNA profiles of gonococcal and meningococcal strains (Versalovic and Lupski, 1995), confirming the highly repetitive nature of *N*grep.

A highly repetitive 24-bp direct repeat was isolated from the genome of *Mycobacterium bovis* (Doran *et al.*, 1993). A *M. bovis* genomic library was hybridized with genomic DNA from the same organism. Recombinant clones that yielded particularly intense hybridization signals were selected as probes for Southern hybridization experiments with *M. bovis* genomic DNA. Clones that generated identical and complex hybridization patterns were selected for characterization. A 24-bp direct repeat (DR) was isolated which clustered 38 times in a 1.4 kb region and was widely distributed throughout the *M. bovis* genome.

Larger Interspersed Repetitive Elements (greater than 50 bp)

Larger repetitive elements that occur primarily in non-coding regions and are found less frequently have been described from distantly related bacteria. One such element from enteric bacteria is the intergenic repeat unit (IRU) (Sharples and Lloyd, 1990) or enterobacterial repetitive intergenic consensus (ERIC) (Hulton *et al.*, 1991). The intact ERIC or IRU sequence is 126 bp in length though partial remnants have been described (Hulton *et al.*, 1991). The ERIC remnants contain the ends of the ERIC element minus the central inverted repeat (Sharples and Lloyd, 1990; Hulton *et al.*, 1991; Versalovic *et al.*, 1993). ERIC-like sequences have been demonstrated throughout the bacterial kingdom (Versalovic *et al.*, 1991). Like the REP sequence, ERIC elements occur in either orientation with respect to transcription and include a conserved inverted repeat. The chromosomal locations of ERIC vary between different strains and species. In analogous chromosomal locations, regardless of whether or not the IRU or ERIC sequence is present, the surrounding sequence is not disturbed, suggesting either a precise insertion or deletion (Sharples and Lloyd, 1990; Hulton *et al.*, 1991; Versalovic *et al.*, 1993). No evidence has been found for classical transposition mechanisms involving the ERIC element.

The *Mycobacterium leprae* genome contains a family of species-specific dispersed repeats named RLEP (Woods and Cole, 1990). RLEP elements possess a 545-bp central domain and flanking sequences of various sizes. Due to the variations in sizes of the terminal extensions, RLEP elements vary in size from 545 bp to at least 1063 bp. A minimum of 28 copies of the RLEP element are dispersed throughout the 2.8 Mb genome of *M. leprae,* accounting for 0.6% of the chromosome. RLEP sequences do not resemble insertion sequences since these elements lack detectable secondary structure and are not flanked by repeated DNA.

A restriction fragment containing the *ops* and *tps* genes of *Myxococcus xanthus* highlighted multiple genomic fragments when used as a probe in hybridization experiments (Fujitani *et al.*, 1991). Three percent of clones from an *M. xanthus* cosmid library hybridized with the same restriction fragment. Six clones containing this repetitive element were sequenced and revealed a consensus sequence with an 87-bp core. One of the Mx-rep elements was located downstream from *M. xanthus rpoD* gene in a position analogous to that of an *E. coli* REP element.

A highly conserved repetitive sequence was found in the radioresistant bacterium *Deinococcus radiodurans* SARK (Lennon *et al.*, 1991). Three percent of *E. coli* colonies containing *D. radiodurans* genomic DNA in the pUC18 vector hybridized with an insert containing the repetitive sequence. The repetitive element (*Dr*-rep) is variable in length ranging from 150 to 192 bp and contains dyad symmetries near each end.

Multiple repetitive elements have been described from the genome of *Mycoplasma pneumoniae.* These elements are larger in size and occur only 8 to 10 times per genome. The elements vary in size from the 150-bp RepMP2 (Wenzel and Herrmann, 1988), the 300-bp RepMP1 (Wenzel and Herrmann, 1988), and 400-bp SDC1 (Colman *et al.*, 1990) to elements that exceed 1 kilobase in size. PCR amplification with a primer (RW3A) complementary to RepMP2 demonstrated the presence of RepMP2-like elements in the genome of *Staphylococcus aureus* (Del Vecchio *et al.*, 1995), suggesting that these elements may also be conserved in diverse bacteria. Remarkably, at least 6% of the 840-kb *M. pneumoniae* genome appears to consist of repetitive DNA elements (Ruland *et al.*, 1990).

Two repeated palindromic elements have been isolated from the genomes of the obligate intracellular bacteria *Brucella abortus* and other *Brucella* species (Halling and Bricker, 1994). The elements, Bru-RS1 and Bru-RS2, are 103 bp and 105 bp in size respectively, and with one exception occur singly (not clustered) at each genomic locus. The elements are dispersed in the 3.25 Mb *Brucella* genome (Michaux *et al.*, 1993) with a copy number exceeding 35. Assuming a typical gene approximates 1000 bp in size, the Bru-RS elements occur once per less than 100 genes. The Bru-RS elements, like REP and ERIC, are palindromic, bounded by nearly perfect inverted repeats, and conserved among *Brucella* species. Bru-RS elements appear to be hotspots for insertion of IS711 (Halling *et al.*, 1993) since several copies of Bru-RS elements are adjacent to a IS711 element (Halling *et al.*, 1993).

Mosaic Repetitive Elements

Interspersed repetitive elements may be clustered in complex mosaic structures that include combinations of different repetitive elements. Bacterial Interspersed

Mosaic Elements (BIME) were described (Gilson *et al.*, 1991a; Gilson *et al.*, 1991b) following sequence analysis of regions containing clusters of REP, or palindromic unit (PU), elements. Approximately 500 of such mosaic elements were estimated in the *E. coli* genome. Such BIME elements were conserved in other Gram-negative enterobacteria such as *Klebsiella pneumoniae* (Bachellier *et al.*, 1993). Each BIME element was exclusively composed of combinations of REP and seven other repetitive motifs. Two major BIME families have been defined based on structural differences (Bachellier *et al.*, 1994). The BIME-1 family is highly conserved in sequence and length (145 bp), consist of two specific REPs, and contains the highly repetitive L motif. The BIME-2 family is variable in length and contains a variable number of two specific REP elements. Different functions for each type of BIME element have been proposed based on the differential affinity of gyrase for each REP element and the presence of the L sequence. Based on the combination of REPs present, BIME-2 elements would be expected to bind to DNA gyrase with greater affinity. The L sequence contains IHF-binding sites and confers IHF binding ability to BIME-1 elements. A subset of BIME elements, known as repetitive IHF-binding palindromic (RIP) elements, is approximately 100 bp in length and includes one L sequence surrounded by individual REP elements (Oppenheim *et al.*, 1993). The L sequence within each RIP element binds specifically with integration host factor (IHF) and may mediate DNA "looping" or rearrangements *in vivo.* Extrapolations suggest that approximately 70 RIP elements are present in the *E. coli* genome in a near random pattern. Such functional diversity based on structural differences in mosaic repetitive elements may have important implications for bacterial genome structure and function.

Other examples of mosaic elements include the REP MP 1 repetitive mosaic element and the BOX element. Rep MP 1 exists as a 300-bp core element flanked by shorter repetitive sequences in the genome of *Mycoplasma pneumoniae* (Forsyth and Geary, 1996). BOX elements are modular in nature and represent the first interspersed repetitive element from a Gram-positive organism (*Streptococcus pneumoniae*) (Martin *et al.*, 1992). Three different subunits were originally defined within BOX elements, namely boxA (59 bp), boxB (45 bp), and boxC (50 bp) (Martin *et al.*, 1992). BOX elements are mosaic combinations of repetitive subunits. Examples of elements with boxB alone, boxA and multiple boxB subunits, and the complete combination of boxA, boxB, and boxC are all present within the *S. pneumoniae* genome. Oligonucleotide probes complementary to BOX subunit sequences demonstrated widespread conservation of the boxA subunit among diverse bacteria (Koeuth *et al.*, 1995). Sequences similar to the boxB and boxC subunits were found only in *S. pneumoniae* (Koeuth et al., 1995). The boxA subunit exists independently of boxB and boxC in diverse bacterial genomes. Organisms unrelated to Gram-positive pathogens such as *E. coli* and *Salmonella typhimurium* contain multiple copies of boxA-like subsequences dispersed in their respective genomes. The BOX subunits appear to have evolved independently of one another within the *S. pneumoniae* genome and in evolutionarily diverse prokaryotic chromosomes.

Repetitive Sequences and Chromosomal Rearrangements

Repetitive elements provide substrates for intrachromosomal recombination events (*see also* Chapter 11). Such recombination events may be important for the production of genetic diversity and may yield biological consequences such as changes in toxin production or pathogenicity. REP elements have been associated with chromosomal duplication events in *Salmonella typhimurium* (Shyamala *et al.*, 1990). Tandem duplications involving the *S. typhimurium hisD* gene were characterized in strains with or without RecA (Shyamala *et al.*, 1990). Joinpoints of duplications were demonstrated within REP in both $RecA^+$ and $RecA^-$ strains. All $RecA^+$ duplications occurred by recombination between REP sequences. Amplification and deletion events involving the *Vibrio cholerae* toxin genes *ctxAB* depended on the integrity of the homologous recombination system (Goldberg and Mekalanos, 1986). Chromosomal rearrangements presumably occurred by unequal crossing over between 2.7-kb RS1 repetitive sequences flanking the *ctxAB* locus.

The presence of a BOX element downstream of the *S. pneumoniae glpF* gene conferred high frequency phase variation (Saluja and Weiser, 1995). Phase variation occurred between transparent and opaque colony phenotypes and was manifested by changes in colony morphology. Transparent pneumococcal variants were able to efficiently colonize the nasopharynx of infant rats (Weiser *et al.*, 1994a; Weiser *et al.*, 1994b). Transformation of a recipient cell lacking the BOX element by donor DNA containing the BOX element downstream of *glpF* increased the frequency of transformation from 1.4×10^{-6} to 2.7×10^{-3}. Presumably the presence of the BOX element affects a putative regulatory sequence downstream of the element which affects the frequency of phase variation and colonization potential. The exact nature of this regulatory sequence is currently unknown, though it is thought to encode a *trans* acting element.

Summary and Conclusions

Bacterial genomes contain a fascinating variety of interspersed repetitive elements. These elements vary in size from 15 bp to several hundred bp, do not encode proteins, and are dispersed at different locations within each bacterial chromosome. Repetitive elements such as REP and ERIC are present in high copy number in enteric bacteria, but similar elements are present in diverse bacteria. Other elements such as RLEP appear to be specific to a particular genus such as *Mycobacterium.* In addition to variations in size and the extent of sequence conservation, relative copy numbers also vary significantly. REP elements are present several hundred times in the *E. coli* genome, whereas the RepMP1 element is present fewer than ten times in the *M. pneumoniae* genome. Several interspersed repetitive elements possess secondary structure by virtue of inverted repeats present within each element (e.g. REP, ERIC, Bru-RS1, Bru-RS2). Other elements lack discernible secondary structure (e.g. *Ng*rep, RLEP).

Though bacterial genomes are smaller and more compact than their eukaryotic counterparts, prokaryotic chromosomes are not simple amalgamations of coding sequences with regulatory elements lying between genes. Bacterial chromosomes must be folded and replicated within intact cells and information contained in interspersed repetitive sequences may provide the "genomic code" necessary for proper chromosomal structure and function *in vivo.* The complete genomic sequences of different bacterial organisms, like that recently obtained from *H. influenzae* Rd (Fleischmann *et al.*, 1995), will provide a wealth of DNA sequence information to analyze repetitive sequence elements, evaluate the roles of repetitive sequences in genome structure, and perform comparative genomics.

Acknowledgments

JV acknowledges the support of the NIH Medical Scientist Training Program at Baylor College of Medicine. JRL acknowledges support from the Departments of Molecular and Human Genetics and Pediatrics at Baylor College of Medicine.

References

Bachellier, S., D. Perrin, M. Hofnung, and E. Gilson. 1993. Bacterial interspersed mosaic elements (BIMEs) are present in the genome of *Klebsiella. Mol. Microbiol.* 7:537–544.

Bachellier, S., W. Saurin, D. Perrin, M. Hofnung, and E. Gilson. 1994. Structural and functional diversity among bacterial interspersed mosaic elements (BIMEs). *Mol. Microbiol.* 12:61–70.

Blaisdell, B. E., K. E. Rudd, A. Matin, and S. Karlin. 1993. Significant dispersed recurrent DNA sequences in the *Escherichia coli* genome. *J. Mol. Biol.* 229:833–848.

Burge, C., A. M. Campbell, and S. Karlin. 1992. Over- and under-representation of short oligonucleotides in DNA sequences. *Proc. Natl. Acad. Sci. USA* 89:1358–1362.

Cardon, L. R., C. Burge, G. A. Schachtel, B. E. Blaisdell, and S. Karlin. 1993. Comparative DNA sequence features in two long *Escherichia coli* contigs. *Nucl. Acids Res.* 21:3875–3884.

Colman, S. D., P. C. Hu, and K. F. Bott. 1990. Prevalence of novel repeat sequences in and around the P1 operon in the genome of *Mycoplasma pneumoniae. Gene* 87:91–96.

Correia, F. F., S. Inouye, and M. Inouye. 1986. A 26-base-pair repetitive sequence specific for *Neisseria gonorrhoeae* and *Neisseria meningitidis* genomic DNA. *J. Bacteriol.* 167:1009–1015.

Del Vecchio, V. G., J. M. Petroziello, M. J. Gress, F. K. McCleskey, G. P. Melcher, H. K. Crouch, and J. R. Lupski. 1995. Molecular genotyping of methicillin-resistant *Staphylococcus aureus* via fluorophore-enhanced repetitive-sequence PCR. *J. Clin. Microbiol.* 33:2141–2144.

Dimri, G. P., K. E. Rudd, M. K. Morgan, H. Bayat, and G. F.-L. Ames. 1992. Physical mapping of repetitive extragenic palindromic sequences in *Escherichia coli* and phylogenetic distribution among *Escherichia coli* strains and other enteric bacteria. *J. Bacteriol.* 174:4583–4593.

Doll, L., S. Moshitch, and G. Frankel. 1993. Poly$(GTG)_5$-associated profiles of *Salmonella* and *Shigella* genomic DNA. *Res. Microbiol.* 144:17–24.

Doran, T. J., A. L. Hodgson, J. K. Davies, and A. J. Radford. 1993. Characterisation of a highly repeated DNA sequence from *Mycobacterium bovis.* FEMS Microbiol. Lett. 111:147–152.

Fleischmann, R. D., M. D. Adams, O. C. R. White, E. F. Kirkness, A. R. Kerlavage, C. J. Bult, J.-F. Tomb, and B. A. Dougherty. 1995. Whole-genome random sequencing and assembly of *Haemophilus influenzae* Rd. *Science* 269:496–512.

Forsyth, M. H. and S. J. Geary. 1996. The repetitive element Rep MP 1 of *Mycoplasma pneumoniae* exists as a core element within a larger, variable repetitive mosaic. *J. Bacteriol.* 178:917–921.

Fujitani, S., T. Komano, and S. Inouye. 1991. A unique repetitive DNA sequence in the *Myxococcus xanthus* genome. *J Bacteriol.* 173:2125–2127.

Gilson, E., J. M. Clement, D. Brutlag, and M. Hofnung. 1984. A family of dispersed repetitive extragenic palindromic DNA sequences in *E. coli. EMBO J.* 3:1417–1421.

Gilson, E., D. Perrin, and M. Hofnung. 1990. DNA polymerase I and a protein complex bind specifically to *E. coli* palindromic unit highly repetitive DNA: implications for bacterial chromosome organization. *Nucleic Acids Research* 18:3941–3952.

Gilson, E., W. Saurin, D. Perrin, S. Bachellier, and M. Hofnung. 1991a. The BIME family of bacterial highly repetitive sequences. *Res. Microbiol.* 142:217–222.

Gilson, E., W. Saurin, D. Perrrin, S. Bachellier, and M. Hofnung. 1991b. Palindromic units are part of a new bacterial interspersed mosaic element (BIME). *Nucl. Acids Res.* 19:1375–1383.

Goldberg, I. and J. J. Mekalanos. 1986. Effect of a *recA* mutation on cholera toxin gene amplification and deletion events. *J. Bacteriol.* 165:723–731.

Halling, S. M., F. M. Tatum, and B. J. Bricker. 1993. Sequence and characterization of an insertion sequence. IS711, from *Brucella ovis. Gene* 133:123–127.

Halling, S. M. and B. J. Bricker. 1994. Characterization and occurrence of two repeated palindromic DNA elements of *Brucella* spp.: Bru-RS1 and Bru-RS2. *Mol. Microbiol.* 14:681–689.

Hermans, P. W. M., D. van Soolingen, and J. D. A. van Embden. 1992. Characterization of a major polymorphic tandem repeat in *Mycobacterium tuberculosis* and its potential use in the epidemiology of *Mycobacterium kansasii* and *Mycobacterium gordonae. J. Bacteriol.* 174:4157–4165.

Higgins, C. F., G.F.L. Ames, W. M. Barnes, J. M. Clement, and M. Hofnung. 1982. A novel intercistronic regulatory element of prokaryotic operons. *Nature* 298:760–762.

Higgins, C. F., R. S. McLaren, and S. F. Newbury. 1988. Repetitive extragenic palindromic sequences, mRNA stability and gene expression: evolution by gene conversion?—a review. *Gene* 72:3–14.

Hulton, C. S. J., C. F. Higgins, and P. M. Sharp. 1991. ERIC sequences: a novel family of repetitive elements in the genomes of *Escherichia coli, Salmonella typhimurium* and other enterobacteria. *Molecular Microbiology* 5:825–834.

Karlin, S., I. Ladunga, and B. E. Blaisdell. 1994. Heterogeneity of genomes: measures and values. *Proc. Natl. Acad. Sci. USA* 91:12837–12841.

Koeuth, T., J. Versalovic, and J. R. Lupski. 1995. Differential subsequence conversation of interspersed repetitive *Streptococcus pneumoniae* BOX elements in diverse bacteria. *Genome Research* 5:408–418.

Lennon, E., P. D. Gutman, H. Yao, and K. W. Minton. 1991. A highly conserved repeated chromosomal sequence in the radioresistant bacterium *Deinococcus radiodurans* SARK. *J. Bacteriol.* 173:2137–2140.

Lewin, B. 1994. *Genes* V (Oxford: Oxford University Press).

Lupski, J. R. and G. M. Weinstock. 1992. Short, interspersed repetitive DNA sequences in prokaryotic genomes. *J Bacteriol.* 174:4525–4529.

Martin, B., O. Humbert, M. Camara, E. Guenzi, J. Walker, T. Mitchell, P. Andrew, M. Prudhomme, G. Alloing, R. Hakenbeck, D. A. Morrison, G. J. Boulnois, and J.-P. Claverys. 1992. A highly conserved repeated DNA element located in the chromosome of *Streptococcus pneumoniae. Nucl. Acids Res.* 20:3479–3483.

Mazel, D., J. Houmard, A. M. Castets, and N. Tandeau de Marsac. 1990. Highly repetitive DNA sequences in cyanobacterial genomes. *J. Bacteriol.* 172:2755–2761.

Michaux, S., J. Paillisson, M. J. Charles-Nurit, G. Bourg, A. Allardet-Servent, and M. Ramuz. 1993. Presence of two independent chromosomes in the *Brucella melitensis* 1.6M genome. *J. Bacteriol.* 175:701–705.

Newbury, S. F., N. H. Smith, E. C. Robinson, I. D. Hiles, and C. F. Higgins. 1987. Stabilization of translationally active mRNA by prokaryotic REP sequences. *Cell* 48:297–310.

Oppenheim, A. B., K. E. Rudd, I. Mendelson, and D. Teff. 1993. Integration host factor binds to a unique class of complex repetitive extragenic DNA sequences in *Escherichia coli. Mol. Microbiol.* 10:113–122.

Ruland, K., R. Wenzel, and R. Herrmann. 1990. Analysis of three different repeated DNA elements present in the P1 operon of *Mycoplasma pneumoniae:* size, number, and distribution on the genome. *Nucl. Acids Res.* 18:6311–6317.

Saluja, S. K. and J. N. Weiser. 1995. The genetic basis of colony opacity in *Streptococcus pneumoniae:* evidence for the effect of box elements on the frequency of phenotypic variation. *Mol. Microbiol.* 16:215–227.

Sharples, G. J. and R. G. Lloyd. 1990. A novel repeated DNA sequence located in the intergenic regions of bacterial chromosomes. *Nucleic Acids Research* 18:6503–6508.

Shyamala, V., E. Schneider, and G.F.L. Ames. 1990. Tandem chromosomal duplications: role of REP sequences in the recombination event at the join-point. *EMBO J.* 9:939–946.

Smith. H. O., J.-F. Tomb, B. A. Dougherty, R. D. Fleischmann, and J. C. Vender. 1995. Frequency and distribution of DNA uptake signal sequences in the *Haemophilus influenzae* Rd genome. *Science* 269:538–540.

Stahl, F. W. 1979. Special sites in generalized recombination. *Ann. Rev. Genet.* 13:7–24.

Stern, M. J., G.F.L. Ames, N. H. Smith, E. C. Robinson, and C. F. Higgins. 1984. Repetitive extragenic palindromic sequences: A major component of the bacterial genome. *Cell* 37:1015–1026.

Tompkins, L. S., N. Troup, A. Labigne-Roussel, and M. L. Cohen. 1986. Cloned, random chromosomal sequences as probes to identify *Salmonella* species. *J. Infect. Dis.* 154:156–162.

Versalovic, J., T. Koeuth, and J. R. Lupski. 1991. Distribution of repetitive DNA sequences in eubacteria and application to fingerprinting of bacterial genomes. *Nucleic Acids Research* 19:6823–6831.

Versalovic, J., T. Koeuth, R. Britton, K. Geszvain, and J. R. Lupski. 1993. Conservation and evolution of the *rpsu-dnaG*-rpoD macromolecular synthesis operon. *Mol. Microbiol.* 8:343–355.

Versalovic, J. and J. R. Lupski. 1995. DNA fingerprinting of *Neisseria* strains by rep-PCR. *Meth. Mol. Cell Biol.* 5:96–104.

Weiser, J. N., R. Austrian, P. K. Sreenivasan, and H. R. Masure. 1994a. Phase variation in pneumococcal opacity: relationship between colonial morphology and nasopharyngeal colonization. *Infect. Immun.* 62:2582–2589.

Weiser, J. N., E. I. Tuomanen, D. R. Cundell, P. K. Sreenivasan, R. Austrian, and H. R. Masure, 1994b. The effect of colony opacity variation on pneumococcal colonization and adhesion. *J Cell Biochem.* 18A:S55.

Wenzel, R. and R. Herrmann. 1988. Repetitive DNA sequences in *Mycoplasma pneumoniae. Nucl. Acids Res.* 16:8337–8350.

Woods, S. A. and S. T. Cole. 1990. A family of dispersed repeats in *Mycobacterium leprae. Mol. Microbiol.* 4:1745–1751.

Yang, Y. and G.F.L. Ames. 1988. DNA gyrase binds to the family of prokaryotic repetitive extragenic palindromic sequences. *Proc. Natl. Acad. Sci. USA* 85:8850–8854.

Yang, Y. and G.F.L. Ames. 1990. The family of repetitive extragenic palindromic sequences: Interaction with DNA gyrase and histonelike protein HU. In *The Bacterial Chromosome.* K. Drlica, ed. pp. 211–225. American Society of Microbiology, Washington, D.C.

6

Chi Sites and Their Consequences

Gerald R. Smith

Chi sites are octameric nucleotide sequences in DNA that stimulate the RecBCD pathway of homologous recombination in *Escherichia coli.* Stimulation is maximal at the Chi site, decreases approximately a factor of two for each 2–3 kb to one side, but is insignificant to the other side of Chi. Chi stimulates recombination by interaction with RecBCD enzyme, which has multiple enzymatic activities and multiple physiological roles in recombination, repair, and replication. Chi appears to be active throughout the enteric bacteria; other nucleotide sequences may similarly interact with RecBCD-like enzymes in other bacteria. This chapter reviews the properties of Chi, its interaction with RecBCD enzyme, and its occurrence on the chromosome of *E. coli* and other organisms.

A Brief History of Chi

Chi was first indicated by mutations that enhance the plaque-size of bacteriophage λ Red$^-$ Gam$^-$ mutants (Henderson and Weil, 1975). Packaging of λ requires concatemeric DNA, which can arise either by rolling circle replication or by recombination. The λ Gam protein inhibits RecBCD enzyme; in its absence RecBCD enzyme blocks rolling circle replication. The λ Red proteins promote homologous recombination; in their absence recombination is limited to that by the *E. coli* RecBCD pathway. This pathway is feeble on λ Red$^-$ Gam$^-$ phage, thus packaging little of the λ DNA and forming only small plaques. (For further discussion of these views, see Smith, 1983.) Better-growing mutants arise in λ Red$^-$ Gam$^-$ stocks propagated on *E. coli* $recB^+C^+D^+$ cells; the responsible mutations, called χ^+, have been found at χA, χB, χC, and χD (Stahl, Crasemann, and Stahl, 1975).

Stahl and his colleagues showed that the χ^+ mutations create a site, called Chi, that enhances homologous recombination by the RecBCD pathway but not by other pathways of recombination (the λ Red and *E. coli* RecE and RecF pathways of homologous recombination and the λ Int pathway of site-specific recombination) (Stahl and Stahl, 1977). Stimulation is maximal at or near Chi and is greater to the left of Chi than to its right (relative to the conventional λ genetic map) (Stahl et al., 1980). Stimulation decreases approximately exponentially to the left, a factor of about 2 per 2-3 kb, and precipitously to the right (Cheng and Smith, 1989) (Figure 6-1a). An active Chi site is largely inactivated when it is inverted in λ (Faulds *et al.*, 1979). Chi's activity depends upon its orientation with respect to the λ *cos* site, at which the λ terminase protein cuts λ DNA to generate the ends of mature (packaged) λ DNA (Kobayashi et al., 1982; Kobayashi, Stahl and Stahl, 1984). Packaging begins at the left end of λ DNA, leaving the right end exposed to a factor (RecBCD enzyme; see section 4) that travels along the DNA, recognizes a properly oriented Chi site, and promotes recombination downstream of it.

Nucleotide Sequence of Chi and its Variants

The nucleotide sequence of Chi was shown to be 5′ G-C-T-G-G-T-G-G 3′ (or its complement or the duplex) by a comparison of the sequences of active and inactive Chi sites and their flanking sequences (reviewed by Smith *et al.*, 1984, with references). Three sites of mutation in λ (χB, χC, and χD) and three in the plasmid pBR322 inserted into λ (χE, χF, and χG) were sequenced from both the active mutant (χ^+) and inactive wild-type (χ°) alleles. In each case the χ^+ allele contains the sequence 5′ G-C-T-G-G-T-G-G 3′, and the χ^+ mutation created one or another of these nucleotides Figure 6-2. A Chi site in the *E. coli lacZ* gene, designated *chi*$^+$*lacZ*, also has this sequence, and a mutation inactivating it occurred within the octamer. As expected from the orientation dependence of Chi in λ, the Chi octamer in each case is oriented in the same direction (left to right as written here and on the conventional λ genetic map). These seven Chi sites are active with *cos* in the wild-type orientation. Chemical mutagenesis of λ Red$^-$ Gam$^-$ $\chi^+ C$ phage produced small-plaque derivatives with a partially or essentially completely inactivated Chi site; in each of nine independent $\chi^- C$ mutations analyzed one of four nucleotides in the Chi octamer was changed (Figure 6-2).

Comparison of the sequences at these seven loci reveals little in common other than the Chi octamer; all analyzed mutations creating or inactivating Chi occur within this octamer (Figure 6-2). Wild-type λ and pBR322 lack the Chi octamer and, as far as one can tell from plaque size and recombination of λ derivatives, lack active Chi sites. Insertion of the Chi octamer flanked by *Hin*dIII linkers into λ creates a Chi hotspot (H. Coleman, C. Roberts, K. Helde, and G. Smith, unpublished observations). Insertion of the octamer into pBR322 creates a site acted upon by RecBCD enzyme and indistinguishable from other Chi sites (Dixon

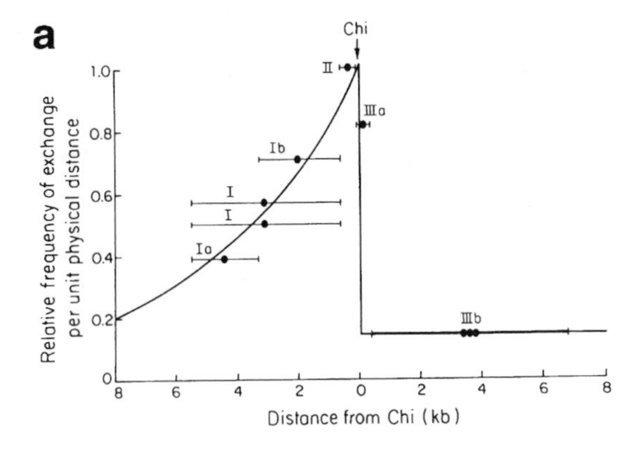

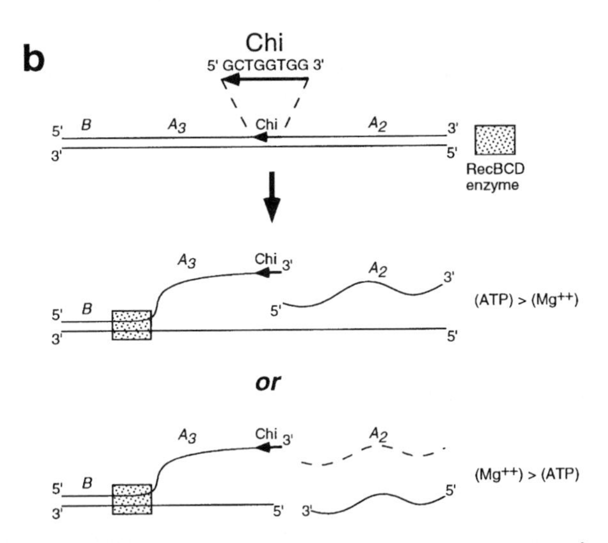

Figure 6-1. (a) Localized stimulation of recombination by Chi in phage λ crosses. I, Ia, *etc.* are genetic intervals bounded by markers located the indicated distance from a Chi site in λ. Solid circles indicate the midpoints of each interval and the frequency of recombinants per physical length of that interval, normalized to interval II = 1. From Cheng and Smith, 1989. (b) Action of purified RecBCD enzyme at Chi. With (ATP) > (Mg^{++}), RecBCD enzyme unwinds ds DNA, nicks the upper strand 4–6 nucleotides to the right of Chi, and continues unwinding (Taylor *et al.*, 1985). With (Mg^{++}) > (ATP), RecBCD enzyme degrades the upper strand up to, but not beyond, Chi (Dixon and Kowalczykowski, 1993); the lower strand is cut near Chi and is unwound beyond Chi (Taylor and Smith, 1995b; information cited in Dixon and Kowalczykowski, 1995). Both conditions produce a ss DNA fragment with a 3′ end near Chi and extending to its left. This fragment is a potential substrate for homologous pairing by the RecA and SSB proteins (see text and Figure 6-3). For simplicity the loop-tail and twin loop DNA structures (Figure 6-3b and c) have been omitted. Reprinted with permission from Smith et al. (1995).

Chi locus	Location of site[a]	Sequence of *l* strand in λ
χ^+B	λ *xis-exo* interval (bp 30529-30536)	5' GGCAGATATA GCTGGTGG TTCAGGCGGC 3'
χ^+C	λ *c*II gene (bp 38481-38488)	5' TCGCAGATCA GCTGGTGG AAGAGGGACT 3'
χ^+D	λ *Q-S* interval (bp 45025-45032)	5' CTTCGTGAAA GCTGGTGG CAGGAGGTCG 3'
χ^+E	pBR322 *tet* gene (bp 983-990)	5' GCGACGCGAG GCTGGTGG CCTTCCCCAT 3'
χ^+F	pBR322 (bp 1492-1499)	5' ACCCGGCTAG GCTGGTGG GGTTGCCTTA 3'
χ^+G	pBR322 (bp 3061-3068)	5' ACAAACCACC GCTGGTGG CGGTGGTTTT 3'
chi[+]*lacZ*	*E. coli lacZ* gene (bp 2229-2222)	5' AATCCATTTC GCTGGTGG TCAGATGCGG 3'

Chi 5' GCTGGTGG 3'

↓↓↓↓ ↓↓

Mutations affecting Chi CTAA CA

A G

Figure 6-2. Nucleotide sequences of Chi sites. Nucleotide sequences are aligned with respect to the 8-bp Chi sequence enclosed in the box. Underlined nucleotides indicate mutations creating active Chi sites; bold nucleotides indicate sites of mutations inactivating Chi sites. Two independent nucleotide changes have created χ^+B: a C→G transversion and a single base-pair deletion of C at the underlined position. The mutations creating χ^+E delete an A between the two underlined nucleotides. For references to the sequence analyses see Smith *et al.* (1984). Modified and reproduced with permission from Smith *et al.* (1984).

[a] The location of the 8-bp Chi sequence is given according to the conventional numbering of the nucleotide sequence of λ (Sanger *et al.*, 1982), pBR322 (Sutcliffe, 1979) and the *E. coli lacZ* gene (Kalnins *et al.*, 1983). The sequence shown for *chi*[+]*lacZ* is that of the *l* strand of λp*lac5*, in which Chi is present in the active orientation; this sequence is complementary to that of the *lacZ* mRNA.

and Kowalczykowski, 1991; A. Taylor and A. Eisen, personal communication). Synthetic DNA containing the Chi octamer and various flanking sequences alters the replication and degradation of plasmids in *E. coli* (see below). Thus, the Chi octamer appears to be necessary and sufficient (with other factors described below) for Chi activity.

Although wild-type λ lacks a fully active Chi hotspot (or Chi octamer) in either orientation, it does contain closely related sequences. There are two copies each of 5′ A-C-T-G-G-T-G-G 3′ and 5′ G-*T*-T-G-G-T-G-G 3′, which have about 5% and 10% respectively of the activity of fully active Chi (Cheng and Smith, 1984). These sequences can account for the low level of recombination of λ χ° Red^- Gam^- by the RecBCD pathway. Notably, λ does not contain the sequence 5′ G-C-T-*A*-G-T-G-G 3′, which has about 40% of the activity of fully active Chi. This sequence appears to be underrepressed in the *E. coli* genome also (see next section).

Distribution of Chi on the *E. coli* Chromosome

Early in the study of Chi, it became clear that Chi was present in the *E. coli* genome, since λ transducing phages such as λ *bio* have active Chi sites (McMilin, Stahl, and Stahl, 1974). By studying λ phage with *E. coli Eco*RI inserts, Malone *et al.* (1978) and Faulds *et al.* (1979) deduced that Chi occurs about once per 5 kb on the average. *E. coli* nucleotide sequences in EcoSeq2 have 378 Chi sites in 1.6 mb, about one Chi per 4.2 kb on the average, or ~1000 Chi sites per genome (Blaisdell *et al.*, 1993). Chi is the tenth most frequent among the 32,768 ($\frac{1}{2} \times 4^8$) possible 8 bp sequences. Random association of nucleotides in 50% G+C DNA, as in *E. coli,* would predict an octamer to occur once per 33 kb. In addition to its higher than random frequency, Chi is preferentially oriented, within analyzed regions, such that in about 85% of the cases recombination would be stimulated in the shorter arc of the chromosome between Chi and the origin of replication (*oriC*); Chi sites are also particularly dense around *oriC*; about 1 per 1.5 kb (Burland *et al.,* 1993).

The basis of these biases is not entirely clear. An attractive possibility is that Chi has been preserved at high density and with its preferential orientation in order to restore a broken replication fork (Kuzminov, Schabtach, and Stahl, 1994) (see below for discussion). However, without the preferential bias an active Chi site would be located, on the average, no farther than twice the distance from the break as it is with the bias. Another possibility is that the bias stems from the preferential orientation of transcription away from *oriC* (Brewer, 1988) and the occurrence of frequently used codons (such as 5′ C-T-G 3′, for leucine) within Chi (Triman, Chattoraj, and Smith, 1982). It has been suggested that the action of the *E. coli* very short patch (VSP) repair system within 5′ C-C-A-G-G 3′ (or related sequences) helps maintain the high frequency of Chi and the low frequency of 5′ G-C-T-A-G-T-G-G 3′ (A. Taylor, personal communication and cited in Lieb and Rehmat 1995; *see* Chapter 12).

Interaction of Chi with RecBCD Enzyme

The first suggestion that Chi interacts with RecBCD enzyme came from the specificity of Chi's stimulation of recombination by the RecBCD pathway (Gillen and Clark, 1974; Stahl and Stahl, 1977); RecBCD enzyme is the only known enzyme specific to that pathway. Further evidence came from special (presumably missense) mutations in *recB* and *recC* that reduce or abolish Chi stimulation, while leaving some activities of the enzyme intact (Schultz, Taylor, and Smith, 1983; Lundblad *et al.*, 1984). A direct demonstration came from the Chi-dependent DNA strand-cleavage by purified RecBCD enzyme and the lack of this activity in extracts of the special *recB* and *recC* mutants but its presence in wild-type extracts (Ponticelli et al., 1985).

RecBCD enzyme, also called exonuclease V (EC 3.1.11.5), is a multifunctional enzyme containing three polypeptides encoded by three genes: *recB* (134 kDa), *recC* (129 kDa), and *recD* (67 kDa); its activities can be broadly classed as DNA unwinding and DNA degradation (reviewed by Taylor, 1988). The enzyme has a high affinity for double-stranded (ds) DNA ends ($K_D \sim 10^{-10}$ *M;* Taylor and Smith, 1995a). In the presence of Mg^{++} and ATP, which it hydrolyzes, the enzyme rapidly moves along the DNA (~300 bp per sec) generating single-stranded (ss) DNA loops and tails (diagrammed in Figure 6-3b and c). The enzyme can also hydrolyze ds or ss linear DNA to oligonucleotides; it has no detectable activity on ds circular DNA and only a weak endonuclease activity on ss circular DNA.

There are two modes of action of RecBCD enzyme at Chi, depending upon the reaction conditions. With (ATP) > (Mg^{++}) (*i.e.*, with little free Mg^{++} not

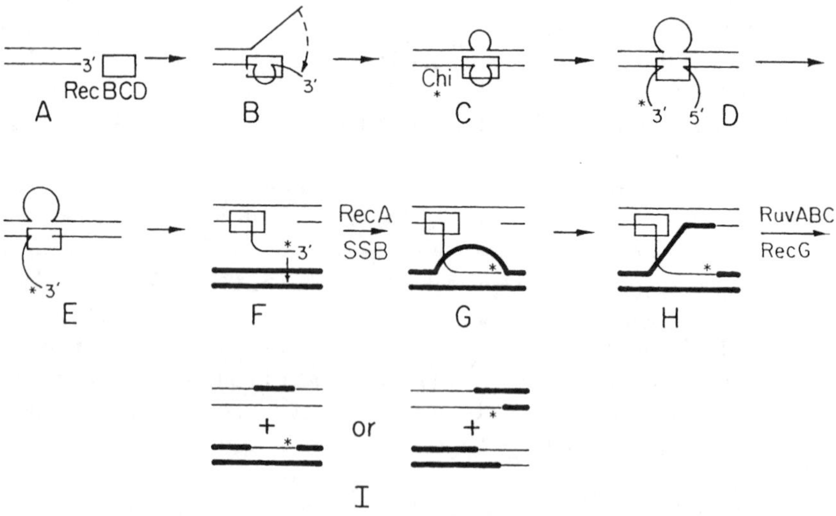

Figure 6-3. Model for Chi stimulated recombination by the RecBCD pathway of *E. coli* acting on phage λ. RecBCD enzyme (open rectangle) initiates DNA unwinding at a ds DNA end of one parent (thin lines) (a). Unwinding produces a loop-tail structure (b); annealing of the ss DNA tails produces a twin-loop structure (c). Upon encountering a properly oriented Chi site (*) RecBCD enzyme cuts one strand to produce linear ss DNA with 5′ G-C-T-G-G-T-G-G 3′ near its 3′ end (d). Continued unwinding elongates this tail (e, f), which is a substrate for homologous pairing with the second parental DNA (thick lines), promoted by RecA and SSB proteins (g). Cleavage of the displaced D-loop by an unspecified enzyme, and a second homologous pairing, produce a Holliday junction (h), which is resolved by some combination of the RuvA, B, C, and RecG proteins into a reciprocal pair of "patch" (l, left) or "splice" (l, right) recombinants. Modified and reproduced with permission from Smith *et al.* (1981).

chelated by ATP) the enzyme's Chi-independent ("general") nuclease activity is low. Under this condition the enzyme unwinds DNA, nicks one strand (that containing 5′ G-C-T-G-G-T-G-G 3′) about five nucleotides to the 3′ side of this sequence, and continues unwinding the DNA (Taylor *et al.*, 1985) (Figure 6-1b, middle). Nicking occurs only during unwinding and only as the enzyme unwinds the DNA from right to left, as written here. Upon cutting at Chi, RecBCD enzyme loses its ability to cut at another Chi site on the same DNA molecule, as well as the ability to unwind a separate DNA molecule (or to cut it at Chi) (Taylor and Smith, 1992).

With (Mg^{++}) > (ATP) (i.e., with excess, unchelated Mg^{++}) the enzyme's general nuclease activity is high. Under this condition the enzyme degrades the DNA, preferentially the strand with a 3′ end at the terminus attacked; degradation proceeds up to Chi and is then attenuated, but DNA unwinding continues (Dixon and Kowalczykowski, 1993). The complementary strand is frequently cut near or within the Chi octamer (Taylor and Smith, 1995b; information cited in Dixon and Kowalczykowski, 1995) (Figure 6-1b, bottom).

These two reactions have both distinctive and common features. Under the first condition [(ATP) > (Mg^{++})], Chi stimulates a nuclease activity and only one strand is cut, whereas under the second condition [(Mg^{++}) > (ATP)], Chi attenuates a nuclease activity and both strands are cut. Under both conditions, however, RecBCD enzyme produces ss DNA with Chi near its 3′ end and extending to the left ("downstream") of Chi (Figure 6-1b); this "Chi tail" is a potent substrate for homologous pairing and strand exchange promoted by RecA and SSB proteins (Dixon and Kowalczykowski, 1991) (Figure 6-3f). Also, under both conditions, the enzyme loses the ability to act at a second Chi site; this property is hypothesized to lead to a single recombinational exchange near each end of a linear DNA fragment bearing Chi sites (Taylor and Smith, 1992) (see next section).

It is unclear which of the two reaction conditions discussed here more nearly reflects that in *E. coli* cells. The behavior inside *E. coli* of mutant RecBCD enzymes, RecBCD-like enzymes from other bacteria (see below), and partially active Chi-like sites closely parallels their behavior outside *E. coli* with (ATP) > (Mg^{++}) (Figure 6-1b, middle) (Smith, 1987; Amundsen *et al.*, 1990). This observation supports the "nick at Chi" reaction inside *E. coli.* A similar study with (Mg^{++}) > (ATP) has not, to my knowledge, been reported with this complete set of mutants and RecBCD-like enzymes (but see Eggleston and Kowalczykowski, 1993; Dixon and Kowalczykowski, 1995 for three examples). The parallel effects of Chi on DNA degradation by RecBCD enzyme both inside and outside *E. coli* supports the "attenuation at Chi" reaction (Figure 6-1b, bottom) (see below).

The change of RecBCD enzyme at Chi has not, to my knowledge, been directly demonstrated, but it has been hypothesized to be the release of the RecD subunit (Thaler *et al.*, 1988). *recD* nonsense and deletion mutants lack RecBCD nuclease and Chi-cutting activities, yet they are recombination-proficient; in *recD* mutants λ Red$^-$ Gam$^-$ χ° recombination occurs at the level of Chi-stimulated recombination in *rec*⁺ cells and is not detectably stimulated by Chi (Chaudhury and Smith, 1984; Amundsen *et al.*, 1986). Cells with the *recD*⁺ gene overexpressed from a plasmid are less susceptible than wild-type cells to reduction of Chi hotspot activity by excess Chi sites on linearized non-homologous DNA or by bleomycin-induced fragmentation of the chromosome (Myers, Kuzminov, and Stahl, 1995, Köppen *et al.*, 1995). These results are consistent with the hypothesis that RecD is released at Chi and that excess RecD restores interaction with Chi.

The preceding observations establish that Chi interacts with RecBCD enzyme, but they leave open the possibility that Chi interacts with another cellular component. Tracy and Kowalczykowski (1996) found that RecA protein preferentially binds to short ss DNA with GT-rich, Chi-like sequences and promotes their pairing with homologous ds DNA. This feature may contribute to Chi's enhancement of recombination by the RecBCD pathway. Chi does not, however, significantly enhance recombination by other pathways in which RecA protein acts (*see* above). Perhaps Chi stimulates RecA-protein-promoted synapsis only when Chi is near the 3′ end of the ss DNA (Figures 6-1b and 6-3f).

Chi's Effects on Recombination

As noted above, Chi was first characterized as a site that stimulates recombination of phage λ during its vegetative growth. In addition to its properties discussed there, Chi acts when it is located opposite a heterology of as much as 4 kb in the non-Chi parent (Stahl and Stahl, 1975). In this case, recombination is stimulated in the region to the left of the heterology. Chi is nearly as active when in one parent as in two; i.e., an active Chi site is dominant (Lam *et al.*, 1974). Chi enhances the level of hybrid DNA (that with one strand from each parent) in mature λ phage (Rosenberg, 1987, 1988; Hagemann and Rosenberg, 1991; Siddiqi, Stahl, and Stahl, 1991) and in total intracellular λ DNA during vegetative crosses (Holbeck and Smith, 1992). This is the result expected if Chi acts before hybrid DNA formation, as inferred from the action of purified RecBCD enzyme on DNA (see below and Figures 6-1 and 6-3).

As expected from the occurrence of Chi at high density in the *E. coli* chromosome and from its stimulation of the *E. coli* RecBCD pathway of recombination (see above), Chi is active in Hfr-mediated conjugational crosses and in phage P1-mediated transductions. The frequency of exchanges is enhanced to the left of a Chi site (as written here) in the λ prophage in conjugation and transduction; with one paradoxical exception, Chi was active only when it was in the donor (Dower and Stahl, 1981). This is the result expected if RecBCD enzyme has access to the linear donor fragment but not to the circular recipient chromosome (Figure 6-4).

In conjugation and transduction most of the selected recombinants have just two exchanges, the minimal number needed to replace a recipient marker with

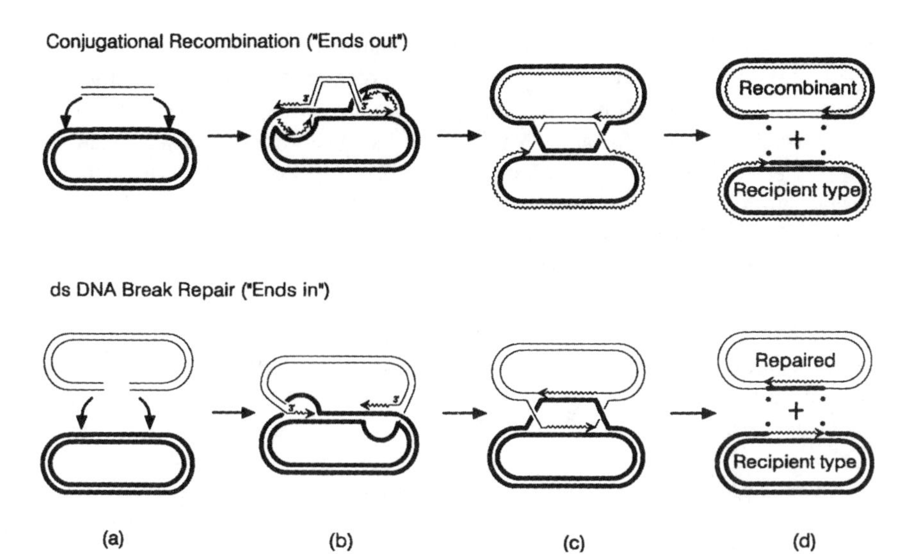

Figure 6-4. Model for conjugational and transductional recombination and for ds DNA break repair. (Top) Each end of a linear DNA fragment introduced during conjugation or transduction (a) undergoes the reactions shown in Figure 6-3 a–g. The 3′ ends of the invading DNA (thin lines) prime DNA replication (wavy lines) (b), which produces two circular chromosomes joined by two Holliday junctions (c). Some combination of the RuvA, B, C, and RecG proteins resolves this joint molecule into a recombinant and a recipient-type molecule (d). Alternative resolutions would produce reciprocals of the types shown or a dimeric circular chromosome, which could be resolved to monomeric chromosomes by *dif* site-specific recombination (Blakely *et al.*, 1991; *see also* Chapters 7 and 10).

(Bottom) The ends of a broken chromosome undergo the reactions outlined above. Replication restores any DNA lost at the ends. Resolution proceeds as above. Note that the ends of the broken DNA point toward each other ("ends in") over the short arc of the chromosome, whereas in the top panel they point away from each other ("ends out"). Reproduced with permission from Smith (1991).

a donor fragment and to restore the circularity of the chromosome (Smith, 1991). In P1-mediated transduction, with ~110kb donor fragments, and in conjugation with donor fragments $\lesssim$ 1 mb the majority of the exchanges appear to occur near the ends of the donor fragment (Smith, 1991; Lloyd and Buckman, 1995). These observations suggested the "long chunk" integration model in which exactly one exchange occurs "downstream" of a Chi site near each donor fragment end (Figure 6-4). This model is supported by the ability of RecBCD enzyme to act at exactly one Chi site (see above).

Transformation with short (<10 kb) linear DNA fragments (gene replacement) occurs at a very low or undetectable frequency in wild-type (*rec*$^+$ *sbc*$^+$) *E. coli* but does occur in *E. coli* mutants expressing the RecF (*recBC sbcBC*) or ‡ (*recD*) pathway (e.g. Jasin and Schimmel, 1984; Winans *et al.*, 1985; Shevell *et al.*, 1988; Russell, Thaler, and Dahlquist, 1989). Inclusion of two properly oriented and positioned Chi sites on the linear fragment enhances gene replacement during transformation about 40-fold in wild-type cells; dual Chi sites also enhance a specialized transduction in which intracellular *Eco*RI restriction enzyme cuts the linear Chi-containing fragment out of infecting λ DNA (Dabert and Smith, in press). Transformation or transduction in this way offers a useful method of "gene targeting" with cloned DNA fragments in wild-type *E. coli* (as opposed to *recBC sbc* or *recD* mutants used previously).

These observations indicate that Chi is an important element in *E. coli* homologous recombination. In its absence recombination following conjugation or transduction in wild-type cells would presumably occur at very low or undetectable frequency. Testing this proposal is thwarted by the ~1000 Chi sites in the *E. coli* genome.

A hotspot of recombination such as Chi increases the ratio of genetic distance to physical distance; in other words, genomic regions containing a hotspot have more recombination per unit physical distance than do regions lacking a hotspot. (In purely genetic studies, however, hotspots go unnoticed unless mutations create or destroy the hotspot.) Chi may be responsible for certain regions of the *E. coli* genome recombining at ~25-fold higher frequency than others; a region near *oriC* with an especially high density of Chi sites (see above) is transduced by P1 at higher-than-average frequency (Newman and Masters, 1980). Although a small part of this high frequency is due to preferential packaging of DNA in the donor cell, most is due to events in the recipient, where Chi is expected to enhance recombination.

Chi's Effects on Replication and DNA Degradation

Replication of certain plasmids produces rolling circles, presumably when a replication fork in the initial θ form breaks. In *E. coli* these plasmids produce high molecular weight DNA (HMW) when RecBCD enzyme is absent, due to mutation, or when the plasmid contains a properly oriented Chi site (Dabert, Ehrlich, and Gruss, 1992; Zaman and Boles, 1994). Apparently, RecBCD enzyme gains access to the end of the rolling circle and degrades it unless Chi is present to protect it. This observation supports the view that Chi attenuates the RecBCD nuclease activity in *E. coli* cells [the (Mg^{++}) > (ATP) condition; above and Figure 6-1b, bottom]. The requirement for RecA protein for HMW formation argues, however, that HMW formation requires not just attenuation of the nuclease activity, but recombination, which may indirectly protect the DNA. This scenario is consistent with the view that Chi stimulates the nicking activity of RecBCD enzyme in *E. coli* cells [the (ATP) > (Mg^{++}) condition; Figure 6-1b, middle]. In this scenario linear DNA that does not recombine is degraded by nucleases, perhaps other RecBCD enzyme molecules.

Linear DNA produced from a λ *cos*-containing plasmid cut with λ terminase protein is also degraded in *E. coli* cells unless the plasmid contains a properly oriented Chi (Kuzminov, Schabtach, and Stahl, 1994). Here, too, protection by Chi requires RecA protein and SSB protein and may involve recombination.

Chi appears to stimulate recombination at a stalled, presumably broken replication fork, perhaps to repair it. The evidence for this has come from a study of *E. coli rnh* mutants, which lack RNase H activity and in which DNA replication can occur in the absence of *oriC* and its activating protein DnaA. Nishitani, Hidaka, and Horiuchi (1993) sought *E. coli* DNA fragments that would allow replication of circular DNA containing the *kan* resistance determinant, which lacks an origin of replication, in *rnh* mutants. Eight unique fragments conferring high frequency Kan^R transformation were isolated; seven arose from an ~280 kb region near the terminus of replication, and five of five tested contained Chi. Mutation of the Chi octamer in one of these fragments, called HotA, eliminated its "hot activity" (Horiuchi *et al.*, 1994). The *kan*-HotA DNA does not appear to replicate as an autonomous circle, as initially anticipated, but rather appears to integrate by homologous recombination into the chromosome at high frequency and to be excised at high frequency. High frequency Kan^R transformation is abolished by a chromosomal deletion corresponding to HotA, and plasmid-like DNA is found at high levels in cell extracts only when the DNA contains HotA with Chi (the "hot activity" assay). The activity of HotA depends upon the Tau (or Tus) protein, which blocks DNA replication at Ter sites located around the terminus of replication. Horiuchi et al. (1994) propose that a replication fork blocked by Tau at Ter frequently breaks, allowing RecBCD enzyme to enter the resultant ds break. RecBCD enzyme then promotes recombination between the duplicated HotA DNA, to excise the *kan*-HotA DNA. They further propose that RecBCD enzyme and Chi sites are components of recombinational repair of broken replication forks in wild-type cells; Kuzminov et al. (1994) and Smith (1991) made similar proposals (see Fig. 6-4).

Chi may stimulate DNA replication directly by providing a 3′ OH primer for replication (see Figs. 6-3g and 6-4). Asai et al. (1994; see also Chapter 7) studied replication of *E. coli* plasmids containing *cos* after induction of λ terminase, which presumably cleaves some but not all of the plasmid DNA. The amount of covalently closed circular plasmid DNA increases several-fold over a few hours. The increase is stimulated by a properly oriented Chi and requires RecA protein and RecBCD enzyme. Henderson and Weil (1975) also noted that Chi increases the amount of replicated λ DNA. These observations support the view that RecBCD-dependent recombination is coupled to DNA replication, which is necessary for faithful ds DNA break repair if nucleotides are lost at the break (Fig. 6-4).

Chi and Chi-like Sites in Other Organisms

The presence of RecBCD-like enzymes, identified as ATP-dependent DNases, in both Gram-positive and Gram-negative bacteria, suggests that Chi or equivalent sites may play important roles in bacteria other than *E. coli.* Chi is active in λ vegetative crosses in *Salmonella typhimurium* (Smith, Roberts, and Schultz, 1986). It is likely active in other enteric bacteria, too, for extracts of most of the genera tested (*Shigella, Citrobacter, Salmonella, Edwardsiella, Proteus, Morganella, Erwinia, Vibrio,* and *Photobacterium*) have Chi-dependent DNA strand-cleavage activity like that in *E. coli* extracts (Schultz and Smith, 1986). Furthermore, cloned genes from the enteric bacteria tested (*Shigella, Citrobacter, Salmonella, Klebsiella, Proteus,* and *Serratia*) complement an *E. coli recBCD* deletion mutation for ATP-dependent nuclease activity, Chi-cutting activity, and Chi hotspot activity in λ vegetative crosses (McKittrick and Smith, 1989; Weichenhan and Wackernagel, 1989). Cloned genes from three *Pseudomonas* isolates, however, complement for ATP-dependent nuclease but not for either Chi activity. Large plaque formation by λ Red^- Gam^- χ° phage on *E. coli recBCD* deletion cells bearing the *Pseudomonas* clones suggests that the encoded RecBCD-like enzyme recognizes a sequence in λ, other than Chi, as a recombination hotspot, but that sequence has not been identified.

The Gram-positive bacterium *Lactococcus lactis* appears to use 5′ G-C-G-C-G-T-G 3′ as a Chi-like site, for rolling circle replication plasmids with this sequence form HMW in this bacterium (Biswas *et al.*, 1995) (see above). Mutation of any one of these 7 bp severely reduces HMW formation, indicating a strict sequence requirement, as for Chi in *E. coli.* Presumably, this sequence interacts with the ATP-dependent nuclease detected in *L. lactis* extracts; a direct interaction has not been reported.

There have been periodic speculations that Chi sites or closely related sequences stimulate recombination in mammals. Kenter and Birshtein (1981) noted Chi and related sequences in the region of recombination underlying immunoglobulin class switching in mice. Copies of the human hypervariable minisatellite DNA vary more frequently than does other DNA; this variation has been attributed to recombination. DNA with multiple copies of the core (5′ A-G-A-G-G-T-G-G-G-C-A-G-G-T-G-G 3′) of the minisatellite sequence confers hyper-recombinogenicity to plasmids in human tissue culture cells (Wahls, Wallace, and Moore, 1990). Two proteins binding to this core sequence have been partially purified (Wahls, Swenson, and Moore, 1991), but their role in promoting recombination or sequence variation is unknown. A subset of this sequence, 5′ G-C-(T/A)-G-G-(T/A)-G-G 3′ occurs at or close to breakpoints of oncogene translocations apparently promoted by aberrant V(D)J immunoglobulin rearrangement (Krowczynska, Rudders, and Krontiris, 1990; Wyatt *et al.*, 1992) and at the breakpoints of spontaneous deletions at various loci (Dewyse and Bradley, 1991). Rüdiger (1995) noted that an especially highly conserved region of the human *Alu* DNA repeat is associated with gene rearrangements and contains the sequence 5′ G-C-T-G-G 3′. RecBCD-like enzymes (i.e., ATP-dependent nucleases) have not, to my knowledge, been reproducibly reported from any eukaryote. If Chi and Chi-like sites have a role in eukaryotic recombination, it may be distinct from that in bacteria. Perhaps DNA with Chi or related sequences assumes a special structure that is recognized by a variety of recombination-promoting enzymes.

Perspective

Elucidating the mechanism of Chi's enhancement of λ Red^- Gam^- plaque size has unexpectedly revealed much about the mechanism of homologous recombination, particularly the role of RecBCD enzyme. Recent work with Chi has shown connections between recombination, replication, and repair, also shown by earlier work with phage. Chi has been a useful point of comparison for studies of eukaryotic recombination hotspots (Stahl, 1979; Smith, 1988, 1994). Further work may use Chi and sites like it to enhance gene targeting in bacteria, and perhaps in eukaryotes as well. Chi is also likely to remain an important element in more fully elucidating the mechanism of homologous recombination in *E. coli.*

Acknowledgments

I am grateful to Sue Amundsen, Joseph Farah, and Andrew Taylor for their helpful comments on this review and to Karen Brighton for processing it. I thank Howard Coleman, Patrick Dabert, Andrew Eisen, Kathy Helde, Chris Roberts, and Andrew Taylor for unpublished observations. Research on recombinational hotspots in my laboratory is supported by grant GM31693 from the National Institutes of Health.

References

Amundsen, S. K., A. F. Taylor, A. M. Chaudhury, and G. R. Smith. 1986. *recD:* The gene for an essential third subunit of exonuclease *V. Proc. Natl. Acad. Sci. USA* 83:5558–5562.

Amundsen, S. K., A. M. Neiman, S. M. Thibodeaux, and G. R. Smith. 1990. Genetic dissection of the biochemical activities of RecBCD enzyme. *Genetics* 126:25–40.

Asai, T., D. B. Bates, and T. Kogoma. 1994. DNA replication triggered by double-stranded breaks in E. coli: dependence on homologous recombination functions. *Cell* 78:1051–1061.

Biswas, I., E. Maguin, S. D. Ehrlich, and A. Gruss. 1995. A 7-base-pair sequence protects DNA from exonucleolytic degradation in *Lactococcus lactis. Proc. Natl. Acad. Sci. USA* 92:2244–2248.

Blaisdell, B. E., K. E. Rudd, A. Matin, and S. Karlin. 1993. Significant dispersed recurrent DNA sequences in the *Escherichia coli* genome. *J. Mol. Biol.* 229:833–848.

Blakely, G., S. Colloms, G. May, M. Burke, and D. Sherratt. 1991. *Escherichia coli* XerC recombinase is required for chromosomal segregation at cell division. *New. Biol.* 3:789–798.

Brewer, B. J. 1988. When polymerases collide: replication and the transcriptional organization of the Escherichia coli chromosome. *Cell* 53:679–686.

Burland, V., G. Plunkett III, D. L. Daniels, and F. R. Blattner. 1993. DNA sequence and analysis of 136 kilobases of the *Escherichia coli* genome: Organizational symmetry around the origin of replication. *Genomics* 16:551–561.

Chaudhury, A. M., and G. R. Smith. 1984. A new class of *Escherichia coli recBC* mutants: Implications for the role of RecBC enzyme in homologous recombination. *Proc. Natl. Acad. Sci. USA,* 81:7850–7854.

Cheng, K. C., and G. R. Smith. 1984. Recombinational hotspot activity of Chi-like sequences. *J. Mol. Biol.* 180:371–377.

Cheng, K. C., and G. R. Smith. 1989. Distribution of Chi-stimulated recombinational exchanges and heteroduplex endpoints in phage lambda. *Genetics* 123:5–17.

Dabert, P., S. D. Ehrlich and A. Gruss. 1992. χ sequence protects against RecBCD degradation of DNA *in vivo. Proc. Natl. Acad. Sci. USA* 89:12073–12077.

Dabert, P. and G. R. Smith. Gene Replacement with linear DNA fragments in wild-type *Escherichia coli* enhancement by Chi sites. *Genetics,* in press.

Dewyse, P. and W. E. C. Bradley. 1991. A very large spontaneous deletion at *aprt* locus in CHO cells: sequence similarities with small *aprt* deletions. *Somatic Cell and Molecular Genetics* 17:57–68.

Dixon, D. A. and S. C. Kowalczykowski. 1991. Homologous pairing in vitro stimulated by the recombination hotspot, Chi. *Cell* 66:361–371.

Dixon, D. A. and S. C. Kowalczykowski. 1993. The recombination hotspot χ is a regulatory sequence that acts by attenuating the nuclease activity of the E. coli RecBCD enzyme. *Cell* 73:87–96.

Dixon, D. A. and S. C. Kowalczykowski. 1995. Role of the *Escherichia coli* recombination hotspot, χ, in RecABCD-dependent homologous pairing. *J. Biol. Chem.* 270:16360–16370.

Dower, N. A. and F. W. Stahl. 1981. Chi activity during transduction-associated recombination. *Proc. Natl. Acad. Sci. USA* 78:7033–7037.

Eggleston, A. K. and S. C. Kowalczykowski. 1993. Biochemical characterization of a mutant recBCD enzyme, the recB^{2109} CD enzyme, which lacks χ-specific, but not non-specific, nuclease activity. *J. Mol. Biol.* 231:605–620.

Faulds, D., N. Dower, M. M. Stahl, and F. W. Stahl. 1979. Orientation-dependent recombination hotspot activity in bacteriophage lambda. *J. Mol. Biol.* 131:681–695.

Gillen, J. R. and A. J. Clark. 1974. The RecE pathway of bacterial recombination. In *Mechanisms in Recombination,* R. F. Grell, ed. pp. 123–126. Plenum Press, New York.

Hagemann, A. T. and S. M. Rosenberg. 1991. Chain bias in Chi-stimulated heteroduplex patches in the lambda *ren* gene is determined by the orientation of lambda *cos*. *Genetics* 129:611–621.

Henderson, D. and J. Weil. 1975. Recombination-deficient deletions in bacteriophage lambda and their interaction with Chi mutations. *Genetics* 79:143–174.

Holbeck, S. L. and G. R. Smith. 1992. Chi enhances heteroduplex DNA levels during recombination. *Genetics* 132:879–891.

Horiuchi, T., Y. Fujimura, H. Nishitani, T. Kobayashi, and M. Hidaka. 1994. The DNA replication fork blocked at the *Ter* site may be an entrance for the RecBCD enzyme into duplex DNA. *J. Bacteriol.* 176:4656–4663.

Jasin, M. and P. Schimmel. 1984. Deletion of an essential gene in *Escherichia coli* by site-specific recombination with linear DNA fragments. *J. Bacteriol.* 159:783–786.

Kalnins, A., K. Otto, U. Ruther, and B. Muller-Hill. 1983. Sequence of the *lacZ* gene of *Escherichia coli. EMBO J.* 2:593–597.

Kenter, A. L. and B. K. Birshtein. 1981. Chi, a promoter of generalized recombination in λ phage, is present in immunoglobulin genes. *Nature* 293:402–404.

Kobayashi, I., M. M. Stahl, and F. W. Stahl. 1984. The mechanism of the Chi-*cos* interaction in RecA-RecBC-mediated recombination in phage lambda. *Cold Spring Harbor Symp. Quant. Biol.* 49:497–506.

Kobayashi, I., H. Murialdo, J. M. Crasemann, M. M. Stahl, and F. W. Stahl. 1982. Orientation of cohesive end site *cos* determines the active orientation of chi sequence in stimulating recA-recBC mediated recombination in phage lambda lytic infections. *Proc. Natl. Acad. Sci. USA* 79:5981–5985.

Köppen, A., S. Krobitsch, B. Thoms, and W. Wackernagel. 1995. Interaction with the recombination hot spot χ *in vivo* converts the RecBCD enzyme of *Escherichia coli* into a χ-independent recombinase by inactivation of the RecD subunit. *Proc. Natl. Acad. Sci. USA* 92:6249–6253.

Krowczynska, A. M., R. A. Rudders, and T. G. Krontiris. 1990. The human minisatellite consensus at breakpoints of oncogene translocations. *Nucl. Acids Res.* 18:1121–1127.

Kuzminov, A., E. Schabtach, and F. W. Stahl. 1994. χ sites in combination with RecA protein increase the survival of linear DNA in *Escherichia coli* by inactivating exoV activity of RecBCD nuclease. *EMBO J.* 13:2764–2776.

Lam, S. T., M. M. Stahl, K. D. McMilin, and F. W. Stahl. 1974. Rec-mediated recombinational hotspot activity in bacteriophage lambda. II. A mutation which causes hotspot activity. *Genetics* 77:425–433.

Lieb, M. and S. Rehmat. 1995. Very short patch repair of T:G mismatches in vivo: Importance of context and accessory proteins. *J. Bacteriol.* 177:660–666.

Lloyd, R. G. and C. Buckman. 1995. Conjugational recombination in *Escherichia coli:* Genetic analysis of recombinant formation in Hfr × F^- crosses. *Genetics* 139:1123–1148.

Lundblad, V., A. F. Taylor, G. R. Smith, and N. Kleckner. 1984. Unusual alleles of *recB* and *recC* stimulate excision of inverted repeat transposons Tn*10* and Tn*5*. *Proc. Natl. Acad. Sci. USA* 81:824–828.

Malone, R. E., D. K. Chattoraj, D. H. Faulds, M. M. Stahl, and F. W. Stahl. (1978).

Hotspots for generalized recombination in the *Escherichia coli* chromosome. *J. Mol. Biol.* 121:473–491.

McKittrick, N. H. and G. R. Smith. 1989. Activation of Chi recombinational hotspots by RecBCD-like enzymes from enteric bacteria. *J. Mol. Biol.* 210:485–495.

McMilin, K. D., M. M. Stahl, and F. W. Stahl. 1974. Rec-mediated hotspot recombinational activity in bacteriophage lambda. I. Hot spot activity associated with *spi* deletions and *bio* substitutions. *Genetics* 77:409–423.

Myers, R. S., A. Kuzminov, and F. W. Stahl. 1995. The recombination hot spot χ activates RecBCD recombination by converting *Escherichia coli* to a *recD* mutant phenocopy. *Proc. Natl. Acad. Sci. USA* 92:6244–6248.

Newman, B. J. and M. Masters. 1980. The variation in frequency with which markers are transduced by phage P1 is primarily a result of discrimination during recombination. *Mol. Gen. Genet.* 180:585–589.

Nishitani, H., M. Hidaka, and T. Horiuchi. 1993. Specific chromosomal sites enhancing homologous recombination in *Escherichia coli* mutants defective in RNase H. *Mol. Gen. Genet.* 240:307–314.

Ponticelli, A. S., D. W. Schultz, A. F. Taylor, and G. R. Smith. 1985. Chi-dependent DNA strand cleavage by RecBC enzyme. *Cell* 41:145–151.

Rosenberg, S. M. 1987. Chi-stimulated patches are heteroduplex, with recombinant information on the phage lambda *r* chain. *Cell* 48:855–865.

Rosenberg, S. M. 1988. Chain-bias of *Escherichia coli* Rec-mediated lambda patch recombinants is independent of the orientation of lambda *cos. Genetics* 120:7–21.

Rüdiger, N. S., N. Gregersen, and M. C. Kielland-Brandt. 1995. One short well conserved region of *Alu*-sequences is involved in human gene rearrangements and has homology with prokaryotic *chi. Nucl. Acids Res.* 23:256–260.

Russell, C. B., D. S. Thaler, and F. W. Dahlquist. 1989. Chromosomal transformation of *Escherichia coli recD* strains with linearized plasmids. *J. Bacteriol.* 171:2609–2613.

Sanger, F., A. R. Coulson, G. F. Hong, D. F. Hill, and G. B. Petersen. 1982. Nucleotide sequence of bacteriophage lambda DNA. *J. Mol. Biol.* 162:729–773.

Schultz, D. W. and G. R. Smith. 1986. Conservation of Chi cutting activity in terrestrial and marine enteric bacteria. *J. Mol. Biol.* 189:585–595.

Schultz, D. W., A. F. Taylor, and G. R. Smith. 1983. *Escherichia coli* RecBC pseudorevertants lacking Chi recombinational hotspot activity. *J. Bacteriol.* 155:664–680.

Shevell, D. E., A. M. Abou-Zamzam, B. Demple, and G. C. Walker. 1988. Construction of an *Escherichia coli* K-12 *ada* deletion by gene replacement in a *recD* strain reveals a second methyltransferase that repairs alkylated DNA. *J. Bacteriol.* 170:3294–3296.

Siddiqi, I., M. M. Stahl, and F. W. Stahl. 1991. Heteroduplex chain polarity in recombination of phage lambda by the Red, RecBCD, RecBC(D$^-$) and RecF pathways. *Genetics* 128:7–22.

Smith, G. R. 1983. General recombination, in *Lambda II.* R. W. Hendrix, J. W. Roberts, F. W. Stahl and R. A. Weisberg, eds. pp. 175–209. Cold Spring Harbor Laboratory, Cold Spring Harbor, New York.

Smith, G. R. 1987. Mechanism and control of homologous recombination. *Escherichia coli. Annu. Rev. Genet.* 21: 179–201.

Smith, G. R. 1988. Homologous recombination sites and their recognition. In *The Recombination of Genetic Material.* pp. 115–154. B. Low, ed. Academic Press, New York.

Smith, G. R. 1991. Conjugational recombination in *E. coli: Myths and mechanisms. Cell* 64:19–27.

Smith, G. R. 1994. Hotspots of homologous recombination. *Experientia* 50:234–241.

Smith, G. R., C. M. Roberts, and D. W. Schultz. 1986. Activity of Chi recombinational hotspots in *Salmonella typhimurium. Genetics* 112:429–439.

Smith, G. R., D. W. Schultz, A. F. Taylor, and K. Triman. 1981. Chi sites, RecBC enzyme, and generalized recombination. *Stadler Genetics Symposium* 13:25–37.

Smith, G. R., S. K. Amundsen, A. M. Chaudhury, K. C. Cheng, A. S. Ponticelli, C. M. Roberts, D. W. Schultz, and A. F. Taylor. 1984. Roles of RecBC enzyme and Chi sites in homologous recombination. Cold Spring Harbor Symp. *Cold Spring Harbor Symp. Quant. Biol.* 49:485–495.

Smith, G. R., S. K. Amundsen, P. Dabert, and A. F. Taylor. 1995. The initiation and control of homologous recombination in *Escherichia coli. Phil. Trans. R. Soc. London* 347:13–20.

Stahl, F. W. 1979. Special sites in generalized recombination. *Annu. Rev. Genet.* 13:7–24.

Stahl, F. W., J. M. Crasemann, and M. M. Stahl. 1975. Rec-mediated recombinational hot spot activity in bacteriophage lambda. III. Chi mutations are site-mutations stimulating Rec-mediated recombination. *J. Mol. Biol.* 94:203–212.

Stahl, F. W. and M. M. Stahl. 1975. Rec-mediated recombinational hot spot activity in bacteriophage lambda. IV. Effect of heterology on Chi-stimulated crossing over. *Mol. Gen. Genet.* 140:29–37.

Stahl, F. W. and M. M. Stahl. 1977. Recombination pathway specificity of Chi. *Genetics* 86:715–725.

Stahl, F. W., M. M. Stahl, R. E. Malone, and J. M. Crasemann. 1980. Directionality and nonreciprocality of Chi-stimulated recombination in phage lambda. *Genetics* 94:235–248.

Sutcliffe. J. G. 1979. Complete nucleotide sequence of the *Escherichia coli* plasmid pBR322. *Cold Spring Harbor Symp. Quant. Biol.* 43:77–90.

Taylor, A. F. 1988. RecBCD enzyme of *Escherichia coli,* in *Genetic Recombination,* R. Kucherlapati and G. R. Smith, eds. pp. 231–263. American Society for Microbiology, Washington, DC.

Taylor, A. F. and G. R. Smith. 1992. RecBCD enzyme is altered upon cutting DNA at a Chi recombination hotspot. *Proc. Natl. Acad. Sci. USA* 89:5226–5230.

Taylor, A. F. and G. R. Smith. 1995. Monomeric RecBCD enzyme binds and unwinds DNA. *J. Biol. Chem.* 270:24451–24458.

Taylor, A. F. and G. R. Smith. 1995b. Strand specificity of nicking of DNA at Chi sites by RecBCD enzyme: modulation by ATP and magnesium levels. *J. Biol. Chem.* 270:24459–24467.

Taylor, A. F., D. W. Schultz, A. S. Ponticelli, and G. R. Smith. 1985. RecBC enzyme nicking at Chi sites during DNA unwinding: Location and orientation dependence of the cutting. *Cell* 41:153–163.

Thaler, D. S., E. Sampson, I. Siddiqi, S. M. Rosenberg, F. W. Stahl, and M. Stahl. 1988. A hypothesis: Chi-activation of RecBCD enzyme involves removal of the RecD subunit, in *Mechanisms and Consequences of DNA Damage Processing*. E. Friedberg and P. Hanawalt, eds. pp. 413–422. Alan R. Liss, New York.

Triman, K. L., D. K. Chattoraj, and G. R. Smith. 1982. Identity of a Chi site of *Escherichia coli* and Chi recombinational hotspots of bacteriophage lambda. *J. Mol. Biol.* 154:393–399.

Wahls, W. P., G. Swenson, and P. D. Moore. 1991. Two hypervariable minisatellite DNA binding proteins. *Nucl. Acids Res.* 19:3269–3274.

Wahls, W. P., L. J. Wallace, and P. D. Moore. 1990. Hypervariable minisatellite DNA is a hotspot for homologous recombination in human cells. *Cell* 60:95–103.

Weichenhan, D. and W. Wackernagel. 1989. Functional analyses of *Proteus mirabilis* wild-type and mutant RecBCD enzymes in *Escherichia coli* reveal a new mutant phenotype. *Mol. Microbiol.* 3:1777–1784.

Winans, S. C., S. J. Elledge, J. H. Krueger, and G. C. Walker. 1985. Site-directed insertion and deletion mutagenesis with cloned fragments in *Escherichia coli. J. Bacteriol.* 161:1219–1221.

Wyatt, R. T., R. A. Rudders, A. Zelenetz, R. A. Delellis, and T. G. Krontiris. 1992. BCL2 oncogene translocation is mediated by a χ-like consensus. *J. Exp. Med.* 175:1575–1588.

Zaman, M. M. and T. C. Boles. 1994. Chi-dependent formation of linear plasmid DNA in exonuclease-deficient *recBCD*+ strains of *Escherichia coli. J. Bacteriol.* 176:5093–5100.

7

Origins of Chromosome Replication

Tokio Kogoma

The first obligatory step for replication of duplex DNA is strand separation. Duplex melting (unwinding) leading to initiation of bacterial chromosome replication normally occurs at a fixed site on the chromosome termed *oriC*. This local duplex unwinding is effected by specific binding of an initiator, DnaA protein encoded by the *dnaA* gene, to 9-bp repeated sequences (DnaA boxes) which are clustered within the *oriC* site. The DnaA box sequence and the amino acid sequence of DnaA protein are evolutionarily well conserved among eubacteria. Furthermore, with a few exceptions, most eubacteria have a gene arrangement in which the *oriC* region is located next to the *dnaA* gene. The arrangements of several genes in the vicinity of *dnaA* are also well conserved. The evolutionary conservation of the DnaA protein and DnaA box sequences suggests that the initiation mechanism for chromosome replication in eubacteria has been largely conserved. In this chapter, the structures of the *oriC* sites for several species of eubacteria are reviewed. At least in *E. coli,* two other normally repressed chromosome replication systems are activated under certain specific conditions. The origins (*oriM* and *oriK*) of these alternative forms of replication are also summarized. The readers are referred to recent reviews for more details of the subjects (Messer and Weigel, 1996; Skarstad and Boye, 1994; Yoshikawa and Ogasawara, 1991; Asai and Kogoma, 1994).

The *oriC* Sites of Enteric Bacteria

The oriC *Structure*

The first origin of chromosome replication that was cloned and sequenced was the *Escherichia coli oriC.* The minimal sequence necessary for the origin activity, i.e. the minimal *oriC,* was determined by introducing deletions from either side of cloned *oriC* fragments (Oka *et al.*, 1980). The *E. coli oriC* sequence (Fig. 7-1) contains several unique motifs: four or five DnaA-binding sites (DnaA boxes), three AT-rich 13-bp direct repeats (13-mers) and 11 GATC sites. In addition, the sequence includes the binding sites for DNA bending proteins such as IHF (integration host factor) and FIS (factor for inversion stimulation) (Fig. 7-1; *see also* Chapter 2). The minimal *oriC* was initially determined to be 245-bp long. Later, a 12-bp AT-rich segment (the AT cluster) at the left end of *oriC* was demonstrated also to be required for the activity (Asai et al., 1990).

Chromosomal DNA fragments capable of autonomous replication in *E. coli* were also cloned from several other members of the family *Enterobacteriaceae.* Sequence analyses revealed that the origin structure is remarkably well conserved among these enterics (Zyskind *et al.*, 1983). The intensive mutational analysis of cloned *E. coli oriC* (Oka *et al.*, 1984) and the comparison of the enteric *oriC* sequences (Zyskind *et al.*, 1983) led to the concept that the origin sequence affords (i) the recognition sites for proteins that interact with it and (ii) the spacer regions that keep the recognition sites in precise arrangement. Thus, the spacing of the recognition sites in *oriC* is crucial for the origin activity. For example, a change in the distance between DnaA boxes R3 and R4 (Fig. 7-1) by a 4-bp insertion or a 5-, 7-, or 8-bp deletion inactivates the origin activity although base substitutions in the same region do not (Hirota *et al.*, 1981). However, insertion or deletion of 10–12 bp (approximately one helical turn) can be tolerated (Woelker and Messer, 1993). This suggests that a specific helical phasing of the R3 and R4 sites for DnaA binding is important for the formation of the initiation complex. Similarly, the distance between the rightmost 13-mer and DnaA box R1 must be strictly maintained: Even 1-bp insertion or deletion has a profound effect on

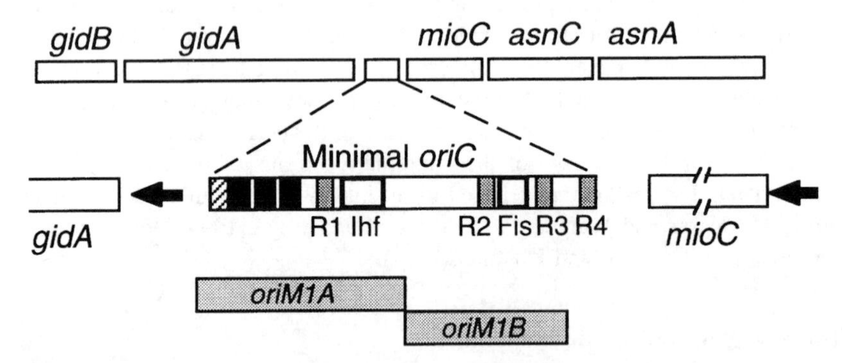

Figure 7-1. The *E. coli* minimal *oriC* and the map of *oriM1* sites. The minimal *oriC* and the neighboring genes are shown along with the positions of the *oriM1A* and *oriM1B* sites within the *oriC*. The arrows denote the position and direction of the promoters of the *mioC* and *gidA* genes. Hatched and solid rectangles are the AT cluster and the 13-mer repeats, respectively. Approximate positions of DnaA boxes (R1, R2, R3, R4) and the binding sites (Ihf, Fis) for IHF and FIS are also shown.

the activity, and any distance changes of two or more basepairs completely inactivate the activity (Hsu *et al.*, 1994). In contrast to the R3-R4 spacing above, 10- and 11-bp insertions are not tolerated in this case.

The Mechanism of Initiation

The availability of the cloned *E. coli oriC* sequence facilitated development of *in vitro* systems to study the initiation mechanism. The initiation reaction begins with binding of DnaA protein (20 to 40 monomers) to DnaA boxes within *oriC* (reviewed by Messer and Weigel, 1996). With the aid of IHF, FIS, and HU (*E. coli* histone-like protein), DnaA binding leads to formation of a compact complex in which duplex DNA wraps around the proteins. Facilitated by the AT-richness of the AT cluster and 13-mer repeats, helical distortion as a result of complex formation induces local duplex unwinding at the 13-mers. DnaA protein may have an active role in this step (Hsu *et al.*, 1994). Transcription around *oriC* (e.g., *mioC* and *gid* transcription; Fig. 7-1) facilitates the strand melting. DnaA protein then guides the DnaB helicase to the melted 13-mer region, forming a prepriming complex. The DnaB helicase enlarges the initial bubble and interacts with the DnaG primase, which lays down RNA primers for DNA polymerase III. Subsequent assembly of dimeric DNA polymerase III holoenzyme (replisomes) at both ends of the enlarged bubble leads to bidirectional replication of the chromosome.

Chromosomal oriC *vs. Cloned* oriC

The analysis of the structure and function of *E. coli oriC* summarized above is based almost exclusively on the studies that utilized *oriC* sites cloned on plasmids. There are some indications that cloned *oriC* sites behave differently from the native *oriC* on the chromosome. For example, it was recently demonstrated that a 2-kb insertion between R3 and R4 which completely destroys the *oriC* activity on plasmids can be tolerated on the chromosome; mutant cells carrying the modification at the native *oriC* site on the chromosome are viable (Bates *et al.*, 1995). Furthermore, R4 can even be deleted from chromosomal *oriC* without loss of cell viability although the modified *oriC* is completely inert on plasmid. These results indicate that the sequence requirement for the origin activity on the chromosome is different from the so-called minimal *oriC* which was determined with cloned *oriC* sites.

The *oriC* Structures of Other Eubacteria

Several *oriC* sites from other eubacteria have been cloned and the nucleotide sequences determined. These origins of replication were cloned primarily by three different approaches. One approach was shotgun cloning of chromosomal

DNA fragments capable of autonomous replication (e.g., *Pseudomonas putida* and *Streptomyces lividans*). The second approach was cloning of autonomously replicating fragments by virtue of the proximity of *oriC* to *dnaA* (e.g., *Bacillus subtilis* and *Spiroplasma citri*). The third was subcloning of autonomously replicating fragments from a larger fragment that spans the earliest replication region after synchronization of DNA replication (e.g., *Caulobacter crescentus*).

Gram-negative Bacteria

The *oriC* sites cloned from two Pseudomonads (*P. aeruginosa* and *P. putida*) exhibit nucleotide sequence similarity, but they are distinct from the enteric *oriC* sites (Yee and Smith, 1990). In fact, these origins can function in both *P. aeruginosa* and *P. putida* but not in *E. coli.* The *Pseudomonas oriC* sequences contain three 13-mer direct repeats and several DnaA boxes, and are located very close to the *dnaA* gene (Fig. 7-2). Although the nucleotide sequences are distinct, the Pseudomonad and enteric *oriC* sites share some common features. For example, the consensus sequence obtained by comparison of the two Pseudomonad *oriC*s shows four shared DnaA boxes, two of which have the same spacing and orientation of the R1 and R4 sites in the enteric *oriC* (Yee and Smith, 1990). Furthermore, the position of the three 13-mers relative to R1 mimics that of the

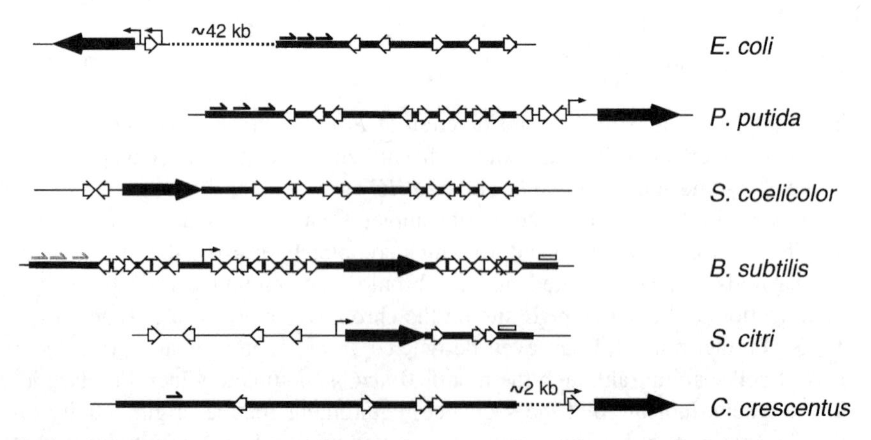

Figure 7-2. The structures of eubacterial *oriC* sites. The *oriC* structures of five eubacteria species are compared to the *E. coli* minimal *oriC*. Thick solid lines denote the *oriC* sites, and thick solid arrows indicate the position and orientation of *dnaA* genes (drawn not to scale). Thick open arrows denote DnaA boxes, i.e., the consensus sequence TTATCCACA and one- or two-base variations from the consensus except for *S. coelicolor* where the consensus is TTGTCCACA. Bent thin arrows, promoters of dnaA; solid and stippled half arrows, 13-mer and 16-mer direct repeats, respectively; open rectangles above the thick lines, AT-rich segments.

enteric *oriC*. These structural features suggest that initiation of chromosome replication in *Pseudomonas* occurs in a manner similar to the initiation in *E. coli.*

Gram-positive Bacteria

The *oriC* site of *Bacillus subtilis* was cloned on an autonomously replicating large DNA fragment that includes two DnaA box regions which flank the *dnaA* gene (Moriya *et al.*, 1992) (Fig. 7-2). Deletion of either DnaA box region destroys the origin activity *in vivo* (Moriya *et al.*, 1992) and *in vitro* (Moriya et al., 1994). Thus, the structure of the *B. subtilis oriC* site is unique in that two regions separated by 1.4 kb (the *dnaA* gene) are essential for initiation. Located upstream of the left DnaA box region are three 16-bp AT-rich direct repeats, reminiscent of the three 13-mers of the *E. coli oriC* site. The function of the 16-mers is not clear, however, because removal of the leftmost 16-mer inactivates the *in vivo* origin activity but not *in vitro.* The right DnaA box region contains a 33-bp, extremely AT-rich (97%) segment which is also required for the origin activity. A study with a recently developed *in vitro* replication system localized the initiation site within the right DnaA box region (Moriya *et al.*, 1994). How these two regions separated by a large distance act to initiate chromosome replication is not known. The distance can be shortened to less than 300 bp without loss of activity by deletion of the intervening sequence including the *dnaA* gene. The elucidation of the initiation mechanism at this unusual *oriC* site awaits further analysis with the *in vitro* replication system. Recently, a 300-bp non-coding region which contains a cluster of DnaA boxes was also located immediately downstream of the *Staphylococcus aureus dnaA* gene (Alonso and Fisher, 1995).

Streptomyces, Gram-positive filamentous bacteria, have genomic DNA of a high G+C content (ca. 70%) relative to the low G+C content of the *B. subtilis* genome. A fragment capable of autonomous replication was cloned from *Streptomyces lividans* chromosomal DNA and sequenced (Zakrzewska-Czerwinska and Schrempf, 1992). Comparison of the sequence to that of the *dnaA* region of the closely related *S. coelicolor* (Calcutt and Schmidt, 1992) places this *S. lividans* sequence downstream of the *dnaA* gene (Fig. 7-2). A 600-bp noncoding region (the *oriC* site) contains numerous DnaA boxes. Interestingly, the consensus sequence (TTGTCCACA) obtained by comparison of these DnaA box sequences has a G at the third position instead of an A in the consensus (TTATCCACA) for the low G+C bacteria such as *E. coli* and *B. subtilis.* This may reflect the high G+C pressure exerted during the course of evolution (Fujita et al., 1990). Despite the overall high G+C content of the genomic DNA, the *oriC* region is relatively AT-rich (46%). However, no obvious AT-rich repeat or AT cluster is discernible. The *dnaA* region of *Micrococcus luteus,* another Gram-positive high G+C bacterium, contains a structure very similar to the *oriC* regions of *S. lividans* and *S. coelicolor* (Fujita *et al.*, 1990). Whether or not the structure confers the capacity for autonomous replication has not been determined.

Mycoplasma

The parasitic, wall-less bacterium *Mycoplasma capricolum* has a small genome (ca. 700 kb) of an extremely low G+C content (*see also* Chapters 41 and 42). The *dnaA* region of this organism was shown to contain a structure similar to the *oriC* region of *B. subtilis,* i.e., the *dnaA* gene flanked by two DnaA box regions (Fujita *et al.*, 1992). Whether or not both regions, like *B. subtilis,* are required for the origin activity is not known. In fact, no fragment containing this structure has yet been isolated as an autonomously replicating entity. Recently, an *oriC* site of *Spiroplasma citri,* a plant pathogen that shares some molecular properties with *M. capricolum,* such as the G+C content (26%), gene structure, and genomic organization, was isolated and shown to have similarity to the *oriC* region of *B. subtilis* (Ye *et al.*, 1994) (Fig. 7-2). The two regions flanking the *dnaA* gene each contain several DnaA boxes. In addition, the left and right DnaA box regions include three AT-rich 9-bp direct repeats and a highly AT-rich segment, respectively, a feature that mimics the *oriC* structure of *B. subtilis.* However, a deletion analysis clearly indicated that, unlike the *B. subtilis oriC,* only the right DnaA box region is sufficient for the origin activity and the left region is dispensable (Renaudin et al., 1995).

Caulobacter

Unlike other eubacteria, *Caulobacter crescentus* (*see also* Chapter 56) undergoes asymmetric cell division to produce a stalker cell and a swarmer cell. Initiation of a next round of chromosome replication occurs only in the stalker cell. It is therefore an interesting question whether the *oriC* site contains information for the asymmetric control of DNA replication. The *oriC* site (*Cori*) of *C. crescentus* may be characterized by the sparse distribution of DnaA boxes and the separation from the *dnaA* gene by 2 kb (Marczynski and Shapiro, 1992; Zweigler and Shapiro, 1994) (Fig. 7-2). It includes an *E. coli*-like 13-mer and a 40-bp highly AT-rich segment. There are several motifs that are unique to the *C. crescentus oriC* site, some of which are required for the origin activity. Whether any of the unique motifs relates to the asymmetric control of initiation remains to be seen.

Alternative Replication Systems

E. coli possesses two other DNA replication systems which are normally repressed but can be activated under certain specific conditions (reviewed by Asai and Kogoma, 1994). One is inducible stable DNA replication (iSDR) which is activated upon the SOS response to DNA damage, and the other is constitutive stable

DNA replication (cSDR) which occurs in *rnhA* mutants lacking RNase HI activity. Other bacteria are yet to be shown to have similar cryptic replication activities.

oriM *sites*

The origins of iSDR are located in the *oriC* and *terC* regions of the *E. coli* chromosome, and designated *oriM1* and *oriM2,* respectively (Fig. 7-3). Although *oriM1* overlaps with *oriC* (Fig. 7-1), the intact *oriC* site is not essential for the OriM1 activity. Thus, DnaA box mutations which completely inactivate *oriC* have no effect on iSDR initiation (Asai *et al.*, 1994b). In fact, each of the two adjacent segments within the minimal *oriC* has origin activity (Asai *et al.*, 1994b) (Fig. 7.1). As expected, DnaA protein is dispensable for the initiation at the *oriM* sites. Instead, the initiation requires homologous recombination proteins such as RecA and RecBCD. Duplex opening for iSDR initiation is very likely to be achieved by formation of a D-loop, an intermediate of homologous recombination, instead of the action of DnaA protein (Asai *et al.*, 1993). It was proposed that upon induction of the SOS response, a double-strand break is introduced at *oriM,* and the resulting ends are processed by RecBCD enzyme to generate single-stranded DNA with a 3′ end, which is then assimilated into an intact *oriM* site by the action of RecA recombinase, yielding a D-loop (Asai *et al.*, 1993). It was recently demonstrated that double-strand breaks artificially introduced into a plasmid trigger plasmid replication in a manner dependent on D-loop formation like iSDR (Asai *et al.*, 1994a). The activity, termed homologous recombination-dependent DNA replication (RDR), can occur in the absence of SOS induction

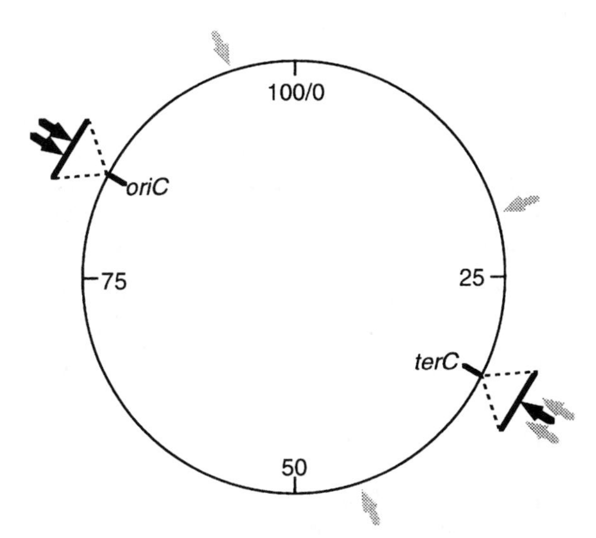

Figure 7-3. The maps of *oriM* and *oriK* sites. Three *oriM* sites (solid arrows) and five *oriK* sites (stippled arrows) are shown on the *E. coli* chromosome.

provided that the linearized plasmid molecules are protected from the RecBCD nuclease activity. The essential roles of RDR in homologous recombination and double-strand break repair have been proposed (Kogoma *et al.*, 1996; Kogoma, 1996). iSDR is envisioned to be a special type of RDR that starts exclusively from fixed sites (*oriM*s) on the chromosome (Asai *et al.*, 1994a).

oriK *sites*

Duplex opening for initiation can also be achieved by R-loop formation. Thus, an R-loop which is generated by invasion of the duplex by an RNA transcript (Hong et al., 1995) is stabilized in *E. coli rnhA* mutants due to the lack of RNase HI activity, giving rise to an initiation site for cSDR. Initiation of cSDR does not require DnaA protein. In fact, activation of cSDR by *rnhA* mutations renders the *oriC* site and *dnaA* gene dispensable for viability (Kogoma and von Meyenburg, 1983). The origins of cSDR, *oriK*s, map in several regions of the *E. coli* chromosome including the *terC* region (de Massy *et al.*, 1984) (Fig. 7-3). Despite considerable efforts, no *oriK* site has been cloned. Mutations in the *recG* gene encoding RecG helicase, which specifically recognizes the Holliday intermediate in homologous recombination and is proposed to be involved in removal of R-loops, were also shown to activate cSDR (Hong *et al.*, 1995). Whether the same *oriK* sites are used both in *rnhA* and *recG* mutants is not known.

Conclusion

The *oriC* sites of eubacteria have several common features. 1) Unlike the replication origins of plasmids and bacteriophages (Kornberg and Baker, 1992), they contain no coding region. 2) They contain one or two clusters of DnaA boxes and one or more of AT-rich elements, though this latter feature is obscure in high G+C bacteria. 3) They are located in the immediate vicinity of the *dnaA* gene, except in *E. coli* (42 kb apart) and *C. crescentus* (2 kb apart). *E. coli oriC* may have suffered transposition and inversion during the course of evolution (Yoshikawa and Ogasawara, 1991), and *C. crescentus oriC* may have acquired an intervening gene by insertion. Since the list of cloned *dnaA* genes is growing (Skarstad and Boye, 1994), additional *oriC* sites will continue to be identified and studied in the near future. The close examination of these *oriC* sites is expected to help elucidate the essential features of the bacterial origins of chromosome replication.

How many origins are present per chromosome? This important question has not been critically addressed except for *E. coli,* where the singleness of the *oriC* site was genetically demonstrated. Deletion of the *oriC* site kills the cell unless an integrated plasmid origin is activated (von Meyenburg and Hansen, 1980) or *oriK* sites are derepressed (Kogoma and von Meyenburg, 1983). In the case of *P. aeruginosa,* two different sequences capable of autonomous replication were isolated. One has sequence similarities to the *P. putida oriC* (see above) and the other is distinct. The latter sequence hybridizes to the *P. aeruginosa* chromosomal DNA fragments which show properties not expected of replication origins (Yee and Smith, 1990). It may therefore represent a cryptic origin of replication which is activated by cloning. *rnhA* mutants, in which several origins (*oriKs*) are simultaneously active, grow poorly particularly in nutritionally rich media because these multiple origins are not regulated (von Meyenburg *et al.*, 1987). This observation suggests that the bacterial chromosome normally has a single replication origin, or, if more than one origin is active, cells must have a regulatory mechanism that coordinates the activity of the multiple origins of replication.

Acknowledgments

I thank T. Asai, E. Boye, N. Ogasawara, and W. Messer for helpful comments on the manuscript. The work from my laboratory was supported by Public Health Service grant GM22092 from the National Institutes of Health and by National Science Foundation grant BIR9218818.

References

Alonso, J. C., and L. M. Fisher. 1995. Nucleotide sequence of the *recF* gene cluster from *Staphylococcus aureus* and complementation analysis in *Bacillus subtilis recF* mutants. *Mol. Gen. Genet.* 246:680–686.

Asai, T., D. B. Bates, and T. Kogoma. 1994a. DNA replication triggered by double-stranded breaks in E. coli: dependence on homologous recombination functions. *Cell* 78:1051–1061.

Asai, T., M. Imai, and T. Kogoma. 1994b. DNA damage-inducible replication of the *Escherichia coli* chromosome is initiated at separable sites within the minimal *oriC*. *J. Mol. Biol.* 235:1459–1469.

Asai, T., and T. Kogoma. 1994. D-loops and R-loops: alternative mechanisms for the initiation of chromosome replication in *Escherichia coli. J. Bacteriol.* 176:1807–1812.

Asai, T., S. Sommer, A. Bailone, and T. Kogoma. 1993. Homologous recombination-dependent initiation of DNA replication from DNA damage-inducible origins in *Escherichia coli. EMBO J.* 12:3287–3295.

Asai, T., M. Takanami, and M. Imai. 1990. The AT richness and *gid* transcription determine the left border of the replication origin of the *E. coli* chromosome. *EMBO J.* 9:4065–4072.

Bates, D. B., T. Asai, Y. Cao, M. W. Chambers, G. W. Cadwell, E. Boye, and T. Kogoma. 1995. The DnaA box R4 in the minimal *oriC* is dispensable for initiation of *Escherichia coli* chromosome replication. *Nucl. Acids Res.* 23:3119–3125.

Calcutt, M. J., and F. J. Schmidt. 1992. Conserved gene arrangement in the origin region of the *Streptomyces coelicolor* chromosome. *J. Bacteriol.* 174:3220–3226.

de Massy, B., O. Fayet, and T. Kogoma. 1984. Multiple origin usage for DNA replication in *sdrA* (*rnh*) mutants of *Escherichia coli* K12: initiation in the absence of *oriC*. *J. Mol. Biol.* 178:227–236.

Fujita, M. Q., H. Yoshikawa, and N. Ogasawara. 1990. Structure of the *dnaA* region of *Micrococcus luteus:* conservation and variations among eubacteria. *Gene* 93:73–78.

Fujita, M. Q., H. Yoshikawa, and N. Ogasawara. 1992. Structure of the *dnaA* and DnaA-box region in the *Mycoplasma capricolum* chromosome: conservation and variations in the course of evolution. *Gene* 110:17–23.

Hirota, Y., A. Oka, K. Sugimoto, K. Asada, H. Sasaki, and M. Takanami. 1981. *Escherichia coli* origin of replication: structural organization of the region essential for autonomous replication and the recognition frame model, in *The Initiation of DNA Replication, ICN-UCLA Symp. Mol. and Cell. Biol.* pp. 1–12. D. S. Ray, ed. Academic Press, New York.

Hong, X., G. W. Cadwell, and T. Kogoma. 1995. *Escherichia coli* RecG and RecA proteins in R-loop formation. *EMBO J.* 14:2385–2392.

Hsu, J., D. Bramhill, and C. M. Thompson. 1994. Open complex formation by DnaA initiation protein at the *Escherichia coli* chromosomal origin requires the 13-mers precisely spaced relative to the 9-mers. *Mol. Microbiol.* 11:903–911.

Kogoma, T. 1996. Recombination by replication. *Cell* 85:625–627.

Kogoma, T., G. W. Cadwell, K. G. Barnard, and T. Asai. 1996. Requirement of the DNA replication priming protein, PriA, for homologous recombination and double-strand break repair. *J. Bacteriol.* 178:1258–1264.

Kogoma, T., and K. von Meyenburg. 1983. The origin of replication, *oriC,* and the *dnaA* protein are dispensable in stable DNA replication (*sdrA*) mutants of *Escherichia coli* K12. *EMBO J.* 2:463–468.

Kornberg, A., and T. A. Baker. 1992. *DNA replication,* 2nd ed. W. H. Freeman and Co., New York.

Marczynski, G. T., and L. Shapiro. 1992. Cell-cycle control of a cloned chromosomal origin of replication from *Caulobacter crescentus. J. Mol. Biol.* 226:959–977.

Messer, W., and C. Weigel. 1996. Initiation of chromosome replication, in *Escherichia coli and Salmonella typhimurium,* 2nd ed. R. C. Neidhardt, J. L. Ingraham, E.C.C. Lin, K. B. Low, B. Magasanik, W. S. Reznikoff, M. Riley, M. Schaechter and H. E. Umbarger eds., Washington, D. C. American Society for Microbiology, pp. 1579–1601.

Moriya, S., T. Atlung, F. G. Hansen, H. Yoshikawa, and N. Ogasawara. 1992. Cloning of an autonomously replicating sequence (*ars*) from the *Bacillus subtilis* chromosome. *Mol. Microbiol.* 6:309–315.

Moriya, S., W. Firshein, H. Yoshikawa, and N. Ogasawara. 1994. Replication of a *Bacillus subtilis oriC* plasmid *in vitro. Mol. Microbiol.* 12:469–478.

Oka, A., H. Sasaki, K. Sugimoto, and M. Takanami. 1984. Sequence organization of replication origin of the *Escherichia coli* K-12 chromosome. *J. Mol. Biol.* 176:443–458.

Oka, A., K. Sugimoto, M. Takanami, and Y. Hirota. 1980. Replication origin of the *Escherichia coli* K-12 chromosome: The size and structure of the minimum DNA segment carrying the information for autonomous replication. *Mol. Gen. Genet.* 178: 9–20.

Renaudin, J., A. Marais, E. Verdin, S. Duret, X. Foissac, F. Laigret, and J. M. Bove. 1995. Integrative and free *Spiroplasma citri oriC* plasmids: expression of the *Spiroplasma phoeniceum* spiralin in *Spiroplasma citri*. *J. Bacteriol.* 177:2870–2877.

Skarstad, K., and E. Boye. 1994. The initiation protein DnaA: evolution, properties and function. *Biochim. Biophys. Acta* 1217:111–130.

von Meyenburg, K., E. Boye, K. Skarstad, L. Koppes, and T. Kogoma. 1987. Mode of initiation of constitutive stable DNA replication in RNase H-defective mutants of *Escherichia coli* K-12. *J. Bacteriol.* 169:2650–2658.

von Meyenburg, K., and F. G. Hansen. 1980. The origin of replication, *oriC,* of *Escherichia coli* chromosome: gene near *oriC* and construction of *oriC* deletion mutations. *ICN-UCLA Symp. Mol. Cell. Biol.* 21:137–159.

Woelker, B., and W. Messer. 1993. The structure of the initiation complex at the replication origin, *oriC,* of *Escherichia coli. Nucl. Acids Res.* 21:5025–5033.

Ye, F., J. Renaudin, J. Bove, and J.-M. Laigret. 1994. Cloning and sequencing of the replication origin (*oriC*) of the *Spiroplasma citri* chromosome and construction of autonomously replicating artificial plasmids. *Curr. Microbiol.* 29:23–29.

Yee, T. W., and D. W. Smith. 1990. *Pseudomonas* chromosomal replication origins: a bacterial class distinct from *Escherichia coli*-type origins. *Proc. Natl. Acad. Sci. USA* 87:1278–1282.

Yoshikawa, H., and N. Ogasawara. 1991. Structure and function of DnaA and the DnaA-box in eubacteria: evolutionary relationship of bacterial replication origins. *Mol. Microbiol.* 5:2589–2597.

Zakrzewska-Czerwinska, J., and H. Schrempf. 1992. Characterization of an autonomously replicating region from the *Streptomyces lividans* chromosome. *J. Bacteriol.* 174:2688–2693.

Zweigler, G. and L. Shapiro. 1994. Expression of *Caulobacter dnaA* as a function of the cell cycle. *J. Bacteriol.* 176:401–408.

Zyskind, J. W., J. M. Cleary, W. S. A. Brusilow, N. E. Harding, and D. W. Smith. 1983. Chromosomal replication origin from the marine bacterium *Vibrio harveyi* functions in *Escherichia coli: oriC* consensus sequence. *Proc. Natl. Acad. Sci. USA* 80:1164–1168.

8

Restriction Modification Systems: Where They Are and What They Do

Elisabeth A. Raleigh and Joan E. Brooks

Definition

This review concentrates on restriction-modification (RM) in the context of bacterial genome evolution and how the systems affect bacterial populations. RM systems regulate the entry of foreign DNA into cells. A model of how the systems work is shown in Figure 8-1. Foreign DNA is *restricted* by a *restriction endonuclease* that recognizes a specific sequence and cleaves the DNA unless the sequence is protected. Typically, a *modification methyltransferase* confers protection, by methylating a particular base within the sequence recognized by the restriction enzyme, thereby rendering it resistant to cleavage.

Alternatively, however, some restriction enzymes recognize a sequence only when it *is* methylated. In this case, methylation of a suitable base confers sensitivity to restriction and protection arises from failure to methylate the relevant sequence. Both sorts of restriction can act to limit the transfer of DNA into cells.

One key feature of RM is that the systems can be effective only if they are variable and fluid within a bacterial population. Modifications made to foreign DNA escaping restriction are epigenetic, i.e., not heritable.

How are they found?

Genetic Methods

The first RM systems were detected by genetic means, as host factors reducing phage viability in *E. coli* (Bertani and Weigle, 1953; Luria and Human, 1952). Restriction also acts to reduce the frequency of other forms of genetic exchange in *E. coli,* including the transfer of chromosomal markers by conjugation or transduction, as well as plasmid transfer (Arber and Linn, 1969; Raleigh, 1992).

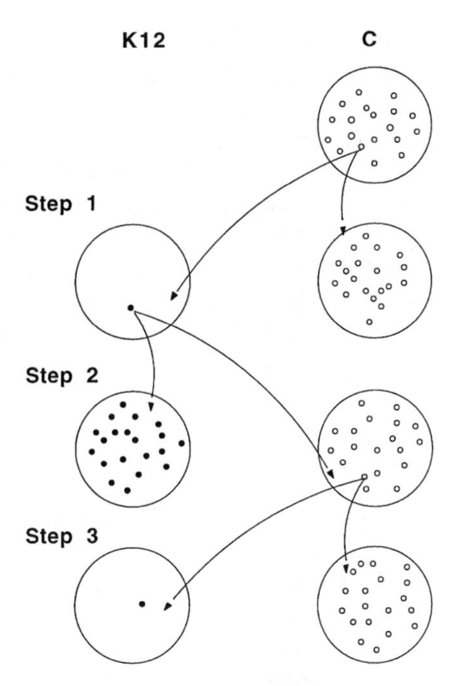

Figure 8-1. Restriction-modification (RM) is exemplified by exchange of DNA between two common laboratory strains of *E. coli,* K-12 and C. K-12 has an RM system, but C does not. Large circles represent petri plates, while small circles represent successful DNA transfers: in this example, colonies of bacteria that have acquired a selectable plasmid by conjugation. Colonies exhibiting K-12 modification are filled circles, while colonies lacking it are open circles.

Step 1. Progeny of an initial plasmid-containing colony of C are transferred at high frequency back to itself, but at low frequency (10^{-2}) to the restricting strain, K-12.

Step 2. Retransfer from K12 back to itself is not restricted and occurs at high frequency; the DNA is said to be K-12-modified, and protected from restriction. Retransfer to the C strain also occurs at high frequency, due to lack of a restriction system in C.

Step 3. The presence of an RM system is definitively demonstrated: all the progeny of plasmids passed through C have lost K-12-specific protection and regained sensitivity to K-12-specific restriction. As before, retransfer from C back to itself is at high frequency.

In chromosomal recombination, restriction also has a profound effect on the size and structure of replacement patches (McKane and Milkman, 1995). However, it is noteworthy that in bacteria naturally competent for transformation, transfer of chromosomal markers is not restricted, either in Gram-positive (Noyer-Weidner and Trautner, 1993) or Gram-negative bacteria (Kahn *et al.*, 1983; Stein, 1991; but see Zawadzki *et al.*, 1995). This is attributed to the transport of single-stranded DNA into the cells, followed by the efficient formation of heteroduplex

DNA, which is insensitive to cleavage. Other forms of gene transfer are still restricted in these organisms (Butler and Gotschlich, 1991; Noyer-Weidner and Trautner, 1993).

New RM systems are still being discovered as researchers develop cloning systems in previously uncharacterized organisms. Sometimes the systems can be circumvented by *in vitro* modification of cloning vectors (e.g. in particular strains of *Nostoc* and *Xanthomonas;* De Feyter and Gabriel, 1991; Moser *et al.*, 1993). In other cases, researchers find it expedient to remove the systems, usually without identifying them. In one particularly industrious effort, five independent RM systems were removed from a single strain of *Streptomyces* (Matsushima et al., 1987).

Biochemical Method

This is useful for systems whose restriction component works well *in vitro* and cuts DNA in a simple site-specific manner. Endonucleolytic activity yielding defined fragments is detected in crude extracts of bacterial cells when used to cleave standard DNA substrates followed by agarose gel electrophoresis (Smith and Marley, 1980). Because of the utility of this type of enzyme, extensive screening programs have been carried out. Thousands of site-specific endonucleases have been found, many from obscure organisms with no available genetic tools for testing biological effects. In most cases no attempt was made to fully characterize the RM system components. An electronic database of endonucleases of this group is accessible (Roberts and Macelis, 1994) and is updated continuously.

Sequence-based Methods

Identification of new systems by homology has recently become possible. In some cases, DNA similarity is sufficient for Southern blots to be useful (Daniel et al., 1988). In addition, amino acid sequence comparisons among RM systems have identified a common architecture for DNA methyltransferases, enabling recognition of methyltransferase coding sequences in DNA (Lauster et al., 1989; Pósfai et al., 1989; Timinskas et al., 1995). This method now may replace molecular biological and genetic methods for precisely locating methyltransferase genes within cloned DNA fragments (e.g. Noyer-Weidner *et al.*, 1994.) With the coming of large-scale sequencing, previously unsuspected RM candidates can be identified. Researchers located not only the two known systems of *Haemophilus influenzae*, Rd (*Hind*II and *Hind*III) but also four other candidate systems (Fleischmann *et al.*, 1995). Of course, the sequence alone cannot predict whether or under what conditions the putative system are active.

How are They Organized?

Both sequence recognition and the molecular organization of RM systems are highly variable. This subject has been well reviewed (Barcus and Murray, 1995; Bickle and Krüger, 1993; Wilson and Murray, 1991) and will only be sketched here.

The Type II Systems

These are the familiar and commercially available restriction enzymes. The modification methyltransferase and endonuclease proteins are specified by separate genes and they act independently. The recognition sequence is four to eight base pairs in length. Although both proteins recognize the same sequence, they do so quite differently and are unrelated in protein sequence. The methyltransferase requires only S-adenosylmethionine (AdoMet) as cofactor. The endonuclease requires only Mg^{++} as cofactor and cuts at a defined position within or adjacent to the recognition site, yielding discrete banding patterns on agarose gel electrophoresis. It was this group that proved easy to identify by screening extracts for sequence-specific endonuclease activity.

Most enzymes identified this way (but not type IIS, see below) recognize palindromic sequences and the endonuclease cleaves within the site. For the few endonucleases examined crystallographically, residues involved in recognition and catalysis are tightly integrated in space and within the protein sequence. No protein sequence similarity has been detected between enzymes that recognize different sequences.

Type IIS Systems

The type IIS endonucleases, which split recognition and catalysis into separate domains, are a fundamentally distinct class of enzyme. A DNA Recognition Domain (DRD) recognizes the sequence, which is asymmetric, and a catalytic domain cleaves at a defined location to one side (Li and Chandrasegaran, 1993; Li *et al.*, 1992). As with other type II enzymes, Mg^{++} is the only cofactor, and modification is carried out by a separate protein. Because of the asymmetry of the site, the sequences of the two strands differ, and these are modified by different methyltransferase activities. These activities are usually found in two separate proteins but sometimes as two domains within one protein (Bickle and Krüger, 1993). None of the type IIS systems has yet been associated with a biological restriction phenotype.

Type I Systems

These, exemplified by *Eco*KI and *Eco*BI, are much more complex than type II systems. DNA recognition, methyl transfer, and DNA cleavage are divided among three proteins, but the action of the three is tightly integrated by protein-protein and protein-cofactor interactions. Target recognition for both methylation and restriction is carried out by the specificity polypeptide, HsdS. The recognition sequences are asymmetric and bipartite, with two specific sequence elements separated by a nonspecific spacer of fixed length. Each specific sequence element is recognized by a separate well-defined DRD of the HsdS protein. Two polypeptides (HsdM and HsdS) and AdoMet are required for modification activity; HsdR is required together with the others for restriction activity. Endonuclease cleavage requires ATP, Mg^{++} and AdoMet and occurs at a large and variable distance from the recognition site. No banding pattern is detected on agarose gels, except under special conditions.

Other Types

Type III enzymes comprise two proteins. Mod, a two-domain protein, both determines specificity via its DRD and carries out methylation. A second polypeptide, Res, and cofactors ATP and Mg^{++} are required in addition to Mod for cleavage activity. The recognition sequences of type III systems are asymmetric and uninterrupted; endonucleolytic cleavage is at a short fixed distance from the recognition sequence. Other ATP-dependent nucleases can interfere with detection of type III enzymes in crude extracts by agarose gel assays.

As more RM systems are found and characterized, new ones with novel subunit structures, cofactor requirements and cleavage patterns are still being identified, such as *Eco*57I (Janulaitis *et al.*, 1992), *Bcg*I (Kong *et al.*, 1994) and *Lla*I (O'Sullivan *et al.*, 1995). It is difficult to accommodate all types into one evolutionary scheme.

Modification-dependent Systems

These are RM systems that only restrict *modified* DNA substrates. DNA is protected by the absence of particular modified sequences. For technical reasons, these are difficult to assay biochemically, so their prevalence and distribution are uncertain, but they could be widespread.

At least four mechanistically distinct classes are known. The best known of these is *Dpn*I from *Streptococcus pneumoniae* (previously called *Diplococcus pneumoniae*). The *Dpn*I endonuclease is encoded by one gene, and the protein looks and acts like a type II endonuclease. However, it will only cleave DNA with the appropriate modified base, N6-methyladenine, in its recognition sequence. Similar enzymes are widely distributed (see below). A methyladenine-dependent enzyme from *Streptomyces bambergensis* (Zotchev *et al.*, 1995) may also be a isoschizomer.

Three mechanistically distinct sequence-specific, modification-dependent systems have been identified in *E. coli* K12 (Bickle and Krüger, 1993), where these systems have been the most extensively studied. Two, Mrr and McrA, are specified by single genes. These are unrelated to each other or to any other known restriction gene. Mrr is the only system known to restrict DNA containing either methyladenine or 5-methylcytosine in particular sequences. McrA restricts DNA with modified cytosine (hydroxymethylcytosine or methylcytosine) *in vivo.* No biochemical characterization of McrA activity has been carried out, but protein sequence similarity to homing endonucleases has been noted (Ferat and Michel, 1993; Gorbalenya, 1994; Shub *et al.*, 1994).

The most extensively characterized of the modification-dependent systems in *E. coli,* McrBC, has been recently reviewed (Raleigh, 1992). Like McrA, it acts on DNA containing modified cytosine. This system acts as a multi-subunit nucleotide-dependent endonuclease. Unlike other nucleotide-dependent nucleases, it displays an absolute requirement for GTP rather than ATP. It is unrelated to other complex restriction systems at the sequence level.

Anecdotal evidence exists for modification-dependent restriction systems in *Acholeplasma laidlawii* (Sladek *et al.*, 1986), *Bacillus thuringiensis* (Macaluso and Mettus, 1991), coryneform bacteria (Vertès *et al.*, 1993), and a number of streptomycetes, most notably *Streptomyces avermitilis* (MacNeil, 1988). A methylation-dependent restriction system has also been found in Archaea (Holmes *et al.,* 1991).

Population Distribution

The distribution among prokaryotes of restriction enzymes of different types and specificities yields a picture of a pool of genes that have circulated with few taxonomic limitations for a very long time.

Taxonomic limitations are few. Restriction systems are ubiquitous among prokaryotes, but their presence or identity are not taxonomically informative. For technical reasons, the data are most complete for type II enzymes, which have been characterized from at least 11 of 13 phyla of Bacteria and Archaea (Olsen *et al.*, 1994; Roberts and Macelis, 1994). The lack of representatives from the two remaining phyla (Chlamydia and Spirochaetes) may well reflect sampling bias in the dataset, since these are relatively difficult to culture. 76% of isolates assigned to phyla in this survey were derived from low-GC gram positive organisms (especially *Bacillus*) or Proteobacteria; this preponderance reflects more intensive study of these taxa than others.

Particular type II specificities may be found throughout the phylogenetic tree. Four of the most common specificities are CGCG, GGCC, CCGG, and GATC, representing together about 12% of all isolates. These are found throughout the prokaryotes (Table 8-1), with all but CCGG found in the extremes of the phylogenetic tree, the Archaea and the Proteobacteria.

Recent evidence suggests that type I systems may also be widely distributed,

Table 8-1. Distribution of selected type II restriction specificities in prokaryotes.

Domain/Phylum	CCGG		GGCC		CGCG		GATC		G^mATC	
	Enzyme	Organism	Enzyme	Organism	Enzyme	Organism	Enzyme	Organism	Enzyme	Organism
Archaea										
1 Euryarchaeota			*Mth*TI	*Methanobacterium thermoformicicum*	*Mvn*I	*Methanococcus vannielli*	*Mth*AI	*Methanobacterium thermoautotrophicum*		
2 Crenarchaeota			*Sua*I	*Sulfolobus acidocaldarius*	*Tha*I	*Thermoplasma acidophilum*				
Bacteria										
3 Thermotogales					*Tma*I	*Thermotoga maritima*				
4 Green nonsulfur bacteria										
*a *Thermus thermophilus*			*Tsp*560I	*Thermus species*			*Tru*II	*Thermus ruber*		
*b *Deinococcus radiodurans*										
5 Cyanobacteria	*Sec*I	*Synechocystis sp.*	*Aca*IV	*Anabaena catenula*	*Sce*I	*Synechococcus cedrorum*				
6 Low G+C Gram Positive	*Bsu*F	*Bacillus subtilis*	*Bsu*RI	*Bacillus subtilis*	*Bsu*6633I	*Bacillus subtilis*	*Bce*I	*Bacillus cereus*	*Dpn*I	*Streptococcus pneumoniae*
7 Fusobacteria			*Fnu*DI	*Fusobacterium nucleatum*	*Fnu*DII	*Fusobacterium nucleatum*	*Fnu*EI	*Fusobacterium nucleatum*		
8 High G+C Gram Positive			*Mch*AI	*Mycobacterium cheloni*	*Bep*I	*Brevibacterium epidermidis*	*Mgo*I	*Mycobacterium gordonae*		
9 Cytophaga/Flexibacter/ Bacteroides	*Fin*II	*Flavobacterium indologenes*	*Fin*SI	*Flavobacterium indoltheticum*	*Fau*BII	*Flavobacterium aureus*				
10 Fibrobacteria										
11 Spirochaetes										
12 Planctomyces/Chlamydia										
13 Proteobacteria (Purple bacteria)	*Hpa*II	*Haemophilus parainfluenzae*	*Hae*III *Ngo*PII	*Haemophilus aegyptius* *Neisseria gonorrhoeae*	*Hin*1056I	*Haemophilus influenzae*	*Nde*II	*Neisseria denitrificans*	*Nmu*DI	*Neisseria mucosa*

*Species indicated with letters were placed in the tree but not grouped in named phyla (Olsen, 1994).

since systems active *in vivo* and related to type I enzymes by sequence have been found in *Mycoplasma* (Dybvig and Yu, 1994) and *Bacillus* (Xu *et al.*, 1995; Ikawa *et al.*, 1980); these belong to a different bacterial phylum than the enteric bacteria (Olsen *et al.*, 1994) where type I enzymes have been well characterized.

It is also clear that the evolution of both type I and type II systems has involved horizontal transfer over large taxonomic distance. For example, even though they were isolated from widely different taxa (Table 8-1), the isoschizomeric enzymes *Mth*TI, *Fnu*DI and *Ngo*PII (GGCC) are very similar; identity with *Ngo*PII is 54% for *Mth*TI (Nolling and de Vos, 1992) and 59% for *Fnu*DI (Wilson and Murray, 1991). Not all GGCC enzymes are related, however; the amino acid sequences of *Hae*III and *Ngo*PII, which were isolated from the same phylum, show no relationship. Also unlike are type I enzymes found as alleles within *E. coli* populations, which may be so dissimilar in sequence as to reflect divergence at the phyletic level; horizontal transfer was invoked to explain this divergence (Murray *et al.*, 1993).

Intrataxon diversity is high

Diversity within species is very great, no matter at what level diversity is examined. *E. coli* has been the most extensively studied. There are at least six distinct mechanistic classes of restriction enzyme within the species (types I, II and III and three modification-dependent types). Sequence specificity is also highly variable. In this species, at least 28 different type II specificities (Janulaitis *et al.*, 1988) and at least 14 type I specificities (Barcus and Murray, 1995; Barcus *et al.*, 1995) have been found. Specificity for restriction of modified DNA must also vary within *E. coli,* because some methyltransferases isolated from *E. coli* (e.g. M.*Eco*47II and M.*Eco*47III) confer sensitivity to the K12 allele of McrBC (Povilenis *et al.*, 1989).

Numerous systems are frequently present in the same cell. Four active systems are found in *E. coli* K12; *Haemophilus influenzae* Rd expresses at least three systems that are biologically active (Roszczyk and Goodgal, 1975) and three more, plus a *dam*-like methyltransferase, are recognizable in the complete sequence (Fleischmann *et al.*, 1995).

Variation

Intrinsic mechanisms for varying specificity

Three mechanisms have been described for changing type I restriction specificity without importing new DNA sequence. A type I system from *Mycoplasma pulmonis* is subject to phase variation by inversion of the DNA, such that one specificity is expressed in only a portion of the population (Dybvig and Yu, 1994). A second mechanism allows variation by changing the number of nonspecific basepairs required between the two specific elements of the recognition site. For *Eco*R124I, an enteric type I system, this variation has been observed to occur with a frequency of about 10^{-7} (Glover *et al.*, 1983). The switch is accomplished by unequal crossing-over within a region encoding a "measuring" segment of the recognition protein, which connects the two DRDs (Price *et al.*, 1989). Finally, in two different instances, new specificities have been created by deletion of one of the two DRDs characteristic of type I enzymes (Abadjieva *et al.*, 1993; Meister *et al.*, 1993). In both instances, the remaining polypeptide was able to dispose the lone DRD in a dimeric arrangement, so that a palindromic repeat of the appropriate sequence element could now be recognized. A model has been proposed for the organization of polypeptide sequence that enables this event (Kneale, 1994).

Varying specificity by exchange

For enzymes that have segregated target recognition into one or more DRDs separate from the catalytic functions, specificity can be varied by domain swaps. Exchange of one DRD for another has been accomplished by *in vitro* methods for modification methyltransferases (Klimasauskas *et al.*, 1991; Mi and Roberts, 1992; Walter *et al.*, 1992) and for the type IIS endonuclease *Fok*I (Kim and Chandrasegaran, 1994). In the latter case, the DNA binding domain of a eukaryotic transcription factor was substituted for the DRD. In these cases, the specific activity of the resultant construct was generally low, and in the latter case the selectivity of the enzyme was poor compared with true restriction enzymes. Presumably pressures for more specific and more efficient action would eventually fine-tune the action of such chimeras in nature.

The modular organization that permits change of specificity by these means very likely results from selection to enable such changes in the real world. The type I enzymes are clearly organized to facilitate combinatorial variation in restriction (and methylation) specificity. The two specific elements of DNA sequence that comprise the bipartite recognition site of type I enzymes are recognized essentially independently by the two highly variable DRDs of the specificity polypeptide. The connector and flanking regions are much more highly conserved than the DRDs—so much so that homologous recombination between genes of different specificity has been observed to yield completely new specificities by pairwise reassortment of DRDs (Fuller-Pace *et al.*, 1984; Gann *et al.*, 1987). These naturally reassorted recognition proteins function quite efficiently *in vivo.*

Evolving New Specificities

Although domain swaps can yield novel recognition sequences for old enzyme frameworks, the evolution of truly new DRDs is an open problem (for a discussion of this issue see Roberts and Halford, 1993). Some progress has been made in identifying relationships between the target recognition domains of methyltransferases of different but related specificities (Gopal et al., 1994; Kumar et al., 1994) but the rules governing de novo construction of a DRD are so far undiscovered.

Conclusion: Open Questions

Although much remains to be clarified concerning the mechanistic details of sequence recognition by restriction enzymes, there are also wider questions. In our view there are four open questions that relate to the role of restriction systems in the population, and to their effects on "the genome."

A major question is how fast and by what mechanisms a population diversifies with respect to its complement of RM systems after it has been homogenized by selection. Clonal replacement, the expansion of a single (presumably fitter) genotype to take over the niche previously occupied by a diverse population, is thought to occur frequently (Levin, 1981; Selander *et al.*, 1987). An epidemiologic series would be of particular value, since new pandemics may represent clonal expansions (Achtman, 1994; Karaolis *et al.*, 1995), and suitable strain collections exist. Both the processes by which new specificities are acquired and the regulatory mechanisms that contribute to the changeover are of interest.

A second, related question concerns the evolutionary pressures driving diversification of restriction enzymes. Population genetic experiments suggest that restriction may be of selective value to the host organism only episodically and transiently (Korona and Levin, 1993; Levin, 1988), so that systems may fall into disuse and require resurrection or new recruitment from other sources. It has also been suggested that restriction genes, like plasmids and transposons, may respond to selection on the genes themselves, favoring retention and spread of the genes directly, rather than through contributions to the competitive success of the organisms carrying them (Kulakauskas *et al.*, 1995; Naito *et al.*, 1995). The predictions of this "selfish DNA" model have not yet been fully explored.

A third question concerns the relationship between modification-dependent enzymes and more conventional restriction. The sequence-specificity of modification-dependent systems sets limits on what sorts of conventional enzymes can be maintained in the same cell. What effect does this have on the behavior of the cell population on the one hand, and on the behavior of circulating restriction genes on the other?

The final question is how these enzymes have shaped the genomes of their hosts. Whatever the pressures driving diversification of restriction systems, these systems do function as gatekeepers, restricting the entrance of foreign DNA and thus limiting the pool of genes in circulation within a population. On the other hand, restriction may facilitate some types of genetic exchange. It has been proposed that restriction usually breaks DNA into pieces conveniently the size of one or a few genes (McClelland, 1988); it also provides double-strand ends, which are recombinogenic (e.g. Myers and Stahl, 1994). Detailed analysis of the molecular consequences of restriction for the outcome of general recombination has only just begun (McKane and Milkman, 1995), but it seems likely that the mosaic structure of the *E. coli* genome (Milkman and Bridges, 1993; *see also* Chapters 17 and 18) results at least in part from recombination acting on restricted fragments. How large a role has restriction played, compared with other forces? Is that role the same in all taxa? Answers to these questions will increase our understanding of prokaryotic genome evolution.

References

Abadjieva, A., J. Patel, M. Webb, V. Zinkevich, and K. Firman. 1993. A deletion mutant of the type IC restriction endonuclease *EcoR*1241 expressing a novel DNA specificity. *Nucl. Acids Res.* 21:4435–4443.

Achtman, M. 1994. Clonal spread of serogroup A meningococci: a paradigm for the analysis of microevolution in bacteria. *Mol. Microbiol.* 11:15–22.

Arber, W., and S. Linn. 1969. DNA modification and restriction. *Annu. Rev. Biochem.* 38:467–500.

Barcus, V. A., and N. E. Murray. 1995. Barriers to recombination: restriction. *Soc. Gen. Microbiol. Symp.* 52:31–58.

Barcus, V. A., A. J. B. Titheradge, and N. E. Murray. 1995. The diversity of alleles at the *hsd* locus in natural populations of *Escherichia coli. Genetics* 140:1187–1197.

Bertani, G., and J. J. Weigle. 1953. Host-controlled variation in bacterial viruses. *J. Bacteriol.* 65:113–121.

Bickle, R. A., and D. H. Krüger. 1993. Biology of DNA restriction. *Microbiol. Rev.* 57:434–450.

Butler, C. A., and E. C. Gotschlich. 1991. High-frequency mobilization of broad-host-range plasmids into *Neisseria gonorrhoeae* requires methylation in the donor. *J. Bacteriol.* 173:5793–5799.

Daniel, A. S., F. V. Fuller-Pace, D. M. Legge, and N. E. Murray. 1988. Distribution and diversity of *hsd* genes in *Escherichia coli* and other enteric bacteria. *J. Bacteriol.* 170:1775–1782.

De Feyter, R., and D. W. Gabriel. 1991. Use of cloned DNA methylase genes to increase the frequency of transfer of foreign genes into *Xanthomonas campestris* pv. *malvacearum. J. Bacteriol.* 173:6421–6427.

Dybvig, K., and H. Yu. 1994. Regulation of a restriction and modification system via DNA inversion in *Mycoplasma pulmonis. Mol. Microbiol.* 12:547–560.

Ferat, J.-L., and F. Michel. 1993. Group II self-splicing introns in bacteria. *Nature* 364:358–361.

Fleischmann, R. E., M. D. Adams, O. White, R. A. Clayton, E. F. Kirkness, A. R. Kerlavage, C. J. Bult, J. F. Tomb, B. A. Dougherty, J. M. Merrick, K. McKenney, G. Sutton,

W. FitzHugh, C. Fields, J. D. Gocayne, J. Scott, R. Shirley, L. Liu, A. Glodek, J. M. Kelley, J. F. Wiedman, C. A. Phillips, T. Spriggs, E. Hedblom, M. D. Cotton, T. R. Utterback, M. C. Hanna, D. T. Nguyen, D. M. Saudek, R. C. Brandon, L. D. Fine, J. L. Fritchman, J. L. Fuhrmann, N. S. M. Geoghagan, C. L. Gnehm, L. A. McDonald, K. V. Small, C. M. Fraser, H. O. Smith, and J. C. Venter. 1995. Whole-genome random sequencing and assembly of *Haemophilus influenzae* Rd. *Science* 269:496–512.

Fuller-Pace, F. V., L. R. Bullas, H. Delius, and N. E. Murray. 1984. Genetic recombination can generate altered restriction specificity. *Proc. Natl. Acad. Sci. USA* 81:6095–6099.

Gann, A.A.F., A.J.B. Campbell, J. F. Collins, A. F. W. Coulson, and N. E. Murray. 1987. Reassortment of DNA recognition domains and the evolution of new specificities. *Mol. Microbiol.* 1:13–22.

Glover, S. W., K. Firman, G. Watson, C. Price, and S. Donaldson. 1983. The alternate expression of two restriction and modification systems. *Mol. Gen. Genet.* 190:65–69.

Gopal, J., M. J. Yebra, and A. S. Bhagwat. 1994. *Dsa*V methyltransferase and its isoschizomers contain a conserved segment that is similar to the segment in *Hha*I methyltransferase that is in contact with DNA bases. *Nucl. Acids Res.* 22:4482–4488.

Gorbalenya, A. E. 1994. Self-splicing group I and group II introns encode homologous (putative) DNA endonucleases of a new family. *Protein Sci.* 3:1117–1120.

Gorbalenya, A. E., and E. V. Koonin. 1991. Endonuclease (R) subunits of type-I and type-III restriction-modification enzymes contain a helicase-like domain. *FEBS Lett.* 291:277–281.

Holmes, M. L., S. D. Nuttall, and M. L. Dyall-Smith. 1991. Construction and use of halobacterial shuttle vectors and further studies on *Haloferax* DNA gyrase. *J. Bacteriol.* 173:3807–3813.

Ikawa, S., S. Takehiko, T. Ando, and H. Saito. 1980. Genetic studies on site-specific endodeoxyribonucleases in *Bacillus subtilis:* multiple modification and restriction systems in transformant of *Bacillus subtilis* 168. *Mol. Gen. Genet.* 177:359–368.

Janulaitis, A., R. Kazlauskiene, L. Lazareviciute, R. Glvonauskaite, D. Steponaviciene, M. Jagelavicius, M. Petrusyte, J. Bitinaite, Z. Veneviciute, E. Kiuduliene, and V. Butkus. 1988. Taxonomic specificity of restriction-modification enzymes. *Gene* 74:229–232.

Janulaitis, A., M. Petrusyte, Z. Maneliene, S. Klimasauskas, and V. Butkus. 1992. Purification and properties of the *Eco*57I restriction endonuclease and methylase—prototypes of a new class (type IV). *Nucl. Acids Res.* 20:6043–6049.

Janulaitis, A., R. Vaisvila, A. Timinskas, S. Klimasauskas, and V. Butkus. 1992. Cloning and sequence analysis of the genes coding for *Eco*57I type IV restriction-modification enzymes. *Nucl. Acids Res.* 20:6051–6056.

Kahn, M. E., F. Barany, and H. O. Smith. 1983. Transformasomes: specialized membranous structures that protect DNA during *Haemophilus* transformation. *Proc. Natl. Acad. Sci. USA* 80:6927–6931.

Karaolis, D. K., R. Lan, and P. R. Reeves. 1995. The sixth and seventh cholera pandemics are due to independent clones separately derived from environmental, nontoxigenic, non-01 *Vibrio cholerae. J. Bacteriol.* 177:3191–3198.

Kim, Y. G., and S. Chandrasegaran. 1994. Chimeric restriction endonuclease. *Proc. Natl. Acad. Sci. USA* 9:883–887.

Klimasauskas, S., J. L. Nelson, and R. J. Roberts. 1991. The sequence specificity domain of cytosine-C5 methylases. *Nucl. Acids Res.* 19:6183–6190.

Kneale, G. G. 1994. A symmetrical model for the domain structure of type I DNA methyltransferases. *J. Mol. Biol.* 243:1–5.

Kong, H., S. E. Roemer, P. A. Waite-Rees, J. S. Benner, G. G. Wilson, and D. O. Nwankwo. 1994. Characterization of *Bcg*I, a new kind of restriction-modification system. *J. Biol. Chem.* 269:683–690.

Korona, R., and B. R. Levin. 1983. Phage-mediated selection and the evolution and maintenance of restriction-modification. *Evolution* 47:556–575.

Kulakauskas, S., A. Lubys, and S. D. Ehrlich. 1995. DNA restriction-modification systems mediate plasmid maintenance. *J. Bacteriol.* 177:3451–3454.

Kumar, S., X. Cheng, S. Klimasauskas, S. Mi, J. Posfai, R. J. Roberts, and G. G. Wilson. 1994. The DNA (cytosine-5) methyltransferases. *Nucl. Acids Res.* 22:1–10.

Lauster, R., T. A. Trautner, and M. Noyer-Weidner. 1989. Cytosine-specific type II DNA methyltransferases: a conserved enzyme core with variable target-recognizing domains. *J. Mol. Biol.* 206:305–312.

Levin, B. R. 1988. Frequency-dependent selection in bacterial populations. *Philos. Trans. R. Soc. Lond. B. Biol. Sci.* 319:459–472.

Levin, B. R. 1981. Periodic selection, infectious gene exchange and the genetic structure of *E. coli* populations. *Genetics* 99:1–23.

Li, L., and S. Chandrasegaran. 1993. Alteration of the cleavage distance of *Fok*I restriction endonuclease by insertion mutagenesis. *Proc. Natl. Acad. Sci. USA* 90:2764–2768.

Li, L., L. P. Wu, and S. Chandrasegaran. 1992. Functional domains in *Fok*I restriction endonuclease. *Proc. Natl. Acad. Sci. USA* 89:4275–4279.

Luria, S. E., and M. L. Human. 1952. A nonhereditary, host-induced variation of bacterial viruses. *J. Bacteriol.* 64:557–559.

Macaluso, A., and A.-M. Mettus. 1991. Efficient transformation of *Bacillus thuringiensis* requires nonmethylated plasmid DNA. *J. Bacteriol.* 173:1353–1356.

MacNeil, D. J. 1988. Characterization of a unique methyl-specific restriction system in *Streptomyces avermitilis*. *J. Bacteriol.* 170:5607–5612.

Matsushima, P., K. L. Cox, and R. H. Baltz. 1987. Highly transformable mutants of *Streptomyces fradiae* defective in several restriction systems. *Mol. Gen. Genet.* 206:393–400.

McClelland, M. 1988. Recognition sequences of type II restriction systems are constrained by the G+C content of host genomes. *Nucl. Acids Res.* 16:2283–2294.

McKane, M., and R. Milkman. 1995. Transduction, restriction and recombination patterns in *Escherichia coli*. *Genetics* 139:35–43.

Meister, J., M. MacWilliams, P. Hubner, H. Jutte, E. Skrzypek, A. Piekarowicz, and T. A. Bickle. 1993. Macroevolution by transposition: drastic modification of DNA recognition by a type I restriction enzyme following Tn5 transposition. *EMBO J.* 12:4585–4891.

Mi, S., and R. J. Roberts. 1992. How M.*Msp*I and M.*Hpa*II decide which base to methylate. *Nucl. Acids Res.* 20:4811–4816.

Milkman, R., and M. M. Bridges. 1993. Molecular evolution of the *Escherichia coli* chromosome. IV. Sequence comparisons. *Genetics* 133:455–468.

Moser, D. P., D. Zarka, and T. Kallas. 1993. Characterization of a restriction barrier and electrotransformation of the cyanobacterium *Nostoc* PCC 7121. *Arch. Microbiol.* 160:229–237.

Murray, N. E., A. S. Daniel, G. M. Cowan, and P. M. Sharp. 1993. Conservation of motifs within the unusually variable polypeptide sequences of type I restriction and modification enzymes. *Mol. Microbiol.* 9:133–143.

Myers, R. S., and F. W. Stahl. 1994. Chi and RecBCD enzyme of *Escherichia coli. Annu. Rev. Genet.* 28:49–70.

Naito, T., K. Kusano, and I. Kobayashi. 1995. Selfish behavior of restriction-modification systems. *Science* 267:897–899.

Nolling, J., and W. M. de Vos. 1992. Characterization of the archaeal, plasmid-encoded Type II restriction-modification system *Mth*TI from *Methanobacterium thermoformicicum* THF: homology to the bacterial *Ngo*PII system from *Neisseria gonorrhoeae. J. Bacteriol.* 174:5719–5726.

Noyer-Weidner, M., and T. A. Trautner. 1993. Methylation of DNA in prokaryotes. *EXS* 64:39–108.

Noyer-Weidner, M., J. Walter, P. A. Terschuren, S. Chai, and T. A. Trautner. 1994. M.phi 3TII: a new monospecific DNA (cytosine-C5) methyltransferase with pronounced amino acid sequence similarity to a family of adenine-N6-DNA-methyltransferases [corrected and republished article originally printed in Nucl. Acids Res. 1994 Oct 11; 22(20):4066–4072]. *Nucl. Acids Res.* 22:5517–5523.

O'Sullivan, D. J., K. Zagula, and T. R. Klaenhammer. 1995. In vivo restriction by *Lla*I is encoded by three genes, arranged in an operon with *llaIM,* on the conjugative *Lactococcus* plasmid pTR2030. *J. Bacteriol.* 177:134–143.

Olsen, G. J., C. R. Woese, and R. Overbeek. 1994. The winds of (evolutionary) change: breathing new life into microbiology. *J. Bacteriol.* 176:1–6.

Pósfai, J., A. S. Bhagwat, G. Pósfai, and R. J. Roberts. 1989. Predictive motifs derived from cytosine methyltransferases. *Nucl. Acids Res.* 17:2421–2435.

Povilenis, P. I., A. A. Luis, R. I. Vaishvila, S. T. Kulakauskas, and A. A. Ianulaitis. 1989. Methyl-cytosine specific restriction in *Escherichia coli* K-12. *Genetika* (USSR) 25:753–755.

Price, C., J. Lingner, T. A. Bickle, K. Firman, and S. W. Glover. 1989. Basis for changes in DNA recognition by the *Eco*R124 and *Eco*R124/3 type I DNA restriction and modification enzymes. *J. Mol. Biol.* 205:115–125.

Raleigh, E. A. 1992. Organization and function of the *mcrBC* genes of *E. coli* K-12. *Mol. Microbiol.* 6:1079–1086.

Roberts, R. J., and S. E. Halford. 1993. Type II restriction endonucleases. In *Nucleases,* S. M. Linn, R. S. Lloyd and R. J. Roberts, eds. pp. 35–88. Cold Spring Harbor, N. Y. Cold Spring Harbor Laboratory Press.

Roberts, R. J., and D. Macelis. 1994. REBASE—restriction enzymes and methylases. *Nucl. Acids Res.* 22:3628–3639.

Roszczyk, E., and S. Goodgal. 1975. Methylase activities from *Haemophilus influenzae* that protect *Haemophilus parainfluenzae* transforming deoxyribonucleic acid from inactivation by *Haemophilus influenzae* endonuclease R. *J. Bacteriol.* 123:287–293.

Selander, R. K., D. A. Caugant, and T. S. Whittam. 1987. Genetic structure and variation in natural populations of *Escherichia coli.* In *Escherichia coli and Salmonella typhimurium Cellular and Molecular Biology,* F. C. Neidhardt, J. L. Ingraham, K. B. Low, B. Magasanik, M. Schaechter, and H. E. Umbarger, eds. pp. 1625–1648. American Society for Microbiology, Washington, D. C.

Shub, D. A., H. Goodrich-Blair, and S. R. Eddy. 1994. Amino acid sequence motif of group I intron endonucleases is conserved in open reading frames of group II introns. *Trends Biochem. Sci.* 19:402–404.

Sladek, R. L., J. A. Nowak, and J. Maniloff. 1986. *Mycoplasma* restriction: identification of a new type of restriction specificity for DNA containing 5-methylcytosine. *J. Bacteriol.* 165:219–225.

Smith, H., and G. Marley. 1980. Purification and properties of *Hind*II and *Hind*III endonucleases from *Haemophilus influenzae* Rd. *Methods Enzymol.* 65:104–108.

Stein, D. C. 1991. Transformation of *Neisseria gonorrhoeae:* physical requirements of the transforming DNA. *Can. J. Microbiol.* 37:345–349.

Timinskas, A., V. Butkus, and A. Janulaitis. 1995. Sequence motif characteristics for DNA [cytosine-N4] and DNA [adenine-N6] methyltransferases. Classification of all DNA methyltransferases. *Gene* 157:3–11.

Vertès, A. A., M. Inui, M. Kobayashi, Y. Kurusu, and H. Yukawa. 1993. Presence of *mrr*- and *mcr*-like restriction systems in coryneform bacteria. *Res. Microbiol.* 144:181–185.

Walter, J., T. A. Trautner, and M. Noyer-Weidner. 1992. High plasticity of multispecific DNA methyltransferases in the region carrying DNA target recognizing enzyme modules. *EMBO J.* 11:4445–4450.

Wilson, G. G., and N. E. Murray. 1991. Restriction and modification systems. *Annu. Rev. Genet.* 25:585–627.

Xu, G., J. Willert, W. Kapfer, and T. A. Trautner. 1995. *Bsu*CI, a type-I restriction-modification system in *Bacillus subtilis. Gene* 157:59.

Zawadzki, P., M. S. Roberts, and F. M. Cohan. 1995. The log-linear relationship between sexual isolation and sequence divergence in *Bacillus* transformation is robust. *Genetics* 140:917–932.

Zotchev, S. B., H. Schrempf, and C. R. Hutchinson. 1995. Identification of a methyl-specific restriction system mediated by a conjugative element from *Streptomyces bambergiensis. J. Bacteriol.* 177:4809–4812.

PART B

Genome Stability

9

Genome Ploidy

Nancy J. Trun

What is Ploidy?

Traditionally, ploidy (euploidy) has been defined in eukaryotic cells. A cell containing only one homologue of each chromosome is haploid. Cells containing two homologues of each chromosome are diploid; three homologues, triploid, and so on. By this convention, most bacteria in general, and *E. coli* in particular, contain one homologue of their single chromosome and are considered to be haploid. However, when the differences in the cell cycles of eukaryotes and prokaryotes are considered, this distinction becomes less clear.

Ploidy and the Cell Cycle

In eukaryotes, cells begin the cell cycle in G1 with completely replicated chromosomes (Pardee, 1989). Once G1 has been completed, they enter S and initiate replication. At the end of S phase, the cells contain completely replicated chromosomes. S is followed by G2, then M, when the chromosomes are segregated and the daughter cells separate.

Several aspects of the eukaryotic cell cycle are particularly relevant to this discussion. Within each of the phases, a distinct set of events takes place. Once the cycle is set in motion, cells go through each phase sequentially, as long as all events in the previous phase have taken place (Murray and Kirschner, 1989). This sequence is monitored by checkpoint events that must be finished before passage to the next phase is sanctioned (Hartwell and Weinert, 1989). Cells only complete the phases in the correct sequence, and, most importantly, the phases are not overlapping. For example, each cell starts S phase with one amount of DNA, and finishes S phase with twice that amount. Before it can initiate DNA replication again, it must go through G2, M and G1. Because of this cell cycle design, telling haploid from diploid can be accomplished with relative ease.

In *E. coli,* the cell cycle (Fig. 9-1) begins after separation of daughter cells, in what has been termed the B phase (Helmstetter, 1987). After this period of growth, cells enter the C phase and replicate their DNA. The remainder of the cell cycle, from termination of replication through division of the daughter cells is known as the D phase. Similar to eukaryotes, once the cycle has begun, prokaryotic cells will continue through each phase sequentially as long as all events, including checkpoint events, have taken place (Donachie and Robinson, 1987). Unlike the eukaryotic cell cycle, however, certain phases of the bacterial cell cycle can be overlapped. *E. coli* chromosomes take a minimum of approximately 40 minutes to replicate, although the cells can divide as fast as every 20 minutes (Helmstetter, 1987). By necessity, the C period can overlap with the B and D periods. Thus, the structure of the chromosome at the beginning of a cell cycle depends on the previous growth rate of the cells.

Ploidy at Different Growth Rates

The influence of the growth rate on the state of the chromosome can be illustrated by considering bacteria at two different rates. When *E. coli* is growing with a generation time of 60 minutes or more, each newborn cell inherits a fully replicated chromosome (Fig. 9-2). Upon initiating a cell cycle, the cell will go through a replication-less B period before it initiates DNA replication on that chromosome

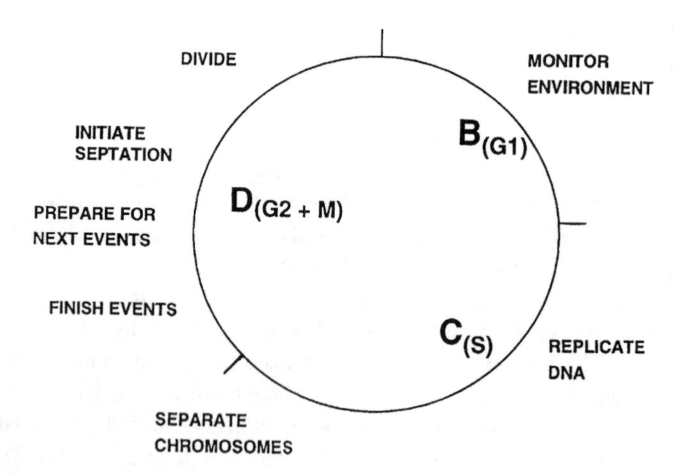

Figure 9-1. The *E. coli* cell cycle. The three phases of the cell cycle with the corresponding eukaryotic phase are indicated. The major events that take place within each phase are also indicated.

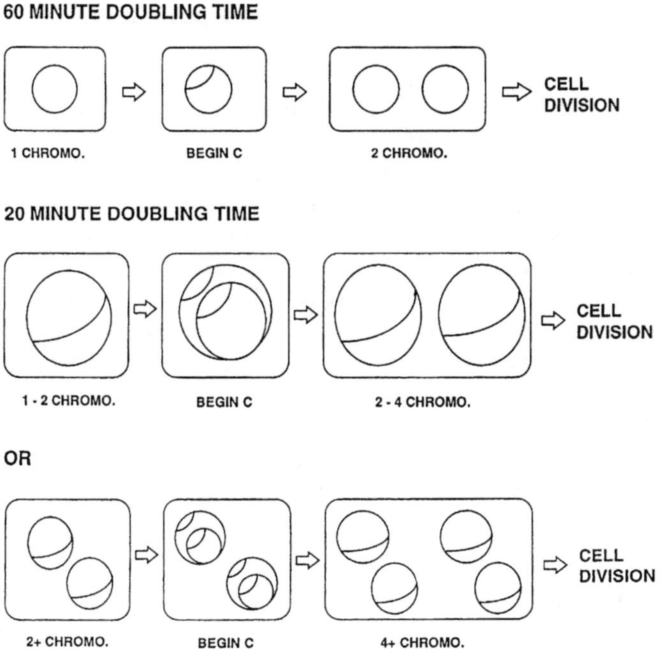

60 MINUTE DOUBLING TIME

1 CHROMO. BEGIN C 2 CHROMO. CELL DIVISION

20 MINUTE DOUBLING TIME

1 - 2 CHROMO. BEGIN C 2 - 4 CHROMO. CELL DIVISION

OR

2+ CHROMO. BEGIN C 4+ CHROMO. CELL DIVISION

Figure 9-2. The state of the chromosome(s) at different growth rates. When the growth rate of the cell exceeds 60 minutes, the cells inherit a chromosome with no replication forks (top panel). When the growth rate of the cells is faster than the DNA replication time, cells inherit chromosomes with replication forks (bottom two panels). Under these conditions, two models for the state of the chromosome exist. A newborn cell could either inherit one multiforked chromosome with a single terminus or several replicating chromosomes. The difference between these two models is the number of termini that a cell inherits. The fact that *E. coli* is genetically haploid leads to two interpretations depending on which model is considered. See text for details.

(Kubitschek and Newman, 1978). Once the cell has gone through a C period, it must pass through D and B period before it can again initiate DNA replication.

If, however, *E. coli* is dividing faster than the time necessary for chromosome replication, it inherits a chromosome that contains replication forks (Helmstetter and Cooper, 1968). In this case, cells will simultaneously be in a B phase and a C phase and will not have gone through an intervening, replication-less D period. At maximal growth rates, the measured amount of DNA per cell is three to four genome equivalents (Cooper and Helmstetter, 1968). The DNA content is expressed as genome equivalents because the exact replication state of these chromosomes is unknown; a cell may contain several complete chromosomes with replication forks or one chromosome with multiple replication forks and only one terminus region (Fig. 9-2). The critical measurement in these cells is the number of termini regions per cell. Thus, in fast growing cells, the distinction between haploid and diploid is blurred. If the cell contains two chromosomes with several replication forks, it is technically a diploid. If, however, the cell contains a chromosome with many replication forks and one terminus region it is a haploid.

Cells deviate from a 40-minute chromosome replication time when the generation time is longer than 60 minutes. Under these conditions, the replication time increases linearly with the generation time until the growth rate approaches 16 to 17 hours (Kubitschek and Newman, 1978; Skarstad *et al.,* 1983). At these very slow growth rates, the C period can occupy as little as 10–15% of the cell cycle (Skarstad *et al.,* 1983). Thus, *E. coli* grows under two conditions that affect the rate of DNA replication and the DNA content of the cells. At generation times under approximately 40 minutes, DNA replication (the C period) occupies virtually 100% of the cell cycle and cells inherit chromosomes that contain replication forks. Once the cell cycle exceeds 60 minutes, the C period occupies a progressively shorter amount of the cell cycle and cells inherit chromosomes without replication forks.

Ploidy at Changing Growth Rates

The effects of growth rate on the chromosome that have been described above concern cells that are in balanced growth at a given rate. Balanced or steady state growth is characterized by the doubling, in mass, of each cell constituent between successive divisions. Because *E. coli* has adapted to accommodate rapidly changing growth rates, how the cells deal with a decrease or increase in rate helps define the mechanisms for regulating these changes.

What happens upon a change in growth rate has been most thoroughly and elegantly described by O. Maaloe and coworkers (Maaloe, 1960; Maaloe and Kjeldgaard, 1966). The most remarkable aspect of a shift is that the synthetic capacities of the cell do not change synchronously. An increase in growth rate results in the immediate increase of both the rate of RNA synthesis and in mass accumulation. Increases in the rates of DNA synthesis and cell division lag somewhat behind. After a short lag, DNA synthesis increases abruptly at first, followed several generations later by an increase in the rate of cell divisions (Kjeldgaard *et al.,* 1958). The result of this order of change is that the amount of DNA per cell overshoots after the change, but in subsequent generations is adjusted down to fit the growth rate of the cell.

Upon a decrease in the growth rate, RNA synthesis and mass accumulation again respond immediately (Kjeldgaard *et al.,* 1958). DNA synthesis and cell divisions remain high for at least one generation. The net result of these changes is that the amount of DNA per cell initially increases, then is subsequently adjusted. Thus, on *any* growth rate change, the cell responds with a temporary increase in the DNA content per cell. This is followed by a readjustment of the DNA synthesis rate to match the new growth rate.

Physical vs. Genetic Ploidy

In the prior discussion, the state of the chromosome was described from physical measurements of the DNA. The concept of haploid and diploid cells however, also makes very specific predictions for genetic experiments. When a cell is haploid, it carries, by definition, only one copy of any given gene. If a cell is diploid then it carries two copies of any given gene. Using this definition, *E. coli* always behaves experimentally as a haploid, even when growing at maximal growth rates (Lederberg and Tatum, 1946).

Genetic haploidy is consistent either with one multi-forked chromosome or with multiple chromosomes. If fast growing cells contain one, multi-forked chromosome with a single terminus, a gene near the origin has a greater chance of being replicated before the second allele is introduced. The first cell division would then result in the two different alleles being segregated into different daughter cells. Because genetic experiments examine the cells 15–20 generations after introduction of the second allele, the two alleles will not remain in the same cell long enough to be detected. On the other hand, if fast growing cells contain several multi-forked chromosomes, then genetic haploidy would indicate that segregation of chromosomes to daughter cells is non random. Regardless of the explanation, genetically *E. coli* behaves as a haploid at fast growth rates, while physically it contains three to four chromosome equivalents.

Control of Ploidy

Studies of how cells control the amount of DNA per cell indicate the existence of several mechanisms that act at different steps in the process. Initiation of DNA replication is controlled at two major steps. First, the amount and activity of the essential, initiator protein DnaA are tightly regulated (von Meyenburg and Hansen, 1987). This insures that initiation of replication can only occur at very specific times in the cell cycle. Second, the newly replicated hemi-methylated origin is sequestered (Campbell and Kleckner, 1990). Thus, access to the newly replicated origin is regulated so that many forks are not initiated in a short period of time. These two mechanisms insure that initiation of DNA replication occurs at the correct time in the cell cycle.

After the replication forks have formed, there is increasing evidence for a second level of control at movement of the replication forks. Within the replication complex that is formed at the origin resides the DnaB helicase that binds to the lagging strand and is essential for DNA replication (LeBowitz and McMacken, 1986). A second helicase, Rep, which binds to the leading strand, also affects replication fork movement (Lane and Denhardt, 1975). In the absence of Rep, the replication forks move 50% slower. Mutations that allow twice as many forks to traverse the chromosome also map to *rep* (N. Trun, unpublished observations). While *rep* is not an essential gene (Bialkowska-Hobrzanska and Denhardt, 1984), deletion of both *rep* and another leading strand helicase, *uvrD,* is lethal to the cells (Taucher-Scholz *et al.,* 1983). It has been shown *in vitro* that Rep and UvrD form heterodimers that have helicase activity (Wong *et al.,* 1993). An intriguing possibility is that the two leading strand helicases, Rep and UvrD, in conjunction with the lagging strand helicase DnaB, regulate movement of the forks around the chromosome. Additionally, control of these helicases, and thus control of movement of the replication forks, may be linked to the growth rate of the cell. Post-initiation control of replication is not unique to *E. coli;* under certain circumstances it has been shown to be an important control point in *B. subtilis* (Levine *et al.,* 1995).

Ploidy in Other Bacteria

While all of the information summarized above comes from studies of *E. coli,* the question of ploidy also has been studied in other bacteria. Ploidy studies in *Azotobacter vinelandii* initially gave apparently contradictory results. Physical studies indicated that *A. vinelandii* is polyploid, having as many as 100 chromosome equivalents per cell (Sadoff *et al.,* 1979). However, genetic studies showed that recessive mutations in various genes are isolated at normal frequencies (Maldonado *et al.,* 1992). If a cell always contains 100 chromosomes, then identification of recessive mutations should be very difficult.

Flow cytometry studies of *A. vinelandii* during different stages of growth indicated an unexpected situation (Maldonado *et al.,* 1994). Cells in early exponential phase have a ploidy number similar to that of *E. coli.* As they progress to stationary phase, the chromosome number increases dramatically. In late exponential cultures, the chromosome number approaches 40 chromosome equivalents per cell, while early stationary phase cultures have 80 chromosome equivalents per cell and late stationary phase cells can have as many as 100 chromosome equivalents. These dramatic changes occur during rapid growth, at slow growth rates, the number of chromosome of *A. vinelandii* equivalents is very similar to that of *E. coli.*

Summary

The DNA content of an *E. coli* cell is not constant. Bacterial cells respond to changes in growth rate quickly and dramatically. These changes insure, first and foremost, that sufficient chromosomes are made for each daughter cell. While these states have been described in physiological studies, the mechanisms regulating the transitions between them have yet to be elucidated. Current knowledge begins to account for the accuracy of DNA replication. However, few experiments address the timing and coordination of the rest of the cell cycle events or the state of the inherited chromosome at different growth rates. Investigating these processes will provide valuable insights into how "simple" bacteria solve these most intriguing fundamental problems.

Acknowledgments

I would like to thank S. Wickner, M. Brennan, and S. Benson for critically reading the manuscript.

References

Bialkowska-Hobrzanska, H., and D. T. Denhardt. 1984. The *rep* mutation. VII. Cloning and analysis of the functional *rep* gene of Escherichia coli K-12. *Gene* 28:93–102.

Campbell, J. L., and N. Kleckner. 1990. *E. coli oriC* and the *dnaA* gene promoter are sequestered from Dam methyltransferase following the passage of the chromosomal replication fork. *Cell* 62:967–79.

Cooper, S., and C. E. Helmstetter. 1968. Chromosome replication and the division cycle of *Escherichia coli* B/r. *J. Mol. Biol.* 31:519–540.

Donachie, W. D., and A. C. Robinson. 1987. Cell division: Parameters values and the process. In *Escherichia coli and Salmonella typhimurium. Cellular and molecular biology.* pp. 1578–1593. F. C. Neidhardt, ed. American Society for Microbiology, Washington, D.C.

Hartwell, L. H., and T. A. Weinert. 1989. Checkpoints: Controls that ensure the order of cell cycle events. *Science* 246:629–634.

Helmstetter, C. E. 1987. Timing of synthetic activities in the cell cycle. In *Escherichia coli and Salmonella typhimurium. Cellular and Molecular Biology,* F. C. Neidhardt, ed., pp. 1594–1605. American Society for Microbiology, Washington, D.C.

Helmstetter, C. E., and S. Cooper. 1968. DNA synthesis during the division cycle of rapidly growing *Escherichia coli* B/r. *J. Mol. Biol.* 31:507–518.

Kjeldgaard, N. O., O. Maaloe, and M. Schaechter. 1958. The transition between different physiological states during balanced growth of *Salmonella typhimurium. J. Gen. Microbiol.* 19:607.

Kubitschek, H. E., and C. N. Newman. 1978. Chromosome replication during the division cycle in slowly growing, steady-state cultures of three *Escherichia coli* B/r strains. *J. Bacteriol.* 136:179–190.

Lane, H. E., and D. T. Denhardt. 1975. The *rep* mutation. IV. Slower movement of replication forks in Escherichia coli *rep* strains. *J. Mol. Biol.* 97:99–112.

LeBowitz, J. H., and R. McMacken. 1986. The *Escherichia coli* DnaB replication protein is a DNA helicase. *J. Biol. Chem.* 261:4738–4748.

Lederberg, J., and E. L. Tatum. 1946. Gene recombination in *Escherichia coli. Nature* 158:558.

Levine, A., S. Autret, and S. J. Seror. 1995. A checkpoint involving RTP, the replication terminator protein, arrests replication downstream of the origin during the Stringent Response in Bacillus subtilis. *Mol. Microbiol.* 15:287–95.

Maaloe, O. 1960. The nucleic acids and the control of bacterial growth. In *Society for General Microbiology Symposium: Bacterial Genetics: Society for General Microbiology,* pp. 272–293.

Maaloe, O., and N. O. Kjeldgaard. 1966. *Control of Macromolecular Synthesis* (New York: W. A. Benjamin).

Maldonado, R., A. Garzon, D. Dean, and J. Casadesus. 1992. Gene dosage analysis in *Azotobacter vinelandii. Genetics* 132:869–878.

Maldonado, R., J. Jimenez, and J. Casadesus. 1994. Changes in ploidy during the *Azotobacter vinelandii* growth cycle. *J. Bacteriol.* 176:3911–3919.

Murray, A. W., and M. W. Kirschner. 1989. Dominoes and clocks: The union of two views of the cell cycle. *Science* 246:614–621.

Pardee, A. B. 1989. G1 events and regulation of cell proliferation. *Science* 246:603–608.

Sadoff, H. L., B. Shimei, and S. Ellis. 1979. Characterization of *Azotobacter vinelandii* deoxyribonucleic acid and folded chromosomes. *J. Bacteriol.* 138:871–877.

Skarstad, K., H. B. Steen, and E. Boye. 1983. Cell cycle parameters of slowly growing Escherichia coli B/r studied by flow cytometry. *J. Bacteriol.* 154:656–62.

Taucher-Scholz, G., M. Abdel-Monem, and H. Hoffman-Berling. 1983. Functions of DNA helicases in *Escherichia coli.* In *UCLA Symposia on Molecular and Cellular Biology: Mechanisms of DNA replication and recombination,* pp. 1–12. N. Cozzarelli, ed. Liss, New York.

Wong, I., M. Amaratunga, and T. M. Lohman. 1993. Heterodimer formation between *Escherichia coli* Rep and UvrD proteins. *J. Biol. Chem.* 268:20386–91.

10

Segregation of the Bacterial Chromosome

Robert A. Britton and James R. Lupski

Bacteria must successfully complete several processes of the cell cycle to ensure that a single cell will become two viable daughter cells. These processes include: 1) replication of the chromosome, 2) segregation of the chromosomes to opposite poles, and 3) cell division. Each of these discontinuous processes must be coordinated, but how this is achieved is poorly understood (*see also* Chapters 7 and 9).

A mechanism exists in *Escherichia coli* that makes sure that each daughter cell receives a single chromosome. Fewer than 0.03% of daughter cells fail to receive a chromosome in *E. coli,* demonstrating that chromosome segregation is an efficient process (Hiraga *et al.,* 1989). For the purpose of this chapter, chromosome segregation refers to the physical separation of the chromosomes after DNA replication has been completed and the movement, or partitioning, of the chromosomes to the ¼ and ¾ points of the cell that follows this physical separation (Rothfield, 1994).

Resolution of Catenanes Created during Replication

The replication of the *E. coli* chromosome is a semi-conservative process that is initiated at a fixed site, *oriC,* and proceeds bi-directionally to the *ter* region, where termination takes place (*see* Chapter 7). Because the chromosome is a covalently closed circular molecule, the newly formed chromosomes remain linked, or catenated, following the termination of DNA replication. Before the chromosomes can be partitioned to their correct positions (¼ and ¾ points of the cell) in the cell prior to cell division they must first be decatenated. Two types of enzymes are capable of removing catenanes *in vitro.* Type I topoisomerases (topoisomerase I and topoisomerase III) are able to separate catenanes that contain a nick or a gap, while type 2 topoisomerases (DNA gyrase and topoisomerase IV) are able to decatenate double-stranded interlocked circles (Schmid and Sawitzke, 1993; Luttinger, 1995). Type II topoisomerases are essential and mutants display phenotypes that would be predicted in partitioning defective mutants (Steck and Drlica, 1984; Luttinger *et al.,* 1991; Kato *et al.,* 1990; Schmid, 1990).

DNA gyrase and topoisomerase IV are the products of the *gyrA, gyrB* and *parC, parE* genes, respectively. The fact that both of these enzymes are essential demonstrates that the functions performed by each are not totally redundant. Mutations in the genes for both DNA gyrase and topoisomerase IV can cause partitioning defects in which the chromosomes do not segregate and remain as a mass in the center of the cell (Steck and Drlica, 1984; Luttinger *et al.,* 1991; Schmid, 1990; Kato *et al.,* 1988; Kato *et al.* 1990). DNA gyrase has been shown to be necessary for the initiation and elongation of DNA replication, which made it difficult to assess its role in segregation. However, studies utilizing plasmid-based systems strongly suggest that topoisomerase IV is responsible for the removal of catenanes generated by DNA replication both *in vivo* and *in vitro* (Adams *et al.,* 1992; Peng and Marians, 1993). Adams *et al.,* (1992) reported that the incubation of topoisomerase IV mutants at the non-permissive temperature yielded an increase of catenated plasmid molecules that were the structure one would predict if these catenanes were produced during DNA replication. A similar accumulation of catenated plasmids was not observed when DNA gyrase was inhibited.

Recently it was demonstrated that topoisomerase IV has 100 times greater activity than DNA gyrase in removing plasmid replicon catenanes generated during DNA replication *in vivo,* providing further evidence for the role of topoisomerase IV in this process (Zechiedrich and Cozzarelli, 1995). Thus, it appears that the role of DNA gyrase is for the unlinking of catenanes formed by site-specific recombination and tangling while topoisomerase IV is necessary for the removal of catenanes formed during DNA replication (Adams *et al.,* 1992). DNA gyrase also has the ability to negatively supercoil DNA, while topoisomerase IV cannot, further demonstrating the functional differences of these type II topoisomerases. One reason for these differences is that topoisomerase IV and DNA gyrase bind DNA differently (Peng and Marians, 1995). It has been proposed that topoisomerase IV is a part of a protein-DNA complex dedicated to the termination of DNA replication (Adams *et al.,* 1992). This complex may exclude the action of other topoisomerases (Adams *et al.,* 1992).

Resolution of Chromosomal Dimers

During the replication of DNA, recombination can occur between the sister chromosomes (sister chromosome exchange). If an odd number of recombinational events occurs, the newly formed chromosomes will be dimers, which must be resolved to monomers before they can be partitioned to their correct position in the cell. *Escherichia coli* utilizes a site-specific recombination system that acts at a site near the terminus of replication to resolve dimers into monomers. Two recombinases that share homology with the λ integrase family, XerC and XerD, carry out this recombinational event (Blakely *et al.,* 1993). XerC was first discovered as an enzyme provided by the host cell capable of resolving plasmid ColE1 multimers to monomers at a 250bp site on the plasmid termed *cer,* while the XerD protein was identified based on its homology to XerC (Blakely *et al.,* 1991; Blakely *et al.,* 1993). Recombination occurs at a site designated *dif* (deletion induced filamentation), which is located near the terminus of replication and is analogous to *cer,* the ColE1 plasmid recombination site (Blakely *et al.,* 1991; Kuempel *et al.,* 1991).

Mutations in *xerC* and deletions of the *dif* sequence are viable but exhibit a chromosome partitioning defect (Blakeley *et al.,* 1991; Kuempel *et al.,* 1991; Blakely *et al.,* 1993). Cells are elongated with abnormally positioned nucleoids, some with a nucleoid in the center of the cell and some with nucleoids in other aberrant positions. When either *dif* or *xerC* mutants are combined with mutations in *recA,* the *par* phenotype is no longer observed, strongly suggesting that the function of XerC/XerD recombination is to resolve dimers that are formed by homologous recombination. Blakely *et al.* (1993) proposed that the requirement for two recombinases ensures that only correctly aligned sites are recombined and that this type of system may be involved in the segregation of most circular chromosomes and replicons.

Zyskind *et al.* (1992) have proposed that *recA* plays a role in the partitioning of chromosomes. Analysis of *recA* mutants by microscopy showed that up to 10% of the cells were anucleate. However, nuclease degradation of the chromosomes may be responsible for these DNA-less cells rather than a partitioning defect (Skarstad and Boye, 1993). Thus, the role of *recA* in chromosome segregation is unclear.

par Genes

Hirota and co-workers were the first to describe partitioning defective mutants (Hirota *et al.,* 1968; Hirota *et al.,* 1971). There have been six *par* alleles described to date; *parA, parB, parC, parD, parE* and *parF* (Hirota *et al.,* 1968; Hirota *et al.,* 1971; Hussain *et al.,* 1987; Kato *et al.,* 1988; Kato *et al.,* 1990; Luttinger *et al.,* 1991). These alleles have centrally located nucleoids in elongated cells at the non-permissive temperature. The *parA* and *parD* alleles were later identified as *gyrA* and *gyrB,* the subunits of DNA gyrase, while the *parC* and *parE* alleles encode the subunits of topoisomerase IV (Hussain *et al.,* 1987; Kato *et al.,* 1989; Kato *et al.,* 1990; Luttinger *et al.,* 1991). The *parB* allele is a mutation in *dnaG,* which encodes the protein primase (Norris *et al.,* 1986; Grompe *et al.,* 1991). The abnormal chromosome segregation phenotype of this allele may be partially due to the activation of the SOS response, which would be activated in response to the single stranded gaps produced by missed priming events on the lagging strand by the mutant primase (Versalovic, 1994). However, the SOS response is not entirely responsible for the *par* phenotype because *lexA3, parB* double mutants still have a partitioning defect (Versalovic, 1994; Versalovic and Lupski, 1997).

muk Genes

Hiraga and co-workers described the isolation of novel mutants that produce increased numbers of anucleate cells (Hiraga *et al.,* 1989). Of the alleles identified (termed *muk* for *mukaku,* which means anucleate), one of the most interesting appears to be *mukB* (Niki *et al.,* 1991). The *mukB* gene encodes an 177 kD protein that is predicted to form a homodimer with a rod-and-hinge structure and has similarity to the heavy chains of the eukaryotic force generating-enzymes myosin and kinesin (Niki *et al.,* 1991; Niki *et al.,* 1992). MukB can bind ATP and GTP, but no associated ATPase or GTPase activities have been reported. MukB also has the ability to bind DNA, but it appears to do so nonspecifically (Niki *et al.,* 1992). A null mutant of *mukB* is viable at 22°C but cannot form colonies at higher temperatures. The segregation defect in the *mukB* null mutant is observed at all temperatures but appears to be more severe at higher temperatures (Niki *et al.,* 1991).

Whether or not MukB is the force-generating enzyme responsible for partitioning chromosomes remains to be proven. It is likely there are other unidentified factors that participate in the movement of chromosomes. Candidates for such proteins would be a stimulator of MukB GTPase or ATPase activity, an accessory factor that would provide specificity for the binding of MukB to DNA, or a filamentous protein polymer that would interact with MukB and act as a "rail" along which MukB could move. (Hiraga *et al.,* 1991; Hiraga, 1992; Rothfield, 1994). One protein in *E. coli* that has been shown to have the ability to form filaments is FtsZ, which is essential for cell division. Interestingly, the *mukB* null allele appears to be inhibited or delayed in cell division at 42°C. However, mutants of *ftsZ* are capable of proper chromosome segregation; therefore FtsZ probably is not the "rail." Whether another protein capable of forming filaments plays a role in the positioning of chromosomes remains to be determined.

Two other recently identified *muk* genes, *mukE* and *mukF,* are likely to be in an operon with the *mukB* gene. As with *mukB,* null mutations in *mukE* and *mukF* are viable at 25°C but are unable to form colonies at higher temperatures and cause an increase in anucleate cells (Yamanaka *et al.,* 1996). Further characterization of these three genes and their protein products may lead to a better understanding of the segregation process.

Passive versus Active Segregation

After the chromosomes have been physically separated, the daughter chromosomes must be partitioned to the quarter positions of the cell prior to division. How this is achieved is largely unknown. In 1963, Jacob *et al.* proposed a passive model in which the daughter chromosomes are attached to the cell envelope and that new synthesis of the cell envelope from the midpoint of the cell provided the force necessary for the partitioning of the nucleoids (Jacob *et al.,* 1963). An extension of this model suggested that *oriC* was the site of DNA attachment to the membrane, thereby linking chromosome replication and segregation (*see also* Chapter 7). However, a specific component of the membrane has not been identified that is involved in the binding of DNA and chromosome partitioning. Although hemimethylated *oriC* has been shown to bind to the cell membrane, it likely does not play a role in chromosome partitioning. Plasmids that contain *oriC* are segregated at random, demonstrating that *oriC* alone cannot serve as the bacterial centromere (Hiraga, 1992). Mutations in the *dam* gene, which are unable to methylate the GATC sites in *oriC,* do not show a partitioning defective phenotype (Vinella *et al.,* 1992; Lobner-Olesen *et al.,* 1994). Although in most models an attachment of the chromosome to the cell membrane is proposed to be important for segregation, the component(s) of the membrane involved and what role this interaction plays in segregation is unknown.

The movement of chromosomes to the ¼ positions of the cell requires post-replication protein synthesis (Donachie and Begg, 1989; Hiraga *et al.,* 1990; van Helvoort and Woldringh, 1994). Cells treated with inhibitors of protein synthesis such as chloramphenicol are able to complete current rounds of replication but the nucleoids are unable to segregate and are found positioned at the midpoint of the cell. In support of the passive model of partitioning it was found that nucleoid movement is associated with an increase in cell length upon recovery from protein synthesis inhibition (van Helvoort and Woldringh, 1994; *see also* Chapter 2). These authors proposed that the nucleoid migrates gradually with growth of the cell envelope and that cells must elongate before segregation can occur. They further suggested this could be achieved by a transient association of the chromosome with the membrane via a transcription, translation, and protein export complex.

However, other evidence suggests that an active partitioning mechanism may occur in *E. coli* (Donachie and Begg, 1989; Hiraga *et al.,* 1990). Resumption of protein synthesis resulted in the rapid movement of nucleoids to the ¼ positions faster than cell length increased (Donachie and Begg, 1989; Hiraga *et al.,* 1990). In addition, inhibitors of cell wall growth such as ampicillin did not affect the movement of the chromosomes (Hiraga *et al.,* 1990). These observations suggest that growth of the cell wall is not a major driving force in the movement of chromosomes. A model in which the segregation of chromosomes is dependent on the synthesis of a new protein(s) specifically involved in the positioning of the chromosomes has been proposed (Hiraga *et al.,* 1990). Clearly, more work is needed to determine if nucleoid migration occurs by a passive or active mechanism.

Summary

Chromosome segregation is an integral part of the cell cycle (Rothfield, 1994; Lobner-Olesen and Kuempel, 1992; Hiraga, 1992; Schmid and von Friesleben, 1996). *E. coli* has developed mechanisms to ensure that daughter chromosomes are physically separated and partitioned both faithfully and efficiently. Topoisomerase IV and XerC/XerD recombination at *dif* ensure that the newly formed chromosomes will be physically separated and monomeric. MukB may prove to be the force-generating enzyme that moves the chromosomes to their proper positions at the cell quarters.

While many advances have been made in the study of chromosome segregation during the last decade, many questions still remain. Why does the cell need two different type II topoisomerases? What other proteins participate with MukB in segregation? How does the cell ensure that the chromosomes migrate to opposite ends of the cell? What role does the membrane play in the movement of the nucleoids? Is there a bacterial equivalent to a centromere and if so, what defines it? The discovery of a centromere in *E. coli* would allow the study of segregation in a plasmid-based system, which was essential to elucidating the functions of topoisomerase IV and the XerC and XerD recombinases. Further investigations will undoubtedly identify new components involved in these processes and bring us closer to an understanding of how a single cell efficiently segregates the newly replicated chromosomes.

Acknowledgments

We would like to thank Jim Sawitzke for the critical reading of this chapter.

References

Adams, D. E., E. M. Shektman, E. L. Zechiedrich, M. B. Schmid, and N. R. Cozzarelli. 1992. The role of topoisomerase IV in partitioning bacterial replicons and the structure of catenated intermediates in DNA replication. *Cell* 71:277–288.

Blakely, G., S. Colloms, G. May, M. Burke, and D. Sherratt. 1991. *Escherichia coli* XerC recombinase is required for chromosomal segregation at cell division. *New Biol.* 3:789–798.

Blakely, G., G. May, R. McCulloch, L. K. Arciszewska, M. Burke, S. T. Lovett, and D. J.

Sherratt. 1993. Two related recombinases are required for site-specific recombination at *dif* and *cer* in *E. coli* K-12. *Cell* 72:351–361.

Donachie, W. D. and K. J. Begg. 1989. Chromosome partition in *Escherichia coli* requires post-replication protein synthesis. *J. Bacteriol.* 171:5405–5409.

Grompe, M., J. Versalovic, T. Koeuth, and J. R. Lupski. 1991. Mutations in the *Escherichia coli dnaG* gene suggest coupling between DNA replication and chromosome partitioning. *J. Bacteriol.* 173:1268–1278.

Hiraga, S. 1992. Chromosome and plasmid partition in *Escherichia coli. Annu. Rev. Biochem.* 61:283–306.

Hiraga, S., H. Niki, R. Imamura, T. Ogura, K. Yamanaka, J. Feng, B. Ezaki, and A. Jaffe. 1991. Mutants defective in chromosome partitioning in *E. coli. Res. Microbiol.* 142:189–194.

Hiraga, S., H. Niki, T. Ogura, C. Ichinose, H. Mori, B. Ezaki, and A. Jaffe. 1989. Chromosome partitioning in *Escherichia coli:* novel mutants producing anucleate cells. *J. Bacteriol.* 171:1496–1505.

Hiraga, S., T. Ogura, H. Niki, C. Ichinose, and H. Mori. 1990. Positioning of replicated chromosomes in *Escherichia coli. J. Bacteriol.* 172:31–39.

Hirota, Y., M. Ricard, and B. Shapiro. 1971. The use of thermosensitive mutants of *E. coli* in the analysis of cell division. *Biomembranes* 2:13–31.

Hirota, Y., A. Ryter, and F. Jacob. 1968. Thermosensitive mutants of *E. coli* affected in the processes of DNA synthesis and cellular division. *Cold Spring Harbor Symp. Quan. Biol.* 33:677–693.

Hussain, K., E. J. Elliot, and G. P. C. Salmond. 1987. The $ParD^-$ mutant of *Escherichia coli* also carries a $gyrA_{am}$ mutation. The complete sequence of *gyrA*. *Mol. Microbiol.* 1:259–273.

Jacob, F., S. Brenner, and F. Cuzin. 1963. On the regulation of DNA replication in bacteria. *Cold Spring Harbor Symp. Quan. Biol.* 28:329–348.

Kato, J., Y. Nishimura, R. Imamura, H. Niki, S. Hiraga, and H. Suzuki. 1990. New topoisomerase essential for chromosome segregation in *E. coli. Cell* 63:393–404.

Kato, J., Y. Nishimura, and H. Suzuki. 1989. *Escherichia coli parA* is an allele of the *gyrB* gene. *Mol. Gen. Genet.* 217:178–181.

Kato, J., Y. Nishimura, M. Yamada, H. Suzuki, and Y. Hirota. 1988. Gene organization in the region containing a new gene involved in chromosome partition in *Escherichia coli. J. Bacteriol.* 170:3967–3977.

Kuempel, P. L., J. M. Henson, L. Dircks, M. Tecklenberg, and D. F. Lim. 1991. *dif,* a *recA*-independent recombination site in the terminus region of the chromosome of *Escherichia coli. New Biol.* 3:799–811.

Lobner-Olesen, A., F. G. Hansen, K. V. Rasmussen, B. Martin, and P. L. Kuempel. 1994. The initiation cascade for chromosome replication in wild-type and Dam methyltransferase deficient *Escherichia coli* cells. *EMBO J.* 13:1856–1862.

Logner-Olesen, A., and P. L. Kuempel. 1992. Chromosome partitioning in *Escherichia coli. J. Bacteriol.* 174:7883–7889.

Luttinger, A. L. 1995. The twisted 'life' of DNA in the cell: bacterial topoisomerases. *Mol. Microbiol.* 15:601–6.

Luttinger, A. L., A. L. Springer, and M. B. Schmid. 1991. A cluster of genes that affects nucleoid segregation in *Salmonella typhimurium. New Biol.* 3:687–697.

Niki, H., R. Imamura, M. Kitaoka, K. Yamanaka, T. Ogura, and S. Hiraga. 1992. *E. coli* MukB protein involved in chromosome partition forms a homodimer with a rod-and-hinge structure having DNA binding and ATP/GTP binding activities. *EMBO J.* 11:5101–5109.

Niki, H., A. Jaffe, R. Imamura, T. Ogura, and S. Hiraga. 1991. The new gene *mukB* codes for a 177 kd protein with coiled-coil domains involved in chromosome partitioning of *E. coli. EMBO J.* 10:183–193.

Norris, V., T. Alliotte, A. Jaffe, and R. D'Ari. 1986. DNA replication termination in *Escherichia coli parB* (a *dnaG* allele) and *parA* and *gyrB* mutants affected in DNA distribution. *J. Bacteriol.* 168:494–504.

Peng, H. and K. J. Marians. 1993. Decatenation activity of topoisomerase IV during *oriC* and pBR322 DNA replication *in vitro. Proc. Natl. Acad. Sci. USA* 90:8571–8575.

Peng, H. and K. J. Marians. 1995. The interaction of *Escherichia coli* topoisomerase IV with DNA. *J. Biol. Chem.* 270:25286–25290.

Rothfield, L. 1994. Bacterial chromosome segregation. *Cell* 77:963–966.

Schmid, M. 1990. A locus affecting nucleoid segregation in *Salmonella typhimurium. J. Bacteriol.* 172:5416–5424.

Schmid, M. B. and U. von Friesleben. 1996. Nucleoid segregation. In Escherichia coli *and* Salmonella. *Cellular and molecular biology,* Second edition, F. C. Neidhardt, R. Curtiss III, J. L. Ingraham, E. C. C. Lin, K. B. Low, B. Magasanik, W. S. Reznikoff, M. Riley, M. Schaechter, and H. E. Umbarger, eds., pp. 1662–1669. ASM Press, Washington, D.C.

Schmid, M. B. and J. A. Sawitzke. 1993. Multiple bacterial topoisomerases: specialization or redundancy? *Bioessays* 15:445–449.

Skarstad, K. and E. Boye. 1993. Degradation of individual chromosomes in *recA* mutants of *Escherichia coli. J. Bacteriol.* 175:5505–5509.

Steck, T. R. and K. Drlica. 1984. Bacterial chromosome segregation: evidence for DNA gyrase involvement in decatenation. *Cell* 36:1081–1088.

van Helvoort, J. M. L. M. and C. L. Woldringh. 1994. Nucleoid partitioning in *Escherichia coli* during steady-state and upon recovery from chloramphenicol treatment. *Mol. Microbiol.* 13:577–583.

Versalovic, J. 1994. Evolution of the macromolecular synthesis operon and analysis of bacterial primase. Ph.D. Thesis. Baylor College of Medicine.

Versalovic, J., and Lupski, J. R. (1997) Missense mutations in the 3′ end of *dna6* gene do not destroy primase activity but confer the chromosome segregation defective (par) phenotype. *Microbiology* 143:585–594.

Vinella, D., A. Jaffe, R. D'Ari, M. Kohiyama, and P. Hughes. 1992. Chromosome partitioning in *Escherichia coli* in the absence of Dam-directed methylation. *J. Bacteriol.* 174:2388–2390.

Yamanaka, K., T. Ogura, H. Niki, and S. Hiraga. 1996. Identification of two new genes, *mukE* and *mukF,* involved in chromosome partitioning in *Escherichia col. Mol. Gen. Genet.* 250:241–251.

Zechiedrich, E. L. and N. R. Cozzarelli. 1995. Roles of topoisomerase IV and DNA gyrase in DNA unlinking during replication in *Escherichia coli. Genes Dev.* 9:2859–2869.

Zyskind, J. W., A. L. Svitil, W. B. Stine, M. C. Biery, and D. W. Smith. 1992. RecA protein of *Escherichia coli* and chromosome partitioning. *Mol. Microbiol.* 6:2525–2537.

11

Chromosomal Rearrangements

George M. Weinstock and James R. Lupski

Repeated sequences provide the potential for altering and destabilizing the genome (reviewed by Petes and Hill, 1988). Multigene families, IS elements and shorter repeated sequences like REP (Higgins *et al.,* 1982; Gibson *et al.,* 1984; Stern *et al.,* 1984; Dimri *et al.,* 1992) and ERIC (Sharples and Lloyd, 1990; Hulton *et al.,* 1991) are dispersed throughout bacterial genomes (reviewed by Riley and Krawiec, 1987; Lupski and Weinstock, 1992; *see also* Chapters 4 and 5). These sequences provide regions of homology for unequal crossing-over events. Such ectopic recombination events can lead to inversions, deletions, or duplications of regions of the chromosome. An inversion results when the repeated sequences involved in the recombination event are in inverse orientation to one another on the chromosome. Deletions and tandem duplications result from ectopic recombination between repeated sequences in the same or tandem orientation (Figure 11-1).

A generally accepted model for the formation of chromosomal duplication in bacteria proposes that, after chromosomal replication, misaligned repeated sequences (e.g. *rrn* operons, IS elements, transposons) act as regions of homology or substrates for homologous recombination, leading to duplication or deletion of the specific region between the repeated sequences (*see* Chapters 4, 5, 20 and 21). Since these repeats are usually far apart on the chromosome, duplications are large and the corresponding deletions may not be recovered since they remove essential genes. A variety of long-sequence repeats probably serve as substrates for homologous recombination at frequencies that reflect the length of the repeat and the degree of sequence similarity (Roth *et al.,* 1996). Support for the idea that most duplications form by recombination between repeated sequences is based on the observation that duplication formation generally is found to be highly dependent on the RecA function, a protein essential for homologous recombination. Additionally, the observation that DNA-damaging treatment increases formation of duplications is not surprising in view of the well-known

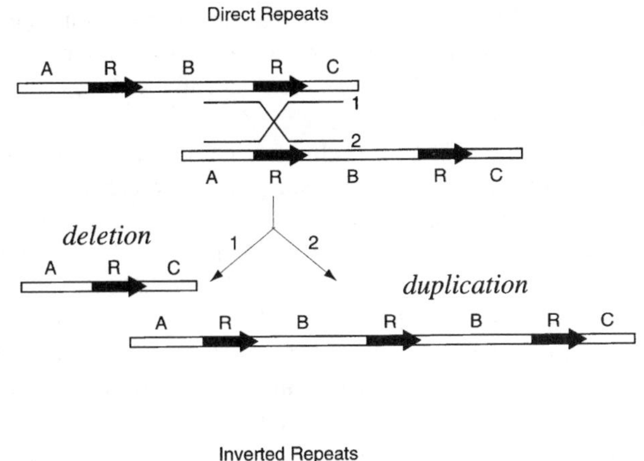

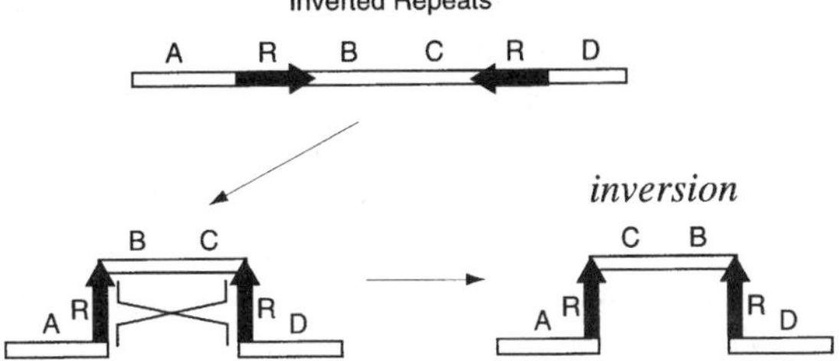

Figure 11-1. The letters A, B, and C represent a unique DNA sequence in the chromosome (open box), while the letter R represents a repeat sequence with the filled arrow showing the orientation on the chromosome. The repeat sequences could be *rrn* operons, *rhs* elements, transposons, IS elements, repeated genes, ERIC, REP, or other short repetitive sequences. The top of the figure shows misalignment of direct repeats and the products of the recombination (thin lines). Crossover 1 results in a deletion while crossover 2 yields a duplication. The bottom of the figure demonstrates recombination between inverted repeats which results in a chromosomal inversion.

stimulation of homologous recombination by DNA damage (Hill and Cambriato, 1973; Hill *et al.,* 1977; Hoffman *et al.,* 1978; Heath and Weinstock, 1991; Heath, 1992).

Duplications are particularly frequent in bacteria and have been postulated to play a role in the evolutionary process (Riley and Anilionis, 1978; *see also* Chapters 17 and 18). Measurements of tandem duplications in bacteria have demonstrated duplications ranging in size up to at least 320 kb (7% of the chromosome) in *Escherichia coli* and up to at least 22% of the *Salmonella typhimurium* chromosome (Anderson and Roth, 1978; Heath and Weinstock, 1991; Hill *et al.,* 1977).

Possibly the first report of bacterial duplications was that of Horiuchi *et al.,* (1963); they reported duplications of the *lac* operon isolated following selection for faster growth on lactose. The basic genetic properties of duplications in bacteria were first outlined by Campbell (1963 and 1965), using duplications of the *gal* region. These and other early work on genetics of duplications have been reviewed by Anderson and Roth (1977), more recently by Petes and Hill (1988), and by Roth *et al.* (1996).

Genetic studies of the haploid unicellular bacterium *S. typhimurium* demonstrate that in an unselected culture, the frequency of duplication of particular loci varies from 3% to approximately 10^{-5}, and an average locus is duplicated in about one in every 1,000 cells. It is estimated that nearly 10% of the cells in an unselected culture carry a duplication of some region of the chromosome. Analysis of chromosomal duplications in the *lac* region of *E. coli* reveals a basal frequency of 0.7%; this confirms the observation that tandem duplications are present at a surprisingly high frequency in bacterial populations (Heath and Weinstock, 1991). Considered together, these estimates suggest that the bacterial chromosome is in a constant state of flux and that duplications are continually acquired and lost (Anderson *et al.,* 1976; Anderson and Roth, 1977, 1978, 1979, 1981; Roth *et al.,* 1996). Duplication and their reversion may be quite common and examples have been documented across species (Lupski *et al.,* 1996).

Tandem chromosomal duplications have also been identified by genetic analysis in *E. coli* where mild ultraviolet (UV) irradiation causes a large increase in duplication frequency (Hill and Cambriato, 1973; Hill *et al.,* 1977). Following UV irradiation, 12% of survivors have duplication of the *lac* region, a 16-fold increase over the basal level (Heath, 1992). If all regions of the bacterial chromosome form duplications of similar size and frequency as the *lac* region, it is likely that every cell surviving mild UV irradiation carries a duplication of one or more portions of its chromosome.

Further studies of different regions of the *E. coli* chromosome indicate that duplications occur at a frequency of 10^{-4} to 10^{-3}, and are stimulated at least 10-fold in most regions by relatively mild UV irradiation (Heath, 1992). The spontaneous tandem duplication frequency of the *metE* locus in an *E. coli* mismatch repair mutant, *mutL,* increased 6-fold over the frequency observed in a wild type strain, indicating that mismatch repair stabilizes the chromosome and maintains gene dosage (Heath, 1992). Physical measurements of *E. coli* duplications by pulsed-field gel electrophoresis (PFGE) reveal that they are very large and range from 140 kb to at least 2100 kb (Weinstock, 1994; Heath, 1992), the larger duplication measuring about ½ of the 4.7 million base pair *E. coli* chromosome (Heath, 1992).

Genetic and physical mapping of bacterial duplication endpoints demonstrates a non-random distribution. The chromosomal region duplicated is flanked by a large sequence repeat in direct orientation. Many different repeated sequences have been shown to be involved in duplication formation. There are seven ribosomal RNA loci in the *E. coli* and *S. typhimurium* chromosomes that provide large, highly homologous sequences for ectopic recombination (*see* Chapter 21). Previous studies have shown that rRNA operons are almost exclusively responsible for duplication formation in the *glyT-purD* region of the *E. coli* chromosome (Hill *et al.,* 1977). Anderson and Roth (1981) have shown that spontaneous duplications between rRNA operons in *S. typhimurium* are present in as many as 3% of the bacterial population. Other large repeated sequences can also be involved in duplication formation (Chumley and Roth, 1980). For example, duplications of the *glyS* region of *E. coli* are known to occur by recombination between two highly conserved loci, *rhsA* and *rhsB* (Lin *et al.,* 1984; *see* Chapter 23). The toxin gene of *Vibrio cholerae* is also amplified by recombination of flanking 2.7 kb repeated sequences (Goldberg and Mekalanos, 1986). On a smaller scale, IS*200* (708 bp) has recently been implicated in the formation of duplication in *S. typhimurium* (Haack and Roth, 1995). Similarly, a 380 bp repeated sequence is involved in the amplification of a *tetR* gene in *Enterococcus faecalis* plasmids (Yagi and Clewell, 1977). Smaller still, the 35 bp REP sequence, present in about 500 copies per chromosome, has been demonstrated to play a role in the formation of a specific duplication of the *his* locus in *S. typhimurium* (Shyamala *et al.,* 1990). Sequences as small as 12 base pairs are responsible for some duplication events leading to the amplification of the *amp*C gene (Edlund and Normark, 1981) and repeated sequences of 6 base pairs have been reported to be responsible for deletions in the *lac* region (Albertini *et al.,* 1982). Interestingly, the seven *E. coli rrn* operons are not used equivalently for duplication formation, suggesting that higher order chromosome structure, or slight variation in the DNA sequence of recombined sequences, may influence the frequency of duplication (Heath, 1992).

Duplication formation has been shown to be influenced by the SOS response (Dimpfl and Echols, 1989) and mismatch repair (Petit *et al.,* 1991), suggesting that it may be a controlled cellular process. Homologous recombination is known to be stimulated by DNA discontinuities such as nicks, gaps and double-stranded breaks caused by DNA damaging agents. Consistent with this, tandem duplications have been shown to be stimulated by DNA damaging treatments such as ultraviolet irradiation (UV), X-rays, mitomycin C and other mutagens (for example, Heath and Weinstock, 1991; Hill and Combriato, 1973; Hoffman *et al.,* 1978).

The high frequency of the formation of duplications, and their sometimes rapid loss through segregation under nonselective conditions, suggest that chromosomal duplication may be a mechanism by which bacteria can amplify particular functions and thereby adapt to stressful conditions in nature without undergoing irreversible changes in their genomes (Sonti and Roth, 1989).

Chromosomal duplications may provide the increased gene dosage of a required allele, provide a novel fusion at the join point, or supply redundant DNA for genetic divergence (Anderson and Roth, 1978). Chromosomal duplications and resultant gene duplication are ubiquitous features of genome evolution and have been viewed as the predominant mechanisms for the evolution of a new gene functions and adaptive responses (Ohno, 1970; Li and Graur, 1991; *see also* Chapters 17, 18, 22 and 23). Chromosomal rearrangements occur frequently and even in closely related strains (Weinstock, 1994). Further studies will likely identify additional genome structural features predisposing to chromosomal rearrangements.

References

Albertini, A. M., M. Hofer, M. P. Calos, J. H. Miller. 1982. On the formation of spontaneous deletions: The importance of short sequence homologies in the generation of large deletions. *Cell* 29:319–328.

Anderson, R. P., C. G. Miller, J. R. Roth. 1976. Tandem duplications of the histidine operon observed following generalized transduction in *Salmonella typhimurium. J. Mol. Biol.* 105:201–218.

Anderson, R. P., J. R. Roth. 1977. Tandem genetic duplications in phage and bacteria. *Ann. Rev. Microbiol.* 31:473–505.

Anderson, R. P., J. R. Roth. 1978. Tandem chromosomal duplications in *Salmonella typhimurium:* Fusion of histidine genes to novel promoters. *J. Mol. Biol.* 119:147–166.

Anderson, R. P., J. R. Roth. 1979. Gene duplication in bacteria: Alteration of gene dosage by sister-chromosome exchanges. *Cold Spring Harbor Symposium of Quantitative Biology* 43:1083–1087.

Anderson, P., J. Roth. 1981. Spontaneous tandem genetic duplications in *Salmonella typhimurium* arise by unequal recombination between rRNA *(rrn)* cistrons. *Proc. Natl. Acad. Sci. USA* 78:3113–3117.

Campbell, A. 1963. Segregates from lysogenic heterogenotes carrying recombinant lambda prophages. *Virology* 20:344–356.

Campbell, A. 1965. The steric effect in lysogenization by bacteriophage lambda. 1. Lysogenization of a partially diploid strain of *Escherichia coli* K-12. *Virology* 27:329–339.

Chumley, F. G., J. R. Roth. 1980. Rearrangement of the bacterial chromosome using Tn*10* as a region of homology. *Genetics* 94:1–14.

Dimpfl, J., H. Echols. 1989. Duplication mutation as an SOS response in *Escherichia coli:* enhanced duplication formaton by a constitutively activated RecA. *Genetics* 123:255–260.

Dimri, G. P., K. E. Rudd, M. K. Morgan, H. Bayat, G. F.-L. Ames. 1992. Physical mapping of repetitive extragenic palindromic sequences in *Escherichia coli* and phylogenetic distribution among *Escherichia coli* strains and other enteric bacteria. *J. Bacteriol.* 174:4583–4593.

Edlund, T., S. Nomark. 1981. Recombination between short DNA homologies causes tandem duplications. *Nature* 292:269–271.

Gilson, E., J. M. Clement, D. Brutlag, M. Hofnung. 1984. A family of dispersed repetitive extragenic palindromic DNA sequences in *E. coli. EMBO J.* 3:1417–142.

Goldberg, I. and J. J. Mekalanos. 1986. Effect of a *recA* gene mutation on cholera toxin gene amplification and deletion events. *J. Bacteriol.* 165:723–731.

Haack, K. R., J. R. Roth. 1995. Recombination between chromosomal IS200 elements supports frequent duplication formation in *Salmonella typhimurium. Genetics* 141:1245–1252.

Heath, J. D. 1992. Control of chromosomal rearrangements in *Escherichia coli* Ph.D. Thesis University of Texas Health Science Center at Houston.

Heath, J. D., G. M. Weinstock. 1991. Tandem duplications of the *lac* region of the *Escherichia coli* chromosome. *Biochimie.* 73:343–352.

Higgins, C. F., G. F. L. Ames, W. M. Barnes, J. M. Clement, M. Hofnung. 1982. A Novel intercistronic regulatory element of prokaryotic operons. *Nature* 298:760–762.

Hill, C. W., G. Combriato. 1973. Genetic duplications induced at very high frequency by ultraviolet irradiation in *Escherichia coli. Molec. Gen. Genet.* 127:197–214.

Hill, C. W., R. H. Grafstrom, B. W. Harnish, B. S. Hillman. 1977. Tandem duplications resulting from recombination between ribosomal RNA genes in *Escherichia coli. J. Mol. Biol.* 116:407–428.

Hoffman, G. R., R. W. Morgan, R. C. Harvey. 1978. Effects of chemical and physical mutagens on the frequency of a large genetic duplication in *Salmonella typhimurium* I. Induction of duplications. *Mutation Research* 52:73–80.

Horiuchi, R., S. Horiuchi, A. Novick. 1963. The genetic basis of hypersynthesis of β-galactosidase. *Genetics* 48:157–169.

Hulton, C. S. J., C. F. Higgins, P. M. Sharp. 1991. ERIC sequences: a novel family of repetitive elements in the genomes of *Escherichia coli, Salmonella typhimurium* and other enterobacteria. *Mol. Microbiol.* 5:825–834.

Li W-H, D. Graur. 1991. *Fundamentals of molecular evolution.* Sinauer Associates, Inc., Sunderland, Massachusetts.

Lin, R. J., M. Capage, C. W. Hill. 1984. A repetitive DNA sequence *rhs,* responsible for duplications within the *Escherichia coli* K-12 chromosome. *J. Mol. Biol.* 177:1–18.

Lupski, J. R., J. R. Roth, G. M. Weinstock. 1996. Chromosomal duplications in bacteria, fruit flies, and humans. *Am. J. Hum. Genet.* 58:21–27.

Lupski, J. R., G. M. Weinstock. 1992. Short, interspersed repetitive DNA sequences in prokaryotic genomes. *J. Bacteriol.* 174:4525–4529.

Ohno, S. 1970. *Evolution by Gene Duplication.* Springer-Verlag, Berlin.

Petes, T. D., C. W. Hill. 1988. Recombination between repeated genes in microorganisms. *Ann. Rev. Genet.* 22:147–168.

Petit. M.-A., J. Dimpfl, M. Radman, H. Echols. 1991. Control of large chromosomal duplications in *Escherichia coli* by the mismatch repair system. *Genetics* 126:327–332.

Riley, M., A. Anilionia. 1978. Evolution of the bacterial genome. *Ann. Rev. Microbiol.* 32:519–560.

Riley, M., S. Krawiec. 1987. Evolutionary history of enteric bacteria, pp. 967–981. In, F. C. Neidhardt, J. L. Ingraham, L. B. Low, B. Magasanik, M. Schaechter, H. E. Umbarger (ed.), *Escherichia coli and Salmonella typhimurium: Cellular and Molecular Biology,* vol. 2. American Society for Microbiology, Washington, D.C.

Roth, J. R., N. Benson, T. Galitski, K. Haack, J. Lawrence, L. Miesel. 1996. Rearrangements of the bacterial chromosome: formation and applications. In: Neidhardt F.C.

(ed) Escherichia coli *and* Salmonella typhimurium*: Cellular and Molecular Biology.* American Society for Microbiology, Washington, D.C. (in press).

Sharples, G. J., R. G. Lloyd. 1990. A novel repeated DNA sequence located in the intergenic regions of bacterial chromosomes. *Nucleic. Acids. Res.* 18:6503–6508.

Shyamala, V., E. Schneider, G. F. L. Ames. 1990. Tandem chromosomal duplications: role of REP sequences in the recombination event at the joinpoint. *EMBO J.* 9:939–946.

Sonti, R. V., J. R. Roth. 1989. Role of gene duplications in the adaption of *Salmonella typhimurium* to growth on limiting carbon sources. *Genetics* 123:19–28.

Stern, M. J., G. F. L. Ames, N. H. Smith, E. C. Robinson, C. F. Higgins. 1984. Repetitive extragenic palindromic sequences: A major component of the bacterial genome. *Cell* 37:1015–1026.

Weinstock, G. M. 1994. Bacterial genomes: Mapping and stability. *ASM News* 60:73–78.

Yagi, Y. and D. B. Clewell. 1977. Identification and characterization of a small sequence located at two sites on the amplifiable tetracycline resistance plasmid pAMalpha1 in *Streptococcus faecalis. J. Bacteriol.* 129:400–406.

12

Mechanism of Avoidance of 5-methylcytosine to Thymine Mutations in Bacteria

Ashok S. Bhagwat and Margaret Lieb

Introduction

It has been known for some time that 5-methylcytosine (5meC) in DNA is a mutation hazard, since deamination of 5meC converts it to thymine, a normal base (Lindahl and Nyberg, 1974). Replication of DNA containing such T:G mismatches results in C to T mutations. Sites of cytosine methylation were first recognized to be hotspots for transition mutations in the *lacI* system of *E. coli* (Coulondre *et al.,* 1978), and this observation has been confirmed in other *E. coli* gene systems and has been extended to human cells. Transition mutations in CpG sequences are the largest class of point mutations found in genetic diseases including cancer (Cooper and Youssoufian, 1988; Greenblatt *et al.,* 1994).

The discovery of the first repair system with the ability to reduce 5meC to T mutations came from the studies of marker effects during phage lambda recombination. Certain amber mutations in the repressor gene *cI* recombine at unexpectedly high frequencies with nearby mutations to give *cI*$^+$phage. Sequencing of the aberrant *cI* mutations revealed that they were C to T transitions resulting in 5′-CTAGG or related sequences. *E. coli* K-12 contains a DNA cytosine methylase (Dcm) that methylates the second cytosine within 5′-CCWGG (W is A or T) sequences. Because the context of the T:G mismatches arising due to deamination of 5meC created by Dcm would be identical to those found in the heteroduplexes arising during recombination, it was suggested that the cause of excess wild-type progeny in the phage crosses was the result of specific mismatch correction targeted to a site of cytosine methylation (Lieb, 1983). The putative mismatch repair system was called "very short patch" (VSP) because repair of the T:G mismatch was not accompanied by corepair of mutations only a few base pairs on either side (Lieb *et al.,* 1986). Mutagenic consequence of 5meC deamination and the antimutagenic effects of VSP repair are schematically shown in Figure

12-1. Subsequently, hints about the existence of similar systems in other bacteria and of analogous systems in mammalian cells have surfaced. We summarize here what is known about these systems and discuss why, despite error correction mechanisms, sites of cytosine methylation remain hotspots for mutations.

A Gene Required for VSP Repair Overlaps *Dcm*

The gene for the methyltransferase, *dcm,* and a gene required for VSP repair, *vsr,* lie at 43′ on the *E. coli* map (Marinus, 1973). The two genes are transcribed counterclockwise from an unidentified promoter upstream of *dcm.* Remarkably, the first seven codons of *vsr* overlap the 3′ end of *dcm* in the +1 register (Dar and Bhagwat, 1993; Sohail *et al.,* 1990). Despite this, *dcm* and *vsr* produce separate gene products.

Mutational analysis of this region identified *dcm* and *vsr* as separate genes. Deletions and point mutations in the region were used to complement chromosomal mutations defective in both methylation and VSP repair activities. Mutants were identified that could complement only the Dcm^- or Vsr^- phenotype (Dar and Bhagwat, 1993; Sohail *et al.,* 1990). In addition, a *dcm-vsr* segment deleted for all of *dcm* except the last seven codons was expressed from a T7 promoter and shown to restore VSP repair activity to bacteria deleted for *vsr.* Although it is clear that Vsr expression does not involve ribosomal frameshifting, the mechanism by which ribosomes initiate translation of *vsr* is not yet clear. The structure of the *dcm-vsr* region is shown in Figure 12-2.

The relationship between *dcm* and *vsr* was further clarified by the analysis of the mutant allele *dcm-6* (Marinus and Morris, 1973) which has a Vsr^- Dcm^- phenotype. Sequencing of *dcm-6* revealed two mutations in *dcm:* a Gln to Arg mutation at codon 26, and a TGG to TGA change at codon 34 (Dar and Bhagwat, 1993). Based on the discovery of the nonsense mutation and on analysis of RNA

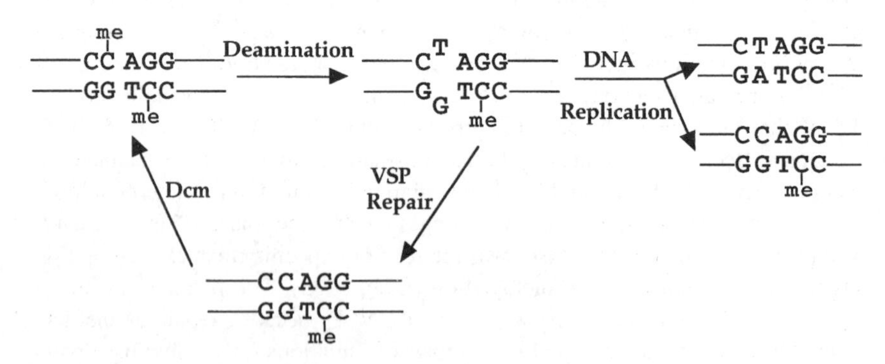

Figure 12-1. Cytosine Methylation, Mutagenesis, and VSP Repair
The life cycle of a Dcm site in *E. coli* is shown.

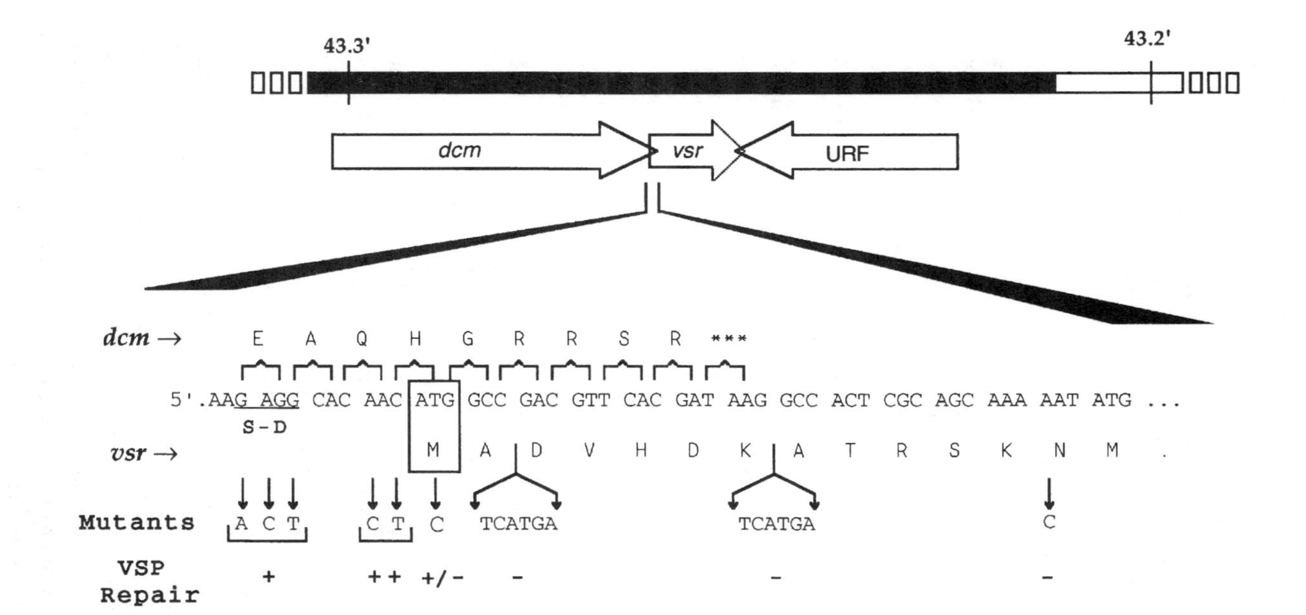

Figure 12-2. Organization of *dcm-vsr* Region and Mutational Analysis of Overlap between *dcm* and *vsr*
The uppermost part of the diagram shows the location of the *dcm* region on the *E. coli* genetic map. Below it, open arrows represent the major open reading frames in this region. An uncharacterized reading frame (URF) overlaps the last four codons of *vsr;* the overlap between *dcm* and *vsr* is shown in detail. The predicted sequences of the two proteins are shown above or below the sequence using the single letter amino acid code. The Shine-Dalgarno (S-D) sequence for *vsr* is underlined. Sequence changes in mutants constructed to analyze the *dcm-vsr* overlap and their VSP repair phenotypes are summarized below the DNA sequence. When mutants contained more than one sequence change the sequence changes are bracketed together. VSP repair activity of the mutants was determined by a phage λ recombination assay (Dar and Bhagwat, 1993). The repair activities of the mutants are shown on a scale where wild-type *vsr* would be "+++" (not shown) and Δ*vsr* would be "–".

in wild-type cells and in cells with the *dcm-6* allele, it has been concluded that *dcm* and *vsr* are transcribed as a single unit and that the nonsense mutation in *dcm* has a polar effect on Vsr expression. A coupling between the expression of the two genes represents a form of biological regulation. Because transcription and translation of *dcm* also leads to the translation of *vsr,* VSP repair is likely to be active whenever Dcm is present in the cell. As Dcm is responsible for the occurrence of T:G mismatches at CCWGG sites, this strategy for Vsr synthesis should minimize C to T mutations at sites of methylation.

Vsr Reduces Mutations at Dcm Sites

Two studies involving *E. coli* grown in rich medium have shown that VSP repair reduces the frequency of mutations at Dcm sites, but does not eliminate them. A *dcm* site in gene *cI* of phage λ is a hotspot for C to T mutation that creates an amber codon. In a dcm$^+$ *vsr*$^-$ background 95% of all amber mutations are at this site (Lieb, 1991)! In a dcm$^+$ *vsr*$^+$ background, the frequency of amber mutations at the Dcm site is reduced 10-fold. However, the Dcm site remains a hotspot and even the overproduction of Vsr does not completely eliminate it. Qualitatively similar results were also obtained using a genetic reversion assay. A mutant kanamycin-resistance gene was constructed such that deamination of 5meC in the CCAGG context results in a change from kanamycin-sensitivity to kanamycin-resistance (Kan^R). The frequency of Kan^R revertants obtained in a Dcm$^+$ Vsr$^-$ strain was 100 times the frequency in a Dcm$^-$ Vsr$^-$ strain (Wyszynski *et al.,* 1994). In a Dcm$^+$ Vsr$^+$ strain, the reversion frequency was still 25-fold higher than the frequency found in the Dcm$^-$ Vsr$^-$ strain.

If VSP repair is designed to avoid 5meC to T mutations, why are Dcm sites hotspots for transition mutations in wild-type *E. coli?* We suggest two related answers to this question. The efficiency of mutation avoidance is likely to be proportional to the time between the deamination event and replication of the strand containing the resultant thymine. In λ crosses, the frequency of recombinants attributable to mismatch repair is increased by agents that reduce the rate of phage DNA synthesis. These include UV irradiation of the host bacteria, or incubation of *dnaBts* bacteria at high temperatures during the cross (Lieb and Rehmat, 1995). When *E. coli* is grown in rich media, its generation time shortens and hence the time available for repair decreases. As a result, some of the mismatches may escape repair. Another consequence of replication is the generation of DNA single strands. If 5meC in single-stranded DNA within replication forks deaminates to thymine, it will be copied by the replication complex fixing the mutation. In this case, VSP repair will not be able to repair the lesion because at no time will it exist as a T:G mismatch. Comparing the ability of VSP repair to suppress 5meC to T mutations in slow growing or stationary phase cells with those grown in rich media would be a way to test these hypotheses.

Vsr is a Sequence-specific and Mismatch-specific Endonuclease

Gene *vsr* codes for a protein 156 amino acids in length. The native Vsr has recently been purified and shown to be a monomer by gel-exclusion chromatography (C. Putnam and J. Tainer, personal communication). However, most biochemical studies of Vsr have been done with protein that was expressed as a fusion protein. In some studies the fused protein was treated with proteases to release Vsr prior to its use. Vsr acts as a sequence-specific endonuclease, and cleaves the phosphodiester linkage preceding the thymine in a T:G mismatch to produce 5′-phosphate and 3′-OH termini (Hennecke *et al.,* 1991). In addition, Vsr recognizes U:G mismatches (Gläsner *et al.,* 1992) and VSP repair can correct such mismatches to C:G (Gabbara *et al.,* 1994). Vsr is unique among enzymes that attack "damaged" bases in that it has no detectable glycosylase activity. Specifically, Vsr is unlike other enzymes that process U:G and T:G mismatches i.e., uracil-DNA glycosylases and the thymine glycosylase from HeLa cells (see below).

Table 12-1 compares the current information on the effect of context on VSP repair. The efficiency of VSP repair of different *cI* mutations was studied in lambda crosses, or purified Vsr was used to compare nicking of substrates contain-

Table 12-1.[a] Effect of Sequence Context on VSP Repair

	Sequence	Relative repair frequency[b]	Relative nicking rate constant[c]
1	CTAGG	1.0	1.0
2	TTAGG	.60	.39
3	ATAGC		.22
4	GTAGG	.32	.36
5	CTAGA	.39	.27
6	CTAGT		.35
7	CTAGC		.57
8	TTAGA		.06
9	TTAGT	.12	
10	TTAGC		.04
11	ATAGT	.04	
12	GTAGT	.02	
13	GTAGC	.03	
14	CTTGG		.68
15	TTTGG		.27
16	CTCGG		.24
17	CTTGC		.16
18	CTGGG		.05
19	CTAAG	.005	

[a]from (Lieb and Bhagwat, 1996)

[b]from (Lieb and Rehmat, 1995)

[c]from (Gläsner et al., 1995)

ing T:G mispairs. There is good agreement between the results obtained by these two different methods (Table 12-1, lines 2, 4 and 5). Although cytosine methylation occurs in both 5′-CCAGG and 5′-CCTGG, Vsr nicks more efficiently when the cytosine is in 5′-CTAGG. Interestingly, nicking was also seen with several sequences that contained four out of five bases from the Dcm recognition sequence. For example, nicking adjacent to the mismatched T was observed in all 5′CTNGG contexts, with efficiency lowest when N=G. Similarly, VSP repair was observed in two four-out-of-five sequences tested (Table 12-1, lines 4 and 5).

The discrepancy between the methylation specificity of Dcm and the sequence specificity of Vsr is curious. *In vivo,* methylation of non-CCWGG sites by *Eco*RII methylase is seen only under conditions of enzyme overproduction. If Dcm has a similar strong preference for CCWGG sites, there would be little methylation of non-canonical sequences, and consequently no need for VSP repair. However, there may be physiological conditions, such as stationary phase, under which Dcm is overproduced in the bacteria, and VSP repair may be needed to deal with such a contingency. VSP repair in replicating bacteria would also help prevent C to T mutations in χ sites (5′GCTGGTGG) that are known to be important for recombination (*see* Chapter 6). VSP repair would also tend to maintain REP sequences of *E. coli* which contain 5′CC (G or T) GA and TC (C or A) GG motifs. It should be noted that REP sequences are abundant in *E. coli, Shigella,* and *Salmonella* species (*see* Chapter 5) and these species also contain *vsr* (see below).

Because of its broad sequence-specificity, VSP repair may act as a mutagenic agent upon certain sequences. It could compete with the long patch repair system creating "four-out-of-five" sites. For example, during replication of 5′CTAG, if the T in the parental strand is mispaired with a G in the newly synthesized strand, VSP repair could "correct" the T to C, creating a CCAG site. Comparison of frequencies of tetranucleotides in the *E. coli* genome of the form 5′-TWGG, 5′-CTWG and their complements with frequencies predicted from the observed frequencies of component di- and tri-nucleotides shows a consistent underrepresentation of these tetranucleotide sequences in the *E. coli* genome (Bhagwat and McClelland, 1992; Merkl *et al.,* 1992). Concurrently, CWGG, CCWG sequences and their complements are significantly overrepresented in the genome. Such biases are not found in *B. subtilis* or *S. cerevisiae,* organisms that do not contain Dcm. There is also a rough correlation between the efficiency with which Vsr nicks various T:G mismatches and the extent of underrepresentation of the corresponding T-containing sequences in the genome (Gläsner *et al.,* 1995).

Other Genes Active in VSP Repair

A Lac reversion assay was used to isolate a mutator in *E. coli* that shows a higher frequency of 5meC to T mutations (Ruiz *et al.,* 1993). The mutation maps to the 43′ region of the chromosome, and DNA sequencing has found no mutation in *vsr* (Ruiz *et al.*, 1993). Therefore, the mutation that causes this phenotype may be in an unidentified gene, or at a site important in *vsr* expression. Mutations in many identified bacterial genes have been screened for their effects on VSP repair. Significant reductions in repair are produced by mutations in *polA, lig, mutL,* and *mutS.* The 5′ to 3′ exonuclease and polymerase activities of polymerase I presumably remove and replace a short stretch of DNA 3′ to the nick produced by Vsr, and DNA ligase completes the repair. Although VSP repair *in vivo* is reduced significantly in their absence, MutS and MutL are dispensable. Nicking by purified Vsr also occurs without the addition of other proteins (Hennecke *et al.*, 1991). If the amount of Vsr in bacteria in *mutS* or *mutL* backgrounds is increased by the introduction of vsr^+ on a multicopy plasmid, repair of a T:G mispair in the 5′CTAGG context is restored to levels found in $mutL^+$ $mutS^+$ bacteria. However, mispairs in the 5′TTAGG and 5′GTAGG contexts require MutS and MutL for optimum repair suggesting a collaboration between MutL/ MutS and Vsr (Lieb and Rehmat, 1995). Collaboration between Vsr and MutL/ MutS is somewhat unexpected, since these proteins compete for the same mismatch. Competition between Vsr and MutS is suggested by the observation that VSP repair is reduced in bacteria in which excess MutS is supplied by a multicopy plasmid (Lieb and Rehmat, 1995).

How might proteins which are essential for long patch repair participate in VSP repair? We would like to propose a "conveyor belt" model of the cooperation between MutS, MutL and Vsr. This is based on the reasonable hypothesis that in addition to binding to T:G mispairs in the appropriate context, Vsr binds with lower affinity to the large excess of other sites in the genomic DNA. MutS and MutL (presumably present at a relatively high concentration) bind to the mismatch, and by a translocation mechanism reel in DNA to which one or more Vsr molecules are non-specifically bound. Vsr, having a higher affinity for T:G in the appropriate context, then replaces MutS and MutL and proceeds to exercise its nicking function. Complete elucidation of the functions of MutS and MutL in VSP repair awaits the development of an *in vitro* system for repair.

VSP Repair in Other Organisms

It was shown that DNA from several enteric bacteria that are closely related to *E. coli* including *Shigella sonnei, Salmonella typhimurium, Salmonella enteritidis, Enterobacter cloacae* and *Klebsiella pneumonia* is methylated within Dcm sites (Gomez-Eichelmann *et al.*, 1991). Southern blot hybridization of these DNAs with probes from *dcm* and *vsr* genes have detected homologues of both the genes in these bacteria (M. Lieb, unpublished results). In addition, neither study has found evidence for the presence of *dcm* or *vsr* in *Serratia marcescens* or in *Yersinia* and *Proteus* species. The homologue of *vsr* from *Shigella* has been cloned into *E. coli* and sequenced. It is nearly identical to its *E. coli* counterpart (Figure 12-3).

```
                                                                        A      L F                         YR V   H
Consensus -------------------D-L--E-RSKNMQ-A-SUN-GTKPEKALRSLL--LGYRYRK-D-SLPG-----TPDIVF--RK-AIFIDG
E.coli    -----------------MADVHDKATRSKNMRAIATRD--TAIEKRLASLLTGQGLAFRVQDASLPG-----RPDFVVDEYRCVIFTHG 65
Shigella  -----------------MADVHDKATRSKNMRAIATRD--TAIEKRLASLLTGQGLAFRVQDASLPG-----SPDFVVDEYRCVIFTHG 65
XorII     -----------------MTDRLSPERRRYLMQQVRSKN--TRPEKAVRSLLHSIAYRFRLHRKDLPG-----TPDIVFPSRRLVLFVHG 65
HpaII     ------------------MDKLTPEQRKKCMK-ASSKNKGTKPELLLAKYLWALGLRYRKNDRSIFG-----TPDLSFKRYKIAIFIDG 65
NaeI      MSDKSSRAAARARAHASGTYPAPLNAGRSRNMQ-ANRRS-GTKPEAALRSALFKLGYRYRKDFLLRLGDGVKVKPDIVFTARKVAVFIDG 88

                   K     TR  F
Consensus CFWH-H-C-IRQ--PKSN-DYWSPKIE-NVERDRRVN--L---GWRJLRVWE--------L-D--AL-ARL------E--------------
E.coli    CFWHHHHCYLFK-VPATRTEFWLEKIGKNVERDRRDISRLQELGWRVLIVWECALRGREKLTD-EALTERLEEWICGEGASAQIDTQGIHLLA 156
Shigella  CFWHHHHCYLFK-VPATRTEFWLEKIGKNVERDRRDISRLQELGWRVLIVWECALRGREKLTD-EALTERLEEWICGEGASAQIDTQGIHLLA 156
XorII     CFWHGHGCRIGQ-LPKSRLDYWSPKIEANRARDQRKEAALAAEGWRVAVVWQC------ELSDLGALEARLRNILDPS-------------- 136
HpaII     EFWHGKDWDIRKYDIKSNKDFWISKIEHNMNRDKKVNDYLISNGWVIFRFWGK-----DVLKN--PEKFSLEIQKAIYERCVR--------- 141
NaeI      CFWHVCPDHGRQ--PTTNEWYWSPKLRRNVERDRTVNQSLTNAGWRVLRVWEH-----EELQD--AVAAVVDTLHHLEHGFDTSAED------ 166
```

Figure 12-3. Sequences of Known and Putative Vsr Homologues

The sequences were aligned using the pattern-induced multi-sequence alignment program (Smith and Smith, 1990). Sequences of the *E. coli vsr* and the putative homologs from *Xor*II, *Hpa*II and *Nae*I systems were from Genbank (accession no.s M32307/X13330, U06424, X51322 and U09581). The *Shigella* sequence is from the Lieb laboratory (unpublished). The *Shigella* sequence is identical to the *E. coli* except for the Arg-49 in *E. coli,* which is changed to Ser. These residues are presented by hollow letters in the figure. Residues that are identical in all five sequences are presented in bold. The consensus sequence represents the residues at different positions that appear most frequently among the homologs. Because of their near identity, *E. coli* and *Shigella* sequences were treated as one in arriving at the consensus. When two residues appeared equally frequently among the homologs, the two choices were indicated one above the other as the consensus.

E. coli Vsr does not share much sequence similarity with any other protein with an identified function. Its sequence can be aligned with the sequences of some predicted proteins in sequence databases (Figure 12-3). Remarkably, the open reading frames for these proteins lie close to cytosine methylase genes, and hence it is likely that these proteins are also involved in the repair of T:G mismatches at sites of cytosine methylation. However, this remains to be demonstrated. In a study done with *E. coli,* the putative *vsr* homolog located near the *Hpa*II methylase gene did not reduce C to T mutations within a CCGG site (Bandaru *et al.,* 1995). A possible explanation of this result is that the *H. parainfluenza* homologue of *vsr* does not interact effectively with the *E. coli* MutL and MutS proteins.

5-methylcytosine is found in the DNA of most bacteria and of many eukaryotes, including some fungi, plants, and vertebrates. It is likely that repair processes to reduce 5meC to T mutations exist in all such organisms. Repair of T:G mispairs to C:G occurs in DNA transfected into mammalian tissue cultures, and a glycosylase isolated from human cells removes thymine from T:G mispairs (Nedderman and Jiricny, 1993). As yet the sequence-specificity of this enzyme has not been studied and its biological role remains to be elucidated. It will be interesting to see if other organisms that contain 5meC employ a VSP repair-like pathway, thymine glycosylase pathway, or some other novel mechanism to avoid 5meC to T mutations.

Acknowledgments

This article was adapted from a recent review (Lieb and Bhagwat, 1996) by the authors.

References

Bandaru, B., *et al.* 1995. *Hpa*II Methyltransferase Is Mutagenic in *Escherichia coli. J. Bacteriol.* 177:2950–2952.

Bhagwat, A. S., and M. McClelland. 1992. DNA mismatch correction by very short patch repair may have altered the abundance of oligonucleotides in the *E. coli* genome. *Nucl. Acids Res.* 20:1663–1668.

Cooper, D. N., and H. Youssoufian. 1988. The CpG dinucleotide and human genetic disease. *Hum. Genet.* 78:151–155.

Coulondre, C., *et al.* 1978. Molecular basis of substitution hotspots in *Escherichia coli.* Nature 274:775–780.

Dar, M. E., and A. S. Bhagwat. 1993. Mechanism of expression of DNA repair gene *vsr,* an *Escherichia coli* gene that overlaps the DNA cytosine methylase gene, *dcm. Mol. Microbiol.* 9:823–833.

Gabbara, S., *et al.* 1994. A DNA Repair Process in *Escherichia coli* Corrects U:G and T:G Mismatches to C:G at Sites of Cytosine Methylation. *Molec. Gen. Genet.* 243:244–248.

Gläsner, W., *et al.* 1992. Enzymatic Properties and Biological Functions of Vsr DNA Mismatch Endonuclease. In *Structural Tools for the Analysis of Protein-Nucleic Acid Complexes,* D. M. J. Lilley, H. Heumann, and D. Suck, eds. pp. 165–173. Birkhäuser Verlag Basel.

Gläsner, W., *et al.* 1995. Substrate Preferences of Vsr DNA Mismatch Endonuclease and their Consequences for the Evolution of the *Escherichia coli* K-12 Genome. *J. Mol. Biol.* 245:1–7.

Gomez-Eichelmann, M. C., *et al.,* 1991. Presence of 5-Methylcytosine in CC(A/T)GG Sequences (Dcm Methylation) in DNAs from Different Bacteria. *J. Bacteriol.* 173:7692–7694.

Greenblatt, M. S., *et al.* 1994. Mutations in the *p53* Tumor Suppressor Gene: Clues to Cancer Etiology and Molecular Pathogenesis. *Cancer Research* 54:4855–4878.

Hennecke, F., *et al.* 1991. The *vsr* gene product of *E. coli* K-12 is a strand- and sequence-specific DNA mismatch endonuclease. *Nature* 353:776–778.

Lieb, M. 1983. Specific Mismatch Correction in Bacteriophage Lambda Crosses by Very Short Patch Repair. *Mol. Gen. Genet.* 191:118–125.

Lieb, M. 1991. Spontaneous mutation at a 5-methylcytosine hotspot is prevented by very short patch (VSP) mismatch repair. *Genetics* 128:23–27.

Lieb, M., *et al.* 1986. Very Short Patch Repair in Phage Lambda: Repair Sites and Length of Repair Tracts. *Genetics* 114:1041–1060.

Lieb, M., and A. S. Bhagwat. 1996. VSP: Reducing the cost of cytosine methylation. *Molec. Microbiol.* 20: in press.

Lieb, M., and S. Rehmat. 1995. Very short patch repair of T:G mismatches in vivo: Importance of context and accessory proteins. *J. Bacteriol.* 177:660–666.

Lindahl, T., and B. Nyberg. 1974. Heat-induced deamination of cytosine residues in deoxyribonucleic acid. *Biochemistry* 13:3405–3410.

Marinus, M. G. 1973. Location of DNA Methylation Genes on the *Escherichia coli* Genetic Map. *Mol. Gen. Genet.* 127:47–55.

Marinus, M. G., and N. R. Morris. 1973. Isolation of Deoxyribonucleic Acid Methylase Mutants of *Escherichia coli* K-12. *J. Bacteriol.* 114:1143–1150.

Markl, R., *et al.* 1992. Statistical evaluation and biological interpretation of nonrandom abundance in the *E. coli* K-12 genome of tetra- and pentanucleotide sequences related to VSP DNA mismatch repair. *Nucl. Acids Res.* 20:1657–1662.

Nedderman, P., and J. Jiricny. 1993. The purification of a mismatch-specific thymine-DNA glycosylase from HeLa cells. *J. Biol. Chem.* 268:21218–21224.

Ruiz, S. M., *et al.* 1993. Isolation and characterization of an *Escherichia coli* strain with a high frequency of C-to-T mutations at 5-methylcytosines. *J. Bacteriol.* 175:4985–4989.

Smith, R. F., and T. F. Smith. 1990. Automatic generation of primary sequence patterns from sets of related protein sequences. *Proc. Natl. Acad. Sci. (USA)* 87:118–122.

Sohail, A., *et al.* 1990. A Gene Required for Very Short Patch repair in E. coli is Adjacent to the DNA Cytosine Methylase Gene. *J. Bacteriol.* 172:4214–4221.

Wyszynski, M., *et al.* 1994. Cytosine Deaminations Catalyzed by DNA Cytosine Methyltransferases are unlikely to be the Major Cause of Mutational Hot-spots at Sites of Cytosine Methylation in *E. coli. Proc. Natl. Acad. Sci. (USA)* 91:1574–1578.

13

Conjugative Transposons

Don B. Clewell

Conjugative transposons are able to move from one bacterial cell to another by a process requiring cell-to-cell contact. Such elements have been found in many bacterial genera but are particularly common among the Gram-positive streptococci and enterococci. They move via an excision/insertion process that involves a non-replicative circular DNA intermediate possessing plasmid-like conjugative properties. Some exhibit a relatively low level of target-site specificity, whereas others can be very specific. The transposons commonly carry antibiotic resistance determinants and, at least in some species (e.g. *Streptococcus pneumoniae* and *Streptococcus pyogenes*), are probably more responsible for the dissemination of these genes than plasmids. Anaerobic, Gram-negative *Bacteroides* strains are also known to carry non-plasmid resistance elements, and certain tetracycline-resistance transposons in this group exhibit drug-inducible transfer. Some strains of lactococci carry conjugative transposons with determinants for nisin production and sucrose metabolism. For recent reviews of these elements see Clewell and Flannagan (1993), Clewell, Flannagan, and Jaworski (1995), Scott and Churchward (1995), and Salyers *et al.* (1995).

The Tn*916*/Tn*1545* Family

The closely related Tn*916* from *E. faecalis* DS16 (Franke and Clewell, 1981; Gawron-Burke and Clewell, 1992) and Tn*1545* from *S. pneumoniae* BM4200 (Courvalin and Carlier, 1986) are among the most intensively studied conjugative transposons, and their mechanism of movement is believed to be similar if not identical. Tn*916* encodes resistance to tetracycline [*tet*(M)]; Tn*1545* also carries *tet*(M) but encodes resistance to erythromycin and kanamycin as well. The complete 18-Kb nucleotide sequence is known in the case of Tn*916;* (Flannagan *et al.,* 1994) and although the larger Tn*1545* (25.3 Kb) has not been entirely sequenced, regions near its ends that have been determined are essentially identical to those of Tn*916* (Trieu-Cuot *et al.,* 1991). Members of the Tn*916*/Tn*1545* family, of which there are now more than 10 known examples, exhibit a broad bacterial host-range extending across more than 50 different species including 24 genera (Clewell, Flannagan and Jaworski, 1995).

Most conjugative transposons have been found originally on the bacterial chromosome; indeed only one, Tn*925,* has been identified on a plasmid (pCF10 in *E. faecalis*). Transposition to low-copy number plasmids can generally be observed if, like the case for Tn*916,* the target site is not highly specific. When cloned on a multicopy plasmid in *E. coli,* Tn*916* is relatively unstable and tends to excise, with the flanking plasmid DNA being spliced together (Gawron-Burke and Clewell, 1984). The excised element is usually lost (segregated) but can also integrate into the chromosome where it is relatively stable. The presence of a single copy of Tn*916* on the *E. coli* chromosome results in a barely detectable level of tetracycline resistance—unlike the case in *E. faecalis,* where a single copy results in a higher and more easily-detectable level of resistance. When the *tet*(M) determinant is replaced by an erythromycin-resistance determinant *(erm),* a single copy is easily monitored on the chromosome of *E. coli,* and intracellular transposition to a low-copy conjugative plasmid such as pOX38 (a derivative of F) can be followed (Clewell *et al.,* 1991).

The elevated excision rate of Tn*916* from a multicopy plasmid in *E. coli* facilitated the isolation of "intermediate" circular DNA which could be introduced into the chromosome of *Bacillus subtilis* protoplasts, where it maintained its conjugative potential (Scott, Kirchman, and Caparon, 1988). Transformation using DNA from *E. coli* has been useful for the introduction of Tn*916* to various bacterial species. Because *E. coli* preparations of plasmid DNA carrying the transposon are likely to also contain a significant amount of intermediate Tn*916* DNA, the latter is probably responsible for transformation (Clewell and Flannagan, 1993).

Mechanism of Movement

Unlike the case for a number of other transposons (*see* Chapter 4) the Tn*916*-related elements do not duplicate their target sequence upon insertion. The target corresponds to an AT-rich region with a non-specific hexanucleotide "core" at its center (Clewell and Flannagan, 1993; Trieu-Cuot *et al.,* 1993). After insertion, the target core can be found at one or the other end of the transposon, where it is referred to as a "coupling sequence." The junction at the opposite end represents a hexanucleotide coupling sequence brought in by the transposon, having served to join the two ends of the incoming element while in its circular intermediate structure.

A current view of how movement occurs holds that excision is initiated via a

6-bp staggered cut across the coupling sequences at each end of the transposon followed by a splicing together of resulting single strand overhangs (Caparon and Scott, 1989; Poyart-Salmeron *et al.,* 1989). Since the coupling sequences are generally different, the junction of the "coupled" ends correspond to a heteroduplex structure. In addition, the flanking DNA from which the transposon excised is also spliced together and may also manifest as a heteroduplex at the 6-bp overlap. Repair processes and/or replication should readily remove the heteroduplexes, although there is some evidence that in the case of the circular intermediate, the heteroduplex may be protected. Assuming that conjugative transfer is a plasmid-like event with only a single strand being transferred and replicated to a duplex circular structure in the recipient cell, the 6-bp coupling segment is likely to become a homoduplex. Scott *et al.* (1994) have published evidence consistent with the transfer of a single strand. They found that only the coupling sequence at the left end of the donor element was present in transconjugants, although this could be found with equal probability at either end of the newly inserted transposon.

Integration is viewed as the opposite of excision, occurring at a target core that is generally different from the coupling sequence of the intermediate. The DNA at each of the transposon contains 26-bp imperfect (20 of 26 identical) inverted repeats (Clewell *et al.,* 1988). These sequences, which are A-rich on one end and T-rich on the other, are thought to facilitate alignment of the circular intermediate with the target sequence. Since there can be many different target cores, and the transposon brings in a coupling sequence, insertions can result in different combinations of coupling sequences flanking the integrated element.

Organization of Tn*916*

The map of Tn*916* shown in Figure 13-1 indicates a number of open reading frames, all but two of which are in the same orientation. Key determinants relating to excision and insertion are located near the left end of the transposon (Senghas *et al.,* 1988; Poyart-Salmeron *et al.,* 1989, 1990; Su and Clewell, 1993). The determinant *int-Tn* is necessary for both integration and excision, whereas *xis-Tn* plays a role only in excision. There are two open reading frames within and in-frame with the *int-Tn* determinant whose products appear to play a role in regulating excision (Clewell, *et al.* 1995). The region between *tet* and the right end is primarily devoted to conjugation functions. Between *tet* and the left end, there is only one reading frame, now designated *traA,* currently known to relate to conjugation. The 5′ ends of this determinant and *xis-Tn* overlap and diverge. The possibility that *traA* may be a key positive regulator of conjugation functions has recently been considered (Clewell, Flannagan, and Jaworski, 1995). In this view, the "trigger event" for Tn*916* excise relates to a coupled expression of *traA* and *xis-Tn,* followed by a TraA-promoted upregulation of operonically

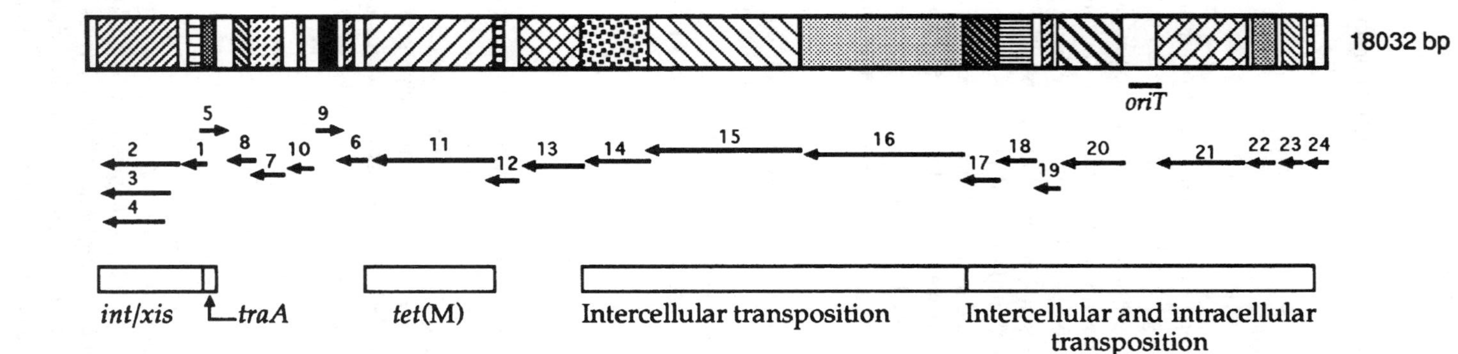

Figure 13-1. Map of Tn*916*. Open reading frames are designated as ORFs 1–24. ORFs 3 and 4 are within ORF2 and in the same frame. ORFs 1 and 2 correspond to *xis-Tn* and *int-Tn,* respectively. ORF5 corresponds to *traA,* which is necessary for conjugative transfer. ORF11 corresponds to *tet*(M). ORFs 14–24 are not required for excision/insertion but are necessary for intercellular transfer; however, mutations in the right portion of this segment also greatly reduce intracellular transposition (see Clewell, *et al.,* 1995). A functional origin of conjugative transfer is indicated by *oriT.* (Modified from Clewell *et al.,* 1995 with permission of the publisher.)

organized conjugation genes under control of the ORF24 promoter near the right end of the transposon.

Differences in Donor Potential

In *E. faecalis* and other Gram-positive bacteria, Tn*916* transfers with widely differing donor potentials. Matings on filter membranes result in transfer frequencies of from less than 10^{-8} to greater than 10^{-4} per donor, depending on the particular isolate. The differences relate to the nature of the coupling sequences (Jaworski and Clewell, 1994) and probably involve an effect on the binding of Int-Tn. It has been reported that Int-Tn has two DNA binding domains, one of which binds to the ends of the transposon and DNA just outside each end (Lu and Churchward, 1994). Since binding would appear to occur across the coupling sequence, differences in these sequences could be easily envisioned to have significant effects on binding and/or the subsequent cleavage event—thus explaining the connection between coupling sequences and donor potentials (see above). Indeed, there is recent biochemical evidence that supports this notion (Lu and Churchward, 1995). Interestingly, the other domain of Int-Tn binds to a 27-nucleotide sequence located within the transposon 72 bp from the right end and repeated in the same orientation 136 bp from the left end. It is therefore conceivable that the two Int-Tn domains could bind to their corresponding sequences at opposite ends of the transposon and facilitate alignment of the two junctions prior to excision.

Multiple Copies and Activation *in trans*

In *E. faecalis* mating experiments, it is not uncommon for transconjugants to acquire more than one copy of Tn*916*. Frequently, half of a population of transconjugants harbors two or as many as six inserts located at different sites (Gawron-Burke and Clewell, 1982; Clewell *et al.*, 1991); and the presence of one element in a recipient has no effect on the ability to take up additional elements. The basis for which multiple insertions are generated is not known, although various suggestions have been offered (Clewell and Flannagan, 1993).

When donor cells have more than one copy, the movement of one transposon appears to activate the movement of others (Flannagan and Clewell, 1991). For example, using a bacterial host deficient in homologous recombination harboring Tn*916* on a nonconjugative plasmid and Tn*916ΔE (erm* substituted for *tet)* on the chromosome, transconjugants selected for the acquisition of *erm* acquire *tet* (unselected) about half the time. This phenomenon can be taken advantage of when performing various genetic (complementation) analyses; and recently it has facilitated the identification of a functional transfer origin *(oriT)* located between ORF20 and ORF21 (Jaworski and Clewell, 1995). For example, when a small segment of DNA containing this region was cloned on a nonconjugative plasmid and placed in an *E. faecalis* strain harboring an intact transposon on the chromosome, selection for transfer of the transposon resulted in unselected co-transfer of the plasmid about half the time. It is conceivable that the product of the nearby ORF23 is responsible for a single stranded cleavage at this *oriT,* since it exhibits significant sequence similarity with the ColE1 MbeA protein involved in nicking at the transfer origin (Boyd, Archer, and Sherratt, 1989).

Complex Conjugative Transposons

Tn*916*-like elements can transpose into other mobile elements such as plasmids, as noted above, and even other conjugative transposons. The transposons Tn*5253* of *S. pneumoniae* (Ayoubi, Kilic, and Vijayakumar, 1991) and Tn*3703* of *S. pyogenes* (Horaud *et al.,* 1991) appear to represent elements where this has occurred in nature. For example, Tn*5253* contains a Tn*916*-like element (designated Tn*5251*) which can be removed leaving a nonhomologous element (designated Tn*5252*) that is able to transpose conjugatively; the latter, however, has a very specific target site, unlike the case for Tn*916*-like elements.

Conjugative Transposons in Bacteroides

The *Bacteroides* elements are generally of a size greater than 65 kb, and most carry a resistance determinant of the *tet*(Q) variety (Speer, Shoemaker, and Salyers, 1992). The resistance encoded by both *tet*(Q) and *tet*(M) is of the ribosome protection type, in contrast to a number of other tetracycline-resistance determinants that involve a drug efflux mechanism. Like the case for the Tn*916* family, movement appears to occur via an excision/insertion process (Bedzyk, *et al.,* 1992). There appears to be a greater degree of target specificity than the case for the Tn*916*-like elements, although insertions do occur at multiple sites. Some examples of *Bacteroides* elements are designated Tc^r ERL, Tc^{r}Em^r 12256, and the cryptic XBU4422 (Salyers and Shoemaker, 1992). A smaller element designated Tn*4399* (9.6 kb) appears able to mobilize nonconjugative plasmids when present *in cis,* although mobilization is greatly enhanced by a chromosome-borne element (Murphy and Malamy, 1993). An *oriT* site has been identified on this element (Murphy and Malamy, 1995). Transfer has been observed between *Bacteroides* strains/species and *E. coli.*

Certain conjugative transposons have been found to activate *in trans* the excision of unlinked elements referred to as NBUs (non-replicating *Bacteroides* units) (Shoemaker *et al.,* 1993). These are smaller (10–12 kb in size) than the conjugative transposons that activate them and can be cryptic. A recent element of this type, designated Tn*4555,* has been found to encode cefoxitin resistance (Smith and Parker, 1993). Intermediate circular DNA forms of NBUs have been identified directly. NBUs can be mobilized to recipient cells and can insert into the recipient chromosome. *oriT* regions have been identified on some of these elements (see Salyers *et al.,* 1995).

The movement of conjugative elements that encode tetracycline resistance has been shown to be greatly stimulated by exposure to subinhibitory concentrations of drug. For example, exposure of cells carrying the element Tc^rEm^r DOT to 0.5 ug/ml of tetracycline can result in up to a 1,000-fold enhancement of conjugative transfer (Stevens *et al.,* 1992).

Concluding Remarks

The conjugative transposons are an interesting and widely diverse group of elements that are ubiquitous in the bacterial world. Like conjugative plasmids, they appear to carry determinants that provide a selective advantage under atypical conditions. The extremely broad host range of elements such as those of the Tn*916* group, with their relatively low level of target specificity, implies that their horizontal transfer has played a significant role in bacterial evolution.

Acknowledgments

Related research conducted in the author's laboratory was supported by National Institutes of Health grant AI10318.

References

Ayoubi, P., A. O. Kilic, and M. N. Vijayakumar. 1991. Tn*5253,* the pneumococcal Ω*(cat tet)* BM6001 element, is a composite structure of two conjugative transposons Tn*5251* and Tn*5252. J. Bacteriol.* 173:1617–1622.

Bedzyk, L. A., N. B. Shoemaker, K. E. Young, and A. A. Salyers. 1992. Insertion and excision of *Bacterioides* conjugative chromosomal elements. *J. Bacteriol.* 174:166–172.

Boyd, A. C., J. A. K. Archer, and D. J. Sherratt. 1989. Characterization of the ColE1 mobilization region and protein products. *Mol. Gen. Genet.* 217:488–498.

Caparon, M. G., and J. R. Scott. 1989. Excision and insertion of the conjugative transposon Tn*916* involves a novel recombination mechanism. *Cell* 59:1027–1034.

Clewell, D. B., and S. E. Flannagan. 1993. The conjugative transposons of Gram positive bacteria, in *Bacterial Conjugation,* D. B. Clewell, ed. pp. 369–393. Plenum Press, New York.

Clewell, D. B., S. E. Flannagan, Y. Ike, J. M. Jones, and C. Gawron-Burke. 1988. Sequence analysis of termini of conjugative transposon Tn*916. J. Bacteriol.* 170:3046–3052.

Clewell, D. B., S. E. Flannagan, and D. D. Jaworski. 1995. Unconstrained bacterial promiscuity: the Tn*916*-Tn*1545* family of conjugative transposons. *Trends in Microbiology* 3:229–236.

Clewell, D. B., S. E. Flannagan, L. A. Zitzow, Y. A. Su, P. He, E. Senghas, and K. E. Weaver. 1991. Properties of conjugative transposon Tn*916,* in, *Genetics and Molecular Biology of Streptococci, Lactococci, and Enterococci.* G. M. Dunny, P. P. Cleary, and L. L. McKay, eds. pp. 39–44. Washington, D.C., Am. Soc. Microbiol.

Clewell, D. B., D. D. Jaworski, S. E. Flannagan, L. A. Zitzow, and Y. A. Su. 1995. The conjugative transposon Tn*916* of *Enterococcus faecalis:* structural analysis. In, *Genetics of Streptococci, Enterococci and Lactococci,* J. J. Ferretti, M. S. Gilmore, and T. R. Klaenhammer, eds. pp. 11–17. Dev. Biol. Stand., Basel, Karger, Vol. 85.

Courvalin, P. and C. Carlier. 1986. Transposable multiple antibiotic resistance in *Streptococcus pneumoniae. Mol. Gen. Genet.* 205:291–297.

Franke, A. E., and D. B. Clewell. 1981. Evidence for a chromosome-borne resistance transposon (Tn*916*) in *Streptococcus faecalis* that is capable of "conjugal" transfer in the absence of a conjugative plasmid. *J. Bacteriol.* 145:494–502.

Flannagan, S. E., and D. B. Clewell. 1991. Conjugative transfer of Tn*916* in *Enterococcus faecalis: trans* activation of homologous transposons. *J. Bacteriol.* 173:7136–7141.

Flannagan, S. E., L. A. Zitzow, Y. A. Su, and D. B. Clewell. 1994. Nucleotide sequence of the 18-kb conjugative transposon Tn*916* from *Enterococcus faecalis. Plasmid* 32:350–354.

Gawron-Burke, C., and D. B. Clewell. 1982. A transposon in *Streptococcus faecalis* with fertility properties. *Nature* 300:281–284.

Gawron-Burke, C., and D. B. Clewell. 1984. Regeneration of insertionally inactivated streptococcal DNA fragments after excision of transposon Tn*916* in *Escherichia coli:* strategy for targeting and cloning of genes from gram-positive bacteria. *J. Bacteriol.* 159:214–221.

Horaud, T., G. de Cespedes, D. Clermont, F. David, and F. Delbos. 1991. Variability of chromosomal genetic elements in streptococci, in, *Genetics and Molecular Biology of Streptococci, Lactococci, and Enterococci.* pp. 16–20. G. M. Dunny, P. P. Cleary, and L. L. McKay, eds. Am. Soc. Microbiol., Washington, D.C.

Jaworski, D. D., and D. B. Clewell. 1994. Evidence that coupling sequences play a frequency-determining role in conjugative transposition of Tn*916* in *Enterococcus faecalis. J. Bacteriol.* 176:3328–3335.

Jaworski, D. D., and D. B. Clewell. 1995. A functional origin of transfer *(oriT)* on the conjugative transposon Tn*916. J. Bacteriol.* 177:6644–6651.

Lu, F., and G. Churchward. 1994. Conjugative transposition: Tn*916* integrase contains two independent DNA binding domains that recognize different DNA sequences. *EMBO J.* 13:1541–1548.

Lu, F., and G. Churchward. 1995. Tn*916* target DNA sequences bind the C-terminal domain of integrase protein with different affinities that correlate with transposon insertion frequency. *J. Bacteriol.* 177:1938–1946.

Murphy, C. G., and M. H. Malamy. 1993. Characterization of a "mobilization cassette" in transposon Tn*4399* from *Bacterioides fragilis. J. Bacteriol.* 175:5814–5823.

Murphy, C. G., and M. H. Malamy. 1995. Requirements for strand- and site-specific cleavage within the *oriT* region of Tn*4399,* a mobilizing transposon from *Bacteroides fragilis. J. Bacteriol.* 177:3158–3165.

Poyart-Salmeron, C., P. Trieu-Cuot, C. Carlier, and P. Courvalin. 1989. Molecular characterization of two proteins involved in the excision of the conjugative transposon Tn*1545;* homologies with other site-specific recombinases. *EMBO J.* 8:2425–2433.

Poyart-Salmeron, C., P. Trieu-Cuot, C. Carlier, and P. Courvalin. 1990. The integration-excision system of the conjugative transposon Tn*1545* is structurally and functionally related to those of lambdoid phages. *Mol. Microbiol.* 4:1513–1521.

Salyers, A. A., and N. B. Shoemaker. 1992. Chromosomal gene transfer elements of the *Bacteroides* group. *Eur. J. Clin. Microbiol. Infect. Dis.* 11:1032–1038.

Salyers, A. A., N. B. Shoemaker, A. M. Stevens, and L. Li. 1995. Conjugative transposons: an unusual and diverse set of integrated gene transfer elements. *Microbiol. Rev.* 59:579–590.

Scott, J. R., and G. G. Churchward. 1995. Conjugative transposition. *Annu. Rev. Biochem.* 49:367–397.

Scott, J. R., F. Bringel, D. Marra, G. Van Alstine, and C. K. Rudy. 1994. Conjugative transposition of Tn*916:* preferred targets and evidence for conjugative transfer of a single strand and for a double-stranded circular intermediate. *Mol. Microbiol.* 11:1099–1108.

Scott, J. R., P. A. Kirchman, and M. G. Caparon. 1988. An intermediate in the transposition of the conjugative transposon Tn*916. Proc. Natl. Acad. Sci. USA* 85:4809–4813.

Senghas, E., J. M. Jones, M. Yamamoto, C. Gawron-Burke, and D. B. Clewell. 1988. Genetic organization of the bacterial conjugative transposon. Tn*916. J. Bacteriol.* 170:245–249.

Shoemaker, N. B., G. Wang, A. M. Stevens, and A. A. Salyers. 1993. Excision, transfer, and integration of NBU1, a mobilizable site-selective insertion element. *J. Bacteriol.* 175:6578–6587.

Smith, C. J., and A. C. Parker. 1993. Identification of a circular intermediate in the transfer and transposition of Tn*4555,* a mobilizable transposon from *Bacteroides* species. *J. Bacteriol.* 175:2682–2691.

Speer, B. S., N. B. Shoemaker, and A. A. Salyers. 1992. Bacterial resistance to tetracycline: mechanisms, transfer, and clinical significance. *Clin. Microbiol. Rev.* 5:387–399.

Stevens, A. M., J. M. Sanders, N. B. Shoemaker, and A. A. Salyers. 1992. Genes involved in production of plasmidlike forms by a *Bacteroides* conjugal chromosomal element share amino acid homology with two-component regulatory systems. *J. Bacteriol.* 174:2935–2942.

Su, Y. A., and D. B. Clewell. 1993. Characterization of the left 4 kb of conjugative transposon Tn*916:* determinants involved in excision. *Plasmid* 30:234–250.

Trieu-Cuot, P., C. Poyart-Salmeron, C. Carlier, and P. Courvalin. 1991. Molecular dissection of the transposition mechanism of conjugative transposons from gram-positive cocci, In *Genetics and Molecular Biology of Streptococci, Lactococci, and Enterococci* pp. 21–27. G. M. Dunny, P. P. Cleary, and L. L. McKay, eds. Am. Soc. Microbiol., Washington, D.C.

Trieu-Cuot, P., C. Poyart-Salmeron, C. Carlier, and P. Courvalin. 1993. Sequence requirements for target activity in site-specific recombination mediated by the Int protein of transposon Tn*1545*. *Mol. Microbiol.* 8:179–185.

14

Phase Variation

Marjan van der Woude, Bruce Braaten, and David Low

Introduction

One way in which bacteria adapt to changes in their environment is by altering the expression states of their cellular surface structures such as fimbriae, flagella, and capsules. In some cases, expression of a structure varies between OFF and ON states, whereas in other cases expression fluctuates between high and low levels. This mechanism, known as phase variation, results in a skewed distribution of phenotypic characteristics within a bacterial population. Bacteria may also undergo antigenic variation in which they display qualitative differences in cell surface molecules that can be detected by specific antisera. As discussed below, phase variation and antigenic variation are not always separate processes. For example, antigenic variation between H1 and H2 flagellar expression in *Salmonella* is regulated by phase variation of H2 (Glasgow *et al.,* 1989). In contrast, fimbrial antigenic variation in *Neisseria gonorrhoeae* can result in fimbrial phase variation (Davies *et al.,* 1994). In this chapter, we focus on four basic mechanisms by which phase variation occurs: homologous recombination, site-specific recombination, slipped strand mispairing at nucleotide repeats, and DNA methylation pattern formation.

Homologous (General) Recombination

Neisseria gonorrhoeae *fimbriae*

The Gram-negative bacterium *Neisseria gonorrhoeae* displays extensive variability of its type IV fimbriae. The first 53 amino acids of the gonococcal major fimbrial subunit are conserved, but the remainder of the protein contains both semi-variable and hypervariable regions (Hagblom *et al.,* 1985). It has been estimated that a single gonococcus has the potential to produce over 10^6 fimbrial variants (Rudel *et al.,* 1992). These fimbrial variants are generated by a RecA-dependent mechanism involving recombination between homologous sequences within fimbrial "silent" loci (*pilS*$_1$, *pilS*$_2$. . ., *pilS*$_n$) and a fimbrial expression locus (*pilE*) (Davies *et al.,* 1994) (Fig. 14-1a). The *pilS* sequences are incomplete fimbrial sequences, lacking a promoter as well as the 5′ end and, in most cases, the 3′ fimbrial gene regions (Haas and Meyer, 1986).

The process of fimbrial antigenic variation results in phase variation if a *pilS* sequence containing a missense mutation(s) preventing fimbrial assembly or a nonsense or frameshift mutation causing fimbrial truncation, is recombined into *pilE* (Bergstrom *et al.,* 1986). (Fig. 14-1a). This switch to a phase OFF fimbrial expression state is reversible since an additional recombination event with an intact, export-competent fimbrial sequence can occur (the frequency of both phase and antigenic variation is about 10^{-4} per cell per generation) (Sparling *et al.,* 1986). In addition, phase variation can also occur if recombination with *pilS* silent copies gives rise to L-fimbrial variants, which contain extra fimbrial sequences at the 3′ end and are not assembled into fimbrial structures (Manning *et al.,* 1991). In such cases, reversion to a fimbriae$^+$ phenotype can occur by deletion of 3′ fimbrial DNA sequences (Fig. 14-1b).

There appear to be two pathways by which recombination between the *pilS* and *pilE* fimbrial loci occur to give rise to both antigenic and phase variants. Gonococci spontaneously lyse in culture, releasing DNA which is transformed into neighboring cells by an efficient uptake mechanism. General recombination between an exogenous *pilS* sequence and *pilE* can result in antigenic variation (Seifert *et al.,* 1988). However, under conditions where uptake of DNA is blocked by mutation, antigenic variation still occurs (Swanson *et al.,* 1990). These results suggest that antigenic variation can also occur by gene conversion (non-reciprocal recombination) since the silent loci are conserved, whereas the expressed copy is lost. The relative contributions of each mechanism to antigenic and phase variation within the human host is not known.

Borrelia hermsii

A spirochete causing relapsing fever, expresses variable major lipoproteins (Vmps) which, like gonococcal fimbriae, undergo extensive antigenic variation. Both silent and expressed *vmp* genes are located on linear plasmids, with the expressed copy having a telomeric location (Barbour, 1993). Activation of silent *vmp* genes can occur by intra- and interplasmid general recombination events which are non-reciprocal and are thus consistent with gene conversions. In addition, antigenic variation can occur by point mutations generated following gene conversion (Restrepo and Barbour, 1994). Although phase variation of Vmps has not been reported, it may occur as a result of conversion of an expression

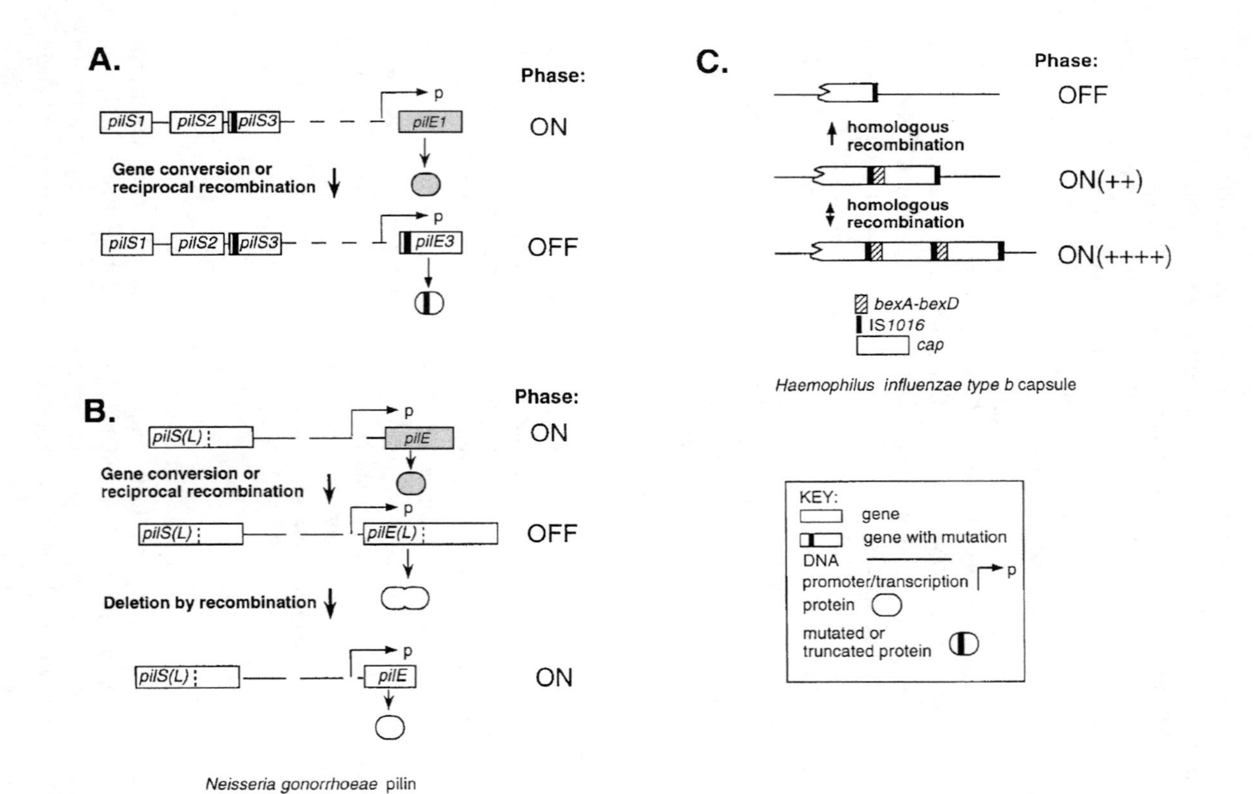

Figure 14-1. Phase Variation by General Recombination

site with a defective *vmp* gene, analogous to fimbrial phase variation in the gonococcus.

Haemophilus influenzae *serotype b capsule*

H. influenzae serotype b strains contain *cap* genes coding for the expression of polyribosylribitol phosphate capsule, which may enhance the virulence of this pathogenic bacterium. The expression of the capsule alternates between moderate and high levels, which presumably (RecA-dependence has not yet been shown) result from a general recombination-mediated amplification of *cap* from two copies to as many as five copies (Roche and Moxon, 1995) (Fig. 14-1c). In addition, an irreversible[1] loss of capsule synthesis can occur in phylogenetic division I strains of *H. influenzae* by loss of a single copy of *bexA,* which is required for capsule synthesis and is located between the *cap* genes.

Site-specific Recombination

Site-specific recombination involves rearrangements between DNA sequences that do not share large regions of homology and is mediated by specific DNA binding proteins which are not part of the general recombination machinery. There are two major classes of site-specific recombination: conservative site-specific recombination (CSSR) and transposition. In CSSR recombination occurs within short regions of sequence identity and strand exchange is reciprocal whereas in transposition there is no homology between recombination sites and recombination is not reciprocal (Craig, 1988; *see also* Chapters 4, 15 and 20). All CSSR systems studied to date, with the exception of Piv of *Moraxella bovis* (see below), fall into two major groups: the Tn*3* resolvase and λ integrase families, which differ in the type of recombination intermediate formed, strand exchange mechanism, and other attributes (Stark *et al.,* 1992; *see also* Chapters 3 and 4).

Conservative Site-specific Recombination

Salmonella *Flagellar Phase Variation (Tn*3 *Resolvase family)*

Salmonella typhimurium has the potential to express two antigenically distinct forms of flagellar protein, H1 and H2. Under conditions where H2 is expressed, H1 is repressed by the rH1 repressor. Conversely, when H2 expression is phase OFF, H1 expression is phase ON (Glasgow *et al.,* 1989). H2 expression is regulated by CSSR by the action of Hin recombinase at two 26 base-pair recombination sites, hixL and hixR, which results in the inversion of a one kilobase-pair

[1]*H. influenzae,* like *N. gonorrhoeae,* is highly transformable. Thus, it is possible that a capsule deficiency caused by *cap* loss could be repaired by recombination with exogenous *cap* DNA sequences.

DNA sequence containing the H2 promoter. Another DNA binding protein, Fis (factor for inversion stimulation), also participates in recombination, binding to two sites within a 60 base-pair recombinational enhancer sequence and stimulating inversion 150-fold. Fis interacts with Hin on a supercoiled DNA substrate to form a complex nucleoprotein structure containing three DNA loops termed an invertasome (Feng *et al.,* 1994; Heichman and Johnson, 1990) (Fig. 14-2).

Members of the Hin recombinase family include *gin* of phage Mu and *cin* of phage P1, which mediate inversions that control tail fiber gene expression and therefore govern phage host range and *pin,* encoded by what appears to be a defective prophage (*e*14) on the *E. coli* chromosome. The Hin, Gin, Cin, and Pin recombinases are highly related at the amino acid level (ca. 60% identity) and cross-complement each other (Glasgow *et al.,* 1989).

E. coli *Type 1 Fimbrial Phase Variation (λ Integrase family)*

Expression of type I fimbriae (also known as mannose-sensitive fimbriae) by *E. coli* is under a reversible OFF-ON phase variation control mechanism which involves the inversion of a 314 base-pair DNA sequence containing the *fim* promoter (Abraham *et al.,* 1985). This inversion is mediated by the FimB recombi-

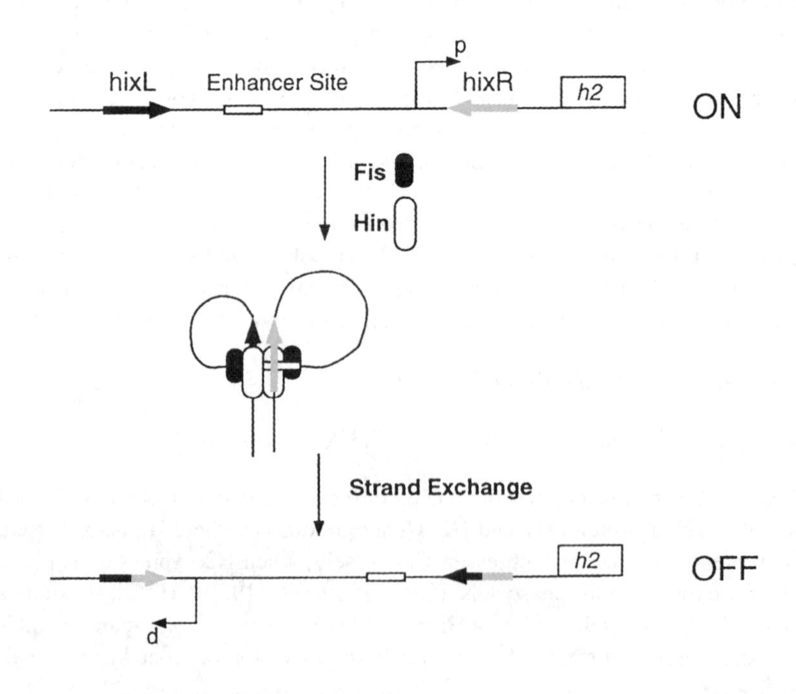

Figure 14-2. Phase Variation by Site-specific Recombination

nase which catalyzes both ON to OFF and OFF to ON switching and by the FimE recombinase which catalyzes ON to OFF switching exclusively (McClain *et al.,* 1991). These recombinases share 50% amino acid identity with the carboxyl-terminal region of phage λ integrase (Int). Notably, phage λ recombination is also asymmetric, requiring Int for integration and both Int and Xis for excision.

Additional factors involved in *fim* inversion include integration host factor (IHF), first identified as a factor required for phage λ integration and excision, histone-like protein H-NS, and leucine-responsive regulatory protein (Lrp). IHF consensus binding sequences are located near the left 9 base-pair inverted repeat flanking the invertible *fim* segment and Lrp binding sites are located within the invertible segment itself. These factors may bend and orient *fim* DNA sequences into a nucleoprotein complex competent for recombination, much like Fis helps form the invertasome in the *Salmonella* flagellar switch (see above). Another member of the λ integrase family is the IncI plasmid "shufflon" system, which involves inversion of multiple DNA segments that regulate conjugation specificity (Komano *et al.,* 1994).

Moraxella *Fimbrial Phase Variation (Piv family)*

A third class of CSSR-dependent phase variation systems occurs in *Moraxella bovis* and *M. lacunata,* causative agents of bovine keratoconjunctivitis and human conjunctivitis, respectively. These bacteria have the potential to express two different type IV fimbriae designated as Q and I, coded for by the *tfpQ* and *tfpI* genes, respectively. Similar to other type IV fimbriae, *tfpQ* and *tfpI* are transcribed by RNA polymerase containing the alternative sigma factor RpoN. Both *tfpQ* and *tfpI* fimbrial gene sequences are contained within a 2.1 kilobase-pair DNA fragment. The *tfpQ* and *tfpI* genes lack fimbrial amino termini and are only expressed when fused to a common amino terminus and promoter. Therefore, inversion of the 2.1 kb DNA fragment results in the alternate expression of either TfpQ or TfpI fimbriae. (Marrs, 1994). Inversion of *Moraxella* fimbrial genes is catalyzed by the Piv recombinase, which does not share homology with either the Hin or λ integrase families of recombinases. Instead, Piv shares sequence similarities with a group of insertion sequence elements including *IS492* from *Pseudomonas atlantica,* which, unlike many transposons, does not have terminal inverted repeats (Glasgow and Lenich, 1994; Marrs *et al.,* 1990) (see below). It is not yet known if other accessory factors are required for inversion stimulation.

Transposition

The marine bacterium *Pseudomonas atlantica* undergoes a reversible transition between mucoid and nonmucoid phenotypes. This phase variation is regulated by the highly specific insertion and excision of an IS*492* element within the *eps* gene controlling extracellular polysaccharide expression (Bartlett *et al.,* 1988).

Reversible expression of the Vi capsular antigen of *Citrobacter freundii* is regulated by the insertion and excision of an IS*1*-like element within the *viaB* gene, required for Vi expression (Ou *et al.,* 1988).

Slipped-strand Mispairing

The expression of outer membrane protein II (P.II) in *N. gonorrhoeae* is under both phase and antigenic variation control mechanisms. There are multiple *opa* genes, each containing an intact promoter and coding for antigenically distinct P.II proteins. The 5′ region of each *opa* gene contains polypyrimidine repeats (CTCTT) coding for a hydrophobic leader sequence. Expression of a particular P.II variant is regulated at the translational level by alteration in the number of CTCTT coding repeats, which shift the AUG start codon in and out of frame with downstream P.II coding sequence (Stern *et al.,* 1986).

Available evidence indicates that alteration in the number of CTCTT coding repeats in *opa* occurs by a slipped-strand mispairing mechanism (SSM) in which mutations arise during DNA replication due to misalignment of the DNA strands (Belland *et al.,* 1989; Levinson and Gutman, 1987) (Fig. 14-3a). This process is RecA-independent, indicating that the general recombination apparatus is not involved. The coding repeats appear to form a triple helix (H-form DNA), which could play a role in SSM (Belland, 1991).

Other phase variation systems under translational control that may be regulated by SSM include LPS biosynthesis in *Haemophilus influenzae,* BvgS expression by *Bordetella pertussis,* and *Neisseria gonorrhoeae* fimbrial adhesin PilC (Jonsson *et al.,* 1992; Stibitz *et al.,* 1989; Weiser, *et al.,* 1989). Expression of LPS epitopes in *H. influenzae* appears to be controlled in part by changes in the number of CAAT repeats within the open reading frame of three *lic* genes involved in LPS regulation. In the case of *lic-1,* the reading frame shifts caused by alteration in the number of CAAT coding repeats result in three different LPS expression states: high, low and off (Weiser *et al.,* 1989).

Regulation of phase variation at the transcriptional level also occurs by alteration of nucleotide repeats in upstream regulatory sequences. For example, in *Mycoplasma hyorhinus* variation in the number of adenosines within a poly(A) tract near the RNA polymerase binding site of the *vlp* gene appears to control expression of variant surface lipoprotein (Vlp) (Fig. 14-b) (Yogev *et al.,* 1991). Other examples include phase variation of *H. influenzae* fimbriae by TA repeat length, phase variation of Opc protein in *N. meningitidis* by poly(C) repeat length, and phase variation of class I outer membrane protein in *N. meningitidis* by poly(G) repeat length (Sarkari *et al.,* 1994; van der Ende *et al.,* 1995; van Ham *et al.,* 1993). In these examples, three levels of gene expression (high, moderate, and off) are effected by alteration in the number of nucleotide repeats. It seems likely that at least for *H. influenzae* fimbriae, SSM is involved in phase variation

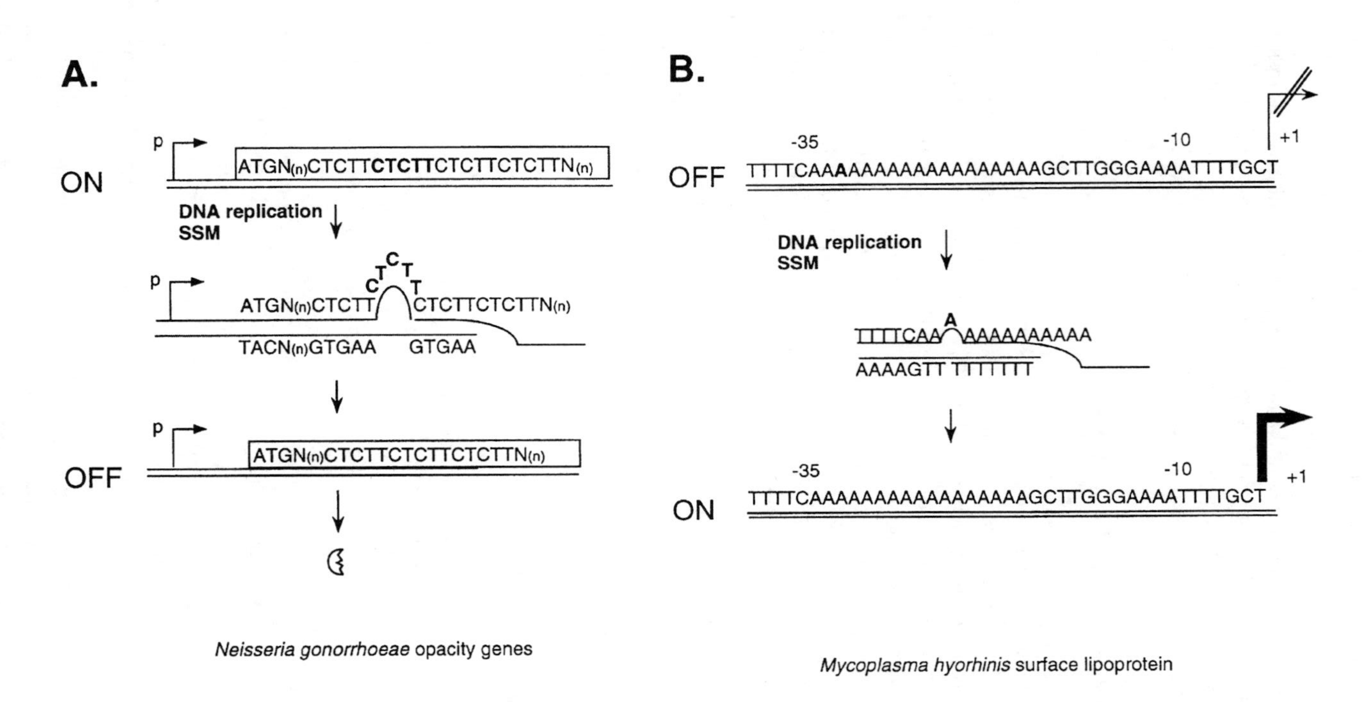

Figure 14-3. Phase Variation by Slipped-strand Mispairing

since fimbrial switching occurs in the absence of RecA and involves loss or gain of integral TA repeats (van Ham *et al.,* 1993). Additional examples of phase variation systems which contain short repeats are described by Robertson and Meyer (Robertson and Meyer, 1992).

Methylation-dependent Phase Variation

All three of the phase variation mechanisms described above involve changes in the primary DNA sequence. In contrast, Pap fimbrial phase variation in *Escherichia coli* is regulated by a methylation-dependent epigenetic mechanism as are a number of other phase variation systems (Blyn *et al.,* 1990; Braaten *et al.,* 1994; Van der Woude and Low, 1994). The transcription of *pap* is regulated by the binding of leucine-responsive regulatory protein (Lrp) to two upstream regulatory regions separated by over 100 base-pairs. Within the DNA of each Lrp binding region a GATC sequence is present (designated as GATC-I and GATC-II) which can be methylated by the deoxyadenosine methylase (Dam) if the GATC sequence is not protected by Lrp binding. In the OFF phase Lrp binds cooperatively and with high affinity around the GATC-II site near the *papBAp* promoter, shutting OFF transcription (Nou *et al.,* 1993; van der Woude *et al.,* 1995). Binding of Lrp around GATC-II forms a DNA methylation pattern characteristic of phase OFF cells: GATC-I is unoccupied and becomes fully methylated by Dam whereas GATC-II is nonmethylated due to protection by Lrp (Fig. 14-4).

After DNA replication (see below), transition to the phase ON state is facilitated by the binding of PapI, a small *pap*-encoded regulatory protein, to the Lrp moiety of Lrp-*pap* DNA complex (Kaltenbach *et al.,* 1995). PapI reduces the affinity of Lrp for the GATC-II region and increases its affinity for sites around GATC-I, resulting in a shift in binding of the Lrp-*pap* nucleoprotein complex to the GATC-I region (Nou *et al.,* 1995). Activation of *pap* transcription appears to occur as a result of binding of Lrp near GATC-I in conjunction with binding of CAP at a site 50 base-pairs upstream of GATC-I.

DNA methylation plays dual roles in Pap phase variation, which was initially suggested by the observation that over- and under-expression of Dam locks cells in the phase OFF state. Methylation of GATC-I in phase OFF cells blocks the transition to the phase ON state by inhibiting binding of Lrp-PapI to the GATC-I region. This effectively locks cells in the phase OFF state until cell division occurs, generating a hemimethylated GATC-I site. Since Lrp-PapI binds with low affinity to a hemimethylated GATC-I sequence, this provides a time window in which translocation of Lrp-PapI to GATC-I can occur. In contrast, methylation of the GATC-II site is required for the phase OFF to ON transition. It is not yet clear how this occurs, but one possibility is that methylation of GATC-II inhibits binding of Lrp near this site, aiding the translocation of Lrp-PapI to GATC-I (Braaten *et al.,* 1994).

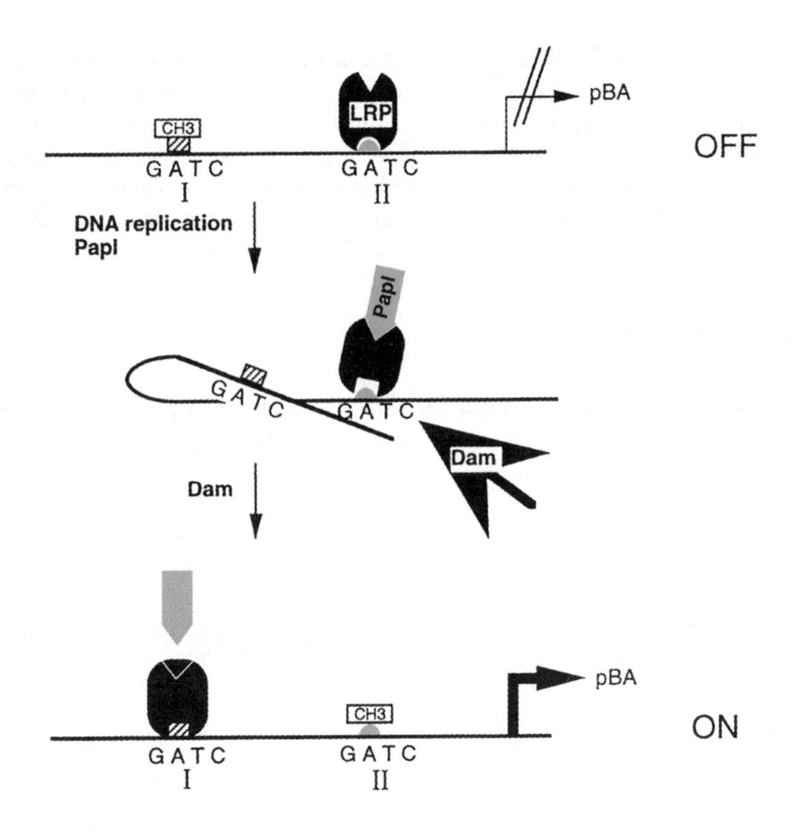

Figure 14-4. Phase Variation by Methylation Pattern Formation

Conclusions

Bacteria have developed a number of phase variation mechanisms to control gene expression. Most systems studied involve the regulation of cell surface structures, although phase variation of a restriction-modification system in *Mycoplasma pulmonis* has recently been described (Dybvig and Yu, 1994). Both the site-specific recombination and methylation-dependent phase variation systems rely on global regulators (e.g., Lrp and Fis) as well as local regulators (e.g., recombinases and PapI). These specialized systems have likely evolved in response to the selective pressures imposed by the varied environments that bacteria live in. Both global and local regulatory networks provide the means by which phase variation switch frequencies can be regulated by environmental signals. For example, type I fimbrial phase switch frequencies are modulated by aliphatic amino acid levels *via* Lrp control, and *pap* fimbrial switch frequencies are regulated by environmental carbon sources via CAP control. In contrast, the general recombination and slipped-strand mispairing mechanisms appear to use nonspecialized host machinery to regulate phase variation. This limits the potential for environmental control of phase variation rates in these systems. It will be important to determine what roles different phase variation systems play in the colonization and transmission of bacteria in different environments.

Acknowledgments

We thank Anna Glasgow and Ian Blomfield for helpful discussions. We are grateful to the National Institutes of Health (grant #2R01 AI23348) for continued support.

References

Abraham, J. M., C. S. Freitag, J. R. Clements, and B. I. Eisenstein. 1985. An invertible element of DNA controls phase variation of type 1 fimbriae of *Escherichia coli. Proc. Natl. Acad. Sci. USA* 82:5724–5727.

Barbour, A. 1993. Linear DNA of *Borrelia* species and antigenic variation. *Trends in Microbiol.* pp. 239–239.

Bartlett, D. H., M. E. Wright, and M. Silverman. 1988. Variable expression of extracellular polysaccharide in the marine bacterium *Pseudomonas atlantica* is controlled by genome rearrangement. *Proc. Natl. Acad. Sci. USA* 85:3923–3927.

Belland, R. J. 1991. H-DNA formation by the coding repeat elements of neisserial *opa* genes. *Mol. Microbiol.* 5:2351–2360.

Belland, R. J., S. G. Morrison, P. van der Ley, and J. Swanson. 1989. Expression and phase variation of gonococcal P.II genes in *Escherichia coli* involves ribosomal frameshifting and slipped-strand mispairing. *Mol. Microbiol.* 3:777–786.

Bergstrom, S., K. Robbins, J. M. Koomey, and J. Swanson. 1986. Piliation control mechanisms in *Neisseria gonorrhoeae. Proc. Acad. Natl. Sci. USA* 83:3890–3894.

Blyn, L. B., B. A. Braaten, and D. A. Low. 1990. Regulation of *pap* pilin phase variation by a mechanism involving differential dam methylation states. *EMBO J.* 9:4045–4054.

Braaten, B. A., X. Nou, L. S. Kaltenbach, and D. A. Low. 1994. Methylation patterns in *pap* regulatory DNA control the pyelonephritis-associated pili phase variation in *E. coli. Cell* 76:577–588.

Craig, N. L. 1988. The mechanism of conservative site-specific recombination. *Ann. Rev. Genet.* 22:77–105.

Dybvig, K., and H. Yu. 1994. Regulation of a restriction and modification system via DNA inversion in *Mycoplasma pulmonis. Mol. Microbiol.* 12:547–560.

Feng, J. A., R. C. Johnson, and R. E. Dickerson. 1994. Hin recombinase bound to DNA: the origin of specificity in major and minor groove interactions. *Science* 263:348–355.

Glasgow, A. C., and A. G. Lenich. 1994. Amino acid sequence homology between Piv, an essential protein in site-specific DNA inversion in *Moraxella lacunata,* and transposases of an unusual family of insertion elements. *J. Bacteriol.* 176:4160–4164.

Haas, R. and T. F. Meyer. 1986. The repertoire of silent pilus genes in Neisseria gonorrhoeae: evidence for gene conversion. *Cell* 44:107–115.

Hagblom, P., E. Segal, E. Billyard, and M. So. 1985. Intragenic recombination leads to pilus antigenic variation in *Neisseria gonorrhoeae. Nature* 315:156–158.

Heichman, K. A. and R. C. Johnson. 1990. The hin invertasome: protein-mediated joining of distant recombination sites at the enhancer. *Science* 249:511–517.

Jonsson, A. B., J. Pfeifer, and S. Normark. 1992. Neisseria gonorrhoeae PilC expression provides a selective mechanism for structural diversity of pili. *Proc. Natl. Acad. Sci. USA* 89:3204–3208.

Kaltenbach, L. S., B. A. Braaten, and D. A. Low. 1995. Specific binding of PapI to Lrp-*pap* DNA complexes. *J. Bacteriol.* 177:6449–6455.

Komano, T., S. R. Kim, T. Yoshida, and T. Nisioka. 1994. DNA rearrangement of the shufflon determines recipient specificity in liquid mating of IncI1 plasmid R64. *J. Mol. Biol.* 243:6–9.

Levinson, G. and G. A. Gutman. 1987. Slipped-strand mispairing: A major mechanism for DNA sequence evolution. *Mol. Biol. Evol.* 4:203–221.

Manning, P. A., A. Kaufmann, U. Roll, J. Pohlner, T. F. Meyer, and R. Haas. 1991. L-pilin variants of *Neisseria gonorrhoeae* MS11. *Mol. Microbiol.* 5:917–926.

Marrs, C. F., F. W. Rozsa, M. Hackel, S. P. Stevens, and A. C. Glasgow. 1990. Identification, cloning, and sequencing of piv, a new gene involved in inverting the pilin genes of Moraxella lacunata. *J. Bacteriol.* 172:4370–4377.

Marrs, C. F. 1994. Type IV pili in the families *Moraxellaceae* and *Neisseriaciae.* In Molecular Genetics of Bacterial Pathogenesis, V. L. Miller *et al.,* eds. pp. 127–143. ASM Press, Washington, D.C.

McClain, M. S., I. C. Blomfield, and B. I. Eisenstein. 1991. Roles of *fimB* and *fimE* in site-specific DNA inversion associated with phase variation of type 1 fimbriae in *Escherichia coli. J. Bacteriol.* 173:5308–5314.

Nou, X., B. Braaten, L. Kaltenbach, and D. Low. 1995. Differential binding of Lrp to two sets of *pap* DNA binding sites, mediated by PapI, regulates Pap phase variation in *Escherichia coli. EMBO J.* 14:5785–5797.

Nou, X., B. Skinner, B. Braaten, L. Blyn, D. Hirsh, and D. Low. 1993. Regulation of pyelonephritis-associated pili phase variation in *Escherichia coli:* binding of the PapI and Lrp regulatory proteins is controlled by DNA methylation. *Mol. Microbiol.* 7:545–553.

Ou, J. T., L. S. Baron, F. A. Rubin, and D. J. Kopecko. 1988. Specific insertion and deletion of insertion sequence *l*-like DNA element causes the reversible expression of the virulence capsular antigen of *Citrobacter freundii* in *Escherichia coli. Proc. Natl. Acad. Sci. USA* 85:4402–4405.

Restrepo, B. I., and A. G. Barbour. 1994. Antigen diversity in the bacterium *B. hermsii* through somatic mutations in rearranged *vmp* genes. *Cell* 78:867–876.

Robertson, B. D. and T. F. Meyer. 1992. Genetic variation in pathogenic bacteria. *Trends in Genetics* 8:422–427.

Roche, R. J. and E. R. Moxon. 1995. Phenotypic variation of carbohydrate surface antigens and pathogenesis of *Haemophilus influenzae* infections. *Trends Microbiol.* 8:304–309.

Rudel, T., J. P. M. van Putten, C. P. Gibbs, R. Haas, and T. F. Meyer. 1992. Interaction of two variable proteins *(pilE* and *pilC)* required for pilus-mediated adherence of *Neisseria gonorrhoeae* to human epithelial cells. *Mol. Microbiol.* 6:3439–3450.

Sarkari, J., N. Pandit, E. R. Moxon, and M. Achtman. 1994. Variable expression of the Opc outer membrane protein of *Neisseria meningitidis* is caused by size variation of a promoter containing poly-cytidine. *Mol. Microbiol.* 13:207–217.

Seifert, H. S., R. S. Ajioka, C. Marchal, P. F. Sparling, and M. So. 1988. DNA transformation leads to pilin antigenic variation in *Neisseria gonorrhoeae. Nature* 336:392–396.

Sparling, P. F., J. G. Cannon, and M. So. 1986. Phase and antigenic variation of pili and outer membrane protein II of *Neisseria gonorrhoeae. J. Infect. Dis.* 153:196–201.

Stark, W. M., M. R. Boocock, and D. J. Sherratt. 1992. Catalysis by site-specific recombinases. *Trends in Genetics* 8:432–438.

Stern, A., M. Brown, P. Nickel, and T. F. Meyer. 1986. Opacity genes in Neisseria gonorrhoeae: control of phase and antigenic variation. *Cell* 47:61–71.

Stibitz, S., W. Aaronson, D. Monack, and S. Falkow. 1989. Phase variation in Bordetella pertussis by frameshift mutation in a gene for a novel two-component system. *Nature* 338:266–269.

Swanson, J., S. Morrison, O. Barrerea, and S. Hill. 1990. Piliation changes in transformation-defective *Neisseria gonorrhoeae: J. Exp. Med.* 171:2131.

van der Ende, A., C. T. Hopman, S. Zaat, B. B. O. Essink, B. Berkhout, and J. Dankert. 1995. Variable expression of Class 1 outer membrane protein in *Neissera meningitidis* is caused by variation in the spacing between the –10 and –35 regions of the promoter. *J. Bacteriol.* 177:2475–2480.

van der Woude, M. W. and D. A. Low. 1994. Leucine-responsive regulatory protein and deoxyadenosine methylase control the phase variation and expression of the *E. coli sfa* and *daa* pili operons. *Mol. Microbiol.* 11:605–618.

van der Woude, M. W., L. S. Kaltenbach, and D. A. Low. 1995. Leucine-responsive regulatory protein plays dual roles as both an activator and a repressor of the *E. coli pap* operon. *Molec. Microbiol.* 17:303–312.

van Ham, S. M., L. van Alphen, F. R. Mooi, and J. P. M. van Putten. 1993. Phase variation of *H. influenzae* fimbriae: Transcriptional control of two divergent genes through a variable combined promoter region. *Cell* 73:1187–1196.

Weiser, J. N., J. M. Love, and R. E. Moxon. 1989. The molecular mechanism of phase variation of *H. influenzae* lipopolysaccharide. *Cell* 59:657–665.

Yogev, D., R. Rosengarten, R. Watson-McKown, and K. S. Wise. 1991. Molecular basis of *Mycoplasma* surface antigenic variation: a novel set of divergent genes undergo spontaneous mutation of periodic coding regions and 5″ regulatory sequences. *EMBO J.* 13:4069–4079.

15

The Dynamic Genome of *Rhizobium*

David Romero, Guillermo Davila, and Rafael Palacios

Introduction

A major, indeed the predominant, biological feature of *Rhizobium* spp. is their ability to establish nitrogen-fixing symbioses with leguminous plants. The understanding of the molecular basis for interaction with plants has been the major thrust for molecular genetics studies of these bacteria. This research has culminated with the identification of a wealth of bacterial genes involved in the symbiotic process, and some of their biochemical and developmental functions (reviewed by Fischer, 1994 and Schultze *et al.*, 1994).

Moreover, the breadth of the present studies has revealed, either for reason or by accident, some unanticipated characteristics of the *Rhizobium* genome. *Rhizobium* spp have a well-deserved fame because of the existence of several megaplasmids in their genome. Calculations based on assumed sizes for the chromosome, indicate that megaplasmids may represent from 25 to 50% of the total genome size (Martínez *et al.*, 1990; *see also* Chapter 74). Many of the genes important to nodulation and nitrogen fixation are encoded in the symbiotic plasmid or p*Sym,* although other megaplasmids carry genes relevant for symbiotic and/ or saprophytic life styles (Brom *et al.*, 1992; Hynes and McGregor, 1990). Here we review information regarding two other prominent features of the *Rhizobium* genome, namely extensive gene reiteration and frequent genome rearrangements.

Gene Reiteration is a Widespread Characteristic

A distinctive characteristic of the *Rhizobium* genome is the frequent finding of gene reiteration. This feature was recognized originally during studies focused on the identification of genes involved in the symbiotic process. Hybridization with some of these cloned genes revealed the presence of additional copies.

Initially considered as a peculiarity of some genes or genomic regions, the list of sequences that present some sort of reiteration has grown steadily over the years.

How frequent is gene reiteration in the genome of *Rhizobium?* One study obtained a rough estimate through the analysis of a random sample of the genome by Southern hybridization (Flores et al., 1987). Under the conditions employed, this technique may detect reiterated sequences larger than 300 bp, with at least 85% similarity. The number of positive occurrences was then extrapolated to the rest of the genome. Assuming a genome size of about 5000 Kb, a typical figure for the genomes of *R. etli* and *R. meliloti* is approximately 200 different reiterated families with a mean repetition of 3.5. Thus, using 300 bp as a lower estimate for the size of a repeated sequence, about 4.2% of the genome is in the form of repeated sequences.

It is important to state that this is a minimum estimate. Short repeated sequences that are present in these genomes, such as REP sequences (see Chapters 5 and 32) were almost certainly excluded from this calculation. Also, many identified reiterated sequences are larger than 300 bp. This frequency of reiteration is well above the one observed in many prokaryotes. Only members of the class *Archaebacteria* (Sapienza and Doolittle, 1982) and genus *Streptomyces* (Hopwood and Kieser, 1990) appear to exhibit such a high frequency. This number is in the range of those observed in some simple eukaryotes, such as yeast.

Where are these repeated sequences located? Available evidence indicates scattered location throughout the genome, both in the chromosome and megaplasmids. Distribution, on the contrary, appears to be non-random. Some repeated families are located only in certain plasmids (see below), while others are exclusive to the chromosome. Reiteration mode is usually dispersed; only the *fla* genes in *R. meliloti* have been shown to be reiterated in tandem (Bergman *et al.*, 1991).

What is the nature of these reiterations? Not surprisingly, several kinds of transposable elements are reiterated (Martínez et al., 1990; *See also* Chapters 31, 32 and 34). However, it is notable to find reiteration in several housekeeping genes, such as the *fla, ftsZ* (Margolin and Long, 1994), *groEL* (Fischer et al., 1993), *rpoN* (Kündig *et al.*, 1993) and citrate synthase (Pardo et al., 1994). Several genes involved solely in the symbiotic process are also reiterated, such as *nifHDK* (Quinto et al., 1985), *fixN* (David et al., 1987), and the *nodD* genes (Schultze *et al.*, 1994). Finally, the list includes reiterated genes with a related, but not identical, function. Examples are the *nodPQ-saa* pair, where *nodPQ* is involved in the sulfation of a modulation factor, while *saa* participates in the sulfation step for cysteine and methionine biosynthesis (Schwedock and Long, 1992). Another case is the *nodM-glmS* pair, where *nodM* synthesizes the glucosamine needed for *nod* factor biosynthesis, while *glmS* elaborates the glucosamine required for cell function (Marie *et al.*, 1992). Thus, extensive reiteration in *Rhizobium* probably was not due to an unusually harsh invasion by transposable elements; reiterated sequences frequently are expressed genes with important consequences for cell physiology.

What are the reasons for extensive reiteration? Large gaps in knowledge preclude a definitive answer, but the data presented above give some hints about the possible reasons. Ignoring the case of transposable elements, which have multiplied by their own, probably for selfish reasons, two functional explanations can be hypothesized. One postulates that reiteration of a specific region might be selected under environmental conditions that demand overexpression of a gene product. The case of the *nifHDK* reiterations in *R. etli* illustrates this point. It is clear that both reiterations cooperate to determine the nitrogen fixation ability of this species (Romero *et al.*, 1988). The second explanation assumes that, upon demand for a slightly different product, reiterations followed by divergence might be selected. The *nodPQ-saa* and *nodM-glmS* cases are clear examples of this. Both explanations make a strong emphasis on the selective value of a single gene reiteration. An alternative viewpoint is that the reiteration of single genes has by itself a slight selective value. Under a dispersed mode of reiteration, stronger selection might be exerted because it allows the multiplication of interactive sets of genes flanked by the original reiterations. This possibility will be discussed later in this chapter.

Genomic Rearrangements Generate Broad Structural Variability

The third unusual characteristic in the genome of *Rhizobium* is the frequent occurrence of genomic rearrangements. Many laboratories have reported the occurrence of genomic rearrangements, practically in all *Rhizobium* species examined (Brom *et al.*, 1991; Djordjevic *et al.*, 1982; Flores *et al.*, 1988; Hahn and Hennecke, 1987; Selbitschka and Lotz, 1991). Two reports (Brom et al., 1991; Flores *et al.*, 1988), elucidating genome instability in *R. etli,* contributed greatly to the demonstration that the *Rhizobium* genome is highly dynamic. The basic approach employed in both studies was the isolation of approximately 300 single-colony derivatives from a single mother culture. These derivatives were screened for the occurrence of genomic rearrangements, either through hybridization with a set of plasmid and chromosomal probes (Flores *et al.*, 1988), or by analysis of megaplasmid profiles (Brom et al., 1991). Both reports arrived at the same surprising conclusion: a high frequency of these derivatives (3 to 6%) showed some sort of genomic rearrangement, implying a high rate of genomic reorganization. The frequency of derivatives carrying rearrangements may increase up to 35% after subculturing for one year under normal laboratory conditions.

Although this experimental approach is not common in bacterial genetics, a recent report that analyzed variation in insertion sequences in *Escherichia coli* maintained in stab cultures (Naas *et al.*, 1994) provides a comparative study. No variation in insertion sequence patterns was observed among derivatives (sample size, 120) from a young stab culture (1 week). In contrast, among derivatives from an old stab culture (30 years) the frequency of variants was 57% (Naas *et al.*, 1994). *Rhizobium* species have not been tested under a similar stressful condition.

Frequencies of variation as high as those observed in *Rhizobium* have been reported only for *Archaebacteria* (Sapienza *et al.*, 1982) and *Streptomyces* (Hopwood and Kieser, 1990). As mentioned before, both genera share with *Rhizobium* the presence of high levels of reiteration in their genomes.

Many different types of genomic rearrangements, including cointegration, translocation, deletion and amplification of specific sequences participate in the frequent generation of structural diversity (see Romero *et al.,* 1995 and references therein). Recombination between reiterated sequences appears to be responsible for such variation. With few exceptions, the mechanisms involved in the generation of rearrangements in *Rhizobium* have been poorly studied. The rest of this chapter focuses on the analysis of rearrangements in the p*Sym* of *R. etli.*

The p*Sym* of *Rhizobium etli:* A Model for the Study of Gene Reiteration and Genome Rearrangements

To understand the forces involved in the shaping of the dynamic genome of *Rhizobium,* it is important to analyze different types of rearrangements in regions of the genome where accurate physical maps are available. We have employed the p*Sym* of *R. etli* CFN42 as a model because there is a detailed physical map that locates both symbiotic genes and reiterated sequences (Girard *et al.,* 1991). Additionally, this plasmid is not required for growth under free-living conditions, thus eliminating phenotypic constraints for the isolation of genomic rearrangements.

All the symbiotic genes identified on this plasmid are located in a sector that spans 1/3 of its size (Fig. 1, innermost circle). Many of them are located in a subsector of 120 Kb whose boundaries are the *nifHDK* reiterations. Clustering of genes involved in the symbiotic process has been observed previously (reviewed by Sanjuan *et al.*, 1993).

This map also contains information about the number and location of reiterated sequences (middle and outermost circles, Fig. 15-1). There are about 49 reiterated elements in the p*Sym.* This is roughly the number to be expected in a molecule 390 Kb in size knowing the density of reiterated sequences in the genome (see above). Some of these elements are internally repeated in the pSym, constituting eleven families of two to three elements each (identified as capital letters on the middle circle (Fig. 15-1). With the exception of the *nifHDK* and *nodD* reiterated families, the nature of the remaining families is still unknown. The distribution of internal repetitions is clearly not random. Fifteen out of 26 elements are located in a sector 45 Kb in size (6 to 8 o'clock sector on the middle circle (Fig. 15-1). Besides internally reiterated sequences, 23 additional regions are repeated in other plasmids present in the strain (lowercase letters on the outermost circle, Fig. 15-1). Most of these sectors are repeated in p*a*, a smaller (150 kb) self-transmissible plasmid. The remaining sectors are repeated on p*b* and p*f.*

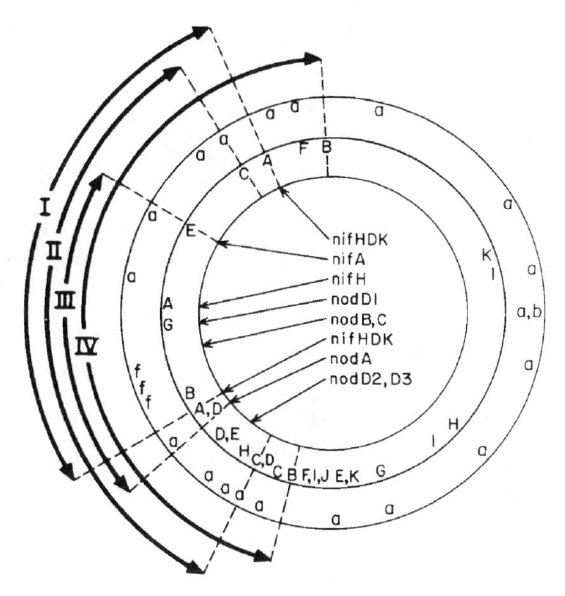

Figure 15-1. Structure and dynamics of the symbiotic plasmid of *R. etli* CFN42. The general structure of the p*Sym* (390 Kb) is presented in three concentrical circles; 1 Kb= 0.93°. The position of some symbiotic genes is indicated by arrows in the internal circle. The next circle indicates the position of reiterated DNA families. Each family is indicated by a letter (A–K). The external circle indicates regions of the p*Sym* that have similarity to other replicons of the strain: p*a,* p*b,* or p*f.* Structures labeled I–IV indicate the four amplicons detected in the p*Sym.* See text for explanation and references.

A global search for regions on the p*Sym* that undergo frequent genomic rearrangements revealed the location of four variable regions (represented by arcs on Fig. 15-1) that we have called amplicons (Flores *et al.*, 1993; Romero *et al.*, 1991; Romero *et al.*, 1995). The term refers to specific DNA regions that have the potential to be amplified or deleted as a unit. Size of the amplicons is large, ranging from 90 to 175kb and they are overlapping. Each amplicon is flanked by repeated sequences in direct orientation; homologous recombination between them generates amplifications and deletions in the p*Sym,* in a similar way as has been found in other microorganisms (Weinstock and Lupski, Chapter 11). Rearrangements fostered by each amplicon occur at high frequencies, that range from 10^{-3} (type I amplicon) to 10^{-4} (type II–IV amplicons). It is notable that amplicons are located exclusively in the region that contains symbiotic genes. No amplicons were observed on the rest of the p*Sym,* despite the presence of repeated sequences (Romero *et al.*, 1991; Romero *et al.*, 1995).

A schematic representation of the different rearrangements observed for the p*Sym* is shown in Fig. 15-2. Besides the events of amplification and deletion discussed above (central part, Fig. 15-2), cointegration between p*Sym*-p*a* and

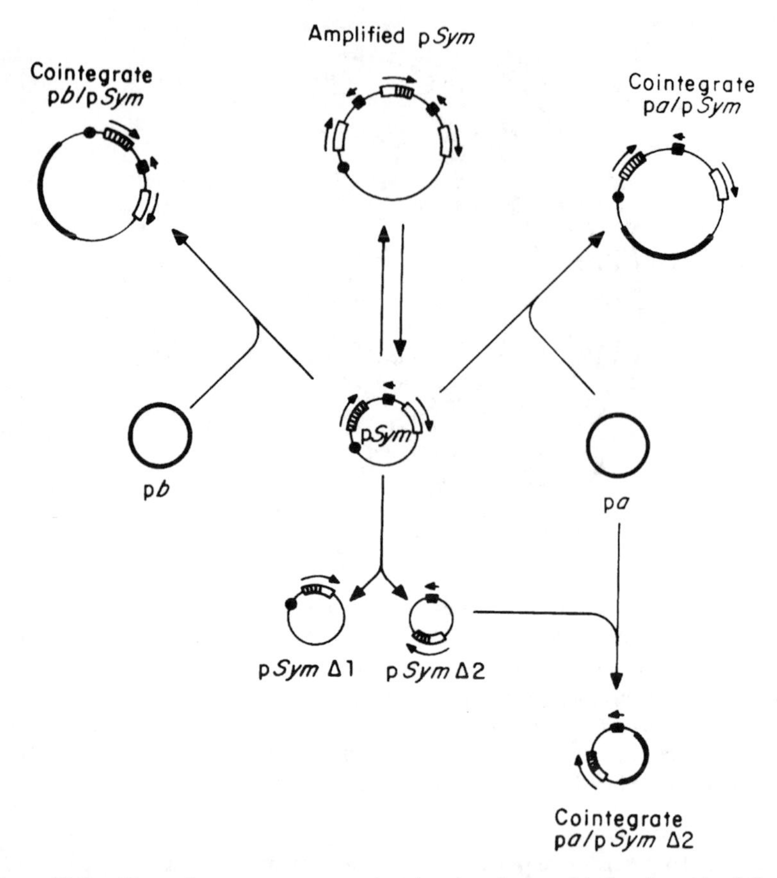

Figure 15-2. Genomic rearrangements that involve the symbiotic plasmid of *R. etli* CFN42. Homologous recombination between the two reiterated *nifHDK* operons of the p*Sym* leads to either amplification of the plasmid or to deletions (p*Sym* Δ1 and p*Sym* Δ2). The p*Sym* can be cointegrated with other replicons, either p*a* or p*b* and some deletions of the p*Sym* can be cointegrated with p*a*. The three elements of the nitrogenase genes family of the p*Sym* are shown as boxes; arrows indicate the relative orientation of the repeats. The origin of replication of the p*Sym* is also indicated (•). See text for explanation and references.

p*Sym*-p*b* also occurs at high frequencies. Additionally, events that translocate the whole or part of the symbiotic zone into p*a* also occur readily (10^{-4}). This last event is generated apparently through the excision of amplicon sequences from the p*Sym,* followed by reinsertion onto p*a* (Romero, D., unpublished data). Conceivably, these rearrangements may occur *via* recombination between reiterations shared in these replicons. Both amplifications and cointegrations are reversible at high frequencies (10^{-2} to 10^{-3}). These results clearly show the potential for generation of broad structural variability through genomic rearrangements.

Although no comparable data exist for other genome regions, we speculate that the dynamic picture of the p*Sym* might be representative of the *Rhizobium* genome. Examples of high frequency deletions and cointegrations have been reported for other regions in the genome besides the p*Sym*. In fact, amplicon structures might be a common feature in *Rhizobium.* One study screened for the occurrence of amplification in other sectors of the genome (Flores *et al.*, 1993). Amplification was found in several of the cases analyzed, indicating the presence of amplicon structures. These were found not only in many of the plasmids present in the strain, but also in the chromosome.

What are the biological consequences of such a dynamic genome? We propose that this genomic plasticity creates an enlarged pool of variants that might be useful for adaptation under the changing environmental conditions found in the soil microenvironment, or for interaction with different plants. There is a growing body of evidence that overexpression of gene products, accrued through amplification, might be important for adaptation (see Sonti and Roth, 1989 and references therein; *see also* Chapter 11). This has lead to the proposal that, in at least some regions of the genome, the position of genes has been shaped as to favor frequent amplification of interactive sets of genes (Sonti and Roth, 1989).

The distribution of amplicons in the p*Sym* of *R. etli* lends some support to this idea, because all the symbiotic genes detected on this plasmid are immersed in amplicon structures. Moreover, we have recently found that amplification of one of these sectors (the type I amplicon) in certain genomic backgrounds leads to a strong increase in nitrogen fixation ability (Romero, unpublished data). Thus, amplifications might be a source for beneficial variants. Amplicons also generate deletions, which are frequently deleterious. However, that outcome is not invariable, since a large deletion in a plasmid of *R. leguminosarum* bv. viceae, previously thought to be cryptic, lead to a strong increase in nitrogen fixation ability (Selbitschka and Lotz, 1991). We believe that the generation of deleterious variants, a byproduct of a dynamic genome, might be compensated by the beneficial effects of some rearrangements.

Research so far has clearly shown the unusual dynamic characteristics of the *Rhizobium* genome. More information is needed in regard to the phenotypic consequences of genomic rearrangements in *Rhizobium,* as well as their mechanisms of formation. This research will likely illuminate the selective forces that have contributed to the shaping of this complex genome.

Acknowledgements

Work from the authors laboratory was supported in part by grants DGAPA-UNAM IN200693 and IN200593, CONACyT 4321, and by The Rockefeller Foundation.

References

Bergman, K., E. Nulty, and L. Su. 1991. Mutations in the two flagellin genes of *Rhizobium meliloti*. *J. Bacteriol.* 173:3716–3723.

Brom, S., A. García de los Santos, M. L. Girard, G. Dávila, R. Palacios, and D. Romero. 1991. High-frequency rearrangements in *Rhizobium leguminosarum* bv. phaseoli plasmids. J. Bacteriol. 173:1344–46.

Brom, S., A. García de los Santos, T. Stepkowski, M. Flores, G. Dávila, D. Romero, and R. Palacios. 1992. Different plasmids of *Rhizobium leguminosarum* bv. phaseoli are required for optimal symbiotic performance. *J. Bacteriol.* 174:5183–5189.

David, M., O. Domergue, P. Pognonec, and D. Kahn. 1987. Transcription patterns of *Rhizobium meliloti* symbiotic plasmid pSym: Identification of nifA-independent fix genes. *J. Bacteriol.* 169:2239–2244.

Djordjevic, M. A., W. Zurkowski, and B. G. Rolfe. 1982. Plasmids and stability of symbiotic properties of *Rhizobium trifolii*. *J. Bacteriol.* 15:560–568.

Fischer, H-H. 1994. Genetic regulation of nitrogen fixation in Rhizobia. *Microbiol. Rev.* 58:352–386.

Fischer, H-M., M. Babst, T. Kaspar, G. Acuña, F. Arigoni, and H. Hennecke. 1993. One member of a *groESL*-like chaperonin multigene family in *Bradyrhizobium japonicum* is coregulated with symbiotic nitrogen fixation genes. *EMBO J.* 12:2901–2912.

Flores, M., S. Brom, T. Stepkowski, M. L. Girard, G. Dávila, D. Romero, and R. Palacios. 1993. Gene amplification in *Rhizobium:* identification and *in vivo* cloning of discrete amplifiable DNA regions (amplicons) from *Rhizobium leguminosarum* bv. phaseoli. *Proc. Natl. Acad. Sci. USA* 90:4932–4936.

Flores, M., V. González, S. Brom, E. Martínez, D. Piñero, D. Romero, G. Dávila and R. Palacios. 1987. Reiterated DNA sequences in *Rhizobium* and *Agrobacterium* spp. *J. Bacteriol.* 169:5782–5788.

Flores, M., V. González, M. A. Pardo, A. Leija, E. Martínez, D. Romero, D. Piñero, G. Dávila, and R. Palacios. 1988. Genomic instability in *Rhizobium phaseoli*. *J. Bacteriol.* 170:1191–1196.

Girard, M. L., M. Flores, S. Brom, D. Romero, R. Palacios, and G. Dávila. 1991. Structural complexity of the symbiotic plasmid of *Rhizobium leguminosarum* bv. phaseoli. *J. Bacteriol.* 173:2411–2419.

Hahn, M., and H. Hennecke. 1987. Mapping of a *Bradyrhizobium japonicum* DNA region carrying genes for symbiosis and an asymmetric accumulation of reiterated sequences. *Appl. Environ. Microbiol.* 53:2247–2252.

Hopwood, D. A., and T. Kieser. 1990. The *Streptomyces* genome. In *The Bacterial chromosome.* K. Drlica and M. Riley, eds. pp. 147–162. American Society for Microbiology, Washington, D.C.

Hynes, M. F., and N. F. McGregor. 1990. Two plasmids other than the nodulation plasmid are necessary for formation of nitrogen fixing nodules by *Rhizobium leguminosarum*. *Mol. Microbiol.* 4:567–574.

Kündig, C., H. Hennecke and M. Göttfert. 1993. Correlated physical and genetic map of the *Bradyrhizobium japonicum* 110 genome. *J. Bacteriol.* 175:613–622.

Marie, C., M. A. Barny and J. A. Downie. 1992. *Rhizobium leguminosarum* has two glucosamine synthetases, GlmS and NodM, required for nodulation and development of nitrogen-fixing nodules. *Mol. Microbiol.* 6:843–51.

Margolin, W., and S. Long. 1994. *Rhizobium meliloti* contains a novel second homolog of the cell division gene *ftsZ*. *J. Bacteriol.* 176:2033–43.

Martínez, E., D. Romero and R. Palacios. 1990. The *Rhizobium* genome. *Crit. Revs. in Plant Sci.* 9:59–93.

Naas, T., M. Blot, W. M. Fitch and W. Arber. 1994. Insertion sequence-related genetic variation in resting *Escherichia coli* K-12. *Genetics* 136:721–730.

Pardo, M. A., J. Lagúnez, J. Miranda, and E. Martínez. 1994. Nodulating ability of *Rhizobium tropici* is conditioned by a plasmid-encoded citrate synthase. *Mol. Microbiol.* 11:315–321.

Quinto, C., H. de la Vega, M. Flores, J. Leemans, M. A. Cevallos, M. A. Pardo, R. Azpiroz, M. L. Girard, E. Calva, and R. Palacios. 1985. Nitrogenase reductase: a functional multigene family in *Rhizobium phaseoli*. *Proc. Natl. Acad. Sci. USA* 82:1170–1174.

Romero, D., S. Brom, J. Martínez-Salazar, M. L. Girard, R. Palacios and G. Dávila. 1991. Amplification and deletion of a *nod-nif* region in the symbiotic plasmid of *Rhizobium phaseoli*. *J. Bacteriol.* 173:2435–2441.

Romero, D., J. Martínez-Salazar, L. Girard, S. Brom, G. Dávila, R. Palacios, M. Flores, and C. Rodríguez. 1995. Discrete Amplifiable Regions (Amplicons) in the Symbiotic Plasmid of *Rhizobium etli* CFN42. *J. Bacteriol.* 177:973–980.

Romero, D., P. W. Singleton, L. Segovia, E. Morett, B. B. Bohlool, R. Palacios, and G. Dávila. 1988. Effect of naturally occurring *nif* reiterations on symbiotic effectiveness in *Rhizobium phaseoli*. *Appl. Environ. Microbiol.* 54:848–850.

Sanjuan, J., S. Luka, and G. Stacey. 1993. Genetic maps of *Rhizobium* and *Bradyrhizobium* species. In *Genetic Maps,* S. O'Brien, ed. pp. 2136–2145. Cold Spring Harbor Laboratory Press, New York, N.Y.

Sapienza, C., and W. F. Doolittle. 1982. Unusual physical organization of the *Halobacterium* genome. *Nature* 295:384–389.

Sapienza, C., M. R. Rose, and W. F. Doolittle. 1982. High-frequency genomic rearrangements involving archaebacterial repeat sequence elements. *Nature* 299:182–185.

Schultze, M., E. Kondorosi, P. Ratet, M. Buiré, and A. Kondorosi. 1994. Cell and molecular biology of *Rhizobium*-plant interactions. *Int. Rev. Cytol.* 156:1–75.

Schwedock, J. and S. Long. 1992. *Rhizobium meliloti* genes involved in sulfate activation: the two copies of *nodPQ* and a new locus, *saa*. *Genetics* 132:899–909.

Selbitschka, W., and W. Lotz. 1991. Instability of cryptic plasmids affects the symbiotic effectivity of *Rhizobium leguminosarum* bv. viceae strains. *Mol. Plant-Microbe Interact.* 4:608–618.

Sonti, R. V., and J. R. Roth. 1989. Role of gene duplications in the adaptation of *Salmonella typhimurium* to growth on limiting carbon sources. *Genetics* 123:19–28.

16

Programmed DNA Rearrangements in Cyanobacteria

James W. Golden

Introduction

Three developmentally programmed DNA rearrangements occur during heterocyst differentiation in the nitrogen-fixing cyanobacterium *Anabaena* sp. strain PCC 7120 (*Anabaena* PCC 7120). All three rearrangements involve the excision of an element from the chromosome by site-specific recombination and each requires an element-encoded recombinase. Each element is harbored in a heterocyst-specific gene that is required for the production of a fully functional heterocyst. Precise excision of the elements is necessary to restore these genes. Elements similar to these *Anabaena* PCC 7120 elements have been found in a variety of heterocystous cyanobacterial strains.

Cyanobacterial microfossils have been dated to over three billion years old (Schopf, 1993). Cyanobacteria are thought to be the evolutionary ancestors of plastids (Douglas, 1994). Their oxygen-evolving photosynthetic apparatus is very similar to that of higher plant chloroplasts. Biological nitrogen fixation requires a microaerobic environment and several mechanisms for meeting this requirement have evolved. Many filamentous cyanobacteria capable of both oxygenic photosynthesis and nitrogen fixation exhibit a striking example of prokaryotic cellular specialization to segregate these two incompatible processes. They produce heterocysts, highly differentiated cells that are responsible for nitrogen fixation (Wolk *et al.*, 1994). In the absence of a combined nitrogen source, approximately 10% of the photosynthetic vegetative cells differentiate into heterocysts at semiregular intervals along the filament to produce a simple multicellular organism composed of two interdependent cell types.

Heterocyst development requires global changes in gene expression, with large numbers of genes turned off and others turned on, while also allowing for "housekeeping" genes to be expressed in both vegetative cells and heterocysts

(Wolk *et al.*, 1994). The genes necessary for the production of nitrogenase (*nif* genes) are expressed only in heterocysts (Elhai and Wolk, 1990). Heterocysts are terminally differentiated in most strains and are unable to undergo further cell division once fully formed; therefore, their genomes will not be passed on to the next generation. Their sole function is to supply the neighboring vegetative cells in the filament with fixed nitrogen.

This paper describes the three programmed DNA rearrangements that occur during heterocyst differentiation. Programmed DNA rearrangements occur in both eukaryotes and prokaryotes (*see also* Chapter 11). In prokaryotes, developmentally regulated genome rearrangements have been found in two types of organisms: filamentous cyanobacteria and *Bacillus* sp. (Haselkorn, 1992). The *Bacillus* rearrangement will only be briefly mentioned in this chapter. During *Bacillus subtilis* sporulation, a 42-kb element named *skin* is deleted from within the *sigK* gene in the terminally differentiating mother cell (Kunkel *et al.*, 1990). The rearrangement is the result of site-specific recombination between two 5-bp direct repeats and requires the *spoIVCA* gene, which is located on the excised element and which shows homology to the Tn*3* resolvase family of site-specific recombinases (Kunkel *et al.*, 1990; Sato *et al.*, 1990; Stark *et al.*, 1992).

Anabaena Rearrangements

During heterocyst differentiation in *Anabaena* PCC 7120, two DNA elements are excised from adjacent nitrogen-fixation operons by site-specific recombination (Apte and Prabhavathi, 1994; Golden *et al.*, 1987; Golden *et al.*, 1985). A recently identified third rearrangement involves the excision of a DNA element from within a hydrogenase gene that is distant from the *nif* gene cluster (Carrasco *et al.*, 1995; Matveyev *et al.*, 1994). The three DNA elements present in the vegetative cell chromosome are shown in Figure 16-1.

nifD *element*

The *nifD* rearrangement is the excision of an 11-kb element from within the *nifD* open reading frame (ORF) and results in the restoration of the complete *nifD* gene, which encodes the alpha subunit of dinitrogenase, and the *nifHDK* operon (Golden *et al.*, 1985; Golden *et al.*, 1991; Haselkorn *et al.*, 1986). Excision occurs by site-specific recombination between two 11-bp direct repeats that border the element (Fig. 16-2) and requires the *xisA* gene, which is present on the element (Brusca *et al.*, 1990; Golden and Wiest, 1988; Lammers *et al.*, 1986). The conservative recombination event excises the 11-kb element as a circular molecule that is neither degraded nor amplified in heterocysts. The *nifD* element has been partially sequenced. Four ORFs have been identified in addition to *xisA* and one of these shows homology with cytochrome P-450 ω-hydroxylases (Lammers et al., 1990).

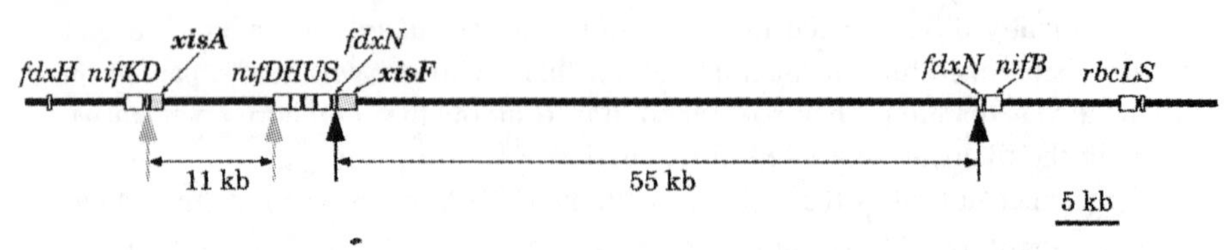

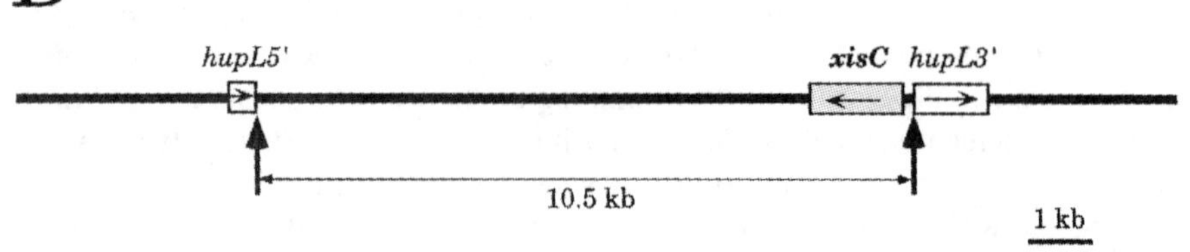

Figure 16-1. The *Anabaena* PCC 7120 programmed DNA rearrangements. (a) The vegetative cell chromosomal region containing the *nifD* and *fdxN* elements. The 11-kb *nifD* element interrupts the *nifHDK* operon, which encodes the nitrogenase polypeptides. The 55-kb *fdxN* element interrupts the *nifB-fdxN-nifS-nifU* operon. The *fdxH* gene encodes a heterocyst-specific ferredoxin and the *rbcLS* operon, which is only expressed in vegetative cells, encodes the large and small subunits of ribulose-1,5-bisphosphate carboxylase. *xisA* and *xisF* encode the *nifD* and *fdxN* element site-specific recombinases, respectively. All genes read from right to left except for *xisA, xisF,* and *rbcLS.* The *nifD* and *fdxN* recombination sites are indicated by gray and black vertical arrows, respectively. (b) The *hupL* region. The 10.5-kb *hupL* element interrupts the *hupL* gene, which may be part of a *hupSL* operon. *xisC* encodes the element's recombinase. Horizontal arrows show the orientation of the genes. Vertical arrows indicate recombination sites.

nifD	11 kb element	GGCA----T-C---**GCCTCATTAGG**-----CAC--AA----C
fdxN	55 kb element	----T-G-----A-T--**TATTC**--AGAA-TTT-C---A----
hupL	10.5 kb element	------G----**CACAGCAGTTATATGG**-------T---G--A

Figure 16-2. Comparison of the recombination sites involved in the *nifD, fdxN,* and *hupL* rearrangements. Nucleotides that are repeated at the borders of the elements are shown. Recombination occurs within the nucleotide sequences shown in bold type. In each case, site-specific recombination between the directly repeated sequences that flank the element results in excision of the element from the heterocyst chromosome.

fdxN *element*

The *fdxN* rearrangement is the excision of a 55-kb element from within the *fdxN* gene (Figure 16-1), which encodes a bacterial-type ferredoxin and is part of the *nifB-fdxN-nifS-nifU* operon (Golden *et al.*, 1988; Mulligan *et al.*, 1988; Mulligan and Haselkorn, 1989). It is the result of a site-specific recombination between directly repeated DNA sequences (Figure 16-2). The crossover occurs within a 5-bp sequence that is part of a larger imperfect repeat (Golden *et al.*, 1987). The deletion of the 55-kb element reforms the *fdxN* open reading frame and allows the *nifB* promoter to express the downstream genes in the *nifB-fdxN-nifU* operon (Golden *et al.*, 1987; Mulligan *et al.*, 1988). Excision of the *fdxN* element can occur independently from excision of the *nifD* element and does not require the *xisA* gene (Golden *et al.*, 1988; Golden and Wiest, 1988). The *xisF* gene, which encodes a site-specific recombinase required for rearrangement, is found at one end of the *fdxN* element (Carrasco *et al.*, 1994). Mapping and DNA sequence data show that the rearrangement is a conservative site-specific recombination that results in the circularization of the element. However, the intact circular product of excision was not detected in DNA isolated from heterocysts, presumably because of random shearing (Golden *et al.*, 1988). No region of the *fdxN* element shows detectable levels of transcription in either vegetative cells or heterocysts under laboratory growth conditions (Golden *et al.*, 1988), and no other genes have been identified on the element.

hupL *element*

A third heterocyst-specific genome rearrangement was recently identified (Matveyev *et al.*, 1994). A 10.5-kb element is excised from within the *hupL* gene by site-specific recombination between 16-bp direct repeats (Carrasco *et al.*, 1995). The recombination sites for the *nifD*, *fdxN,* and *hupL* rearrangements show no sequence similarities (Fig. 16-2), suggesting that the *hupL* rearrangement requires its own site-specific recombinase. A gene, *xisC,* which is presumed to encode a recombinase, was identified at one end of the element. The excised circular element can be detected in heterocyst DNA.

Southern analysis of genomic DNA fragments separated by pulsed-field gel electrophoresis showed that the *hupL* element is located between 0.4 and 0.9 Mb on the *Anabaena* PCC 7120 chromosome, which is over 700 kb from the *nif* gene cluster (Carrasco *et al.*, 1995; Matveyev *et al.*, 1994).

A *hupS* gene is present upstream of the interrupted *hupL* gene (Carrasco *et al.*, 1995). *hupS* and *hupL* encode the small and large subunits of [NiFe] uptake hydrogenases. In diazotrophs, uptake hydrogenases function to utilize molecular hydrogen, which is produced as a byproduct of nitrogen fixation, for the energy-conserving reduction of electron acceptors. The *Anabaena* PCC 7120 hydrogenase would be required to improve the efficiency of nitrogen fixation in heterocysts, and the putative *hupSL* operon may be expressed only in heterocysts. Therefore, the *hupL* element could be carried in the vegetative cell genome without detriment. Excision of the *nifD* and *fdxN* elements is required for nitrogen fixation (Carrasco *et al.*, 1994; Golden and Wiest, 1988). The *hupL* rearrangement is probably not required for nitrogen fixation. Failure to produce hydrogenase should not eliminate heterocyst function, but should only degrade the efficiency of nitrogen fixation.

Recombinases

XisA

XisA is necessary and sufficient for rearrangement of the *nifD* element in *Anabaena* PCC 7120 and of an artificial substrate in *E. coli* (Brusca *et al.*, 1990; Golden and Wiest, 1988; Lammers et al., 1986). However, XisA does not show significant similarity to the integrase or resolvase families of site-specific recombinases (Stark *et al.*, 1992). The *xisA* gene is 240 bp from the *nifK* proximal end of the *nifD* element (Fig. 16-1). A 354 amino acid 41.6-kDa protein is predicted to be made from the second of two in-frame start codons. It is not clear which of the two start codons is used *in vivo.* Although deletion of the upstream start codon allows production of recombinase activity in *Anabaena* PCC 7120 and *E. coli* (Brusca *et al.*, 1990), the amino acid sequence encoded by the region between the two start codons can be aligned with the XisC recombinase (Carrasco *et al.*, 1995) suggesting that it has a conserved function. XisA protein binds to a sequence that overlaps the upstream start codon. Recombinase and DNA binding activity can be assayed from *xisA* expression clones in which either of the two start codons have been destroyed by site-directed mutagenesis, suggesting that both start sites may be functional (Brusca and Golden, unpublished).

XisF

XisF encodes the site-specific recombinase responsible for the excision of a 55-kb element from within the *fdxN* gene during heterocyst differentiation (Carrasco *et al.*, 1994). A 1.6-kb ORF was found within the element at the *nifS*-proximal end of the element. The predicted XisF recombinase is 58.5 kDa and is 27% identical to the *B. subtilis* SpoIVCA recombinase. The N-terminal third of the XisF and SpoIVCA polypeptides show similarity to the resolvase family of site-specific recombinases.

xisF was shown to be required for excision of the *fdxN* element from the heterocyst chromosome by site-directed gene inactivation. The *xisF* mutant strain, KSR9, forms morphologically normal heterocysts but cannot grow on media lacking a source of combined nitrogen. KSR9 heterocysts fail to excise the *fdxN* element, but excise the *nifD* element normally. These experiments show that *xisF* is necessary for the *fdxN* rearrangement and that the DNA rearrangement is necessary for nitrogen fixation but not for morphological heterocyst differentiation. These results mirror those obtained with *xisA* (Golden and Wiest, 1988), and show that the *fdxN* element must be precisely excised from the chromosome to allow the correct expression of the *nifB-fdxN-nifS-nifU* operon.

XisF may not be sufficient for site-specific recombination. Expression of *xisF* in *E. coli* from a strong promoter caused site-specific recombination and rearrangement of the pJG1A substrate, but only at a low frequency (Carrasco *et al.*, 1994). *xisF* on a conjugal expression vector was moved into *Anabaena* PCC 7120 to complement the site-directed mutation in KSR9. Constitutive expression of *xisF* complemented the *xisF* mutation in KSR9 heterocysts, but failed to cause rearrangement of the *fdxN* element in vegetative cells, suggesting that other factors may be involved in cell-type specificity. The additional factor could regulate the activity of *xisF,* or it could act similarly to IHF or FIS by participating in the assembly of a higher-order nucleoprotein complex required for efficient recombination (Finkel and Johnson, 1992; Landy, 1989).

The *Anabaena* PCC 7120 *fdxN* element and the *Bacillus skin* element have several similar features. The amino acid sequence similarity between their recombinases, XisF and SpoIVCA respectively, indicates that the elements are evolutionarily related. They both excise from the chromosome by conservative site-specific recombination that restores the coding sequences within which the elements reside. The genes that are interrupted by the elements are not required for vegetative growth but they are required in the terminally differentiated cell type that undergoes rearrangement: the *sigK* gene in *Bacillus* mother cells and the *nifB-fdxN-nifS-nifU* operon in heterocysts. For both elements, failure to excise from the chromosome results in a nonfunctional cell.

XisC

The *xisC* gene is a 1.5-kb ORF predicted to encode a 60.2-kDa protein and is located 115 bp inside one border of the *hupL* element. Comparison of the predicted XisC amino acid sequence with GenBank sequences identified apparent homology to a single protein: XisA, the *Anabaena* PCC 7120 *nifD* element recombinase. Although XisC is 25% longer than XisA, the sequences are 61% similar and 43% identical. All recombinases that have been studied *in vitro,* and many others identified genetically, can be assigned to one of two families: the resolvase family or the integrase family of recombinases (Stark *et al.*, 1992). XisC and XisA appear to represent a novel class of site-specific recombinases since they do not show similarity to either of these families or to other proteins.

Regulation

The *nifD, fdxN,* and *hupL* rearrangements all share similar features of their regulation. The elements remain quiescent in the vegetative cell chromosome and are passed on to daughter cells during cell division. Their excision is tightly coupled to development. All three rearrangements occur during the late stages of heterocyst differentiation, at about the same time that the nitrogen-fixation genes begin to be transcribed (Carrasco *et al.*, 1995; Golden *et al.*, 1985; Golden *et al.*, 1991). The elements must be precisely and efficiently excised from the chromosome to restore the open reading frame and operon within which they reside when the organism requires the expression of those genes. All three elements encode their own site-specific recombinase and remain stoichiometric with the chromosome after excision. It is not known how the rearrangements are coordinately regulated. Although several genes have been identified that play a role in the regulation of heterocyst development, they affect early stages of development, well before the time of the DNA rearrangements and transcription of the *nif* genes (Buikema and Haselkorn, 1993; Wolk *et al.*, 1994).

Although a clear picture has yet to emerge, some of the factors potentially involved in the regulation of the *nifD* rearrangement have been identified, however nothing is currently known about the *cis* and *trans*-acting factors that may regulate the *fdxN* or *hupL* rearrangements. Two DNA-binding factors, NtcA (BifA) and Factor-2, interact at multiple adjacent sites located approximately 200 bp upstream of the *xisA* gene and 30 bp from one of the recombination sites. It is not yet known what role these interactions play in the excision of the *nifD* element. The DNA-binding sites for these factors were determined and this information was used to clone the *Anabaena* PCC 7120 *ntcA* (*bifA*) gene. The phenotype of a null mutant indicates an important role for *ntcA* in heterocyst development and nitrogen metabolism.

Initial attempts to express *xisA* in vegetative cells using a shuttle vector failed, although the same constructs could be expressed in *E. coli.* This suggested the presence of a negative regulatory element. Shuttle vectors containing the *E. coli tac* consensus promoter fused to various 5′ deletions of the *xisA* gene were constructed and conjugated into *Anabaena* PCC 7120 (Brusca *et al.*, 1990). Some of the expression plasmids resulted in excision of the *nifD* element, but only if a region upstream of *xisA* was deleted. A strain lacking the *nifD* element was obtained as a result of these experiments; it showed normal growth and heterocyst development.

The *xisA* regulatory region was used to identify trans-acting factors responsible for repression of *xisA* in vegetative cells. A DNA-binding factor was partially purified from *Anabaena* PCC 7120 vegetative cell extracts that interacted with *xisA* upstream sequences (Chastain *et al.*, 1990). Competition experiments showed that this protein also bound to the upstream sequences of the *glnA, rbcL,* and *nifH* genes. DNase footprinting and deletion analysis of the *xisA* upstream region mapped the binding to three adjacent sites in a 66-bp region. This protein, originally called VF1, is now known to be similar to NtcA, which is a global regulator involved in nitrogen control in *Synechococcus* sp. strain PCC 7942 (Frias *et al.*, 1993; Vega-

Palas *et al.*, 1992). The *Anabaena* PCC 7120 NtcA binding sites can be aligned to produce the consensus sequence 5′TGT(N_{9-10})ACA 3′.

The *Anabaena* PCC 7120 *ntcA* gene was cloned using a novel genetic selection strategy (Wei *et al.*, 1993). The deduced *Anabaena* PCC 7120 NtcA amino acid sequence shows 77% identity to *Synechococcus* sp. strain PCC 7942 NtcA (Frias *et al.*, 1993). NtcA belongs to the Crp family of prokaryotic regulatory proteins. The *Anabaena* PCC 7120 *ntcA* gene is expressed in both vegetative cells and heterocysts, but shows an interesting burst of transcription early in heterocyst development (Wei *et al.*, 1993).

An *ntcA* mutant has been constructed in *Anabaena* PCC 7120 (Frias *et al.*, 1994; Wei *et al.*, 1994). The resulting strain required ammonia for growth; it failed to grow in media lacking combined nitrogen or containing nitrate as the sole source of nitrogen. The *ntcA* mutant showed the unexpected phenotype of being defective for heterocyst differentiation. The *ntcA* mutant fails to excise the *nifD* and *fdxN* elements, but this is presumably an indirect consequence of the earlier block in heterocyst differentiation.

A second DNA-binding protein, Factor-2, was identified that interacts with *xisA* (Ramasubramanian *et al.*, 1994). Factor-2 is detected only in vegetative cell extracts, and may therefore inhibit *xisA* expression. Factor-2 can be resolved from NtcA by heparin-Sepharose chromatography and was present in extracts from the *ntcA* mutant strain. The Factor-2 binding site on *xisA* was localized to a 68-bp region that showed considerable overlap with the NtcA binding sites.

Transcription and DNA rearrangement of the *nifHDK* operon both occur late during heterocyst differentiation, about 18–24 hours after induction, suggesting that the regulation of these events might be coupled. They are coincident with the appearance of morphologically mature heterocysts and nitrogenase activity in induced wild-type filaments. It was thought that the elements may derive their developmental regulation from the operon in which they reside, so that transcription of the operon triggers excision of the element.

This hypothesis was tested by separately blocking each of the events, transcription or rearrangement, by an appropriate genetic manipulation, and then determining if the other event occurred normally. The results showed that heterocyst specific transcription and DNA rearrangement of the *nifHDK* operon are independent of one another (Golden et al., 1991). Northern analysis of the *xisA* mutant strain DW12-2.2, which cannot excise the *nifD* 11-kb element or fix nitrogen, showed that the *nifH* gene is still transcribed on unrearranged chromosomes. The *nifK* gene was not transcribed in DW12-2.2, indicating that its expression is dependent on the *nifH* promoter and the excision of the 11-kb element from the operon. A 1.68-kb DNA fragment containing the *nifH* promoter was deleted from the chromosome to produce the mutant strain LW1. LW1 formed heterocysts, but did not grow on nitrogen-free medium and showed no transcription through *nifD*. Southern analysis of LW1 showed normal excision of the 11-kb element from the *nifHDK* operon indicating that transcription from the *nifH* promoter is not required for the developmental regulation of the DNA rearrangement.

Distribution of the Elements

The *nifD* element appears to be widespread among heterocystous cyanobacteria, although it is absent from a few strains (Apte and Prabhavathi, 1994; Carrasco and Golden, 1995). The element has not been found in any nonheterocystous unicellular or filamentous strain including those capable of nitrogen fixation. The *fdxN* element is less common; it is present in only three of nine *Anabaena* and *Nostoc* strains tested (Carrasco and Golden, 1995). The distribution of the *hupL* element is similar to that of the *fdxN* element (Carrasco and Golden, unpublished data).

Much of the data for the occurrence of the elements is based on Southern analysis in which probes from *Anabaena* PCC 7120 were used to show that the *nifHDK* operon is not contiguous (Apte and Prabhavathi, 1994). In some cases a probe containing a portion of the *Anabaena* PCC 7120 *nifD* element was used to confirm the presence of similar sequences. Southern analysis of *Nostoc* sp. strain 7801 and a cultured isolate from a symbiotic association with the water fern *Azolla caroliniana* showed hybridization to an *xisA* probe and also showed evidence for rearrangement in samples that contained heterocysts (Meeks *et al.*, 1988).

Rearrangement of the *nifD* and *fdxN* elements has been directly demonstrated in several strains. The *Anabaena variablis nifD* gene contains an 11-kb element that excises from the chromosome during heterocyst differentiation (Brusca *et al.*, 1989). The element is flanked by 11-base-pair direct repeats that are identical to the repeats present at the ends of the *nifD* element in *Anabaena* PCC 7120. *A. variabilis* element contains an *xisA* gene that can complement a defective *Anabaena* PCC 7120 *xisA* gene. However, *A. variabilis* does not contain the equivalent of the *Anabaena* PCC 7120 *fdxN* element.

Southern analysis of *Nostoc* sp. strain MAC showed the presence of a *nifD*-like element and analysis of spontaneous Fox^+ revertant clones identified one rare clone, isolate R1, that had lost the element in all cells (Meeks *et al.*, 1994). DNA samples from other revertant clones, which still carry the *nifD*-like element, showed faint bands on Southern blots that were interpreted as representing excision of the element in heterocysts.

Nostoc MAC also contains an *fdxN*-like element, and *Anabaena cylindrica* contains *nifD*-like and *fdxN*-like elements (Carrasco and Golden, 1995). Southern analysis of vegetative cell and heterocyst DNA from *A. cylindrica* and a Het^+ revertant of *Nostoc* MAC (isolate R2) showed rearrangement of both elements in heterocysts. The presence of these elements in different genera suggests significant selective pressure to retain them; however, what benefit they may provide the organism remains an enigma.

Acknowledgments

This work was supported by Public Health Service grant GM36890 from the National Institutes of Health.

References

Apte, S. K., and N. Prabhavathi. 1994. Rearrangements of nitrogen fixation (*nif*) genes in the heterocystous cyanobacteria. *J. Biosci.* 19(5):579–602.

Brusca, J. S., C. J. Chastain, and J. W. Golden. 1990. Expression of the *Anabaena* sp. strain PCC 7120 *xisA* gene from a heterologous promoter results in excision of the *nifD* element. *J. Bacteriol.* 172(7):3925–3931.

Brusca, J. S., M. A. Hale, C. D. Carrasco, and J. W. Golden. 1989. Excision of an 11-kilobase-pair DNA element from within the *nifD* gene in *Anabaena variabilis* heterocysts. *J. Bacteriol.* 171(8):4138–4145.

Buikema, W. J., and R. Haselkorn. 1993. Molecular genetics of cyanobacterial development. *Annu. Rev. Plant Physiol. Plant Mol. Biol.* 44:33–52.

Carrasco, C. D., and J. W. Golden. 1995. Two heterocyst-specific DNA rearrangements of *nif* operons in *Anabaena cylindrica* and *Nostoc* sp. strain Mac. *Microbiology,* in press.

Carrasco, C. D., K. S. Ramaswamy, T. S. Ramasubramanian, and J. W. Golden. 1994. *Anabaena xisF* gene encodes a developmentally regulated site-specific recombinase. *Genes Dev.* 8:74–83.

Carrasco, C. D., J. A. Simon, and J. W. Golden. 1995. Programmed DNA rearrangement of a cyanobacterial *hupL* gene in heterocysts. *Proc. Natl. Acad. Sci. USA* 92:791–795.

Chastain, C. J., J. S. Brusca, T. S. Ramasubramanian, T.-F. Wei, and J. W. Golden. 1990. A sequence-specific DNA-binding factor (VF1) from *Anabaena* sp. strain PCC 7120 vegetative cells binds to three adjacent sites in the *xisA* upstream region. *J. Bacteriol.* 172(9):5044–5051.

Douglas, S. E. 1994. Chloroplast origins and evolution, in *The molecular biology of cyanobacteria,* pp. 91–118. D. A. Bryant, eds. Kluwar Academic Publishers, Dordrecht.

Elhai, J., and C. P. Wolk. 1990. Developmental regulation and spatial pattern of expression of the structural genes for nitrogenase in the cyanobacterium *Anabaena. EMBO J.* 9(10):3379–3388.

Finkel, S. E., and R. C. Johnson. 1992. The Fis protein: it's not just for DNA inversion anymore. *Mol. Microbiol.* 6(22):3257–3265.

Frias, J. E., E. Flores, and A. Herrero. 1994. Requirement of the regulatory protein NtcA for the expression of nitrogen assimilation and heterocyst development genes in the cyanobacterium *Anabaena* sp. PCC 7120. *Mol. Microbiol.* 14(4):823–832.

Frias, J. E., A. Merida, A. Herrero, J. Martin-Nieto, and E. Flores. 1993. General distribution of the nitrogen control gene *ntcA* in cyanobacteria. *J. Bacteriol.* 175(17):5710–5713.

Golden, J. W., C. D. Carrasco, M. E. Mulligan, G. J. Schneider, and R. Haselkorn. 1988.

Deletion of a 55-kilobase-pair DNA element from the chromosome during heterocyst differentiation of *Anabaena* sp. strain PCC 7120. *J. Bacteriol.* 170:5034–5041.

Golden, J. W., M. E. Mulligan, and R. Haselkorn. 1987. Different recombination site specificity of two developmentally regulated genome rearrangements. *Nature (London)* 327:526–529.

Golden, J. W., S. J. Robinson, and R. Haselkorn. 1985. Rearrangement of nitrogen fixation genes during heterocyst differentiation in the cyanobacterium *Anabaena. Nature (London)* 314:419–423.

Golden, J. W., L. L. Whorff, and D. R. Wiest. 1991. Independent regulation of *nifHDK* operon transcription and DNA rearrangement during heterocyst differentiation in the cyanobacterium *Anabaena* sp. strain PCC 7120. *J. Bacteriol.* 173(22):7098–7105.

Golden, J. W., and D. R. Wiest. 1988. Genome rearrangement and nitrogen fixation in *Anabaena* blocked by inactivation of *xisA* gene. *Science* 242:1421–1423.

Haselkorn, R. 1992. Developmentally regulated gene rearrangements in prokaryotes. *Annu. Rev. Genet.* 26:113–130.

Haselkorn, R., J. W. Golden, P. J. Lammers, and M. E. Mulligan. 1986. Developmental rearrangement of cyanobacterial nitrogen-fixation genes. *Trends in Genetics:* 2:255–259.

Kunkel, B., R. Losick and P. Stragier. 1990. The *Bacillus subtilis* gene for the development transcription factor sigma K is generated by excision of a dispensable DNA element containing a sporulation recombinase gene. *Genes Dev.* 4(4):525–535.

Lammers, P. J., J. W. Golden, and R. Haselkorn. 1986. Identification and sequence of a gene required for a developmentally regulated DNA excision in Anabaena. *Cell* 44:905–911.

Lammers, P. J., S. McLaughlin, S. Papin, C. Trujillo-Provencio and A. J. Ryncarz II. 1990. Developmental rearrangement of cyanobacterial *nif* genes: nucleotide sequence, open reading frames, and cytochrome P-450 homology of the *Anabaena* sp. strain PCC 7120 *nifD* element. *J. Bacteriol.* 172(12):6981–6990.

Landy, A. 1989. Dynamic, structural, and regulatory aspects of lambda site-specific recombination. *Annu. Rev. Biochem.* 58:913–949.

Matveyev, A. V., E. Rutgers, E. Soderback, and B. Bergman. 1994. A novel genome rearrangement involved in heterocyst differentiation of the cyanobacterium *Anabaena* sp. PCC 7120. *FEMS Microbiol. Let.* 116:201–208.

Meeks, J. C., E. L. Campbell, and P. S. Bisen. 1994. Elements interrupting nitrogen fixation genes in cyanobacteria: presence and absence of a *nifD* element in clones of *Nostoc* sp. strain Mac. *Microbiology* 140:3225–3232.

Meeks, J. C., C. M. Joseph, and R. Haselkorn. 1988. Organization of the *nif* genes in cyanobacteria in symbiotic association with *Azolla* and *Anthoceros. Arch. Microbiol.* 150:61–71.

Mulligan, M. E., W. J. Buikema, and R. Haselkorn. 1988. Bacterial-type ferredoxin genes in the nitrogen fixation regions of the cyanobacterium *Anabaena* sp. strain PCC 7120 and *Rhizobium meliloti. J. Bacteriol.* 170:4406–4410.

Mulligan, M. E., and R. Haselkorn. 1989. Nitrogen-fixation (*nif*) genes of the cyanobacterium *Anabaena* sp. strain PCC 7120: the *nifB-fdxN-nifS-nifU* operon. *J. Biol. Chem.* 264(32):19200–19207.

Ramasubramanian, T. S., T.-F. Wei, and J. W. Golden. 1994. Two *Anabaena* sp. strain PCC 7120 DNA-binding factors interact with vegetative cell- and heterocyst-specific genes. *J. Bacteriol.* 176(5):1214–1223.

Sato, T., Y. Samori, and Y. Kobayashi. 1990. The *cisA* cistron of *Bacillus subtilis* sporulation gene *spoIVC* encodes a protein homologous to a site-specific recombinase. *J. Bacteriol.* 172(2):1092–1098.

Schopf, J. W. 1993. Microfossils of the Early Archean Apex chert: new evidence of the antiquity of life. *Science* 260:640–646.

Stark, W. M., M. R. Boocock, and D. J. Sherratt. 1992. Catalysis by site-specific recombinases. *Trends Genet.* 8(12):432–439.

Vega-Palas, M. A., E. Flores, and A. Herrero. 1992. NtcA, a global nitrogen regulator from the cyanobacterium *Synechococcus* that belongs to the Crp family of bacterial regulators. *Mol. Microbiol.* 6(13):1853–1859.

Wei, T.-F., T. S. Ramasubramanian, and J. W. Golden. 1994. *Anabaena* sp. strain PCC 7120 *ntcA* gene required for growth on nitrate and heterocyst development. *J. Bacteriol.* 176(15):4473–4482.

Wei, T.-F., T. S. Ramasubramanian, F. Pu, and J. W. Golden. 1993. *Anabaena* sp. strain PCC 7120 *bifA* gene encoding a sequence-specific DNA-binding protein cloned by in vivo transcriptional interference selection. *J. Bacteriol.* 175(13):4025–4035.

Wolk, C. P., A. Ernst and J. Elhai. 1994. Heterocyst metabolism and development, in *The molecular biology of cyanobacteria,* D. A. Bryant, eds. pp. 769–823. Kluwar Academic Publishers, Dordrecht.

PART C

Genome Evolution

17

Evolution of the *E. coli* Genome

Ulfar Bergthorsson and Howard Ochman

Evolution in bacterial genomes occurs at two levels: (1) alterations of individual nucleotides resulting from point mutations, or from recombinational events that do not incorporate novel sequences, and (2) changes in chromosome structure involving the insertion, deletion or translocation of DNA. The evolution of chromosome organization has received much less attention than point mutational evolution due to both the lack of convenient techniques to study large-scale changes in bacterial genomes at the population level and the relative paucity of quantitative theory pertaining to the rates and patterns of genome rearrangements.

Much of the initial work on genome evolution in bacteria applied classical genetics procedures and yielded several insights into the rates and patterns of chromosome change: (1) interspecies comparisons of linkage maps of *Escherichia coli* K-12 and *Salmonella enterica* serovar Typhimurium LT2 established that the organization of bacterial chromosomes is highly conserved (Riley and Krawiec, 1987; *see also* Chapters 18 and 22); (2) rates of spontaneous rearrangements, such as duplications and deletions, are fairly high, orders of magnitude greater than the nucleotide substitution rate, but most of these changes are unstable (Anderson and Roth, 1977, 1981; Starlinger, 1977; Drake, 1991); (3) only certain portions of the chromosome can tolerate the stable maintenance of large-scale inversions (Segall *et al.*, 1988; Mahan *et al.*, 1990; François *et al.*, 1990).

The recent advent of pulsed-field gel electrophoresis (PFGE; *see* Chapters 24–26 and 28) has greatly facilitated the analysis of the physical structure of bacterial chromosomes (Fonstein and Haselkorn, 1995). Unlike conventional genetic techniques, these procedures can be used to examine strains or species that are not amenable to genetic manipulation, and to assess the extent of variation and patterns of chromosome change within and among closely-related species.

Comparisons between *E. coli* and *Salmonella*

The evolutionary dynamics of bacterial genomes has been assessed through both inter- and intra-species comparisons. The most comprehensive information concerning the evolution of bacterial chromosomes has been obtained through alignments of the genetic maps of *E. coli* K-12 and *S. enterica* serovar Typhimurium LT2 (Sanderson, 1971; Riley and Krawiec, 1987; *see also* Chapter 22). Although these two enteric species diverged an estimated 120 to 160 million years ago (Ochman and Wilson, 1987), their chromosomes are almost identical in length, and the order and spacing of genes is largely congruent, leading to what Charlebois and St. Jean (1995) have called "a paradigm of stability." However, there is also some evidence that large-scale genomic rearrangements occurred after the divergence of these species: their chromosomes differ with respect to a large inversion encompassing 10% of the genome and numerous regions that are unique to one of the species (Riley and Anilionis, 1978; Riley and Krawiec, 1987). These regions, termed "loops," arise from the addition or deletion of DNA in one of the species, and were originally identified as chromosomal segments of more than 0.6 minutes required to restore alignment of the two genetic maps. Taken together, almost 15% of the chromosomal DNA of these species resides in loops, which appear to be randomly distributed throughout the chromosome.

Some species-specific genes of *E. coli* K-12 and Typhimurium LT2 are known to be located in loops (Riley and Sanderson, 1990; Ochman and Groisman, 1994). For example, the *lac* operon and *arg*F gene map to 8′ on the *E. coli* chromosome and, based on their phylogenetic distribution and atypical base composition, are thought to have been acquired horizontally (Van Vliet *et al.*, 1988). Loops are also known to contain genes, such as *bgl* and *spe*C, that are present only in some strains of *E. coli* (Riley and Krawiec, 1987); and in *S. enterica.* Mills et al. (1995) have recently defined a 40-kb region required for invasion of serovar Typhimurium into host tissue. This region, which is not present in *E. coli* K-12, maps to 59′ on the Typhimurium LT2 chromosome, and many of the genes within the region have a low GC content suggesting acquisition by horizontal transfer.

Despite the alterations that have occurred since the divergence of *E. coli* and *S. enterica,* the overall conservation in chromosomal organization is particularly surprising given the high rate of chromosomal alterations in laboratory populations, the ability of bacteria to acquire exogenous DNA, and the amount of time since the divergence of *E. coli* and *S. enterica.*

Variation in the Physical Structure of Bacterial Genomes

Chromosomal variation among enteric species has also been studied by PFGE and the construction of low-resolution physical maps. Comparisons of the physical maps of *E. coli, S. enterica* serovar Typhimurium LT2 and serovar Enteritidis revealed that Enteritidis harbors an inversion involving 18% of the chromosome (Liu *et al.*, 1993a). This inversion contains the terminus and is congruent with one detected from comparisons of *E. coli* K-12 and Typhimurium LT2, but is more extensive. Liu et al. (1993a, 1993b, 1994; Liu and Sanderson, 1995a, 1995b) have applied the homing endonuclease I-*Ceu*I to evaluate variation in the physical maps of *E. coli* K-12 and several additional serovars of *S. enterica.* I-*Ceu*I recognizes a 19-bp target sequence situated in bacterial *rrn* operons (Marshall and Lemieux, 1992), generating seven fragments in *E. coli* and *S. enterica* ranging from 44 to 2,460 kb in size (Liu *et al.*, 1993b). When the I-*Ceu*I maps were compared among serovars of *S. enterica,* and to *E. coli,* it appeared that the distribution of *rrn* loci is well conserved, with the exception of serovar Paratyphi C where one *rrn* locus has shifted in position. According to these I-*Ceu*I digests, the chromosome lengths of different serovars ranges from 4.7 to 4.9 Mb, all slightly larger than the estimated length of the *E. coli* K-12 chromosome. In a survey of Typhimurium serovars from the SARA collection (Beltran *et al.*, 1991), there was very little variation in I-*Ceu*I restriction patterns, with 15 out of 17 isolates yielding identical patterns (Liu and Anderson, 1995b), whereas the chromosomes of some other serovars of *S. enterica,* notably Typhi Ty2, and Paratyphi A and C, harbor large rearrangements, some of which are produced by recombination at *rrn* operons (Liu and Sanderson, 1995a, 1995b; *see also* Chapter 22).

Using PFGE, the chromosome size of *Shigella flexneri* 2a was identical to that of laboratory strains of *E. coli,* and there is a general similarity between the physical maps of these species, as expected from their close genetic relatedness (Ochman *et al.*, 1983; Okada *et al.*, 1991). Some of the disparities between the *S. flexneri* 2a and *E. coli* K-12 chromosomal maps include an inversion spanning about 10% of the chromosome located roughly between 30′ and 40′ on the genetic map, and a 30-kb region present on the *E. coli* K-12 chromosome that includes the *lac* operon is absent from *S. flexneri* (Okada *et al.*, 1991).

The use of PFGE to study variation in chromosome structure within *E. coli* has mainly focused on closely related lineages derived from laboratory strains. Several studies have constructed physical maps of derivatives of *E. coli* K-12 and the resulting estimates of chromosome lengths ranged from 4.5 to 4.7 megabases (Smith *et al.*, 1987; Daniels, 1990; Heath, 1992; Perkins *et al.*, 1992, 1993). Some strains harbor inversions, duplications, or deletions, and a disproportionately large amount of the differences mapped to the region surrounding the replication terminus, presumably due to higher levels of recombination in that section (Perkins et al., 1993; Louarn, 1994). However, laboratory strains are probably not representative of genetic variation in the species at large, and many of the observed changes in chromosome size and organization are thought to have resulted from mutagenic treatment in the laboratory environment (Perkins *et al.*, 1993).

Using DNA renaturation procedures, Brenner et al. (1972) detected substantial variation in genome sizes in laboratory and clinical isolates of two serotypes of *E. coli.* The genome sizes of these strains were estimated to be in the range of 2.29×10^9 to 2.97×10^9 daltons, corresponding to 3.7 to 4.8 megabases. While some of this variation could be attributable to plasmids, many strains had smaller genomes than an isolate of *E. coli* K-12 known to lack plasmids, implying that these differences are due in part to variation in the amount of chromosomal DNA.

This variation within *E. coli,* as revealed by DNA hybridization procedures, stands in sharp contrast to the conservation observed in comparisons of the *E. coli* K-12 and *S. enterica* sv. Typhimurium LT2 linkage maps. To examine the diversity in genome size in natural populations of *E. coli,* we have recently employed PFGE to analyze isolates of *E. coli* of known genetic and genealogical relationships (Bergthorsson and Ochman, 1995). The strains used for these studies were selected from the ECOR collection (Ochman and Selander, 1984), and represent the major subgroups within *E. coli* as defined by multilocus enzyme electrophoresis (Selander *et al.*, 1987; Herzer *et al.*, 1990). Genome sizes, as estimated from summing the sizes of restriction fragments generated by digestion with *Bln*I and *Not*I, ranged from roughly 4.65 to 5.3 megabases, a wider range than that detected among enteric species (Figure 17-1). Based on these samples, estimates of genome size for natural isolates of *E. coli* were, in general, larger than those of laboratory strains, and there were also significant differences in average genome sizes among the major subgroups of *E. coli.* The difference between laboratory strains and natural isolates is probably not due to reduction in genome size subsequent to the cultivation of laboratory strains, but rather from the phylogenetic ancestry of the strains: the laboratory strain *E. coli* K-12 types to subgroup A which includes the strains with the smallest genomes, and its genome size is similar to two natural isolates in that group, ECOR 4 and 13.

In our sample of ECOR strains, there is variation in the number and sizes of plasmids harbored by each strain and considering only large plasmids (those over 40 kb), individual isolates contain up to three different plasmids with sizes approaching 200 kb. The cumulative size of large plasmids per isolate was in one case 250 kb. The effect of these plasmids on the genome size estimates based on PFGE depends on the specific enzyme restriction sites in a plasmid. Not all these plasmids contained restriction sites for *Bln*I or *Not*I, and only about 25% of them contain recognition sites for both enzymes. The total range in genome size is largely unaffected by the presence of plasmids. For example, the strains with the largest genome sizes—ECOR 38 and 40—both harbor a 60 kb plasmid with a *Bln*I restriction site. Therefore, the variation in genome size as measured by PFGE principally reflected the differences in chromosomal DNA content.

Since duplications of parts of the *E. coli* chromosome are usually unstable, the variation in genome size among natural isolates is most likely due to the cumulative effects of deletions and the acquisition of DNA through lateral pro-

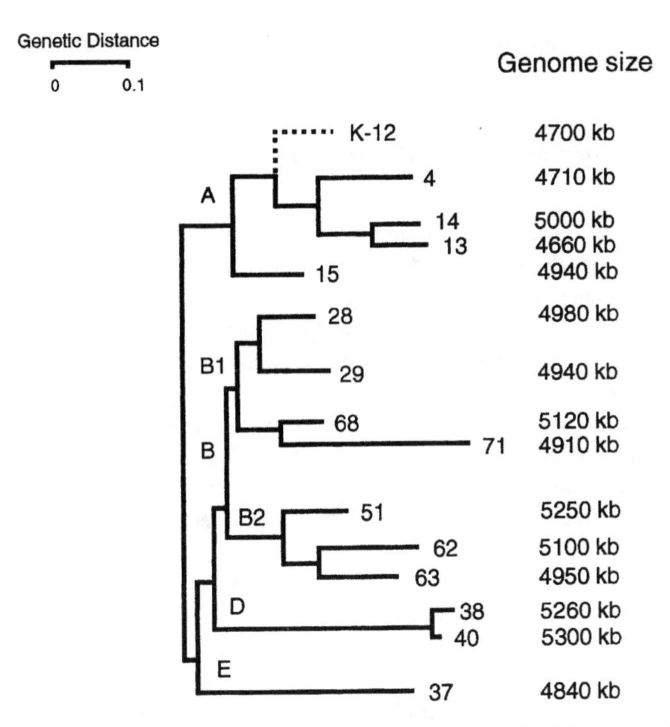

Figure 17-1. Genome sizes and evolutionary relationship of 15 isolates of *E. coli.* Numbers at the branches are the ECOR strain designations and letters refer to *E. coli* subgroups. Phylogenetic relationships are based on allelic variation at 38 polymorphic enzyme loci (Herzer et al., 1991), and the position of *E. coli* K-12 on this tree is based on similarities in its electrophoretic profile to that of ECOR 2. The column to the right includes the genome sizes of each strain as estimated by PFGE. There is a significant difference in mean genome size among the four major subgroups A, B1, B2, and D (Analysis of variance, $F = 7.5$, $p < 0.01$; Kruskal-Wallis, $H = 8.3$, $p < 0.05$).

cesses. The precise origins and nature of the sequences contributing to the variation in genome size in natural isolates of *E. coli* are still unclear, but preliminary physical maps of these strains indicate that the differences are distributed throughout the chromosome. Some of the variation is likely due to mobile genetic elements such as plasmids, prophages, IS elements, and transposons (*see also* Chapters 3, 4 and 20) which are known to be highly variable in their occurrences in natural populations (Silver *et al.*, 1980; Nyman *et al.*, 1983; Hartl et al., 1986; Bisercic and Ochman, 1993); however, these extrachromosomal elements rarely constitute hundreds of kilobases. Some natural uropathogenic strains of *E. coli* are known to harbor pathogenicity islands—clusters of chromosomal genes required for virulence—which can be upwards of 190 kb (Blum *et al.*, 1994), and the large 220-kb virulence plasmids of *Shigella* and enteroinvasive *E. coli* are known to integrate into the host chromosome, thereby both modulating virulence and stabilizing the plasmid (Zagaglia *et al.*, 1991).

Base Composition and Horizontal Transfer

Base composition is relatively homogenous over the entire bacterial chromosome, but can vary greatly among species (Sueoka, 1961, 1988; Muto and Osawa, 1987). Among bacteria, GC contents range from 25% in Mycoplasma (*see* Chapters 39–42) to 75% in Micrococcus, and it is generally thought that these differences in genomic base composition arise from biases in the mutation rates among species; *i.e.* the A/T → G/C and G/C → A/T are not equal. The effect of these mutational biases would be most apparent at the nucleotide positions under the least selective constraints: GC contents at the third position of codons are 10% in *Mycoplasma,* 58% in *E. coli,* and 95% in *Streptomyces,* whose chromosomal base composition are 25%, 51%, and 73% GC respectively (Muto and Osawa, 1987).

Because the GC contents of genes within a given bacterial species fall within a fairly narrow range, atypical base composition is often taken as evidence for the acquisition of a gene from a non-related species of a differing GC content (Aoyama *et al.*, 1994; Ochman and Groisman, 1994; Ochman and Lawrence, 1995). By examining the GC contents and codon usage patterns of sequenced genes from *E. coli,* it has been suggested that at least 6% (Whittam and Ake, 1993) or as much as 16% (Médigue *et al.*, 1991) of the *E. coli* genome has been acquired through horizontal transfer, and a similar estimate has been derived for *S. enterica* (Ochman and Lawrence, 1996). These numbers probably underestimate the total amount of genes that have been horizontally transferred to *E. coli* and *S. enterica* since gene transfer can occur from species of similar base composition and acquired genes are subjected to mutational biases which gradually change their base composition to that of their new host species (Ochman and Lawrence, 1996).

Once the original base composition of a gene is known, it is possible to analyze the change in GC content on an evolutionary timescale (Sueoka, 1993). Aoyama et al. (1994) have established that *cps* genes in *E. coli* K-12 and *S. enterica* Serovar Typhimurium LT2 were independently acquired from the same source, but at different times. By applying certain assumptions about mutation rates, mutational pressure, and the number of generations per year in natural populations of *E. coli,* they estimated that the *cps* cluster in *E. coli* K-12 was acquired some 32 million years ago. Ochman and Lawrence (1996) used an alternative approach to estimate the rate of GC content change of the *spa* genes in *Salmonella* and *Shigella.* The *spa* genes are required for bacterial entry into intestinal host cells and were independently acquired by *Salmonella* and *Shigella* from a species of low GC content. The GC content of the *spa* genes in *Shigella* is 34%, but the homologous genes in *S. enterica* have a GC content of 46% compared to 52% for the whole chromosome. The distribution of these genes in *Salmonella, Shigella* and *E. coli* indicate that the *spa* genes have been in *S. enterica* for at least 140 million years, while *Shigella* acquired the virulence plasmid with the *spa* genes 20 million years ago. Assuming that the GC content of the *spa* genes in *Shigella* are close to that of the donor species, it is possible to estimate the rate at which the GC content of the *spa* genes in *S. enterica* has approached that of their new host. Over the course of evolution within *Salmonella* there has been a 12% increase in GC content of the *spa* genes in *S. enterica,* and the change was most pronounced in third position of codons, with an increase of 23% in GC content. Both of these previous studies support the view that base composition of bacterial genes is principally the result of species-specific biases in mutation rates.

Concluding Remarks

Despite the conservation in genome size and gene order in *E. coli* and *S. enterica,* the bacterial chromosome is fairly dynamic—with genes frequently gained by lateral processes and lost through deletions—thus suggesting that overall genome size may reflect selective constraints operating on the bacterial chromosome. The potential for constraints on chromosome organization arises from several sources, such as: (1) gene dosage depending on the distance from the origin of replication (Schmid and Roth, 1987); (2) the need for the terminus to be opposing the origin of replication to minimize chromosomal replication rates (Riley and Sanderson, 1990); (3) maintaining an advantageous orientation of transcription relative to direction of replication (Brewer, 1990); (4) selection against extraneous DNA, and the streamlining of the genome to promote replication rates; and (5) local effects of superhelical domains on gene expression (Charlebois and St. Jean, 1995). The individual contribution of each constraint, as well as the cumulative effect of these constraints, on genome organization await further investigation.

References

Anderson, R. P., and J. R. Roth. 1977. Tandem genetic duplications in phage and bacteria. *Ann. Rev. Microbiol.* 31:473–505.

Anderson, R. P., and J. R. Roth. 1981. Spontaneous tandem genetic duplications in *Salmonella typhimurium* arise by unequal recombination between rRNA (*rrn*) cistrons. *Proc. Natl. Acad. Sci. USA* 78:3113–3117.

Aoyama, K., A. M. Haase, and P. R. Reeves. 1994. Evidence for effect of random genetic drift on G+C content after lateral transfer of fucose pathway genes to *Escherichia coli* K-12. *Mol. Biol. Evol.* 6:829–838.

Beltran, P., S. A. Plock, N. H. Smith, T. S. Whittam, D. C. Old, and R. K. Selander.

1991. Reference collection of strains of the *Salmonella typhimurium* complex from natural populations. *J. Gen. Microbiol.* 137:601–606.

Bergthorsson, U., and H. Ochman. 1995. Heterogeneity of genome sizes among natural isolates of *Escherichia coli. J. Bacteriol.* 177:5784–5789.

Bisercic, M., and H. Ochman. 1993. Natural populations of *Escherichia coli* and *Salmonella typhimurium* harbor the same classes of insertion sequences. *Genetics* 133:449–454.

Blum, G., M. Ott, A. Lischewski, A. Ritter, H. Imrich, H. Tschäpe, and J. Hacker. 1994. Excision of large DNA regions termed pathogenicity islands from tRNA-specific loci in the chromosome of *Escherichia coli* wild-type pathogen. *Infect. Immun.* 62:606–614.

Brenner, D. J., G. R. Fanning, F. J. Skerman, and S. Falkow. 1972. Polynucleotide sequence divergence among strains of *Escherichia coli* and closely related organisms. *J. Bacteriol.* 109:953–965.

Brewer, B. J. 1990. Replication and the transcriptional organization of the *Escherichia coli* chromosome. In *The Bacterial Chromosome,* K. Drlica and M. Riley, eds. pp. 61–83. American Society for Microbiology, Washington, D.C.

Charlebois, R. L., and A. St. Jean. 1995. Supercoiling and map stability in the bacterial chromosome. *J. Mol. Evol.* 41:15–23.

Daniels, D. L. 1990. The complete *Avr*II restriction map of the *Escherichia coli* genome and comparisons of several laboratory strains. *Nucl. Acids Res.* 18:2649–2651.

Drake, J. W. 1991. A constant rate of spontaneous mutation in DNA-based microbes. *Proc. Natl. Acad. Sci. USA* 88:7160–7164.

Fonstein, M., and R. Haselkorn. 1995. Physical mapping of bacterial genomes. *J. Bacteriol.* 177:3361–3369.

François, V., J. Louarn, J. E. Rebello, and J.-M. Louarn. 1990. Replication termination, nondivisible zones, and structure of the *Escherichia coli* chromosome. In *The Bacterial Chromosome,* K. Drlica and M. Riley, eds. pp. 351–359. American Society for Microbiology, Washington, D.C.

Hartl, D. L., M. Medhora, L. Green, and D. E. Dykhuizen. 1986. The evolution of DNA sequences in *Escherichia coli. Phil. Trans. R. Soc. Lond. B* 312:191–204.

Heath, J. D., J. D. Perkins, B. Sharma, and G. M. Weinstock. 1992. *Not*I genomic cleavage map of *Escherichia coli* K-12 strain MG1655. *J. Bacteriol.* 174:558–567.

Herzer, P. J., S. Inouye, M. Inouye, and T. S. Whittam. 1990. Phylogenetic distribution of branched RNA-linked multicopy single-stranded DNA among natural isolates of *Escherichia coli. J. Bacteriol.* 172:6175–6181.

Liu, S.-L., A. Hessel, and K. E. Sanderson. 1993a. The *Xba*I-*Bln*I-*Ceu*I genomic cleavage map of *Salmonella enteritidis* shows an inversion relative to *Salmonella typhimurium* LT2. *Mol. Microbiol.* 10:655–664.

Liu, S.-L., A. Hessel, and K. E. Sanderson. 1993b. Genomic mapping with I-*Ceu*I, an intron-encoded endonuclease specific for genes for ribosomal RNA, in *Salmonella* spp., *Escherichia coli,* and other bacteria. *Proc. Natl. Acad. Sci. USA* 90:6874–6878.

Liu, S.-L., A. Hessel, H.-Y.M. Cheng, and K. E. Sanderson. 1994. The *Xba*I-*Bln*I-*Ceu*I genomic cleavage map of *Salmonella paratyphi B. J. Bacteriol.* 176:1014–1024.

Liu, S.-L., and K. E. Sanderson. 1995a. Rearrangements in the genome of the bacterium *Salmonella typhi. Proc. Natl. Acad. Sci. USA* 92:1018–1022.

Liu, S.-L., and K. E. Sanderson. 1995b. I-*Ceu*I reveals conservation of the genome of independent strains of *Salmonella typhimurium. J. Bacteriol.* 177:3355–3357.

Louarn, J., F. Cornet, V. François, J. Patte, and J. M. Louarn. 1994. Hyperrecombination in the terminus region of the *Escherichia coli* chromosome: Possible relation to nucleoid organization. *J. Bacteriol.* 176:7524–7531.

Mahan, M. J., A. M. Segall, and J. R. Roth. 1990. Recombination events that rearrange the chromosome: Barriers to inversion. In *The Bacterial Chromosome,* K. Drlica and M. Riley, eds. pp. 341–350. American Society for Microbiology, Washington, D.C.

Marshall, P., and C. Lemieux. 1992. The I-*Ceu*I endonuclease recognizes a sequence of 19 base pairs and preferentially cleaves the coding strand of the *Chlamydomonas moewusii* chloroplast large subunit rRNA gene. *Nucl. Acids Res.* 20:6401–6407.

Médigue, C., T. Rouxel, P. Vigier, A. Hénaut, and A. Danchin. 1991. Evidence for horizontal gene transfer in *Escherichia coli* speciation. *J. Mol. Biol.* 222:851–856.

Mills, D. M., V. Bajaj, and C. A. Lee. 1995. A 40-kb chromosomal fragment encoding *Salmonella typhimurium* invasion genes is absent from the corresponding region of the *Escherichia coli* K-12 chromosome. *Mol. Microbiol.* 15:749–759.

Muto, A., and S. Osawa. 1987. The guanine and cytosine content of genomic DNA and bacterial evolution. *Proc. Natl. Acad. Sci. USA* 84:166–169.

Nyman, K., H. Ohtsubo, D. Davison, and E. Ohtsubo. 1983. Distribution of insertion element IS*1* in natural isolates of *Escherichia coli. Mol. Gen. Genet.* 189:516–518.

Ochman, H., and R. K. Selander. 1984. Standard reference strains of *Escherichia coli* from natural populations. *J. Bacteriol.* 157:690–693.

Ochman, H., and A. Wilson. 1987. Evolution in bacteria: Evidence for a universal substitution rate in cellular genomes. *J. Mol. Evol.* 26:74–86.

Ochman, H., and E. A. Groisman. 1994. The origin and evolution of species differences in *Escherichia coli* and *Salmonella typhimurium.* In *Molecular Ecology and Evolution: Approaches and Applications,* B. Schierwater, B. Streit, G. P. Wagner and R. DeSalle, eds. pp. 479–493. Birkhäuser, Basel.

Ochman, H., and J. Lawrence. 1996. Phylogenetics and the amelioration of bacterial genomes. In *Escherichia coli* and *Salmonella typhimurium.* Volume 2, F.C. Neidhardt ed. (Washington D.C.: American Society for Microbiology), in press.

Ochman, H., T. S. Whittam, D. A. Caugant, and R. K. Selander. 1983. Enzyme polymorphism and genetic population structure in *Escherichia coli* and *Shigella. J. Gen. Microbiol.* 129:2715–2726.

Okada, N., C. Sasakawa, T. Tobe, K. A. Talukder, K. Komatsu, and M. Yoshikawa. 1991. Construction of a physical map of the chromosome of *Shigella flexneri* 2a and the direct assignment of nine virulence-associated loci identified by Tn5 insertions. *Mol. Microbiol.* 5:2171–2180.

Perkins, J. D., J. D. Heath, B. R. Sharma, and G. M. Weinstock. 1992. *Sfi*I genomic cleavage map of *Escherichia coli* K-12 strain MG1655. *Nucl. Acids Res.* 20:1129–1137.

Perkins, J. D., J. D. Heath, B. R. Sharma, and G. M. Weinstock. 1993. *Xba*I and *Bln*I genomic cleavage maps of *Escherichia coli* K-12 strain MG1655 and comparative analysis of other strains. *J. Mol. Biol.* 232:419–445.

Riley, M., and A. Anilionis. 1978. Evolution of the bacterial genome. *Ann. Rev. Microbiol.* 32:519–560.

Riley, M., and S. Krawiec. 1987. Genome organization. In Escherichia coli *and* Salmonella typhimurium: *Cellular and Molecular Biology,* F. C. Neidhardt, J. L. Ingraham, K. B. Low, B. Magasinik, M. Schaechter and H. E. Umbarger, eds. pp. 967–981. American Society for Microbiology, Washington, D.C.

Riley, M., and K. E. Sanderson. 1990. Comparative genetics of *Escherichia coli* and *Salmonella typhimurium.* In *The Bacterial Chromosome,* K. Drlica and M. Riley, eds. pp. 85–95. American Society for Microbiology, Washington, D.C.

Sanderson, K. E. 1971. Genetic homology in the Enterobacteriaceae. *Adv. Genet.* 16:35–51.

Schmid, M. B., and J. R. Roth. 1987. Gene location affects expression level in *Salmonella typhimurium. J. Bacteriol.* 169:2872–2875.

Segall, A., M. J. Mahan, and J. R. Roth. 1988. Rearrangement of the bacterial chromosome: forbidden inversions. *Science* 241:1314–1318.

Selander, R. K., D. A. Caugant, and T. S. Whittam. 1987. Genetic structure and variation in natural populations of *Escherichia coli.* In Escherichia coli *and* Salmonella typhimurium: *Cellular and Molecular Biology,* F. C. Neidhardt, J. L. Ingraham, K. B. Low, B. Magasinik, M. Schaechter and H. E. Umbarger, eds. pp. 1625–1648. American Society for Microbiology, Washington, D.C.

Silver, R. P., W. Aaronson, A. Sutton, and R. Schneerson. 1980. Comparative analysis of plasmids and some metabolic characteristics of *Escherichia coli* K1 from diseased and healthy individuals. *Infect. Immun.* 29:200–206.

Smith, C. L., J. G. Econome, A. Schutt, S. Klco, and C. R. Cantor. 1987. A physical map of the *Escherichia coli* K12 genome. *Science* 236:1448–1453.

Starlinger, P. 1977. DNA rearrangements in prokaryotes. *Ann. Rev. Genet.* 11:103–126.

Sueoka, N. 1961. Variation and heterogeneity of base composition of deoxyribonucleic acids: a compilation of old and new data. *J. Mol. Biol.* 3:31–40.

Sueoka, N. 1988. Directional mutation pressure and neutral molecular evolution. *Proc. Natl. Acad. Sci. USA* 85:2653–2657.

Sueoka, N. 1993. Directional mutation pressure, mutator mutations, and dynamics of molecular evolution. *J. Mol. Evol.* 37:137–153.

Van Vliet, F., A. Boyen, and N. Glansdorff. 1988. On interspecies gene transfer: the case of the *arg*F gene of *Escherichia coli. Ann. Inst. Pasteur/Microbiol.* 139:493–496.

Whittam, T. S., and S. E. Ake. 1993. Genetic polymorphisms and recombination in natural populations of *Escherichia coli.* In *Mechanisms of Molecular Evolution,* N. Takahata and A. G. Clark, eds. pp. 223–245. Japan Scientific Societies Press, Tokyo.

Zagaglia, C., M. Casalino, B. Colonna, C. Conti, A. Calconi, and M. Nicoletti. 1991. Virulence plasmids of enteroinvasive *Escherichia coli* and *Shigella flexneri* integrate into a specific site on the host chromosome: Integration greatly reduces expression of plasmid-carried virulence genes. *Infect. Immun.* 59:792–799.

18

E. coli Genes: Ancestries and Map Locations

Monica Riley

Introduction

With the advent of improved sequencing methods, information on the genomes and genes of many bacteria is being accumulated rapidly. We can ask whether any of this information throws new light on the ancestries of bacterial genes and chromosomes. Sequence-related proteins seem likely to have descended from common ancestor genes; therefore, methodical examination of sequence similarities could identify evolutionarily related groups of proteins and their genes. Also, the locations of genes on bacterial chromosomes might reflect ancient mechanisms of evolution.

Information Content of Bacterial Genomes

Data on the information content of the genome and knowledge of its gene products are the most comprehensive for *E. coli.* More than half of the genes and gene products of the cell have been described (Bairoch and Boeckman 1993, Riley and Labedan, 1995). It has been possible to categorize the type of gene product (enzyme, regulator, RNA, etc.) (Table 18-1) for 1,933 gene products, and to categorize the cellular functions (Table 18-2) of 2,013 proteins (Riley 1990, Riley and Labedan, 1995 and unpublished work).

One can ask if all the genetic information in bacterial genomes is unique or if some is redundant. Bacteria often have multiple sets of ribosomal RNA genes, presumably to meet through a gene dosage effect the massive needs for ribosomal components (*see also* Chapter 21). In addition, in *E. coli* many sets of multiple enzymes exist whose functions are very nearly redundant. The enzymes of any one set have slightly different properties and specificities, but they catalyse essentially the same reactions. In the small molecule metabolism of *E. coli* there

Table 18-1. Kinds of Gene Products in E. coli*

Type of Gene Product	Number of Genes
Enzymes, leader sequences	1,087
Transport	278
Regulators	235
RNA	106
Structural components	110
Factors	70
Carriers	17
Total	1,933

*excluding orfs and genes known only by phenotype.

are 71 enzymatic functions for which two, three, or in one case four proteins exist. For the 71 functions there are 157 separate proteins involved in altogether 176 pairs of similar proteins (Labedan & Riley, 1995a, and unpublished data). The pairs of enzymes are usually active under different environmental conditions or are sensitive to different inhibitory molecules, or they have different ranges of substrate specificity, are induced to be formed, or are repressed by different molecules. Therefore, the enzymes are not strictly redundant; rather, they provide a breadth of responses to a wide range of environmental and biochemical conditions. It will be interesting to see if bacteria with smaller genomes differ from *E. coli* in that they do not contain multiple enzymes, only a unique set of one enzyme for each reaction (*see also* Chapter 37).

Table 18-2. Functions of E. coli *proteins*

Category	Number of genes	Percent
Small molecule metabolism		
Degradation and energy metabolism	364	18
Central intermediary metabolism	160	8
Broad regulatory functions	57	3
Biosynthesis		
amino acids, polyamines	115	6
purines, pyrimidines, nucleosides, nucleotides	32	2
cofactors, prosthetic groups	108	5
fatty acids	24	1
Large molecule metabolism		
Synthesis and modification	411	20
Degradation	73	4
Cell envelop	174	9
Cell processes		
transport	291	14
other, e.g. cell division, chemotaxis, mobility, osmotic adaptation, detoxification, cell killing	86	4
Miscellaneous	118	6

Evolutionary History of Genes for Multiple Enzymes

Different modes of acquisition of genetic information may have operated over evolutionary time to expand the capabilities of a small primitive genome. One mode is the duplication and divergence of genes; another is the acquisition by lateral transmission of genetic material, transferring from one organism to another; and another is convergent evolution from more than one ancestral gene. Of the 176 pairs of related multiple enzymes in *E. coli,* the sequence is known for both members of 105 pairs, allowing a query about their sequence relatedness. Of the 105, 63 pairs have significantly related sequences as judged by their PAM scores (an inverse measure of sequence similarity (Gonnet et al., 1992), whereas 42 pairs have no evident sequence relatedness (Labedan and Riley 1995a). On this basis it seems likely that somewhat over half of the multiple enzymes in *E. coli* have their origins in duplication and divergence of an ancestral gene, whereas somewhat less than half have their origins either in horizontal transfer of a gene from another organism or in convergent evolution from two different genetic starting points. All modes of expanding the capability of a genome may have been utilized simultaneously or the mechanisms of evolutionary expansion may have shifted over evolutionary time.

Evolutionary History of Individual *E. coli* Genes and Proteins

To look at the evidence for duplication of ancestral genes, all sequenced *E. coli* proteins were examined for sequence similarities among themselves (Labedan and Riley, 1995a, 1995b). Homologous pairs could have arisen by gene duplication within the *E. coli* chromosome, but some would have duplicated and diverged long before *E. coli* became a defined biological entity. In some cases, the common ancestral gene for two sequence-related *E. coli* proteins is an ancestral *E. coli* gene; in other cases the ancestor could date back to early days of evolution of primordial cells.

All pairs of *E. coli* protein sequences were collected that were related by two criteria: their sequences aligned for at least 100 amino acids, and their sequences were related by a PAM score (Gonnet *et al.*, 1992) of less than 200. We identified 3,574 such pairs from among the 2,548 *E. coli* K-12 protein sequences in release 33 of the SwissProt database (Bairoch and Boeckman, 1993) (Labedan and Riley, 1995a, and unpublished data). There are families of sequence-related proteins (or modals of proteins) all members of which might have descended from one original ancestor gene. Many *E. coli* proteins are members of families. Altogether there were 352 families among the 3,574 sequence-related pairs.

We asked whether the 3,574 pairs of sequence-related proteins were often related by function and we found that functional relatedness was observed far above chance couplings. On inspection, 1,345 of the 1,523 pairs (90.3%) were composed of two proteins that had been assigned to the same broad functional class, such as two transport elements, or two regulatory proteins.

Many *E. coli* gene products can be classified by type of gene product (Table 18-1) and by a more narrowly defined physiological function (Table 18-2). The physiological functions of all *E. coli* proteins have been classified in a hierarchical scheme of functional categories (Riley and Labedan, 1996, similar to Riley, 1993). Not all members of the 3,574 pairs could be assessed for function. There were 1,579 protein pairs in which both members of the pair could be labeled with a narrowly defined physiological function. Together, the proteins of the 1,579 pairs belong to 103 of the possible 118 physiological functions. When the physiological assignments of the protein pairs were examined, 430 of 1,579 pairs (27.2%) carried the identical assignment, a frequency far above chance for any pair having the same of 103 possible assignments.

The main conclusion from examining sequence-relatedness among *E. coli* proteins is that we can identify evolutionary siblings. Nearly half of all currently known *E. coli* proteins have a sequence-related partner in *E. coli* and often these are functionally related. Thus, gene duplication and divergence seems to have played a major role in the generation of contemporary proteins of *E. coli.* The degree of sequence similarity between the *E. coli* pairs ranged widely, from almost identical to much less related, indicating that there are both ancient and recent divergences in the genes of the *E. coli* cell of today. Sequences of even distantly related proteins seem to retain detectable similarities. There are constraints for functioning proteins that prevent divergence of the amino acid sequences beyond a certain point, preventing further change. The relatively weak sequence similarities we collected may represent common ancestry dating back perhaps as early as a primordial progenote.

Evolutionary History of the *E. coli* Chromosome

We can go further than sequence analysis and ask whether another mechanism for genome expansion besides duplication of genes and acquisition by horizontal transfer existed in the past. At some early time in the course of evolution, did the *E. coli* chromosome as a whole double, providing the opportunity for divergence and specialization of the replicate genes? Was the duplication of entire genomes a mechanism of evolution (Sparrow and Nauman, 1976, Wallace and Morowitz, 1973, Hopwood, 1967), or did all ancestral duplications deal only with smaller amounts of genetic material such as single genes? Twenty years ago a hypothesis was put forward that the *E. coli* genome may have doubled one or two times, leaving ancestrally related sibling genes disposed at intervals of ¼ or ½ of the chromosome (Zipkas and Riley, 1975). At the time it seemed that the then-mapped functionally related genes were more likely to be located at 90° or 180° intervals on the map than would be expected by chance. However, data on the map positions and functions of *E. coli* genes were sparse at the time compared to the data of today. The question about total chromosome doubling in *E. coli* can be asked again today in a number of different ways.

To ask about map locations of related genes using the data available today, one can first ask if the genes for the 44 pairs of functionally related and sequence-related multiple enzymes discussed above are more likely than chance to be positioned at distances of 25 or 50 map units apart on the 100 map unit map. Inspection of map data shows this is not true (Figure 18-1). The answer is "no," they do not tend to occupy such positions.

Next, one can expand the query to address *all* functionally related or sequence-related, mapped *E. coli* genes, not just those for multiple enzymes. One can ask whether the mapped genes for functionally related gene products (related by any of several definitions) (Riley and Labedan, 1996) are separated on the chromosome by 25 or 50 map units. This question can be asked in three different ways.

(1) Are the genes for enzymes that catalyse similar types of reactions located on the chromosome in clusters separated by 25 or 50 map units? When one looks at map positions of genes for enzymes having the same first three numbers of the Enzyme Commission classification scheme (Webb, 1992), the answer is "no." Figure 18-2 illustrates this approach and shows map positions for those EC numbers beginning

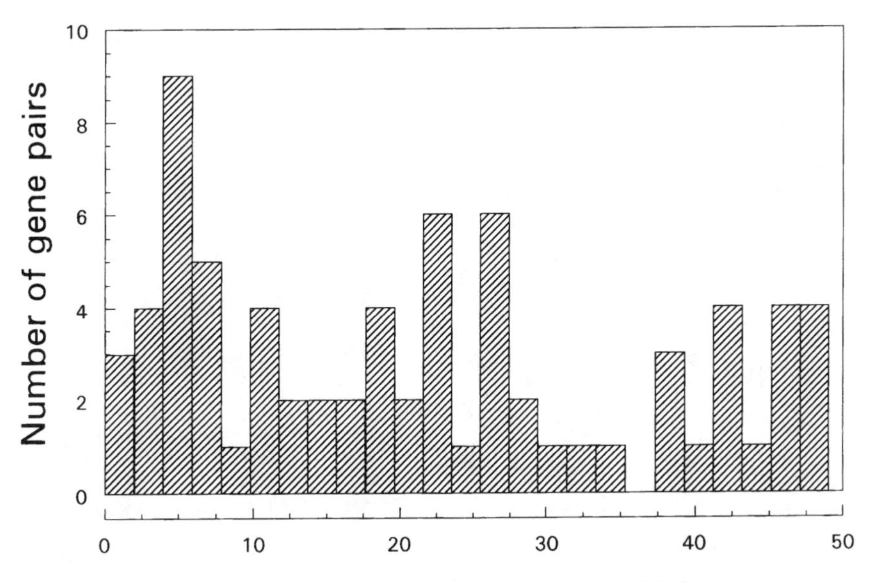

Figure 18-1. Map distances between pairs of genes for multiple enzymes. The shorter arc on the circle is shown, expressed in centisomes (Rudd, 1993).

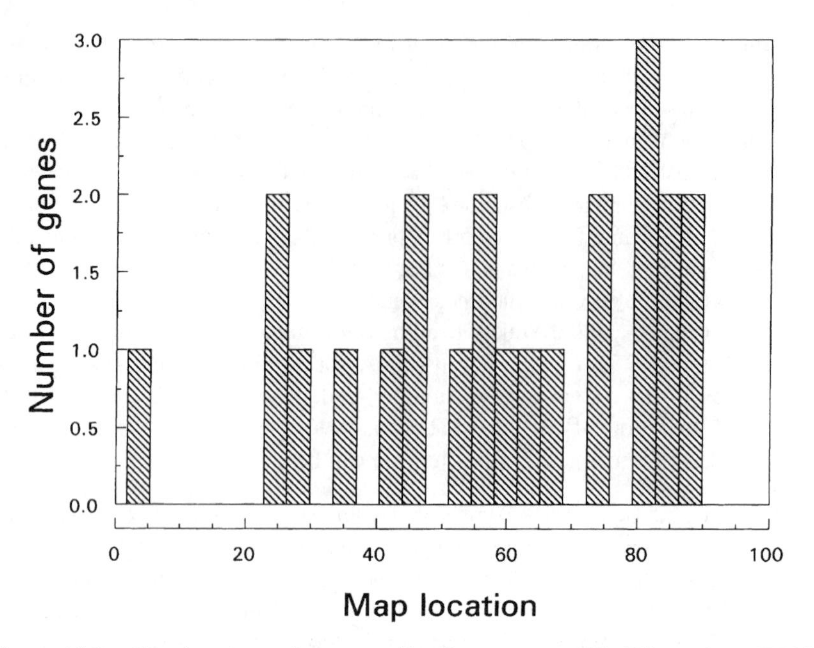

Figure 18-2. Map locations of genes coding for enzymes with EC numbers (Webb, 1992) of 1.1.1-, representing dehydrogenases that act on the CH-OH group of donor molecules using NAD+ or NADP+ as the acceptor.

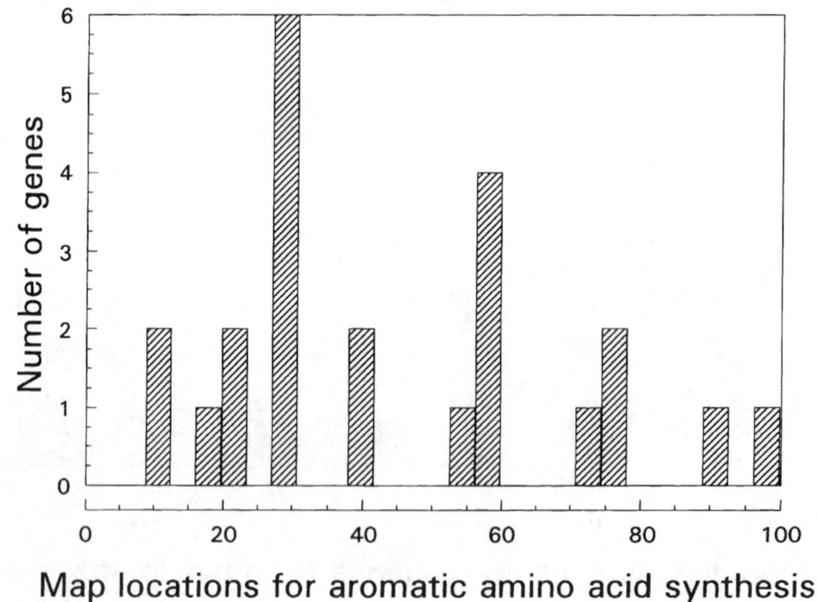

Figure 18-3. Map locations of genes coding for enzymes of aromatic amino acid biosynthesis.

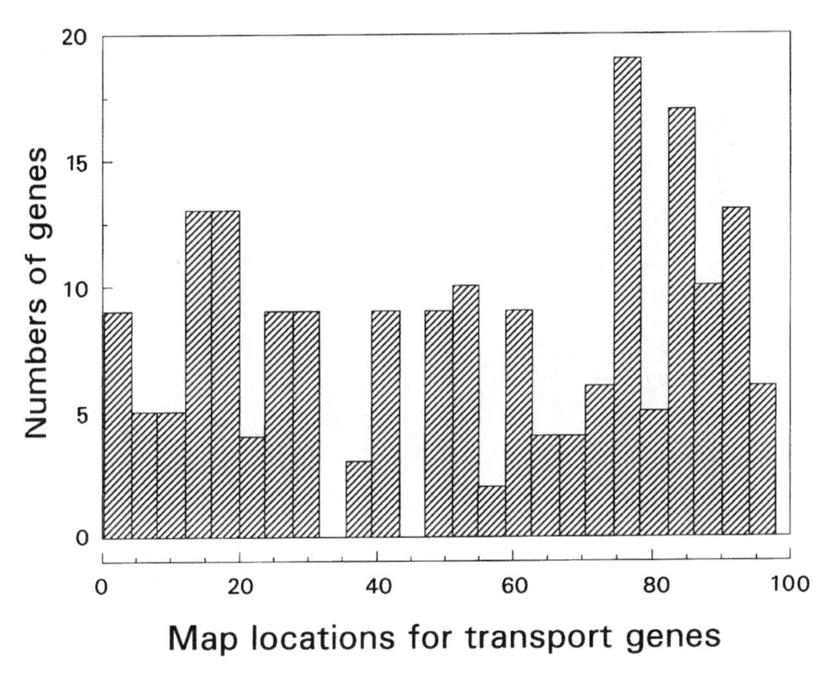

Figure 18-4. Map locations of genes coding for transport functions.

with 1.1.1.-, referring to dehydrogenases that act on the CH-OH group of donor molecules using NAD+ or NADP+ as the acceptor.

(2) Next, one can ask if genes for similar cell functions such as metabolic pathways are located in this pattern. Again, the answer is "no." Figure 18-3 shows an example, the locations of genes for biosynthesis of aromatic amino acids.

(3) Next, one can ask if gene products for a type of cellular function such as transport or regulation are located in this pattern. Again the answer is "no." The distribution of map locations of genes for transport functions is shown in Fig. 18-4.

Finally, one can ask if sequence-related pairs of genes tend to be 25 or 50 map units apart. A list of 2,329 pairs of sequence-related proteins was used to test the chromosome-doubling hypothesis once again. Distance in map positions between the genes of all sequence-related pairs was determined and the distribution is displayed in Fig. 18-5. No tendency to peak at 25 or 50 map units is evident.

With vastly more data available to test the idea than was available 20 years ago, it seems now that the notion of genome doubling is not upheld, at least not in its original form. Since there are not regular patterns of map locations of sequence-related or functionally related genes, it seems likely that evolution by

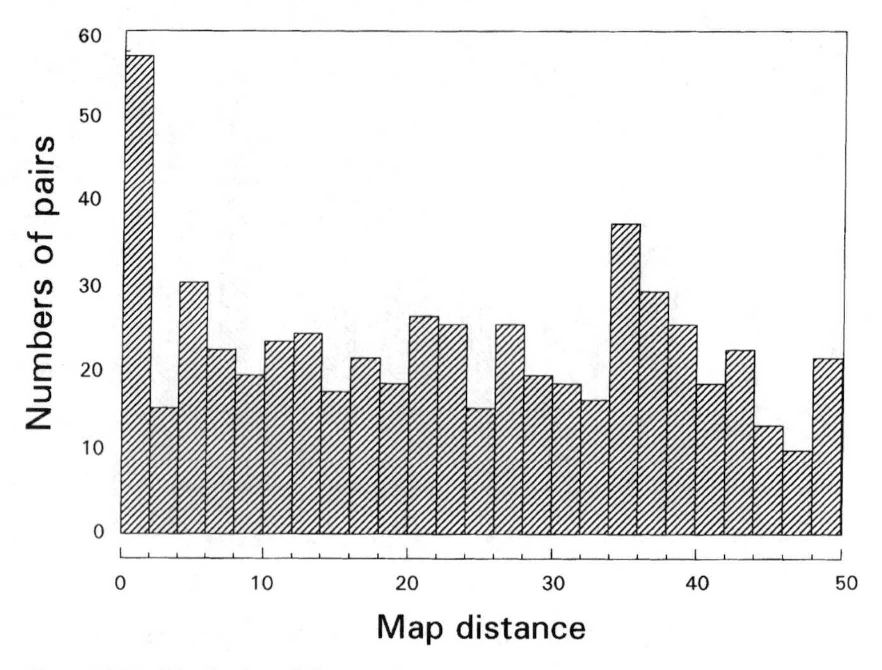

Figure 18-5. Distribution of distances between map locations of 2329 homologous pairs of *E. coli* genes (Labedan and Riley, 1995a).

replication of ancestral sequences used a process of individual gene duplications rather than whole chromosome duplications. These data do not prove that genome doublings did not occur. It is conceivable that locations of modern genes have no relation to ancient gene locations. Diverging genes may have translocated to positions on the chromosome not constrained in respect to distance apart. Looking at available genetic and physical maps of other bacteria, one notes that although related bacteria strongly maintain the same gene order, there is no universal gene order for bacteria as a whole and genetic maps of members of unrelated bacterial families are quite different (O'Brien, 1990). It is possible that contemporary bacterial gene map locations do not reflect evolutionary history of an early ancestral chromosome.

Looking Ahead

Within the year, the entire sequence of the *E. coli* chromosome will have been determined (*see also* Chapter 38). At that point, a complete analysis of sequence similarities will provide us with an idea of the minimum number of ancestral genes that were needed to generated the *E. coli* chromosome as we know it. Determination of the functions of *all* the gene products will enable us to see the totality of genetic functions that are necessary to create, maintain, and reproduce the life of this versatile free-living single-cell organism.

Acknowledgements

The data sets of sequence-related *E. coli* genes were developed with Dr. Bernard Labedan.

References

Bairoch, A., and B. Boeckman. 1993. The SWISS-PROT protein sequence data bank, recent developments. *Nucl. Acids Res.* 21:3093–3096.

Gonnet, G. H., M. A. Cohen, and S. A. Benner. 1992. Exhaustive matching of the entire protein sequence database. *Science* 256:1443–1445.

Hopwood, D. A. 1967. Genetic analysis and genome structure in *Streptomyces coelicolor. Bacteriol. Rev.* 31:373–403.

Labedan, B., and M. Riley. 1995a. Gene products of *Escherichia coli:* Sequence comparisons and common ancestries. *Mol. Biol. Evol.* 12:980–987.

Labedan, B., and M. Riley. 1995b. Widespread protein sequence similarities: Origins of *Escherichia coli* genes. *J. Bacteriol.* 177:1585–1588.

O'Brien, S. J. 1990. *Genetic Maps: Locus Maps of Complex Genomes,* 5th ed., Cold Spring Harbor Laboratory Press, Cold Spring Harbor, N.Y.

Riley, M. 1993. Functions of the gene products of *Escherichia coli. Microbiol. Rev.* 57:862–952.

Riley, M., and B. Labedan. 1996. *E. coli* gene products: physiological functions and common ancestries, in *Escherichia coli* and *Salmonella,* 2nd ed., F. Neidhardt, III Curtis, J. Ingraham, E.C.C. Lin, K. B. Low, B. Magasanik, W. Reznikoff, M. Riley, M. Schaechter, and H. E. Umbarger, eds. American Society for Microbiology, Washington, D.C.

Rudd, K. E. 1993. Maps, genes, sequences and computers: An *Escherichia coli* case study. *ASM News* 59:335–341.

Sparrow, A. H., and A. F. Nauman. 1976. Evolution of Genome Size by DNA Doublings. *Science* 192:524–529.

Wallace, D. C., and H. J. Morowitz. 1973. Genome Size and Evolution. *Chromosoma* 40:121–126.

Webb, E. C. 1992. *Enzyme Nomenclature.* New York: Academic Press, Inc.

Zipkas, D., and M. Riley. 1975. Proposal concerning mechanism of evolution of the genome of *Escherichia coli. Proc. Natl. Acad. Sci. USA* 72:1354–1358.

19

Prokaryotic Genome-Wide Comparisons and Evolutionary Implications

Samuel Karlin and Jan Mrázek

Abstract

Bacterial similarity relationships are inferred using sequence information derived from large aggregates of genomic sequences. Comparisons within and between species sample sequences are based on the vector of dinucleotide relative abundance values, referred to as the genomic signature. Recent studies have demonstrated that the dinucleotide relative abundance values (profiles) of different DNA sequence samples (sample size ≥50kb) from the same organism are generally much more similar to each other than they are to profiles from other organisms, and that closely related organisms generally have more similar profiles than do distantly related organisms. These highly stable DNA-doublet profiles suggest that there may be genome-wide factors such as functions of replication and repair machinery that impose limits on the compositional and structural patterns of a genomic sequence. The genomic signatures of all prokaryotic genomes with available non-redundant DNA of at least about 100 kb were compared. These include 21 proteobacteria, 10 Gram-positives, three cyanobacteria, three archaea, and three unclassified sequences. Among specific results, the genomic signature of thermophilic Archaea deviates substantially from the signature of halophilic Archaea. *Anabaena* sequences are relatively close to the Gram-positives *L. lactis* and *S. aureus*. Gram-positives divide into at least five subgroups. The dinucleotide TA is almost universally under-represented; GC is pervasively over-represented in γ and β-proteobacterial genomes; AT is high in α-proteobacteria, and CG is low in many thermophiles. Interpretations center on DNA structures (e.g., base-step stacking energies, conformational tendencies) and context-dependent mutational biases.

Introduction

Prokaryotic molecular evolutionary relationships have been predominantly derived from sequence comparisons among 16S ribosomal RNA genes (e.g., Woese, 1987; Olsen *et al.*, 1994). The principal prokaryotic subdivision separates eubacteria and archaebacteria. The current major categories of available eubacterial sequences consist of the Gram-negative purple proteobacteria and the Gram-positive bacteria. The archaea include principally extreme thermophiles, extreme halophiles, and methanogens.

Conventional methods of phylogenetic reconstruction from sequence information employ only similarity or dissimilarity assessments of aligned homologous genes or regions (e.g., Nei, 1987; Felsenstein, 1991; Lake, 1994). Difficulties intrinsic to this approach include: (1) alignments of distantly related long sequences (e.g., complete genomes) are generally not feasible; (2) different phylogenetic reconstructions (trees) may result for the same set of organisms based on analysis of different protein, gene, or noncoding sequences (attempts are made to overcome this by "averaging" over many proteins); (3) resultant trees may be highly dependent on details of the alignment algorithm employed; (4) the often-made assumption of constant rates of evolution on the various branches of the tree or at different sites within a sequence may be violated (the problem of unequal rate effects); (5) problems of chimeric origins, recombination, inversions, transpositions, and lateral transfer between distantly related organisms; and (6) tree construction derived from aligned sequences cannot apply to organisms for which similar gene sequences are largely unavailable (e.g., for bacteriophages or diverse eukaryotic viruses).

Relative Abundance Values and Distance Measures

Our methods address the problem of inferring genomic relationships by using nucleotide sequence information derived from entire genomes rather than individual loci. Comparisons within and between species sample sequences are based on *dinucleotide relative abundance values* (Karlin *et al.*, 1994a; Karlin and Ladunga, 1994; Karlin and Burge, 1995; Blaisdell *et al.*, 1996). Dinucleotide contrasts are assessed through the odds ratio $\rho_{xy} = f_{XY}/f_X f_Y$, where f_X denotes the frequency of nucleotide X, and f_{XY} is the frequency of dinucleotide XY in the sequence under study. For double-stranded DNA sequences, a symmetrized version ρ^*_{XY} is computed from frequencies of the sequence concatenated with its inverted complementary sequence. These dinucleotide relative abundances to a great extent factor out the effect of the genomic G+C content.

We measure the distance between two sequences f and g (from different organisms or from different regions of a single genome) by the dinucleotide relative abundance distance calculated as $\delta^*(f,g) = 1/16 \sum_{ij}|\rho^*_{ij}(f) - \rho^*_{ij}(g)|$, where the sum extends over all dinucleotides. The $\delta^*(f,g)$ distance can be interpreted as the average difference between the dinucleotide relative abundance values of the species f and g. A multi-dimensional analysis (partial ordering) comparing the vectors of dinucleotide relative abundance values between genomes is described in Karlin *et al.*, (1994b).

Dinucleotide relative abundance profiles are equivalent to the "general designs" derived from biochemical nearest-neighbor frequency analysis that were evaluated extensively two and three decades ago in samples of genomic DNA from many organisms (e.g., Josse *et al.*, 1961; Russell *et al.*, 1976; Russell and Subak-Sharpe, 1977). It was observed that the set of dinucleotide relative abundance values is essentially the same in most organisms for bulk genomic DNA, and also essentially the same for DNA fractions of differing sequence complexity (renaturation rate fractions), for euchromatin versus heterochromatin, and for distinct base-compositional (density gradient) fractions of nuclear DNA (Russell *et al.*, 1976). Our recent studies have demonstrated that the dinucleotide relative abundance profiles of different sequence samples of DNA from the same organism are generally much more similar to each other than to sequence samples from other organisms and that related organisms generally have more similar dinucleotide relative abundance values than distantly related organisms (Karlin and Ladunga, 1994; Karlin and Burge, 1995), Table 19-1. These observations seem to indicate that there are factors and mechanisms that impose limits upon compositional and structural variation of any particular genome.

Dinucleotide Relative Abundance Distance Orderings among Prokaryotic Genomes

All prokaryotic genomes having aggregate nonredundant available DNA of at least about 100 kb were compared via δ^*-distances. Several independent samples of 50 kb lengths from each genome were formed. The average δ^*-distances with respect to all samples within and between genomes are displayed in Table 19-2. The explicit bacterial representations and abbreviations are explained in the legend to Table 19-2. The sequences of *A. tumefaciens* were derived from chromosomal sources or from the Ti plasmid. These are separated in Table 19-2. The δ^*-distance between these collections is close, 0.039. In four examples (agrtu, strli, chltr, halha), available DNA was just short of 100 kb, and we constructed two partly overlapping 50 kb samples. We highlight several results suggested by the δ^*-distance evaluations. As an aid for our interpretations, Table 19-1 provides δ^*-distances in terms of several familiar examples (cf. Karlin and Ladunga, 1994; Karlin and Cardon, 1994).

(1) *δ^*-distances within and between bacterial genomes.* The average within bacterial genome δ^*-distances (diagonal of Table 19-2) range

Table 19-1. Examples of dinucleotide relative abundance distances (multiplied by 1000) within and between disjoint eukaryotic and prokaryotic genomic collections based on 50kb sequence samples.

Comparison	# of samples	Mean δ*	δ* range
Within *S. cerevisiae*	10	22	11–38
Within *E. coli*	32	26	6–70
Within Human	10	48	16–96
E. coli vs. *S. typhimurium*	32×8	40	15–87
Bovine vs. Pig	7×5	45	20–80
Mouse vs. Rat	10×10	46	16–90
Human vs. Mouse	10×10	57	23–125
Human vs. Chicken	10×10	70	39–131
D. melanogaster vs. *B. mori*	10×6	77	43–113
E. coli vs. *B. subtilis*	32×32	83	53–121
C. elegans vs. *S. cerevisiae*	10×10	88	57–124
D. melanogaster vs. *C. elegans*	10×10	91	55–137
Human vs. Trout	10×2	94	81–104
Human vs. Sea urchin	10×5	123	94–166
Human vs. *S. cerevisiae*	10×10	134	100–175
E. coli vs. *M. genitalium*	32×12	159	114–230
Human vs. *D. melanogaster*	10×10	172	120–251
Human vs. *E. coli*	10×32	229	168–289
E. coli vs. *Sulfolobus sp*	32×2	233	193–255

Sequence samples of about 50 kb size were generated for each species and compared by means of δ*-distances. The table indicates the mean and range of δ*-distances of all the pairwise comparisons within or between species. The eukaryotic sequence samples were generated from the longest available contiguous sequences. The number of samples was limited to 10. For description of prokaryotic samples, see Species listings to Table 19-2.

from 17 to 43 (all δ*-distances are multiplied by 1000) and δ*-distances between genomes range from 32 to 309 (one 23 value between two *Streptomyces* genomes). The bacterial genomes are arranged with respect to standard bacterial groupings (cf. Karlin *et al.*, 1995) and generally show moderate relatedness in δ*-distances among members of each group (usual range about 30–80) and generally consistent distances with respect to members of different groups.

(2) *Are the archaea monophyletic?* On the basis of rRNA gene comparisons (Woese, 1987; Woese *et al.*, 1990), the archaea are deemed monophyletic. This conclusion is consistent for certain protein comparisons, e.g., the elongation factor EF-1α and EF-2G families (Iwabe *et al.*, 1989; Creti *et al.*, 1994), and the eukaryotic and archaea RecA-like sequences of *rad51, dmc*1, *rad*A (Sandler *et al.*, 1996, Brendel *et al.* 1997). However, different protein comparisons challenge the monophyletic character of the archaea. For example, bacterial relationships based on comparison of the 70 kd heat shock proteins (*E. coli*

Table 19-2. The average δ-distances (multiplied by 1000) for ~50 kb samples within (bold face, diagonal) and between (nondiagonal) genomes.*

	esc co	sal ty	kle pn	ser ma	vib ch	yer en	hae in	azo vi	pse ae	pse pu	bor pe	alc eu	nei go	nei me	agr tu	agr Ti	rhi le	rhi me	bra ja	rho ca	rho sp	myx xa
escco	**27**	40	58	48	49	64	63	107	90	82	71	94	116	82	88	70	117	137	132	135	149	142
salty		**28**	46	53	68	72	75	121	104	107	89	112	107	71	78	68	106	124	124	126	142	169
klepn			**34**	62	70	84	90	118	105	110	84	101	137	98	66	65	87	108	110	113	122	172
serma				**25**	81	94	81	110	99	100	61	81	118	89	71	75	98	115	110	115	146	161
vibch					**22**	54	66	103	81	69	86	97	134	100	112	78	132	146	137	165	146	118
yeren						**32**	64	130	109	94	99	114	138	94	135	105	152	172	159	184	171	158
haein							**28**	156	133	125	122	141	110	69	131	113	165	184	180	171	192	175
azovi								**33**	43	65	81	68	195	176	79	67	80	89	66	115	63	62
pseae									**23**	53	82	74	173	154	75	54	86	98	81	122	80	69
psepu										**24**	70	58	176	157	102	77	114	126	109	148	115	81
borpe											**23**	37	166	142	82	79	83	103	83	117	110	128
alceu												**35**	187	165	87	83	84	95	74	124	99	111
neigo													**26**	52	145	141	173	179	192	184	207	225
neime														**35**	124	117	153	167	173	162	188	207
agrtu															**(11)**	39	35	57	61	61	85	133
agrTi																**24**	56	73	70	89	86	115
rhile																	**27**	32	37	62	64	136
rhime																		**19**	39	80	63	144
braja																			**30**	87	56	120
rhoca																				**19**	98	155
rhosp																					**39**	100
myxxa																						**18**

Groups: C_{1a} (escco–serma); C_{1b} (vibch–haein); C_1 (escco–haein); C_2 (azovi–psepu); B_1 (borpe–alceu); B_2 (neigo–neime); A_1 (agrtu–braja); A_2 (rhoca–rhosp); D (myxxa)

	bac su	sta au	lac la	str pn	bac st	str co	str li	myc le	myc tu	cor gl	syn sp	ana sp	syn sq	myc ge	chl tr	the sp	hal ha	met th	sul sp
escco	82	91	118	146	98	201	202	83	101	96	96	103	150	159	162	242	206	201	232
salty	91	104	130	157	84	210	211	107	115	122	111	112	152	181	156	238	208	225	224
klepn	80	109	130	157	98	215	216	115	128	127	109	119	155	190	146	231	222	227	228
serma	93	131	143	176	93	222	221	111	112	120	110	143	165	191	192	251	209	234	265

64	55	82	106	122	190	192	90	126	68	65	71	146	132	124	212	211	168	191	vibch
99	67	104	122	140	232	234	106	137	114	101	78	132	141	151	234	240	172	193	yeren
91	100	112	143	121	237	238	129	154	123	103	102	127	143	158	226	253	212	231	haein
98	114	120	124	149	121	121	86	60	74	93	137	216	209	150	220	150	166	208	azovi
74	103	93	101	128	137	141	89	79	61	62	112	196	182	125	199	169	157	191	pseae
88	79	112	132	136	158	159	70	86	42	76	114	192	164	160	233	180	154	215	psepu
112	112	148	168	128	194	192	80	89	102	117	148	206	204	197	272	185	202	248	borpe
110	123	148	168	141	179	177	88	95	87	112	160	224	211	197	271	181	191	244	alceu
140	166	174	205	70	255	262	186	186	177	154	166	166	221	195	249	244	277	268	neigo
113	129	137	166	68	240	244	156	167	159	133	126	126	181	164	225	237	240	233	neime
70	148	132	156	82	175	174	122	91	109	96	164	183	225	140	203	171	226	245	agrtu
58	110	99	121	86	161	161	93	79	83	73	126	169	199	120	196	166	190	212	agrTi
101	161	154	170	110	163	161	136	103	123	121	180	213	255	154	224	154	239	255	rhile
117	175	166	182	119	154	157	155	120	133	132	192	232	267	155	226	138	252	255	rhime
116	157	158	168	137	149	147	135	102	125	126	183	236	258	171	242	147	227	247	braja
115	196	179	203	121	201	199	168	126	150	142	211	214	274	185	236	211	273	293	rhoca
121	160	147	153	150	116	120	126	104	122	119	176	232	253	148	214	146	200	210	rhosp
122	121	125	116	192	111	114	95	94	70	98	143	235	186	144	211	155	144	186	myxxa
30	99	80	110	93	185	187	118	121	77	52	111	147	167	109	172	205	180	203	bacsu
	41	82	95	160	184	185	93	126	81	88	56	145	125	126	205	209	145	161	staau
		31	44	153	189	190	134	150	83	57	54	129	110	74	131	232	122	140	lacla
			23	186	183	187	145	146	108	85	70	145	116	71	117	227	104	109	strpn
				29	222	229	147	140	144	119	175	165	234	162	216	206	253	249	bacst
					26	23	132	116	145	176	203	285	264	196	247	79	218	228	strco
						(13)	133	117	147	181	206	287	263	202	251	79	221	239	strli
							17	55	88	117	136	193	187	180	258	139	162	208	mycle
								18	96	125	164	211	225	185	244	124	186	232	myctu
									26	56	98	183	146	129	200	182	149	202	corgl
										31	86	149	139	94	159	212	149	177	synsp
											26	120	94	100	165	246	138	144	anasp
												27	159	173	157	295	171	188	synsq
													43	152	179	309	137	172	mycge
														(13)	92	230	174	122	chltr
															32	266	148	121	thesp
																(19)	237	253	halha
																	31	106	metth
																		20	sulsp
P_{1a}	P_{1b}	P_{1c}		P_2	P_3		P_4		P_5	S_1	S_2	S_3	M			Z			
P_1																			

Species listings of Table 19-2.

Groupings of bacterial species. The C_1 group corresponds roughly to γ-proteobacteria with primary habitat in vertebrate hosts. Group C_2 are also γ-proteobacteria of primary soil habitat. Group B is rather diverse, classified mainly as β-proteobacterial types. The α-proteobacteria are covered in the A_1 (bacteria interacting primarily with plants, e.g. rhizobia) and A_2 group (free living photosynthetic bacteria)

The Gram-positive is divided into five subgroups where $[P_1,P_2]$ are composed from low C+G content, whereas $[P_3,P_4,P_5]$ are groups of high C+G content, S_1,S_2,S_3 consist of different cyanobacteria. The Z group includes archaebacterial types. The other examples are unclassified.

List of abbreviations.

C_{1a} escco *Escherichia coli* (1.6Mb, 32 samples)
C_{1a} salty *Salmonella typhimurium* (407kb, 8)
C_{1a} klepn *Klebsiella pneumoniae* (195kb, 4)
C_{1a} serma *Serratia marcescens* (83kb, 2)
C_{1b} vibch *Vibrio cholerae* (96kb, 2)
C_{1b} yeren *Yersinia enterocolitica* (111kb, 2)
C_{1b} haein *Haemophilus influenzae* (1.8Mb, 37)
C_2 azovi *Azotobacter vinelandii* (133kb, 3)
C_2 pseae *Pseudomonas aeruginosa* (298kb, 6)
C_2 psepu *Pseudomonas putida* (191kb, 4)
B_1 borpe *Bordetella pertussis* (113kb, 2)
B_1 alceu *Alcaligenes eutrophus* (113kb, 2)
B_2 neigo *Neisseria gonorrhoeae* (76kb, 2)
B_2 neime *Neisseria meningitidis* (69kb, 2)
A_1 agrtu *Agrobacterium tumefaciens* (62kb, 2 overlaping)
A_1 AgrTi *A. tumefaciens* plasmid Ti (106kb, 2)
A_1 rhile *Rhizobium leguminosarum* (99kb, 2)
A_1 rhime *Rhizobium meliloti* (246kb, 5)
A_1 braja *Bradyrhizobium japonicum* (113kb, 2)
A_2 rhoca *Rhodobacter capsulatus* (116kb, 2)
A_2 rhosp *Rhodobacter sphaeroides* (92kb, 2)
D myxxa *Myxococcus xanthus* (69kb, 2)
P_{1a} bacsu *Bacillus subtilis* (1.6Mb, 32)
P_{1b} staau *Staphylococcus aureus* (298kb, 6)
P_{1c} lacla *Lactococcus lactis* (242kb, 5)
P_{1c} strpn *Streptococcus pneumoniae* (97kb, 2)
P_2 bacst *Bacillus stearothermophilus* (116kb, 2)
P_3 strco *Streptomyces coelicolor* (135kb, 3)
P_3 strli *Streptomyces lividans* (64kb, 2 overlapping)
P_4 mycle *Mycobacterium leprae* (1.1Mb, 22)
P_4 myctu *Mycobacterium tuberculosis* (108kb, 2)
P_5 corgl *Corynebacterium glutamicum* (107kb, 2)
S_1 synsp *Synechococcus sp* (186kb, 4)
S_2 anasp *Anabaena sp* (89kb, 2)
S_3 synsq *Synechocystis sp* (1.0Mb, 20)
M mycge *Mycoplasma genitalium* (580kb, 12)
chltr *Chlamydia trachomatis* (60kb, 2 overlaping)
thesp *Thermus aquaticus+thermophilus* (119kb, 2)
Z halha *Halobacterium halobium* (60kb, 2 overlaping)
Z metth *Methanobacterium thermoautotrophicum* (113kb, 2)
Z sulsp *Sulfolobus acidocaldarius+solfataricus* (99kb, 2)

Sequence samples for *H. influenzae* and *M. genitalium* consist of their complete genomes. *E. coli* sequence sample is a 1.6 Mb contig centered at oriC. Other sequence samples are aggregates of GenBank entries at least 2.5 kb long. Shorter sequences were included only for *Thermus* sp and *Sulfolobus* sp in order to increase the aggregate amount of sequences. Major sequence redundancies were removed. The aggregate sequences were divided into nonoverlapping samples of equal sizes about 50 kb (ranging from 35 to 60 kb depending on the available amount of sequences). Two overlapping 50 kb samples were used when less than 70 kb aggregate sequences were available. The within-species distances assessed from two overlapping samples are shown in parentheses in Table 19-2.

DnaK homologues) place the *Halobacteria* closer to the *Streptomyces* of high G+C Gram-positive bacteria than to Gram-negative or eukaryotic species (Gupta and Golding, 1993; Gupta and Singh, 1994; Karlin, 1994). Results also contesting the pure monophyletic hypothesis from other protein sequence comparisons occur in Benachenhou-Lahfa *et al.* (1993) in terms of glutamate dehydrogenase sequences, and Brown *et al.* (1994) for glutamine synthetase. Rivera and Lake (1992) split the prokaryotes into eubacteria, halobacteria, and eocytes. With respect to genomic signature comparisons, the closest to the *Halobacterium halobium* (halha) are the *Streptomyces* sequences (strco, strli) of δ*-distances 79,79, respectively, "moderately related," and the next more distant by a factor of about 2 are the *M. tuberculosis* and *M. leprae* sequences, δ* (halha, myctu) = 124, δ* (halha, mycle) = 139. The distance of halha to the archaebacterial sequences of *Sulfolobus sp.* (sulsp) and *M. thermoautotrophicum* (metth) are very distant, 253 and 237 respectively. Thus, the monophyletic postulate is not supported by the δ*-distance data.

(3) *What is close to the Sulfolobus genomic sequences?* The bacterial sequences at hand are distantly related or very distant from sulsp, closest δ*-distances being sulsp to metth 106, to strpn 109, and to thesp 121. The closest to metth is again strpn 104. Virtually all other distances to sulsp and metth are more than 150. The thermophile sequences of thesp show δ*-distances to virtually all (one exception) bacterial sequences ≥100 (mostly ≥200) and are very distant to halha (δ* = 266). The closest to thesp is chltr (δ* = 92) and next closest is strpn (δ* = 117). For δ*- distance comparisons, the thermophiles tend to be closer to vertebrate eukaryotes than to proteobacterial sequences (Karlin and Cardon, 1994).

(4) *Relations of cyanobacterial genomes to proteobacterial and Gram-positive sequences.* A relatively close δ*-distance is observed between *Anabaena sp.* (anasp) and the Gram-positive collections of lacla (δ* = 54) and staau (δ* = 56), about equal to the distance between human and mouse (see Table 19-1). *Anabaena* is moderately related to vibch and yeren. *S. aureus* is also moderately close to vibch and yeren (δ* = 55,67). The *Anabaena* versus *Synechococcus* sequences are weakly related (δ* = 86), whereas the genomic collections of *Synechococcus sp.* and *Anabaena* versus *Synechocystis sp.* are distant (δ* = 149, 120). Synsq distances to all proteobacterial and Gram-positive sequences are almost always ≥ 150 and often ≥ 200. Clearly, the three cyanobacterial genomes are not a coherent group.

(5) *How coherent is the enterobacterial group?* The enterobacteria *E. coli, S. typhimurium, K. pneumoniae,* and *S. marcescens* are considered to be

closely to moderately related. Concordantly, the dinucleotide relative abundance distances vary between 40 and 62, where the smallest between species distance is δ* (escco, salty) = 40. The distance of klepn to escco is 58 and almost equally close to klepn is *A. tumefaciens,* δ* = 66. The non-enteric C_{1b} proteobacterial sequences [vibch, yeren] constitute a relatively coherent pair, with mutual δ*-distance 54. *H. influenzae* (haein, with complete genome available) is the most deviant among the C_1 group with corresponding δ*-distances 63 to 90.

Different groups of genes may exhibit very different degrees of similarity both in terms of protein and genomic comparisons as illustrated in comparing *S. flexneri* with *E. coli.* They share many common genes with very high amino acid identity. These include the ten *E. coli* genes *rpoS, gnd, hns, recA, rol, glpX, rfbA, rfbB, rfbC,* and *rfbD,* whose homologues in *S. flexneri,* show ≥90% amino acid identity, respectively. For example, the RecA and Hns proteins match perfectly and for RfbA, the alignment score is 90%. The aggregate length of these genes is 8970 bp in *E. coli* and 8711 bp in *S. flexneri.* The δ*-distance between the DNA of their coding regions has the low value of 29. The *S. flexneri* (shifl) coding DNA of these genes compared to all other available *S. flexneri* genomic sequences (about 82 kb) shows δ*-distance 46, and surprisingly from total comparisons δ* (shifl, escco) = 68 (data not shown).

(6) *How coherent are the β- and α-proteobacteria?* Although the B group sequences are classified as β-proteobacteria, with respect to δ*-distances, these sequences divide into two subgroups, B_1 = [borpe, alceu] and B_2 = [neigo, neime]. Neime is persistently closer to the sequences of C_1 than is neigo. Both neigo and neime are moderately related to *B. stearothermophilus* (δ* = 70, 68), an outlier among Gram-positive (see also Table 19-3). On the other hand, the B_1 group sequences might reasonably be joined to the C_2 group. The *Rhizobium* sequences in the A_1 group (see Table 2) are substantially homogeneous with mutual δ*-distances 32–39, moderately close to agrtu (δ* = 35–61).

(7) *How should the bacterium M. xanthus be classified?* Myxxa is currently labeled as a member of the δ-proteobacterial class. The closest to myxxa are the soil bacteria azovi, δ* = 62, pseae, δ* = 69, and corgl = δ* = 70. *M. xanthus,* a proficient bacterial predator, secretes many digestive enzymes and through swarming behavior absorbs prey bacterial lipids, peptides, nucleic acids, etc. It is conceivable that *M. xanthus* may have acquired DNA and genes from other (especially soil) bacteria in this fashion. From this perspective, genomic accretions concomitant to *M. xanthus* predation activities may be an important factor underlying the expansion of the *M. xanthus* genome (cf. Karlin *et al.*, 1995).

Table 19-3. Extremes of dinucleotide relative abundance values (displayed in symbolic form) in diverse bacterial sequences.

		G+C	CG	GC	TA	AT	CC/GG	TT/AA	TG/CA	AG/CT	AC/GT	GA/TC
C	escco	51%		+	-			(+)				
	salty	52%		+				(+)				
	klepn	56%		++		+		(+)			-	
	serma	60%		++	--	(+)		+				
	erwch	54%		+	-	(+)		(+)		-		
	shifl	41%		+	(-)							
	vibch	43%		+	-							
	yeren	45%		+								
	actpl	37%		+								
	haein	38%		++	-			+				
	azovi	65%			--	(+)						
	pseae	64%			--							
	psepu	60%			--							
	psefl	60%		(+)	--							
	psesy	56%		(+)	--							
B	borpe	66%		+	--	++						
	alceu	65%		+	--	++						
	neigo	52%	++	+	--	+++				--		
	neime	47%	+	+	-			++		-		
A	agrtu	56%	(+)		--	+		(+)			(-)	
	agrTi	52%			--							
	rhile	59%	+		--	++					-	
	rhime	61%	+		--	++					(-)	+
	braja	62%	+		--	++						(+)
	rhoca	66%			---	+++		+			-	
	rhosp	65%			--	++					(-)	+
D	myxxa	66%			---							

E	helpy	39%		++	-			++		-	
P	bacsu	44%		+	--			+		-	
	staau	34%									
	lacla	35%	(-)		-						
	strpn	39%	-		-						
	strmu	38%	--	++	-						
	bacst	51%	++	+	--			++	-		
	cloac	31%	---	+							
	clope	29%	---	+			+				
	strco	71%			---						+
	strli	70%			---						+
	strgr	71%			--						+
	mycle	58%			-						
	myctu	64%			--						
	corgl	55%			--						
S	synsp	53%			--			(+)			
	anasp	42%									
	synsq	48%	-		-		++	++			
	mycge	32%	---		-	-		(+)			
	borbu	32%	-	+			+	(+)		(-)	
	chltr	39%			-					(-)	
	thesp	67%	-	(-)	--	(-)	+	+		(-)	
Z	halha	62%	+		--						++
	metth	49%	--	-	-		(+)				
	sulsp	36%	--						(+)		

See Text for interpretation of the symbols.

Bacteria which were not included in Table 19-2, but are included in Table 19-3: *Erwinia chrysanthemi* (erwch), *Shigella flexneri* (shifl), *Actinobacillus pleuropneumoniae* (actpl), *Pseudomonas fluorescens* (psefl), *Pseudomonas syringae* (psesy), *Helicobacter pylori* (helpy), *Streptococcus mutans* (strmu), *Clostridium acetobutylicum* (cloac), *Clostridium perfringens* (clope), *Streptomyces griseus* (strgr), and *Borrelia burgdorferi* (borbu).

(8) *Diversity of Gram-positive.* The division of Gram-positive into five or six subgroups based on δ^*-distances is reasonably concordant with results of protein comparisons derived from RecA sequences (cf. Karlin *et al.*, 1995).

(9) *Chlamydia trachomatis* is generally considered primitive. On the basis of δ^*-distances, chltr is moderately related to the P_{1c} group at δ^*-distances 71, 74 and to thesp, synsp, anasp ($\delta^* = 92, 94, 100$), whereas generally other distances exceed 140.

Recently, several complete bacterial genomes were made available including *Mycoplasma pneumoniae* (Himmelreich *et al.* 1996), the archaeon *Methanococcus jannaschii* (Bult *et al.* 1996), and about 700kb of another archaeon *Pyrobaculum aerophilum* (J. H. Miller, personal communication). Analysis of these genomes confirmed genome-wide homogeneity of the genome signature (dinucleotide relative abundance values). The average within-species δ^*-distances based on 100kb sequence samples were $\delta^*=31$ in *M. pneumoniae,* $\delta^*=24$ in *M. jannaschii,* and $\delta^*=17$ in *P. aerophilum,* and the maximum within-species δ^*-distances did not exceed 54 in *M. pneumoniae* and *M. jannaschii,* and 32 in *P. aerophilum.* Among the relations of these new genomes to other prokaryotes, the following results stand out:

(a) The genus *Mycoplasma* is highly heterogeneous with the δ^*-distance between the closest related species (Himmelreich et al. 1996) *Mycoplasma pneumoniae* and *Mycoplasma genitalium* 97, only weakly related.

(b) The archaeon *Methanococcus jannaschii* is distant from all other bacteria (all $\delta^*>125$) and mostly very distant ($\delta^*>200$) from classical proteobacteria.

(c) *Pyrobaculum aerophilum* exhibits no dinucleotide relative abundance biases. This archaebacterium is distant to very distant from all other bacteria but marginally closer to classical proteobacteria than is *M. jannaschii* (δ^*-distances 137–204).

Extremes of Dinucleotide Relative Abundances

From statistical theory and data experience, a dinucleotide relative abundance may be conservatively described as significantly low if $\rho^*_{XY} \leq 0.78$ and significantly high if $\rho^*_{XY} \geq 1.23$ [Karlin and Cardon, 1994]. We distinguish symbolically levels of dinucleotide relative abundances as follows: *over-representation:* extremely high, +++, $\rho^*_{XY} \geq 1.50$; very high, ++, $1.30 \leq \rho^*_{XY} < 1.50$; significantly high, +, $1.23 \leq \rho^*_{XY} < 1.30$; marginally high, (+), $1.20 \leq \rho^*_{XY} < 1.23$; *under-representation:* extremely low, – – –, $\rho_{XY} \leq 0.50$; very low, – –, $0.50 < \rho^*_{XY} \leq 0.70$; significantly low, –, $0.70 < \rho^*_{XY} \leq 0.78$; marginally low, (–), $0.78 < \rho^*_{XY}$

≤ 0.81. The symbols marginally low (–) and marginally high (+) are shown only if the same dinucleotide is significantly high or low in another species of the same group (except for the singular cases mycge, borbu, chltr, thesp, halha, metth, and sulsp). The extremes are displayed symbolically in Table 19-3. There are clear trends, as follows:

(1) The dinucleotide TA is broadly under-represented in bacterial sequences. Among the exceptions are yeren and actpl of the C_{1b} group, sulsp, and borbu. In the low normal range are the clostridium genomes (cloac, clope) and anasp. TA suppression is also pervasive in eukaryotic sequences although not in animal mitochondrial genomes (Karlin and Cardon, 1994).

(2) Sequences of the proteobacterial groups C_1 and B are persistently over-represented in GC dinucleotides. Also, many of the low G+C Gram-positive sequences show high ρ^*_{GC} values.

(3) CG under-representations are prominent in the low C+G Gram(+) sequences of lacla, strpn, strmu, cloac, and clope, in many thermophiles including metth, sulsp, and thesp, and in mycoplasma and borbu sequences. At the other extreme, CG over-representations occur at bacst, in the halophile halha, and also in the groups B_2 and A_1. For eukaroytes, CG suppression is familiar in vertebrates, many diverse protist genomes, dicot (but generally not monocot) plants, animal mitochondrial sets, and almost all vertebrate small viral genomes (Karlin and Burge, 1995).

(4) The dinucleotide AT is strongly over-represented in the proteobacterial A and B_1 groups.

(5) *S. aureus* carries all dinucleotide relative abundances in the random range (Table 3). Also, the cyanobacterium *Anabaena* sp. sequences show no dinucleotide biases. Apart from TA avoidance, no other compositional biases occur in *Mycobacterium* sequences.

These trends in extremes over the different classes are consistent with the δ^*-distances.

Genomic Signature and DNA Structure

Collectively, the dinucleotide relative abundances $\{\rho^*_{XY}\}$ give each DNA genome a signature that is generally constant throughout the genome (Karlin and Ladunga, 1994; Karlin and Cardon, 1994; Blaisdell *et al.*, 1996). Average tri- and tetranucleotide relative abundance distances between sequences are highly correlated with dinucleotide δ^*-distances (Karlin and Ladunga, 1994). More generally, the structure of longer oligonucleotides can to a significant extent be predicted on the basis of nearest neighbor (dinucleotide) base pair interactions (e.g., Hunter, 1993). Philips *et al.* (1987) have concluded from a Markov-chain study of *E. coli* sequences that "constraints affecting oligonucleotide frequencies occur at the trinucleotide level or lower." Consistent with these predictions are the observation of the distribution of dinucleotides separated by 0, 1, 2, or more other nucleotides. Although values for contiguous base pairs as seen in Table 19-3 can be highly biased, those for space 1 or more tend to be quite random (Karlin and Burge, 1995). These observations suggest that the double-stranded base steps (dinucleotides) embody determinants of DNA structure/function of primary importance.

What are possible mechanisms underlying the genome signature determinations? DNA structural configurations appear to be principally determined by base-step arrangements reflecting dinucleotide stacking energies, DNA modifications, context dependent mutation biases, and mechanisms attendant to chromosomal replication, repair, segregation, and recombination processes (Breslauer *et al.*, 1986; Delcourt and Blake, 1991; Calladine and Drew, 1992; Hunter, 1993; Karlin and Burge, 1995). Certain base steps are associated with an intrinsic curvature, which can lead to bending and supercoiling (e.g., the wedge model of DNA curvature, Bolshoy *et al.*, 1991). In addition, there appear to be biases in replication and repair efficiency and fidelity depending on neighboring base context (Kunkel, 1992). Many DNA repair enzymes recognize shapes and lesions in DNA structures rather than specific sequences (Echols and Goodman, 1991; Kunkel, 1992). We hypothesize that differences between organisms in replication and repair machinery largely maintain the homogeneity of the whole genome of an organism (the genome signature). We indicate a possible example. The dinucleotide relative abundance values of temperate dsDNA phages are very close to their hosts, filamentous and ssDNA phages are moderately to distantly related to their hosts, and lytic dsDNA phages are generally distant from their hosts, with phage T7 being farther than phage T4 (Blaisdell *et al.*, 1996). This gradient in similarity to the host parallels the decline in the extent to which the phage uses the complete replication and repair machinery of the host and the duration of such use (Blaisdell *et al.*, 1996).

The analysis of dinucleotide relative abundance values in phylogenetic analyses has the following advantages: (1) It does not depend on finding the homologous genes in the sequences compared. (2) It does not require a prior alignment and is unaffected by the presence of gaps and large rearrangements in the sequence. (3) The genomic relative abundance distances can use the entire available genome sequence data for the organisms. The average δ^*-distances for multiple samples of 50 kb or more between genomes almost always substantially exceed within-genome distances. Translation of sequence similarities into evolutionary relatedness will always be tentative as the underlying assumptions about mutation rates, selective forces, and gene transfer events are uncertain. The results on the δ^*-distances within and between genomes for prokaryotes and eukaryotes are overwhelmingly robust although the underlying mechanisms are hardly understood.

However, the biochemical nearest neighbor analyses (Russell *et al.*, 1976; Russell and Subak-Sharpe, 1977) might be used to investigate the effects of altered replication and repair factors and context-dependent mutation tendencies such as evaluating dinucleotide relative abundances in the mutator strain of *E. coli* (Cox and Yanofsky, 1967). Further progress in our understanding of genomic evolution and phylogenetic relationships will require synthesis of sometimes conflicting results from rRNA, protein, and genome-wide sequence comparisons.

Acknowledgments

We thank Drs. B. E. Blaisdell, V. Brendel, C. Burge, and A. Campbell for valuable discussions and comments on the manuscript.

Supported in part by NIH Grants 2R01GM10452-31, 5R01HG00335-07 and NSF Grant DMS 9403553.

References

Benachenhou-Lahfa, N., P. Forterre, and B. Labedan. 1993. Evolution of glutamate dehydrogenase genes: Evidence for two paralogous protein families and unusual branching patterns of the archaebacteria in the universal tree of life. *J. Mol. Evol.* 36:335–346.

Blaisdell, B. E., A. M. Campbell, and S. Karlin. 1996. Similarities and dissimilarities of phage genomes. *Proc. Natl. Acad. Sci. USA* 93:5854–5859.

Bolshoy, A., P. McNamara, R. E. Harrington, and E. N. Trifonov. 1991. Curved DNA without A-A: experimental estimation of all 16 DNA wedge angles. *Proc. Natl. Acad. Sci. USA* 88:2312–2316.

Breslauer, K. J., R. Frank, H. Blöcker, and L. A. Marky. 1986. Predicting DNA duplex stability from the base sequence. *Proc. Natl. Acad. Sci. USA* 83:3746–3750.

Brown, J. R., Y. Masuchi, F. T. Robb, and W. F. Doolittle. 1994. Evolutionary relationships of bacterial and archaeal glutamine synthetase genes. *J. Mol. Evol.* 38:566–576.

Bult, C. J., O. White, G. J. Olsen, L. Zhou, R. D. Fleischmann, G. G. Sutton, J. A. Blake, L. M. Fitzgerald, R. A. Clayton, J. D. Gocayne, *et al.* 1996. Complete genome sequence of the methanogenic archaeon, *Methanococcus jannaschii.* Science 273:1058–1073.

Calladine, C. R., and H. R. Drew. 1992. *Understanding DNA.* Academic Press, San Diego.

Cox, E. C., and C. Yanofsky. 1967. Altered base ratios in the DNA of an Escherichia coli mutator strain. *Proc. Natl. Acad. Sci. USA* 58:1895–1902.

Creti, R., E. Ceccarelli, M. Bocchetta, A. M. Sanangelantoni, O. Tiboni, P. Palm, and P. Cammarano. 1994. Evolution of translational elongation factor (EF) sequences: Reliability of global phylogenies inferred from EF-1α (Tu) and EF-2 (G) proteins. *Proc. Natl. Acad. Sci.* USA 91:3255–3259.

Delcourt, S. G., and R. D. Blake. 1991. Stacking energies in DNA. *J. Biol. Chem.* 266:15160–15169.

Dickerson, R. E. 1992. DNA structure from A to Z. *Methods Enzymol.* 211:67–111.

Echols, H., and M. F. Goodman. 1991. Fidelity mechanisms in DNA replication. *Annu. Rev. Biochem.* 60:477–511.

Felsenstein, J. 1991. *PHYLIP (Phylogeny Inference Package), Version 3.4.* University of Washington, Seattle.

Gupta, R. S., and G. B. Golding. 1993. Evolution of HSP70 gene and its implication regarding relationships between Archaebacteria, Eubacteria and Eukaryotes. *J. Mol. Evol.* 37:573–582.

Gupta, R. S., and B. Singh. 1994. Phylogenetic analysis of 70 kD heat shock protein sequences suggests a chimeric origin for the eukaryotic cell nucleus. *Current Biol.* 4:1104–1114.

Himmelreich, R., H. Hilbert, H. Plagens, E. Pirkl, B.-C. Li, and R. Herrmann. 1996. Complete sequence analysis of the genome of the bacterium *Mycoplasma pneumoniae.* Nucl. Acids Res. 24:4420–4449.

Hunter, C. A. 1993. Sequence-dependent DNA structure. The role of base stacking interactions. *J. Mol. Biol.* 230:1025–1054.

Iwabe, N., K. Kuma, M. Hasegawa, S. Osawa, and T. Miyata. 1989. Evolutionary relationship of archaebacteria, eubacteria, and eukaryotes inferred from phylogenetic trees of duplicated genes. *Proc. Natl. Acad. Sci. USA* 86:9355–9359.

Josse, J., A. D. Kaiser, and A. Kornberg. 1961. Enzymatic synthesis of deoxyribonucleic acid VIII. Frequencies of nearest neighbor base sequences in deoxyribonucleic acid. *J. Biol. Chem.* 263:864–875.

Karlin S. 1994. Statistical studies of biomolecular sequences: Score based methods. *Philos. Trans. Roy. Soc. London Ser. B* 344:391–401.

Karlin, S., E. Mocarski, and G. A. Schachtel. 1994a. Molecular evolution of herpesviruses: Genomic and protein sequence comparisons. *J. Virol.* 68:1886–1902.

Karlin, S., and L. R. Cardon. 1994. Computational DNA sequence analysis. *Ann. Rev. Microbiology* 48:619–654.

Karlin, S., and I. Ladunga. 1994. Comparisons of eukaryotic genomic sequences. *Proc. Natl. Acad. Sci. USA* 91:12832–12836.

Karlin, S., Ladunga, I., and B. E. Blaisdell. 1994b. Heterogeneity of genomes: Measures and values. *Proc. Natl. Acad. Sci. USA* 91:12837–12843.

Karlin, S., and C. Burge. 1995. Dinucleotide relative abundance extremes: A genomic signature. *TIG* 11:283–290.

Karlin, S., G. Weinstock, and V. Brendel. 1995. Bacterial classifications derived from RecA protein sequence comparisons. *J. Bacteriol.* 177:6881–6893.

Kunkel, T. A. 1992. Biological asymmetries and the fidelity of eukaryotic DNA replication. *Bioessays* 14:303–308.

Lake, J. A. 1994. Reconstructing evolutionary trees from DNA and protein sequences: Paralinear distances. *Proc. Natl. Acad. Sci. USA* 91:1455–1459.

Nei, M. 1987. *Molecular Evolutionary Genetics.* Columbia Univ. Press, New York.

Olsen, G. J., C. R. Woese, and R. Overbeek. 1994. The winds of evolutionary change: Breathing new life into microbiology. *J. Bacteriol.* 176:1–6.

Phillips, J. P., J. Arnold, and R. Ivarie. 1987. The effect of codon usage on the oligonucleotide composition of the E. coli genome and identification of over and under represented sequences by Markov chain analysis. *Nucl. Acids Res.* 15:2627–2638.

Rivera, M. C., and J. A. Lake. 1992. Evidence that eukaryotes and eocyte prokaryotes are immediate relatives. *Science* 257:74–76.

Russell, G. J., P.M.B. Walker, R. A. Elton, and J. H. Subak-Sharpe. 1976. Doublet frequency analysis of fractionated vertebrate nuclear DNA. *J. Mol. Biol.* 108:1–28.

Russell, G. J., and J. H. Subak-Sharpe. 1977. Similarity of the general designs of protochordates and invertebrates. *Nature* 266:533–535.

Sandler, S. J., L. H. Satin, H. S. Samra, and A. J. Clark. 1996. *RecA* genes from three archaean species with putative protein products similar to Rad51 and Dmc1 proteins of the yeast *Saccharomyces cerevisiae. Nucl. Acids Res.*, 24:2125–2132.

Woese, C. R. 1987. Bacterial evolution. *Microbiol. Rev.* 51:221–271.

Woese, C. R., O. Kandler, and M. L. Wheelis. 1990. Towards a natural system of organisms: proposal for the domains Archaea, Bacteria, and Eucarya. *Proc. Natl. Acad. Sci. USA* 87:4576–4579.

20

Insertion Sequences and their Evolutionary Role

Michael Syvanen

Many bacterial insertion sequences produce no clear phenotype other than their mutator activity; thus, it has been proposed that insertion sequences function to produce genetic variability (Cohen, 1976; Nevers and Saedler, 1977; Reanney, 1976). This selection hypothesis maintains that the trait of genetic variability is the function insertion sequences serve for the organisms that harbor them, and that these sequences are fixed by natural selection each time that a mutation induced by these elements is selected.

The selection hypothesis is generally accepted by microbiologists as an explanation for bacterial insertion sequences found in association with antibiotic resistance genes, many transposons, and the large resistance plasmids (*see also* Chapter 4). The correlation between the highly selected genes and the presence of tightly linked insertion sequences provides a convincing case.

While insertion sequences in plasmids seem to be highly selected, it is more difficult to argue that insertion sequences in the bacterial chromosome serve any useful purpose. In this case, as with transposable elements in eukaryotic genomes, it may not be too unreasonable to hypothesize that the insertion sequences act as genomic parasites or selfish DNA (Doolittle and Sapienza, 1981). Indeed, the selection hypothesis has been strongly resisted by those less familiar with microorganisms. For example, Mayr (1982) criticizes the selection hypothesis as a "teleological answer [where] seemingly functionless DNA is stored up in order to have it available in future time of need," and Hartl *et al.* (1986) wonders "how some future benefit accruing to such elements could be the driving force behind their evolution." While Mayr (1982) objects to the notion that a complex element could be maintained because it helps its host evolve, Hartl *et al.* (1986) seem to question both how insertion sequences could now contribute to the adaptive evolution of the hosts in which they reside, and how they could have evolved up to the point where they could be selected for their adaptive contribution. The selection hypothesis does not attempt to answer the latter question. It makes no statement about how these sequences arose and does not claim that they have served the same, if any, function throughout evolutionary history. The hypothesis does claim that these elements have contributed to adaptive evolution at least once in the past, and may still possess mutator activities with the potential to induce variants that can be fixed in the future. This hypothesis is not teleological. Stanley (1979) stressed a view of evolution as a race among evolving species; those species that give rise to successful variants at the highest rate will be favored. It seems reasonable to assume that a species that has given rise to successful variants will in some cases carry along with it the mutational mechanism that gave rise to that variant. Such a mechanism may give rise to future variants. If the benefits of carrying these variants outweigh the disadvantage of carrying these mutators, we have a simple, non-teleological explanation for the maintenance of insertion sequences based on their function as mutators. I have argued elsewhere (Syvanen, 1984) that genomic insertion sequences may possibly be maintained during evolutionary stasis by their ability to replicate themselves and spread within genomes, but that they occasionally become fixed through the beneficial mutations they cause. This is a version of the selection hypothesis. As with any evolutionary hypothesis, evidence in its favor is mostly indirect.

Direct Selection

There are a number of examples in which direct selection for mutants results in insertion sequence-induced mutations. There are the early cases of IS1 and IS5 induced mutations for activation of the cryptic bgl operon, IS1 induced reversion of promoterless *gal* mutants, and the finding that IS10 carrying strains of *Escherichia coli* win in chemostat competition experiments with non-IS10 carrying strains because of a beneficial mutation that IS10 elicits, as has been reviewed previously (Syvanen, 1984; Arber 1993; Blot, 1994). Recent examples include the Van der Ploeg *et al.* (1995) demonstration that mutants of *Xanthobacter autotrophicus,* which are resistant to killing by monobromoacetate, arise because of an insertion sequence that causes overproduction of a haloacid dehalogenase. And Hall and Xu (1992) have shown that IS186 activates the cryptic *asc* operon of *E. coli.* A variety of mechanisms by which insertion sequences activate gene expression is documented; these usually entail activation of transcription and consequent expression. Most of these findings involve transposition of bacterial insertion sequences which activate genes by providing transcriptional signals. These observations have led to the argument that part of the function of bacterial insertion sequences is to induce such regulatory mutations. This is a major basis for the selection hypothesis.

Environmental Stress

Another line of evidence supporting an evolutionary role for insertion sequences comes from experiments based on the environmental stress hypothesis. This hypothesis states that *if* environmental stress triggers insertion-sequence-mediated mutations and induces new genetic adaptations that aid survival, then it may be argued that insertion sequences *function* to provide mutations in times of need. The first formulation of the environmental stress hypothesis was the genomic stress hypothesis (McClintock, 1984), which postulated that DNA damage directly stimulates transposition. This notion seems quite plausible, since transposons use elaborate regulatory mechanisms to control their movement, and analogous regulatory mechanisms in bacteriophages and bacteria are known to respond to environmental stimulations such as UV light. However, despite a number of surveys, DNA damaging agents do not appear to stimulate bacterial insertion sequence movement. [One possible exception is the report by Kuan and Tessman (1991, 1992), indicating that Tn5 transposition responds to the UV-stimulated SOS repair system. This remains an isolated example.]

A more promising variant of the environmental stress hypothesis involves stress caused by nutrient deprivation. The idea that nutrient deprivation may stimulate insertion sequence movement derives from two different lines of evidence: (1) the long-term stab studies and (2) adaptive mutation.

Long-term stab studies

A number of observations have led many to believe that transposition events are higher in cells stored in long-term stabs than in exponentially growing cells. We may presume that cells stored thusly exist primarily in stationary phase. The conclusion that transposition rates are higher under such conditions has been supported by casual observations of workers who have found that strains harboring insertion sequences frequently pick up unselected secondary insertions during storage in room-temperature stabs. In the most thorough characterization of this observation, Naas *et al.* (1994, 1995) examined a variety of rearrangements induced by the native insertion sequences in *E. coli* strain K-12. In particular, they surveyed a strain of W3110 that had been stored in a room-temperature stab for 30 years. From 118 single colonies recovered, they found 51 different genotypes with respect to the number of different ISs and their chromosomal locations. Most of the variation is seen with IS5 and IS30 and in decreasing amounts with other ISs. In these long-term stabs, rearrangements are the rule, although ancestral loci can be identified and occasionally recovered. This work makes a good case for stationary-phase growth activating movement. However, there are a number of unanswered questions. It is difficult to compare the transposition frequency from growing cells (measured in events per generation) with that from stationary phase cells (measured in events per unit time).

In the reports cited above, Naas *et al.* estimated a mutation frequency of $2\text{-}8 \times 10^{-6}$ per cell per hour based on a minimal replacement tree of the 51 different genotypes. There are a number of problems with these estimates. One problem is the uncertainty concerning the number of generations that occur during storage (the fact that so many different genotypes were found argues that this number is low, since more fit individuals would otherwise overrun the population). Another problem is the uncertainty concerning whether any of the mutations are selected or not. Finally, a comparable frequency from exponentially grown cells was not available. These issues were addressed but not resolved. There seems to be little doubt that transposition events do in fact occur in stationary cells at a relatively high rate, and this is probably significant, given that so many other cellular processes are turned off under these conditions. The results are suggestive, but not conclusive.

Adaptive Mutation

Another body of observations supporting the notion that nutrient deprivation stimulates IS movement is seen in the work described as adaptive mutation. It should be noted, however, that the mechanism responsible for increased activity is very likely unrelated to the mechanism responsible for the results seen in the long-term stab studies. The observations on adaptive mutation were initiated by Cairns (1990), in which he showed that starvation of Lac$^-$ *E. coli* on a minimal lactose medium-stimulated reversion of the mutant to Lac$^+$. Initially, this was presented as support for a Lamarkian-environmentally directed mutation hypothesis, though later Mittler and Lenski (1990) appear to have established that spontaneous mutations in general are stimulated by prolonged cell starvation. The simplest mechanism for adaptive mutation would be that the nutrient-deprived cells are stimulated to produce mutations at random, but that the cells with these mutations generally die unless growth occurs. The advantage of these adaptive mutation studies is that quantitation of mutation frequencies (and hence comparison of starved cells to growing cells) are possible. So, for example, Foster (1993) has shown that the mutation frequency from growing cells is clearly lower (in mutations per cell *per generation*) than from starved cells (in mutations per cell *per hour*).

The relevance of adaptive mutation studies to the current subject is that these studies have been extended to include a variety of insertion sequence induced events. Hall (1988) showed that precise excision of an IS*103* in the bgl operon occurred predominantly while the cells were starving on a salicin containing medium, rather than during prior growth. Shapiro (1989) showed that deletions induced by phage Mu occurred preferentially in resting bacteria, and Shapiro and Higgins (1988) have shown that spontaneous phage Mu *dlac* insertions occurred late during colony formation. This later experiment measured transposition events. By measuring phage Mu excision and deletion, Mittler and Lenski (1990) showed that mutagenesis was general, occurring not only in the genes needed for adaptation. Foster and Cairns (1994) have also documented that phage Mu excision is stimulated during stationary phase growth, and have concurred with the Mittler and Lenski result.

The mechanisms employed that stimulate precise excision and Mu transposition during starvation remain unknown, but it is very likely that the bacteria itself regulates this adaptive response, possibly independently of any IS-mediated regulation. The adaptive response is not as general as was initially believed. Radicella *et al.* (1995) and Galitski and Roth (1995) have shown that the adaptive response is dependent upon the conjugal transfer mediated by the F plasmid. In addition, it has also been shown that adaptive mutation is dependent on a functioning *recABCD* pathway (Harris *et al.*, 1994; Rosenberg, 1994). These recent studies implicating conjugal transfer were performed by measuring reversion of point mutations, but are likely to apply to the earlier results on insertion sequences as well. This follows because most of the studies employed cells that were F^+. [And in the Mu *dlac* study (Shapiro and Higgins, 1988), the only one to employ F^- cells, many of the insertion events must have occurred during colony growth; therefore, this may not be an example of adaptive mutation.] In addition, in an earlier study, we documented an F-factor stimulation of precise excision of Tn5 and Tn10 that seems relevant to the current subject (Syvanen *et al.* 1986). In these studies, we showed that conjugal transfer from cell to cell stimulated both recA-dependent recombination and precise excision. Though the DNA involved in the excision event need not transfer, entry of DNA into the recipient triggered stimulation by some unknown mechanism. These studies were coincidentally carried out on stationary phase cells, though at the time, we did not interpret these results as adaptive mutation. If adaptive mutation involving insertion sequences seen in these experiments require conjugal transfer, then we would have to conclude that this phenomena is probably unrelated to the results from the long-term stab studies, since those were performed with F^- *E. coli.*

Although evidence suggests that nutrient deprivation causes increased insertion sequence movement in long-term storage in stabs and during starvation of F^+ cells on plates, the mechanisms responsible for the increased movement likely differ.

IS Distribution: Evolutionary Conclusions

The abundance and distribution of insertion sequences in the genomes of *E. coli* and *S. typhimurium* from a variety of different strains have been characterized (Stanley *et al.*, 1992; Sawyer *et al.*, 1987; Sawyer and Hartl, 1986; Hall *et al.*, 1989). These results show a great variability in both number and type of ISs in closely related strains, but it has not been possible, using population genetic theory, to distinguish whether these elements were selected on the basis of their beneficial nature, or whether they behave as genomic parasites. In a few examples, a case can be made that copy number is likely regulated and hence prevents an undesirable genomic load.

Even though studies on the distribution of ISs in differential strains and species have not yielded conclusions about whether or not they are selected at the level of the whole organism, these studies have yielded interesting evolutionary and phylogenetic information. Lawrence *et al.* (1989, 1992) compared the DNA sequence for three different ISs cloned from a variety of enteric bacterial species. They reported that the amount of IS3 sequence divergence correlated well with the amount of divergence among chromosomal genes. That is, the IS3 gene tree was congruent with the "species" tree. At the same time, they found incongruities between the IS1 and IS30 gene trees compared to normal chromosomal gene trees. A simple explanation for this latter result is that IS1 and IS30 are frequently involved in interspecies horizontal gene transfer. These same studies found that IS3 displayed much less than the expected divergence when they were sampled from different strains of *E. coli*, again indicating frequent horizontal interstrain transfers and subsequent gene conversion events. At the same time that a given insertion sequence shows unusual sequence conservation, the number of any given insertion sequences and their location is extremely variable among the different strains of *E. coli* (Sawyer et al., 1987). The situation with *S. typhimurium,* on the other hand, is quite different. First, this enteric has only one IS—IS200—and not the great diversity as seen with *E. coli.* In addition, the relationships among the different *Salmonella* strains, as judged by IS200 fingerprint information, are relatively congruent with other chromosomal markers (Stanley *et al.*, 1992, 1993) (*see also* Chapter 22).

A summary of the studies on the distribution of insertion sequences illustrates the great variability in the properties of the different elements. In addition, these studies seem to show that many different species of enteric bacteria share among themselves their insertion sequences but that *Salmonella* is not a member of the club. This is paradoxical because there seem to be numerous claims of horizontal gene transfer involving other genes for the *Salmonella* (reviewed by Syvanen, 1994).

References

Arber, W. 1993. Evolution of prokaryotic genomes. *Gene* 135:49–56.

Blot, M. 1994. Transposable elements and adaptation of host bacteria. *Genetics* 93:5–12.

Cairns, J. 1990. Causes of mutation and mu excision. *Nature* 345:213.

Cohen, S. N. 1976. Transposable genetic elements and plasmid evolution. *Nature* 263:731–8.

Doolittle, W. F., and C. Sapienza. 1980. Selfish genes, the phenotype paradigm and genome evolution. *Nature* 284:601–3.

Foster, P. L. 1993. Adaptive mutation: the uses of adversity. *Annual Review of Microbiology* 47:467–504.

Foster, P. L., and J. Cairns. 1994. The occurrence of heritable Mu excisions in starving cells of *Escherichia coli. EMBO Journal* 13:5240–4.

Galitski, T. and J. R. Roth. 1995. Evidence that F plasmid transfer replication underlies apparent adaptive mutation. *Science* 268:421–3.

Hall, B. G. 1988. Adaptive evolution that requires multiple spontaneous mutations. I. Mutations involving an insertion sequence. *Genetics* 120:887–97.

Hall, B. G., L. L. Parker, P. W. Betts, R. F. DuBose, S. A. Sawyer, and D. L. Hartl. 1989. IS103, a new insertion element in *Escherichia coli:* characterization and distribution in natural populations. *Genetics* 121:423–31.

Hall, B. G., and L. Xu. 1992. Nucleotide sequence, function, activation, and evolution of the cryptic asc operon of *Escherichia coli* K12. *Molecular Biology and Evolution* 9:688–706.

Harris, R. S., S. Longerich, and S. M. Rosenberg. 1994. Recombination in adaptive mutation. *Science* 264:258–60.

Hartl, D. L., M. Medhorn, L. Green, and D. E. Dykhuizen. 1986. The evolution of DNA sequences in *E. coli. Philos. Trans. R. Soc. London B. Biol. Sci.* 312:191–204.

Kuan, C. T., and I. Tessman. 1992. Further evidence that transposition of Tn5 in *Escherichia coli* is strongly enhanced by constitutively activated RecA proteins. *Journal of Bacteriology* 174:6872–7.

Kuan, C. T., S. K. Liu, and I. Tessman. 1991. Excision and transposition of Tn5 as an SOS activity in *Escherichia coli. Genetics* 128:45–57.

Lawrence, J. G., H. Ochman, and D. L. Hartl. 1992. The evolution of insertion sequences within enteric bacteria. *Genetics* 131:9–20.

Lawrence, J. G., D. E. Dykhuizen, R. F. DuBose, and D. L. Hartl. 1989. Phylogenetic analysis using insertion sequence fingerprinting in *Escherichia coli. Molecular Biology and Evolution* 6:1–14.

McClintock, B. 1984. The significance of responses of the genome to challenge. *Science* 226:792–801.

Mayr, E. 1982. *The growth of biological thought.* Harvard U. Press, Cambridge, MA.

Mittler, J. E., and R. E. Lenski. 1990. New data on excisions of Mu from *E. coli* MCS2 cast doubt on directed mutation hypothesis. *Nature* 344:173–5.

Naas, T., M. Blot, W. M. Fitch, and W. Arber. 1994. Insertion sequence-related genetic variation in resting *Escherichia coli* K-12. *Genetics* 136:721–730.

Naas, T., M. Blot, W. M. Fitch, and W. Arber. 1995. Dynamics of IS-related genetic rearrangements in resting *Escherichia coli* K-12. *Molecular Biology and Evolution* 12:198–207.

Nevers, P., and H. Saedler. 1977. Transposable genetic elements as agents of gene instability and chromosomal rearrangements. *Nature* 268:109–15.

Radicella, J. P., P. U. Park, and M. S. Fox. 1995. Adaptive mutation in *Escherichia coli:* a role for conjugation. *Science* 268:418–20.

Reanney, D. 1976. Extrachromosomal elements as possible agents of adaptation and development. *Bacteriological Reviews* 40:552–90.

Rosenberg, S. M., S. Longerich, P. Gee, and R. S. Harris. 1994. Adaptive mutation by deletions in small mononucleotide repeats. *Science* 265:405–7.

Sawyer, S., and D. Hartl. 1986. Distribution of transposable elements in prokaryotes. *Theoretical Population Biology* 30:1–16.

Sawyer, S. A., D. E. Dykhuizen, R. F. DuBose, L. Green, T. Mutangadura-Mhlanga, D. F. Wolczyk, and D. L. Hartl. 1987. Distribution and abundance of insertion sequences among natural isolates of *Escherichia coli. Genetics* 115:51–63.

Shapiro, J. A., and N. P. Higgins. 1989. Differential activity of a transposable element in *Escherichia coli* colonies. *Journal of Bacteriology* 171:5975–86.

Stanley, J., N. Baquar, and E. J. Threlfall. 1993. Genotypes and phylogenetic relationships of *Salmonella typhimurium* are defined by molecular fingerprinting of IS200 and 16S rrn loci. *Journal of General Microbiology* 139:1133–40.

Stanley, J., A. Burnens, N. Powell, N. Chowdry, and C. Jones. 1992. The insertion sequence IS200 fingerprints chromosomal genotypes and epidemiological relationships in *Salmonella heidelberg. Journal of General Microbiology* 138:2329–36.

Stanley, S. M. 1979. A theory of evolution above the species level. *Proc. Natl. Acad. Sci. USA* 72:646–650.

Syvanen, M., J. D. Hopkins, T. J. Griffin, T. Y. Liang, K. Ippen-Ihler, and R. Kolodner. 1986. Stimulation of precise excision and recombination by conjugal proficient F plasmids. *Molecular and General Genetics* 203:1–7.

Syvanen, M. 1984. The evolutionary implications of mobile genetic elements. *Annual Review of Genetics* 18:271–93.

Syvanen, M. 1994. Horizontal gene transfer: Evidence and possible consequences. *Annual Review of Genetics* 28:237–61.

van der Ploeg, J., M. Willemsen, G. van Hall, and D. B. Janssen. 1995. Adaptation of *Xanthobacter autotrophicus* GJ10 to bromacetate due to activation and mobilization of the haloacetate dehalogenase gene by insertion element IS1247. *Journal of Bacteriology* 177:1348–56.

21

Multiplicity of Ribosomal RNA Operons in Prokaryotic Genomes

Thomas M. Schmidt

Introduction

While the majority of genes in prokaryotic microorganisms are present at a single locus on the chromosome, as many as 13 copies of each of the ribosomal RNA genes (*rrn*) have been detected. However, it is difficult to correlate *rrn* copy number with any known selective advantage. Multiple copies may be key to the development of resistance to antibiotics that interact with ribosomal RNAs through duplication and subsequent mutations of the duplicated gene. It has also been suggested that multiple copies of the ribosomal RNA genes are required for rapid cellular growth. In addition to the potential physiological implications of multiple *rrn* operons, their repeated and highly conserved gene order and sequence make them viable targets for homologous recombination, which could lead to chromosomal rearrangements. In this chapter, the involvement of *rrn* loci in chromosome structure and organismal fitness is considered through an examination of the organization and distribution of rRNA operons in prokaryotic chromosomes.

Organization and Transcription of *rrn* Genes

Clustering of *rrn* genes into an operon is common (Figure 21.1a). Based on a phylogenetic comparison of the linkage of *rrn* genes in Bacteria, Archaea and Eukarya, the occurrence of rRNA genes in an operon is most likely the ancestral state (Pace and Burgin, 1990). Since equimolar amounts of the 16S, 23S, and 5S rRNAs are present in a functional ribosome, it is tempting to speculate that the assembly of these genes into an operon is required to maintain balanced gene expression. However, alternative organizations also exist with partial or no linkage

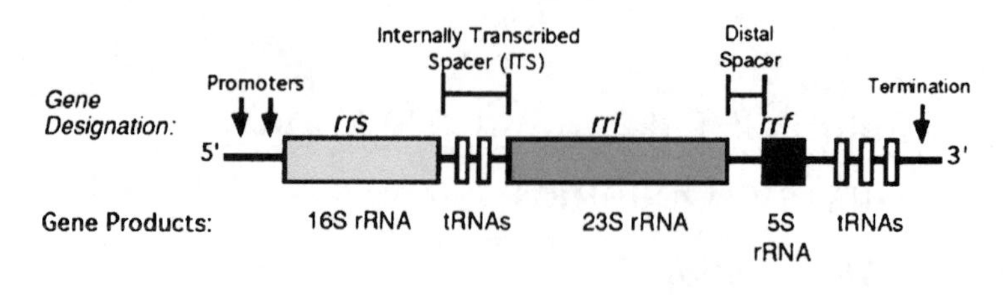

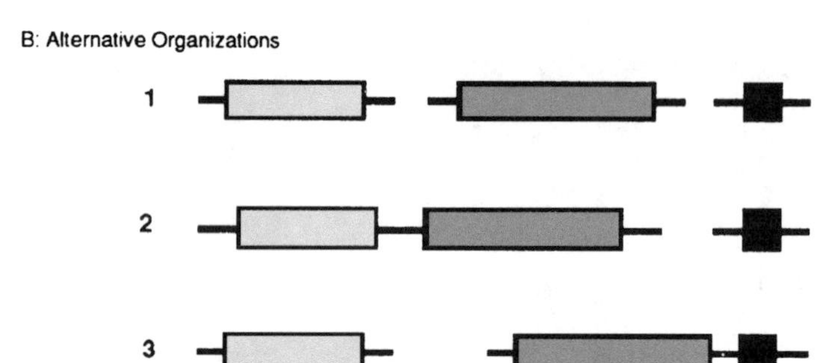

Figure 21-1. Chromosomal organizations of *rrn* genes. The most commonly observed *rrn* operon contains coding regions for all of the ribosomal RNAs and some tRNAs (A). Other arrangements have also been observed (B). For example, the *rrn* genes in *Thermoplasma acidophilum* are separated on the chromosome without any apparent linkage (B1) (Ree and Zimmermann 1990). In *Leptospira interogans, rrf* is separated from *rrs* and *rrl* (B2) (Fukunaga *et al.*, 1990) while *rrs* is transcribed from an isolated transcription unit in *Thermus thermophilus* (B3) (Winder and Rooney 1970).

among *rrn* genes (Fig. 21.1b), indicating that there is not a strict requirement for physical organization as an operon.

Transcription of *rrn* operons generally proceeds away from the origin of replication and is tightly regulated in response to growth rate and the stringent response (Wagner, 1994). The *rrn* transcript is processed by RNases to yield functional rRNAs and transfer RNAs (tRNAs). One of the regions cleaved from the primary transcript is located between the gene encoding the 16S rRNA (*rrs*) and the 23S rRNA (*rrl*), it is known as the internally transcribed spacer (ITS). Genes encoding tRNAs are commonly found in the ITS and in the region distal to the 5S rRNA-encoding gene (*rrf*). In a few organisms, the ribosomal RNA genes are interrupted by introns. Introns are present in *rrl* of several Archaea and there is a 713 bp intron in *rrs* of *Pyrobaculum aeorphilum* (Burggraf *et al.*, 1993). The structural rRNA genes can also be interrupted by elements that are later excised without subsequent ligation of the rRNA. These sequences are known as intervening sequences and have been observed in *rrl* of *Salmonella* spp. (Burgin *et al.*, 1990) and *Leptospira* spp. (Ralph *et al.*, 1993).

Occurrence and Phylogenetic Distribution of Multiple *rrn* Operons

Considerable variation exists in the number of *rrn* operons in prokaryotes (Table 21-1), ranging from a single *rrn* operon in some species of *Mycoplasma* (Amikan *et al.*, 1982) *Mycobacteria* (Bercovier *et al.*, 1986) and *Rickettsia* (Anderson *et al.*, 1995) to as many as 13 rRNA operons in *Clostridium beijerinckii* (Wilkinson and Young 1995). In the Archaea, the highest reported copy number for *rrn* operons is four in *Methanococcus vaniellii* (Jarsch *et al.*, 1983) and in several strains of the extreme halophiles (Sanz *et al.*, 1988). There is tremendous variation found in the number of copies of *rrn* genes in eukaryotes, with as many as tens of thousands of tandemly repeated 18S and 28S rRNA genes in copepods (Wyngaard *et al.,* 1995). Although the variation in prokaryotic *rrn* number is less dramatic

Table 21-1. Multiplicity of rRNA operons in phylogenetically diverse prokaryotes.

Phylogenetic Group	Genus species	rRNA operon copy number	Reference
Euryarchaeota	*Haloarcula californiae*	4	Sanz *et al.*, 1988
	Methanobacterium thermoautotrophicum	2	Neumann *et al.*, 1983
	Methanococcus vannielii	4	Jarsch *et al.*, 1983
	Nantronobacterium magadii	1	Lodwick *et al.*, 1991
	Thermoplasma acidophilum	1	Ree & Zimmerman, 1990
	Thermococcus celer	1	Neumann *et al.*, 1983
Creanarchaeota	*Pyrobaculum aerophilum*	1	Burggraf *et al.*, 1993
	Sulfolobus acidocaldarius	1	Neumann *et al.*, 1983
Alpha Proteobacteria	*Rhizobium meliloti*	3	Honeycutt *et al.*, 1993
	Rhodobacter capsulatus	4	Fonstein and Haselkorn, 1993
	Rhodobacter sphaeroides	3	Choudhary *et al.*, 1994
	Rickettsia prowazekii	1	Anderson *et al.*, 1995
Gamma Proteobacteria	*Escherichia coli*	7	Elwood and Nomura, 1980
	Haemophilus influenzae	6	Fleischmann *et al.*, 1995
	Pseudomonas aeruginosa	4	Hartmann *et al.*, 1986
Delta Proteobacteria	*Campylobacter jejuni*	3	Kim *et al.*, 1993
	Helicobacter pylori	3	Taylor *et al.*, 1992
	Myxococcus xanthus	4	He *et al.*, 1994
High G+C Gram positive	*Mycobacterium phlei*	2	Bercovier *et al.*, 1986
	Mycobacterium tuberculosis	1	Bercovier *et al.*, 1986
	Streptomyces ambofaciens	4	Pernodet *et al.*, 1989
	Streptomyces lividans TK21	6	Suzuki *et al.*, 1988
Low G+C Gram Positive	*Acholeplasma laidlawii*	2	Amikan *et al.*, 1982
	Bacillus subtilis	10	Loughney *et al.*, 1982
	Clostridium perfringes	9	Canard and Cole, 1989
	Clostridium beijerinckii	13	Wilkinson and Young, 1995
	Lactobacillus plantarum	5	Chevallier *et al.*, 1994
	Listeria monocytogenes	6	Michel and Cossart, 1992

than in eukaryotes, the differences must be considered in terms of the relative size of prokaryotic genomes and the tendency for most prokaryotic genes to be present in a single copy.

When multiple *rrn* operons are present in a prokaryotic chromosome, they are generally located within the half of the chromosome that contains the origin of replication (*see* chapter 7). This positioning results in a relative increase in the number of *rrn* loci in rapidly dividing cells through the gene dosage effect. For the rRNA operons of *E. coli,* this translates into the difference between an average of 7.9 copies per genome at a doubling time of 100 min. and an average of 9.5 copies per genome at a doubling time of 24 min. The difference is exaggerated to 12.4 and 36 *rrn* operons per cell, respectively, when the effect of multiple chromosomes per cell is considered (Bremer and Dennis, 1987).

The lack of convergence on an optimal number of copies of rRNA genes in prokaryotes is striking. It is possible that the number of *rrn* copies is determined exclusively by an organism's evolutionary history. However, if *rrn* copy number was strictly limited by an organism's evolutionary history, a strong concordance between phylogeny and *rrn* copy number would be expected. Within each of the phylogenetic assemblages presented in Table 21.1 there is considerable variation in the number of *rrn* operons. The apparent lack of phylogenetic conservation indicates that evolutionary history does not define strictly the number of chromosomal copies of the ribosomal RNA genes, suggesting that *rrn* copy number has been established in response to environmental adaptation.

Potential Consequences of *rrn* Operon Multiplicity

Growth Rate

Multiple copies of *rrn* genes may be needed to support the high concentrations of ribosomes required by rapidly dividing cells. Evidence for a causal relationship between *rrn* copy number and growth rate is found in studies of the bacterial genus *Mycobacterium.* Strains of the genus *Mycobacterium* are separated phylogenetically into two groups which are congruent with the phenotypic division of the genus into fast- and slow-growing isolates. The fast-growing organisms possess two *rrn* operons per genome, while the slow growers contain one (Bercovier *et al.*, 1986). However, *E. coli* strains do not require their full complement of seven ribosomal RNA operons to reach maximal growth rates. Inactivation of two of the seven rRNA operons does not result in a concomitant reduction in maximal growth rate. This has been attributed to increased expression of the remaining intact operons (Condon *et al.*, 1993). In terms of absolute growth rates, the extreme thermophile *Thermococcus celer* achieves a maximal doubling time of less than 50 min. (Zillig 1992) with a single *rrn* copy, thus surpassing the maximal growth rates of many organisms with multiple rRNA operons. Taken together, these results suggest that there is not a simple relationship between the number of copies of ribosomal RNA genes and maximal growth rate.

Recently, the capacity for rapid response to an upshift in nutrients was correlated with the number of intact *rrn* operons in *E. coli* (Condon *et al.*, 1995). If multiple copies of ribosomal RNA genes provide a selective advantage in rapidly fluctuating environments, organisms with multiple operons might be expected to dominate those environments. This raises the intriguing possibility of explaining the distribution of microbial populations based on a favorable match between *rrn* copy number and environmental stability.

Recombination

Hill and colleagues (1990) reviewed mechanisms for generating chromosomal rearrangements through unequal cross-over between *rrn* operons resulting in duplications, deletions, transpositions, and inversions. Inversions of the genome between *rrn* loci have been documented in populations of *E. coli* (Hill and Harnish, 1981) and *Salmonella typhimurium* (Liu and Sanderson, 1995). Inversions in *E. coli* were stable when the geometric relationships between *rrn* operons and the origin of replication were preserved, whereas disruption of the relative placement rendered inversions unstable. The gene dosage effect for genes located near the origin of replication may have provided selection against inversions that disrupted the relative placement of *rrn* operons.

When multiple ribosomal RNA operons are present, there are frequently sequence variations among operons. The greatest variability is generally present in the upstream regions of the operons as well as within the spacer regions, although some variation within the structural genes has also been observed. These sequence variations may play a role in reducing the rate of recombination between operons.

Despite the potential for duplication and deletion of rRNA operons through recombination, there is little evidence that there is any variation in *rrn* copy number within a species. In an expansive study to examine chromosomal structure in the human pathogen *Listeria monocytogenes,* Bruce and colleagues (1995) examined the distribution of *rrn* operons in 1,346 isolates of *L. monocytogenes.* Genomic DNA digested with EcoR1 and probed with rRNA transcripts from *E. coli* revealed 50 distinct patterns indicative of variation of the genome structure, but there was no evidence of alteration in *rrn* copy number. In regards to potential variations of *rrn* copy number within a population, a series of publications described the variation of *rrn* operon copy number in *Bacillus subtilus* (Widom *et al.*, 1988). However, a later study demonstrated that the reported alteration of *rrn* copy number was due to the masking of restriction sites resulting from partially purified DNA (Waterhouse and Glover, 1993).

Perspective

Due to evolutionary constraints on the ribosomal RNAs, the genes that encode them are recognizable in every form of life so far examined and form the basis of a phylogenetic framework into which all of cellular life can be placed (Woese *et al.*, 1990). The *rrn* genes are usually organized into an operon which is likely the ancestral arrangement. Multiple copies of *rrn* operons are frequently present on the bacterial chromosome despite the general tendency towards single copies of most genes. Variation of *rrn* copy number among phylogenetically related organisms suggests that evolutionary history does not play a definitive role in establishing the number of *rrn* operons in prokaryotes, and leaves open the possibility that *rrn* copy number is of adaptive significance in bacterial life histories. Conservation of gene order and sequence in the *rrn* operons makes them likely sites for homologous recombination. The frequency and significance of genomic rearrangements resulting from recombination amongst *rrn* genes remain to be determined, but the consequences of multiple *rrn* genes in terms of organismal fitness are beginning to be understood.

Acknowledgments

Research in the author's laboratory is supported by grants from the Department of Energy (DEFG0393ER61684) and the Environmental Protection Agency (CR822014010), along with funds from the NSF Center for Microbial Ecology (BIR9120006). The author acknowledges Daniel H. Buckley and Bonnie J. Bratina for their thoughtful reviews of this manuscript.

References

Amikan, D., S. Razin, and G. Glaser. 1982. Ribosomal RNA genes in Mycoplasma. *Nucleic Acids Research* 10:4215–4221.

Anderson, S.G.E., A. Zomorodipour, H. H. Winkler, and C. G. Kurland. 1995. Unusual organization of the rRNA genes in *Rickettsia prowazekii*. *J. Bacteriol.* 177:4171–4175.

Bercovier, H., O. Kafri, and S. Sela. 1986. *Mycobacteria* possess a surprisingly small number of ribosomal RNA genes in relation to the size of their genome. *Biochem. Biophys. Res. Commun.* 136:1136–1141.

Bremer, H., and P. P. Dennis. 1987. Modulation of chemical composition and other parameter of the cell by growth rate. In Escherichia coli *and* Salmonella typhimurium: *Cellular and Molecular Biology*. F. C. Niedhardt, J. L. Ingraham, K. B. Low, B. Magasanik, M. Shaechter, H. E. Umbarger. American Society for Microbiology, Washington, D.C.

Bruce, J. L., R. J. Hubner, E. M. Cole, C. I. McDowell, and J. A. Webster. 1995. Sets of EcoRI fragments containing ribosomal RNA sequences are conserved among different strains of *Listeria monocytogenes*. *Proc. Natl. Acad. Sci. USA* 92:5229–5233.

Burggraf, S., N. Larsen, C. R. Woese, and K. O. Stetter. 1993. An intron within the 16S ribosomal RNA gene of the archaeon *Pyrobaculum aerophilum*. *Proc. Natl. Acad. Sci. USA* 90:2547–2550.

Burgin, A. B., K. Parodos, D. J. Lane, and N. R. Pace. 1990. The excision of intervening sequences from Salmonella 23S ribosomal RNA. *Cell* 60:405–414.

Canard, B., and S. T. Cole. 1989. Genome organization of the anaerobic pathogen *Clostridium perfringens*. *Proc. Natl. Acad. Sci. USA* 86:6676–6680.

Chevallier, B., J.-C. Hubert, and B. Kammerer. 1994. Determination of chromosome size and number of rrn loci in *Lactobacillus plantarum* by pulsed-field gel electrophoresis. *FEMS Letters* 120:51–56.

Choudhary, M., C. Mackenzie, K. S. Nereng, E. Sodergren, G. M. Weinstock, and S. Kaplan. 1994. Multiple chromosomes in bacteria: structure and function of chromosome II of *Rhodobacter sphaeroides* 2.4.1T. *J. Bacteriol.* 176:7694–7702.

Condon, C., S. French, C. Squires, and C. L. Squires. 1993. Depletion of functional ribosomal RNA operons in Escherichia coli causes increased expression of the remaining intact copies. *EMBO* 12:4305–4315.

Condon, C., D. Liveris, C. Squires, I. Schwartz, and C. L. Squires. 1995. rRNA operon multiplicity in *Escherichia coli* and the physiological implications of *rrn* inactivation. *J. Bacteriol.* 177:4152–4156.

Ellwood, M., and M. Nomura. 1980. Deletion of a ribosomal ribonucleic acid operon in *Escherichia coli*. *J. Bacteriol.* 143:1077–1080.

Fleischmann, R. D. *et al.* 1995. Whole-genome random sequencing and assembly of *Haemophilus influenzae* Rd. *Science* 269:496–512.

Fonstein, M., and R. Haselkorn. 1993. Chromosomal structure of *Rhodobacter capsulatus* strain SB1003: Cosmid encyclopedia and high-resolution physical and genetic map. *Proc. Natl. Acad. Sci. USA* 90:2522–2526.

Fukunaga, M., T. Masuzawa, N. Okuzako, I. Mifuchi, and Y. Yanagihara. 1990. Linkage of Ribosomal RNA Genes in *Leptospira*. *Microbiol. Immunol.* 34:565–573.

Hartmann, R. K., H. Y. Toschka, N. Ulbrich, and V. A. Erdmann. 1986. Genomic organization of rDNA in *Pseudomonas aeruginosa*. *FEBS Letters* 195:187–193.

He, Q., C. Hongwu, A. Kuspa, Y. Cheng, D. Kaiser, and L. J. Shimkets. 1994. A physical map of the Myxococcus xanthus genome. *Proc. Natl. Acad. Sci. USA* 91:9584–9587.

Hill, C. W., and B. W. Harnish. 1981. Inversions between ribosomal RNA genes of *Escherichia coli*. *Proc. Natl. Acad. Sci. USA* 78:7069–7072.

Hill, C. W., S. Harvey, and J. A. Gray. 1990. Recombination between rRNA genes in *Escherichia coli* and *Salmonella typhimurium*. In *The Ribosome,* W. E. Hill, A. Dahlberg, R. A. Garrett, P. B. Moore, D. Schlessinger, J. R. Warner, eds. pp. 335–339. American Society for Microbiology. Washington, D.C.

Honeycutt, R. J., M. McClelland, and B.W.S. Sobral. 1993. Physical map of the genome of *Rhizobium meliloti* 1021. *J. Bacteriol.* 175:6945–6952.

Jarsch, M., J. Altenbuchner, and A. Böck. 1983. Physical organization of the genes for ribosomal RNA in *Methanococcus vannielii. Molecular and General Genetics* 189:41–47.

Kim, N. W., R. Lombardi, H. Bingham, E. Hani, H. Louie, D. Ng, and V. L. Chan. 1993. Fine mapping of the three rRNA operons on the updated genomic map of *Campylobacter jejuni* TGH9011 (ATCC 43431). *J. Bacteriol.* 175:7468–7470.

Liu, S.-L., and K. E. Sanderson. 1995. I-*Ceu*I reveals conservation of the genome of independent strains of *Salmonella typhimurium. J. Bacteriol.* 177:3355–3357.

Lodwick, D., H.N.M. Ross, J. A. Walker, J. W. Almond, and W. D. Grant. 1991. Nucleotide Sequence of the 16S Ribosomal TNA Gene from the Haloalkaliphilic Archaeon (Archaebacterium) *Natronobacterium magadii,* and the Phylogeny of Halobacteria. *System. Appl. Microbiol.* 14:352–357.

Loughney, K., E. Lund, and J. E. Dahlberg. 1982. tRNA genes are found between the 16S and 23S rRNA genes in *Bacillus subtilis. Nucl. Acids Res.* 10:1607–1624.

Michel, E., and P. Cossart. 1992. Physical map of *Listeria monocytogenes* chromosome. *J. Bacteriol.* 174:7098–7103.

Neumann, H., A. Gierl, J. Tu, J. Leibrock, D. Staiger, and W. Zillig, 1983. Organization of the Genes for Ribosomal RNA in Archaebacteria. *Mol. Gen. Genet.* 192:66–72.

Pace, N. R., and A. B. Burgin. 1990. Processing and evolution of the rRNAs. In *The Ribosome.* W. E. Hill, A. Dahlberg, R. A. Garrett, P. B. Moore, D. Schlessinger, J. R. Warner, eds. pp. 417–425. American Society for Microbiology. Washington, D.C.

Pernodet, J.-L., F. Boccard, M.-T. Alegre, J. Gagnat, and M. Guerineau. 1989. Organization and nucleotide sequence analysis of a ribosomal RNA gene cluster from *Streptomyces ambofaciens. Gene* 79:33–46.

Ralph, D., M. McClelland, J. Welsh, G. Baranton, and P. Perolat. 1993. *Leptospria* species categorized by arbitrarily primed polymerase chain reaction (PCR) and by mapped restriction polymorphisms in PCR-amplified rRNA genes. *J. Bacteriol.* 175:973–981.

Ree, H. K., and R. A. Zimmermann. 1990. Organization and expression of the 16S, 23S and 5S ribosomal RNA genes from the archaebacterium *Thermoplasma acidophilum. Nucleic Acids Research* 18:4471–4478.

Sanz, J. L., I. Marin, L. Ramirez, J. P. Abad, C. L. Smith, and R. Amils. 1988. Variable rRNA gene copies in extreme halobacteria. *Nucleic Acids Research* 16:7827–7832.

Suzuki, Y., Y. Ono, A. Nagata, and T. Yamada. 1988. Molecular cloning and characterization of an rRNA operon in *Streptomyces lividans* TK21. *J. Bacteriol.* 169:1631–1636.

Taylor, D. E., M. Eaton, N. Chang, and S. M. Salama. 1992. Construction of a Helicobacter pylori genome map and demonstration of diversity at the genome level. *J. Bacteriol.* 174:6800–6806.

Wagner, R. 1994. The regulation of ribosomal RNA synthesis and bacterial cell growth. *Arch. Microbiol.* 161:100–109.

Waterhouse, R. N., and L. A. Glover. 1993. Differences in the hybridization pattern of *Bacillus subtilis* genes coding for rRNA depend on the method of DNA preparation. *Appl. Environ. Microbiol.* 59:919–921.

Widom, R. L., E. D. Jarvis, G. LaFauci, and R. Rudner. 1988. Instability of rRNA operons in *Bacillus subtilis. J. Bacteriol.* 170:605–610.

Wilkinson, S. R., and M. Young. 1995. Physical map of the *Clostridium beijerinckii* (formerly *Clostridium acetobutylicum*) NCIMB 8052 chromosome. *J. Bacteriol.* 77:439–448.

Winder, F. G., and S. A. Rooney. 1970. Effects of Nitrogenous Components of the Medium on the Carbohydrate and Nucleic Acid Content of *Mycobacterium tuberculosis* BCG. *J. Gen. Microbiol.* 63:29–39.

Woese, C. R., O. Kandler, and M. L. Wheelis. 1990. Towards a natural system of organisms: Proposal for the domains Archaea, Bacteria, and Eucarya. *Proc. Natl. Acad. Sci. USA* 87:4576–4579.

Wyngaard, G. A., I. A. McLaren, M. M. White, and J.-M. Sévigny. 1995. Unusually high numbers of ribosomal RNA genes in copepods (Arthropoda:Crustacea) and their relationship to genome size. *Genome* 38:97–104.

Zillig, W. 1992. The order Thermococcales, in *The Prokaryotes,* A. Balows, H. G. Trüper, M. Dworkin, W. Harder and K.-H. Schleifer, eds. pp. 702–706. Springer Verlag, New York.

22

Genome Evolution in the Salmonellae

Kenneth E. Sanderson, Shu-Lin Liu, Andrew Hessel, and M. McClelland

SUMMARY The genus *Salmonella,* composed of many species that occupy different habitats, diverged from *Escherichia* over 100 million years ago (*see also* Chapter 17). *S. typhimurium* LT2 has been studied genetically for many years. We have used four major methods to collect data from pulsed-field gel electrophoresis (*see* Chapters 24–26) on the structure of the genome of several species of *Salmonella:* 1) The endonuclease I-*Ceu*I to determine an *rrn* operon skeleton; 2) Tn*10*s, which are inserted in specific genes for transduction into new species, and physical mapping of the genes; 3) Double digestion following excision of fragments in agarose; 4) Mud-P22s (*see also* Chapter 27) as linking probes.

Genome changes can be classified into three types: 1) Base pair changes; These occur in homologous genes of the species, but at low frequency. 2) Intracellular genome rearrangements. The genome of most species of *Salmonella* shows few rearrangements, except for an inversion over the terminus region which is found in many species. In addition, inversions and transpositions due to homologous recombination between the *rrn* operons for rRNA are common in wild-type strains of species which cause enteric fevers, such as *S. typhi* and *S. paratyphi* A and C. 3) Lateral transfer. This commonly adds large blocks of DNA to the genome of species, and may be a major force in evolution.

The objective of this review is to discuss the types of genetic changes which are observed to have occurred during the evolution of different species of *Salmonella.* We can also speculate on which of these are most relevant to the phenotypic differences which are observed between these species. We will focus on data obtained by study of the entire genome by methods including pulsed-field gel electrophoresis and others.

The Genus Salmonella

Salmonella are non-spore-forming Gram-negative rods of the family *Enterobacteriaceae.* The division of the genus *Salmonella* into over 2,300 serovars is based on the somatic and flagellar antigens; these serovars have been called species (Kauffman, 1966). Molecular genetic evidence indicates close relationships of all these serovars (Crosa *et al.*, 1973) and some authors now call all strains *Salmonella enterica,* with a separate indication of the serovar (e.g., *S. enterica* sv. Typhimurium) (Le Minor and Popoff, 1987); in this report we arbitrarily retain the use of different species names (e.g., *S. typhimurium*). Many species grow in a large number of vertebrate or invertebrate hosts, and many of these species cause gastroenteritis (food poisoning) in humans; *S. typhimurium* is an example. A few species are host specific; *S. typhi,* causal organism of typhoid fever in humans, is one of these.

The determination of gene order in chromosomes is an important part of genetic analysis (Fonstein and Haselkorn, 1995; *see also* Chapter 28). Practically all genetic studies in *Salmonella* until recently have been done with strain LT2 of *S. typhimurium.* Edition I of the linkage map of *S. typhimurium* LT2 used phage P22-mediated transduction and F-factor mediated conjugation to determine the linkages of genes and the overall structure of the genome (Sanderson and Demerec, 1965), further editions up to edition VII (Sanderson and Roth, 1988) incorporated some molecular data but the basic map was derived from genetic analysis.

Methods of Genome Analysis

Recently, low resolution maps of *S. typhimurium* LT2 have been constructed using pulsed-field gel electrophoresis for the enzymes *Xba*I (Liu *et al.*, 1993; Liu and Sanderson, 1992), *Bln*I (=*Avr*II) (Wong and McClelland, 1992); and *Ceu*I (Liu *et al.*, 1993); a partial map has been constructed for *Spe*I (S.-L. Liu, unpublished data (*see* Chapter 78)). The locations of genes on these physical maps, shown in edition VIII (Sanderson *et al.*, 1995), agrees well with the original linkage maps. Utilizing methods devised to construct a physical map (which shows the positions of restriction sites for rare-cutting enzymes) and methods which also produce a genetic map (the physical positions of genes), we have studied the genomes of *S. enteritidis* (Liu *et al.*, 1993), *S. typhi* (Liu and Sanderson, 1995b, 1995d; *see also* Chapter 27), *S. paratyphi* A (Liu and Sanderson, 1995a), *S. paratyphi* B (Liu *et al.*, 1995) and *S. paratyphi* C (Hessel *et al.*, 1995). The following methods were used, some of which are new, while others are modifications of methods described by others: (*see also* Chapter 24).

(1) I-*Ceu*I analysis for determination of the "*rrn* skeleton." The endonuclease I-*Ceu*I (hereafter called *Ceu*I) is encoded by a class I mobile intron

which is inserted into the *rrl* gene for the large subunit rRNA (23S-rRNA) in the chloroplast DNA of *Chlamydomonas eugamatos* (Marshall and Lemieux, 1991). *Ceu*I is specific for and cuts in a 19-bp sequence in the *rrl* gene (Marshall *et al.*, 1994). Because rRNA sequences are strongly conserved, this 19bp sequence is present in all seven *rrl* genes of enteric bacteria, but at no other site so far detected (Liu *et al.*, 1993). Because *Ceu*I cleaves only *rrn* operons and because the *rrn* skeleton (the number and locations of *rrn* operons) is highly conserved in enteric bacteria, related wild-type strains usually yield very similar fingerprints. For example, seventeen *S. typhimurium* strains from the SARA set (a set of *Salmonella* strains established by R. K. Selander by multi-locus enzyme electrophoresis, MLEE) gave very similar *Ceu*I digests indicating that the *rrn* skeleton of their genomes is almost identical (Liu and Sanderson, 1995c). In addition, analysis by partial *Ceu*I digestion indicates fragments that are adjacent on the chromosome, and showed that the order of *Ceu*I fragments in the chromosome, which is *Ceu*I-ABCDEFG in *S. typhimurium* LT2 and *E. coli* K-12, appears to be the same in all strains tested from the SARA set (Liu and Sanderson, 1995c).

(2) Use of Tn*10*s which are inserted in specific genes for transduction into new species, and physical mapping of the genes. Tn*10* has sites for the rare-cutting enzymes *Xba*I and *Bln*I; physical mapping of these sites locates the map position of genes which have Tn*10* insertions. Tn*10* which is inserted into a specific gene (e.g., *trp*::Tn*10,* an insertion into a gene for tryptophan synthesis) of *S. typhimurium* can be transduced by phage 22 into other species of *Salmonella* with selection for tetracycline resistance. The transductant usually results from homologous recombination rather than transposition, and this is confirmed if the transductant has the phenotype of the original strain (i.e., tryptophan requirement for growth). The physical location of the Tn*10,* and thus of the gene, was determined through the location of new *XbalI* and *Bln*I sites for *S. entertidis* (60 genes) (Liu *et al.*, 1993), *S. typhi* (82 genes) (Liu and Sanderson, 1995d), *S. paratyphi* A (78 genes), *S. paratyphi* B (61 genes) (Liu *et al.*, 1994), and *S. paratyphi* C (50 genes) (Hessel *et al.*, 1995). The distance between gene pairs (in kb) in different but related species sometimes shows dramatic differences. Strains which do not adsorb P22 because they lack appropriate LPS can be made sensitive to P22 by a cosmid which carries the *rfc* and *rfb* genes of *S. typhimurium* (Neal *et al.*, 1993). If the chromosome of the recipient strain has diverged from that of *S. typhimurium* by nucleotide substitution so that homologous recombination is rare, *mutS, L,* or *U* mutations can be transferred into the recipient by transduction

using linked Tn*10* mutations; this reduces mismatch repair and dramatically increases the frequency of homologous recombination (Zahrt *et al.*, 1994).

(3) Double digestion. Following PFGE separation of fragments from endonuclease digestion, agarose blocks containing individual fragments can be excised from the gel, redigested with a second enzyme, and re-electrophoresed (Liu *et al.*, 1993). The resulting fragments, which are often small, may not be visualized by ethidium bromide but require end-labelling with ^{32}P using T7 polymerase or Klenow fragment and autoradiography. Fragments as small as 1 kb can be detected. Thus the location of restriction sites of the second enzyme within a specific fragment from the first enzyme can be determined; the order of use of the enzymes can be reversed.

(4) Mud-P22 as linking probes. Mud-P22 is an excision-defective derivative of bacteriophage P22 (Youderian *et al.*, 1988; *see also* Chapter 27). Lysogens were constructed with Mud-P22 inserted at over 40 different sites in the *S. typhimurium* chromosome (Benson and Goldman, 1992). The DNA from P22 phage induced from these strains is greatly amplified for ca. 200 kb in the region of phage insertion; this DNA provides linking probes which enables determination of adjacent fragments (Liu and Sanderson, 1992; Wong and McClelland, 1992).

Genome Changes in Salmonella

We describe three types of genetic changes which can occur during evolution of bacteria, discuss which have occurred during evolution of different species of *Salmonella,* and speculate on which may be the most important in the phenotypic differences between species.

(1) Base pair changes. These are probably due primarily to replication errors. Homologous genes in different species of *Salmonella* appear to have only a low percentage of base pair changes, because Salmonella shows high DNA-DNA reassociation (usually > 80%) and low delta Tm_e (usually about 1°C) in studies with different species; this suggests bp changes of an average of only about 1% (Crosa *et al.*, 1973). In addition, nucleotide sequencing of homologous genes of different species within sub-genus I of *Salmonella* (which includes all the human-pathogenic species usually studied) indicates low divergence (Boyd *et al.*, 1994), except in a few highly polymorphic genes such as *rfb* (Reeves, 1993), *fliC,* and *fliB* (Li *et al.*, 1994). Homologous genes of members of two different genera, *S. typhimurium* and *E. coli* K-12, differ by an average of 16% bp changes (Sharp, 1991); a high

proportion of these changes are silent substitutions leading to no amino acid change, and many of the amino acid changes lead to non-detectable difference in protein function. It is not clear if these bp changes lead to important differences in phenotype, either between different genera or between different species of *Salmonella.*

(2) Intracellular genomic rearrangements. These include inversions, translocations, and duplications, in which large blocks of DNA exchange within the cell; no new genetic material is added, but the existing genes are rearranged. Gene order in the chromosomes of *E. coli* K-12 and *S. typhimurium* LT2 is strongly conserved (Krawiec and Riley, 1990; Sanderson, 1976), even though the genera diverged over 100 million years before the present (Ochman and Wilson, 1987, *see also* Chapters 17 and 18). In conformity with this, there are few genomic rearrangements which distinguish most species and strains of the genus *Salmonella.* The order of genes in *S. paratyphi* B (Liu *et al.*, 1994), and in *S. enteritidis* (Liu *et al.*, 1993) is very similar to *S. typhimurium* LT2. In addition, the *rrn* skeleton of the chromosome, and hence probably the total gene order, determined from partial digestion with *Ceu*I, is very similar in seventeen independent strains of *S. typhimurium* (Liu and Sanderson, 1995c). In contrast to this stability of these chromosomes during evolution, chromosomes of enteric bacteria cells in culture frequently rearrange; duplications of segments of the chromosome occur at high frequencies (10^{-2} to 10^{-5}) (Anderson and Roth, 1978; Hill and Harnish, 1981), and some inversions, especially those with endpoints in the *rrn* operons, are common (Hill and Gray, 1988). This high frequency of rearrangements, combined with stability of the chromosome during evolution, indicates that huge selective pressures must exist to maintain the order of genes. The following explanations for elimination of cells with rearranged chromosomes, and hence conservation of genome structure, have been advanced: 1) In cells with rearrangements involving *oriC* (the origin of replication) or Ter (the terminus), the normal 180°C separation of *oriC* and Ter may be interrupted (Hill and Harnish, 1981); b) displacement of genes outside the inversion segment relative to *oriC* affects the average gene dosage of these genes during replication, and hence their gene expression (Schmid and Roth, 1987); c) highly transcribed genes are almost always replicated in the same direction.

However, though most genomic rearrangements seldom survive in evolution, two classes of chromosomal rearrangements are found in *Salmonella* spp. The first class is inversions and transpositions due to homologous recombination between the *rrn* operons. These are found in wild-type strains of species such as *S. typhi* and *S. paratyphi* A and

C, whose growth is restricted to humans, where they cause enteric fever. Whereas in *S. typhimurium* LT2 the order of *Ceu*I fragments is ABCDEFG, in *S. typhi* Ty2 the order is AGCEFDB; thus the fragments are rearranged, though the genes on specific *Ceu*I fragments are similar (Liu and Sanderson, 1995b, 1995d). Other wild-type strains of *S. typhi* have different orders, such as CFDE, and CEDF, etc. (Liu and Sanderson, 1995b), but the order found in *S. typhimurium* LT2 is seldom detected. Ten strains of *S. paratyphi* A have been analysed, and all have an inversion of about half the chromosome, postulated to be due to recombination between *rrnH* and *rrnG* (Liu and Sanderson, 1995a). Strains of *S. paratyphi* C have several different orders of the *Ceu*I fragments (Hessel *et al.*, 1995).

The second category of genetic rearrangement involves inversions over Ter, the terminus region of the chromosome. This is the only region in which inversions in wild-type strains of *Salmonella* are common (other than inversions involving *rrn* genes). Inversions overlap the Ter region in the following species, when compared with *S. typhimurium* (the size of the inversion in kb is shown); *E. coli*, 480 kb (Casse *et al.*, 1972; Sanderson and Hall, 1970); *S. enteritidis,* 750 kb (Liu *et al.*, 1993), *S. typhi,* 500 kb; *S. paratyphi* C, 1700 kb (Hessel *et al.*, 1995). No inversion was detected in *S. paratyphi* B (Liu *et al.*, 1994). All these inversions differ by at least one endpoint. The gene order in *S. typhimurium* may be considered the ancestral order, because all inverted orders could be obtained in single events from the *S. typhimurium* order, whereas more complex events would be required from other orders, but this may simply reflect high frequency of inversions in this region. Inversions at many points in the genome are detected in experimental studies in *E. coli* K-12 and *S. typhimurium* LT2, though some end-points are much less common than others (Mahan *et al.*, 1990). Two types of recombination occur at high frequency in *E. coli* K-12 in the Ter region: *recA*-dependent homologous recombination, resulting in deletions; site-specific recombination at the *dif*-site in the Ter region (Louarn *et al.*, 1994a, 1994b). The role of these recombination events in inversion in wild type strains is unknown. Surprisingly, in experimental studies in *E. coli* K-12, inversions with endpoints in the regions just outside TerA and TerC, called non-divisible zones, are seldom found, but in natural populations of *Salmonella* and *E. coli* we find this to be the only region in which inversions are common; the rules governing survival of such inversions in natural populations must differ from laboratory studies in some undefined way.

(3) Addition of new genes through lateral transfer. The first two mechanisms involve vertical transfer of genes, with mutations due to bp

changes (Anderson and Roth, 1978) or rearrangements (Baumler and Heffron, 1995), but mechanism 3 involves entry of non-homologous genes, presumably by classical genetic transfer methods, and integration of these genes into the chromosome. This results in segments or "loops" of DNA in one strain for which there is no homologue in closely related species or genera. Loops of DNA distinguishing *E. coli* and *S. typhimurium* were noted by Riley and colleagues (Krawiec and Riley, 1990; Riley and Sanderson, 1990) and others (Mills *et al.*, 1995). Chromosomes of different species of *Salmonella* also differ by loops of this type. A segment of DNA with the *nanH* gene of *S. typhimurium* LT2 is absent from all other species of sub-genus I (Hoyer *et al.*, 1992), having entered by lateral transfer, perhaps from *Clostridium.* Genetic maps of *Salmonella* show that intervals between the same gene pair in different species may be very different. For example, in *S. typhi* the *mel-poxA* interval is 120 kb longer than *S. typhimurium* and other species; this loop was found to carry the *viaB* gene for the Vi antigen, missing from most other *Salmonella.* "Loops" of species-specific DNA of this type, probably due to lateral transfer, are common, and detected in several species of *Salmonella,* though the resolution of the method can detect only inserts greater than 20 to 40 kb. In addition, the I-*Ceu*I-D fragment, is 136 kb in *S. typhi,* but only 92 kb in other species, indicating an insertion of about 44 kb in this region (Liu and Sanderson, 1995d). *S. typhimurium* has recently been shown to have an insertion, in the interval near *glyS* at 3800 kb of DNA of about 8 kb containing the *lpfABCDE* genes for fimbriae; this DNA is missing from *S. typhi* and many other species (Baumler and Heffron, 1995). Lateral (also called horizontal) transfer resulting in "loops" of species-specific DNA is probably the most important genetic mechanism resulting in species adaptation in Salmonella.

Variation in the Restriction map, but Conservation of the Genetic Map

Even though the gene order in the chromosomes of most species of Salmonella is very similar, the numbers and locations of *Xba*I and *Bln*I sites in the chromosomes are very different, for *S. typhimurium* (Liu *et al.*, 1993), *S. enteritidis* (Liu *et al.*, 1993) and *S. paratyphi* B have 24, 16, and 19 *Xba*I sites and 16, 12 and 10 *Bln*I sites, respectively. The fingerprints of these species with these enzymes suggest major differences, yet the general chromosome structure and the gene order is very similar. Thus, variations in fingerprint pattern are of value in determining if different strains are clones, and in doing "outbreak" analysis, but most enzymes do not tell much about the similarity of genetic maps. *Ceu*I, because it cleaves only sites in the *rrn* operons, reveals the relatively-conserved *rrn* skeleton of the chromosome. As expected, *Ceu*I sites in the *rrn* operons and *Xba*I and *BInI* sites in the glt-tRNA genes in the *rrn* operons are fully conserved in all these species, but the *Bln*I site present in *rrs* (16S-rRNA) genes of some *Salmonella* species are missing from all seven *rrn* operons of *S. typhi.*

References

Anderson, R. P., and J. R. Roth. 1978. Gene duplication in bacteria: alteration of gene dosage by sister chromosome exchange. *Cold Spring Harbor Symp. Quantit. Biol.* 43:1083–1087.

Baumler, A. J., and F. Heffron. 1995. Identification and sequence analysis of *lpfABCDE,* a putative fimbrial operon of *Salmonella typhimurium. J. Bacteriol.* 177:2087–2097.

Benson, N. R., and B. S. Goldman. 1992. Rapid mapping in *Salmonella typhimurium* with Mud-P22 prophages. *J. Bacteriol.* 174:1673–1681.

Boyd, E. F., K. Nelson, F. S. Wang, T. S. Whittam, and R. K. Selander. 1994. Molecular genetic basis of allelic polymorphism in malate dehydrogenase (*mdh*) in natural populations of *Escherichia coli* and *Salmonella enterica. Proc. Natl. Acad. Sci. USA* 91:1280–1284.

Casse, F., M.-C. Pascal, and M. Chippaux. 1972. Comparison between the chromosome maps of *Escherichia coli* and *Salmonella typhimurium:* length of the inverted segment in the *trp* region. *Mol. Gen. Genet.* 51:253–275.

Crosa, J. H., D. J. Brenner, W. H. Ewing, and S. Falkow. 1973. Molecular relationships among the Salmonellae. *J. Bacteriol.* 115:307–315.

Fonstein, M., and R. Haselkorn. 1995. Physical mapping of bacterial genomes. *J. Bacteriol.* 177:3361–3369.

Hessel, A., S.-L. Liu, and K. E. Sanderson. 1995. The chromosome of *Salmonella paratyphi* C contains an inversion and is rearranged relative to *S. typhimurium* LT2. Abstr., *Ann. Mtng., Amer. Soc. Microbiol.* 95th Annual Meeting: (Abstract).

Hill, C. W., and J. A. Gray. 1988. Effects of chromosomal inversion on cell fitness in *Escherichia coli* K-12. *Genetics* 119:771–778.

Hill, C. W., and B. W. Harnish. 1981. Inversions between ribosomal RNA genes of *Escherichia coli. Proc. Natl. Acad. Sci. USA* 78:7069–7072.

Hoyer, L. L., A. C. Hamilton, S. M. Stenbergen, and E. R. Vimr. 1992. Cloning, sequencing, and distribution of the *Salmonella typhimurium* LT2 sialidase gene, *nanH,* provides evidence for interspecies gene transfer. *Mol. Microbiol.* 6:873–884.

Kauffman, F. 1966. *The Bacteriology of Enterobacteriaceae.* Williams and Wilkens, Baltimore, MD.

Krawiec, S., and M. Riley. 1990. Organization of the bacterial genome. *Microbiol. Rev.* 54:502–5349.

Le Minor, L., and M. Y. Popoff. 1987. Designation of *Salmonella enterica* sp. nov. as the type and only species of the genus *Salmonella. Int. J. Syst. Bacteriol.* 37:465–468.

Li, J., K. Nelson, A. C. McWhorter, T. S. Whittam, and R. K. Selander. 1994. Recombinational basis of serovar diversity in Salmonella enterica. *Proc. Natl. Acad. Sci. USA* 91:2552–2556.

Liu, S.-L., A. Hessel, H.-Y.M. Cheng, and K. E. Sanderson. 1994. The *Xba* I-*Bln* I-*Ceu* I genomic cleavage map of *Salmonella paratyphi* B. *J. Bacteriol.* 176(4):1014–1024.

Liu, S.-L., A. Hessel, and K. E. Sanderson. 1993. Genomic mapping with I-*Ceu*I, an intron-encoded endonuclease, specific for genes for ribosomal RNA, in *Salmonella* spp., *Escherichia coli,* and other bacteria. *Proc. Natl. Acad. Sci. USA* 90:6874–6878.

Liu, S.-L., A. Hessel, and K. E. Sanderson. 1993. The *Xba* I-*Bln* I-*Ceu* I genomic cleavage map of *Salmonella typhimurium* LT2 determined by double digestion, end-labelling, and pulsed-field gel electrophoresis. *J. Bacteriol.* 175:4104–4120.

Liu, S.-L., A. Hessel, and K. E. Sanderson. 1993. The *Xba*I-*Bln* I-*Ceu* I genomic cleavage map of *Salmonella enteritidis* shows an inversion relative to *Salmonella typhimurium* LT2. *Mol. Microbiol.* 10:655–664.

Liu, S.-L., and K. E. Sanderson. 1992. A physical map of the *Salmonella typhimurium* LT2 genome made by using *Xba* I analysis. *J. Bacteriol.* 174(5):1662–1672.

Liu, S.-L., and K. E. Sanderson. 1995a. The chromosome of *Salmonella paratyphi* A is inverted by recombination between *rrnH* and *rrnG. J. Bacteriol.* 177:6585–6792.

Liu, S.-L., and K. E. Sanderson. 1995b. Rearrangements in the genome of the bacterium *Salmonella typhi. Proc. Natl. Acad. Sci. USA* 92:1018–1022.

Liu, S.-L., and K. E. Sanderson. 1995c. I-*Ceu*I reveals conservation of the genome of independent strains of *Salmonella typhimurium. J. Bacteriol.* 177:3355–3357.

Liu, S. L., and K. E. Sanderson. 1995d. The genomic cleavage map of *Salmonella typhi* Ty2. *J. Bacteriol.* 177:5099–5107.

Louarn, J., F. Cornet, V. Francois, J. Patte, and J.-M. Louarn. 1994a. Hyperrecombination in the terminus region of the *Escherichia coli* chromosome: possible relation to nucleoid organization. *J. Bacteriol.* 176:7524–7531.

Louarn, J., F. Cornet, V. Francois, J. Patte, and M.-J. Louarn. 1994b. *Dif* activity in the cell cycle. Mapping the region where *dif* must lie, and comparison with the terminal recombination domain. *J. Bacteriol.* 176:7524–7531.

Mahan, M. J., A. M. Segall, and J. R. Roth. 1990. Recombination events that rearrange the chromosome: barriers to inversion, *The Bacterial Chromosome.* In K. Drlica and M. Riley, eds. p. 341–349. American Society for Microbiology, Washington, D.C.

Marshall, P., T. B. Davis, and C. Lemieux. 1994. The I-*Ceu*I endonuclease: purification and potential role in the evolution of *Chlamydomonas* group I introns. *Eur. J. Bacteriol.* 220:855–859.

Marshall, P., and C. Lemieux. 1991. Cleavage pattern of the homing endonuclease encoded by the fifth intron in the chloroplast subunit rRNA-encoding gene of *Chlamydomonas eugamatos. Gene* 104:1241–1245.

Mills, D.M.V., V. Balaj, and C. A. Lee. 1995. A 40 kilobase chromosomal fragment encoding *Salmonella typhimurium* invasion genes is absent from the corresponding region of the *Escherichia coli* chromosome. *Mol. Microbiol.* 15:749–759.

Neal, B. L., P. K. Brown, and P. R. Reeves. 1993. Use of *Salmonella* phage P22 for transduction in *Escherichia coli. J. Bacteriol.* 175:7115–7118.

Ochman, H., and A. C. Wilson. 1987. Evolutionary history of enteric bacteria, *Escherichia coli* and *Salmonella typhimurium:* cellular and molecular biology. p. 1649–1654. In F. C. Neidhardt, American Society for Microbiology, Washington, D.C.

Reeves, P. R. 1993. Evolution of *Salmonella* O antigen variation by interspecific gene transfer on a large scale. *Trends Genet.* 9:17–22.

Riley, M., and K. E. Sanderson. 1990. Comparative genetics of *Escherichia coli* and *Salmonella typhimurium, The Bacterial Chromosome* p. 85–95. In K. Drlica and M. Riley (eds), American Society for Microbiology, Washington, D.C.

Sanderson, K. E. 1976. Genetic relatedness in the family *Enterobacteriaceae. Ann. Rev. Microbiol.* 30:327–349.

Sanderson, K. E., and M. Demerec. 1965. The linkage map of *Salmonella typhimurium. Bacteriol. Rev.* 51:897–913.

Sanderson, K. E., and C. A. Hall. 1970. F-prime factors of *Salmonella* and an inversion between *S. typhimurium* and *Escherichia coli. Genetics* 64:214–228.

Sanderson, K. E., A. Hessel, and K. E. Rudd. 1995. The genetic map of *Salmonella typhimurium* LT2, edition VIII. *Microbiol. Revs.* 59:241–303.

Sanderson, K. E., and J. R. Roth. 1988. Linkage map of *Salmonella typhimurium,* edition VII. *Microbiol. Rev.* 52:485–532.

Schmid, M., and J. R. Roth. 1987. Gene location affects expression level in *Salmonella typhimurium. J. Bacteriol.* 169:2872–2875.

Sharp, P. M. 1991. Determinants of DNA sequence divergence between *Escherichia coli* and *Salmonella typhimurium:* codon usage, map position, and concerted evolution. *J. Mol. Evol.* 33:23–33.

Wong, K. K., and M. McClelland. 1992. A *Bln*I restriction map of the *Salmonella typhimurium* LT2 genome. *J. Bacteriol.* 174:1656–1661.

Youderian, P., P. Sugiono, K. L. Brewer, N. P. Higgins, and T. Elliott. 1988. Packaging specific segments of the Salmonella genome with locked-in Mud-P22 prophages. *Genetics* 118:581–592.

Zahrt, T. C., G. C. Mora, and S. Maloy. 1994. Inactivation of mismatch repair overcomes the barrier to transduction between *Salmonella typhimurium* and *Salmonella typhi. J. Bacteriol.* 176:1527–1529.

23

Structure and Evolution of *Escherichia coli Rhs* Elements

Charles W. Hill

Bacteria commonly host accessory genetic elements, a multifarious category that includes prophages, insertion sequences, transposons and episomes (Campbell, 1981; *see also* Chapters 3, 4, and 20). Such sequences are generally characterized by their ability to move from genome to genome, their great variation among independent strains, and their non-essential nature. The *Rhs* elements of *Escherichia coli* are accessory elements with some exceptional properties. Although the biological function of the *Rhs* elements is not known, circumstances suggest they are under strong selection in nature, and that they may have considerable evolutionary significance to the *E. coli* that possess them. This chapter summarizes their structure and natural distribution, and it considers their possible origins and contributions to *E. coli* evolution.

Structure

The Prototypical Rhs Element

The *Rhs* elements are genetic composites of discrete components. Some components are highly conserved, while others are divergent or even unrelated in sequence, yet retain common features. A prototypical *Rhs* structure is shown in Fig. 23-1. No individual *Rhs* element has exactly the structure depicted, but all maintain most features. The different components of these composites are discussed below, but the most prominent feature, the core-ORF, requires preliminary comment. Its protein product is predicted to be in excess of 155 kDa, and it is encoded by two distinct genetic components: the *Rhs* core itself and a core-extension (Feulner *et al.*, 1990). While the core sequences are conserved, the core-extensions of different elements are dissimilar. In fact, of the various *Rhs* components, only the core shows clear sequence homology for all elements. These 3.7 kb *Rhs* cores provide some of the most significant sequence repetition

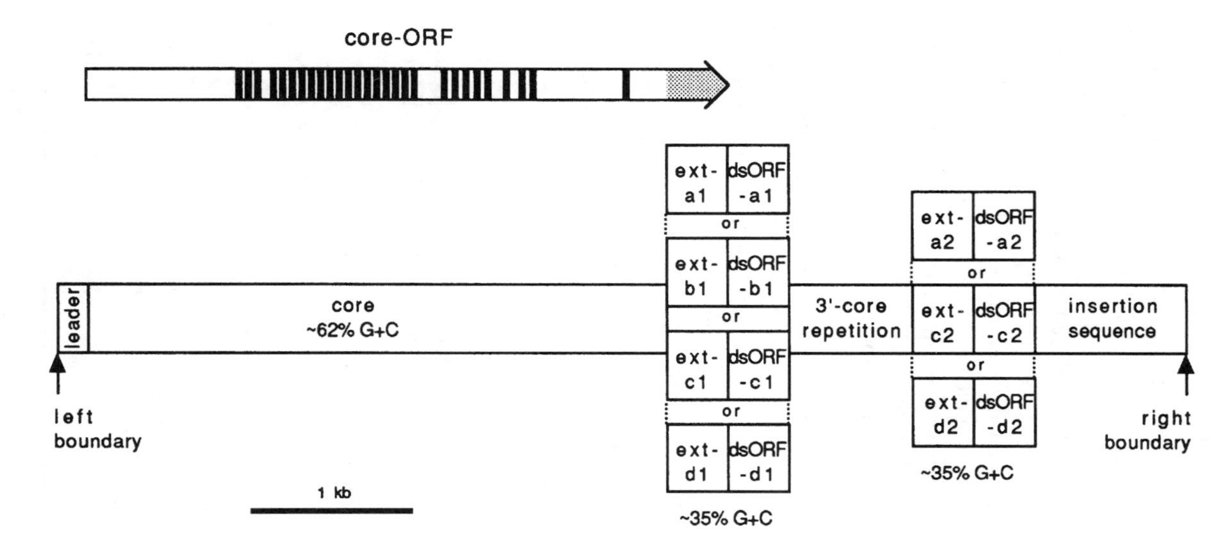

Figure 23-1. Structure components of an *Rhs* element. Each *Rhs* component is discussed in the text. No known *Rhs* element has precisely the arrangement shown, but all share most features. The *Rhs* core is the only highly conserved component common to all elements. The core-ORF is comprised of the core and adjacent core-extension, and it encodes a mosaic protein in excess of 155 kDa. Positions of the 28 repetitions of a peptide motif within the core-ORF are shown as vertical black bars.

in the *E. coli* chromosome, second in extent only to the rRNA operons. Like the rRNA operons and other large repetitions, the cores provide homology for RecA-dependent recombination, leading to chromosomal rearrangement (Hill *et al.*, 1990). In fact, amplification of the three-minute interval between two elements was the basis for their name: *R*earrangement *h*ot*s*pot (Lin *et al.*, 1984).

Rhs *Element Definition and Location*

Not all strains of *E. coli* have *Rhs* elements, but many strains such as K-12 have multiple copies. The five *Rhs* elements of K-12 range in size up to 9.6 kb, and collectively they comprise 37.8 kb or 0.8% of the genome. Their chromosomal locations are listed in Table 23.1. Included in the table are *RhsF* and *RhsG* (Hill *et al.*, 1995) which are not present in strain K-12. The boundaries of the elements have been defined through sequence comparison of genomic DNA with the DNA of a reference strain lacking the element. In each case, the reference strain has an alternate sequence, but the size of this alternate sequence varies widely from 10 bp in the case of *RhsC* (Feulner *et al.*, 1990) to 3336 bp in the case of *RhsE* (N. McCann and C. W. Hill, unpublished).

A general property of accessory elements is some sort of sequence repetition at the element boundaries. This repetition is either necessary for, or a consequence of, site-specific insertion (see Chapter 4). Such terminal repetition has not been observed for the *Rhs* elements. For a given element, the left boundary does not share significant sequence similarity to its right boundary (either direct or inverted), nor does the end of one element show any sequence similarity to the end of other elements. *Rhs* elements have not been observed to transpose in the laboratory (Zhao *et al.*, 1993), and their positional stability in natural *E. coli* populations (Hill *et al.*, 1995) (see below) suggests that their movement might be an exceedingly rare event.

Rhs *Subfamilies*

The *Rhs* elements are divided into subfamilies, according to sequence divergence of the cores. The cores of the *RhsA-B-C-F* subfamily differ from those of the

Table 23-1. Location of Rhs *elements relative to the* E. coli *K-12 map*

		Alternate sequence	
Element	Minutes	Size	Reference strain
RhsA	80.9	32 bp	ec50 or ec55
RhsB	77.8	807 bp	ec32
RhsC	15.8	10 bp	ec55
RhsD	11.3	224 bp	ec39
RhsE	32.8	3336 bp	ec40
RhsF	89.2	12bp	K-12
RhsG	unknown	unknown	

RhsD-E subfamily by 22% at the nucleotide level, while divergence within subfamilies is limited to about 4%. A survey of a collection of independent *E. coli* strains has revealed that at least one additional subfamily exists. A portion of a new element, *RhsG,* was isolated from strain ec50 and based on the sequence of a 471 bp segment, *RhsG* diverges 23% from *RhsD* and 27% from *RhsA* (Hill et al., 1995). *RhsG* is not present in K-12, and its map location has not been established.

Rhs *Core-ORFs*

The predicted proteins made from the respective *Rhs* core-ORFs are mosaics, each sharing a long, conserved segment joined to a shorter unique segment at the C-terminus. The initiation codon for the core-ORF lies precisely at the left end of the 3.7 kb core, but the ORF extends 139 or more codons beyond the core's right end through the core-extension (Fig. 23-1). Overall, the predicted proteins are extremely hydrophilic, although a hydrophobic region near the N-terminus may provide a membrane spanning function (Sadosky *et al.*, 1991). A striking feature of the core sequence is the presence of a peptide motif that is repeated 28 times. The motif can be written xxGxxxRYxYDxxGRL(I or T)xxxx and in the most regular region, it occurs with a rough periodicity of 21 amino acids (shown by small black bars in Fig. 23-1). A similar motif is seen in a *Bacillus subtilis* wall-associated protein, where it appears 31 times (Foster, 1993). These motifs in turn resemble motifs associated with carbohydrate binding by other bacterial cell surface proteins (Wren, 1991; von Eichel-Streiber *et al.*, 1992). These considerations have led us to speculate that the core-ORF protein is expressed on the cell surface, and that it allows the cell to bind a macromolecular substrate(s) in the environment. Recently, we have obtained experimental evidence for the association of the core-ORF protein with the cell membrane when artificially expressed from a *lac* promoter (C. H. Sandt and C. W. Hill, unpublished).

Downstream-ORFs

In every case studied, the core-extension is followed by a much shorter ORF called the downstream ORF. While the various downstream ORFs are generally dissimilar in sequence, most have a candidate signal-peptide for transport across the cytoplasmic membrane; the only downstream ORF that lacks an apparent signal sequence is in the K-12 *RhsA* element. In two cases, signal peptide activity has been verified experimentally (Hill *et al.*, 1994). Drawing on analogies with colicin expression (Luria and Suit, 1987), we suggest that the function of these downstream-ORFs is to aid in the transport of the large core-ORF protein to the cell surface. The sequence of the core-ORF does not predict a signal peptide of its own.

Leader Segments

The leader regions at the left end of the elements (Fig. 23.1) contain the presumed promoters for core-ORF expression. They vary from 112 bp to 425 bp and, with the exception of the *RhsA-RhsC* pair, show little sequence similarity. However, candidate promoters can be found within each leader by inspection. Understanding *Rhs* expression has been severely hampered by the fact that the core-ORFs are not expressed in strain K-12 under conditions of routine laboratory cultivation. Our laboratory has used a variety of approaches in efforts to detect expression. The most comprehensive efforts involved protein fusions designed to reveal promoter activity in the *Rhs* leader regions. The constructs used both β-galactosidase and alkaline phosphatase as reporters fused to the N-termini of the *Rhs* core-ORFs. No significant expression was detected (Hill *et al.*, 1994). We suspect that either expression requires a yet undiscovered environmental signal, or that strain K-12 has suffered the loss of a function or system necessary for expression.

Insertion Sequences

All *Rhs* elements include one or more insertion sequences within their composite structures. These components are designated as insertion sequences because of their homology with known insertion sequences, but the versions in the *Rhs* elements appear very inert, and frequently are obviously defective. A sequence called H-rpt (Zhao *et al.*, 1993) appears in *RhsB, RhsC, RhsE* and *RhsF.* Highly divergent, degenerate copies appear in *RhsA* and *RhsE.* H-rpt is a homolog of ISAS2 of *Aeromonas salmonicida* (Gustafson *et al.*, 1994) and a related sequence appears in *Salmonella enterica* (Xiang *et al.*, 1994). An IS*1* homolog is found in *RhsF* (Zhao and Hill, 1995).

Partial Core-Repetitions

A feature common to *RhsA, RhsC, RhsD* and *RhsF* is the presence of one or two repetitions of the 3′-end of the core (Fig. 23.1). The sizes of the core fragments vary, but they extend to the 3′-end and are followed by novel core-extensions and dsORFs. Our view is that they were once parts of primary cores that were displaced by the introduction of new core-extensions; i.e. they are debris of evolution.

Origins of the Rhs Elements

Correlation of Rhs *Elements with* E. coli *Population Structure*

As mentioned above, an expected character of accessory elements is the ability to move from genome to genome. Experimental efforts to detect *Rhs* element movement, or even the movement of subcomponents have had consistently negative results (Zhao *et al.*, 1993). However, comparison of natural *E. coli* isolates has shown that variation does exist in nature, consistent with our categorization of *Rhs* elements as genetic elements. The ECOR reference collection of independent *E. coli* (*see also* Chapter 17) was scored for both the presence of *Rhs* elements (defined by their location) and for the association of specific core-extensions with individual elements. The focus of this analysis was the *RhsA-B-C-F* subfamily; the analysis of the *RhsD-E* and *RhsG* subfamilies is ongoing. It was found that many natural *E. coli* lack *Rhs* elements altogether, but that those which have them tend to have several. More importantly, there is strong correlation between the elements (Hill *et al.*, 1995) and the clonal structure of the ECOR collection as defined by multilocus enzyme electrophoresis (MLEE) (Herzer *et al.*, 1990). This correlation is at several levels (Table 23-2). First, 23 of the 72 strains have no *Rhs* elements. These Rhs° strains include all 15 group B2 and 8 strains from group D (the ec35-41 clade and the ec44-47 clade). Thus, the Rhs° genotype is confined to three clades of the ECOR phylogeny. Second, phylogenetically related Rhs^+ strains tend to have the same array of elements. Whereas *RhsA* and *RhsC* are widespread among Rhs^+ strains, *RhsB* and *RhsF* show clear specificity for clonal groups. Notably, *RhsB* is in 23 of 25 group A strains, but is absent from all 13 group B1* strains. *RhsF* is present in only 4 of 25 group A, but is present in all 16 group B1. The third level of correlation regards shuffling of core-extensions. Ext-a1, the primary extension of K-12 *RhsA,* is also associated with *RhsA* in 21 of 25 group A strains and in all three strains of the ec70–72 cluster, but this extension is absent from the *RhsA* element present in the 13 group B1* strains and from the remainder of the collection. Ext-b1, the primary extension of K-12 *RhsB,* is associated with *RhsB* in all 23 of the group A strains that contain *RhsB.* Ext-b1 is also present in all 13 group B1* strains despite the absence of *RhsB* in these strains. Here it is associated with

Table 23-2. Rhs *elements and core-extensions in the ECOR reference collection*

Group[a]	Presence of *Rhs* elements	*RhsA-B-C-F* subfamily	Core-extension associations[b]		
			ext-a1	ext-b1	ext-a2
ECOR group A	Rhs+	A, B, C	*RhsA*	*RhsB*	*RhsA*
ECOR group B1*	Rhs+	A, C, F	absent	*Rhs*A	*RhsA*
ec70-71 clade	Rhs+	A, C, F	*RhsA*	absent	*RhsA*
ECOR group B2	Rhs°				
ec35-41 clade	Rhs°				
ec44-47 clade	Rhs°				
ec49-50 clade	Rhs+	B, C, F	absent	absent	*RhsF*

[a]Groups refer to the MLEE phylogeny determined for the ECOR collection (Herzer et al., 1990) except that group B1* refers to the original MLEE group B1 after removal of the ec70, 71, 72 clade. The ec35-41 clade includes ec35, 36, 38, 39, 40 and 41 from MLEE group D. The laboratory strain K-12 belongs to MLEE group A.

[b]Assocations designated are the most common found within the groups; see text for exceptions.

RhsA, and more detailed work has shown that it is the primary extension of *RhsA* in ec32. Ext-a2, a secondary extension of K-12 *RhsA,* is associated with *RhsA* throughout groups A and B1. However, in the clade of ec46, 49 and 50, it is associated with *RhsF,* and more detailed work has shown that it is the primary extension of *RhsF* in ec50. Clearly, there has been considerable shuffling of *Rhs* extensions between elements, but the shuffling has occurred in a way that has not scrambled the correlation with the ECOR clonality.

Sequence Conservation of Rhs *Core-Extensions*

To explore further the correlations between *Rhs* elements and the *E. coli* population structure (Table 23-2), we have tested for sequence divergence in examples of various core-extensions. Comparative sequencing of various genes by many laboratories has shown that genetic variation of 1–4% or greater is generally observed between strains of different clonal groups, while genetic variation within groups is usually much less (summarized in Hill *et al.*, 1995). As described above, the ext-b1/dsORF-b1 segment is associated with *RhsB* in group A strains, but with *RhsA* in group B1*. In the context of the genetic variation of other genes, a 1% or greater divergence of the ext-b1/dsORF-b1 structures in the two settings could be expected. However, sequencing four of these structures from group A strains and three from group B1* has shown that all are identical throughout the entire 950 bp (Hill *et al.*, 1995). These results are statistically significant and indicate that the ext-b1 structures are recently related. Ext-b1 appears to have entered the two clonal groups (into *RhsB* of the group A lineage and into *RhsA* of the group B1* lineage) before the apparent population sweeps that caused the groups to be prevalent among existing *E. coli,* but long after the last common ancestor of the groups. Similar high sequence conservation has been seen for ext-a1 which has been sequenced for 9 group A strains and for ec70 and ec72 (C. W. Hill, unpublished). This conservation indicates introduction of ext-a1 into the population long after group A divergence from other groups.

Origins and Evolution of Rhs *Elements*

The results summarized above concerning the structures of *Rhs* elements, the sequence divergence of *Rhs* core subfamilies, the shuffling of core-extensions between elements, the correlation of elements and their core-extensions with the *E. coli* population structure and the near absolute sequence conservation of core-extensions is suggestive of a remarkable course for *Rhs* evolution. The following progression is offered to draw together these observations. The cores are GC-rich (~62% G+C), and the core subfamilies are ≥22% divergent. Therefore, the cores likely evolved in a GC-rich genetic background. The core-extensions (and associated dsORFs) are AT-rich (<40% G+C). They likely evolved independently of the cores in an AT-rich background. In most cases, the various extensions are so divergent from each other that it is unclear whether they even had a common origin. At some less remote time, the cores and extensions became joined and entered the *E. coli* lineage, but there seems little basis for favoring one of these events as preceding the other. The elements became established at a limited number of locations (Table 23-1), most likely before divergence of the ECOR clonal groups that are defined by MLEE analysis. A single ancestor of the *RhsA-B-C-F* core subfamily may have entered the *E. coli* lineage because the sequence divergence within the core subfamily is sufficiently small (<5%) that it could have occurred entirely since the *E. coli-Salmonella* split. Shuffling of core extensions (or the introductions of core-extensions from more exotic sources) has occurred very recently, as judged by the near absolute sequence conservation of extensions.

Taken in the aggregate, the complex correlations of the *Rhs* elements and especially the near absence of genetic variation in the core-extensions is very difficult to explain without evoking direct natural selection on the *Rhs* elements as a major causal agent for the observed *E. coli* population structure. Greater knowledge of the actual function of the core-ORF protein and of its control is essential to understanding the importance of these novel elements to *E. coli* biology.

Acknowledgments

This work was supported by. Public Health Service research grant GM16329 from the National Institutes of Health.

References

Campbell, A. 1981. Evolutionary significance of accessory DNA elements in bacteria. *Annu. Rev. Microbiol.* 35:55–83.

Feulner, G., J. A. Gray, J. A. Kirschman, A. F. Lehner, A. B. Sadosky, D. A. Vlazny, J. Zhang, S. Zhao, and C. W. Hill. 1990. Structure of the *rhsA* locus from *Escherichia coli* K-12 and comparison of *rhsA* with other members of the *rhs* multigene family. *J. Bacteriol.* 172:446–456.

Foster, S. J. 1993. Molecular analysis of three major wall-associated proteins of *Bacillus subtilis* 168: Evidence for processing of the product of a gene encoding a 258 kDa precursor two-domain ligand-binding protein. *Mol. Microbiol.* 8:299–310.

Gustafson, C. E., S. Chu, and T. J. Trust. 1994. Mutagenesis of the paracrystalline surface protein array of *Aeromonas salmonicida* by endogenous insertion elements. *J. Mol. Biol.* 237:452–463.

Herzer, P. J., S. Inouye, M. Inouye, and T. S. Whittam. 1990. Phylogenetic distribution of branched RNA-linked multicopy single-stranded DNA among natural isolates of *Escherichia coli. J. Bacteriol.* 172:6175–6181.

Hill, C. W., S. Harvey, and J. A. Gray. 1990. Recombination between rRNA genes in *Escherichia coli* and *Salmonella typhimurium,* in *The Bacterial Chromosome,* K. Drlica and M. Riley, eds. American Society for Microbiology, Washington, D.C., pp. 335–340.

Hill, C. W., C. H. Sandt, and D. A. Vlazny. 1994. *Rhs* elements of *Escherichia coli:* a family of genetic composites each encoding a large mosaic protein. *Mol. Microbiol.* 12:865–871.

Hill, C. W., G. Feulner, M. S. Brody, S. Zhao, A. B. Sadosky, and C. H. Sandt. 1995. Corelation of *Rhs* elements with *Escherichia coli* population structure. *Genetics* 141:15–24.

Lin, R.-J., M. Capage, and C. W. Hill. 1984. A repetitive DNA sequence, *rhs,* responsible for duplications within the *Escherichia coli* K-12 chromosome. *J. Mol. Biol.* 177:1–18.

Luria, S. E., and J. L. Suit. 1987. Colicins and Col plasmids, in Escherichia coli *and* Salmonella typhimurium: *Cellular and Molecular Biology,* F. C. Neidhardt ed. pp. 1615–1624. American Society for Microbiology, Washington, D. C.

Sadosky, A. B., J. A. Gray, and C. W. Hill. 1991. The *RhsD-E* subfamily of *Escherichia coli* K-12. *Nucl. Acids Res.* 19:7177–7183.

von Eichel-Streiber, C., M. Sauerborn, and H. K. Kuramitsu. 1992. Evidence for a modular structure of the homologus repetitive C-terminal carbohydrate-binding sites of *Clostridium difficile* toxins and *Streptococcus mutans* glucosyltransferases. *J. Bacteriol.* 174:6707–6710.

Wren, B. W. 1991. A family of clostridial and streptoccoal ligand-binding proteins with conserved C-terminal repeat sequences. *Mol. Microbiol.* 5:797–803.

Xiang, S.-H., M. Hobbs, and P. R. Reeves. 1994. Molecular analysis of the *rfb* gene cluster of a group D2 *Salmonella enterica* strain: evidence for intraspecific gene transfer in O antigen variation. *J. Bacteriol.* 176:4357–4365.

Zhao, S., C. H. Sandt, G. Feulner, D. A. Vlazny, J. A. Gray, and C. W. Hill. 1993. *Rhs* elements of *Escherichia coli* K-12: Complex composites of shared and unique components that have different evolutionary histories. *J. Bacteriol.* 175:2799–2808.

Zhao, S., and C. W. Hill. 1995. Reshuffling of *RHs* to create a new element. *J. Bacteriol.* 177:1393–1398.

SECTION 2

Strategies for Genome Analysis

PART A

Physical Mapping Strategies

24

Physical Mapping and Fingerprinting of Bacterial Genomes using Rare Cutting Restriction Enzymes

Michael McClelland, K. K. Wong, and Ken Sanderson

Introduction

Pulsed-field gel electrophoresis (PFGE) has allowed the separation of DNA fragments of five million base pairs or more, which is about the same size or greater than most bacterial genomes. When a bacterial genome is cut into a few discrete fragments, then PFGE is an ideal tool for the construction of physical maps (see also Chapters 25 and 26).

In general, cleavage agents with larger target sequences should produce larger DNA fragments. However, no restriction enzymes are known that have target sites longer than 8 bp and such sites would occur in a typical DNA more frequently than once every 65kb, on average. Fortunately, DNA sequences are not random, which causes certain restriction sites to be very rare in certain genomes. Statistical tools that have aided researchers in choosing the restriction enzymes that are most likely to generate large fragments in the genome in question. The most important observations for eubacteria have been the exceptional rarity of CTAG in many genomes and the profound effect of high G+C or A+T content on the frequency of restriction cleavage sites (McClelland *et al.*, 1987). Consequently, there are enough 6 bp and 8 bp restriction enzymes that cleave most bacterial genome into 5 to 20 fragments to construct a preliminary map of the genome. Recently, an intron endonuclease from the ribosomal gene of a *Chlamydomonas* chloroplast (I-*Ceu*I) (Dujon *et al.*, 1989; Gauthier *et al.*, 1991; Marshall *et al.*, 1994) has been shown to cleave *rrl* (23S rRNA) genes in some bacteria (Liu *et al.*, 1993a; Honeycutt *et al.*, 1993). Conservation of the cleavage site among bacteria indicates this enzyme will probably cleave most bacterial *rrl* genes. This tool allows *rrl* genes to be mapped and provides a convenient rare cutter that generally cleaves only once or a few times in the genome, depending on the number of *rrl* loci.

Table 24-1. Restriction enzymes used in physical mapping of Archaebacterial and Eubacterial genomes

Species (and strain)	Genome size (or other)	G+C content	Restriction Enzyme	Number of sites	References
ARCHAEBACTERIA					
Haloferax mediterranei	Taxon.		*Pac*I		Lopez Garcia *et al.*, 1993
Haloferax mediterranei	2900+490+	^60%	*Pac*I	4	Lopez Garcia *et al.*, 1992
	320+130		*Bam*HI	38	
Halobacterium	2000+600	58–68%	*Afl*II	23–21	Hackett *et al.*, 1994
salinarium NRC-1			*Ase*I	27–25	
Haloferax volcanii DS2	2920+690+442+86+64		*Dra*I	17–23	Charlebois *et al.*, 1991
Halobacterium sp. GRB	2038+305+90+37+18				St. Jean *et al.*, 1994
Methanobacterium	1623		*Not*I	6	Stettler and Leisinger, 1992
thermoautotrophicum Marburg			*Pme*I	14	Stettler *et al.*, 1995
			*Nhe*I	19	
Methanobacterium wolfei	1729		*Not*I	8	Stettler *et al.*, 1995
			*Nhe*I	9	
Methanococcus voltae	1880	30%	*Sfi*I	0	Sitzmann *et al.*, 1991
			*Not*I	0	
			*Bss*HII	0	
			*Rsr*II	3	
			*Pvu*I	4	
			*Eag*I	5	
			*Bcl*I	6	
			*Bgl*II	6	
			*Bam*HI	11	
			*Apa*I	16	
			*Sac*II	17	
			*Sma*I	18–22	
			*Nae*I	20–25	
			*Nar*I	30	
			*Mbo*I	50–60	

Sulfolobus acidocaldarius	3100	40%	*Not*I	2	Yamagishi and Oshima, 1990
Sulfolobus acidocaldarius 7	2760		*Bss*HII		Kondo *et al.*, 1993
			*Eag*I		
			*Not*I		
			*Rsr*II		
Thermococcus celer Vu13	1890		*Nhe*I	5	Noll, K. M., 1989
			*Spe*I	5	
			*Xba*I	12	
EUBACTERIA					
Acholeplasma oculi ISM1499 1633			*Apa*I		Tigges and Minion, 1994
			*Bss*HII		
			*Eag*I		
			*Sma*I		
Acinetobacter baumannii	Epidem.		*Apa*I	>20	Grundmann *et al.*, 1995;
					Crowe *et al.*, 1995
					Mulin *et al.*, 1995;
					Seifert *et al.*, 1995
					Webster *et al.*, 1995
Acinetobacter baumannii	Epidem.		*Apa*I	>12	Marcos *et al.*, 1995
			*Sma*I		
Acinetobacter calcoaceticus	Epidem.	38–47%	*Sma*I	<30	Allardet-Servent *et al.*, 1989
Agrobacterium tumefaciens			*Swa*I	<20	Jumas Bilak *et al.*, 1995
Agrobacterium vitis	Epidem.		*Sfi*I		Schultz *et al.*, 1993
			*Spe*I		
			*Xba*I		
Anabaena sp. strain PCC 7120	6370		*Sph*I	5	Bancroft *et al.*, 1989
			*Sst*II	5	
			*Pst*I	9	
			*Avr*II	19	
			*Sal*I	21	
Aquifex pyrophilus	4700		*Swa*I	7	Shao *et al.*, 1994
			*Spe*I	21	

Table 24-1. Continued

Species (and strain)	Genome size (or other)	G+C content	Restriction Enzyme	Number of sites	References
Azospirillum species			*Xba*I		Eid *et al.*, 1995
Azotobacter vinelandii	4700		*Spe*I		Maldonado *et al.*, 1994
			*Swa*I		
Bacillus anthracis			*Not*I		Harrell *et al.*, 1995
B. cereus ATCC 14579			*Sfi*I		
B. mycoides ATCC 6462			*Sma*I		
Bacillus cereus	5485-6270	36%	*Not*I	10-14	Carlson *et al.*, 1992
ATCC 10876, ATCC 11778, ATCC 14579					
Bacillus cereus	5700	36%	*Not*I	11	Kolsta *et al.*, 1990
			*Apa*I	>30	
			*Nae*I	>30	
			*Nar*I	>30	
			*Sac*II	>30	
Bacillus subtilis 168			*Not*I		Itaya, 1993
			*Sfi*I		
			I-*Ceu*I	10	Toda and Itaya 1995
Bacillus subtilis	4165	43%	*Sfi*I	24	Amjad *et al.*, 1991
			*Not*I	^50	
Bacillus thuringiensis	5400		*Not*I	16	Carlson and Kolsto, 1993
Bacillus sp.	3700	43.7%	*Asc*I	17	Sutherland *et al.*, 1993
strain C-125			*Sse*83871	18	
			*Sfi*I	^20	
			*Pac*I	>50	
			*Not*I	>70	

Bartonella (*Rochalimaea*) species	1500–2000	39–40%	*Not*I	2–5	Roux and Raoult, 1995
			*Sma*I	12–34	
			*Eag*I	28–38	
Bartonella bacilliformis	1600		I-*Ceu*I	2	Krueger *et al.*, 1995
			*Sfi*I	4	
			*Not*I	6	
Bifidobacterum species			*Spe*I		Roy *et al.*, 1996
			Xba		
Bordetella bronchiseptica			*Spe*I		Gueirard *et al.*, 1995
			*Xba*I	>20	
Bordetella pertussis	3750	66%	*Pme*I	1	Stibitz and Garletts, 1992
			*Pac*I	2	Khattak *et al.*, 1992
			*Spe*I	16	
			*Xba*I	25	
			*Dra*I	30	
			*Ase*I	40	
			*Ssp*I	80	
	Epidem.		*Dra*I		Beall *et al.*, 1995
			*Spe*I		
			*Xba*I		
Bordetella pertussis	Epidem.		*Xba*I	20	Syedabubakar *et al.*, 1995
Borrelia afzelii	948		*Csp*I	2	Ojaimi *et al.*, 1994
			*Bss*HII	3	
			I-*Ceu*I	3	
			*Sac*II	3	
			*Sgr*AI	3	
			*Eag*I	4	
			*Mlu*I	4	
			*Sma*I	13	
			*Apa*I	14	

Table 24-1. Continued

Species (and strain)	Genome size (or other)	G+C content	Restriction Enzyme	Number of sites	References
Borrelia burdorferi			*Asc*I	0	Davidson *et al.*, 1992
212	946(L)		*Not*I	1	Old *et al.*, 1992
Sh-2-82	952	30%	*Sgr*AI	2–3	Casjens and Huang, 1993
			*Rsr*II	1	
			*Srf*I	1	
			*Sfi*I	3	
			*Ksp*I	4	
			*Pvu*I	4	
			*Sac*II	3–4	
			*Sse*83871	4	
			*Spl*I	5	
			*Mlu*I	5–6	
			*Bss*HII	4, 7	
			*Eag*I	5–7	
			*Nae*I	6–7	
			*Apa*I	12–14	
			*Sma*I	11–14	
			*Ehe*I	13	
			*Kpn*I	>15	
			*Nhe*I	>15	
			*Pst*I	>15	
			*Xho*I	16	
			*Bam*HI	>18	
			*Bgl*I	>20	
			*Sst*I	>20	
			*Cla*I	>20	
			*Pvu*II	>40	
			*Bcl*I	>40	

Borrelia burdorferi	Epidem.		*Mlu*I	>5	Kolbert *et al.*, 1995; Picken *et al.*, 1995
Borrelia garinii	953		*Csp*I	0	Ojaimi *et al.*, 1994
			*Sgr*AI	3	
			I-*Ceu*I	3	
			*Eag*I	6	
			*Sma*I	7	
			*Bss*HII	8	
			*Mlu*I	9	
			*Apa*I	11	
Borrelia, 21 strains	935–955				Casjens *et al.*, 1995
Borrelia species	Epidem.		*Apa*I		Busch *et al.*, 1995
			*Ksp*I		
			*Sma*I		
			*Xho*I		
			*Mlu*I		Busch *et al.*, 1996
Borrelia species	Epidem.		*Mlu*I		Will *et al.*, 1995
Bradyrhizobium japonicum 110	8700		*Pac*I	2	Kundig *et al.*, 1993
			*Pme*I	5	
			*Swa*I	9	
Bradyrhizobium japonicum USDA 424	5379		*Ase*I		Sobral *et al.*, 1991
			*Spe*I		
			*Ssp*I		
			*Xba*I		
Branhamella catarrhalis	Epidem.		*Spe*I	>10	Klingman *et al.*, 1995
			*Nhe*I		
Brucella melitensis 16M	2100+1150	59%	*Pac*I	9	Michaux *et al.*, 1993
			*Spe*I	26	
			*Xho*I	32	
			*Xba*I	45	

Table 24-1. Continued

Species (and strain)	Genome size (or other)	G+C content	Restriction Enzyme	Number of sites	References
Burkholderia cepacia ATCC 25416 (basonym *Pseudomonas cepacia*)	3650+3170+1070+200		I-*Ceu*I	6	Rodley *et al.*, 1995
			*Pac*I	11	
			*Pme*I	11	
			*Swa*I	11	
			*Spe*I	38	
Burkholderia cepacia	Epidem.		*Xba*I		Ouchi *et al.*, 1995; Valcin *et al.*, 1996
Burkholderia cepacia	Epidem.		*Spe*I		Liu *et al.*, 1995
Burkholderia cepacia	Epidem.		*Xba*I		Pitt *et al.*, 1996
			*Spe*I		
Caldocellum saccharolyticum	2780	34%	*Sst*II	21	Borges and Bergquist, 1996
			*Nhe*I	23	
Campylobacter coli JCM 2529T	2000		*Apa*I		Matsuda *et al.*, 1991
			*Sal*I		
			*Sma*I		
Campylobacter coli strain UA417	1700	28–38%	*Not*I	0	Yan and Taylor, 1991
			*Sal*I	7	
			*Sma*I	11	
Campylobacter coli	1700		*Sal*I	6–7	Taylor *et al.*, 1992
Campylobacter coli	Epidem.		*Sma*I	9–13	Stanley *et al.*, 1995
Campylobacter fetus	Epidem.		*Sal*I	<15	Fujita *et al.*, 1992; Fujita *et al.*, 1995
			*Sma*I	<20	
Campylobacter fetus subsp. *fetus* ATCC 27374	1160		*Sal*I	4–14	Salama *et al.*, 1995
			*Sma*I	4–14	
			*Not*I	4–14	
Campylobacter jejuni	1700	32–35%	*Sal*I	6	Taylor *et al.*, 1992
			*Nci*I	14	Nuijten *et al.*, 1990
			*Bss*HII	15	

Campylobacter jejuni	Epidem.		*Sma*I	7–10	Owens *et al.*, 1995
Campylobacter jejuni	Epidem.		*Sma*I	^10	Gibson *et al.*, 1995
			*Kpn*I	>15	
Campylobacter jejuni JCM 2013	1900		*Apa*I		Matsuda *et al.*, 1991
			*Sal*I		Matsuda *et al.*, 1995
			*Sma*I		
Campylobacter laridis JCM2530T	1590–1700		*Apa*I		Matsuda *et al.*, 1991
			*Sal*I		Matsumoto *et al.*, 1992
			*Sma*I		
Campylobacter upsaliensis ATCC 43954	2000		*Bss*HII	10	Bourke *et al.*, 1995
			*Nar*I	8	
			*Sal*I	5	
Carnobacterium divergens	3200		*Sma*I	^20	Daniel, 1995
Caulobacter crescentus	3800–4000	67%	*Ase*I	13	Ely and Gerardot, 1988; Ely *et al.*, 1990
			*Dra*I	33–35	Steinman and Ely, 1990
			*Spe*I	17–23	Dingwall and Shapiro, 1989
Chlamydia trachomatis serovar L2	1045	45%	*Not*I	4	Birkelund and Stephens, 1992
			*Sgr*AI	4	
			*Sse*83871	17	
Chromatium vinosum	3670		*Spe*I	13	Gaju *et al.*, 1995
			*Ase*I	24	
Chlorobium limicola		51–58%	*Swa*I	>7	Mendez Alvarez *et al.*, 1995
Chlorobium tepidum			*Pac*I	2	Naterstad *et al.*, 1995
			*Xba*I	6	
Clostridium beijerinckii (formerly *C. acetobutylicum*)	6700				Wilkinson and Young, 1995
Clostridium botulinum type A	4039		*Ksp*I		Lin and Johnson, 1995
			*Mlu*I		
			*Nae*I		
			*Nru*I		
			*Rsr*II		
			*Sma*I		
			*Xho*I		

Table 24-1. Continued

Species (and strain)	Genome size (or other)	G+C content	Restriction Enzyme	Number of sites	References
Clostridium difficile	Epidem.		*Sma*I	5–10	Chachaty *et al.*, 1994
			*Nru*I	6–12	
Clostridium difficile	Epidem.		*Sma*I	>10	Samore *et al.*, 1996;
					Talon *et al.*, 1995
Clostridium perfringens	3070–3650		*Nru*I	1–6	Canard *et al.*, 1992
Seven strains			*Fsp*I	3–17	
			*Sma*I	4–8	
			*Mlu*I	5–7	
			*Apa*I	9–13	
Clostridium perfringens	3600	25%	*Not*I	0	Canard and Cole, 1989
			*Mlu*I	5	
			*Nru*I	8	
			*Sac*II	10	
			*Apa*I	12	
			*Sma*I	12	
			*Fsp*I	16	
Clostridium perfringens			I-*Ceu*I	1	Katayama *et al.*, 1996
Corynebacterium diphtheriae	Epidem.		*Not*I		Hogg *et al.*, 1996
			*Sma*I	>20	
			*Sfi*I		
Corynebacterium glutamicum	2640–2880		*Not*I	40	Nikolskii *et al.*, 1992
ATCC 13032			*Sfi*I	43	
Corynebacterium diphtheriae			*Sfi*I	43	De Zoysa *et al.*, 1995
Enterobacter aerogenes	Epidem.		*Xba*I		Neuwith *et al.*, 1996
Enterobacter cloacae	Epidem.		*Not*I		Haertl and Bandlow, 1993
			*Xba*I		
			*Apa*I	>20	
			*Nhe*I	>20	
			*Spe*I	>20	

Enterococcus faecalis OG1	2825		*Sfi*I	5	Murray *et al.*, 1993
			*Asc*I	9	
			*Not*I	15	
			*Sma*I		
Enterococcus faecalis	Epidem.		*Sma*I		Murray *et al.*, 1990; Coque and Murray, 1995
					Tomayo *et al.*, 1995
Enterococcus faecalis	Epidem.		*Sma*I		Kuhn *et al.*, 1995
			*Apa*I		
Enterococcus faecalis 226 NWC			*Sma*I		Salzano *et al.*, 1992
Enterococcus faecium	Epidem.		*Sma*I		Green *et al.*, 1995;
					Handwerger *et al.*, 1995
					Plessis *et al.*, 1995;
					Barbier *et al.*, 1996
Enterococcus faecium	Epidem.		*Apa*I		Morris *et al.*, 1995
			*Sma*I		
Escherichia coli K12	4595	50%	*Not*I	22	Smith *et al.*, 1987
			*Sfi*I	28	
Escherichia coli K-12 strain MG1655	Epidem.		*Avr*II	13	Daniels, 1990
			*Sfi*I	31	Perkins *et al.*, 1992
			*Not*I		Heath *et al.*, 1992
Escherichia coli	Epidem.		*Not*I		Arbeit *et al.*, 1990
Escherichia coli 06	Epidem.		*Xba*I		Blum *et al.*, 1991
Escherichia coli	Epidem.		*Not*I		Honda *et al.*, 1995
			*Xba*I		
Escherichia coli	Epidem.		*Not*I		Russo *et al.*, 1995
			*Sfi*I		
			*Xba*I		
Escherichia coli	Epidem.		*Xba*I		Kern *et al.*, 1994;
					Krause *et al.*, 1996
					Meng *et al.*, 1995
Escherichia coli	Epidem.				Johnson *et al.*, 1995

Table 24-1. Continued

Species (and strain)	Genome size (or other)	G+C content	Restriction Enzyme	Number of sites	References
Escherichia coli	4660–5300		*Bln*I	11–23	Bergthorsson and Ochman, 1995
			*Not*I	11–28	
Flavimonas oryzihabitans	Epidem.				Liu *et al.*, 1996
Flavobacterium meningosepticum					Sader *et al.*, 1995
Haemophilus influenzae	1980–2100	37–39%	*Not*I	0–1	Kauc *et al.*, 1989
serotype b, strain Eagan			*Sfi*I	0	Butler and Moxon, 1990
Rd			*Nar*I	?	Lee *et al.*, 1989
			*Rsr*II	4	
			*Sma*I	16	
			*Eag*I	16–28	
			*Apa*I	21	
			*Nae*I	28	
Haemophilus influenzae Sb	2028–2045		*Not*I		Kauc *et al.*, 1996
			*Rsr*II		
			*Sma*I		
			*Srf*I		
Haemophilus parainfluenzae	2340	39%	*Nae*I	0	Kauc and Goodgal, 1989
			*Sfi*I	0	
			*Not*I	7	
			*Rsr*II	10	
			*Apa*I	18	
			*Eag*I	>3	
			*Xma*I	>3	
			*Sac*II	>30	
Helicobacter pylori	Epidem.		*Not*I	–>20	Salama *et al.*, 1995
			*Nru*I	–>20	
Klebsiella pneumoniae	Epidem.		*Spe*I		Gouby *et al.*, 1994
			*Xba*I		

Lactobacillus plantarum	2700–2905	*Sfi*I	23–25	Daniel, 1995
five strains		*Not*I	<30	
		*Swa*I	<30	
Lactococcus	2000–2700			Davidson *et al.*, 1995
Lactococcus lactis	Epidem.	*Sma*I	>10	Tanskanen *et al.*, 1990
29 strains of subsp. *lactis* and subsp. cremoris		*Not*I	<10	
		*Sfi*I	<10	
Lactococcus lactis		*Apa*I		Gireesh *et al.*, 1992
subsp. lactis LM0230		*Bgl*I		
		*Bss*HII		
		*Nci*I		
		*Sal*I		
		*Sma*I		
Lactococcus lactis		*Apa*I		Le Bourgeois *et al.*, 1989
		*Sma*I		
Lactococcus lactis	2420	*Not*I	3	Le Bourgeois *et al.*, 1992
subsp. lactis IL1403		*Sma*I	23	
		*Apa*I	31	
Lactococcus lactis		*Apa*I		Prevots *et al.*, 1994
Legionella bozemanni	Epidem.	*Not*I	^4	Luck *et al.*, 1995
		*Sfi*I		
Legionella micdadei	Epidem.	*Not*I		Luck *et al.*, 1995
		*Sfi*I	>10	
Legionella pneumophila	Epidem.	*Sfi*I	>6	Pruckler *et al.*, 1995
Legionella pneumophila Philadelphia 1		*Not*I		Bender *et al.*, 1991
Legionella pneumophila	Epidem.	*Not*I		Lueck *et al.*, 1991
Legionella pneumophila	Epidem.	*Bss*HII	>30	Marrie *et al.*, 1995
		*Sal*I	>30	
		*Spe*I	>30	

Table 24-1. Continued

Species (and strain)	Genome size (or other)	G+C content	Restriction Enzyme	Number of sites	References
Legionella pneumophila	Epidem.		*Not*I		Luck *et al.*, 1995
			*Sfi*I		
			*Apa*I		
			*Eag*I		
			*Nae*I		
			*Sac*II		
			*Sma*I		
Leptospira biflexa serovar patoc I	3500		*Not*I	12	Taylor *et al.*, 1991
			*Sal*I	15	
			*Mlu*I	16	
			*Apa*I	23	
			*Sma*I	24	
			*Sac*II	29	
			*Xba*I	>30	
			*Pvu*I	>30	
Leptospira interrogans	4700+350	34–40%	*Asc*I	3–4	Baril *et al.*, 1992
			*Sfi*I	3–4	Zuerner, 1991
			*Srf*I	4–5	Taylor *et al.*, 1991
			*Sgr*AI	8	Zuerner *et al.*, 1993
			*Sse*I	9	
			*Not*I	10–13	
			*Kpn*I	17	
			*Bgl*I	18	
			*Apa*I	20	
			*Sac*II	22	
			*Sma*I	22	
			*Mlu*I	27	
			*Xba*I	>30	

Leptotrichia buccalis	Epidem.		*Sma*I		Schwartz *et al.*, 1995
Leuconostoc oenos	Epidem.		*Apa*I		Lamoureux *et al.*, 1993
41 strains			*Not*I		
			*Sfi*I		
Listeria monocytogenes	3150	38%	*Not*I	8	Michel and Cossart, 1992
LO28 (serovar 1/2c)			*Sse*8387I	8	
			*Apa*I	30–40	
Listeria monocytogenes					Michel, 1993;
					Loncarevic *et al.*, 1996
Listeria monocytogenes	Epidem.		*Sma*I		Moore and Datta, 1994
			*Asc*I		Proctor *et al.*, 1995
Listeria monocytogenes	Epidem.		*Apa*I		Louie *et al.*, 1996
			*Sma*I		
Micrococcus Sp. Y-1	4061	70%	*Eco*RI		Park *et al.*, 1994
			*Hpa*I		
			*Ssp*I		
			*Spe*I		
			*Xba*I		
Moraxella (Branhamella) catarrhalis	1940		*Sma*I	9	Furihata *et al.*, 1995
ATCC25238			*Not*I	10	
Mycobacterium avium	Epidem.		*Ase*I	<30	Slutsky *et al.*, 1994
Mycobacterium BCG	Epidem.		*Asn*I	>25	Zhang *et al.*, 1995
			*Dra*I	>20	
			*Spe*I	>20	
			*Xba*I	>20	
Mycobacterium bovis	Epidem.		*Dra*I	>20	Feizbadi *et al.*, 1996
			*Spe*I	>30	
			*Vsp*I	>30	
			*Xba*I	>30	

Table 24-1. Continued

Species (and strain)	Genome size (or other)	G+C content	Restriction Enzyme	Number of sites	References
Mycobacterium tuberculosis H37Rv	4400		*Dra*I	>20	Philipp *et al.*, 1995
			*Asn*I	>20	
Mycobacterium tuberculosis	Epidem.		*Dra*I	>20	Griffith *et al.*, 1995
Mycobacterium tuberculosis	Epidem.		*Dra*I		Olson *et al.*, 1995
			*Xba*I	>20	
Mycoplasma capricolum	1155		*Bam*HI		Miyata *et al.*, 1993
Mycoplasma fermentans (incognitus strain)			*Bam*H	12	Huang *et al.*, 1995
			*Bgl*I	10	
			*Mlu*I	7	
Mycoplasma gallisepticum S6	1054		I-*Ceu*I	2	Gorton *et al.*, 1995
			*Eag*I	7	
			*Sma*I	8	
Mycoplasma genitalium	600	33%	*Apa*I	3	Colman *et al.*, 1990
			*Xho*I	7	
			*Sma*I	8	
Mycoplasma hominis five strains	704–825	29%	*Not*I	0	Ladefoged *et al.*, 1992
			*Sfi*I	0	
			*Apa*I	1–2	
			*Sal*I	1–3	
			*Xho*I	4–5	
			*Bam*HI	7–11	
			*Sma*I	13–16	
Mycoplasma mobile ATCC 43663	780		*Apa*I	2	Bautsch, 1988
			*Mlu*I	3	
			*Bam*HI	6	
			*Nru*I	7	

Mycoplasma mycoides	1040–1280		*Apa*I	2	Pyle *et al.*, 1988
subsp. mycoides			*Bss*HII	2	Pyle *et al.*, 1990
strains (PG1, KH3J, Gladysdale, V5 and Y)			*Sma*I	3	
and strain PG50			*Kpn*I	3	
			*Sal*I	3–4	
			*Xho*I	3–4	
			*Bgl*I	4–6	
			*Bam*HI	9	
Mycoplasma pneumoniae	795–775	40%	*Not*I	2	Krause and Lee, 1991
			*Sfi*I	2	Krause and Mawn, 1990
			*Apa*I	13	
			*Not*I	2	
Myxococcus xanthus DK1622	9200				He *et al.*, 1994
Myxococcus xanthus	9454		*Ase*I	16	Chen *et al.*, 1991
			*Spe*I	21	
Neisseria gonorrhoeae FA1090	2219	50%	*Not*I	0	Dempsey *et al.*, 1991, 1994
			*Sfi*I	0	
			*Nhe*I	16	
			*Spe*I	17	
			*Pac*I	21?	
			*Bgl*II	30	
Neisseria gonorrhoeae			*Bgl*II		Poh *et al.*, 1995
IB-2 and IB-6 serovar strains			*Spe*I		
Neisseria gonorrhoeae IA-6	Epidem.		*Xba*I	7	Xia *et al.*, 1995
			*Nhe*I	8	
Neisseria gonorrhoeae	Epidem.		*Spe*I		Li and Dillon, 1995; Ng *et al.*, 1995
Neisseria gonorrhoeae	Epidem.		*Spe*I	>14	Poh *et al.*, 1995
			*Nhe*I		
Neisseria gonorrhoeae	Epidem.		*Spe*I	>14	Poh *et al.*, 1996
			*Bgl*II		

Table 24-1. Continued

Species (and strain)	Genome size (or other)	G+C content	Restriction Enzyme	Number of sites	References
Neisseria meningitidis Z2491	2226		*Bgl*II	26	Dempsey *et al.*, 1995
			*Nhe*I	21	
			*Spe*I	21	
			*Sgf*I	10	
			*Pac*I	22	
			*Pme*I	22	
Neisseria meningitidis serogroup B	Epidem.		*Spe*I	>15	Yakuba and Pennington, 1995
Neisseria meningitidis	Epidem.		*Sfi*I		Edmond *et al.*, 1995
Nocardia opaca MR11			*Asn*I		Kalkus *et al.*, 1990
			*Spe*I		
			*Xba*I		
Ochrobactrum anthropi	Epidem.		*Xba*I	>20	van Dijck *et al.*, 1995
Pediococcus pentosaceus	1200		*Sma*I	^20	Daniel, 1995
Pediococcus acidilactici	1560		*Not*I	^20	Daniel, 1995
Planctomyces limnophilus DSM 3776T	5200		*Pac*I	5	Ward Rainey *et al.*, 1996
			*Pme*I	10	
			*Swa*I	17	
Proteus mirabilis	4200	39%	*Sfi*I	10	Allison and Hughes, 1991
			*Not*I	12	
Pseudomonas aeruginosa	Epidem.				Ojeniya *et al.*, 1993
	Epidem.		*Spe*I		Grundmann *et al.*, 1995
	Epidem.	67.2%	*Xba*I	<30	Seifert *et al.*, 1995; Barth *et al.*, 1995 Bennekov *et al.*, 1996

	Epidem.		*Dra*I		Talon *et al.*, 1996, 1995
	Epidem.		I-*Ceu*I		Mahenthiralingam *et al.*, 1996
	Epidem.		*Soe*I		Lau *et al.*, 1995
Pseudomonas aeruginosa PAO	5862		*Spe*I	37	Ratnaningsih *et al.*, 1990
Pseudomonas aeruginosa	5900		*Dpn*I	15*	Roemling and Tuemmler, 1989
PAO (DSM 1707)			*Spe*I	38	Roemling *et al.*, 1991
			*Dpn*I		Liao *et al.*, 1996
Pseudomonas aeruginosa C	6500		I-*Ceu*I	4	Schmidt *et al.*, 1996
			*Swa*I	5	
			*Pac*I	8	
			*Spe*I	47	
Pseudomonas cepacia (see *Burkholderia*)					
Pseudomonas	Epidem/Taxon.	67.2%	*Asn*I		Grothues and Tuemmler, 1991
32 species			*Dra*I		
			*Spe*I		
			*Xba*I		
Pseudomonas			*Sep*I		Holloway *et al.*, 1992
P. aeruginosa, P. putida and *P. solanacearum*					
Pseudomonas fluorescens	6630		*Pac*I	15	Rainey and Bailey, 1996
SBW25			*Spe*I	53	
			*Xba*I	71	
Rhizobium galegae	5852		*Asn*I		Huber *et al.*, 1994
			*Spe*I		
Rhizobium	Taxon.		*Xba*I		Haukka and Lindstrom, 1994
			*Spe*I		
			*Pac*I	<20	
			*Bfr*I	<20	
			*Cla*I	<20	
			*Dra*I	<20	

Table 24-1. Continued

Species (and strain)	Genome size (or other)	G+C content	Restriction Enzyme	Number of sites	References
Rhizobium loti			*Spe*I		Sullivan *et al.*, 1995
			*Swa*I		
Rhizobium meliloti 1021	3400+1700+1400		*Swa*I	6	Honeycutt *et al.*, 1993
			*Pac*I	4	Sobral *et al.*, 1991
			I-*Ceu*I		Sobral *et al.*, 1991
			*Pme*I		
			*Ase*I		
			*Spe*I		
			*Ssp*I		
			*Xba*I		
Rhodobacter capsulatus	3,600+134		*Spe*I	3	Fonstein *et al.*, 1992
			*Sca*I	14	
			*Xba*I	16	
			*Ase*I	25	
			*Dra*I	25	
Rhodobacter sphaeroides 2.4.1	3,046+914+five plasmids		*Sna*BI	10	Suwanto and Kaplan, 1992
			*Spe*I	16	
			*Ase*I	17	
			*Dra*I	>25	
Rhodobacter sphaeroides	2973	65%	*Sna*BI	4	Mackenzie *et al.*, 1995
			*Spe*I	4	
			*Ase*I	8	
			*Dra*I	10	
Rhodococcus sp. R312	6440		*Dra*I	15	Bigey *et al.*, 1995
			*Asn*I	24	

Rickettsia akari	Epidem.	*Bss*HII	>2	Eremeeva *et al.*, 1995
		*Eag*I	>7	
		*Sma*I	>15	
Rickettsia massiliae	1370	*Eag*I	15–20	Roux and Raoult, 1993
		*Sma*I	20–30	
		*Bss*HII		
R. helvetica	1397	*Eag*I	15–20	Roux and Raoult, 1993
		*Sma*I	20–30	
		*Bss*HII		
R. bellii	1660	*Eag*I	15–20	Roux and Raoult, 1993
		*Sma*I	20–30	
		*Bss*HII		
Rickettsiella grylli	2100	*Not*I	7	Frutos *et al.*, 1989
		*Sfi*I	6	
R. melolonthae	1720	*Not*I	8	Frutos *et al.*, 1989
		*Sfi*I	4	
Salmonella species		*Bln*I	>10	Murase *et al.*, 1995
S. typhimurium, S. thompson,		*Xba*I	>15	
S. enteritidis				
Salmonella species		*Avr*II		Thong *et al.*, 1995
S. typhimurium, S. typhi,		*Spe*I		Thong *et al.*, 1996
S. enteritidis				
		*Xba*I		
Salmonella enteritica	Epidem.	*Not*I		Liebisch and Schwarz, 1996
		*Spe*I		
		*Xba*I		
Salmonella enteritica	Epidem.	*Not*I	>20	Olson *et al.*, 1995
Salmonella enteritica	Epidem.	*Spe*I		Powell *et al.*, 1995

Table 24-1. Continued

Species (and strain)	Genome size (or other)	G+C content	Restriction Enzyme	Number of sites	References
Salmonella enteritidis strain SSU7998	4600		I-*Ceu*I	7	Liu *et al.*, 1994
			*Bln*I	12	
			*Xba*I	16	
Salmonella enteritidis	Epidem.		*Not*I	>20	Suzuki *et al.*, 1995
			*Xba*I	>10	
Salmonella typhi	Epidem.		*Xba*I	>12	McKenna *et al.*, 1995
			*Spe*I	>12	
			*Sfi*I	>30	
Salmonella typhimurium LT2	4807+90		*Xba*I	24	Liu and Sanderson, 1992
			I-*Ceu*I	7	Liu *et al.*, 1993
			*Bln*I	12	Wong and McClelland, 1992
Salmonella paratyphi B			I-*Ceu*I	7	Liu *et al.*, 1994
			*Bln*I	10	
			*Xba*I	19	
Shigella flexneri 2a (YSH6000)	4590		*Not*I	19	Okada *et al.*, 1991
			*Sfi*I	>30	
			*Sse*8387I	>30	
Shigella sonnei	Epidem.		*Xba*I	<25	Liu *et al.*, 1991
Spiroplasma citri	1780		*Sal*I	8	Ye *et al.*, 1992
			*Apa*I	9	
			*Bss*HII	9	
			*Sma*I	17	
			*Sst*II	14	
Spiroplasma species	940-2220		*Not*I	8	Carle *et al.*, 1995
Staphylococcus aureus 50 oxacillin-resistant strains	Epidem.		*Sma*I		Schwarzkopf *et al.*, 1993
MRSA					Nada *et al.*, 1996; Sanches *et al.*, 1995

Staphylococcus aureus	Epidem.		*Apa*I		Maslow *et al.*, 1995
Staphylococcus aureus	Epidem.		*Sma*I		Pantucek *et al.*, 1996;
					Prevost *et al.*, 1995
Staphylococcus aureus					De Lancastre *et al.*,
					1994; Wanger *et al.*, 1992
					Bannerman *et al.*, 1995;
					Hu *et al.*, 1995
					Leonard *et al.*, 1995;
					Talon *et al.*, 1995
					Cookson *et al.*, 1996;
					Couto *et al.*, 1995
					Hartstein *et al.*, 1995;
					Kluytmans *et al.*, 1995
					Teixeira *et al.*, 1995
Staphylococcus aureus	Epidem.				Prevost *et al.*, 1991;
					Jorgensen *et al.*, 1996
Staphylococcus epidermidis	Epidem.		*Sma*I		Lina *et al.*, 1995
Stenotrophomonas maltophilia	Epidem.		*Dra*I	5–>10	Yao *et al.*, 1995
(*Xanthomonas*)	Epidem.		*Dra*I		Fabe *et al.*, 1995
			*Xba*I		
Streptococcus mutans	2200	36–38%	*Sfi*I	0	Okahashi *et al.*, 1990
strain MT8148			*Sma*I	0?	
			*Not*I	10	
			*Apa*I	>15	
			*Nae*I	>30	
			*Sac*II	>30	
Streptococcus mutans		36–38%	*Asc*I	5	Hantman *et al.*, 1993
GS-5			*Not*I	10	
			*Apa*I	18	
			*Rsr*II	21	
			*Sma*I	24	
			*Eag*I	>30	
			*Nae*I	>30	

Table 24-1. Continued

Species (and strain)	Genome size (or other)	G+C content	Restriction Enzyme	Number of sites	References
Streptococcus mutans	Epidem.		*Not*I		Kiska and Macrina, 1994
Streptococcus pneumoniae	2270	39%	*Not*I	0	Gasc *et al.*, 1991
Formerly *Diplococcus*			*Rsr*II	0	
			*Sfi*I	0	
			*Sma*I	20	
			*Apa*I	22	
			*Sac*II	29	
Streptococcus pneumoniae	2500–2700		*Pac*I	6–7	McDougal *et al.*, 1995
			*Sma*I	11–16	
Streptococcus pneumoniae	Epidem.		*Apa*I	>10	Herman *et al.*, 1995; Lefevre *et al.*, 1995
			*Sma*I		Gasc *et al.*, 1995
Streptococcus pneumoniae	Epidem.		*Eag*I		Moreno *et al.*, 1995
			*Sma*I	>10	
Streptococcus pneumoniae	Epidem.		*Apa*I		Moissenet *et al.*, 1995
Streptococcus (group A)	Epidem.		*Apa*I		Shah *et al.*, 1995
			*Sfi*I		
			*Sma*I	>10	
Streptococcus (group A)	Epidem.		*Sma*I	>7	Stanley *et al.*, 1996
Streptococcus pyogenes (serotype M)			*Sma*I	>10	Stanley *et al.*, 1995
Streptococcus pyogenes	Epidem.		*Sma*I	^6	Chaussee *et al.*, 1996
			*Sfi*I	^5	
Streptococcus pyogenes	Epidem.		*Sfi*I	>4	Musser *et al.*, 1995
Streptococcus pyogenes	Epidem.		*Sfi*I	5–10	Upton *et al.*, 1995

Streptococcus salivarius subsp. thermophilus			*Apa*I		Le Bourgeois *et al.*, 1989
			*Sma*I		
Streptococcus sorbinus strain 6715	2200	36–38%	*Sma*I	0?	Okahashi *et al.*, 1990
			*Sfi*I	0	
			*Not*I	6	
			*Apa*I	>15	
			*Nae*I	>30	
			*Sac*II	>30	
Streptomyces avermitilis					Irnich and Cullich, 1993
Streptomyces 59 strains representing eight species	Taxon.	69–78%	*Ase*I	3–15	Beyazova *et al.*, 1993
			*Dra*I	5–>20	
Streptomyces coelicolor A3(2)	8000(L)		*Dra*I	8	Kieser *et al.*, 1992
	Epidem.		*Ase*I	17	Duchene *et al.*, 1994
			*Ssp*I	25	
Streptomyces griseus 2247	7800		*Ase*I	15	Lezhava *et al.*, 1995
			*Dra*I	9	
Streptomyces lividans 66	8000(L)		*Dra*I	7	Leblond *et al.*, 1993
			*Ase*I	14–16	Lin *et al.*, 1993
Streptomyces lividans					Irnich and Cullich, 1993
Synechococcus sp. strain PCC 7002	2700		*Not*I	21	Chen and Widger, 1993
			*Sal*I	37	
			*Sfi*I	27	
			*Asc*I		
Synechocysti sp. Strain PCC 6803	3600		*Asc*I	6	Kotani *et al.*, 1994
			*Mlu*I	25	Churin *et al.*, 1995
			*Spl*I	30	
Taylorella equigenitalis	Epidem.		*Apa*I		Miyazawa *et al.*, 1995
			*Not*I		

Table 24-1. Continued

Species (and strain)	Genome size (or other)	G+C content	Restriction Enzyme	Number of sites	References
Thermus thermophilus HB8	1740	69%	*Ase*I	0	Borges and Bergquist, 1993
			BspHI	0	
			*Pac*I	0	
			*Pme*I	0	
			*Sal*I	0**	
			*Swa*I	0	
			*Hpa*I	6	
			*Mun*I	7	
			*Nde*I	12	
			*Bst*BI	<40	
			*Dra*I	<40	
			*Eco*RI	<40	
			*Eco*RV	<40	
			*Hpa*I	<40	
			*Ssp*I	<40	
			*Xba*I	<40	
Thermus thermophilus HB27	1820		*Eco*RI		Tabata *et al.*, 1993
			*Ssp*I		
			*Mun*I		
Treponema denticola ATCC 33520	3000		*Asc*I	4	MacDougall and Saint Girons, 1995
			*Not*I	21	
			*Srf*I	9	
Ureaplasma urealyticum 960T	900		*Apa*I	1	Cocks *et al.*, 1989
			*Sma*I	1	
			*Bss*HII	6	
			*Bgl*I	7	
			*Asp*718	8	

Vibrio cholerae	Epidem.	*Not*I	>18	Evins *et al.*, 1996; Kurazono *et al.*, 1996
Vibrio cholerae O1	Epidem.	*Not*I	>20	Desmarchelier *et al.*, 1996
Vibrio cholerae O139	Epidem.	*Cpo*I	7	Dalsgaard *et al.*, 1996
Vibrio cholerae O139	Epidem.	*Sfi*I	>20	Gyobu *et al.*, 1995; Bhadra *et al.*, 1995
		*Not*I	>20	
Vibrio cholerae 569B	3200	I-*Ceu*I	7	Majumder *et al.*, 1996
		*Sfi*I	22	
		*Not*I	37	
Vibrio vulnificus	Epidem.	*Sfi*I		Buchrieser *et al.*, 1995
		*Srf*I		
Xanthomonas campestris strains	Epidem./Taxon.	*Spe*I	>10	Egel *et al.*, 1991
		*Xba*I	>10	
Xanthomonas maltophilia	Epidem.	*Spe*I	>5	Laing *et al.*, 1995
Yersinia enterocolitica	Epidem./Taxon.	*Xba*I	^33	Buchrieser *et al.*, 1994
Sixty strains		*Not*I	^36	
Yersinia pestis		*Spe*I		Iteman *et al.*, 1993
Yersinia ruckeri strains	4460–4770	*Not*I	35–43	Romalde *et al.*, 1991
	Epidem.	*Sfi*I	41–47	

Epidem. = application to epidemiology. Taxon. = application to taxonomy. L = linear genome

** = endogenous methylation inhibits this enzyme. * = endogenous methylation creates sites for this enzyme.

Here we present a list of the enzymes that have been used in bacterial mapping (Table 24-1) and briefly review some of the strategies that have been used to enhance cleavage specificities. Table 24-1 also lists publications that utilize the fingerprint of restriction products as an epidemiological tool to relate bacterial strains.

Other recent review include Hillier and Davidson, 1995, Mendez Alvarez *et al.*, 1995, Tenover *et al.*, 1995, and Fonstein and Haselkorn, 1995.

Estimating Average Fragment Sizes Generated by Restriction Enzymes

The distribution of bases in a naturally occurring DNA sequence is not random (e.g., Nussinov, 1980; Nussinov, 1981). If DNA sequence data are available, one can predict which restriction enzymes will cut the DNA least frequently. To do statistical analyses for prediction of frequency of target sites for a given genome, it is necessary to first generate a database containing the sequenced regions of the genome(s) in question. Sequences of closely related species are also useful. The database is then used to extract mono-, di-, and trinucleotide frequencies (see also Chapter 19). The larger the database available, the closer the mono-, di-, and trinucleotide frequencies are to their respective probabilities of occurrence in the genome in question (e.g., Phillips *et al.*, 1987; Phillips *et al.*, 1987).

The sequences that have been entered for the organism (or group of organisms) can be assembled from the GenBank database. Mono-, di-, and trinucleotide frequencies can be calculated by using sequence analysis programs such as MacVector, Oxford Molecular Group. These frequencies are used to calculate the estimated frequency of a restriction enzyme target site as follows. For a 6-base-pair recognition site, the frequency of the sequence $X_1X_2X_3X_4X_5X_6$ (where Xn are specific bases in the recognition sequence) can be calculated from di- and trinucleotide frequencies in the database by the following formula:

$$p(X_1X_2X_3)\cdot p(X_4X_5X_6)\cdot p(X_3X_4X_5)\cdot p(X_4X_5X_6)$$
$$/p(X_2X_3)\cdot p(X_3X_4)\cdot p(X_5X_6)$$

Probably the best estimator considers the frequencies of more subsets of trinucleotides (Karlin and Cardon, 1994; see Chapter 19) including those of the form $p(X_1N_0X_2X_3)$ where N_0 is an unspecified base separating the three specified bases.

Data from PFGE experiments using enzymes that cut at rare sequences, as calculated in this manner, show that most of the predictions are quite accurate or sometimes underestimate the sizes of the observed fragments (e.g., Sobral *et al.*, 1990; Sobral *et al.*, 1991). Thus, the frequencies of di- and trinucleotides in finite-length sequences are not necessarily equal to the actual frequencies of these di- and trinucleotides in the genome. One of the sources of inaccuracy in these methods is that regions of the genome vary in their sequence organization and by adding the frequency of di- or trinucleotides from disparate regions of the genome, one can mask significant differences.

Enzymatic Methods to Alter Restriction Enzyme Specificity

Many of these strategies are summarized in Nelson and McClelland, (1992).

*Dpn*I-Based Strategies

The restriction enzyme *Dpn*I (5′-G^{m}ATC-3), cuts double-stranded DNA molecules efficiently if both strands are N^{6}-methylated at adenine residues (McClelland *et al.*, 1984). The vast majority of species do not methylate their genomes at 5′- G^{m}ATC-3′. By using *Dpn*I in combination with various methyltransferases, effective recognition sequences of eight to twelve base pairs (bp) have been demonstrated (McClelland *et al.*, 1984; Hanish and McClelland, 1991).

Cross-Protection

Restriction enzymes are sensitive to methylation at most bases within their target site (Nelson *et al.*, 1993). This fact can be used to generate large DNA fragments by a 'cross-protection' process (Nelson *et al.*, 1984), in which a defined subset of restriction target sites is blocked from cleavage at partly overlapping methylase/restriction endonuclease sites by prior methylation. For example, modification by M.*Fnu*DII or M.*Bep*I (5′-^{m}CGCG-3′) blocks *Not*I cleavage at overlapping sites (5′-*CGCG*GCCGC-3′, which is equivalent to 5′-GCGGC*CGCG*-3′) and increases the apparent specificity of *Not*I digestion about two-fold (Qiang *et al.*, 1990). Other combinations have been demonstrated (Nelson *et al.*, 1993).

Competition reactions

Competition reactions between methylases and endonucleases are an alternative to restriction-enzyme-limited or Mg^{2+}-limited partial digestions (Albertsen *et al.*, 1989), as the latter strategies do not always work well if the DNA substrate is embedded in agarose blocks. In competitions, partial cleavage of DNA with a restriction endonuclease is achieved by using a methylase that has the same target site specificity. Competition reactions are exemplified by the combination of *Not*I and M.*Bsp*RI (5′-GG^{m}CC-3′). The competition is controlled because the restriction enzyme and methylase are used in a specific ratio, both in excess of the amount required to completely methylate or cleave the DNA sample. This results in a uniform partial digest (Hanish and McClelland, 1990).

Other Strategies

Besides strategies that rely on the joint use of modification and methyltransferases and restriction endonucleases, there are other methods for generating large DNA fragments. One is to create synthetic DNA-cleaving reagents by combining a sequence-specific DNA-binding domain with a DNA-cleaving moiety (Moser and Dervan, 1987). In this strategy, sequence-specific cleavage of double helical DNA is accomplished by binding of modified homopyrimidine EDTA oligonucleotides that form a triple helix structure and, in the presence of Fe(II) and dithiothreitol (DTT), cleave one (or both) strands of the 'double-stranded' DNA at a specific site. The practical limitation of this strategy, so far, is that cleavage is a low efficiency reaction and well-defined ends are not generated. The *trp* repressor gene from *E. coli* has also been chemically converted into a site-specific endonuclease by covalently attaching it to the 1,10-phenanthroline-copper complex (Chen and Sigman, 1987; Pan *et al.*, 1994); once again, however, cleavage is not very efficient.

A second general method for producing large DNA molecules is to block a restriction modification methylase site with a DNA binding protein, triple helix or D-loop structure. One example makes use of the high specificities of repressor proteins (Koob *et al.*, 1988; Koob and Szybalski, 1990). The site-specific *lac* repressor, which has an effective recognition sequence of 20 bp, blocks methylation at the *lac* repressor binding site. When the repressor is removed, the cognate restriction cuts only at the selectively protected site. This has been demonstrated for *E. coli* and yeast by cutting their respective genomes at a single introduced target site (Koob and Szybalski, 1990).

An alternative strategy, RecA-assisted restriction endonuclease (RARE) cleavage, is based on the ability of RecA protein to pair an oligonucleotide to its homologous sequence in duplex DNA and to form a three-stranded complex. This complex is protected from *Eco*RI methylase; after methylation and RecA protein removal, *Eco*RI restriction enzyme cleavage was limited to the site previously protected from methylation (Ferrin and Camerini-Otero, 1991).

Intron-encoded Endonucleases

The intron-encoded endonuclease I-*Ceu*I (hereafter called *Ceu*I) (Colleaux *et al.*, 1986; Dujon *et al.*, 1989; Gauthier *et al.*, 1991; Marshall *et al.*, 1994) is encoded by a Class I mobile intron which is inserted into the *rrl* gene for the large subunit ribosomal RNA (23S rRNA) in the chloroplast DNA or *Chlamydomonas eugamatos. Ceu*I is specific for in a 19 bp sequence in the *rrl* gene (Dujon *et al.*, 1989; Gauthier *et al.*, 1991; Marshall *et al.*, 1994). Because rRNA sequences are strongly conserved, this 19 bp sequence is present in the genomes of many bacteria as well as in chloroplasts and mitochondria of eukaryotes; due to the large number of bases in the recognition site, it is predicted not to occur at other sites in the bacterial genome. *Ceu*I sites are present in all seven *rrl* genes for 23S rRNA in *S. typhimurium* (Liu *et al.*, 1993 (see Chapter 78)) and in the other enteric bacteria we have tested so far (Honeycutt *et al.*, 1993), but at no other locations. The 19 bp sequence for *Ceu*I recognition and cleavage is found in the *rrn* operons of a large number of prokaryotes, but not in eukaryotes (Liu *et al.*, 1993). Thus, the use of I-*Ceu*I in PFGE physical mapping allows the determination of the number, location and orientation of *rrn* loci on the physical map of a eubacterium (Honeycutt *et al.*, 1993).

Because of the specificity of *Ceu*I in cleaving only ribosomal RNA (rRNA) cistrons, and because the number and locations of these cistrons are highly conserved in enteric bacteria (Krawiec and Riley, 1990), we expect the number and size of *Ceu*I fragments obtained from digestion of the DNA (the fingerprint) of related bacteria to be very similar. On the other hand, the cleavage sites for class II restriction endonucleases are usually only 6 bp or 8 bp and are present everywhere in the genome, including less well conserved genes. These sites should be frequently gained or lost due to random nucleotide mutation during evolution, and thus should produce different fingerprints even for related species. Related wild type strains of *Salmonella typhimurium* yield similar *Ceu*I fingerprints. Interestingly however, partial *Ceu*I analysis of three independent strains of *S. typhi* reveals substantial genomic rearrangements rather than the stability seen in *S. typhimurium*. Thus genomic rearrangements are rare in *S. typhimurium* but surprisingly common in *S. typhi* (Liu and Sanderson, 1995; *see also* Chapter 22). Fingerprints generated by digestion with *Xba*I, a type II restriction endonuclease with a 6 bp site, are much more variable, providing little direct genomic information. Though data from Type II restriction endonucleases such as *Xba*I, *Bln*I, *Not*I and others cannot be used for genomic comparisons in the same direct ways as for *Ceu*I. the data from these enzymes is still valuable for construction of genomic cleavage maps (see also Chapters 22, 77 and 78).

Transposons Carrying Rare Cleavage Sites

One way to utilize endonucleases that cleave at rare sites, particularly those that cleave at sites that are so rare that they do not naturally occur in the genome, is to put their recognition sites in transposons and insert them into the genome. At the simplest level, one can take advantage of the fact transposons often differ dramatically from the rest of the genome in their DNA statistics (mono-, di- and trinucleotide frequencies), presumably reflecting their recent origin in phylogenetically distant species (see also Chapt. 19). The first example that took advantage of such a circumstance involved the rare *Not*I site that occurs in Tn*5* which resulted in a new site in the *E. coli* genome wherever it was integrated (McClelland *et al.*,1987; Heath *et al.*,1992). Another example is Tn*10*, which serendipitously carries an *Xba*I and two *Bln*I sites that contain the rare sequence CTAG, which are exceptionally rare in many bacteria, particularly enterobacteria. This fact was particularly fortuitous because this transposon had been used extensively for genetic mapping in *Salmonella*. Thus, it was very easy to construct a genetic/

physical map for *S. typhimurium* by relying on the cleavage of one fragment and the production of two new fragments in every Tn*10* genetic mapping strain (Wong and McClelland, 1992; Wong and McClelland, 1993; Liu *et al.*, 1993a *see* Chapter 22). Maps have now been completed for *S. enteritidis* (Liu *et al.*, 1993b) and *S. paratyphi* B (Liu *et al.*, 1994) and *S. typhi* (Liu and Sanderson, 1995) using this strategy (see also Chapters 77 and 78).

Similarly, Tn*5*-233, which has been used of genetic mapping in *Rhizobium*, fortuitously contains the rare *Spe*I site which proved useful in our mapping of the *Rhizobium meliloti* genome (Honeycutt *et al.*, 1993; *see* Chapter 74). A derivative of Tn*5*-751 carrying a *Swa*I site was used to inactivate and map genes in *Pseudomonas cepacia* (Cheng and Lessie, 1994).

In general, one cannot rely on fortuitous events, and therefore we constructed a series of transposons based on the broad host range Tn*5*. This transposon carries the rare recognition sites for M.*Xba*I-*Dpn*I (TCTAGATCTAGA), which is rare in most bacterial genomes, *Swa*I (ATTTAAAT) and *Pac*I (TTAATTAA), which are rare in (G+C)-rich genomes, *Not*I (GCGGCCGC) and *Sfi*I (GGCCN$_5$GGCC), which are rare in (A+T)-rich genomes, and *Bln*I (CCTAGG), *Spe*I (ACTAGT), and *Xba*I (TCTAGA), (Wong and McClelland, 1992). Tn*5*(pulsed field mapping), or Tn*5*(pfm) constructs are valuable tools for pulsed-field mapping of gram-negative bacterial genomes by assisting in the generation of physical maps and restriction fragment catalogs. A long range restriction map with rare cutting enzymes *Bln*I and *Xba*I has been constructed for *Salmonella typhimurium* LT2 (Wong and McClelland, 1992a; Liu and Sanderson, 1992; Liu *et al.*, 1993; *see* Chapter 78). By taking advantage of the *Bln*I restriction map of *S. typhimurium* LT2, and the presence of *Bln*I sites in both Tn*5*(pfm) (Wong and McClelland, 1992b) and Tn*10* transposons, we were able to construct a set of Pulsed-field gel electrophoresis mapping strains. Different combinations of mapped Tn*5* (pfm) insertions and Tn*10* insertions were selected and transduced sequentially into the same strain by selecting the corresponding Tc and Km markers. Each constructed strain produced specific *Bln*I "drop-out" fragments in the range 190 kb to 380 kb. The drop-out fragments were clearly resolved by pulsed field gel electrophoresis and purified for subsequent manipulation. Suwanto and Kaplan, (1992) have constructed a Tn*5* derivative that contains an *Ase*I, *Dra*I and *Sna*BI recognition site [rare in (G+C)-rich genomes]. A Tn*5* derivative containing the 18 bp I-*Sce*I site, delivered from a RP4-mobilizable, RK6-derived suicide vector has been developed (Jumas Bilak *et al.*, 1995). A Tn*10* derivative with an I-*Sce*I site has also been constructed (Bloch *et al.*, 1996).

Another strategy uses rare restriction sites within vectors bearing cloned sequences that undergo homologous recombination with the genome of the species from which they were derived. For example, Le Bourgeois *et al.*, 1992, used this strategy to map genes in *L. lactis* using a vector carry sites for *Apa*I, *Not*I and *Sma*I. Genes were localized and oriented in *Anabaena* using a similar strategy (Kuritz *et al.*, 1993; see also Chapt. 29).

Over the next few years other vector systems that integrate into genomes may be devised that contain other rare cleavage sites, such as targeted sites for the intronases I-*Ppo*I (Muscarella *et al.*, 1989), I-*Sce*I (Colleaux *et al.*, 1986), I-*Sce*II (Delahodde *et al.*, 1989), I-*Tev*I, I-*Tev*II (Bell-Pedersen *et al.*, 1990), and I-*Cre*I (Durrenberger and Rochaix, 1991).

New Technologies

While PFGE is still the preferred method for separating large DNA molecules in physical mapping (see Chapters 25 and 26), other methods are being developed that may be more efficient. In particular, David Schwartz, the inventor of PFGE, has devised such a method (Wang *et al.*, 1995). Fluorescence microscopy can be used to produce high-resolution restriction maps rapidly by directly imaging restriction digestion cleavage events occurring on single deproteinized DNA molecules. Ordered maps are then constructed by noting fragment order and size, using several optically based techniques. Soon even cleavage may not even be necessary to physically map a genome. Peptide Nucleic Acids (PNAs) bind sequence specifically and tenatiously to DNA (Demidov *et al.*, 1995). When modified with an optically visible moiety these could act as landmarks for fluorescence microscope mapping.

There has been controversy about the relative utility of PFGE restriction versus other methods such as the Arbitrarily primed PCR methods (Welsh and McClelland, 1990; Williams *et al.*, 1990; *see* Chapter 33) for epidemiology in bacteria. A number of papers compare PFGE to other methods (e.g., Collier *et al.*, 1996; Hojo *et al.*, 1995; Kersulyte *et al.*, 1995; Liu *et al.*, 1995; Louie *et al.*, 1996; Pruckler *et al.*, 1995; van Belkum *et al.*, 1995; Webster *et al.*, 1996). Improvements, such as automated laser fluorescence (Webster *et al.*, 1996) and neural networks (Carson *et al.*, 1995), should help analyse of DNA fingerprints, regardless of how they are generated.

Acknowledgments

The work was supported by grants R01AI34829 (McClelland) and R29AI32644 (John Welsh) from the National Institute of Allergy and Infectious Diseases, and by an Operating grant from the Natural Sciences and Engineering Research Council of Canada (Sanderson).

References

Albertsen, H. M., D. Le Paslier, H. Adberrahim, J. Dausset, H. Cann, and D. Cohen. 1989. Improved control of partial DNA restriction enzyme digest in agarose using limiting concentrations of Mg++. *Nucl. Acids Res.* 17:808–808.

Allardet-Servent, A., N. Bouziges, M. J. Carles-Nurit, G. Bourg, A. Gouby, and M. Ramuz. 1989. Use of low-frequency-cleavage restriction endonucleases for DNA analysis in epidemiological investigations of nosocomial bacterial infections. *J. Clin. Microbiol.* 27:2057–61.

Allison, C., and C. Hughes. 1991. Closely linked genetic loci required for swarm cell differentiation and multicellular migration by *Proteus mirabilis*. *Mol. Microbiol.* 5:1975–1982.

Amjad, M., J. M. Castro, H. Sandoval, J. J. Wu, M. Yang, and D. J. Henner. Piggot P. J. 1991. An *Sfi*I restriction map of the *Bacillus subtilis* 168 genome. *Gene*. 101:15–21.

Arbeit, R. D., M. Arthur, R. Dunn, C. Kim, R. K. Selander, and R. Goldstein. 1990. Resolution of recent divergence among *Escherichia coli* from related lineages the application of pulsed field electrophoresis to molecular epidemiology. *J. Infect. Dis.* 161:230–235.

Bancroft, I., C. P. Wolk, and E. V. Oren. 1989. Physical and genetic maps of the genome of the heterocyst-forming cyanobacterium *Anabaena* sp. strain PCC 7120. *J. Bacteriol.* 171:5940–8.

Bannerman, T. L., G. A. Hancock, F. C. Tenover and J. M. Miller. 1995. Pulsed-field gel electrophoresis as a replacement for bacteriophage typing of *Staphylococcus aureus. J. Clin. Microbiol.* 33:551–555.

Barbier, N., P. Saulnier, E. Chachaty, S. Dumontier and A. Andremont. 1996. Random amplified polymorphic DNA typing versus pulsed-field gel electrophoresis for epidemiological typing of vancomycin-resistant enterococci. *J. Clin. Microbiol.* 34: 1096–1099.

Baril, C., J. L. Herrmann, C. Richaud, D. Margarita, and I. Saint Girons. 1992. Scattering of the rRNA genes on the physical map of the circular chromosome of *Leptospira interrogans* serovar *icterohaemorrhagiae J. Bacteriol.* 174:7566–7571.

Barth, A. L. and T. L. Pitt. 1995. Auxotrophic variants of *Pseudomonas aeruginosa* are selected from prototrophic wild-type strains in respiratory infections in patients with cystic fibrosis. *J. Clin. Microbiol.* 33:37–40.

Bautsch, W. 1988. Rapid physical mapping of the *Mycoplasma mobile* genome by two-dimensional field inversion gel electrophoresis techniques. *Nucl. Acids Res.* 16:11461–7.

Beall, B., P. K. Cassiday and G. N. Sanden. 1995. Analysis of *Bordetella pertussis* isolates from an epidemic by pulsed-field gel electrophoresis. *J. Clin. Microbiol.* 33:3083–3086.

Bell-Pedersen, D., S. Quirk, J. Clyman, and M. Belfort. 1990. Intron mobility in phage T4 is dependent upon a distinctive class of endonucleases and independent of DNA sequence encoding the intron core: mechanistic and evolutionary implications. *Nucl. Acids Res.* 18:3763.

Beltan, P., S. A. Plock, N. H. Smith, T. S. Whitam, D. C. Old, and R. K. Selander. 1991. Reference collections of strains of the *Salmonella typhimurium* complex from natural sources. *J. Gen. Microbiol.* 137:601–606.

Bender, L., M. Ott, A. Debes, U. Rdest, J. Heesemann, and J. Hacker. 1991. Distribution expression and long-range mapping of legiolysin genelly-specific DNA sequences in *Legionellae. Infection & Immunity.* 59:3333–3336.

Bennekov, T., H. Colding, B. Ojeniyi, M. W. Bentzon and N. Hoiby. 1996. Comparison of ribotyping and genome fingerprinting of *Pseudomonas aeruginosa* isolates from cystic fibrosis patients. *J. Clin. Microbiol.* 34:202–204.

Bergthorsson, U. and H. Ochman. 1995. Heterogeneity of genome sizes among natural isolates of *Escherichia coli. J. Bacteriol.* 177:5784-b5789.

Beyazova, M., and M. P. Lechevalier. 1993. Taxonomic utility of restriction endonuclease fingerprinting of large DNA fragments from *Streptomyces* strains. *Int. J. System. Bacteriol.* 43:674–682.

Bhadra, R. K., S. Roychoudhury, R. K. Banerjee, S. Kar, R. Majumdar, S. Sengupta, S. Chatterjee, G. Khetawat and J. Das. 1995. Cholera toxin (CTX) genetic in *Vibrio cholerae* O139, *Microbiology* 141:1977–1983.

Bigey, F., G. Janbon, A. Arnaud and P. Galzy. 1995. Sizing of the *Rhodococcus* sp. R312 genome by pulsed-field gel electrophoresis. Localization of genes involved in nitrile degradation. *Antonie Van Leeuwenhoek.* 68:173–179.

Birkelund, S., and R. S. Stephens. 1992. Construction of physical and genetic maps of *Chlamydia trachomatis* serovar L2 by pulsed-field gel electrophoresis. *J. Bacteriol.* 174:2742–7.

Bloch, C. A., C. K. Rode, V. H. Obreque, and J. Mahillon. 1996. Purification of *Escherichia coli* chromosomal segments without cloning. *Biochem. Biophys. Res. Commun.* 223:104–111.

Blum, G., M. Ott, A. Cross, and J. Hacker. 1991. Virulants determinants of *Escherichia coli* O6 extraintestinal isolates analyzed by southern hybridizations and DNA long range mapping techniques. *Microbial PathoGenesis.* 10:127–136.

Borges, K. M., and P. L. Bergquist. 1993. Genomic restriction map of the extremely thermophilic bacterium *Thermus thermophilus* HB8. *J. Bacteriol.* 175:103–110.

Borges, K. M., and P. L. Bergquist. 1993. Pulsed field gel electrophoresis study of the genome of *Caldocellum saccharolyticum. Current Microbiol.* 27:15–19.

Bourke, B., P. Sherman, H. Louis, E. Hani, P. Islur and V. L. Chan. 1995. Physical and genetic map of the genome of *Campylobacter upsaliensis. Microbiology.* 141:2417–2424.

Boyd, E. F., F.-S. Wang, P. Beltran, S. A. Plock, K. Nelson, and R. K. Selander. 1993. *Salmonella* reference collection B (SARB): strains of 37 serovars of subspecies I. *J. Gen. Microbiol.* 139:1125–1132.

Buchrieser, C., S. D. Weagant, and C. W. Kaspar. 1994. Molecular characterization of *Yersinia enterocolitica* by pulsed-field gel electrophoresis and hybridization of DNA fragments to ail and pYV probes. *Appl. & Environ. Microbiol.* 60:4371–4379.

Buchrieser, C., V. V. Gangar, R. L. Murphree, M. L. Tamplin and C. W. Kaspar. 1995. Multiple *Vibrio vulnificus* strains in oysters as demonstrated by clamped homogeneous electric field gel electrophoresis. *Appl. Environ. Microbiol.* 61:1163–1168.

Busch, U., C. Hizo Teufel, R. Boehmer and V. Preacmursic. 1996. Differentiation of *Borrelia burgdorferi* sensu lato strains isolated from skin biopsies and tick by pulsed-field gel electrophoresis. *Rocz. Akad. Med. Bialymst.* 41:51–58.

Busch, U., C. Hizo Teufel, R. Boehmer, V. Fingerle, H. Nitschko, B. Wilske and V. Preac

Mursic. 1996. Three species of *Borrelia burgdorferi* sensu lato (*B. burgdorferi* sensu stricto, *B. afzelii*, and *B. garinii*) identified from cerebrospinal fluid isolates by pulsed-field gel electrophoresis and PCR. *J. Clin. Microbiol.* 34:1072–1078.

Butler, P. D., and E. R. Moxon. 1990. A physical map of the genome of *Haemophilus influenzae* type B. *J. Gen. Microbiol.* 136:2333–2342.

Canard, B., and S. T. Cole. 1989. Genome organization of the anaerobic pathogen. *Clostridium perfringens.* Proc. Natl. Acad. Sci. USA. 86:6676–80.

Canard, B., B. Saint-Joanis, and S. T. Cole. 1992. Genomic diversity and organization of virulence genes in the pathogenic anaerobe *Clostridium perfringens. Mol. Microbiol.* 6:1421–9.

Carle, G. F., M. Frank and M. V. Olson. 1986. Electrophoretic separation of large DNA molecules by periodic inversion of the electric field. *Science* 232:65–68.

Carle, P., F. Laigret, J. G. Tully and J. M. Bove. 1995. Heterogeneity of genome sizes within the genus Spiroplasma. *Int. J. Syst. Bacteriol.* 45:178–181.

Carlson, C. R., A. Gronstad, and A. B. Kolsto. 1992. Physical maps of the genomes of three *Bacillus cereus* strains. *J. Bacteriol.* 174:3750–3756.

Carlson, C. R., and A. B. Kolsto. 1993. A complete physical map of a *Bacillus thuringiensis* chromosome. *J. Bacteriol.* 175:1053–60.

Carson, C. A., J. M. Keller, K. K. McAdoo, D. Wang, B. Higgins, C. W. Bailey, J. G. Thorne, B. J. Payne, M. Skala and A. W. Hahn. 1995. *Escherichia coli* O157:H7 restriction pattern recognition by artificial neural network. *J. Clin. Microbiol.* 33:2894–2898.

Casjens, S. and W. M. Huang. 1993. Linear chromosomal physical and genetic map of *Borrelia burgdorferi*, the Lyme disease agent. *Mol. Microbiol.* 8:967–980.

Casjens, S., M. Delange, H. L. Ley, P. Rosa, W. M. Huang. 1995. Linear chromosomes of Lyme disease agent spirochetes: genetic diversity and conservation of gene order. *J. Bacteriol.* 177:2769–2780.

Chachaty, E., P. Saulnier, A. Martin, N. Mario, and A. Andremont. 1994. Comparison of ribotyping, pulsed-field gel electrophoresis and random amplified polymorphic DNA for typing *Clostridium difficile* strains. *FEMS Microbiol. Lett.* 122:61–68.

Charlebois, R. L., L. C. Schalkwyk, J. D. Hofman, and W. F. Doolittle. 1991. Detailed physical map and set of overlapping clones covering the genome of the *Archaebacterium haloferax volcanii* DS2. *J. Mol. Biol.* 222:509–524.

Chaussee, M. S., J. Liu, D. L. Stevens and J. J. Ferretti. Genetic and phenotypic diversity among isolates of *Streptococcus pyogenes* from invasive infections. *J. Infect. Dis.* 173:901–908.

Chen, C.-H. and D. D. Sigman. 1987. Chemical conversion of a DNA-binding protein into a site-specific nuclease. *Science* 237:1197–1201.

Chen, H., A. Kuspa, I. M. Keseller, and L. J. Shimkets. 1991. Physical map of the *Myxococcus xanthus* chromosome. *J. Bacteriol.* 173:2109–2115.

Chen, X., and W. R. Widger. 1993. Physical genome map of the unicellular *Cyanobacterium synechococcus* sp. strain PCC 7002. *J. Bacteriol.* 175:5106–16.

Cheng, H. P. and T. G. Lessie. 1994. Multiple replicons constituting the genome of *Pseudomonas cepacia* 17616. *J. Bacteriol.* 176:4034–42.

Churin, Y. N., In. N. Shalak, T. Boerner, and S. V. Shestakov. 1995. Physical and genetic map of the chromosome of the unicellular cyanobacterium *Synechocystis* sp. strain PCC 6803. *J. Bacteriol.* 177:3337–3343.

Cocks, B. G., L. E. Pyle, and L. R. Finch. 1989. A physical map of the genome of *Ureaplasma urealyticum* 960T with ribosomal RNA loci. *Nucl. Acids Res.* 17:6713–9.

Colleaux, L., L. d'Auriol, M. Betermier, G. Cottarel, A Jacquier, F. Galibert, and B. Dujon, B. 1986. Universal code equivalent of a yeast mitochondrial intron reading frame is expressed in *E. coli* as a specific double strand endonuclease. *Cell* 44:512–533.

Collier, M. C., F. Stock, P. C. DeGirolami, M. H. Samore and C. P. Cartwright. 1996. Comparison of PCR-based approaches to molecular epidemiologic analysis of Clostridium difficile. *J. Clin. Microbiol.* 34:1153–1157.

Colman, S. D., P. C. Hu, W. Litaker, and K. F. Bott. 1990. A physical map of the *Mycoplasma genitalium* genome. *Mol. Microbiol.* 4:683–687.

Cookson, B. D., P. Aparicio, A. Deplano, M. Struelens, R. Goering and R. Marples. 1996. Inter-centre comparison of pulsed-field gel electrophoresis for the typing of methicillin-resistant *Staphylococcus aureus. J. Med. Microbiol.* 44:179–184.

Coque, T. M. and B. E. Murray. 1995. Identification of *Enterococcus faecalis* strains by DNA hybridization and pulsed-field gel electrophoresis [letter]. *J. Clin. Microbiol.* 33:3368–3369.

Couto, I., J. Melo Cristino, M. L. Fernandes, T. Garcia, N. Serrano, M. J. Salgado, A. Torres Pereira, I. S. Sanches and H. de Lencastre. 1995. Unusually large number of methicillin-resistant *Staphylococcus aureus* clones in a Portuguese hospital. *J. Clin. Microbiol.* 33:2032–2035.

Crowe, M., K. J. Towner and H. Humphreys. 1995. Clinical and epidemiological features of an outbreak of acinetobacter infection in an intensive therapy unit. *J. Med. Microbiol.* 43:55–62.

Dalsgaard, A., M. N. Skov, O. Serichantalergs and P. Echeverria. 1996. Comparison of pulsed-field gel electrophoresis and ribotyping for subtyping of *Vibrio cholerae* O139 isolated in Thailand. *Epidemiol. Infect.* 117:51–58.

Daniel, P. 1995. Sizing of the *Lactobacillus plantarum* genome and other lactic Acid bacterial species by Transverse Alternating Field Electrophoresis. *Current Microbiol.* 30:243–246.

Daniels, D. L. 1990. The complete *Avr*II restriction map of the *Escherichia coli* genome and comparisons of several laboratory strains. *Nucl. Acids Res.* 18:2649–2652.

Davidson, B. E., J. Macdougall, and I. Saint Girons. 1992. Physical map of the linear chromosome of the bacterium *Borrelia burgdorferi* 212 a causative agent of Lyme disease and localization of rRNA genes. *J. Bacteriol.* 174:3766–3774.

Davidson, B. E., N. Kordias, N. Baseggio, A. Lim, M. Dobos and A. J. Hillier. 1995. Genomic organization of Lactococci. *Dev. Biol. Stand.* 85:411–422.

De Lancastre, H., I. Couto, I. Santos, J. Melo-Cristino, A. Torres-Pereira, and A. Tomasz.

1994. Methicillin-resistant *Staphylococcus aureus* disease in a Portuguese hospital: Characterization of clonal types by a combination of DNA typing methods. *European J. Clin. Microbiol. & Infectious Dis.* 13:64–73.

De Zoysa, A., A. Efstratiou, R. C. George, M. Jahkola, J. Vuopio Varkila, S. Deshevoi, G. Tseneva and Y. Rikushin. 1995. Molecular epidemiology of *Corynebacterium diphtheriae* from northwestern Russia and surrounding countries studied by using ribotyping and pulsed-field gel electrophoresis. *J. Clin. Microbiol.* 33:1080–1083.

Delahodde, A., V. Goguel, A. M. Becam, F. Creusot, J. Perea, J. Banroques, and C. Jacq. 1989. Site specific DNA endonuclease and RNA maturase activities of two homologous intron-encoded proteins from yeast mitochondria. *Cell* 56:431–441.

Dempsey, J. A., A. B. Wallace and J. G. Cannon. 1995. The physical map of the chromosome of a serogroup A strain of *Neisseria meningitidis* shows complex rearrangements relative to the chromosomes of the two mapped strains of the closely related species *N. gonorrhoeae. J. Bacteriol.* 177:6390–6400.

Dempsey, J. A.F., and J. G. Cannon. 1994. Locations of genetic markers on the physical map of the chromosome of *Neisseria gonorrhoeae* FA 1090. *J. Bacteriol.* 176:2055:2060.

Dempsey, J.A.F., W. Litaker, A. Madhure, T. L. Snodgrass, and J. G. Cannon. 1991. Physical map of the chromosome of *Neisseria gonorrhoeae* FA 1090 with locations of genetic markers including OPA and PIL genes. *J. Bacteriol.* 173:5476–5486.

Desmarchelier, P. M., F. Y. Wong and K. Mallard. 1995. An epidemiological study of *Vibrio cholerae* O1 in the Australian environment based on rRNA gene polymorphisms. *Epidemiol. Infect.* 115:435–446.

Dingwall, A., and L. Shapiro. 1989. Rate, origin, and bidirectionality of *Caulobacter* chromosome replication as determined by pulsed-field gel electrophoresis. *Proc. Natl. Acad. Sci.* USA. 86:119–23.

Dingwall, A., L. Shapiro, and B. Ely. 1990. Analysis of bacterial genome organization and replication using pulsed field gel electrophoresis. *Methods* 1:160–168.

Duchene, A-M., H. M. Kieser, D. A. Hopwood, C. J. Thompson, and P. Mazodier. 1994. Characterization of two *groEL* genes in *Streptomyces coelicolor* A3. *Gene* 144:97–101.

Dujon, B., M. Belfort, R. A. Butow, C. Jacq, C. Lemieux, P. S. Perlman, and V. M. Vogt. 1989. Mobile introns: definition of terms and recommended nomenclature. *Gene* 82:115–118.

Durrenberger, F. and J. D. Rochaix. 1991. Chloroplast ribosomal intron of *Chlamydomonas reinhardtii*: in vitro self-splicing DNA endonuclease activity, and in vivo mobility. *EMBO J.* 10:3495–3501.

Edmond, M. B., R. J. Hollis, A. K. Houston and R. P. Wenzel. 1995. Molecular epidemiology of an outbreak of meningococcal disease in a university community. *J. Clin. Microbiol.* 33:2209–2211.

Edwards, M. F. and Stocker, B. A. 1988. Construction of delta-*aroA his* delta-*pur* strains of *Salmonella typhi. J. Bacteriol.* 170:3991–3995.

Egel, D. S., J. H. Graham, and R. E. Stall. 1991. Genomic relatedness of *Xanthomonas campestris* strains causing diseases of citrus. *Appl. & Environ. Microbiol.* 57:2724–2730.

Eid, M. M. and J. E. Sherwood. 1995. Pulsed-field gel electrophoresis fingerprinting for identification of *Azospirillum* species. *Curr. Microbiol.* 30:127–131.

Ely, B., and C. J. Gerardot. 1988. Use of pulsed-field-gradient gel electrophoresis to construct a physical map of the *Caulobacter crescentus* genome. *Gene.* 68:323–33.

Ely, B., T. W. Ely, C. J. Gerardot, and A. Dingwall. 1990. Circularity of the *Caulobacter crescentus* chromosome determined by pulsed-field gel electrophoresis. *J. Bacteriol.* 172:1262–6.

Eremeeva, M., N. Balayeva, V. Ignatovich and D. Raoult. 1995. Genomic study of *Rickettsia akari* by pulsed-field gel electrophoresis. *J. Clin. Microbiol.* 33:3022–3024.

Eremeeva, M., N. Balayeva, V. Roux, V. Ignatovich, M. Kotsinjan and D. Raoult. 1995. Genomic and proteinic characterization of strain S, a rickettsia isolated from *Rhipicephalus sanguineus* ticks in Armenia. *J. Clin. Microbiol.* 33:2738–2744.

Erickson, J. W., and G. Altman. 1979. A search for patterns in the nucleotide sequence of the MS2 genome. *J. Math. Biol.* 7:219–230.

Evins, G. M., D. N. Cameron, J. G. Wells, K. D. Greene, T. Popovic, S. Giono Cerezo, I. K. Wachsmuth and R. V. Tauxe. 1995. The emerging diversity of the electrophoretic types of *Vibrio cholerae* in the Western Hemisphere. *J. Infect. Dis.* 172:173–179.

Fabe, C., P. Rodriguez, P. Cony Makhoul, P. Parneix, C. Bebear and J. Maugein. 1996. [Molecular typing by pulsed field gel electrophoresis of *Stenotrophomonas maltophilia* isolated in a department of hematology]. *Pathol. Biol. Paris.* 44:435–441.

Fasman, G. D. (ed.) 1976. *CRC Handbook of Biochemistry and Molecular Biology.* CRC Press.

Feizabadi, M. M., I. D. Robertson, D. V. Cousins and D. J. Hampson. 1996. Genomic analysis of *Mycobacterium bovis* and other members of the *Mycobacterium tuberculosis* complex by isoenzyme analysis and pulsed-field gel electrophoresis. *J. Clin. Microbiol.* 34:1136–1142.

Ferrin, L. J., and R. D. Camerini-Otero. 1991. Selective cleavage of human DNA: RecA-assisted restriction endonuclease (RARE) cleavage. *Science* 254:1494–7.

Fonstein, M. and R. Haselkorn. 1995. Physical mapping of bacterial genomes. *J. Bacteriol.* 177:3361–3369.

Fonstein, M., S. Zheng, and R. Haselkorn. 1992. Physical map of the genome of *Rhodobacter capsulatus* SB 1003. *J. Bacteriol.* 174:4070–4077.

Frutos, R., M. Pages, M. Bellis, G. Roizes, and M. Bergoin. 1989. Pulsed-field gel electrophoresis determination of the genome size of obligate intracellular bacteria belonging to the genera *Chlamydia, Rickettsiella*, and *Porochlamydia. J. Bacteriol.* 171:4511–3.

Fujita, M., S. Fujimoto, T. Morooka and K. Amako. 1995. Analysis of strains of *Campylobacter fetus* by pulsed-field gel electrophoresis. *J. Clin. Microbiol.* 33:1676–1678.

Fujita, M., T. Morooka, S. Fujimoto, T. Moriya and K. Amako. 1995. Southern blotting analyses of strains of *Campylobacter fetus* using the conserved region of *sapA. Arch. Microbiol.* 164:444–447.

Furihata, K., K. Sato and H. Matsumoto. 1995. Construction of a combined NotI/SmaI physical and genetic map of *Moraxella (Branhamella) catarrhalis* strain ATCC25238. *Microbiol. Immunol.* 39:745–751.

Gaju, N., V. Pavon, I. Marin, I. Esteve, R. Guerrero, and R. Amils. 1995. Chromosome map of the phototropic anoxygenic bacterium *Chromatium vinosum. FEMS Microbiol. Lett.* 126. 241–247.

Gasc, A. M., L. Kauc, P. Barraille, M. Sicard, and S. Goodgal. 1991. Gene localization size and physical map of the chromosome of *Streptococcus pneumoniae. J. Bacteriol.* 173:7361–7367.

Gasc, A. M., P. Geslin and M. Sicard.1995. Relatedness of penicillin-resistant *Streptococcus pneumoniae* serogroup 9 strains from France and Spain. *Microbiology* 141:623–627.

Gauthier, A., M. Turmel, and C. Lemieux. 1991. A group I intron in the chloroplast large subunit rRNA gene of *Chlamydomonas eugametos* encodes a double strand endonuclease that cleaves the homing site of this intron. *Curr. Genetics* 19:43–47.

Gibson, J. R., C. Fitzgerald and R. J. Owen. 1995. Comparison of PFGE, ribotyping and phage-typing in the epidemiological analysis of *Campylobacter jejuni* serotype HS2 infections. *Epidemiol. Infect.* 115:215–225.

Gireesh, T., B. E. Davidson, and A. J. Hillier. 1992. Conjugal transfer in *Lactococcus lactis* of a 68-kilobase-pair chromosomal fragment containing the structural gene for the peptide bacteriocinnisin. *Appl. & Environ. Microbiol.* 58:1670–1676.

Gorton, T. S., M. S. Goh and S. J. Geary. 1995. Physical mapping of the *Mycoplasma gallisepticum* S6 genome with localization of selected genes. *J. Bacteriol.* 177:259–263.

Gouby, A., C. Neuwirth, G. Bourg, N. Bouziges, M. J. Carles-Nurit, E. Despaux, E., and M. Ramuz. 1994. Epidemiological study by pulsed-field gel electrophoresis of an outbreak of extended-spectrum beta-lactamase-producing *Klebsiella pneumoniae* in a geriatric hospital. *J. Clin. Microbiol.* 32:301–305.

Green, M., E. R. Wald, B. Dashefsky, K. Barbadora and R. M. Wadowsky. 1996. Field inversion gel electrophoretic analysis of *Legionella pneumophila* strains associated with nosocomial legionellosis in children. *J. Clin. Microbiol.* 34:175–176.

Green, M., K. Barbadora S. Donabedian and M. J. Zervos. 1995. Comparison of field inversion gel electrophoresis with contour-clamped homogeneous electric field electrophoresis as a typing method for *Enterococcus faecium. J. Clin. Microbiol.* 33:1554–1557.

Griffith, D. E., J. L. Hardeman, Y. Zhang, R. J. Wallace and G. H. Mazurek. 1995. Tuberculosis outbreak among healthcare workers in a community hospital. *Am. J. Respir. Crit. Care. Med.* 152:808–811.

Grothues, D., and B. Tuemmler. 1991. New approaches in genome analysis by pulsed field gel electrophoresis application to the analysis of *Pseudomonas* spp. *Mol. Microbiol.* 5:2763–2776.

Grundmann, H., C. Schneider, D. Hartung, F. D. Daschner and T. L. Pitt. 1995. Discriminatory power of three DNA-based typing techniques for *Pseudomonas aeruginosa. J. Clin. Microbiol.* 33:528–534.

Grundmann, H., C. Schneider, H. V. Tichy, R. Simon, I. Klare, D. Hartung, and F. D. Daschner. 1995. Automated laser fluorescence analysis of randomly amplified polymorphic DNA: a rapid method for investigating nosocomial transmission of *Acinetobacter baumannii. J. Med. Microbiol.* 43:446–451.

Gueirard, P., C. Weber, A. Le Coustumier and N. Guiso. 1995. Human *Bordetella bronchiseptica* infection related to contact with infected animals: persistence of bacteria in host. *J. Clin. Microbiol.* 33:2002–2006.

Gyobu, Y., S. Hosorogi and T. Shimada. 1995. [Molecular epidemiological study on *Vibrio cholerae* O139]. *Kansenshogaku. Zasshi.* 69:501–505.

Hackett, N. R., Y. Bobovnikova, and N. Heyrovska. 1994. Conservation of chromosomal arrangement among three strains of the genetically unstable Archaeon *Halobacterium salinarium. J. Bacteriol.* 176:7711–7718.

Haertl, R., and G. Bandlow. 1993. Epidemiological fingerprinting of *Enterobacter cloacae* small-fragment restriction endonuclease analysis and pulsed field gel electrophoresis of genomic restriction fragments. *J. Clin. Microbiol.* 31:128–133.

Handwerger, S. and J. Skoble. 1995. Identification of chromosomal mobile element conferring high-level vancomycin resistance in *Enterococcus faecium. Antimicrob. Agents Chemother.* 39:2446–2453.

Hanish, J., and M. McClelland. 1991. Enzymatic cleavage of a bacterial chromosome at a transposon-inserted rare site. *Nucl. Acids Res.* 19:829–832.

Hantman, M. J., S. Sun, P. J. Piggot, and L. Daneo-Moore. 1993. Chromosome organization of *Streptococcus mutans* GS-5. *J. Gen. Microbiol.* 139:67–77.

Harrell, L. J., G. L. Andersen and K. H. Wilson. 1995. Genetic variability of *Bacillus anthracis* and related species. *J. Clin. Microbiol.* 33:1847–1850.

Hartstein, A. I., C. L. Phelps, R. Y. Kwok and M. E. Mulligan. 1995. In vivo stability and discriminatory power of methicillin-resistant *Staphylococcus aureus* typing by restriction endonuclease analysis of plasmid DNA compared with those of other molecular methods. *J. Clin. Microbiol.* 33:2022–2026.

Haukka, K., and K. Lindstrom. 1994. Pulsed-field gel electrophoresis for genotypic comparison of *Rhizobium* bacteria that nodulate leguminous trees. *FEMS Microbiol. Lett.* 119:215–220.

He, Q., H. Chen, A. Kuspa, Y. Cheng, D. Kaiser, and L. J. Shimkets. 1994. A physical map of the *Myxococcus xanthus* chromosome. *Proc. Natl. Acad. Sci.* USA 91:9584–9587.

Heath, J. D., J. D. Perkins, B. Sharma, and G. M. Weinstock. 1992. *Not*I genomic cleavage map of *Escherichia coli* K-12 strain MG1655. *J. Bacteriol.* 174:558–67.

Hermans, P. W., M. Sluijter, T. Hoogenboezem, H. Heersma, A. van Belkum and R. de Groot. 1995. Comparative study of five different DNA fingerprint techniques for molecular typing of *Streptococcus pneumoniae* strains. *J. Clin. Microbiol.* 33:1606–1612.

Hesse, S., N. Low, B. Molokwu, A. H. Uttley, A. E. Jephcott, A. Turner and A. L. Pozniak. 1995. Molecular typing as an epidemiological tool in the study of sexual networks [letter]. *Lancet* 346:976–977.

Hillier, A. J., and B. E. Davidson, 1995. Pulsed field gel electrophoresis. *Methods Mol. Biol.* 46:149–164.

Hogg, G. G., J. E. Strachan, L. Huayi, S. A. Beaton, P. M. Robinson and K. Taylor. 1996.

Non-toxigenic *Corynebacterium diphtheriae* biovar gravis: evidence for an invasive clone in a south-eastern Australian community. *Med. J. Aust.* 164:72–75.

Hojo, S., J. Fujita, K. Negayama, T. Ohnishi, G. Xu, Y. Yamaji, H. Okada and J. Takahara. 1995. [DNA fingerprinting by arbitrarily primed polymerase chain reaction (AP-PCR) for methicillin-resistant *Staphylococcus aureus*]. *Kansenshogaku. Zasshi.* 69:506–510.

Holloway, B. W., M. D. Escuadra, A. F. Morgan, R. Saffery, and V. Krishnapillai. 1992. The new approaches to whole genome analysis of bacteria. *FEMS Microbiol. Lett.* 100:101–105.

Honda, M., T. Kodama, Y. Ishida, K. Takei and M. Matsuda. 1995. Molecular subtyping of clinical isolates of *Escherichia coli* from patients with lower urinary tract infections in Japan. *Cytobios.* 81:55–60.

Hone, D. M., A. M. Harris, S. Chatfield, G. Dougan, and M. Levine. 1991. Construction of genetically defined double *aro* mutants of *Salmonella typhi. Vaccine* 9:810–815.

Honeycutt, R. J., M. McClelland, and B.W.S. Sobral. 1993. Physical map of the genome of *Rhizobium meliloti* 1021. *J. Bacteriol.* 175:6945–6952.

Hu, L., A. Umeda, S. Kondo, and K. Amako. 1995. Typing of *Staphylococcus aureus* colonising human nasal carriers by pulsed-field gel electrophoresis. *J. Med. Microbiol.* 42:127–132.

Huang, Y., J. A. Robertson and G. W. Stemke. 1995. An unusual rRNA gene organization in *Mycoplasma fermentans* (incognitus strain). *Can. J. Microbiol.* 41:424–427.

Huber, I., and S. Selenska-Pobell. 1994. Pulsed-field gel electrophoresis-fingerprinting, genome size estimation and *rrn* loci number of *Rhizobium galegae. J. Appl. Bacteriol.* 77:528–533.

Irnich, S., and J. Cullum. 1993. Random insertion of Tn4560 in *Streptomyces lividans* and *Streptomyces avermitilis. Biotechnology Lett.* 15:895–900.

Itaya, M. 1993. Physical mapping of multiple homologous genes in the *Bacillus subtilis* 168 chromosome: Identification of ten ribosomal RNA operon loci. *BioScience Biotechnology & Biochemistry* 57:1611–1614.

Iteman, I., A. Guiyoule, A.M.P. De Almeida, I. Guilvout, G. Baranton, and E. Carniel. 1993. Relationship between loss of pigmentation and deletion of the chromosomal iron-regulated IRP2 gene in *Yersinia pestis*: Evidence for separate but related events. *Infection & Immunity* 61:2717–2722.

Jensen, A. E., N. F. Cheville, D. R. Ewalt, J. B. Payeur and C. O. Thoen. 1995. Application of pulsed-field gel electrophoresis for differentiation of vaccine strain RB51 from field isolates of *Brucella abortus* from cattle, bison, and elk. *Am. J. Vet. Res.* 56:308–312.

Johnson, J. M., S. D. Weagant, K. C. Jinneman and J. L. Bryant. 1995. Use of pulsed-field gel electrophoresis for epidemiological study of *Escherichia coli* O157:H7 during a food-borne outbreak. *Appl. Environ. Microbiol.* 61:2806–2808.

Jorgensen, M., R. Givney, M. Pegler, A. Vickery and G. Funnell. 1996. Typing multidrug-resistant *Staphylococcus aureus*: conflicting epidemiological data produced by genotypic and phenotypic methods clarified by phylogenetic analysis. *J. Clin. Microbiol.* 34:398–403.

Jumas Bilak, E., C. Maugard, S. Michaux Charachon, A. Allardet Servent, A. Perrin, D. O'Callaghan and M. Ramuz. 1995. Study of the organization of the genomes of *Escherichia coli, Brucella melitensis* and *Agrobacterium tumefaciens* by insertion of a unique restriction site. *Microbiology* 141:2425–2432.

Kalkus, J., M. Reh, and H. G. Schlegel. 1990. Hydrogen autotrophy of *Nocardia opaca* strains is encoded by linear megaplasmids. *J. Gen. Microbiol.* 136:1145–1152.

Katayama, S., B. Dupuy, G. Daube, B. China and S. T. Cole. 1996. Genome mapping of *Clostridium perfringens* strains with I-CeuI shows many virulence genes to be plasmid-borne. *Mol. Gen. Genet.* 251:720–726.

Kauc, L., A. Michalowska and M. Gnas Wiernicka. 1996. Physical maps of the chromosomes of *Haemophilus influenzae* Sb. *Acta Microbiol. Pol.* 45:19–30.

Kauc, L., and S. H. Goodgal. 1989. A physical map of the chromosome of *Haemophilus parainfluenzae. Gene.* 83:377–380.

Kauc, L., M. Mitchell, and S. H. Goodgal. 1989. A physical map of the chromosome of *Haemophilus influenzae. J. Bacteriol.* 171:2474–9.

Kern, W. V., E. Andriof, M. Oethinger, P. Kern, J. Hacker, and R. Marre. 1994. Emergence of fluoroquinolone-resistant *Escherichia coli* at a cancer center. *Antimicrobial Agents & Chemotherapy.* 38:681–687.

Kersulyte, D., M. J. Struelens, A. Deplano and D. E. Berg. 1995. Comparison of arbitrarily primed PCR and macrorestriction (pulsed-field gel electrophoresis) typing of *Pseudomonas aeruginosa* strains from cystic fibrosis patients. *J. Clin. Microbiol.* 33:2216–2219.

Kieser, H. M., T. Kieser, and D. A. Hopwood. 1992. A combined genetic and physical map of the *Streptomyces coelicolor* A32 chromosome. *J. Bacteriol.* 174:5496–5507.

Kiska, D. L., and F. L. Macrina, F. L. 1994. Genetic analysis of fructan-hyperproducing strains of *Streptococcus mutans. Infection & Immunity* 62:2679–2686.

Klingman, K. L., A. Pye, T. F. Murphy and S. L. Hill. 1995. Dynamics of respiratory tract colonization by *Branhamella catarrhalis* in bronchiectasis. *Am. J. Respir. Crit. Care. Med.* 152:1072–1078.

Kluytmans, J., W. van Leeuwen, W. Goessens, R. Hollis, S. Messer, L. Herwaldt, H. Bruining, M. Heck, J. Rost, N. van Leeuwen, and et al. 1995. Food-initiated outbreak of methicillin-resistant *Staphylococcus aureus* analyzed by pheno- and genotyping. *J. Clin. Microbiol.* 33:1121–1128.

Kolbert, C. P., D. S. Podzorski, D. A. Mathiesen, A. T. Wortman, A. Gazumyan, I. Schwartz and D. H. Persing. 1995. Two geographically distinct isolates of *Borrelia burgdorferi* from the United States share a common unique ancestor. *Res. Microbiol.* 146:415–424.

Kolsto, A. B., A. Gronstad, and H. Oppegaard. 1990. Physical map of the *Bacillus cereus* chromosome. *J. Bacteriol.* 172:3821–3825.

Kondo, S., A. Yamagishi, and T. Oshima. 1993. A physical map of the sulfur-dependent Archaebacterium *Sulfolobus acidocaldarius* 7 chromosome. *J. Bacteriol.* 175:1532–1536.

Koob, M. and W. Szybalski 1990. Cleaving yeast and *Escherichia coli* genomes at a single site. *Science* 250:271–273.

Koob, M., E. Grimes, and W. Szybalski. 1988. Conferring operator specificity on restriction endonucleases. *Science* 241:1084–1086.

Kotani, H., T. Kaneko, T. Matsubayashi, S. Sato, M. Sugiura, and S. Tabat. 1994. A physical map of the genome of a unicellular cyanobacterium *Synechocystis* sp. strain PCC6803. *DNA Res.* 1:303–307.

Krause, D,C., and K. K. Lee. 1991. Juxtaposition of the genes encoding *Mycoplasma pneumoniae* cytadherence-accessory proteins HMW1 and HMW3. *Gene* 107:83–90.

Krause, D., and C. B. Mawn. 1990. Physical analysis and mapping of the *Mycoplasma pneumoniae* chromosome. *J. Bacteriol.* 172:4790–4797.

Krause, U., F. M. Thomson Carter and T. H. Pennington. 1996. Molecular epidemiology of *Escherichia coli* O157:H7 by pulsed-field gel electrophoresis and comparison with that by bacteriophage typing. *J. Clin. Microbiol.* 34:959–961.

Krawiec, S. and M. Riley. 1990. Organization of the bacterial genome. *Microbiol. Rev.* 54:502–539.

Krueger, C. M., K. L. Marks and G. M. Ihler. 1995. Physical map of the *Bartonella bacilliformis* genome. *J. Bacteriol.* 177:7271–7274.

Kuhn, I., L. G. Burman, S. Haeggman, K. Tullus and B. E. Murray. 1995. Biochemical fingerprinting compared with ribotyping and pulsed-field gel electrophoresis of DNA for epidemiological typing of enterococci. *J. Clin. Microbiol.* 33:2812–2817.

Kundig, C., H. Hennecke, and M. Gottfert. 1993. Correlated physical and genetic map of the *Bradyrhizobium japonicum* 110 genome. *J. Bacteriol.* 175:613–22.

Kurazono, H., S. Yamasaki, O. Ratchtrachenchai, G. B. Nair and Y. Takeda. 1996. Analysis of *Vibrio cholerae* O139 Bengal isolated from different geographical areas using macrorestriction DNA analysis. *Microbiol. Immunol.* 40:303–305.

Kuritz, T., A. Ernst, T. A. Black, and C. P. Wolk. 1993. High-resolution mapping of genetic loci of *Anabaena* PCC 7120 required for photosynthesis and nitrogen fixation. *Mol. Microbiol.* 8:101–10.

Ladefoged, S. A., and G. Christiansen. 1992. Physical and genetic mapping of the genomes of five *Mycoplasma hominis* strains by pulsed field gel electrophoresis. *J. Bacteriol.* 174:2199–2207.

Laing, F. P., K. Ramotar, R. R. Read, N. Alfieri, A. Kureishi, E. A. Henderson and T. J. Louie. 1995. Molecular epidemiology of *Xanthomonas maltophilia* colonization and infection in the hospital environment. *J. Clin. Microbiol.* 33:513–518.

Lamoureux, M., Prevost, H., Cavin, J. F., and Divies, C. 1993. Recognition of *Leuconostocoenos* strains by the use of DNA restriction profiles. *Appl. Microbiol. & Biotechnol.* 39:547–552.

Lau, Y. J., P. Y. Liu, B. S. Hu, J. M. Shyr, Z. Y. Shi, W. S. Tsai, Y. H. Lin and C. Y. Tseng. 1995. DNA fingerprinting of *Pseudomonas aeruginosa* serotype O11 by enterobacterial repetitive intergenic consensus-polymerase chain reaction and pulsed-field gel electrophoresis. *J. Hosp. Infect.* 31:61–66.

Le Bourgeois, P., M. Lautier, M. Mata, and P. Ritzenthaler. 1992. Physical and genetic map of the chromosome of *Lactococcus lactis* subsp. lactis IL1403. *J. Bacteriol.* 174:6752–62.

Le Bourgeois, P., M. Lautier, M. Mata, and P. Ritzenthaler. 1992. New tools for the physical and genetic mapping of *Lactococcus* strains. *Gene* 111:109–114.

Le Bourgeois, P., M. Mata, and P. Ritzenthaler. 1989. Genome comparison of *Lactococcus* strains by pulsed-field gel electrophoresis. *FEMS Microbiol. Lett.* 50:65–9.

Leblond, P., M. Redenbach, and J. Cullum. 1993. Physical map of the *Streptomyces lividans* 66 genome and comparison with that of the related strain *Streptomyces coelicolor* A3-2. *J. Bacteriol.* 175:3422–3429.

Lee, J. J., H. O. Smith, and R. J. Redfield. 1989. Organization of the *Haemophilus influenzae* Rd genome. *J. Bacteriol.* 171:3016–24.

Lefevre, J. C., M. A. Bertrand and G. Faucon. 1995. Molecular analysis by pulsed-field gel electrophoresis of penicillin-resistant *Streptococcus pneumoniae* from Toulouse, France. *Eur. J. Clin. Microbiol. Infect. Dis.* 14:491–497.

Leonard, R. B., J. Mayer, M. Sasser, M. L. Woods, B. R. Mooney, B. G. Brinton, P. L. Newcomb Gayman and K. C. Carroll. 1995. Comparison of MIDI Sherlock system and pulsed-field gel electrophoresis in characterizing strains of methicillin-resistant *Staphylococcus aureus* from a recent hospital outbreak. *J. Clin. Microbiol.* 33:2723–2727.

Lezhava, A., T. Mizukami, T. Kajitani, D. Kameoka, M. Redenbach, H. Shinkawa, O. Nimi and H. Kinashi. 1995. Physical map of the linear chromosome of *Streptomyces griseus. J. Bacteriol.* 177:6492–6498.

Li, H. and J. A. Dillon. 1995. Utility of ribotyping, restriction endonuclease analysis and pulsed-field gel electrophoresis to discriminate between isolates of *Neisseria gonorrhoeae* of serovar IA-2 which require arginine, hypoxanthine or uracil for growth. *J. Med. Microbiol.* 43:208–215.

Liao, X., I. Charlebois, C. Ouellet, M. J. Morency, K. Dewar, J. Lightfoot, J. Foster, R. Siehnel, H. Schweizer, J. S. Lam, R. E. Hancock and R. C. Levesque. 1996. Physical mapping of 32 genetic markers on the *Pseudomonas aeruginosa* PAO1 chromosome. *Microbiology* 142:79–86.

Liebisch, B. and S. Schwarz. 1996. Molecular typing of *Salmonella enterica* subsp. enterica serovar Enteritidis isolates. *J. Med. Microbiol.* 44:52–59.

Lilleengen, K. 1948. Typing *Salmonella typhimurium* by means of bacteriophage. *Acta Pathol. Microbiol. Scandin. Suppl.* 77:11–125.

Lin, W. J., and E. A. Johnson. 1995. Genome analysis of *Clostridium botulinum* type A by pulsed-field gel electrophoresis. *Appl. Environ. Microbiol.* 61:4441–4447.

Lin, Y-S., H. M. Kieser, D. A. Hopwood, and C. W. Chen. 1993. The chromosomal DNA of *Streptomyces lividans* 66 is linear. *Mol. Microbiol.* 10:923–933.

Lina, B., F. Forey, J. D. Tigaud, and J. Fleurette. 1995. Chronic bacteraemia due to *Staphylococcus epidermidis* after bone marrow transplantation. *J. Medical Microbiol.* 42:156–160.

Liu, P. Y-F. Y-J. Lau, B-S. Hu, J-M. Shyr, Z-Y. Shi, W-S. Tsai, Y-H. Lin, and C-Y. Tseng. 1995. Analysis of clonal relationships among isolates of *Shigella sonnei* by different molecular typing methods. *J. Clin. Microbiol.* 33:1770–1783.

Liu S.-L and K. E. Sanderson. 1993a. Genomic mapping with I-*Ceu*I, an intron-encoded endonuclease, specific for genes for ribosomal RNA, in *Salmonella* spp., *Escherichia coli*, and other bacteria. *Proc. Natl. Acad. Sci.* USA 90:6874–6878.

Liu, S.-L and K. E. Sanderson. 1995. Rearrangements in the genome of the bacterium *Salmonella typhi. Proc. Natl. Acad. Sci.* USA 92:1018–1022.

Liu, P. Y., Y. J. Lau, B. S. Hu, J. M. Shyr, Z. Y. Shi, W. S. Tsai, Y. H. Lin, and C. Y. Tseng. 1995. Analysis of clonal relationships among isolates of Shigella sonnei by different molecular typing methods. *J. Clin. Microbiol.* 33:1779–1783.

Liu, P. Y., Z. Y. Shi, Y. J. Lau, B. S. Hu, J. M. Shyr, W. S. Tsai, Y. H. Lin, and C. Y. Tseng. 1996. Epidemiological typing of *Flavimonas oryzihabitans* by PCR and pulsed-field gel electrophoresis. *J. Clin. Microbiol.* 34:68–70.

Liu, P. Y., Z. Y. Shi, Y. J. Lau, B. S. Hu, J. M. Shyr, W. S. Tsai, Y. H. Lin, and C. Y. Tseng. 1995. Comparison of different PCR approaches for characterization of *Burkholderia (Pseudomonas) cepacia* isolates. *J. Clin. Microbiol.* 33:3304–3307.

Liu, S-L., A. Hessel, and K. E. Sanderson. 1993. The *Xba*I-*Bln*I-*Ceu*I genomic cleavage map of *Salmonella enteritidis* shows an inversion relative to *Salmonella typhimurium* LT2, *Mol. Microbiol.* 10:655–664.

Liu, S-L., A. Hessel, H-Y.M. Cheng, and K. E. Sanderson. 1994. The *Xba*I-*Bln*I-*Ceu*I genomic cleavage map of *Salmonella paratyphi B. J. Bacteriol.* 176:1014–1024.

Liu, S.-L., A. Hessel, and K. E. Sanderson. 1993. The *Xba*I-*Bln*I-*Ceu*I genomic cleavage map of *Salmonella typhimurium* LT2, determined by double digestion, end-labelling, and pulsed-field gel electrophoresis. *J. Bacteriol.* 175:4104–4120.

Liu, S. L. and K. E. Sanderson. 1995. Genomic cleavage map of *Salmonella typhi* Ty2. *J. Bacteriol.* 177:5099–5107.

Liu, S. L. and K. E. Sanderson. 1995. I-CeuI reveals conservation of the genome of independent strains of *Salmonella typhimurium. J. Bacteriol.* 177:3355–3357.

Liu, S. L. and K. E. Sanderson. 1995. Rearrangements in the genome of the bacterium *Salmonella typhi. Proc. Natl. Acad. Sci. U.S.A.* 92:1018–1022.

Liu, S. L. and K. E. Sanderson. 1995. The chromosome of *Salmonella paratyphi* A is inverted by recombination between *rrnH* and *rrnG. J. Bacteriol.* 177:6585–6592.

Liu, S. L. and K. E. Sanderson. 1992. A physical map of the *Salmonella typhimurium* LT2 genome made by using *Xba*I analysis. *J. Bacteriol.* 174:1662–1672.

Loncarevic, S., W. Tham and M. L. Danielsson Tham. 1996. The clones of *Listeria monocytogenes* detected in food depend on the method used. *Lett. Appl. Microbiol.* 22:381–384.

Lopez Garcia, P., J. P. Abad and R. Amils. 1993. Genome analysis of different *Haloferax mediterranei* strains using pulsed field gel electrophoresis. *Systematic & Appl. Microbiol.* 16:310–321.

Lopez Garcia, P., J. P. Abad, C. Smith, and R. Amils. 1992. Genomic organization of the halophilic *Archaeon haloferax mediterranei* physical map of the chromosome. *Nucl. Acids Res.* 20:2459–2464.

Louie, M., P. Jayaratne, I. Luchsinger, J. Devenish, J. Yao, W. Schlech and A. Simor.

1996. Comparison of ribotyping, arbitrarily primed PCR, and pulsed-field gel electrophoresis for molecular typing of *Listeria monocytogenes*. *J. Clin. Microbiol.* 34:15–19.

Luck, P. C., J. Kohler, M. Maiwald and J. H. Helbig. 1995. DNA polymorphisms in strains of *Legionella pneumophila* serogroups 3 and 4 detected by macrorestriction analysis and their use for epidemiological investigation of nosocomial legionellosis. *Appl. Environ. Microbiol.* 61:2000–2003.

Luck, P. C., J. H. Helbig, H. J. Hagedorn and W. Ehret. 1995. DNA fingerprinting by pulsed-field gel electrophoresis to investigate a nosocomial pneumonia caused by *Legionella bozemanii* serogroup 1. *Appl. Environ. Microbiol.* 61:2759–2761.

Luck, P. C., J. H. Helbig, V. Drasar, N. Bornstein, R. J. Fallon and M. Castellani Pastoris. 1995. Genomic heterogenicity amongst phenotypically similar *Legionella micdadei* strains. *FEMS. Microbiol. Lett.* 126:49–54.

Luck, P. C., R. J. Birtles, and J. H. Helbig. 1995. Correlation of MAb subgroups with genotype in closely related *Legionella pneumophila* serogroup 1 strains from a cooling tower. *J. Med. Microbiol.* 43:50–54.

Lueck, C. P., L. Bender, M. Ott, J. H. Helbig, and J. Hacker. 1991. Analysis of *Legionella pneumophila* serogroup 6 strains isolated from hospital warm water supply over a three-year period by using genomic long-range mapping techniques and monoclonal antibodies. *Appl. & Environ. Microbiol.* 57:3226–3231.

MacDougall, J. and I. Saint Girons. 1995. Physical map of the *Treponema denticola* circular chromosome. *J. Bacteriol.* 177:1805–1811.

MacKenzie, C., M. Chidambaram, E. J. Sodergren, S. Kaplan and G. M. Weinstock. 1995. DNA repair mutants of *Rhodobacter sphaeroides*. *J. Bacteriol.* 177:3027–3035.

Mahenthiralingam, E., M. E. Campbell, J. Foster, J. S. Lam and D. P. Speert. 1996. Random amplified polymorphic DNA typing of *Pseudomonas aeruginosa* isolates recovered from patients with cystic fibrosis. *J. Clin. Microbiol.* 34:1129–1135.

Majumder, R., S. Sengupta, G. Khetawat, R. K. Bhadra, S. Roychoudhury and J. Das. 1966. Physical map of the genome of *Vibrio cholerae* 569B and localization of genetic markers. *J. Bacteriol.* 178:1105–1112.

Maldonado, R., J. Jimenez, and J. Casadesus. 1994. Changes of ploidy during the *Azotobacter vinelandii* growth cycle. *J. Bacteriol.* 176:3911–3919.

Marcos, M. A., M. T. Jimenez de Anta and J. Vila. 1995. Correlation of six methods for typing nosocomial isolates of *Acinetobacter baumannii*. *J. Med. Microbiol.* 42:328–335.

Marrie, T. J., W. Johnson, S. Tyler, G. Bezanson, D. Haldane, S. Burbridge and J. Joly. 1995. Potable water and nosocomial Legionnaires' disease—check water from all rooms in which patient has stayed. *Epidemiol. Infect.* 114:267–276.

Marshall, P., T. B. Davis, and C. Lemieux. 1994. The I-*Ceu*I endonuclease: purification and potential role in the evolution of *Chlamydomonas* group I introns. *Eur. J. Bacteriol.* 220:855–859.

Maslow, J. N., S. Brecher, J. Gunn, A Durbin, M. A. Barlow and R. D. Arbeit. 1995. Variation and persistence of methicillin-resistant *Staphylococcus aureus* strains among individual patients over extended periods of time. *Eur. J. Clin. Microbiol. Infect. Dis.* 14:282–290.

Matsuda, M., K. Matsumoto, C. Kaneuchi, T. Masaoka, F. Akahori, and T. Yamada. 1991. DNA analysis of the thermophylic *Campylobacter* by pulsed field gel electrophoresis. *Physico-Chemical Biol.* 35:49–54.

Matsuda, M., M. Tsukada, M. Fukuyama, Y. Kato, Y. Ishida, M. Honda, and C. Kaneuchi, 1995. Detection of genomic variability among isolates of *Campylobacter jejuni* from chickens by crossed-field gel electrophoresis. *Cytobios.* 82:73–79.

Matsumoto, K., M. Matsuda, and C. Kaneuchi. 1992. Analysis of chromosome-sized DNA from the bacterial genome of thermophilic *Campylobacter lardis* by pulsed field gel electrophoresis and physical mapping. *Microbios.* 71:7–14.

McClelland, M. and M. Nelson. 1987. Enhancement of the Apparent Cleavage Specificities of Restriction Endonucleases: Applications to Megabase Mapping of Chromosomes, in Chirikjian, J. G. (ed.). *Gene Amplification and Analysis*, vol. 5. Restriction Endonucleases and Methylases, Elsevier, Amsterdam.

McClelland, M., J. Hanish, M. Nelson, and Y. Patel. 1988. A single type of buffer for all restriction endonucleases. *Nucl. Acids Res.* 15:364–364.

McClelland, M., L. Kessler, and M. Bittner. 1984. Site-specific cleavage of DNA at 8- and 10-base-pair-sequences. *Proc. Natl. Acad. Sci.* USA 81:983–987.

McClelland, M., M. Nelson, and C. R. Cantor. 1985. Purification of *Mbo*II methylase (GAAG^{m}A) from *Moraxella bovis*: site specific cleavage of DNA at nine and ten base pair sequences. *Nucl. Acids Res.* 13:7171–7182.

McClelland, M., R. Jones, Y. Patel, and M. Nelson. 1987. Restriction endonucleases for pulsed field mapping of bacterial genomes. *Nucl. Acids Res.* 15:5985–6005.

McDougal, L. K., J. K. Rasheed, J. W. Biddle and F. C. Tenover. 1995. Identification of multiple clones of extended-spectrum cephalosporin-resistant *Streptococcus pneumoniae* isolates in the United States. *Antimicrob. Agents Chemother.* 39:2282–2288.

McKenna, A. J., J. A. Bygraves, M. C. Maiden and I. M. Feavers. 1995. Attenuated typhoid vaccine *Salmonella typhi* Ty21a: fingerprinting and quality control. *Microbiology* 141:1993–2002.

Mendez Alvarez, S., V. Pavon, I. Esteve, R. Guerrero and N. Gaju. 1995. Genomic heterogeneity in *Chlorobium limicola*: chromosomic and plasmidic differences among strains. *FEMS. Microbiol. Lett.* 134:279–285.

Mendez Alvarez, S., V. Pavon, I. Esteve, R. Guerrero and N. Gaju. 1995. Analysis of bacterial genomes by pulsed field gel electrophoresis. *Microbiologia* II:323–336.

Meng, J., S. Zhao, T. Zhao and M. P. Doyle. 1995. Molecular characterisation of *Escherichia coli* O157:H7 isolates by pulsed-field gel electrophoresis and plasmid DNA analysis. *J. Med. Microbiol.* 42:258–263.

Meng, X., K. Benson, K. Chada, E. J. Huff and D. C. Schwartz. 1995. Optical mapping of lambda bacteriophage clones using restriction endonucleases. *Nat. Genet.* 9:432–438.

Michaux, S., J. Paillisson, M. J. Carles-Nurit, G. Bourg, A. Allardet-Servent, and M. Ramuz. 1993. Presence of two independent chromosomes in the *Brucella melitensis* 16M genome. *J. Bacteriol.* 175:701–5.

Michel, E. 1993. Mapping of bacterial chromosomes application to *Listeria monocytogenes. Methods in Mol. & Cellular Biol.* 4:81–86.

Michel, E., and P. Cossart. 1992. Physical map of the *Listeria monocytogenes* chromosome. *J. Bacteriol.* 174:7098–7103.

Miyata, M., L. Wang, and T. Fukumura. 1993. Localizing the replication origin region on the physical map on the *Mycoplasma capricolum* genome. *J. Bacteriol.* 175:655–660.

Miyazawa, T., M. Matsuda, Y. Isayama, T. Samata, Y. Ishida, S. Ogawa, K. Takei, M. Honda and M. Kamada. 1995. Genotyping of isolates of *Taylorella equigenitalis* from thoroughbred brood mares in Japan. *Vet. Res. Commun.* 19:265–271.

Moissenet, D., L. Guet, M. Valcin, E. N. Garabedian, P. Geslin, A. Garbarg Chenon and H. Vu Thien. 1996. [Molecular epidemiology of pneumococci with decreased susceptibility to penicillin isolated in a Parisian pediatric hospital]. *Pathol. Biol. Paris.* 44:423–429.

Moore, M. A., and A. R. Datta. 1994. DNA fingerprinting of *Listeria monocytogenes* stains by pulsed-field gel electrophoresis. *Food Microbiol.* 11:31–38.

Moreno, F., C. Crisp, J. H. Jorgensen and J. E. Patterson. 1995. The clinical and molecular epidemiology of bacteremias at a university hospital caused by pneumococci not susceptible to penicillin. *J. Infect. Dis.* 172:427–432.

Morris, J. G., Jr., D. K. Shay, J. N. Hebden, R. J. McCarter, Jr., B. E. Perdue, W. Jarvis, J. A. Johnson, T. C. Dowling, L. B. Polish and R. S. Schwalbe. 1995. Enterococci resistant to multiple antimicrobial agents, including vancomycin. Establishment of endemicity in a university medical center. *Ann. Intern. Med.* 123:250–259.

Moser, H. E., and P. B. Dervan. 1987. Sequence-specific cleavage of double helical DNA by triple helix formation. *Science* 238:645–650.

Mulin, B., D. Talon, J. F. Viel, C. Vincent, R. Leprat, M. Thouverez and Y. Michel Briand. 1995. Risk factors for nosocomial colonization with multiresistant *Acinetobacter baumannii. Eur. J. Clin. Microbiol. Infect. Dis.* 14:569–576.

Murase, T., T. Okitsu, R. Suzuki, H. Morozumi, A. Matsushima, A. Nakamura and S. Yamai. 1995. Evaluation of DNA fingerprinting by PFGE as an epidemiologic tool for *Salmonella* infections. *Microbiol. Immunol.* 39:673–676.

Murray, B. E., K. V. Singh, J. D. Heath, B. R. Sharma, and G. M. Weinstock. 1990. Comparison of genomic DNA's of different Enterococcal isolates using restriction endonucleases with infrequent recognition sites. *J. Clin. Microbiol.* 28:2059–2063.

Murray, B. E., K. V. Singh, R. P. Ross, J. D. Heath, G. M. Dunny, and G. M. Weinstock. 1993. Generation of restriction map of *Enterococcus faecalis* OGI and investigation of growth requirements and regions encoding biosynthetic function. *J. Bacteriol.* 175:5216–5223.

Muscarella, D. E. and V. M. Vogt. 1989. A mobile group I intron in the nuclear rDNA of *Physarum polycephalum. Cell* 56:443–454.

Musser, J. M., V. Kapur, J. Szeto, X. Pan, D. S. Swanson and D. R. Martin. 1995. Genetic diversity and relationships among *Streptococcus pyogenes* strains expressing serotype MI protein: recent intercontinental spread of a subclone causing episodes of invasive disease. *Infect. Immun.* 63:994–1003.

Nada, T., S. Ichiyama, Y. Osada, M. Ohta, K. Shimokata, N. Kato and N. Nakashima.

1996. Comparison of DNA fingerprinting by PFGE and PCR-RFLP of the coagulase gene to distinguish MRSA isolates. *J. Hosp. Infect.* 32:305–317.

Naterstad, K., A. B. Kolsto, R. Sirevag. 1995. Physical map of the genome of the green phototrophic bacterium *Chlorobium tepidum. J. Bacteriol.* 177:5480–5484.

Nelson, M., and M. McClelland. 1992. Use of DNA methyltransferase-endonuclease enzyme combinations for megabase mapping of chromosomes. *Method. Enzymol.*, 216:279–303

Nelson, M., C. Christ, and I. Schildkraut. 1984. Alteration of apparent restriction endonuclease recognition specificities by DNA mathylases. *Nucl. Acids Res.* 12:5165–5173.

Nelson, M., E. Raschke, and M. McClelland. 1993. Effect of Site-Specific Methylation on Restriction Endonucleases with DNA Modification Methyltransferases. *Nucl. Acids Res.* 21:3139–3154.

Neuwirth, C., E. Siebor, J. Lopez, A. Pechinot and A. Kazmierczak. 1996. Outbreak of TEM-24-producing *Enterobacter aerogenes* in an intensive care unit and dissemination of the extended-spectrum beta-lactamase to other members of the family enterobacteriaceae. *J. Clin. Microbiol.* 34:76–79.

Ng, L. K., M. Carballo and J. A. Dillon. 1995. Differentiation of *Neisseria gonorrhoeae* isolates requiring proline, citrulline, and uracil by plasmid content, serotyping, and pulsed-field gel electrophoresis. *J. Clin. Microbiol.* 33:1039–1041.

Nikolskii, Yu, V. Fonstein, M. Gering, Kh. Rostova, G. Yu, and N. K. Yankovskii. 1992. Mapping of genetic markers of *Corynebacterium glutamicum* ATCC13032 with respect to the elements of the genome physical map. *Genetika.* 28:38–46.

Noll, K. M. 1989. Chromosome map of the thermophilic *Archaebacterium thermococcus celer. J. Bacteriol.* 171:6720–5.

Nuijten, P.J.M., C. Bartels, Pluym N.M.C. Bleumink, W. Gaastra, and B.A.M. Van Der Zeijst. 1990. Size and physical map of the *Campylobacter jejuni* chromosome. *Nucl. Acids Res.* 18:6211–6214.

Nussinov, R. 1980. Some rules in the ordering of nucleotides in the DNA. *Nucl. Acids Res.* 8:4545–4562.

Nussinov, R. 1981. The universal dinucleotide asymmetry rules in DNA and amino acid codon choice. *J. Mol. Evol.* 17:237–244.

Ojaimi, C., B. E. Davidson, I. Saint Girons, and I. G. Old. 1994. Conservation of gene arrangement and an unusual organization of rRNA genes in the linear chromosomes of the Lyme disease Spirochaetes *Borrelia burgdorferi, B. garinii* and *B. afzelii. Microbiology* 140:2931–2940.

Ojeniyi, B., U. S. Petersen, and N. Hoiby. 1993. Comparison of genome fingerprinting with conventional typing methods used on *Pseudomonas aeruginosa* isolates from cystic fibrosis patients. *APMIS (Acta Pathologica Microbiologica et Immunologica Scandinavica).* 101:168–175.

Okada, N., C. Sasakawa, T. Tobe, K. A. Talukder, K. Komatsu, and M. Yoshikawa. 1991. Construction of a physical map of the chromosome of *Shigella flixneri* 2A and the direct assignment nine virulence-associated loci identified by Tn*5* insertions. *Mol. Microbiol.* 5:2171–2180.

Okahashi, N. I., C. Sasakawa, N. Okada, M. Yamada, M. Yoshikawa, M. Tokuda, I. Takahashi, and T. Koga. 1990. Construction of a *Not*I restriction map of the *Streptococcus mutans* genome. *J. Gen. Microbiol.* 136:2217–2224.

Old, I. G., J. MacDougall, I. Saint Girons, and B. E. Davidson. 1992. Mapping of genes on the linear chromosome of the bacterium *Borrelia burgdorferi*: possible locations for its origin of replication. *FEMS Microbiol. Letts.* 78:245–50.

Olsen, J. E., M. N. Skov, D. J. Brown, J. P. Christensen and M. Bisgaard. 1996. Virulence and genotype stability of *Salmonella enterica* serovar Berta during a natural outbreak. *Epidemiol. Infect.* 116:267–274.

Olson, E. S., K. J. Forbes, B. Watt and T. H. Pennington. 1995. Population genetics of *Mycobacterium tuberculosis* complex in Scotland analysed by pulsed-field gel electrophoresis. *Epidemiol. Infect.* 114:153–160.

Ouchi, K., M. Abe, M. Karita, T. Oguri, J. Igari and T. Nakazawa. 1995. Analysis of strains of *Burkholderia (Pseudomonas) cepacia* isolated in a nosocomial outbreak by biochemical and genomic typing. *J. Clin. Microbiol.* 33:2353–2357.

Owen, R. J., K. Sutherland, C. Fitzgerald, J. Gibson, P. Borman and J. Stanley. 1995. Molecular subtyping scheme for serotypes HS1 and HS4 of *Campylobacter jejuni. J. Clin. Microbiol.* 33:872–877.

Pan, C. Q., R. Landgraf, and D. Sigman. 1994. DNA-binding proteins as site-specific nucleases. [Review] *Mol. Microbiol.* 12:335–42.

Pantucek, R., F. Gotz, J. Doskar, and S. Rosypal. 1996. Genomic variability of *Staphylococcus aureus* and the other coagulase-positive Staphylococcus species estimated by macrorestriction analysis using pulsed-field gel electrophoresis. *Int. J. Syst. Bacteriol.* 46:216–222.

Park, J. H., J-C. Song, J-C., M. H. Kim, D-S. Lee, and C-H. Kim. 1994. Determination of genome size and a preliminary physical map of an extreme alkaliphile, *Micrococcus* sp. Y-1, by pulsed-field gel electrophoresis. *Microbiology.* 140:2247–2250.

Patel, Y., E. Van Cott, G. G. Wilson, and M. McClelland. 1990. Cleavage at the twelve-base-pair sequence 5′-TCTAGATCTAGA-3′ using M.*Xba*I (TCTAGmA) methylation and *Dpn*I (GmA/TC) cleavage. *Nucl. Acids Res.* 18:1603–1607.

Perkins, J. D., J. D. Heath, B. R. Sharma, and G. M. Weinstock. 1992. *Sfi*I genomic cleavage map of *Escherichia coli* K-12 strain MG1655. *Nucl. Acids Res.* 20:1129–37.

Philipp, W. J., S. Poulet, K. Eiglmeier, L. Pascopella, V. Balasubramanian, B. Heym, S. Bergh, B. R. Bloom, W. R. Jacobs, Jr. and S. T. Cole. 1996. An integrated map of the genome of the tubercle bacillus, *Mycobacterium tuberculosis* H37Rv, and comparison with *Mycobacterium leprae. Proc. Natl. Acad. Sci. U.S.A.* 93:3132–3137.

Phillips, G. J., J. Arnold, and R. Ivarie. 1987. Mono- through hexanucleotide composition of the *Escherichia coli* genome: a Markov chain analysis. *Nucl. Acids Res.* 15:2611–2626.

Phillips, G. J., J. Arnold, and R. Ivarie. 1987. The effect of codon usage on the oligonucleotide composition of the *E. coli* genome and identification of over- and underrepresented sequences by Markov chain analysis. *Nucl. Acids Res.* 16:9185–9198.

Picken, R. N., Y. Cheng, D. Han, J. A. Nelson, A. G. Reddy, M. K. Hayden, M. M.

Picken, F. Strle, J. K. Bouseman and G. M. Trenholme. 1995. Genotypic and phenotypic characterization of *Borrelia burgdorferi* isolated from ticks and small animals in Illinois. *J. Clin. Microbiol.* 33:2304–2315.

Pitt, T. L., M. E. Kaufmann, P. S. Patel, L. C. Benge, S. Gaskin and D. M. Livermore. 1996. Type characterisation and antibiotic susceptibility of *Burkholderia (Pseudomonas) cepacia* isolates from patients with cystic fibrosis in the United Kingdom and the Republic of Ireland. *J. Med. Microbiol.* 44:203–210.

Plessis, P., T. Lamy, P. Y. Donnio, F. Autuly, I. Grulois, P. Y. Le Prise and J. L. Avril. 1995. Epidemiologic analysis of glycopeptide-resistant *Enterococcus* strains in neutropenic patients receiving prolonged vancomycin administration. *Eur. J. Clin. Microbiol. Infect. Dis.* 14:959–963.

Poh, C. L., G. K. Loh and J. W. Tapsall. 1995. Resolution of clonal subgroups among *Neisseria gonorrhoeae* IB-2 and IB-6 serovars by pulsed-field gel electrophoresis. *Genitourin. Med.* 71:145–149.

Poh, C. L., Q. C. Lau and V. T. Chow. 1995. Differentiation of *Neisseria gonorrhoeae* IB-3 and IB-7 serovars by direct sequencing of protein IB gene and pulsed-field gel electrophoresis. *J. Med. Microbiol.* 43:201–207.

Poh, C. L., V. Ramachandran and J. W. Tapsall. 1996. Genetic diversity of *Neisseria gonorrhoeae* IB-2 and IB-6 isolates revealed by whole-cell repetitive element sequence-based PCR. *J. Clin. Mircobiol.* 43:292–295.

Powell, N. G., E. J. Threlfall, H. Chart, S. L. Schofield and B. Rowe. 1995. Correlation of change in phage type with pulsed field profile and 16S rrn profile in *Salmonella enteritidis* phage types 4, 7 and 9a. *Epidemiol. Infect.* 114:403–411.

Prevost, G., B. Pottecher, M. Dahlet, M. Bientz, J. M. Mantz, and Y. Piemont. 1991. Pulsed field gel electrophoresis as a new epidemiological tool for monitoring methicillin-resistant *Staphylococcus aureus* an intensive care unit. *J. Hospital Infection* 17:255–270.

Prevost, G., P. Couppie, P. Prevost, S. Gayet, P. Petiau, B. Cribier, H. Monteil and Y. Piemont. 1995. Epidemiological data on *Staphylococcus aureus* strains producing synergohymenotropic toxins. *J. Med. Microbiol.* 42:237–245.

Prevots, F., E. Remy, M. Mata, and P. Ritzenthaler. 1994. Isolation and characterization of large lactococcal phage resistance plasmids by pulsed-field gel electrophoresis. *FEMS Microbiol. Lett.* 117:7–13.

Proctor, M. E., R. Brosch, J. W. Mellen, L. A. Garrett, C. W. Kaspar and J. B. Luchansky. 1995. Use of pulsed-field gel electrophoresis to link sporadic cases of invasive listeriosis with recalled chocolate milk. *Appl. Environ. Microbiol.* 61:3177–3179.

Proctor, R. A., P. van Langevelde, M. Kristjansson, J. N. Maslow and R. D. Arbeit. 1995. Persistent and relapsing infections associated with small-colony variants of *Staphylococcus aureus. Clin. Infect. Dis.* 20:95–102.

Pruckler, J. M., L. A. Mermel, R. F. Benson, C. Giorgio, P. K. Cassiday, R. F. Breiman, C. G. Whitney and B. S. Fields. 1995. Comparison of *Legionella pneumophila* isolates by arbitrarily primed PCR and pulsed-field gel electrophoresis: analysis from seven epidemic investigations. *J. Clin. Microbiol.* 33:2872–2875.

Pyle, L. E., and L. R. Finch. 1988. A physical map of the genome of *Mycoplasma mycoides* subspecies mycoides Y with some functional loci. *Nucl. Acids Res.* 16:6027–39.

Pyle, L. E., T. Taylor, and L. R. Finch. 1990. Genomic maps of some strains within the *Mycoplasma mycoides* cluster. *J. Bacteriol.* 172:7265–8.

Qiang, B.-q., M. McClelland, S. Podar, A. Spokauskas, and M. Nelson. 1990. The apparent specificity of *Not*I (5′-GCGGCCGC-3′) is enhanced by M.*Fnu*DII or M.*Bep*I methyltransferases (5′-mCGCG-3′): cutting bacterial chromosomes into a few large pieces. *Gene* 88:101–105.

Rainey, P. B. and M. J. Bailey. 1996. Physical and genetic map of the *Pseudomonas fluorescens* SBW25 chromosome. *Mol. Microbiol.* 19:521–533.

Ratnaningsih, E., S. Dharmsthiti, V. Krishnapillai, A. Morgan, M. Sinclair, M., and B. W. Holloway. 1990. A combined physical and genetic map of *Pseudomonas aeruginosa* PAO. *J. Gen. Microbiol.* 136:2351–7.

Riley, M. and K. E. Sanderson. 1990. Comparative genetics of *E. coli* and *S. typhimurium*, p. 85–95. In K. Drlica, and M. Riley (ed.), The bacterial chromosome. American Society for Microbiology, Washington, D. C.

Rodley, P. D., U. Romling and B. Tummler. 1995. A physical genome map of the *Burkholderia cepacia* type strain. *Mol. Microbiol.* 17:57–67.

Roemling, U., and B. Tuemmler. 1991. The impact of two-dimensional pulsed field gel electrophoresis techniques for the consistent and complete mapping of bacterial genomes refined physical map of *Pseudomonas aeruginosa* PAO. *Nucl. Acids Res.* 19:3199–3206.

Roemling, U., D. Grothues, W. Bautsch, and B. Tummler. 1989. A physical genome map of *Pseudomonas aeruginosa* PAO. *EMBO J.* 8:4081–9.

Romalde, J. L., I. Iteman, and E. Carniel. 1991. Use of pulsed field gel electrophoresis to size the chromosome of the bacterial fish pathogen *Yersinia ruckeri. FEMS Microbiol. Lett.* 84:217–226.

Roux, V. and D. Raoult. 1995. Inter- and intraspecies identification of *Bartonella (Rochalimaea)* species. *J. Clin. Microbiol.* 33:1573–1579.

Roux, V., and D. Raoult. 1993.Genotypic identification and phylogenetic analysis of the spotted fever group *Rickettsiae* by pulsed field gel electrophoresis. *J. Bacteriol.* 175:4895–4904.

Roy, D., P. Ward and G. Champagne. 1996. Differentiation of bifidobacteria by use of pulsed-field gel electrophoresis and polymerase chain reaction. *Int. J. Food Microbiol.* 29:11–29.

Russo, T. A., A. Stapleton, S. Wenderoth, T. M. Hooton and W. E. Stamm. 1995. Chromosomal restriction fragment length polymorphism analysis of *Escherichia coli* strains causing recurrent urinary tract infections in young women. *J. Infect. Dis.* 172:440–4445.

Sader, H. S., R. N. Jones and M. A. Pfaller. 1995. Relapse of catheter-related *Flavobacterium meningosepticum* bacteremia demonstrated by DNA macrorestriction analysis. *Clin. Infect. Dis.* 21:997–1000.

Salama, S. M., E. Newnham, N. Chang and D. E. Taylor. 1995. Genome map of *Campylobacter fetus* subsp. fetus ATCC 27374. *FEMS. Microbiol. Lett.* 132:239–245.

Salama, S. M., Q. Jiang, N. Chang, R. W. Sherbaniuk and D. E. Taylor. 1995. Characterization of chromosomal DNA profiles from *Helicobacter pylori* strains isolated from sequential gastric biopsy specimens. *J. Clin. Microbiol.* 33:2496–2497.

Salzano, G., F. Villani, O. Pepe, E. Sorrentino, G. Moschetti, and S. Coppola. 1992. Conjugal transfer of plasmid-borne bacteriocin production in *Enterococcus faecalis* 226 NWC. *FEMS Microbiol. Lett.* 99:1–5.

Samore, M. H., L. Venkataraman, P. C. DeGirolami, R. D. Arbeit and A. W. Karchmer. 1996. Clinical and molecular epidemiology of sporadic and clustered cases of nosocomial *Clostridium difficile* diarrhea. *Am J. Med.* 100:32–40.

Sanches, I. S., M. Ramirez, H. Troni, A. Abecassis, M. Padua, A. Tomasz and H. de Lencastre. 1995. Evidence for the geographic spread of a methicillin-resistant *Staphylococcus aureus* clone between Portugal and Spain. *J. Clin. Microbiol.* 33:1243–1246.

Sanderson, K. E. 1976. Genetic relatedness in the family Enterobacteriaceae. *Annu. Rev. Microbiol.* 30:327–349.

Sanderson, K. E., A. Hessel and K. E. Rudd. 1995. Genetic map of *Salmonella typhimurium*, edition VIII. *Microbiol. Rev.* 59:241–303.

Schmidt, K. D., B. Tummler and U. Romling. 1996. Comparative genome mapping of *Pseudomonas aeruginosa* PAO with *P. aeruginosa* C, which belongs to a major clone in cystic fibrosis patients and aquatic habitats. *J. Bacteriol.* 178:85–93.

Schulz, T. F., C. Bauer, D. Lorenz, R. Plapp, and K. W. Eichhorn. 1993. Studies on the evolution of *Agrobacterium vitis* as based on genomic fingerprinting and its element analysis. *Systematic & Appl. Microbiol.* 16:322–329.

Schwartz, C. D., L. C. Smith, M. Baker, and M. Hsu. 1989. ED: pulsed electrophoresis instrument. *Nature* 342:575–576.

Schwartz, D. N., B. Schable, F. C. Tenover and R. A. Miller. 1995. *Leptotrichia buccalis* bacteremia in patients treated in a single bone marrow transplant unit. *Clin. Infect. Dis.* 20:762–767.

Schwarzkopf, A., H. Karch, H. Schmidt, W. Lenz, and J. Heesemann. 1993. Phenotypical and genotypical characterization of epidemic clumping factor-negative oxacillin-resistant *Staphylococcus aureus. J. Clin. Microbiol.* 31:2281–2285.

Seifert, H. and P. Gerner Smidt. 1995. Comparison of ribotyping and pulsed-field gel electrophoresis for molecular typing of *Acinetobacter* isolates. *J. Clin. Microbiol.* 33:1402–1407.

Seifert, H., W. Richter and G. Pulverer. 1995. Clinical and bacteriological features of relapsing shunt-associated meningitis due to *Acinetobacter baumannii. Eur. J. Clin. Microbiol. Infect. Dis.* 14:130–134.

Shah, R., M. Green, K. A. Barbadora, W. C. Wagener, B. Schwartz, R. R. Facklam and E. R. Wald. 1995. Comparison of M and T type antigen testing to field inversion gel electrophoresis in the differentiation of strains of group A streptococcus. *Pediatr. Res.* 38:988–992.

Shao, Z., W. Mages, and R. Schmitt. 1994. A physical map of the hyperthermophilic bacterium *Aquifex pyrophilus* chromosome. *J. Bacteriol.* 176:6776–6780.

Sitzmann, J., and A. Klein. 1991. Physical and genetic map of the *Methanococcus voltae* chromosome. *Mol. Microbiol.* 5:505–514.

Slaughter, S., M. K. Hayden, C. Nathan, T. C. Hu, T. Rice, J. Van Voorhis, M. Matushek, C. Franklin and R. A. Weinstein. 1996. A comparison of the effect of universal use of gloves and gowns with that of glove use alone on acquisition of vancomycin-resistant enterococci in a medical intensive care unit. *Ann. Intern. Med.* 125:448–456.

Slutsky A. M., R. D. Arbeit, T. W. Barber, J. Rich, C. Fordham Von Reyn, W. Pieciak, M. A. Barlow, J. N. Maslow. 1994. Polyclonal infections due to *Mycobacterium avium* complex in patients with AIDS detected by pulsed-field gel electrophoresis of sequential clinical isolates. *J. Clin. Microbiol.* 32:1773–1778.

Smith, C. L., J. G. Econome, A. Schutt, S. Klco, and C. R. Cantor. 1987. A physical map of the *Escherichia coli* K12 genome. *Science.* 236(4807), 1448-53.

Smith, J. J., L. C. Offord, M. Holderness and G. S. Saddler. 1995. Genetic diversity of *Burkholderia solanacearum* (synonym Pseudomonas solanacearum) race 3 in Kenya. *Appl. Environ. Microbiol.* 61:4263–4268.

Sobral, B.W.S., R. J. Honeycutt, A. G. Atherly, and M. McClelland. 1990. Analysis of the rice (*Oryza sativa L.*) genome using pulsed-field gel electrophoresis and rare-cutting restriction endonucleases. *Plant Mol. Biol. Rep.* 8:252–274.

Sobral, B.W.S., R. J. Honeycutt, A. G. Atherly, and M. McClelland. 1991. Electrophoretic separation of the three *Rhizobium meliloti* replicons. *J. Bacteriol.* 173:5173–5180.

Sobral, B.W.S., R. J. Honeycutt, and A. G. Atherly. 1991. The genomes of the family *Rhizobiaceae* size stability rarely cutting restriction endonucleases. *J. Bacteriol.* 173:704–709.

St. Jean, A., B. A. Trieselmann, and R. L. Charlebois, R. L. 1994. Physical map and set of overlapping cosmid clones representing the genome of the *Archaeon halobacterium* sp. GRB. *Nucl. Acids Res.* 22:1476–1483.

Stanley, J., D. Linton, K. Sutherland, C. Jones and R. J. Owen. 1995. High-resolution genotyping of *Campylobacter coli* identifies clones of epidemiologic and evolutionary significance. *J. Infect. Dis.* 172:1130–1134.

Stanley, J., D. Linton, M. Desai, A. Efstratiou and R. George. 1995. Molecular subtyping of prevalent M serotypes of *Streptococcus pyogenes* causing invasive disease. *J. Clin. Microbiol.* 33:2850–2855.

Stanley, J., M. Desai, J. Xerry, A. Tanna, A. Efstratiou and R. George. 1996. High-resolution genotyping elucidates the epidemiology of group A streptococcus outbreaks. *J. Infect. Dis.* 174:500–506.

Steinman, H. M., and B. Ely. 1990. Copper-zinc superoxide dismutase of *Caulobacter crescentus* cloning sequencing and mapping of the gene and periplasmic location of the enzyme. *J. Bacteriol.* 172:2901–2910.

Stettler R., G. Erauso, T. Leisinger. 1995. Physical and genetic map of the *Methanobacterium wolfei* genome and its comparison with the updated genomic map of *Methanobacterium thermoautotrophicum* Marburg. *Archives of Microbiol.* 163:205–210.

Stettler, R., and T. Leisinger. 1992. Physical map of the methanobacterium *Thermoautotrophicum marburg* chromosome. *J. Bacteriol.* 174:7227–7234.

Stibitz, S., and T. L. Garletts. 1992. Derivation of a physical map of the chromosome of *Bordetella pertussis tohama* I. *J. Bacteriol.* 174:7770–7777.

Sullivan, J. T., H. N. Patrick, W. L. Lowther, D. B. Scott, C. W. Ronson. 1995. Nodulating strains of *Rhizobium loti* arise through chromosomal symbiotic gene transfer in the environment. *Proc. Natl. Acad. Sci.* U.S.A. 92:8985–8989.

Sutherland, K. J., M. Hashimoto, T. Kudo, and K. Horikoshi. 1993. A partial physical map for the chromosome of Alkalophilic *Bacillus-sp* strain C-125. *J. Gen. Microbiol.* 139:661–667.

Suwanto, A., and S. Kaplan. 1992. Chromosome transfer in *Rhodobacter sphaeroides* HFR formation and genetic evidence for two unique circular chromosomes. *J. Bacteriol.* 174:1135–1145.

Suzuki, Y., M. Ishihara, M. Matsumoto, S. Arakawa, M. Saito, N. Ishikawa and T. Yokochi. 1995. Molecular epidemiology of *Salmonella enteritidis.* An outbreak and sporadic cases studied by means of pulsed-field gel electrophoresis. *J. Infect.* 31:211–217.

Syedabubakar, S. N., R. C. Matthews, N. W. Preston, D. Owen and V. Hillier. 1995. Application of pulsed field gel electrophoresis to the 1993 epidemic of whooping cough in the UK. *Epidemiol. Infect.* 115:101–113.

Tabata, K., T. Kosuge, T. Nakahara, and T. Hoshino. 1993. Physical map of the extremely thermophilic bacterium *Thermus thermophilus* HB27 chromosome. *FEBS Letts* 331:81–85.

Talon, D., C. Rouget, V. Cailleaux, P. Bailly, M. Thouverez, F. Barale and Y. Michel Briand. 1995. Nasal carriage of *Staphylococcus aureus* and cross-contamination in a surgical intensive care unit: efficacy of mupirocin ointment. *J. Hosp. Infect.* 30:39–49.

Talon, D., G. Capellier, A. Boillot and Y. Michel Briand. 1995. Use of pulsed-field gel electrophoresis as an epidemiologic tool during an outbreak of *Pseudomonas aeruginosa* lung infections in an intensive care unit. *Intensive Care Med.* 21:996–1002.

Talon, D., P. Bailly, M. Delmee, M. Thouverez, B. Mulin, M. Iehl Robert, V. Cailleaux and Y. Michel Briand. 1995. Use of pulsed-field gel electrophoresis for investigation of an outbreak of *Clostridium difficile* infection among geriatric patients. *Eur. J. Clin. Microbiol. Infect. Dis.* 14:987–993.

Talon, D., V. Cailleaux M. Thouverez and Y. Michel Briand. 1996. Discriminatory power and usefulness of pulsed-field gel electrophoresis in epidemiological studies of *Pseudomonas aeruginosa. J. Hosp. Infect.* 32:135–145.

Tanskanen, E. I., D. L. Tulloch, A. J. Hillier, and B. E. Davidson. 1990. Pulsed field gel electrophoresis of *Sma*I digests of lactococcal genomic DNA A novel method of strain identification. *Appl. & Environ. Microbiol.* 56:3105–3111.

Taylor, D. E., M. Eaton, W. Yan, and N. Chang. 1992. Geneome maps of *Campylobacter jejuni* and *Campylobacter coli. J. Bacteriol.* 174:2332–2337.

Taylor, K. A., A. G. Barbour, and D. D. Thomas. 1991. Pulsed field gel electrophoretic analysis of leptospiral DNA. *Infection & Immunity.* 59:323–329.

Teixeira, L. A., C. A. Resende, L. R. Ormonde, R. Rosenbaum, A. M. Figueiredo, H.

de Lencastre and A. Tomasz. 1995. Geographic spread of epidemic multiresistant *Staphylococcus aureus* clone in Brazil. *J. Clin. Microbiol.* 33:2400–2404.

Tenover, F. C., R. D. Arbeit, R. V. Goering, P. A. Mickelsen, B. E. Murray, D. H. Persing and B. Swaminathan. 1995. Interpreting chromosomal DNA restriction patterns produced by pulsed-field gel electrophoresis: criteria for bacterial strain typing. *J. Clin. Microbiol.* 33:2233–2239.

Thong, K. L., A. M. Cordano, R. M. Yassin and T. Pang. 1996. Molecular analysis of environmental and human isolates of *Salmonella typhi. Appl. Environ. Microbiol.* 62:271–274.

Thong, K. L., M. Passey, A. Clegg, B. G. Combs, R. M. Yassin and T. Pang. 1996. Molecular analysis of isolates of *Salmonella typhi* obtained from patients with fatal and nonfatal typhoid fever. *J. Clin. Microbiol.* 34:1029–1033.

Thong, K. L., S. Puthucheary, R. M. Yassin, P. Sudarmono, M. Padmidewi, E. Soewandojo, I. Handojo, S. Sarasombath and T. Pang. 1995. Analysis of *Salmonella typhi* isolates from Southeast Asia by pulsed-field gel electrophoresis. *J. Clin. Microbiol.* 33:1938–1941.

Thong, K. L., Y. F. Ngeow, M. Altwegg, P. Nagvaratnam and T. Pang. 1995. Molecular analysis of *Salmonella enteritidis* by pulsed-field gel electrophoresis and ribotyping. *J. Clin. Microbiol.* 33:1070–1074.

Tigges, E., and F. C. Minion. 1994. Physical map of the genome of *Acholeplasma oculi* ISM1499 and construction of a Tn4001 derivative for macrorestriction chromosomal mapping. *J. Bacteriol.* 176:1180–1183.

Toda, T., and M. Itaya. 1995. I-CeuI recognition sites in the *rrn* operons of the *Bacillus subtilis* 168 chromosome: inherent landmarks for genome analysis. *Microbiology.* 141:1937–1945.

Tomayko, J. F. and B. E. Murray. 1995. Analysis of *Enterococcus faecalis* isolates from intercontinental sources by multilocus enzyme electrophoresis and pulsed-field gel electrophoresis. *J. Clin. Microbiol.* 33:2903–2907.

Upton, M., P. E. Carter, G. Orange and T. H. Pennington. 1996. Genetic heterogeneity of M type 3 group A streptococci causing severe infections in Tayside, Scotland. *J. Clin. Microbiol.* 34:196–198.

Upton, M., P. E. Carter, M. Morgan, G. F. Edwards and T. H. Pennington. 1995. Clonal structure of invasive *Streptococcus pyogenes* in Northern Scotland. *Epidemiol. Infect.* 115:231–241.

Valcin, M., D. Moissenet, A. Sardet, G. Tournier, A. Garbarg Chenon and H. Vu Thien. 1996. *Pseudomonas (Burkholderia) cepacia* in children with cystic fibrosis: epidemiological investigation by analysis or restriction fragment length polymorphism. *Pathol. Biol. Paris.* 44:442–446.

van Belkum, A., J. Kluytmans, W. van Leeuwen, R. Bax, W. Quint, E. Peters, A. Fluit, C. Vandenbroucke Grauls, A. van den Brule, H. Koeleman and et al. 1995. Multicenter evaluation of arbitrarily primed PCR for typing of *Staphylococcus aureus* strains. *J. Clin. Microbiol.* 33:1537–1547.

van Dijck, P., M. Delmee, H. Ezzedine, A. Deplano and M. J. Struelens. 1995. Evaluation of pulsed-field gel electrophoresis and rep-PCR for the epidemiological analysis of *Ochrobactrum anthropi* strains. *Eur. J. Clin. Microbiol. Infect. Dis.* 14:1099–1102.

Wang Y-K., E. J. Huff, D. C. Schwartz. 1995. Optical mapping of site-directed cleavages on single DNA molecules by the RecA-assisted restriction endonuclease technique. *Proc. Natl. Acad. Sci.* USA 92:165–169.

Wanger, A. R., S. L. Morris, C. Ericsson, K. V. Singh, and M. T. Larocco. 1992. Latex agglutination-negative methicillin-resistant *Staphylococcus aureus* recovered from neonates; epidemiological features and comparison of tying methods. *J. Clin. Microbiol.* 30:2583–2588.

Ward Rainey, N., F. A. Rainey, E. M. Wellington and E. Stackebrandt. 1996. Physical map of the genome of *Planctomyces limnophilus*, a representative of the phylogenetically distinct planctomycete lineage. *J. Bacteriol.* 178:1908–1913.

Webster, C. A., K. J. Towner, H. Humphreys, B. Ehrenstein, D. Hartung and H. Grundmann. 1996. Comparison of rapid automated laser fluorescence analysis of DNA fingerprints with four other computer-assisted approaches for studying relationships between *Acinetobacter baumannii* isolates. *J. Med. Microbiol.* 44:185–194.

Welsh, J. and McClelland, M. (1990). Fingerprinting genomes using PCR with arbitrary primers. *Nucl. Acids Res.* 18:7213–7218.

Wilkinson, S. R., and M. Young. 1995. Physical map of the *Clostridium beijerinckii* (formerly *Clostridium acetobutylicum*) NCIMB 8052 chromosome. *J. Bacteriol.* 177:439–448.

Will, G., S. Jauris Heipke, E. Schwab, U. Busch, D. Rossler, E. Soutschek, B. Wilske and V. Preac Mursic. 1995. Sequence analysis of *ospA* genes shows homogeneity within *Borrelia burgdorferi sensu stricto* and *Borrelia afzelii* strains but reveals major subgroups within the *Borrelia garinii* species. *Med. Microbiol. Immunol. Berl.* 184:73–80.

Williams, J. G., Kubelik, A. R., Livak, K. J., Rafalski, J. A., and Tingey, S. V. (1990). DNA polymorphisms amplified by arbitrary primers are useful as genetic markers. *Nucl. Acids Res.* 18:6531–6535.

Wilson, K. J., A. Sessitsch, J. C. Corbo, K. E. Giller, A. D. Akkermans and R. A. Jefferson. 1995. beta-Glucuronidase (GUS) transposons for ecological and genetic studies of rhizobia and other gram-negative bacteria. *Microbiology* 141:1691–1705.

Wong, K. K., and M. McClelland. 1992. A *Bln*I restriction map of the *Salmonella typhimurium* LT2 genome. *J. Bacteriol.* 174,1656–1661.

Wong, K. K., and M. McClelland. 1992. Dissection of the *Salmonella typhimurium* genome by use of a Tn*5* derivative carrying rare restriction sites. *J. Bacteriol.* 174:3807–3811.

Xia, M., W. L. Whittington, K. K. Holmes, F. A. Plummer, and M. C. Roberts. 1995. Pulsed-field gel electrophoresis for genomic analysis of *Neisseria gonorrhoeae*. *J. Infect. Dis.* 171:455–458.

Yakubu, D. E. and T. H. Pennington. 1995. Epidemiological evaluation of *Neisseria meningitidis* serogroup B by pulsed-field gel electrophoresis. *FEMS. Immunol. Med. Microbiol.* 10:185–189.

Yamagishi, A., and T. Oshima. 1990. Circular chromosomal DNA in the sulfur-dependent Archaebacterium *Sulfolobus acidocaldarius. Nucl. Acids Res.* 18:1133–1136.

Yan, W., and D. E. Taylor. 1991. Sizing and mapping of the genome of *Campylobacter coli* strain UA417R using pulsed field gel electrophoresis. *Gene* 101:117–120.

Yao, J. D., J. M. Conly and M. Krajden. 1995. Molecular typing of *Stenotrophomonas (Xanthomonas) maltophilia* by DNA macrorestriction analysis and random amplified polymorphic DNA analysis. *J. Clin. Microbiol.* 33:2195–2198.

Ye, F., F. Laigret, J. C. Whitley, C. Citti, L. R. Finch, P. Carle, J. Renaudin, and J. M. Bove. 1992. A physical and genetic map of the *Spiroplasma citri* genome. *Nucl. Acids Res.* 20:1559–65.

Zhang, Y., R. J. Wallace, Jr. and G. H. Mazurek. 1995. Genetic differences between BCG substrains. *Tuber. Lung. Dis.* 76:43–50.

Zinder, N. and J. Lederberg. 1992. Genetic exchange in *Salmonella. J. Bacteriol.* 64:679–699.

Zuerner, R. L. 1991. Physical map of chromosomal and plasmid DNA comprising the genome of *Leptospira interrogans. Nucl. Acids Res.* 19:4857–4860.

Zuerner, R. L., J. L. Herrmann, and I. Saint Girons. 1993. Comparison of genetic maps for two *Leptospira interrogans* serovars provides evidence for two chromosomes and intraspecies heterogeneity. *J. Bacteriol.* 175:5445–51.

25

One-dimensional Pulsed-field Gel Electrophoresis

Ute Römling, Karen Schmidt, and Burkhard Tümmler

Introduction

Bacterial genome mapping can be performed at the level of genomic DNA itself (physical mapping) or by following the pattern in which portions of the genome are passed on to the progeny (genetic linkage mapping). Their relatively small genome of 0.5–10 Megabase pairs makes bacteria an appropriate target for the study of genome organization by pulsed-field gel electrophoresis techniques (PFGE; Schwartz and Cantor, 1984). For the construction of a low-resolution restriction map, the bacterial genome is cleaved with rare-cutting restriction endonucleases (see Chapt. 24) and the fragments generated are separated by PFGE. Ordering of fragments by an appropriate strategy results in an anonymous physical genome map that is converted into a genetic map by the assignment of genes to particular regions. This chapter describes the fingerprinting and macrorestriction mapping of bacterial genomes by one-dimensional PFGE techniques.

PFGE techniques

Pulsed-field gel electrophoresis separates DNA molecules of up to 10 Megabase pairs in size. PFGE utilizes at least two alternative fields that do not coincide. Migration of the DNA molecules through the agarose gel is governed by the strength of the electric field, the pulse time, field angle and shape, temperature and ionic strength, and type and concentration of the agarose used (Gemmill, 1991). Optimum size-dependent separation of fragments is usually achieved by the appropriate selection of ramps of pulse time intervals. Although a variety of PFGE techniques have been developed in recent years (TAFE, Gardiner *et al.*, 1986; CFGE, Southern *et al.*, 1987; ROFE, Ziegler and Volz), contour clamped homogeneous electric field electrophoresis (CHEF, Chu *et al.*, 1986) and programmable, autonomously-controlled electrophoresis (PACE, Clark *et al.*, 1986) are almost exclusively used now. CHEF generates uniform electric fields across the gel by a hexagonal array of electrodes of predetermined ('clamped') electric potential at a field angle of 120°. The potential of the electrodes can be individually varied in the PACE system so that all possible field configurations can be created. CHEF and PACE are preferred for bacterial genome analysis because of the high resolution and sharpness of bands in the whole size range from 1 to 10,000 kbp and the straight migration path and almost identical electrophoretic mobility in all lanes (Römling *et al.*, 1994).

Pulse times and electric field strength are the prime parameters which affect the size range of molecules separated by PFGE and need to be optimized on a case-by-case basis. Electrophoretic mobility of linear DNA molecules decreases with size, but the situation becomes more complex at the borders of the low and high molecular weight range. For high molecular weight DNA the inversion point is located close to the compression zone of the gel and shows up by band broadening. Larger molecules move ahead of smaller ones after the inversion point of minimum mobility has been passed. Pulse time dependent inversion of the relative mobility is also seen for linear DNA molecules smaller than 50 kbp, because of the superposition of continuous and alternative electric field effects (example shown in Fig. 26-3; *see* Chapt. 26). These phenomena impede the unequivocal assignment of DNA fragments run under different PFGE separation conditions.

An increase of pulse time and/or electric field results in larger molecules being resolved. Electric fields of 3 V/cm or more, however, may lead to degradation and trapping of Megabase-sized DNA molecules in the agarose matrix (Gemmill, 1991). Faint or fuzzy bands representing the largest fragments are indicative of this phenomenon. Hence, whereas fields of 4–7 V/cm are typically employed for separation in the 10 to 1,000 kb size range, Megabase-sized DNAs should be separated with large pulse times at fields below 2 V/cm.

Topology of Bacterial Genomes

Bacteria show a variety of genomic configurations (*see* Chapter 1). Most species carry one circular chromosome and one or more circular plasmids, but some species have more than one chromosome or harbour a linear chromosome. PFGE can be used to resolve the number, size and topology of the entities of the bacterial genome.

Chromosomes

Unsheared bacterial DNA is prepared by inclusion of intact bacteria into agarose blocks prior to cell lysis. In the initial PFGE experiments the chosen pulse times and ramps should separate linear DNA molecules in the range of the expected chromosomal size. Under these conditions, linear chromosomes or linearized chromosomes will enter the gel and be resolved, whereas linear DNA larger than 10 Mb and open circular molecules will remain trapped in the agarose block. A subsequent PFGE run with different pulse times will discriminate between the pulse-time dependent migration characteristics of linear DNA molecules and the invariant band position of a circle. The size of linear chromosomes is determined from a calibration curve that is constructed from the relative mobilities of size markers separated in adjacent lanes of the gel. The size determination of circular chromosomes requires the linearization of the DNA prior to PFGE. Agarose-trapped DNA samples are exposed to irradiation with X- or gamma rays (Beverley, 1989; Rodley *et al.*, 1995), or to digestion with minute amounts of a restriction endonuclease that cleaves DNA frequently. Separation by PFGE results in a smear of high molecular weight bands with the upper boundaries representing the size of the linearized chromosome cleaved at a single site. Alternatively, a chromosome may be cleaved at a unique restriction site that has been inserted by mutagenesis (Jumas-Bilak *et al.*, 1995; *see also* Chapts. 24 and 29).

Plasmids

Plasmids are not reliably detected by these procedures. Small plasmids (< 10 kb) rapidly diffuse out of agarose plugs and large circular plasmids may not migrate during PFGE. The isolation should be optimized on a case-to-case basis. One may try the Eckhardt method (Eckhardt, 1978), i.e., gel electrophoresis of the bacterial suspensions gently lysed directly in the gel slots. Alternatively, plasmid minipreparation (Lennon and DeCicco, 1991) and subsequent PFGE or conventional agarose gel electrophoresis may be used to isolate the plasmids.

Subsequently, one may assign each genomic entity to be either a self-replicating chromosome or an extrachromosomal plasmid. Chromosomes can be defined as replicons encoding ribosomal RNA genes and other essential genes. Southern blot hybridization of the PFGE-separated linearized genome with a ribosomal *rrn* probe should in principle identify the *rrn*-encoding replicons, but the background caused by multiply-cleaved chromosomes may yield ambiguous signals in the autoradiogram. An alternative is the digestion of the agarose-embedded DNA with I-*Ceu*I that recognizes a conserved 19–24 bp large consensus sequence in the 23S rDNA gene (Liu *et al.*, 1993a). Genomic digests with 1-*Ceu*I allow evaluation of the number of *rrn* operons in the bacterial strain (*see also* Chapt. 24). A chromosome with *n rrn* operons harbouring the recognition sequence is cleaved into *n* I-*Ceu*I fragments, if it is circular and into *n* + *1* I-*Ceu*I fragments, if it is linear. Figure 1 shows the separation of I-*Ceu*I digests of *Pseudomonas aeruginosa* strain C.

Construction of a Macrorestriction Map by One-dimensional PFGE

The practical procedure involves the encapsulation of bacteria in agarose plugs, cell lysis and digestion with protease and detergent, cleavage of the unsheared genomic DNA with restriction endonucleases. PFGE and (optional) blotting and hybridization (Römling *et al.*, 1994).

Choice of Restriction Endonuclease

For fast, economic, and reliable physical mapping of the bacterial genome of interest, one should choose at least two restriction endonucleases, which cleave the genomic DNA into an informative, but still resolvable number of five to 40 fragments. Criteria for the selection of restriction enzymes are the length of the recognition sequence of the enzyme and the GC content, degree of methylation and codon usage of the bacterial species to be typed. The reader is referred to Chapter 24, which provides detailed information about rare cutting enzymes that are useful for macrorestriction mapping of bacterial genomes.

Identification of Fragment Number

The range of separation allowing the identification of all fragments depends on the genome size and the number of restriction sites (compare Figs. 25-1–25-3). In order to uncover double or multiple bands and to allow resolution of all fragments, the conditions for electrophoresis should be separately optimized for at least three size ranges: 500 kbp to 50 kbp; 50 kbp to 15 kbp with pulse times from 1 to 4 s (1.5% agarose gels) and below 15 kbp (conventional 1% agarose gel). Electrophoresis conditions providing high resolution are particularly critical for separating fragments in the intermediate size range between 10 and 50 kb. As already mentioned above, the mobility of these fragments is governed by a superposition of continuous and pulsed field gel electrophoresis and consequently fragment inversion can occur by the selection of different pulse times (see Figure 26-3; *see* Chapter 26 as an example).

Assembly of the Physical Map by One-dimensional PFGE

Macrorestriction fragments can be ordered by applying different criteria of combinatorial analysis. The following paragraphs give a representative, although not exhaustive overview of one-dimensional mapping techniques:

Ordering by Mass Law

If the genome contains only few sites for the respective restriction endonuclease, fragment order may be inferred from fragment size of complete or partial

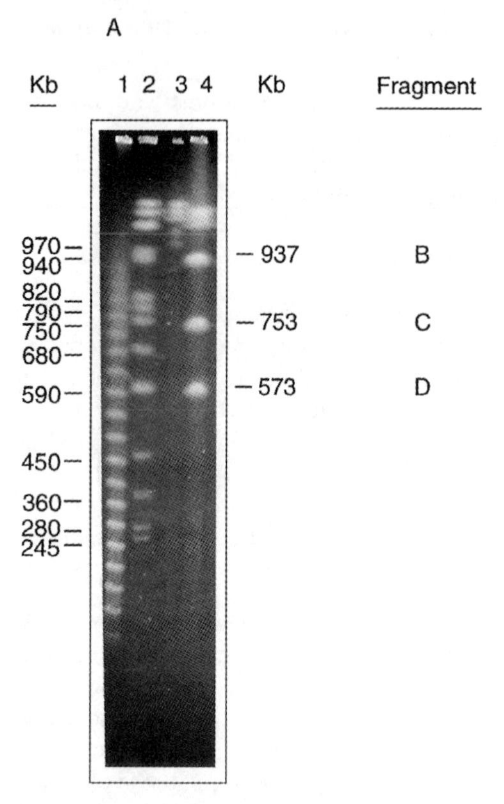

Figure 25-1. PFGE separation of size markers and of an I-*Ceu*I digestion of *Pseudomonas aeruginosa* strain C. Lanes (1) λ oligomers; (2) *Saccharomyces cerevisiae* X-2180/1B chromosomes; (3) *Candida albicans* CBS562 chromosomes; (4) I-*Ceu*I digest of *P. aeruginosa* strain C; (5) *Schizosaccharomyces pombe* chromosomes.

A. CHEF (6 V/cm) separation for fragments in the size range of 50–1,000 kb. CHEF was conducted in a Bio-Rad DR™II cell for 52 h on an 1.5% agarose gel in 0.5 TBE buffer at 10°C. Pulse times were linearly increased from 40 to 120 s.

single and double digestions separated on the same gel (Cocks *et al.*, 1989). An elegant modification of this approach is the methylation of some, but not all restriction recognition sites. The subsequent comparison of restriction fragment patterns of methylated and unmethylated DNA identifies adjacent macrorestriction fragments that carry the overlapping recognition sequence for the methylase (Bautsch, 1993).

Hybridization with Gene Probes

PFGE-separated single and double complete digestions are blotted onto membranes and hybridized with gene probes. This procedure stepwise resolves the

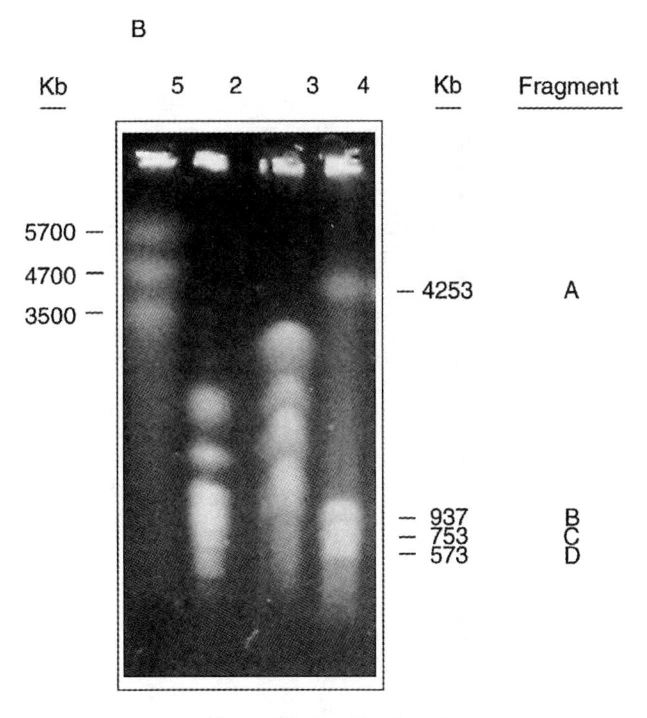

Figure 25-1. Continued

B. CHEF (1.3 V/cm) separation of DNA molecules larger than 2,000 kb on an 0.6% agarose gel in 0.5 TBE buffer at 10°C. Pulse times were linearly increased from 550 s to 4,750 s for 120 h.

fragment order on the bacterial chromosome. If the enzymes cleave at only a few sites, Southern analysis of multiple and/or partial digestions may still result in an informative and interpretable fragment pattern (Canard and Cole, 1989).

Hybridization with (Macro)restriction Fragments

PFGE-separated restriction digestions of enzyme A are probed with a gel-eluted (macro)restriction fragment of enzyme B and vice versa (Allardet-Servent *et al.*, 1991). This approach identifies neighbouring restriction fragments, but not necessarily the fragment order, if the probe recognizes more than two fragments.

The major inherent difficulty of the mapping strategies a, b, and c is the correct placement of small fragments. Linkage to a small fragment may escape notice because of misleading cross-hybridization and/or a poor resolution in the respective molecular weight range. Another difficulty of Southern analysis is the ambiguity caused by repetitive DNA (Römling and Tümmler, 1991).

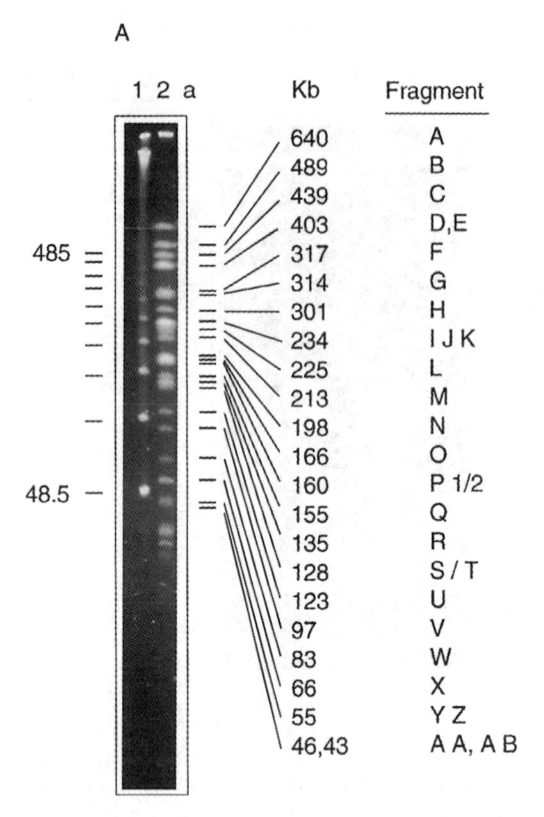

Figure 25-2. Optimized separation of *Spe*I digests of the 6,516 kb large *P. aeruginosa* strain C chromosome. Lanes (1) λ oligomers; (2) *Spe*I digest of strain C, (2a) ethidium bromide strain, (2b), (2c) autoradiograms of the blotted gel (2a) of ^{32}-P end-labeled *Spe*I fragments exposed for (2b) 1 h, (2c) 7 days; (3) λ *Bst*E II digest.

A. Optimized separation of fragments 50–640 kb in size. CHEF (7.5 V/cm) was conducted on a 1% agarose gel in 0.5 TBE buffer at 10°C. Pulse times were increased in two linear ramps of from 1 to 6 s for 6 h and 1 to 35 s for 13 h.

Hybridization with Linking Clones

Linking clones span the recognition site for a rarely cleaving enzyme. A linking clone detects two adjacent macrorestriction fragments and therefore constitutes an end-specific probe. Representative libraries of genomic linking clones can be easily isolated from plasmid libraries of bacterial genomic DNA by a PFGE-based method (Römling and Tümmler, 1993). After digestion of the plasmid library with the infrequently cleaving restriction endonuclease of interest, the linearized linking clone DNAs are separated from undigested circular plasmid

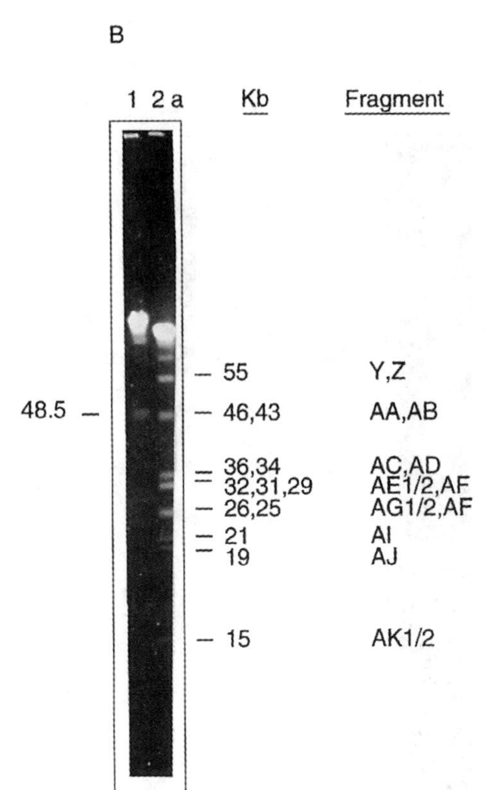

Figure 25-2. Continued

B. Optimized separation of fragments 20–60 kb in size by CHEF (1% agarose gel, 0.5 TBE buffer, 10°C, 6 V/cm, one linear ramp from 1–3 s for 36 h in 1 s increments).

DNAs by pulsed-field polyacrylamide gel electrophoresis, recircularized and introduced into host cells.

Physical mapping by hybridization with linking clones is reliable, but rather tedious, because it is necessary to analyze at least 10 n random linking clones to have a > 95% probability of obtaining clones spanning each of the n rare-cutter sites. A nonrandom distribution of linking clones will demand an even more extensive screening because sequences that are difficult to clone and the size selection of the vector to accommodate foreign DNA lead to an underrepresentation of some linking fragments.

Hybridization of Partially Digested Linear DNA with an End-specific Probe

Fragment contigs of linear chromosomes can be reconstructed from the hybridization of partial digestion kinetics with a labelled end-specific probe. Circular

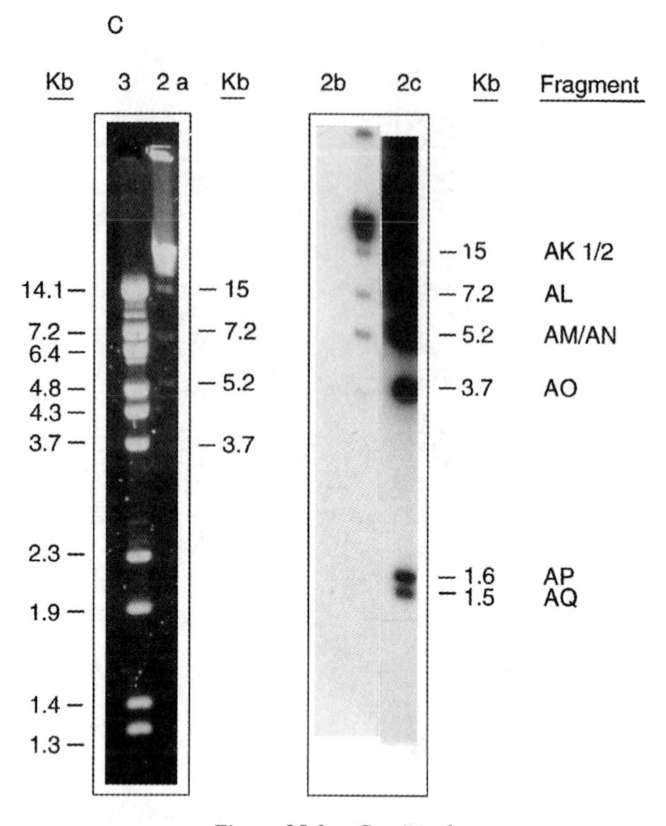

Figure 25-2. Continued

C. Separation of fragments smaller than 15 kb in size by conventional 1% agarose gel electrophoresis in 1 × TBE buffer for 15 h at 2 V/cm.

chromosomes may be mapped by this approach after introduction of an unique restriction site into the genome by e.g. transposon insertion mutagenesis and subsequent linearization at this site (Jumas-Bilak *et al.*, 1995; *see also* Chapters 24 and 29).

Insertion or Deletion Mutagenesis

Transposon mutants can be mapped by the shift of fragment size on the PFGE gel (Ely and Gerardot, 1988). An alternative is the insertional mutagenesis with a transposon carrying a natural (Liu *et al.*, 1993b) or an engineered (Wong and McClelland, 1992; Cheng and Lessie, 1994) rare-cutter site. Map localization of transposon mutants is of general interest for genotype-phenotype studies in bacteria, i.e., the dissection of the genetic elements that are involved in a particular

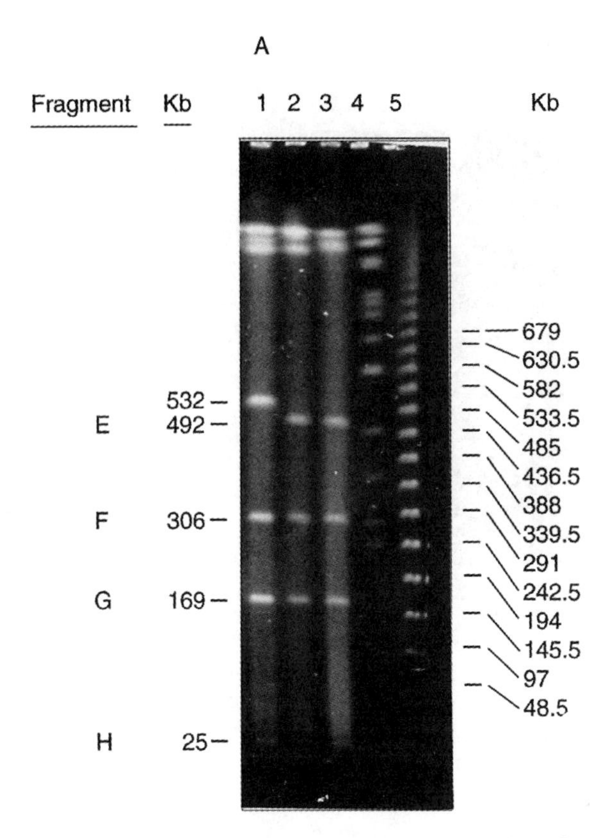

Figure 25-3. Optimized CHEF separation for comparative mapping of *Pac*I sites in the chromosomes of three members of *P. aeruginosa* clone C (strains SG17M, SG50M, B6470) (Römling et al., 1995). Lanes (1), (2), (3), *Pac*I digests of the chromosomes of strain (1) SG17, (2) SG50M, (3) B6470; lane 4, *Saccharomyces cerevisiae* X-2180/b chromosomes; (5), λ oligomers; (6), *Schizosaccharomyces pombe* chromosomes; (7) *Candida albicans* CBS562 chromosomes.

A. Separation of *Pac*I fragments 50–850 kb in size (1.5% agarose gel, 0.5 TBE buffer, 10°C, 6 V/cm, one linear ramp from 10–120 s for 47 h).

pathway, network or phenotype. A limitation of this approach is the possibility of deletions and rearrangements associated with the transposition events.

Gene-directed mutagenesis successively eliminates the recognition site for a rare-cutter enzyme (Itaya and Tanaka, 1991). A cassette which carries a selectable marker flanked by the endogeneous host sequence surrounding the restriction site is introduced into the chromosome by homologous recombination. Two adjacent fragments become a linking fragment and are identified by the change of size and the presence of the foreign marker gene. Mapping by gene disruption has some practical limitations. Mutants will only be obtained if the rare-cutter

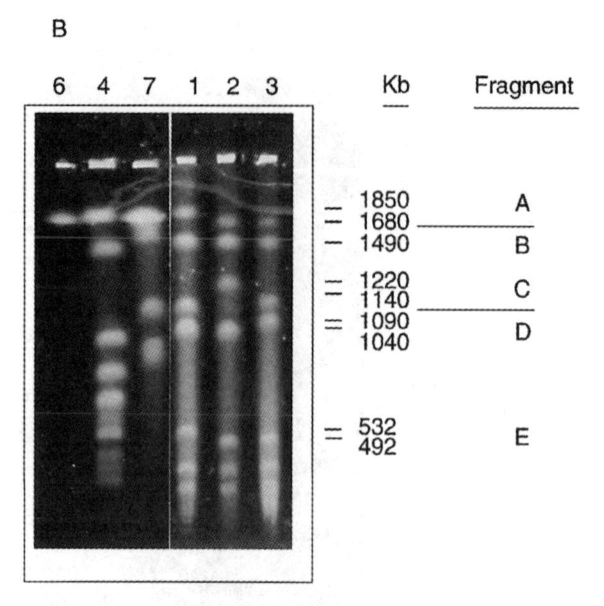

Figure 25-3. Continued

A. Separation of *Pac*I fragments 50–850 kb in size (1.5% agarose gel, 0.5 TBE buffer, 10°C, 6 V/cm, one linear ramp from 10–120 s for 47 h).

B. Separation of *Pac*I fragments 850–1,650 kb in size (0.6% agarose gel, 0.5 TBE buffer, 10°C, 1.3 V/cm, one linear ramp from 500–1,000 s for 160 h).

site does not reside in genes that are essential for growth, competence or recombination.

Combination of Genetic and Physical Mapping

If genetic linkage data are available, mapping of mutants with known genomic alterations may be helpful to resolve equivocal mapping data (Chen et al., 1991). Complementation or knock-outs of a genetically mapped locus can be used for physical mapping if the mutagenesis leads to changes in fragment sizes or hybridization signals.

All one-dimensional mapping techniques have their particular pro's and con's. Complete mapping may be jeopardized by the biology of the organism (methods d, f, g), unfavourable distribution of restriction sites (methods a–c, e–g), hard-to-cleave sites (methods a–e) or sequence repeats (b, c). Fragment order in critical regions like clusters of tiny fragments flanked by large fragments should be elucidated on a case-by-case basis by a combination of various approaches. The inherent problems of mapping by one-dimensional PFGE are overcome by the two-dimensional PFGE techniques that are described in Chapter 26.

Macrorestriction Fingerprinting of Bacterial Genomes

PFGE separation of macrorestriction digests allows the identification of bacteria at all taxonomic levels, particularly at the infrasubspecies level of a clone (Grothues and Tümmler, 1991; Römling *et al.*, 1995). The macrorestriction fragment pattern is diagnostic for a particular strain and provides an estimate of the genomic relationships among strains. Typing of bacteria by macrorestriction genome fingerprinting provides data about the history and the genetic diversity of bacterial clones.

For diagnostic strain-to-strain comparisons the bacterial genome should be digested to yield an informative, but still resolvable number of 20 to 60 fragments. A PFGE instrument should be employed that allows the simultaneous analysis of multiple samples. Size markers and restriction digestions of reference strains should by applied to the outermost and the center lanes in order to correct for distortions across gels.

The 'relatedness' between macrorestriction fingerprints of different bacterial strains is most easily recognized by eye as long as only closely related patterns from a single gel need to be identified. In all other cases reference restriction digests have to be positioned on every gel. Since electric field strength and pulse time have a profound influence on the mobility of DNA molecules during PFGE, visual comparison of fingerprints from PFGE gels run under different conditions is unreliable and should be avoided. Computer-assisted data processing and evaluation are helpful. The software calculates the number of band matches between each pair of fragment pattern. Stepwise combinations of the most similar gel patterns into clusters results in a dendrogram which is the common format to visualize the relatedness of fingerprints (Römling *et al.*, 1994).

Macrorestriction genome fingerprinting is generally applicable to the typing of bacteria, and hence it has become a standard procedure, particularly in clinical microbiology and population genetics (Arbeit *et al.*, 1990; Römling *et al.*, 1994; Tenover *et al.*, 1995).

References

Allardet-Servent, A., M.-J. Carles-Nuirt, G. Bourg, S. Michaux, and M. Ramuz. 1991. Physical map of the *Brucella melitensis* 16M chromosome. *J. Bacteriol.* 173:2219–2224.

Arbeit, R. D., M. Arthur, R. D. Dunn, C. Kim, R. K. Selander, and R. Goldstein. 1990. Resolution of recent evolutionary divergence among *Escherichia coli* from related lineages: the application of pulsed field gel electrophoresis to molecular epidemiology. *J. Infect. Dis.* 161:230–235.

Bautsch, W. 1993. A *Nhe*I macrorestriction map of the *Neisseria meningitidis* B1940 genome. *FEMS Microbiol. Lett.* 107:191–197.

Beverley, S. M. 1989. Estimation of circular DNA size using gamma-irradiation and pulsed field gel electrophoresis. *Anal. Biochem.* 177:110–114.

Canard, B., and S. Cole. 1989. Genome organization of the anaerobic pathogen *Clostridium perfringens. Proc. Natl. Acad. Sci. USA* 86:6676–6680.

Chen, H., A. Kuspa, I. M. Keseler, and L. J. Shimkets. 1991. Physical map of the *Myxococcus xanthus* chromosome. *J. Bacteriol.* 173:2109–2115.

Cheng, H.-P., and T. G. Lessie. 1994. Multiple replicons constituting the genome of *Pseudomonas cepacia* 17616. *J. Bacteriol.* 176:4034–4042.

Chu, G., D. Vollrath, and R. W. Davies. 1986. Separation of large DNA molecules by contour-clamped homogeneous electric fields. *Science* 234:1582–1585.

Clark, S. M., E. Lai, B. W. Birren, and L. Hood. 1986. A novel instrument for separating large DNA molecules with pulsed homogeneous electric fields. *Science* 241:1203–1205.

Cocks, B. G., L. E. Pyle, and L. R. Finch. A physical map of the genome of *Ureaplasma urealyticum* 960^T with ribosomal RNA loci. *Nucl. Acids Res.* 17:6713–6719.

Eckhardt, T. 1978. A rapid method for the identification of plasmid desoxyribonucleic acid in bacteria. *Plasmid* 1:584–588.

Ely, B., and C. J. Gerardot. 1988. Use of pulsed-field gradient gel electrophoresis to construct a map of the *Caulobacter crescentus* genome. *Gene* 68:323–333.

Gardiner, K., W. Laas, and D. Patterson. 1986. Fractionation of large mammalian DNA restriction fragments using vertical pulsed-field gradient gel electrophoresis. *Somatic Cell Genet.* 12:185–195.

Gemmill, R. M. 1991. Pulsed field gel electrophoresis. *Adv. Electrophoresis* 4:1–48.

Grothues, D. and B. Tümmler. (1991). New approaches in genome analysis by pulsed-field gel electrophoresis: application to the analysis of *Pseudomonas* species. *Mol. Microbiol.* 5:2763–2776.

Itaya, M., and T. Tanaka. 1991. Complete physical map of the *Bacillus subtilis* 168 chromosome constructed by a gene-directed mutagenesis method. *J. Mol. Biol.* 220:631–648.

Jumas-Bilak, E., C. Maugard, S. Michaux-Charachon, A. Allardet-Servent, A. Perrin, D. O'Callaghan, and M. Ramuz. 1995. Study of the organization of the genomes of *Escherichia coli, Brucella melitensis* and *Agrobacterium tumefaciens* by insertion of a unique restriction site. *Microbiology* 141:2425–2432.

Lennon, E., and B. T. DeCicco. 1991. Plasmids of *Pseudomonas cepacia* strains of diverse origins. *Appl. Environ. Microbiol.* 57:2345–2350.

Liu, S. L., A. Hessel, and K. E. Sanderson. 1993a. Genomic mapping with I-*Ceu*I, an intron-coded endonuclease specific for genes of ribosomal RNA, in *Salmonella* spp., *Escherichia coli*, and other bacteria. *Proc. Natl. Acad. Sci. (USA)* 90:6874–6878.

Liu, S.-L., A. Hessel, and K. E. Sanderson. 1993b. The *Xba*I-*Bln*I-*Ceu*I genomic cleavage map of *Salmonella typhimurium* LT2 determined by double digestion, end-labelling, and pulsed-field gel electrophoresis. *J. Bacteriol.* 175:4104–4120.

Rodley, P. D., U. Römling, and B. Tümmler. 1995. A physical genome map of the *Burkholderia cepacia* type strain. *Mol. Microbiol.* 17:57–67.

Römling, U., J. Greipel, and B. Tümmler. 1995. Gradient of genomic diversity in the *Pseudomonas aeruginosa* chromosome. *Mol. Microbiol.* 17:323–332.

Römling, U., T. Heuer, and B. Tümmler. 1994. Bacterial genome analysis by pulsed field gel electrophoresis techniques. *Adv. Electrophoresis.* 7:353–406.

Römling, U., and B. Tümmler. 1991. The impact of two-dimensional pulsed-field gel electrophoresis techniques for the consistent and complete mapping of bacterial genomes: refined physical map of *Pseudomonas aeruginosa* PAO. *Nucleic Acids. Res.* 19: 3199–3206.

Römling, U., and B. Tümmler. 1993. Comparative mapping of the *Pseudomonas aeruginosa* PAO genome with rare-cutter linking clones or two-dimensional pulsed-field gel electrophoresis protocols. *Electrophoresis* 14:283–289.

Schwartz, D. C., and C. R. Cantor. 1984. Separation of yeast chromosome-sized DNAs by pulsed field gradient gel electrophoresis. *Cell* 37:67–75.

Southern, E. M., R. Anand, W.R.A. Brown, and D. S. Fletcher. 1987. A model for the separation of large DNA molecules by crossed field gel electrophoresis. *Nucl. Acids Res.* 15:5925–5942.

Tenover, F. C., R. D. Arbeit, R. V. Goering, P. A. Mickelsen, B. E. Murray, D. H. Persing, and B. Swaminathan. 1995. Interpreting chromosomal DNA restrictions patterns produced by pulsed-field gel electrophoresis: criteria for bacterial strain typing. *J. Clin. Microbiol.* 33:2233–2239.

Wong, K. K., and M. McClelland. 1992. Dissection of the *Salmonella typhimurium* genome by use of a Tn*5* derivative carrying rare restriction sites. *J. Bacteriol.* 174:3807–3811.

Ziegler, A., and A. Volz. 1991. ROFE, In *Methods in Molecular Biology*, Vol. 12, Pulsed Field Gel Electrophoresis, pp. 63–72. M. Burmeister and L. Ulanovsky, eds. Humana Press, Totowa NJ.

26

Two-dimensional Pulsed-field Gel Electrophoresis

Ute Römling, Karen Schmidt, and Burkhard Tümmler

Introduction

Two key techniques were responsible for the rapid progress in bacterial genome mapping during the last decade: the identification of infrequently cleaving restriction endonucleases which produce large DNA of defined size (see Chapt 24), and the development of pulsed field gel electrophoresis (PFGE) to separate high molecular weight DNA (see Chapt 25). As pointed out in Chapter 25, the relatively small bacterial genome size (0.5–10 Megabase pairs) makes these organisms ideal objects for the construction of low-resolution physical maps by both one-dimensional and two-dimensional combinatorial analyses. The creation of distinct "spots" enables physical genome analysis by two-dimensional gel electrophoresis as has already been applied to plasmid analysis by conventional continuous field gel electrophoresis some twenty years ago (Villems *et al.*, 1978). The macrorestriction map is assembled from a few two-dimensional PFGE gels of partial-complete single digestions or complete double digestions (Bautsch, 1988). The fragment order can be established without any genetic data and without using hybridization methods, which is particularly advantageous for the mapping of poorly characterized genomes. This chapter describes the essentials of bacterial genome mapping by two-dimensional PFGE.

Physical mapping by two-dimensional PFGE

Principles of the methodology

The fragment order is established by evaluation of gels displaying partial to complete digestions or pairs of reciprocal double digestions (Bautsch, 1988;

Römling and Tümmler, 1991). Two-dimensional mapping requires an optimal quality of the agarose-embedded DNA and optimal separation conditions. The PFGE instrument should generate a straight non-distorted migration path. We routinely use contour clamped homogeneous electric field electrophoresis (CHEF, Chu *et al.*, 1986) for two-dimensional PFGE. To save money and time, one should not run more than two gels for a particular gel combination. To achieve this goal, the quality of the prepared DNA and the conditions for restriction digestion and electrophoretic fragment separation need to be optimized.

Partial to complete mapping: In this technique, a partial restriction digest is first separated by PFGE in one dimension, then redigested to completion with the same enzyme, and subsequently resolved in the second orthogonal dimension (Fig. 26-1). The genomic DNA is partially digested with one enzyme and then (optionally) end-labeled in the agarose plug with ^{32}P-nucleotides by Klenow DNA polymerase (Römling *et al.*, 1994). The partial digest is separated by PFGE, the

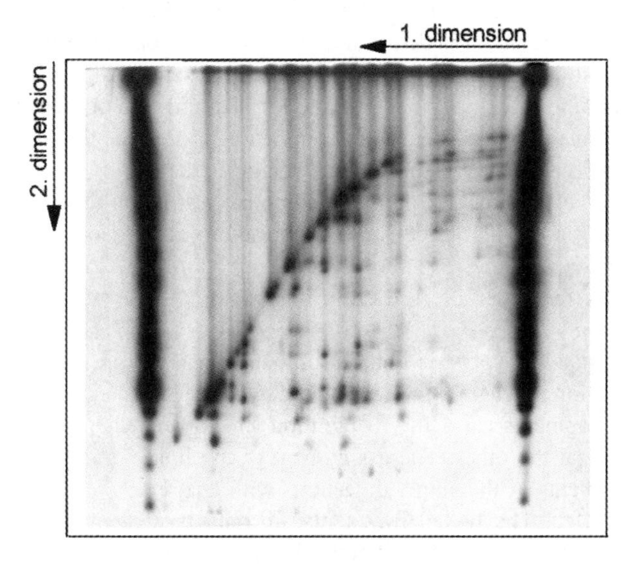

Figure 26-1. Two-dimensional PFGE analysis of a partial/complete *Spe*I restriction digest of the *Pseudomonas aeruginosa* strain C12 chromosome. Autoradiogram of the blotted ^{32}P-end-labeled partial digestion fragments (exposure time: 22 days). After partial digestion of the agarose-embedded DNA with 0.033 U *Spe*I for 90 min and end-labeling, the fragments were separated by CHEF (6 V/cm, 1.5% agarose). Pulse times were increased in three linear ramps, 5–100 s for 37 h, 10–40 s for 12 h, 70–100 s for 13 h. The whole lane was cut out, redigested to completion and separated in the second dimension (6 V/cm, 1.5% agarose). Pulse times were linearly increased in three ramps, 8–50 s for 24 h, 12–25 s for 22 h, 1–15 s for 14 h. Complete *Spe*I digests were applied as size markers to the outermost lanes.

entire lane is cut out, and the gel-separated fragments are redigested to completion with the original enzyme. The lane is oriented in the second agarose gel with the width of the lane of the first dimension becoming the height of the second dimension. The complete digest is separated in the second dimension by PFGE, and the two-dimensional gel is stained with ethidium bromide. Subsequently, the gel is blotted for autoradiography. All fragments which are generated by complete digestion of the initial partial digestion fragments are visualized by ethidium bromide staining, with the fluorescence intensity being proportional to their length. In contrast, the autoradiogram identifies the two terminal fragments of each partial digestion fragment with comparable intensity, irrespective of size (Fig. 26-1). Hence, comparative evaluation of gel stain and autoradiogram will facilitate the determination of the fragment order.

The second dimension resolves the fragment composition of the partial digestion fragments in one straight line (Römling *et al.*, 1989; Römling and Tümmler, 1991). Fragments are identified by comparison with complete digests separated in both outermost lanes of the second dimension. The molecular weight of a partial digestion fragment is determined from its position in the first dimension. If partial digestion fragments accidentially have the same size, other criteria, e.g. mass law and intensity, may be referred to for identification of the corresponding spots. Fragments that were cleaved to completion prior to the first run are located on the diagonal and serve as molecular weight markers.

Reciprocal gels: In this technique, a complete restriction digest with enzyme A is separated in the first dimension, redigested to completion with enzyme B, and then separated in the second orthogonal direction (Fig. 26-2). On a separate series of gels the order of restriction digestions is reversed (Bautsch, 1988; Bihlmaier *et al.*, 1991).

The corresponding spots on both two-dimensional gels are identified by their identical position of fragment order, ranked by increasing electrophoretic mobility. Linked fragments carry the recognition site for enzyme A at one end and for enzyme B at the other. The assignment of the linked fragments to the two parental fragments of the single digestions with enzyme A or B establishes the fragment overlap. The linked fragments are only present in the separation of double digests and cannot be detected in single digests. The separate single digestion products from each enzyme and the double digest are separated on the outermost lanes of the second dimension of the two-dimensional gel. Fragments of the first digest that do not contain the recognition sequence for the second enzyme will be located on the diagonal of the two-dimensional gel and can be identified by comparison with the respective single digest. On the reciprocal gel the audioradiogram of the end-labeled fragments will not reveal the internal neighbours of the linking fragments, i.e. the fragments not digested by the second enzyme; comparison of the autoradiogram with the more complex ethidium bromide stain will again facilitate the elucidation of fragment order.

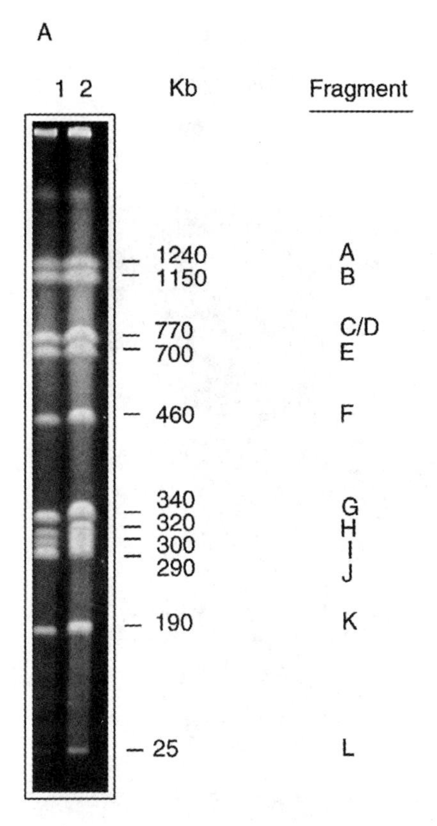

Figure 26-2. Separation of a complete *Pac*I-*Swa*I-*Spe*I triple digest of the *Pseudomonas aeruginosa* strain C17 chromosome by two-dimensional PFGE.

A. Separation of *Pac*I-*Swa*I digestion in the first dimension. Agarose-embedded cell densities were 5×10^9/ml (lane 1) and 1×10^{10}/ml (lane 2). CHEF was conducted on 1% agarose gels in 0.5 TBE buffer at 3.5 V/cm. Pulse times were increased in two linear ramps from 40–80 s for 40 h and then 60–340 s for 40 h.

Practical Procedures

Physical mapping by one-dimensional PFGE and hybridization analysis (see Chapter 25) is a stepwise, laborious and time-consuming procedure, but rather robust. In contrast, only a few experiments are necessary if one performs two-dimensional mapping. The procedure is rapid, but sensitive to technical flaws and hence experimental conditions need to be carefully selected and optimized on a case-by-case basis. The following recommendations and precautionary measures summarize our experience to generate informative two-dimensional gels. For detailed protocols the reader should consult a recent review (Römling *et al.*, 1994).

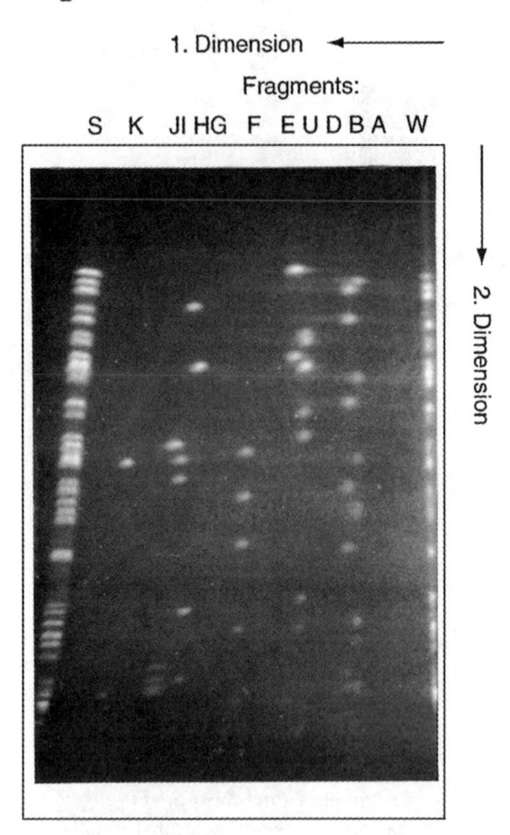

Figure 26-2. Continued

B. After complete digestion with *Spe*I, the triple digest was separated in the second orthogonal dimension by CHEF (1.5% agarose, 0.5 TBE, 6 V/cm). Pulse times were linearly increased in four ramps, 8–45 s in 20 h, 12–25 s for 22 h, 1–14 s for 14 h, 1–4 s for 4 h. A triple digest was applied to the outermost lane (s). (w) indicates the position of the well.

Selection of the Restriction Endonucleases

Each restriction enzyme should cleave the chromosome into an informative, but still resolvable number of five to forty fragments (see Chapt. 24). If a more refined map is desirable, several enzymes may be chosen. At least two restriction enzymes should be employed for mapping in order to prove the consistency of each macrorestriction map. It is often hard to establish fragment order for islands of small fragments or for alternating sequences of vary large and small fragments. In these cases a second or third restriction enzyme will facilitate the mapping of regions harbouring unfavourable spacings of sites for the first enzyme, and vice versa.

Separation Conditions

Ramps and pulse time should be chosen to achieve optimal resolution. Short running times are not an important criterion. Electrophoretic mobility decreases with size, but one should bear in mind that paradoxical migration behavior might occur in the high and low molecular weight range. The inversion point of minimal mobility is located close to the compression zone, and hence larger fragments migrate ahead of smaller ones in the Megabase size range beyond the inversion point. Pulse times have also profound influence on the relative mobility of fragments smaller than 50 kb (Fig. 26-3) which impedes the interpretation of two-dimensional gels. Since pulse times and ramps are usually different in the two dimensions, fragment inversion may occur particularly in the sensitive range from 5 to 25 kb (Fig. 26-3).

Optimization of a Partial Digest

Agarose plugs containing a cell concentration of 2.5×10^9 or 5×10^9 cells/ml should be subjected to digestion kinetics experiments (t > 45 min) over a wide range of enzyme concentrations. Digestion conditions depend on the enzyme, the supplier and the DNA template. Since slight variations of enzyme activity may lead to irreproducible results, it is important to keep one tube of enzyme for the partial digestion experiments and to carry out the optimization speedily. The appropriate partial digest is the optimum compromise between resolution

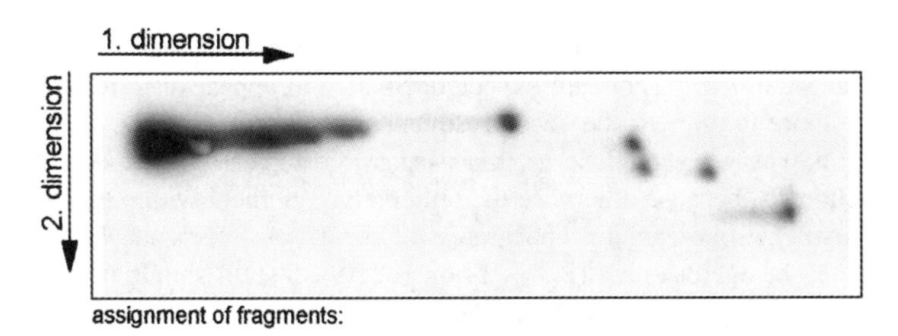

Figure 26-3. Field-dependent shifts of mobility of small macrorestriction fragments. Autoradiogram of the two-dimensional electrophoretic analysis of ^{32}P-end-labeled *Spe*I fragments below 15 kb of the *Pseudomonas aeruginosa* strain C chromosome. The assignment of the fragments is indicated. After digestion of the agarose-embedded DNA with *Spe*I and radioactive end-labeling the fragments were separated by CHEF (6 V/cm, 1.5% agarose, 0.5 × TBE buffer). Pulse times were linearly increased from 0.1–2 s for 36 h. The lane was cut out and rerun in the second dimension by continuous field gel electrophoresis; running conditions were 1 V/cm, 1.5% agarose, 1 × TBE, 18 h.

and information. More fragments will always show up in the two-dimensional gel than expected from the first dimension, due to multiple co-migrating bands (Fig. 26-1). The upper separation limit of the first dimension of the CHEF run should correspond to the sum of the molecular weights of the two largest fragments.

End-labeling of DNA in an Agarose Plug

The restricted chromosomal DNA embedded in a plug of final size (6 × 2 × 1 mm) ready for PFGE, is incubated for 30 min at room temperature with 1 U Klenow DNA polymerase and 2 μCi α-dNTP (> 3000 Ci/mmol). For optimal results neither the incubation time, nor the enzyme concentration should be increased. If the signal intensity is too low, the amount of radioactivity should be increased. Prior to two-dimensional PFGE, end-labeling should be checked by autoradiography of an one-dimensional PFGE gel. Faint signals should be visible after an one day exposure. Optimal signal-to-noise ratios should be achieved by an one week exposure. Linking fragments or end-fragments are detected in two-dimensional gels by end-labeling (Fig. 26-1). Klenow polymerase incorporates nucleotides only at 5′ protruding ends. thus, it is prudent to use the dNTP complementing the first base of the restriction enzyme site. When restriction enzymes create blunt ends or 5′ recessive end, nucleotides are incorporated throughout the fragment. This procedure is useful for increasing the sensitivity of fragment detection when comparing spots on autoradiographs to those on the ethidium bromide stained gels.

Two-dimensional gels

1.5% agarose gels should be used for both directions whenever possible, in particular when many spots are expected. Spots will appear distorted when the agarose concentration of the second dimension is lower than that of the first dimension. The gel should be as thin as possible for effective cooling during electrophoresis, because otherwise the temperature gradient within the gel gives rise to a rocket-like running appearance of the DNA fragments. For the first dimension, the agarose plug (2.5×10^9 or 5×10^9 cells/ml) should be as high as it is wide and should be loaded into the bottom of the gel slot in order to create clear spots in the second dimension. A second plug is loaded into an adjacent slot for quality control. To enhance resolution, the whole length of the gel should be used to separate the restriction digest in the first dimension of PFGE. After the run the lane intended to be used in the second dimension is cut out with a scalpel. The DNA embedded in the gel slice is digested overnight in a tub with about 60 units restriction enzyme per ml. The enzyme concentration may need to be adjusted (empirical range 10–100 U/ml) and incubation times can be extended to 48 h. According to our experience all rare-cutting restriction enzymes except for *Swa*I cleave DNA to completion in Ultra Pure Agarose (Life Technologies); otherwise 1% low-melting agarose can be used for the first dimension. After restriction digestion the gel slice is affixed to the gel for the second dimension whereby the gel slice is turned by 90° so that the previous width becomes the height of the gel. Standards are applied to the outermost lanes and the second dimension is run. DNA is visualized by ethidium bromide staining and autoradiography (if applicable). For partial-complete digests, a total digest is used as standard. A double digest and a single digest using the second enzyme are normally loaded in two adjacent lanes in reciprocal gels.

Interpretation of Gels

The major advantage of two-dimensional PFGE mapping is that all the necessary information resides in a few gels. However, this also represents the greatest challenge for the interpreter. Theoretical examples of maps with very few restriction sites are easily solved, but in reality the gels are much more complex. A correct interpretation has to take into account all of the available experimental evidence. There are many ways to assemble the pieces of the puzzle, but we recommend to observe the following guidelines:

Evaluation of Partial/Total Gels (Fig. 26-1):

(1) One should start with the composition of the partial digestion fragment of the lowest molecular weight.
(2) The sum of the fragments in one lane should be compared with the size of the original partial digestion fragment. This size may be estimated by taking the completely digested fragments from the diagonal as size standards.
(3) The comparison of the autoradiogram with the ethidium bromide stained gel aids to resolve the order of the fragments.
(4) Unequivocal fragment orders from smaller partial digest fragments may be used to interpret a partial digestion fragment with a more complex pattern.
(5) Ellipsoids instead of circles may indicate the presence of more than one partial digestion fragment in close proximity. Corresponding spots may be assigned by their size-dependent intensities and the sum of fragment lengths which must be equivalent to the size of the partial digestion fragment.
(6) Linkage to a small fragment is indicated by slight deviations of the spot from the diagonal.
(7) Isolated links that are not observed in any further partial digestion fragment are suggestive for the existence of a plasmid.

(8) The interpretation of two-dimensional gels in the direction of the second dimension (from left to right in Fig. 26-1) analyses the composition of partial digestion fragments in the order of increasing molecular weight. Complementary evaluation of the gel in the direction of the first dimension uncovers the molecular weight increments of successive neighbours of a particular fragment. This redundancy, namely that fragment links are represented by multiple partial digestion fragments, is valuable to confirm the fragment order. Moreover, intensity variations of partial fragments are a criterion for the susceptibility of the terminal restriction recognition sites to cleavage.

Evaluation of Reciprocal Gels (Fig. 26-2):

(1) First, cut and uncut fragments should be distinguished based on the comparison of one-dimensional single and double digests.

(2) Then fragments should be ordered by decreasing molecular weight in both reciprocal gels. The analysis based on the absolute migration distance is not possible due to the slight distortions and different migration behavior which may occur in different gels.

(3) The comparison of fragment order between the gels will assist in identifying identical fragments.

(4) Next, the two linking fragments in each lane should be identified, if possible, with the help of the corresponding autoradiogram.

(5) The fragment sizes should be added up for each original fragment, in order to confirm fragment size and number.

(6) One should begin deriving the fragment order by first selecting a linking fragment of one original fragment in either of the two gels and then identifying the identical fragment in the second remaining gel. Then the linking fragment from the other end in the second gel should be identified on the first gel, and so forth. By the end of the assignment chain, the gaps between the linking fragments are filled with the fragments not cleaved by the other enzyme. Fragments not cleaved by the second enzyme are found on the diagonal.

(7) After the first round of interpretation, some spots may be difficult to interpret. The major problems arise from plasmids, small fragments and non-resolved fragment overlaps. Strong intensity is evidence for the overlap of two or more fragments. Large plasmids may stick in the well when not cleaved by the first enzyme. One should check for inconsistent genome size if one adds fragment sizes from different restriction digests. Small fragments may accidentially run out of the gel or may not be visible because of low signal intensity. In some

cases the slight deviation of a large fragment from the diagonal is indicative of a link to such a small fragment. These small fragments should be detected by cloning or by separation of end-labeled fragments using pulse times between one to five seconds. When the pulse times for the separation in the two dimensions are very different, fragment order by relative mobility may not be conserved for fragments smaller than 50 kb. Two-dimensional gels of total/total digests with different pulse times (see example in Fig. 26-3) may resolve the running behavior and hence identify the fragments.
Finally, one should merge the information gained from the reciprocal and partial/total gels and prove the consistency of the physical map. In most cases fragment order is reliably established. Any remaining gaps and/or ambiguities, i.e. the exact fragment order in clusters of tiny fragments or in regions with large and very small fragments, should be solved on a case-to-case basis.

Examples of maps that were constructed from the evaluation of two-dimensional PFGE gels are shown in Chapter 73 of this monograph. Instructive examples of the primary data and their interpretation can be found in the original publications by Bihlmaier *et al.*, (1991), Römling and Tümmler (1991) and Schmidt *et al.* (1996).

Concluding Remarks

Analysis of bacterial genomes by two-dimensional PFGE yields reliable and consistent maps. Complete maps can be constructed within a few weeks time, but the procedures demand manual and intellectual skills. Rigorous optimization of conditions for the digestion and separation of the macrorestriction fragments is crucial for the generation of informative two-dimensional CHEF gels. The wealth of information to be extracted from a few experiments is the major advantage of two-dimensional mapping compared to all the step-by-step analyses in one dimension. All primary data necessary for map construction may be stored in a few files which could make two-dimensional gels and autoradiograms easily accessible to peers and electronic libraries, but in contrast to DNA sequencing the practical procedures and the evaluation of data will probably remain refractory to automation.

References

Bautsch, W. 1988. Rapid physical mapping of the *Mycoplasma mobile* genome by two-dimensional field inversion gel electrophoresis techniques. *Nucl. Acids Res.* 16:11461–11467.

Bihlmaier, A. U. Römling, T. F. Meyer, B. Tümmler, and C. P. Gibbs. 1991. Physical and genetic map of the *Neisseria gonorrhoeae* strain MS11-N198 chromosome. *Mol. Microbiol.* 5:2529–2539.

Chu, G., D. Vollrath, and R. W. Davies. 1986. Separation of large DNA molecules by contour-clamped homogeneous electric fields. *Science* 234:1582–1585.

Römling, U., D. Grothues, W. Bautsch, and B. Tümmler. 1989. A physical genome map of *Pseudomonas aeruginosa* PAO. *EMBO J.* 8:4081–4089.

Römling, U., T. Heuer, and B. Tümmler. 1994. Bacterial genome analysis by pulsed field gel electrophoresis techniques. *Adv. Electrophoresis* 7:353–406.

Römling, U., and B. Tümmler. 1991. The impact of two-dimensional pulsed-field gel electrophoresis techniques for the consistent and complete mapping of bacterial genomes: refined physical map of *Pseudomonas aeruginosa* PAO. *Nucl. Acids Res.* 19:3199–3206.

Schmidt, K. D., B. Tümmler, and U. Römling. 1996. Comparative genome mapping of *Pseudomonas aeruginosa* PAO with *P. aeruginosa* C, which belongs to a major clone in cystic fibrosis patients and aquatic habitats. *J. Bacteriol.* 178:85–93.

Villems, R., C. J. Duggleby, and P. Broda. 1978. Restriction endonuclease mapping of DNA using in situ digestion in two-dimensional gels. *FEBS Lett.* 89:267–270.

27

Use of Bacteriophage Mu-P22 Hybrids for Genome Mapping

Matthew Lawes and Stanley R. Maloy

Mu*d*-P22 derivatives are hybrids between phage Mu and phage P22 that can be inserted at essentially any desired site on the *Salmonella* chromosome (Youderian *et al.*, 1988; Benson and Goldman, 1992). Induction of Mu*d*-P22 insertions yields phage particles that, as a population, carry chromosomal DNA from the region between 150–250 Kb on one side of the insertion. Thus, phage lysates from a representative set of Mu*d*-P22 insertions into the *S. typhimurium* chromosome yield an ordered library of DNA that provides powerful tools for the genetic and physical analysis of the *Salmonella* genome (*see* Chapter 22). Although Mu*d*-P22 has not yet been used in other species, this approach should be applicable in a variety of other bacteria as well.

Mu and P22

Phage Mu

Mu is a phage that replicates by transposition (Symonds *et al.*, 1987). Mu has a high transposition frequency and a low bias in selecting its transposition target site, so insertions can be easily recovered in any non-essential gene on the chromosome (Groisman, 1991). Transposition of Mu requires the Mu *A* and *B* gene products which act in *trans*, and sites at the left end (MuL) and right end (MuR) of the Mu genome which acts in *cis* (Symonds *et al.*, 1987).

Deletion derivatives of Mu have been constructed (designated Mu*d*), some of which lack the Mu *A* and *B* genes that are required in *trans* for transposition, but retain the MuL and MuR sequences required in *cis* (Groisman, 1991). Such derivatives are unable to transpose unless the Mu *A* and *B* gene products are provided in *trans*. Delivery systems have been developed to allow a single transposition of the Mu*d* from the donor DNA by providing the Mu *A* and *B* gene products in *trans*. If the Mu *A* and *B* gene products are only provided transiently, no subsequent transposition can occur so the resulting Mu*d* insertions are stable (Hughes and Roth, 1988).

Phage P22

Phage P22 is a temperate *Salmonella* phage that is very easy to grow and very stable during extended storage (Susskind and Botstein, 1978; Poteete, 1988; Lawes and Maloy, 1993; Maloy et al., 1996). In many ways, P22 is quite similar to phage λ. Under conditions that favor lysogeny, P22 can integrate into the host genome at a specific attachment site via a site-specific recombination mechanism, catalyzed by the integrase protein (Int). Upon induction, excision of the prophage from the chromosome requires both the integrase and excisionase (Xis) proteins. In contrast to phage λ, packaging of P22 DNA into phage particles occurs by a headful mechanism.

Mu*d*-P22

Mu*d*-P22 is a hybrid between Mu and P22 phage (Figure 27-1). One end of each Mu*d*-P22 contains 500 bp of the right end of Mu (MuR) and the other end contains 1000 bp from the left end of Mu (MuL). The central portion of Mu*d*-P22 contains the P22 genes required for replication and phage morphogenesis. However, Mu*d*-P22 derivatives are deleted for the P22 *attL, int, xis* and *abc* genes at the left side of the prophage map, and *sieA*, gene *9, conABC,* and *attR* at the right end of the prophage map. The deletion of *attL, attR, int* and *xis* prevents the integration or excision of Mu*d*-P22 via site-specific recombination. Thus, when integrated into the chromosome, the Mu*d*-P22 is "locked-in." Deletion of the *abc* genes prevents a productive switch from bi-directional replication to rolling circle replication (Poteete, 1988). Deletion of *sieA* and *conABC* prevent exclusion of superinfecting P22 phage. Thus, Mu*d*-P22 lysogens are still amenable to further genetic analysis using P22. Gene *9* encodes the tail fiber protein required for adsorption to recipient cells. Although phage that lack tail fibers are not infectious, tail protein can be extracted from cells containing a plasmid which expresses gene *9* constitutively and these "tails" will assemble onto P22 phage heads *in vitro*. The resulting "tailed" phage particles are fully infectious. Alternatively, a plasmid that expresses gene *9* can be introduced into the Mu*d*-P22 containing strain prior to induction so that tail fibers are provided in *trans* during the lytic development of the induced Mu*d*-P22 prophage.

Isolation of chromosomal Mud-P22 insertions

Although it is possible to isolate chromosomal Mu*d*-P22 insertions by transposition if the Mu *A* and *B* gene products are provided in *trans* (Youderian *et al.*, 1988; Higgins and Hillyard, 1988), the frequency of transposition is much less

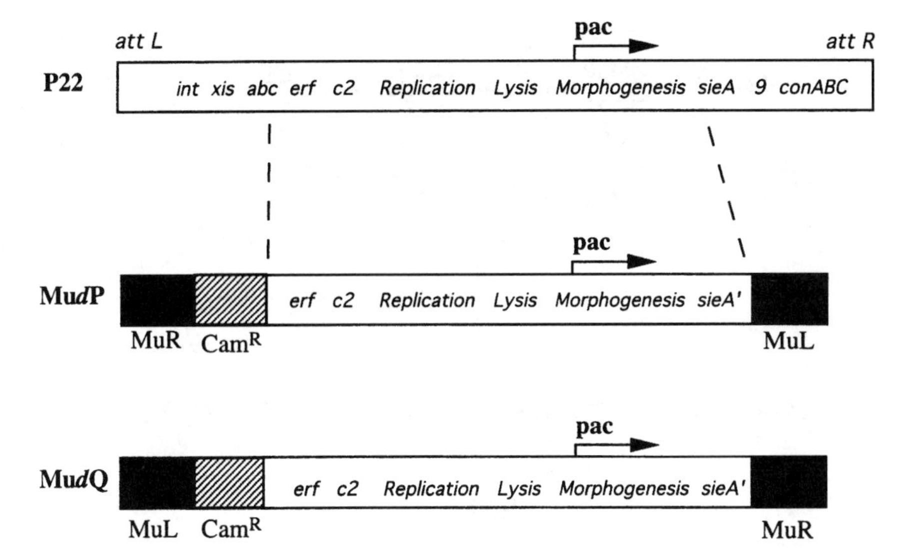

Figure 27-1. Structure of the Mu*d*-P22 derivatives Mu*d*P and Mu*d*Q. The central portion of Mu*d*-P22 contains the P22 genome from the *erf* gene through part of the *sieA* gene. The opposite ends of Mu*d*-P22 contain approximately 500 bp of the right end of Mu (MuR) and 1000 bp from the left end of Mu (MuL). Since the orientation of the P22 segment within the Mu ends is opposite for Mu*d*P relative to Mu*d*Q, a Mu*d*P replacement of a Mu*d* insertion will package flanking DNA *either* clockwise *or* counterclockwise relative to genetic loci on the *S. typhimurium* genetic map, while the Mu*d*Q replacement of the same insertion will package genomic DNA in the opposite direction. Both Mu*d*P and Mu*d*Q are 32.4 Kb in size. A chloramphenicol resistance (Cam^R) gene is located adjacent to one end of Mu*d*-P22.

than for the smaller Mu*d* derivatives. Therefore, chromosomal Mu*d*-P22 insertions are usually isolated by replacement of another chromosomal Mu*d* insertion via recombination *in vivo* (Figure 27-2). Replacement of a chromosomal Mu*d* insertion with both a Mu*d*P and a Mu*d*Q will yield an isogenic pair of strains that package the chromosomal DNA in opposite directions (Figure 27-2).

*Induction of Mu*d*-P22 lysogens*

DNA-damaging agents lead to inactivation of the P22 *c2* repressor, inducing the switch from lysogeny to lytic growth. For example, addition of 2 μg/ml Mitomycin C to a midlog phase culture of a P22 lysogen results in lysis and release of phage particles within three hours. After excision, P22 normally circularizes, then initiates bi-directional replication until about twenty copies of the phage genome are accumulated. In contrast, since "locked-in" prophages like Mu*d*-P22 cannot excise, DNA replication begins from the P22 origin and extends beyond the ends of P22 into adjacent chromosomal DNA in both directions. This results

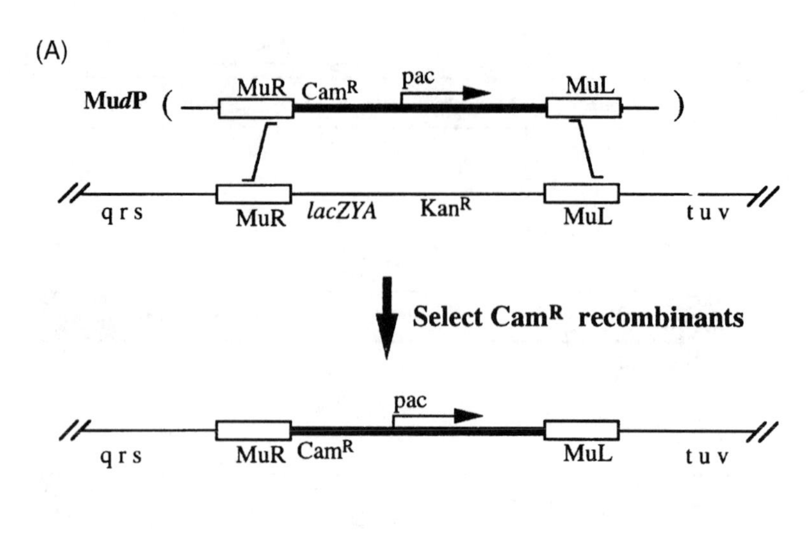

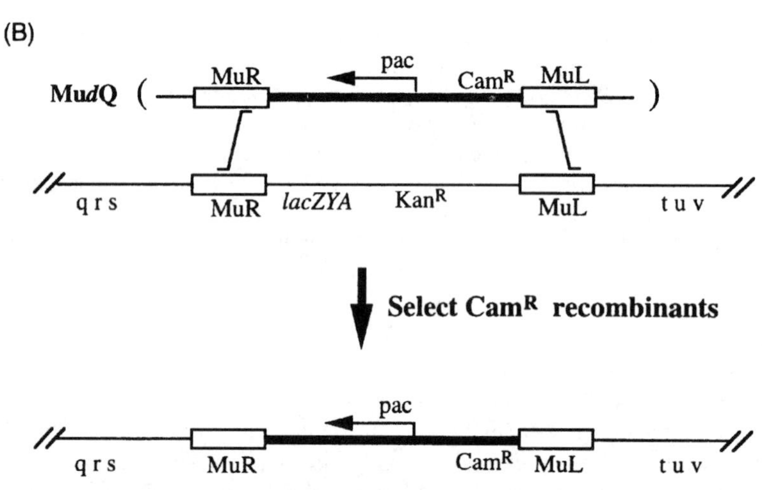

Figure 27-2. Replacement of a chromosomal Mu*d* insertion with both Mu*d*P and a Mu*d*Q will yield an isogenic pair of strains that package the chromosomal DNA in opposite directions. Mu*d*-P22 insertions which package in a clockwise orientation are designated (A), and those that package in a counterclockwise orientation are designated (B), following the convention adopted for orientation of Tn*10* insertions (Chumley *et al.*, 1979). Since the packaging orientation of a Mu*d*-P22 is determined by the orientation of the Mu*d* element which was replaced, for a particular Mu*d* insertion, it is possible to find *either* a set of Mu*d*P(A) Mu*d*Q(B) *or* a set of Mu*d*P(B) Mu*d*Q(A) replacements, but not both sets. Any Mu*d* insertion that shares homology with Mu*d*-P22 at the flanking MuL and MuR sequences can be replaced via homologous recombination with Mu*d*P or Mu*d*Q.

in an "onion-skin" like amplification of the chromosome surrounding the "locked-in" prophage (Figure 27-3).

While DNA replication is proceeding, expression of proteins required for phage morphogenesis also occurs. Once sufficient "Pac" nuclease (encoded by gene *3*) has accumulated, the enzyme will cut a specific DNA sequence within gene *3*, called the *pac* site. After cutting the DNA, the nuclease remains bound to the DNA and feeds it to the portal of the phagehead, where about 44 Kb of linear double-stranded DNA are stuffed into the capsid. Once the head is full, a second, nonspecific, cut is made by the nuclease to release the DNA, and assembly of the phage particle is completed. The end of DNA is then fed to another capsid and another headful is packaged. After about three to five consecutive, sequential "headfuls" of DNA are packaged, the enzyme dissociates from the DNA (Figure 27-4). Only the initial cleavage by the nuclease for the first headful of DNA packaged is sequence dependent, all subsequent cuts occur at nonspecific sequences at the end of the DNA protruding from the capsid.

Since an induced Mu*d*-P22 cannot excise from the chromosome, the first 44 Kb headful of DNA packaged will extend from the *pac* site through the right end of P22 and the flanking end of Mu (MuL for Mu*d*P or MuR for Mu*d*Q), then into adjacent chromosomal DNA. The second and subsequent headfuls will exclusively contain sequential 44 Kb of chromosomal DNA. Although amplification is bi-directional, packaging proceeds in only one direction. Since Mu*d*-P22 encodes the wild-type P22 gene *3* nuclease, the overwhelming majority of packaging events are initiated from the P22 *pac* site and not from pseudo-*pac* sites within the bacterial chromosome. Thus the population of DNA molecules encapsidated upon induction of a specific Mu*d*-P22 insertion is highly enriched for the 3–5 min of chromosomal DNA adjacent to one side of the Mu*d*-P22 insertion.

Using Mu*d*-P22 Insertions for Genetic and Physical Analysis of Chromosomal DNA

The phage particles obtained from a Mu*d*-P22 insertion carry a specific 3–5 min region of adjacent chromosomal DNA. Thus, these phage particles provide a "specialized transducing lysate" for any specific region of the *S. typhimurium* chromosome. A set of Mu*d*-P22 insertions spaced every 3–5 min around the *S. typhimurium* chromosome is available (Benson and Goldman, 1992; Maloy *et al.*, 1996). This ordered collection of Mu*d*-P22 insertions allows the rapid genetic or physical mapping of genes on the *S. typhimurium* chromosome (*see also* Chapter 22).

Rapid genetic mapping

The ordered set of "specialized" transducing lysates can be used for mapping the location of new mutations on the *S. typhimurium* chromosome by transduction.

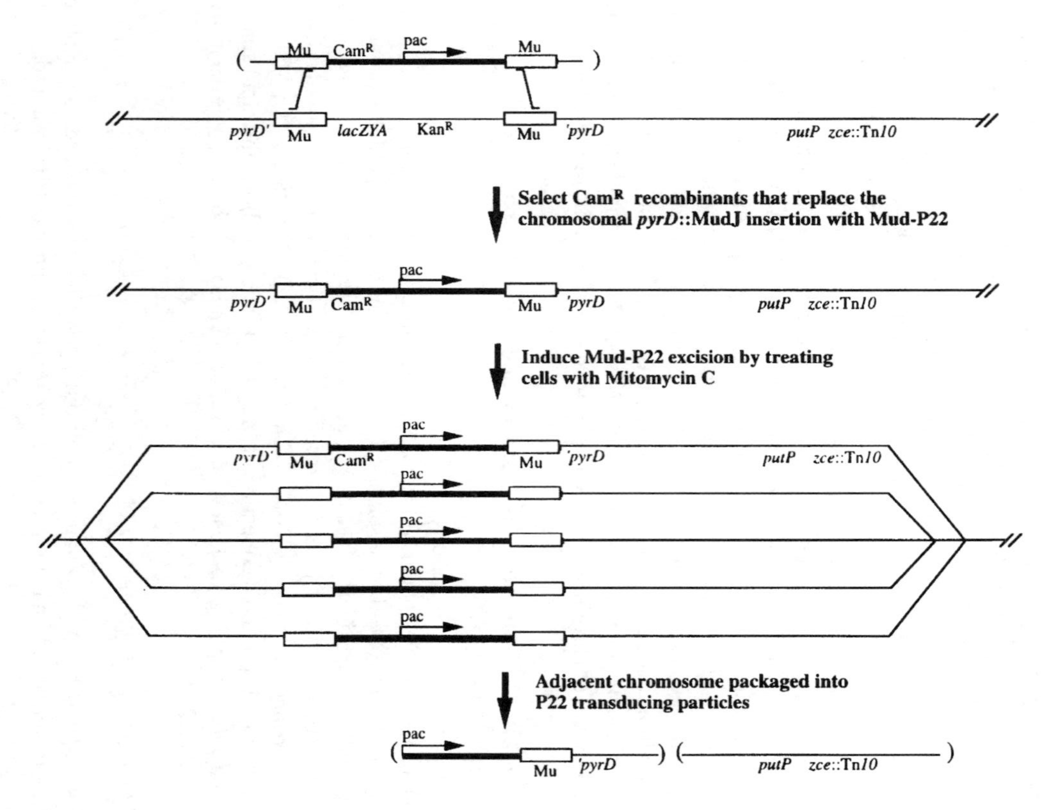

Figure 27-3. Substitution of a Mu*d*J insertion for Mu*d*-P22 and subsequent induction and packaging of chromosomal DNA. An example showing the isolation of a *pyrD*::Mu*d*-P22 at 24 min on the *S. typhimurium* chromosome and the subsequent induction of the Mu*d*-P22 producing phage particles carrying the adjacent DNA (including the *putP* gene at about 25 min). A donor lysate of P22 is prepared on a strain containing F′ *zzf*::Mu*d*P or F′ *zzf*::Mu*d*Q and the lysate used to transduce a Mu containing recipient strain to Cam^R. The desired Cam^R recombinants lose the antibiotic resistance of the original Mu*d* insertion (e.g. replacement of Mu*d*J results in a Kan^S phenotype). Induction of the *pyrD*::Mu*d*-P22 insertion with Mitomycin C results in amplification of the adjacent chromosomal DNA, and packaging of 3–5 headfuls of DNA to one side of the *pac* site into P22 particles.

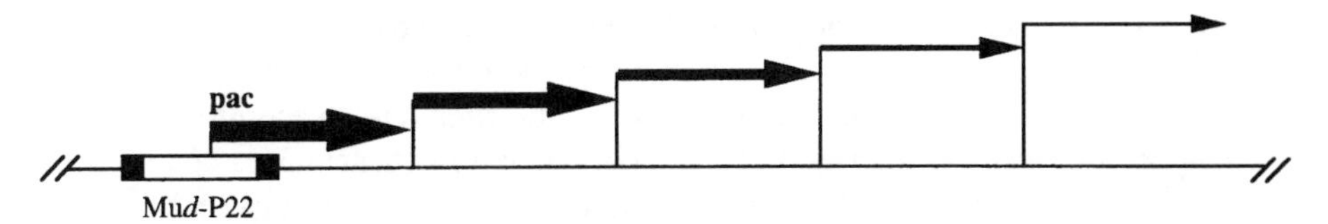

Figure 27-4. Headful packaging of adjacent chromosomal DNA. Each phage head contains about 44 Kb of double stranded linear DNA. The P22 gene *3* protein initially cuts the DNA at the *pac* site within the Mu*d*-P22. One phage head is filled with DNA, then the DNA is cut by the endonuclease. The next head is filled with DNA beginning with the resulting end. These reactions continue processively until 3–5 phage heads are filled with DNA, however, fewer phage particles are packaged with increasing distance from the *pac* site.

Recombinational repair of a chromosomal mutation by a particular Mu*d*-P22 lysate will only occur if the corresponding wild-type allele is packaged into the transducing particles (i.e., located within 3–5 min in the proper orientation from the Mu*d*-P22 insertion). The entire set of Mu*d*-P22 lysates can be quickly tested by spot transduction of the mutant recipient with the Mu*d*-P22 transducing lysates ordered as an array in a microtiter dish, and selecting for repair of the mutant phenotype. For this strategy to work, the recipient strain must be P22 sensitive and the mutant phenotype must be counterselectable, so that only those recipient cells which are transduced to wild-type will form colonies on the selective plates. Some mutant phenotypes are inherently counterselectable. For example, it is possible to select for repair of auxotropic mutations on minimal medium (Benson and Goldman, 1992), or temperature conditional lethal mutations at the non-permissive temperature (Gupta *et al.*, 1993). In addition, certain transposons carry genes with a counterselectable phenotype. For example, the tetracycline resistance gene present on transposon Tn*10* makes enteric bacteria sensitive to certain lipophilic chelating agents (Bochner *et al.*, 1980; Maloy and Nunn, 1981), and the *Bacillus subtilis sacB* (secretory levansucrase) gene present on the Mu*d*-SacI transposon makes Gram negative enteric bacteria sensitive to 5% sucrose (Lawes and Maloy, 1995). It is possible to directly select for tetracycline sensitivity due to loss of a Tn*10* insertion, or for sucrose resistance due to loss of a Mu*d*SacI insertion. Consequently, if a mutation does not have a counterselectable phenotype, it is straightforward to isolate a closely linked Tn*10* or Mu*d*SacI insertion whose location can be mapped by transduction versus the collection of Mu*d*-P22 lysates (Maloy *et al.*, 1996).

An alternative strategy is to generate mutations with Mu*d* insertions and then to replace each insertion with Mu*d*-P22, or to isolate Mu*d*-P22s linked to the gene of interest. Induction of the resulting Mu*d*-P22 yields a "specialized" transducing lysate which can be used to transduce an ordered set of auxotrophs, selecting for repair of the auxotropic mutation minimal media (Higgins and Hillyard, 1988). Since most of the DNA in the transducing particles is from the

3–5 min region adjacent to the Mu*d*-P22 insertion, only those auxotrophic mutations that are near the Mu*d*-P22 insertion will be repaired at high frequency. Thus, the approximate location of the Mu*d*-P22 insertion (and the corresponding Mu*d* insertion) can be inferred from the transduction frequency. This approach is conceptually similar to the use of directed Hfr formation to determine the location of Tn*10* insertions (Chumley *et al.*, 1979).

Cloning, restriction mapping, and DNA sequencing

Mu*d*-P22 lysates can also be used as a highly enriched source of DNA for cloning, restriction mapping, or sequencing genes from the *S. typhimurium* chromosome. If the approximate map location of a mutation is known, then a closely linked Mu*d*-P22 with the appropriate packaging direction can be induced and DNA extracted from the phage used to clone the gene of interest (Higgins and Hillyard, 1988).

The DNA can also be directly used for restriction mapping (Youderian *et al.*, 1988; Higgins and Hillyard, 1988) or DNA sequencing without an intermediate cloning step. For example, Hughes et al. (1993) isolated 6 Mu*d*J insertions located at different positions in the *nadC* gene, replaced each of the insertions with a Mu*d*-P22, then isolated DNA from each of the Mu*d*-P22 lysates. The DNA was directly sequenced using primers specific to one of the ends of the Mu (MuL for Mu*d*P and MuR for Mu*d*Q) to read from the flanking Mu end into the adjacent *nadC* DNA.

DNA hybridization

DNA prepared from the representative set of Mu*d*-P22 lysates also provides an ordered array of chromosomal DNA fragments for physical mapping by DNA hybridization. A single hybridization filter, with an ordered pattern of spots of DNA from a representative set of Mu*d*-P22 insertions, can be probed with smaller chromosomal fragments, cloned genes, or cloned cDNA corresponding to differentially expressed genes (Wong and McClelland, 1992; Wong *et al.*, 1994; Libby *et al.*, 1994; Wong and McClelland, 1994). Applications of this approach are analogous to those used for the Kohara library from *E. coli* (Kohara *et al.*, 1987).

DNA purified from a particular Mu*d*-P22 lysate can be used to probe a Southern blot. For example, Liu and Sanderson (1994) used DNA purified from Mu*d*-P22 lysates to probe digests of *S. typhimurium* total chromosomal DNA separated by pulsed field gel electrophoresis (see Chapters 24–26). This approach facilitated the compilation of a correlated physical-genetic map of *S. typhimurium* (Sanderson *et al.*, 1995; see Chapter 78 of this monograph).

Use of Mu*d*-P22 in Other Species

Mu*d*-P22 can be used in any strain of *Salmonella* that is sensitive to P22. Although Mu*d*-P22 has only been used in *Salmonella* thus far, with some modifications this system should be applicable in a variety of other bacteria as well. For example, a cosmid that carries the *S. typhimurium* lipopolysaccharide genes renders some Gram negative bacteria sensitive to P22 (Neal *et al.*, 1993). Even if Mu does not transpose in the bacterium, Mu*d*-P22 insertions could be obtained by mating an F′ carrying Mu*d*-P22 into the bacterium and selecting for homologous recombination between a transposon present both on the F′ and the chromosome (such as Tn*10*) or by recombination between any gene on the F′ and the corresponding chromosomal gene (Zahrt *et al.*, 1994; Zahrt and Maloy, submitted). Alternatively, it should be possible to develop analogous approaches using phage specific for any particular bacterium.

Summary

Mu*d*-P22 derivatives are hybrids between phage Mu and phage P22 that as a population can package 150–250 Kb of DNA from any specific region of the *Salmonella* chromosome. An ordered set of Mu*d*-P22 insertions spaced every 3–5 min around the chromosome allows the rapid genetic mapping, physical mapping, cloning, and sequencing of genes from *S. typhimurium*. Thus, Mu*d*-P22 derivatives provide powerful tools for the genetic and physical analysis of the *Salmonella* genome, and maybe a useful tool for the genetic and physical analysis of other bacterial species as well.

Acknowledgments

This work was supported by the NIH grant GM34715.

References

Benson, N. R., and B. S. Goldman. 1992. Rapid mapping in *Salmonella typhimurium* with Mu*d*-P22 prophages. *J. Bacteriol.* 174:1673–1681.

Bochner, B., H. C. Huang, G. L. Schrevin, and B. N. Ames. 1980. A positive selection for loss of tetracycline resistance. *J. Bacteriol.* 143:926–933.

Chumley, F. G., R. Menzel, and J. R. Roth. 1979. Hfr formation directed by Tn*10*. *Genetics* 91:639–655.

Groisman, E. A. (1991). *In vivo* genetic engineering with bacteriophage Mu. *Meth. Enzymol.* 204:180–212.

Gupta, S. D., K. Gan, M. B. Schmid, and H. C. Wu. 1993. Characterization of a temperature-sensitive mutant of *Salmonella typhimurium* defective in apolipoprotein N-acetyltransferase. *J. Biol. Chem.* 268:16551–16556.

Higgins, N. P., and D. Hillyard. 1988. Primary structure and mapping of the *hupA* gene of *Salmonella typhimurium. J. Bacteriol.* 170:5751–5758.

Hughes, K. T., A. Dessen, J. P. Gray, and C. Grubmeyer. 1993. The *Salmonella typhimurium nadC* gene: sequence determination by use of Mu*d*-P22 and purification of quinolate phosphoribosyltransferase. *J. Bacteriol.* 175:479–486.

Hughes, K. T., and J. R. Roth. 1988. Transitory *cis* complementation: a method for providing transposition functions to defective transposons. *Genetics* 119:9–12.

Kohara, Y., K. Akiyama, and K. Isono. 1987. The physical map of the whole *E. coli* chromosome: application of a new strategy for rapid analysis and sorting of a large genomic library. *Cell.* 50:495–508.

Lawes, M. C., and S. R. Maloy. 1993. Genetics of DNA injection by phages λ and P22. *Current Topics in Mol. Genet. (Life Sci. Adv.)* 1:133–146.

Lawes, M. C., and S. R. Maloy. 1995. Mu*d*SacI, a transposon with strong selectable and counterselectable markers: use for rapid mapping of chromosomal mutations in *Salmonella typhimurium. J. Bacteriol.* 177:1383–1387.

Libby, S. J., W. Goebel, A. Ludwig, N. Buchmeier, F. Bowe, F. C. Fang, D. G. Guiney, J. G. Songer, and F. Heffron. 1994. A cytolysin encoded by *Salmonella* is required for survival within macrophages. *Proc. Natl. Acad. Sci. (USA)* 91:489–493.

Liu, S. L., and K. E. Sanderson. 1992. A physical map of the *Salmonella typhimurium* LT2 genome made by using *Xba*I analysis. *J. Bacteriol.* 174:1662–1672.

Maloy, S. R., and W. D. Nunn. 1981. Selection for loss of tetracycline resistance by *Escherichia coli. J. Bacteriol.* 145:1110–1112.

Maloy, S. R., V. J. Stewart, and R. K. Taylor. 1996. *Genetic analysis of pathogenic bacteria: A laboratory manual.* (Cold Spring Harbor, New York: Cold Spring Harbor Laboratory Press).

Neal, B. L., P. K. Brown, and P. R. Reeves. 1993. Use of *Salmonella* phage P22 for transduction in *Escherichia coli. J. Bacteriol.* 175:7115–7118.

Poteete, A. R. 1988. Bacteriophage P22. In *The Bacteriophages*, vol. 2, R. Calendar, ed. Plenum Press, pp. 647–682. New York.

Sanderson, K. E., A. Hessel, and K. E. Rudd. 1995. Genetic map of *Salmonella typhimurium*, edition VIII. *Microbiol. Rev.* 59:241–303.

Smith, C. M., W. H. Koch, S. B. Franklin, P. L. Foster, T. A. Cebula, and E. Eisenstadt. 1990. Sequence analysis and mapping of the *Salmonella typhimurium* LT2 *umuDC* operon. *J. Bacteriol.* 172:4964–4978.

Susskind, M.M. and D. Botstein. 1978. Molecular genetics of bacteriophage P22. *Microbiol. Rev.* 42:385–413.

Symonds, N., A. Toussaint, P. van de Putte, and M.M. Howe. (eds) (1987). *Phage Mu.* Cold Spring Harbor, New York: Cold Spring Harbor Laboratory.

Wong, K.K., and M. McClelland. 1992. A *BlnI* restriction map of the *Salmonella typhimurium* LT2 genome. *J. Bacteriol.* 174:1656–1661.

Wong, K.K., and M. McClelland. 1994. Stress-inducible gene of *Salmonella typhimurium* identified by arbitrarily primed PCR of RNA. *Proc. Natl. Acad. Sci. (USA).* 91:639–643.

Wong, K.K., R.M. Wong, K.E. Rudd, and M. McClelland. 1994. High-resolution restriction map for a 240-kilobase region spanning 91 to 96 minutes on the *Salmonella typhimurium* LT2 chromosome. *J. Bacteriol.* 176:5729–5734.

Youderian, P., P. Sugiono, K. L. Brewer, N. P. Higgins, and T. Elliott. 1988. Packaging specific segments of the *Salmonella* chromosome with locked-in Mu*d*-P22 prophages. *Genetics* 118:581–592.

Zahrt, T.C., G.C. Mora, and S. Maloy. 1994. Inactivation of mismatch repair overcomes the barrier to transduction between *Salmonella typhimurium* and *Salmonella typhi. J. Bacteriol.* 176:1527–1529.

28

Encyclopedias of Bacterial Genomes

Michael Fonstein and Robert Haselkorn

Introduction

Summarizing recent studies of bacterial chromosomes, participants in the First International Symposium on Small Genomes in Paris in 1993 declared the birth of a new discipline, bacterial genomics. The subject of genomics can be defined as the study of integral genome structures, integral genome properties and the evolution of genomes. Genome encyclopedias (ordered sets of overlapping clones containing the entire genome), together with Pulsed/Field gel electrophoresis (PFGE)-related methods (see Chapters 24 to 26), and genome sequencing (see Chapts in section 2 of this monograph) are the specific tools of genomics.

Bacterial genomes range in size from 600 kb for *Mycoplasma genitalium* (Colman *et al.*, 1990) to 9.5 Mb for *Myxococcus xanthus* (Chen *et al.*, 1991; *see* Chapter 1). Chromosomes of such sizes can be conveniently explored by a number of physical approaches. At least 90 such studies were completed in the last eight years (summarized in Cole and Saint Girons, 1994, Fonstein and Haselkorn, 1995 and this volume), of which 15 involved genome encyclopedia constructions.

Most of the terms associated with genome encyclopedias are more or less self-explanatory, but some have to be defined. A stored set of DNA fragments, amplified as a result of insertion in a self-propagating cloning vector, and comprising almost all of the genome of an organism is traditionally called a *gene library*. Gene libraries are usually used for repetitive searches for individual genes. Depending on the gene and organism of interest, different vectors and search strategies have been applied to construct and screen these libraries. A specific version of a gene library aimed at the study of whole genomes rather than particular genes is called a *genome encyclopedia* or an *ordered gene library*. Cloned fragments can be linked in a sequence reconstructing the original genome in such a library. The following characteristics are commonly used in descriptions of genome encyclopedias.

The term *N-fold*, or N-hit gene library is used regarding the *representivity* of the library, where **N** is the number of genome equivalents calculated as the number of clones multiplied by the insert size and divided by the genome size. A *contig* is an uninterrupted group of linked or ordered inserts. A *miniset* is an ordered group of clones covering the genome of interest with minimal overlaps. A *gap* in a gene encyclopedia is an area of the genome not represented in the encyclopedia due to unclonability or for statistical reasons. The latter is called a *statistical gap*. This chapter is devoted to the construction of genome encyclopedias together with a brief tabulated survey of completed projects.

Construction of Genome Encyclopedias

The construction of genome encyclopedias can be divided into several stages. First, a representative gene library with the minimal number of rearranged clones has to be made. Second, members of this library have to be grouped by detection of their overlaps. An overall high-resolution restriction map is usually constructed fusing restriction maps of the individual clones following encyclopedia assembly, and known cloned genes are located by hybridization or simple comparison of restriction maps. Total genome sequencing can be considered as the final stage of this process.

Construction of Gene Libraries Used to Assemble Encyclopedias

The two major elements determining the choice of a cloning strategy are the stability of the hybrid clones and the optimal insert size. To minimize the number of clones to be analyzed, essential for all schemes of encyclopedia assembly, larger inserts have to be used. On the other hand, one-step restriction mapping and high-resolution gene mapping is easier to perform with smaller inserts.

A variety of vectors, namely YACs, λ and P1 vectors, cosmids, and Ori F plasmids (BACs) are now available for the cloning of large DNA fragments. λ vectors were used in early encyclopedia projects (*S. cerevisiae* (Olson *et al.*, 1986) and *E. coli* (Kohara *et al.*, 1987). The propagation of hybrid λ phages is not affected by the expression of cloned genes to the same extent as the propagation of plasmid vectors. Though no systematic study of this sort has been carried out, a number of *Bacillus* and *Streptomyces* genes that could be cloned in λ derivatives were shown to be unclonable or unstable in various plasmid systems (Fonstein, unpublished personal communications). On the other hand, seven gaps in Kohara's *E. coli* encyclopedia, constructed using λ vectors, were closed by using the low copy number cosmid pOU61 (Knott *et al.*, 1989) and even the relatively high copy cosmid pJB8 (Birkenbihl and Vielmetter, 1989). λ vectors are also less convenient than plasmids for restriction mapping due to their higher vector insert size ratio and the more laborious methods required for DNA extraction.

Another group of vectors used for encyclopedia constructions are cosmids.

Such vectors were used for nine of the fifteen bacterial encyclopedias described to date (Table 28-1). More recent modifications of cosmids harbor the λ Ori providing constant copy number, independent of the size of the insert. This property greatly reduces the selective advantages of deletion variants common for ColE-type cosmids. The insert size for most cosmids used for encyclopedia projects is about 40 kb. This size is within the limits of resolution of ordinary agarose electrophoresis and permits reliable restriction mapping without slow and expensive PFGE.

Yeast Artificial Chromosomes (YACs) permit the cloning of megabase DNA fragments, which is essential for eukaryotic genome projects. However, splitting the bacterial genome into 4–9 pieces is not obviously a convenience. When YAC4 was used for *B. subtilis*, known for the unstable propagation of its cloned DNA in *E. coli*, and for *Myxococcus xanthus* (Azevedo *et al.*, 1993; Kuspa *et al.*, 1989), much smaller 100-kb inserts were prepared. In the first work, 10% of the YACs were found to be rearranged by hybridization with genetically mapped genes and removed from further study. An even higher proportion of YACs carrying deletions and scrambled inserts (75%) were found in the mapping of the Huntington disease locus (Baxendale *et al.*, 1993). Instability and complicated extraction and restriction mapping of YACs make them less popular in bacterial projects, where their main advantage, huge insert size, is not vital.

P1-derived vectors were successfully used as supplementary systems in many eukaryotic projects (for example, Hoheisel *et al.*, 1993) but not enough experience has been accumulated using P1 in bacterial projects.

BACs, a cloning system developed specifically for genome encyclopedia construction (Shizuya *et al.*, 1992) is getting triumphant reviews in its first applications (Rouqueir *et al.*, 1994; Woo *et al.*, 1994). These vectors, based on the low-copy-number origin of replication of the F-factor, have all modern conveniences such as T7/SP6 promoters for genome walking, rarely-cut restriction sites flanking the insert and λ *cos* and P1 *lox* sites for fast restriction mapping. 100- to 200-kb inserts were cloned in several BAC libraries. No instabilities or rearrangements were found in the studies focused on this aspect of the cloning. This system provides no means of size selection for 100-kb inserts, except physical separation prior to cloning. Cloning of smaller DNA fragments in BACs can be accomplished as with other cosmids, with all the advantages of λ packaging. Summarizing this brief survey, either BACs or λ-Ori cosmids such as Lorist 6 (Gibson *et al.*, 1987) appear to be the most reliable and convenient tools for encyclopedia construction at this time.

The size of the library is the other critical factor for encyclopedia construction, regardless of the vector. Ten-hit libraries are generally sufficient for these projects, since the <0.5% of a genome that is statistically unrepresented in a library of this size is usually less than is absent due to cloning biases. To evaluate these biases, small samples from the libraries are usually analyzed by hybridization or restriction mapping.

Table 28-1. Published Encyclopedias of Bacterial Genomes

Organism	Year	Genome size, Kb	Cloning vector, **cloning site**	Number of clones, *number of contigs,* **% of coverage.** *size of the miniset*	Methods of alignment and mapping	Physical and genetic markers	References
E. coli K12	1976		pLC(ColE1) **poly (dA-dT)**	2,200 *about 600 (mapped),* **20–50%**	Gene complementation, protein expression, hybridization with cloned genes, hybridization with Kohara miniset.	310 genes indexed in the gene/protein database, 282 fts mutants scattered in 50 locations.	(Clarke and Carbon, 1975; (Clarke and Carbon, 1979; (Neidhardt *et al.*, 1989; (Nishimura *et al.*, 1992)
E. coli W3110	1987	4,703	EMBL4, λ2001 ***Sau*3A, *Eco*RI**	3400 *7* **99%** 381	High resolution mapping of 1025 phages with computer assembly produced 70 groups covering 94% of the genome, followed by genome walking with another 2344 clones.	About 8000 restriction sites. 426 genes mapped based on their restriction maps derived from sequences, eventually most of GenBank database was incorporated.	(Dimri *et al.*, 1992; (Kohara *et al.*, 1987; (Medigue *et al.*, 1993)
E. coli K12 MG1655	1987	4,700	Charon 28, 40 ***Eco*RI, *Bam*HI, *Hind*III**	2000 *68 (18–200 kb)* **67%**	Restriction pattern comparison, supported with hybridization with eluted MRFs.	About 20 genes placed by lambda complementation of growth requirements and colorimetric assays.	(Daniels and Blattner, 1987)
E. coli 803	1988	4,736	pTM, Lorist B, pOU61 ***Sau*3A,*Taq*I, shearing**	2512 (6 libraries) *58 (40–300 kb)* **90%**	*Hinf*II fingerprinting and alignment by genetically mapped genes (19 contigs aligned).		(Knott *et al.*, 1989)
E. coli BHB2600	1989	4,700	pJB8 ***Sau*3A**	600 *12 (60–1100 kb)* >95%	Hybridization of gridded sets with whole cosmids verified by restriction pattern analysis. Ordering of the contigs by cloned genes.	26 genes and 6 *rrn*s.	(Birkenbihl and Vielmetter, 1989)

Continued

Table 28-1. Continued

Organism	Year	Genome size, Kb	Cloning vector, **cloning site**	Number of clones, *number of contigs,* **% of coverage.** *size of the miniset*	Methods of alignment and mapping	Physical and genetic markers	References
E. coli W3110	1989	4,500	pHC79 ***Eco*RI, *Sau*3A**	1300 (325 assigned) **75%**	pLC plasmids from the collection constructed by Clarke and Carbon were used as primary probes; 31 islands used to start cosmid walking.	70 (52?) genes out of 90 tested.	(Tabata *et al.*, 1989)
Mycoplasma pneumoniae	1988	809	pcosRW2, λ ZAP ***Eco*RI**	>1000 *1* **100%** *34 cosmids, 2 phages and a plasmid.*	Cosmid walking.	10 genes and 4 repeated elements.	(Wenzel and Herrmann, 1989; (Wenzel *et al.*, 1992)
Myxococcus xanthus	1989	9,454	YAC4 ***Eco*RI**	409 (111 mean insert) *60* **95–99%?**	Alignment of the *Eco*RI maps, with 4-fragment match the initial criterion. Probed genes linked the rest.	18 probes (no major housekeeping genes).	(Kuspa *et al.*, 1989)
Haloferax volcanii DS2	1991	4,140*	Lorist M ***Mlu*I**	About 4,000 (2,000 used in landmark analyses) *7* **96%**	Landmark (pattern) analysis, cosmid walking using cosmid pools. Restriction mapping was done by double and partial digests.	Six semi-frequent cutters (900 sites). Over 100 genes mapped by transformation and hybridization.	(Charlebois *et al.*, 1989; (Charlebois *et al.*, 1991; (Cohen *et al.*, 1992)
*Rhodobacter capsulatus*SB1003	1992	3,700	Lorist 6, λ DASH ***Sau*3A**	1836 *2* **>99%** *192*	Genome walking from uniformly dispersed points (previously mapped genes). Hybridization with eluted MRFs, and further fine mapping to resolve ambiguities.	48 loci (about 300 genes), *Eco*RV map with 670 sites was recently expanded to nearly 3000 *Bam*HI, *Eco*RI and *Hin*dIII sites.	(Fronstein *et al.*, 1995; (Fronstein *et al.*, 1992)

B. subtilis 168	1993	4165	YAC4 ***Eco*RI**	305 (>80 kb inserts) *4* **>98%** *59*	65 cloned genes produced 31 contigs (40% of the map), which were then linked by genome walking.	65 genes used as primary probes. 10 *rrn* operons and several tRNA clusters mapped in good agreement with genetic data.	(Azevedo *et al.*, 1993)
Mycobacterium leprae	1993	Lorist 6 ***Sau*3A**	1000 *4 contigs (380, 400, 800 and 1200 kb)*	Fingerprinting using sequencing gels, then genome walking from the ends of primary contigs.	38 probes (genes and antigenic determinants) located 45 genes, but only one *rrn*).		(Eiglmeier *et al.*, 1993)
Helicobacter pylori NCTC11638	1994	1,730	Lorist 6 ***Sau*3A**	752 *3* **>99%** *68*	Genome walking, hybridization with eluted MRFs used for primary cosmid linkage. Comparisons of the cosmid restriction patterns added to resolve ambiguities.	25 genes, 1 *rrn* operon and a pair of split 23S and 16S rRNA genes. Orientation of transcription of 8 genes was established.	(Bukanov and Berg, 1994)
Desulfovibrio vulgaris	1994	1,720	λ2001 ***Sau*3A**	879 *87* **>100%** (sum of the contigs is much larger than expected genome size) *151*	*Hinf*I fingerprinting (50 bands from the vector and an average of 16 from each insert).	8 genes (5 for electron transport). 2 genes were not found.	(Deckers and Voordouw, 1994)
Mycoplasma genitalium ATCC 33530	1994	578	pcos RW2, λGEM2 ***Sau*3A** ***Eco*RI**	548 *1* **>99%** *20*	Genome walking, hybridization with eluted MRFs.	MgPa repeated operons	(Lucier *et al.*, 1994)
Rhodobacter sphaeroides 2.4.1.	1994	900**	pLA2917, pJRD215 **not described**	92 (chromosome specific) *4* **85%** *46*	Genome walking, hybridization with eluted MRFs. Comparisons of the cosmid restriction patterns were added to resolve ambiguities.	8 loci (about 20 genes and insertions), 40 *Ssp*I sites and 2 *rrn* operons.	(Chouhary *et al.*, 1994)

*Chromosome comprises 2920 kb, the rest is four plasmids.

**There is another 3-Mb chromosome in this organism.

Cloning efficiencies of the vectors described above make prokaryotic library construction a routine and simple task. However, the next stage of encyclopedia construction, the clone assembly, is a different story, and clone instability or a few scrambled inserts can slow down a project for months. The fact that a library has all the overlapping clones needed to assemble an uninterrupted set is not sufficient. The ability to find them depends on the method of assembly. A principal factor describing this ability, which influences the necessary size of the analyzed library, is the minimal detectable overlap (MDO). Its role is discussed below.

Encyclopedia Assembly

Encyclopedia assembly, or the ordering of members of a gene library, is accomplished by establishing the overlaps between them. The minimal detectable overlap (MDO) is the portion of the insert required for detection of linkage related to the whole insert. This parameter describes the screening process and defines the minimal size of the library. The difference in sizes sufficient to assemble a certain number of contigs is 10-fold between cases of 20% and 80% MDOs (Branscomb *et al.*, 1990). Estimates predict ten gaps (rather, undetected links) in a 7-fold library in the case of 20% MDO, whereas 60 gaps will be found in the same library in the case of 50% MDO (Daniels, 1990). There are two major ways to detect overlaps. One is based on the similarity of DNA sequences revealed by hybridization or PCR; the other, on the similarity of restriction maps.

Assembly via Comparison of Restriction Patterns

Assembly based on a comparison of restriction patterns or "fingerprints" was used in the initial stages of the *S. cerevisiae* mapping project (Olson et al., 1986). *Eco*RI plus *Hin*dIII restriction patterns of 5,000 λ clones were assigned to 680 contigs. With an average of eight restriction fragments per hybrid clone, the MDO for this study was actually >70%. To increase the resolution of fragment size determination and thereby reduce the MDO in mapping *C. elegans* (Coulson *et al.*, 1986), cosmid clones were cleaved with *Hin*dIII, end-labeled and redigested with *Sau*3A. These fragments were separated using sequencing gels, significantly increasing the accuracy and thus the reliability of the established overlaps. Still, the MDO was about 50% in this work. This stage of the genome project of *C. elegans* ended with 860 contigs ranging from 35 to 350 kb. Eventually, both the *S. cerevisiae* and the *C. elegans* projects were finished successfully by genome walking procedures.

Direct fingerprinting was applied to several prokaryotic projects. 879 λ clones representing the 1720-kb genome of *Desulfovibrio vulgaris* were grouped in 87 contigs (Deckers and Voordouw, 1994). The same approach produced 68 contigs for *E. coli* MG1655 (Daniels and Blattner, 1987). The *Mycobacteria leprae* project was a more successful example. Pattern comparison with sequencing gels aligned 1000 cosmids into only four large contigs (Eiglmeier *et al.*, 1993). The *Hin*fII fingerprinting of cosmids from an *E. coli* 803 library produced 58 contigs, covering four out of seven gaps in the Kohara library. A more directed search and a low copy number vector were used to close the remaining three gaps (Knott *et al.*, 1989).

So-called landmark analysis was used to group cosmids from a library of *Haloferax volcanii* DS2 around "semi-rare" restriction sites (Charlebois *et al.*, 1989; Charlebois *et al.*, 1991). To characterize them, cosmids were digested with a frequent-cutting enzyme together with a semi-rare one. Two restriction fragments surrounding a semi-rare site produce a unique fingerprint for most chromosomal locations. 59 contigs were further linked by cosmid walking and PFGE into seven contigs.

The MDO is much higher for fingerprinting than for methods based on sequence similarities. Small overlaps, which are the most valuable, are missing in these fingerprinting alignments. However, fingerprinting comparison is insensitive to most DNA repeats, which often confuse hybridization alignments.

Assembly by Comparison of Restriction Maps

1025 λ clones from an *E. coli* W3110 library were assembled in 70 contigs covering 94% of the genome simply by comparing restriction maps of the cloned DNA fragments (Kohara et al., 1987). The criterion of having at least five common fragments in an overlap was applied to establish linkage, which corresponded to a 10–25% MDO. As calculated in Daniels (1990), this cut-off value produces less than one false overlap per thousand. Cosmid walking with another 2344 clones produced the final version of the Kohara map with only seven minor gaps. Similar mapping of YAC clones covering the 9454-kb chromosome of *Myxococcus xanthus* (Kuspa *et al.*, 1989) linked 409 YACs into 60 contigs.

Linking by Hybridization

The assembly of most prokaryotic encyclopedias has been based on hybridization schemes, which can be considered as multi-point genome walking. Evenly spaced or simply random hybrid clones are chosen as the starting points of this procedure. Cloned inserts or their ends are used to probe a gene library. New inserts found in these tests extend the chromosomal region covered by aligned clones. Such contigs can grow by subsequent hybridization cycles until they meet one another or leave a gap in the screened library. To simplify clone handling, clones are usually grouped in 96-well plates. Up to 2500 clones can be gridded onto 6 Petri plate-size filters using automated (Bentley *et al.*, 1992) or manual (Fonstein *et al.*, 1992) devices. The position of a positive signal on a filter determines the address of a clone in the stored collection. No additional purification cycles, as in traditional screening by hybridization, have to be done to get the next walking probe. Such an approach was first used to build the encyclopedia for *Mycoplasma pneumonia* (Wenzel and Herrmann, 1989). The 1.7-Mb genome of *Helicobacter pylory* was covered by 60 walking steps grouped in five sets (Bukanov and Berg, 1994). 300 steps (with large redundancy) were used in the complete mapping of the 3.7-Mb genome of *R. capsulatus* (Fonstein and Haselkorn, 1993). 764 probes were used to map the 14-Mb genome of *S. pombe* (Hoheisel *et al.*, 1993). A thoughtful selection of probes for hybridization cycles can significantly reduce the number of hybridizations required. Screening a cosmid library of a 4-Mb genome, starting with a dozen evenly spaced cosmids, will reveal about 30% of the gridded cosmids as positives. It is practical to select new probes for the next several cycles of hybridization among negative clones. When about 5% of the clones remain negative, most of these will be false negatives and hybridization with transcripts from their ends will most likely position then inside already constructed contigs. At this point, it is better to choose probes from the ends of the contigs. Pooling the probes can drastically reduce the number of hybridizations to be performed by increasing the informational content of each experiment.

An important factor for encyclopedia assembly via hybridization is the size of the probe. Excision of the end fragments of an insert, used in the first chromosome walking protocols, was by-passed by using T7-, T3- or SP6-specific promoters flanking the cloning site of the vector, to prepare transcripts from the ends of the inserts. Insert ends can also be labeled by primer extension with *Taq* polymerase starting from the ends of the vector (Mizukami *et al.*, 1993). The latter procedure reduces probe sizes from 5 to 15 kb for *in vitro* transcription to 500 bp, reducing the chance of hitting repeated elements, which causes ambiguities in the encyclopedia assembly.

Oligo Fingerprinting

Hybridization with a set of 11- and 12-nt oligos was used to supplement cosmid walking in Hoheisel *et al.* (1993). A detailed discussion of the optimization of this approach can be found in Fu *et al.* (1992). Due to technical difficulties, it is not a ready-to-use method, but one can expect substantial technological improvement, as a by-product of the development of "sequencing by hybridization."

Encyclopedias Derived from Genetic Maps

The first organized clone library was the Clarke-Carbon collection of *E. coli* DNA fragments (Clarke and Carbon, 1976). 310 clones selected from 2,200 ColEI plasmids harboring 10- to 14-kb inserts of the *E. coli* genome were ordered by genetic complementation, hybridization with cloned genes and protein overexpression. They were positioned according to the *E. coli* genetic map.

518 other clones were later mapped by hybridization with the Kohara miniset (Nishimura et al., 1992). The YAC encyclopedia of *B. subtilis* was assembled by assigning its members to loci on the genetic map of this organism (Azevedo *et al.*, 1993). However, the accumulation of genetic data that made it possible to apply this strategy is unique for these two organisms, so this approach will most likely be limited to them.

STS Fingerprinting

The Sequence Tagged Sites (STS) fingerprinting used in a variety of eukaryotic projects is a combination of both hybridization and fingerprinting approaches. Multi-dimensional pooling strategies (Barillot *et al.*, 1991) reduce unimaginable numbers of necessary experiments to simply vast in such PCR screening of YAC libraries in the Human Genome Project. Such fingerprinting can be performed in a variety of versions, but one of them looks especially promising for prokaryotic projects. 960 3-kb subclones were prepared from a P1 clone, carrying a 100-kb insert of *Drosophila* DNA, grown in 96-well plates and pooled in three dimensions (Yoshida *et al.*, 1993). Walking was performed using PCRs originating with primers derived from the ends of the vector and from 6 STSs. Walking steps in this screening immediately determine the most outward clones without any mapping, simply by choosing the longest PCR products. Sequencing of the ends of such clones produced primer sequences for the next cycles. Totally, 40 region-specific primers were synthesized and 2400 PCR reactions were performed to combine an ordered set covering this P1 clone. If the recently introduced long-range PCR becomes more reliable, the same small number of experiments can be used to order a 2-Mb genome cloned in a cosmid vector. Moreover, most of the steps in this approach can easily be automated.

Restriction Mapping of Clones of the Encyclopedia

This stage of encyclopedia construction is required to resolve mapping bifurcations, to verify overlaps found by hybridization, and to establish exact distances between elements of the physical map. A high-resolution restriction map is also important for many applications of the genome encyclopedia. λ-terminase cosmid mapping is one of the most convenient among numerous ways of performing restriction analysis. This variant of the Smith-Birnstiel scheme is based on polar end-labeling after λ terminase cleavage followed by partial digestion. Introduced in Rackwitz *et al.* (1985), this method has been used routinely since the key enzyme, λ terminase, became commercially available from reliable sources.

After a decade of scientific development, it has become apparent that the resolution provided by even very high-resolution genome encyclopedias is often insufficient for many genome studies. Large-scale sequencing projects are now thought to be able to close this gap. They are often started as extensions of previous encyclopedia projects. However, at least for microbial genomes smaller than 2 Mb, such sequencing can be done without pre-existing encyclopedias (Fleischmann *et al.*, 1995). The informational content of these projects is greatly increased by the accumulation of published sequences. Due to this, at least 50% of the ORFs found in total genome sequencing can be related to known functions (Borodovsky *et al.*, 1994). However, genome sequencing does not eliminate the demand for encyclopedias, because ordered sets of clones are needed for biological studies, such as systematic construction of deletion strains, studies of gene expression and protein function. These studies can not be accomplished solely with the printout of a shotgun-derived sequence.

References

Azevedo, V., E. Alvarez, E. Zumstein, G. Damiani, V. Sgaramella, S. D. Ehrlich, and P. Serror, 1993. An ordered collection of *Bacillus subtilis* DNA segments cloned in yeast artificial chromosomes. *Proc. Natl. Acad. Sci. U.S.A.*, 90:6047–51.

Barillot, E., B. Lacroix, and D. Cohen, 1991. Theoretical analysis of library screening using a N-dimensional pooling strategy. *Nucl. Acids Res.* 19:6241–7.

Baxendale, S., M. E. MacDonald, R. Mott, F. Francis, C. Lin, S. F. Kirby, M. James, G. Zehetner, H. Hummerich, J. Valdes, and et, *al.*, 1993. A cosmid contig and high resolution restriction map of the 2 megabase region containing the Huntington's disease gene. *Nat Genet,* 4:181–6.

Bentley, D. R., C. Todd, J. Collins, I. Dunham, S. Hassock, A. Bankier, and Giannelli, 1992. The development and application of automated gridding for efficient screening of yeast and bacterial ordered libraries. *Genomics,* 12:534–541.

Birkenbihl, R. P. and W. Vielmetter. 1989. Cosmid-derived map of *E. coli* strain BHB2600 in comparison to the map of strain W3110. *Nucleic Acids Res,* 17:5057–69.

Borodovsky, M., E. V. Koonin, and K. E. Rudd. 1994. New Genes in Old Sequence: Strategy for Finding Genes in the Bacterial Genome. *Genome Mapping & Sequencing,* 25.

Branscomb, E., T. Slezak, R. Pae, D. Galas, A. V. Carrano, and M. Waterman. 1990. Optimizing restriction fragment fingerprinting methods for ordering large genomic libraries. *Genomics,* 8:351–66.

Bukanov, N. O. and D. E. Berg. 1994. Ordered cosmid library and high-resolution physical-genetic map of *Helicobacter pylori* strain NCTC11638. *Molecular Microbiology,* 11:509–23.

Charlebois, R. L., J. D. Hofman, L. C. Schalkwyk, W. L. Lam, and W. F. Doolittle. 1989. Genome mapping in halobacteria. *Can J Microbiol,* 35:21–9.

Charlebois, R. L., L. C. Schalkwyk, J. D. Hofman, and W. F. Doolittle. 1991. Detailed physical map and set of overlapping clones covering the genome of the archaebacterium *Haloferax volcanii* DS2. *J Mol Biol,* 222:509–24.

Chen, H. W., A. Kuspa, I. M. Keseler, and L. J. Shimkets, 1991. Physical map of the *Myxococcus xanthus* chromosome. *J Bacteriol,* 173:2109–15.

Choudhary, M., C. Mackenzie, K. S. Nereng, E. Sodergrn, G. M. Weinstock, and S. Kaplan. 1994. Multiple chromosomes in bacteria: structure and function of chromosome II of *Rhodobacter sphaeroides* 2.4.1. *Journal of Bacteriology* 176:7694–7702.

Clarke, L. and J. Carbon. 1975. Biochemical construction and selection of hybrid plasmids containing specific segments of the *Escherichia coli* genome. *Proc. Natl. Acad. Sci. USA,* 72:4361–5.

Clarke, L. and J. Carbon. (1976). A colony bank containing synthetic ColE1 hybrid plasmids representive of the entire *E. coli* genome. *Cell,* 9:91–99.

Clarke, L. and J. Carbon. 1979. Selection of specific clones from colony banks by suppression or complementation tests. *Methods in Enzymology,* 68:396–408.

Cohen, A., W. L. Lam, R. L. Charlebois, W. F. Doolittle, and L. C. Schalkwyk, 1992. Localizing genes on the map of the genome of *Haloferax volcanii*, one of the *Archaea. Proc Natl Acad Sci USA,* 89:1602–6.

Cole, S. T. and I. Saint Girons. 1994. Bacterial genomics. *FEMS Microbiology Reviews,* 14:139–60.

Colman, S. D., P. C. Hu, W. Litaker, and K. F. Bott, 1990. A physical map of the *Mycoplasma genitalium* genome. *Mol Microbiol, 4:* 683–7.

Coulson, A., J. Sulston, S. Brenner, and J. Karn. 1986. Toward a physical map of the genome of the nematode *Caenorhabditis elegans. Proc. Natl. Acad. Sci. USA,* 83:7821–7825.

Daniels, D. L. 1990. Constructing Encyclopedias of Genomes. In *The Bacterial Chromosome*. K. Drlica and M. Riley, eds. (Washington, D.C.: American Society for Microbiology), pp. 43–52.

Daniels, D. L. and Blattner, F. R. 1987. Mapping using gene encyclopedias. *Nature,* 325:831–2.

Deckers, H. M. and G. Voordouw. 1994. Identification of a large family of genes for putative chemoreceptor proteins in an ordered library of the *Desulfovibrio vulgaris* Hildenborough genome. *J. Bacteriol.,* 176:351–8.

Dimri, G. P., K. E. Rudd, M. K. Morgan, H. Bayat, and G. F. Ames. 1992. Physical mapping of repetitive extragenic palindromic sequences in *Escherichia coli* and phylogenetic distribution among *Escherichia coli* strains and other enteric bacteria. *J. Bacteriol.,* 174:4583–93.

Eiglmeier, K., N. Honore, S. A. Woods, B. Caudron, and S. T. Cole. 1993. Use of an ordered cosmid library to deduce the genomic organization of *Mycobacterium leprae. Mol Microbiol,* 7:197–206.

Fleischmann, R. D., M. D. Adams, O. White, R. C. Clayton, E. F. Kirkness, 1995. Whole-genome random sequencing and assembly of *Haemophilus influenzae Rd. Science,* 269:496–511.

Fonstein, M. and R. Haselkorn, 1993. Chromosomal structure of *Rhodobacter capsulatus* strain SB1003: cosmid encyclopedia and high-resolution physical and genetic map. *Proc Natl Acad Sci USA,* 90:2522–6.

Fonstein, M. and R. Haselkorn, 1995. Physical mapping of bacterial genomes. *J. Bacteriol,* 177:3361–3369.

Fonstein, M., E. G. Koshy, T. Nikolskaya, P. Mourachov, and R. Haselkorn, 1995. Refinement of the high-resolution physical and genetic map of *Rhodobacter capsulatus* and genome surveys using blots of the cosmid encyclopedia. *EMBO Journal,* 14:1827–1841.

Fonstein, M., S. Zheng, and R. Haselkorn, 1992. Physical map of the genome of *Rhodobacter capsulatus* SB 1003. *J Bacteriol,* 174:4070–7.

Fu, Y. X., W. E. Timberlake, and J. Arnold. 1992. On the design of genome mapping experiments using short synthetic oligonucleotides. *Biometrics,* 48:337–59.

Gibson, T. J., A. Rosenthal, and R. H. Waterston. 1987. Lorist6, a cosmid vector with *Bam*HI, *Not*I, *Sca*I and *Hin*dIII cloning sites and altered neomycin phosphotransferase gene expression. *Gene,* 53:283–6.

Hoheisel, J. D., E. Maier, R. Mott, L. McCarthy, A. V. Grigoriev, L. C. Schalkwyk, D. Nizetic, F. Francis, and H. Lehrach. 1993. High resolution cosmid and P1 maps spanning the 14 Mb genome of the fission yeast *S. pombe. Cell,* 73:109–20.

Knott, V., D. J. Blake, and G. G. Brownlee. 1989. Completion of the detailed restriction map of the *E. coli* genome by the isolation of overlapping cosmid clones. *Nucleic Acids Res,* 17:5901–12.

Kohara, Y., K. Akiyama, and K. Isono. 1987. The physical map of the whole *E. coli* chromosome: application of a new strategy for rapid analysis and sorting of a large genomic library. *Cell,* 50:495–508.

Kuspa, A., D. Vollrath, Y. Cheng, and D. Kaiser. 1989. Physical mapping of the *Myxococcus xanthus* genome by random cloning in yeast artificial chromosomes. *Proc. Natl. Acad. Sci. U.S.A.,* 86:8917–21.

Lucier, T. S., P.-Q. Hu, S. N. Petterson, X.-Y. Song, L. Miller, K. Heitzman, K. F. Bott, C. A. Hutchinson III, and P.-C. Hu. 1994. Construction of an ordered genomic library of *Mycoplasma genetalium. Gene,* 150:27–34.

Medigue, C., A. Viari, A. Henaut, and A. Danchin. 1993. Colibri: a functional data base for the *Escherichia coli* genome. [Review]. *Microbiological Reviews,* 57:623–54.

Mizukami, T., W. I. Chang, I. Garkavtsev, N. Kaplan, D. Lombardi, T. Matsumoto, O. Niwa, A. Kounosu, M. Yanagida, T. G. Marr, et al., 1993. A 13 kb resolution cosmid map of the 14 Mb fission yeast genome by nonrandom sequence-tagged site mapping. *Cell,* 73:121–32.

Neidhardt, F. C., D. B. Appleby, P. Sankar, M. E. Hutton, and T. A. Phillips. 1989. Genomically linked cellular protein databases derived from two-dimensional polyacrylamide gel electrophoresis. *Electrophoresis,* 10:116–22.

Nishimura, A., K. Akiyama, Y. Kohara, and K. Horiuchi. 1992. Correlation of a subset of the pLC plasmids to the physical map of *Escherichia coli* K-12. *Microbiol. Rev.,* 56:137–51.

Olson, M. V., J. E. Dutchik, M. Y. Graham, G. M Brodeur, C. Helms, M. Frank, M. MacCollin, R. Scheinman, and T. Frank. 1986. Random-clone strategy for genomic restriction mapping in yeast. *Proc. Natl. Acad. Sci. U.S.A.,* 83:7826–30.

Rackwitz, H. R., G. Zehetner, H. Murialdo, H. Delius, J. H. Chai, A. Poustka, A. Frischauf, and H. Lehrach. 1985. Analysis of cosmids using linearization by phage lambda terminase. *Gene,* 40:259–66.

Rouqueir, S., M. A. Batzer, and D. Giorgi. 1994. Application of bacterial artificial chromosomes to the generation of contiguous physical maps: a pilot study of human ryanodine receptor gene (RYR1) region. *Analyt. Biochem.,* 217:205–209.

Shizuya, H., B. Birren, U. J. Kim, V. Mancino, T. Slepak, Y. Tachiiri, and M. Simon. 1992. Cloning and stable maintenance of 300-kilobase-pair fragments of human DNA in *Escherichia coli* using an F-factor-based vector. *Proc. Natl. Acad. Sci. U.S.A.,* 89:8794–7.

Tabata, S., A. Higashitani, M. Takanami, K. Akiyama, Y. Kohara, Y. Nishimura, A. Nishimura, S. Yasuda, and Y. Hirota. 1989. Construction of an ordered cosmid collection of the *Escherichia coli* K-12 W3110 chromosome. *J. Bacteriol.,* 171:1214–8.

Wenzel, R. and R. Herrmann. 1989. Cloning of the complete *Mycoplasma pneumoniae* genome. *Nucl. Acids Res.,* 17:7029–43.

Wenzel, R., E. Pirkl, and R. Herrmann. 1992. Construction of an EcoRI restriction map of *Mycoplasma pneumoniae* and localization of selected genes. *J. Bacteriol.,* 174:7289–96.

Woo, S.-S., J. Jiang, B. S. Gill, A. H. Paterson, and R. A. Wing, 1994. Construction and characterization of a bacterial artificial chromosome library of *Sorghum bicolor. Nucl. Acids Res.,* 22:4922–4931.

Yoshida, K., M. P. Strathmann, C. A. Mayeda, C. H. Martin, and M. J. Palazzolo, 1993. A simple and efficient method for constructing high resolution physical maps. *Nucl. Acids Res.,* 21:3553–3562.

29

Localizing Genes by the Introduction of Rare Restriction Sites

Tanya Kuritz and C. Peter Wolk

Genetic mapping of bacterial chromosomes began in the 1940s with experiments on conjugal and bacteriophage-mediated transfer of genes in *Escherichia coli* (see Miller, 1972). Pulsed-field gel electrophoresis (PFGE; *see also* Chapters 24 to 26) later enabled physical mapping of a broad range of strains independent of their conjugal proficiency or the availability of transducing phages (Schwartz *et al.*, 1983). Physical genomic maps are now available for over 30 different bacterial taxa (*see* section 3 of this monograph), and in certain instances have been aligned with the maps obtained by more classical methods (e.g., Bachman, 1990; Holloway *et al.*, 1994).

Initially, the positions of genes on physical maps were assigned by Southern hybridization to large restriction fragments that were obtained when genomic DNA was cut by rarely-cutting restriction endonucleases. The choice of endonucleases for any particular bacterial species is related to the GC content of the genome of that species (McClelland *et al.*, 1987; see also Chapt. 24). The assumption that bases are randomly distributed permits a prediction of the average expected number of fragments into which a specific restriction endonuclease should cut a chromosome of a particular size and GC content. However, rarely cutting enzymes frequently cut less often than predicted (Stibitz and Garletts, 1992). The number of restriction endonucleases usable for physical mapping of whole chromosomes is limited by the availability of enzymes that can yield sharp bands upon digestion of agarose-immobilized DNA (Tigges and Minion, 1994a); those suitable for PFGE are often designated as such by commercial suppliers. The endonucleases used for mapping prior to 1990 (Smith and Condemine, 1990) have been augmented by newly discovered enzymes (Liu *et al.*, 1993, Lui *et al.*, 1994; *see also* Chapter 24).

Chromosomally derived PFGE fragments range in size from very small to more than 1 Mb; average lengths of 200 to 300 kb are convenient for mapping. Several means are available to map physically within those fragments, and to show the orientation of the mapped genes relative to the chromosome as a whole. The basic principle is to introduce rare restriction sites within or close to a gene that is being mapped, to map those sites, and to orient the gene relative to those sites (Smith *et al.*, 1989). Upon digestion at a newly introduced rare site, the parental fragment is cleaved into two new fragments. Measurement of the lengths of the new fragments defines the distance of the restriction site, and thus the associated gene, from the ends of the parental fragment, but does not determine to which end of the parental fragment the gene is closer. By introducing and mapping, within or close to the gene, a site for a second rarely-cutting restriction endonuclease whose parental fragments do not exactly overlap those of the first, one can uniquely position the sites, and thus the gene, relative to the ends of both parental fragments. Because the sizing of fragments by PFGE has inherent problems (see below), the positions determined within each of the parental fragments may not precisely superimpose; it is therefore advisable, when possible, to use a third enzyme, and to average the localizations defined by digestion with the different enzymes used. To associate sites with genes normally entails the addition of DNA to the genome, in the form of an integrated plasmid or transposon. When the length of the sequence added approximates or exceeds the resolution of the system of PFGE apparatus, power regime and agarose used, the length of the DNA inserted should be taken into consideration when the position of the gene is calculated.

The precision of a physical map and, therefore, of the localization of a gene within that map, depends on the accuracy of the determination of the sizes of DNA fragments and on the resolution of the PFGE system. The amount of DNA loaded affects the mobility of the bands and thus changes the apparent sizes of fragments (Doggett *et al.*, 1992). Mobility is affected less and less as the amount of DNA that is loaded per lane diminishes. However, visualization, especially for small fragments, can become more of a problem. Each rung of a lambda ladder differs by 48.5 kb from its neighboring rungs, so that bands sized by use of a lambda ladder are resolved to less than 50 kb, although the resolution varies depending on the proximity of the sized fragments to the compression zone of the gel (Bancroft and Wolk, 1988). For (slowly run) large gels with sharp bands, the resolution may be as precise as ca. 5 kb (Kuritz *et al.*, 1993). In our experience, faster migration of bands (permitted by Gold Agarose, FMC, Rockland, Maine) results in poorer resolution than does slower migration (e.g., using SeaKem GTG Agarose, FMC). For other parameters of running PFGE experiments, *see* Chapters 25 and 26.

Introduction of Rare Restriction Sites by Single Recombination

This technique is based on the introduction of rare restriction sites via cell-mediated *rec*-dependent homologous recombination between a cloned sequence and a corresponding sequence in the chromosome (Smith *et al.*, 1989; Stibitz and Garletts, 1992; Kuritz *et al.*, 1993). The cloned sequence, its orientation known, is introduced within a non-replicating ("suicide") vector (Penford and Pemberton, 1992) that contains rare restriction sites used for mapping of the chromosome, and also bears a selectable genetic marker (usually one that confers antibiotic resistance). A conditionally (non-)replicating (Biswas *et al.*, 1993) or an *ori*$^-$ (Hasan *et al.*, 1994) vector can also be used. Upon recombination between the cloned sequence and a homologous sequence in the genome, the vector sequence that bears the rare restriction sites is concomitantly inserted into the chromosome (Fig. 29-1a–d). Although recombination results in duplication of the cloned sequence, the vector portion is normally present in only one copy. The insertion, and thereby the gene, can be localized, and oriented relative to the chromosome as a whole, by digesting with appropriate rarely cutting restriction endonucleases, sizing the parent and daughter fragments, blotting a resulting PFGE gel, and probing the blot with a vector fragment that hybridizes distal to the sites that have been cut (Fig. 29-1e). The orientation can also be determined by comparing the PGFE patterns that result from digestion of chromosomes that contain a sequence recombined in two different orientations (Stibitz and Garletts, 1992). The legitimacy of recombination should be confirmed by Southern hybridization prior to PFGE analysis. Single recombination with a suicide vector bearing a homology should normally be suitable even for the localization of essential genes so long as at least one end of the corresponding operon, including its regulatory regions, is included in the clone, so that an intact copy of the transcriptional unit remains in the recombinant organism. Some cyanobacteria, at least, have 10–15 copies of the chromosome per cell (Thiel, 1994). As long as at least one wild type copy of an essential gene remains, other copies of this gene could bear an insertion.

Introduction of Rare Restriction Sites by Transposon Mutagenesis

Certain transposons have very low site specificity for insertion (Berg, 1989). If a transposon bears restriction site(s) that have been mapped, or a restriction site present nowhere else in the genome, its site of insertion can be mapped (Smith *et al.*, 1989; Jumas-Bilak *et al.*, 1995). Transposon mutagenesis, because of its experimental simplicity, has become the most frequently used means to introduce rare restriction sites (Fig. 29-1a,e,f): with Tn*5* and its derivatives for *E. coli* (Perkins *et al.*, 1993; Jumas-Bilak *et al.*, 1995), *Salmonella typhimurium* (Wong and McClelland, 1992; Liu *et al.*, 1993), *Anabaena* PCC 7120 (Wolk *et al.*, 1991), *Caulobacter crescentus* (Ely and Gerardot, 1988), *Rhizobium meliloti* (Honeycutt *et al.*, 1993), *Rhodobacter sphaeroides* (Choudhary *et al.*, 1994), *Brucella melitensis*, and *Agrobacterium tumefaciens* (Jumas-Bilak *et al.*, 1995); with Tn*10* for *S. typhimurium* (Benson and Goldman, 1992) and *E. coli* (Perkins

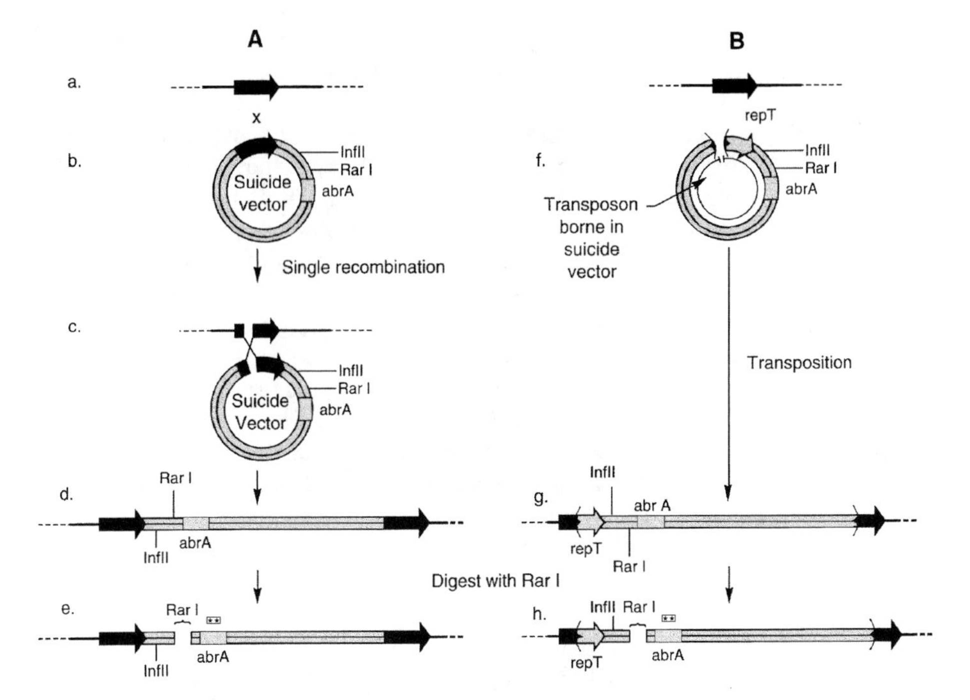

Figure 29-1. Introduction of rare restriction sites via (a) single recombination or (b) transposon mutagenesis.

A region of the genome (a), here shown bearing a gene of interest (heavy black arrow), is insertionally altered in either of two ways. (a) A portion of the region is cloned into a suicide vector that bears rare or very infrequent restriction sites *Rar*I and *Inf*II whose sites in a chromosome have been mapped, and that also contains a selectable marker such as an antibiotic resistance gene (*abrA*). When the resulting plasmid (b) undergoes homologous recombination with the chromosome (c), the vector is inserted into the bacterial chromosome, resulting in a duplication of the cloned portion of the gene (d). Alternatively, a transposon (f) is used that bears sites for *Rar*I and *Inf*II, and the *abrA* gene. As a result of transposition, the transposon is inserted within the same chromosomal region (g). New sites for *Rar*I and *Inf*II, introduced into the chromosome by the inserted plasmid or the transposon, are mapped, indicating the position of the chromosomal region. A fragment (box with two asterisks) of the plasmid vector or of the transposon that lies distal to the sites at which *Rar*I and *Inf*II cut the vector (e) or the transposon (h) is hybridized to a Southern blot of a pulsed-field gel electrophoretogram derived from restriction of the insertionally altered chromosome with *Rar*I or *Inf*II. Which new daughter fragments, derived from homologous recombination or transposition, bear that portion of the vector or the transposon is thereby shown, and demonstrates the orientation of the vector or the transposon relative to the chromosome as a whole. The corresponding orientation of the gene within the chromosomal region is then defined by the relative orientation of the gene and either the vector (as determined by the recombinant plasmid) or the transposon. In the latter case, the orientation may be determined physiologically if the transposon bears a reporter gene (*repT*), or by sequencing outward from one end or the other of the transposon.

et al., 1993); and with Tn*4001* for *Mycoplasma gallisepticum* (Tigges and Minion, 1994a) and *Acholeplasma oculi* (Tigges and Minion, 1994b).

Because transposons that have been used for mutagenesis often bear sites for rarely cutting restriction endonucleases, transposon-generated mutations that are already available are often amenable to mapping (Benson and Goldman, 1992; Perkins *et al.*, 1993; Liu *et al.*, 1993; Honeycutt *et al.*, 1993). If a gene of interest is essential, and therefore normally not subject to transposon mutagenesis, it may nonetheless by mapped by linkage to a transposon (Benson and Goldman, 1992). For *S. typhimurium* 109, insertions were localized on both the physical and the genetic map (Lui *et al.*, 1993; Wong and McClelland, 1992). Localization of a hundred different transpositional insertions within or adjacent to specific developmental genes of *Myxococcus xanthus* showed the clustering (or non-clustering) of specific groups of mutations, and pointed to the existence of previously undescribed genes (Chen *et al.*, 1991). Previously collected transposon-derived mutations were localized to the different replicons of *R. meliloti* (Honeycutt *et al.*, 1993) and in *Anabaena* sp. (Kuritz *et al.*, 1993). The functions of loci to which transposons are mapped are sometimes unknown (Honeycutt *et al.*, 1993; Tigges and Minion, 1994a,b). Mutagenesis of *Mycoplasma gallisepticum* with a Tn*4001* derivative bearing multiple rare restriction sites was used to link restriction fragments during the construction of a physical map (Tigges and Minion, 1994a).

An inserted transposon can be oriented relative to the chromosome by digesting with a rarely cutting restriction endonuclease that has a site within the transposon, followed by blotting a resulting PFGE gel, and probing the blot with a transposon fragment that hybridizes to one side or the other of the rare restriction site (see Fig. 29-1h and the discussion of Fig. 29-1e, above). If the transposon is present within a gene, the gene can then be oriented relative to the chromosome by orienting it relative to the transposon. The latter orientation is possible, for example, for certain derivatives of Tn*5* by sequencing outward from one end or another of the transposon (Kuritz *et al.*, 1993). Alternatively, if the transposon bears a reporter gene (*repT* in Fig. 29-1b), activation of the reporter under conditions that the gene should be activated implies that the reporter gene is oriented parallel to the interrupted gene (Wolk *et al.*, 1991; Kuritz *et al.*, 1993).

Introduction of Rare Restriction Sites by Recombination of DNA Fragments Carrying Them, into other Genomic Restriction Sites

Upon polyethylene glycol-mediated transformation of *Saccharomyces cerevisiae* or electroporation of recipient cells of the slime mold, *Dictyostelium discoideum*, with the restriction endonuclease, *Bam*HI, together with a DNA sequence bearing two *Bam*HI ends, transformants of the yeast or of *D. discoideum* were isolated in which the sequence was incorporated within a chromosomal *Bam*HI site (Schiestl and Petes, 1991; Kuspa *et al.*, 1992; Kuspa and Loomis, 1992, 1994).

This approach, termed restriction enzyme mediated integration, has not to our knowledge been used with any bacterium, but has the potential, in theory, to serve as an alternative means to mutagenize and to introduce rare restriction sites into a bacterial chromosome.

Conclusion

Genes are often localized to large restriction fragments of wild-type DNA by hybridization to blots of pulsed-field gels. Alternatively, homologous recombination or transposon mutagenesis can be used to introduce rare restriction sites within, or close to, those genes, and the introduced sites then mapped. A great increase in the precision of localization of genes in a physical map can thereby be achieved, and the orientation of those genes determined, at the price of little additional experimental manipulation.

Acknowledgments

TK is indebted to Dr. Curtis C. Travis for support and for the opportunity for independence. Mapping work in the laboratory of CPW is supported by the U.S. Department of Energy under grant DE-FG01-90ER20021.

References

Bachman, B. 1990. Linkage map of *Escherichia coli* K-12, edition 8. *Microbiol. Rev.* 54:130–197.

Bancroft, I., and C. P. Wolk. 1988. Pulsed homogeneous orthogonal field gel electrophoresis (PHOGE). *Nucl. Acids Res.* 16:7405–7418.

Benson, N. R., and B. S. Goldman. 1992. Rapid mapping in *Salmonella typhimurium* with Mu*d*-P22 prophages. *J. Bacteriol.* 174:1673–1681.

Berg, D. E. 1989. Transposon Tn*5*. In *Mobile DNA,* Berg, D. E., and M. M. Howe, eds. pp. 185–210 Am. Soc. Microbiol., Washington, D. C.

Biswas, I., A. Gruss, S. D. Ehrlich, and E. Maguin. 1993. High-efficiency gene inactivation and replacement system from Gram-positive bacteria. *J. Bacteriol.* 175:3628–3635.

Chen, S., A. Kuspa, I. M. Keseler, and L. J. Shimkets. 1991. Physical map of the *Myxococcus xanthus* chromosome. *J. Bacteriol.* 173:2109–2115.

Choudhary, M., C. Mackenzie, K. S. Nereng, E. Sodergren, G. M. Weinstock, and S. Kaplan. 1994. Multiple chromosomes in bacteria: structure and function of chromosome II of *Rhodobacter sphaeroides* 2.4.1^{T}. *J. Bacteriol.* 176:7694–7702.

Ely, B., and C. J. Gerardot. 1988. Use of pulsed-field gradient gel electrophoresis to construct a physical map of the *Caulobacter crescentus* genome. *Gene* 68:323–333.

Daniels, D. L. 1990. The complete *Avr*II restriction map of the *Escherichia coli* genome and comparisons of several laboratory strains. *Nucl. Acids Res.* 18:2649–2651.

Doggett, A. N., C. L. Smith, and C. R. Cantor. 1992. The effect of DNA concentration on mobility in pulsed field gel electrophoresis. *Nucl. Acids Res.* 20:859–864.

Hasan, N., M. Koob, and W. Szybalski. 1994. *Escherichia coli* genome targeting, I. Cre-*lox*-mediated in vitro generation of *ori*⁻ plasmids and their in vivo chromosomal integration and retrieval. *Gene* 150:51–56.

Holloway, B. W., U. Römling, and B. Tümmler. 1994. Genomic mapping of *Pseudomonas aeruginosa* PAO. *Microbiology* 140:2907–2929.

Honeycutt, R. J., M. McClelland, and B.W.S. Sobral. 1993. Physical map of the genome of *Rhizobium meliloti* 1021. *J. Bacteriol.* 175:6945–6952.

Jumas-Bilak, E., C. Maugard, S. Michaux-Charachon, A. Allardet-Servent, A. Perrin, D. O'Callaghan, and M. Ramuz. 1995. Study of the organization of the genomes of *Escherichia coli, Brucella melitensis* and *Agrobacterium tumefaciens* by insertion of a unique restriction site. *Microbiology* 141:2425–2432.

Kuritz, T., A. Ernst, T. A. Black, and C. P. Wolk. 1993. High-resolution mapping of genetic loci of *Anabaena* PCC 7120 required for photosynthesis and nitrogen fixation. *Mol. Microbiol.* 8:101–110.

Kuspa, A., and W. F. Loomis. 1992. Tagging developmental genes in *Dictyostelium* by restriction enzyme-mediated integration of plasmid DNA. *Proc. Natl. Acad. Sci. USA* 89:8803–8807.

Kuspa, A., and W. F. Loomis. 1994. REMI-RFLP mapping in the *Dictyostelium* genome. *Gene* 138:665–674.

Kuspa, A., D. Maghakian, P. Bergesch, and W. F. Loomis. 1992. Physical mapping of genes to specific chromosomes in *Dictyostelium discoideum. Genomics* 13:49–61.

Liu, S.-L., A. Hessel, and S. K. Sanderson. 1993. The *Xba*I-*Bln*I-*Ceu*I genomic cleavage map of *Salmonella typhimurium* LT2 determined by double digestion, end labelling, and pulsed-field gel electrophoresis. *J. Bacteriol.* 175:4104–4120.

Liu, S.-L., A. Hessel, H.-Y.M. Cheng, and K.E. Sanderson. 1994. The *Xba*I-*Bln*I-*Ceu*I genomic cleavage map of *Salmonella paratyphi* B. *J. Bacteriol.* 176:1014–1024.

McClelland, M., R. Jones, Y. Patel, and M. Nelson. 1987. Restriction endonucleases for pulsed field mapping of bacterial genomes. *Nucl. Acids Res.* 15:5985–6005.

Miller, L. *Experiments in Microbial Genetics*. 1972. Cold Spring Harbor Laboratory, Cold Spring Harbor, New York.

Penford, R. J., and J. M. Pemberton. 1992. An improved suicide vector for construction of chromosomal insertion mutations in bacteria. *Gene* 118:145–146.

Perkins, J. D., J. D. Heath, B. R. Sharma, and G. M. Weinstock. 1993. *Xba*I and *Bln*I genomic cleavage maps of *Escherichia coli* K-12 strain MG1655 and comparative analysis of other strains. *J. Mol. Biol.* 232:419–445.

Schiestl, R. H., and T. D. Petes. 1991. Integration of DNA fragments by illegitimate recombination in *Saccharomyces cerevisiae. Proc. Natl. Acad. Sci. USA* 88:7585–7589.

Schwartz, D. C., W. Saffran, J. Welsh, R. Haas, M. Goldenberg, and C. R. Cantor. 1983.

New techniques for purifying large DNAs and studying their properties and packaging. *Cold Spring Harbor Symp. Quant. Biol.* 47:189–195.

Smith, C. L., and G. Condemine. 1990. New approaches for physical mapping of small genomes. *J. Bacteriol.* 172:1167–1172.

Smith, C. L., G. Condemine, S.-Y. Cheng, E. McGary, and S. Chang. 1989. Insertion of rare cutting sites nearby genes allows rapid physical mapping: localization of the *E. coli map* locus. *Nucl. Acids Res.* 17:817.

Stibitz, S., and T. L. Garletts. 1992. Derivation of a physical map of the chromosome of *Bordetella pertussis* Tohama I. *J. Bacteriol.* 174:7770–7777.

Thiel, T. 1994. Genetic analysis of cyanobacteria. *In* Bryant, D. A., ed. *The Molecular Biology of Cyanobacteria*, pp. 581–611. Kluwer Acad. Publ., Dordrecht, The Netherlands.

Tigges, E., and F. C. Minion. 1994a. Physical map of *Mycoplasma gallisepticum. J. Bacteriol.* 176:4157–4159.

Tigges, E., and F. C. Minion. 1994b. Physical map of the genome of *Acholeplasma oculi* ISM1499 and construction of a Tn*4001* derivative for macrorestriction chromosomal mapping. *J. Bacteriol.* 176:1180–1183.

Wolk, C. P., Y. Cai, and J.-M. Panoff. 1991. Use of a transposon with luciferase as a reporter to identify environmentally responsive genes in a cyanobacterium. *Proc. Natl. Acad. Sci. USA* 88:5355–5359.

Wong, K. K., and M. McClelland. Dissection of the *Salmonella typhimurium* genome by use of a Tn*5* derivative carrying rare restriction sites. *J. Bacteriol.* 174:3807–3811.

30

Towards a Cosmid-derived Physical Map of the *Synechococcus* PCC 7002 Genome

Hrissi Samartzidou, Fadi Abdi, Wesley Ford, and William R. Widger

Introduction

The generation of an overlapping clone library that represents the correct gene order of an organism of interest is a powerful tool that can be used to characterize a genome (*see also* Chapter 28). Once created, this library is useful for the systematic analysis of the genome including a comparative study of global relationships among genomes. The tremendous amount of information generated by genomics will change the way we contemplate biological problems including phylogenetic information, evolution and possibly the origins of life.

An overlapping cosmid map (contig map) isolated from cloned DNA elements (CDEs) is essential to study global regulation of cellular activity and will lead to the understanding of positional effects of gene location on expression for example (Sankar *et al.*, 1993). To this end, a set of overlapping cosmids are being assembled to form a contiguous minimal set covering the entire genome of the unicellular, non-nitrogen fixing cyanobacterium, *Synechococcus* PCC 7002. This cyanobacterium has been one of several used in the identification and manipulation of genes involved in oxygenic photosynthesis (Schluchter and Bryant, 1992; Schluchter *et al.*, 1993; Widger, 1991; Tan *et al.*, 1994). One salient feature of this organism is the ease of isolation of intact genomic DNA, which aided in the assembly of the physical genome map (Chen and Widger, 1993). This map was composed from pulsed-field-gel (PFG) separated large restriction fragments generated using rare cutting restriction enzymes *Not*I, *Sal*I, *Asc*I and *Sfi*I (see Chapter 83). The overlapping cosmid map was assembled and correlated with the physical map by extensive hybridization to Southern blotted genomic DNA.

Results

Cosmid library

The cosmid library was generated using the Super-cos vector (Stratagene). Size-selected, *Sau*3AI-generated, genomic DNA fragments were ligated into the *Bam*HI site of the vector. The ligated samples were packaged using giga pack gold packaging extracts and used to infect NM544 host cells (Stratagene). The recombinants were selected on kanamycin and 600 individual recombinants were picked and cultured separately in liquid culture. Frozen glycerol stocks were stored at –80°C. Six groups of 96 cosmids were labeled A–F, such that a cosmid name i.e. D20, would define its grid position. Single colonies were grown up in 50 ml of LB augmented with 35 μg/ml of kanamycin and cosmid DNA was isolated. Dot blot filters were generated from isolated cosmid DNA arranged in a 12 × 8 matrix (gridded filters), of which twenty sets were initially made. Hybridization of specific probes to these filters allowed the identification of overlapping cosmids. Field inversion electrophoresis (Lai *et al.*, 1991) on 30 cm long gels was used to separate *Eco*RI restriction fragments from each cosmid, their sizes were determined by extrapolation compared to the 1 Kbp ladder (BRL) and entered into a computer. A program was used to log in all *Eco*RI fragment sizes and these sizes were compared from cosmid to cosmid (Widger and Xavier, unpub. results). This program, resembling the contig7 program (Sulston *et al.*, 1988) and written in C using Xwindows motif running on a Sun IPC workstation, was successful in selecting cosmids with similar fragment sizes. Computer selected cosmids which were postulated to be overlapping were further analyzed by field inversion electrophoresis.

Gel electrophoresis

The level of resolution of DNA fragments on the agarose gel used dictates the accuracy of the fragment size data generated. Longer gels allow a greater number of fragment sizes to be distinguished which in turn leads to a more accurate estimation of fragment size. Using the rationale proposed by Sulston (Coulson *et al.*, 1986; Sulston *et al.*, 1988), genome complexity is a function of genome size and the uniqueness of *Eco*RI restriction fragment sizes within a genome. Fragment sizes can only be determined according to the resolution level of the gel. A 30 cm gel resolves to ≅1 mm if bands are compared on the same gel. The gel resolution increases if rapid field inversion electrophoresis is employed (Lai *et al.*, 1991) since most of the gel or about 250 mm can be used for separation. The identification of three apparently equal sized fragments on differing cosmids is sufficiently unique to declare the cosmids overlapping. In practice, hybridization data with at least four comigrating fragments were used to align cosmids. As more cosmids are added to the contig, a unique fragment order emerges and the self-consistency of the ordering of the *Eco*RI fragment adds confidence that a contig is correct. Occasionally, when cosmids contain a few large *Eco*RI fragments, double digestion with *Bam*HI and *Eco*RI yields sufficient data to determine cosmid overlaps.

Assembly strategies

The overall stratagem used for the building of a contig from cosmids was similar to that reported for *E. coli* (Birkenbihl and Vielmetter, 1989). The assembly process included a) the isolation of approximately 600 cosmids to ensure a 99.5% coverage of the genome, b) an initial screening of the cosmids by *Eco*RI digestion followed by FIGE to select a clean set of cosmids without aberrations, c) recording the *Eco*RI restriction fragment sizes from all the cosmids for comparison by computer analysis, d) the generation of gridded cosmid dot blots for hybridization to establish preliminary contig membership and d) electrophoresis of selected clones to confirm the overlap of *Eco*RI restriction fragments. Because no genetic map was available for *Synechococcus* PCC 7002, the location of contigs on the physical map required hybridization of DNA fragments arising from the contig to a Southern blot of *Not*I, *Sal*I and doubly digested genomic DNA. The degeneracy in fragment sizes arising from either *Not*I or *Sal*I could be overcome by double digestion (Chen and Widger, 1993). The hybridization patterns of all three digestions were found to be almost unique and can resolve most positions on the genome. Spatial relationships among contigs, resolved by hybridization to genome blots, determined the extent of chromosomal walking needed to connect contigs. Two types of probes were used to group cosmids to contigs, the first was to hybridize large *Not*I and *Sal*I fragments to the filters described above. The second method was to hybridize specific DNA fragments from individual cosmids or genes to the filters, to generate small groups of overlapping cosmids. Each method was found to have its advantages and disadvantages. Using *Not*I fragments as probes, many cosmids were selected and several contigs were built, but, on occasion, a contig selected by a large *Not*I fragment did not hybridize to the predicted *Not*I and *Sal*I fragments when small probes were used with Southern blots of PFGE fragments. There are no reports of small repeated DNA sequences in the genome of PCC 7002, although, we have seen several probes selecting two differing sets of cosmids, suggesting some degree of repeated sequence occurrence. Hybridization of probes from opposite ends of a cosmid resolved any ambiguities about the genome position arising from repeated sequences.

The most successful strategy for grouping cosmids into contigs was to select many random probes, containing either known genes or small fragments from random cosmids to screen the cosmid filters, thus rapidly amassing many independent contigs. Ideally, probes chosen should be about 100 Kbp apart on the physical genome map to maximize the initial coverage, however, this is not always possible. DNA fragments of end cosmids of a contig located on the physical map by hybridization were used for chromosome walking and connecting adjacent contigs. Walking was used to close gaps and was not used as a general method to increase the size of a contig.

The small contigs were positioned on the physical map by hybridization to *Not*I, *Sal*I and doubly digested genomic DNA, separated by pulsed field electrophoresis. Three terms are used to describe cosmid positions, floating, tacked or nailed. A floating contig is associated within a large pulsed field fragment, but the location and orientation within the PFG fragment is not known. A tacked cosmid is located within a fragment and its position is known, however, its orientation is unknown. Tacking was done by the identification of a unique site such as a *Not*I or *Sal*I site in the contig. In the case of nailed cosmids both position and orientation data are known. Floating contigs become tacked or nailed if a rare restriction site is identified within the cosmid, which corresponds to one found on the physical map.

A critical aspect on contig assembly is the creation of a self-consistent *Eco*RI restriction pattern as individual cosmids are added to the contig. This generates a reliable fine structure map and identifies cosmids that may have repeated sequences or are rearranged.

Assembly of a 420 Kbp contig encompassing NB-370

Cosmids belonging to the second largest *Not*I fragment in PCC 7002 (*see* Chapter 83) were first identified by hybridization of the entire NB-370 fragment to the gridded library filters. The identification of contigs within this group of cosmids was first done by comparison of fragment sizes, followed by hybridization. Figure 30-1 shows the overlapping cosmid map starting at position 800 Kbp on the genome (see section 3, 38) and extends to 1200 Kbp. The vertical dashed lines are the positions of *Eco*RI sites that have a definite order. Numbers between the dashed lines represent *Eco*RI fragment sizes in Kbp and in several instances several sizes are presented. These fragments are common among the cosmids in

---

▶

Figure 30-1. The overlapping cosmid map covering the entire second largest *Not*I fragment found in *Synechococcus* PCC 7002, NB-370. The *Eco*RI restriction fragment sizes are indicated in small numbers while vertical dashed lines represent *Eco*RI sites that are of known order. Many fragments are common to all overlapping cosmids and their order is presently unknown. Solid vertical lines represent *Not*I (N) or *Sal*I (S) sites. The size of *Eco*RI fragments cut with either *Not*I or *Sal*I and the corresponding N-E or S-E fragments are unique to the genome and act as tagged sites. The *Eco*RI restriction pattern of several of the cosmids placed on the map is shown in Figure 30-2.

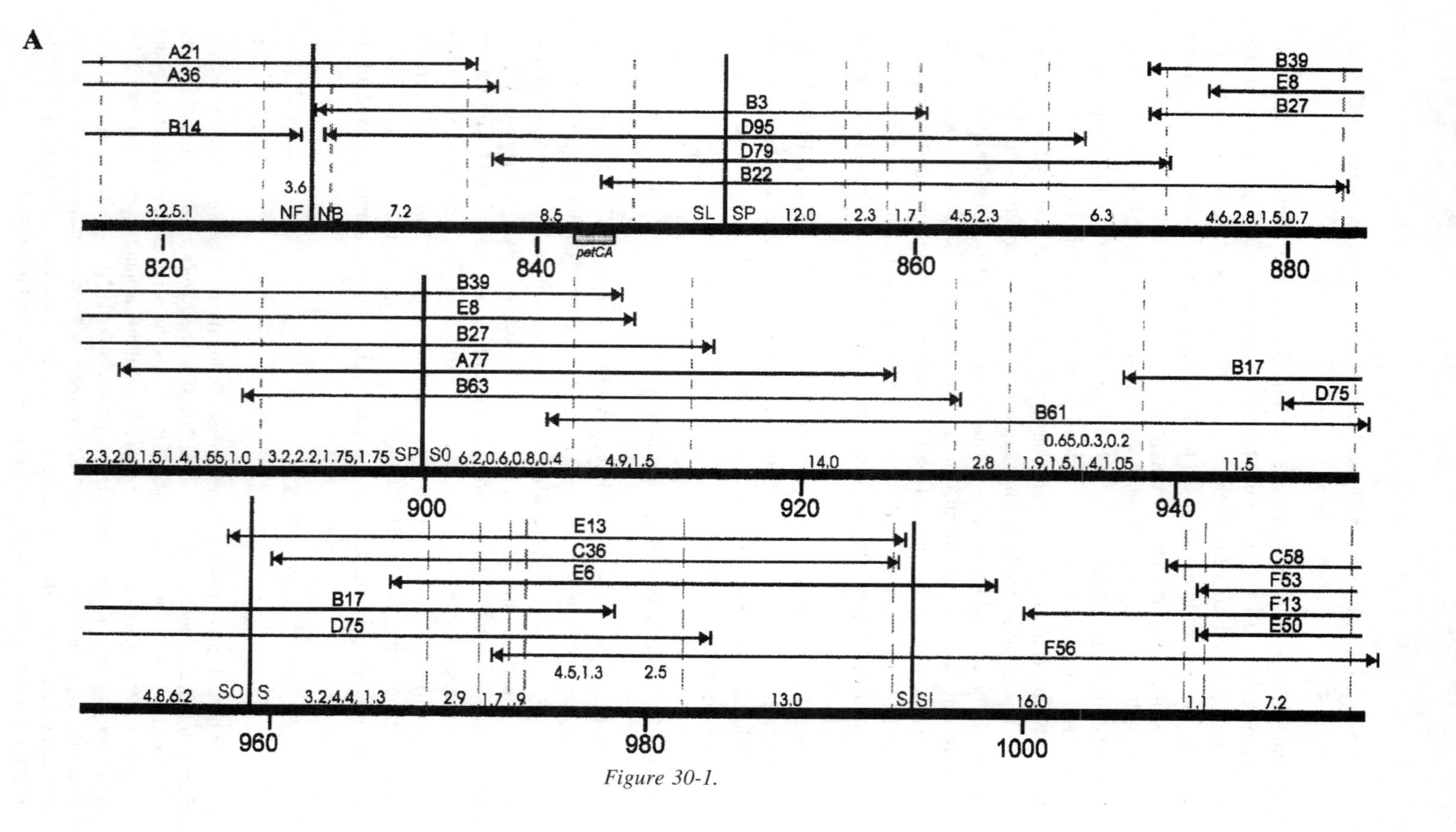
A
A21
A36
B14
B3
D95
D79
B22
B39
E8
B27
3.6
3.2,5.1
NF NB
7.2
8.5
SL SP
12.0
2.3
1.7
4.5,2.3
6.3
4.6,2.8,1.5,0.7
820
840
petCA
860
880
B39
E8
B27
A77
B63
B17
D75
B61
0.65,0.3,0.2
2.3,2.0,1.5,1.4,1.55,1.0
3.2,2.2,1.75,1.75
SP S0
6.2,0.6,0.8,0.4
4.9,1.5
14.0
2.8
1.9,1.5,1.4,1.05
11.5
900
920
940
E13
C36
E6
B17
D75
C58
F53
F13
E50
F56
4.5,1.3
2.5
4.8,6.2
SO S
3.2,4.4, 1.3
2.9
1.7
.9
13.0
S SI
16.0
1.1
7.2
960
980
1000

Figure 30-1.

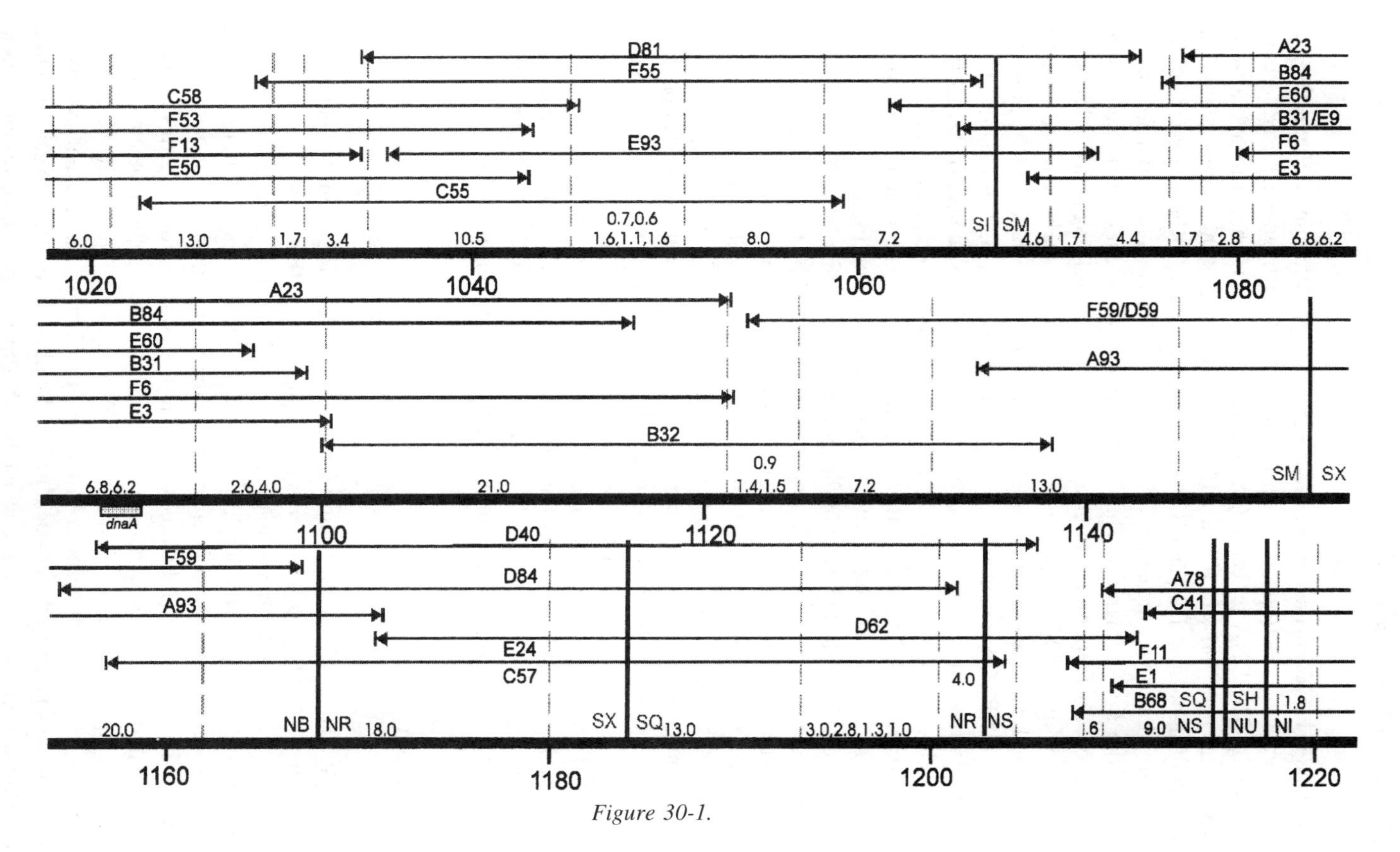
B
D81
F55
A23
B84
E60
B31/E9
F6
E3
C58
F53
F13
E50
E93
C55
SI
SM
1020
1040
1060
1080
A23
B84
E60
B31
F6
E3
F59/D59
A93
B32
SM
SX
dnaA
1100
1120
1140
D40
F59
D84
A93
D62
E24
C57
A78
C41
F11
E1
B68
SQ
SH
NS
NU
NI
NB
NR
SX
SQ
NR
NS
1160
1180
1200
1220

Figure 30-1.

the contig, however, their exact order has not been determined. Figure 30-2 is a representative electrophoresis gel (*Eco*RI restriction fragments) used to build the contig. The DNA probe *pet*CA (Widger, 1991) was shown to hybridize to NB-370 near a *Not*I site (Chen and Widger, 1993) and this probe selected an 8.5 Kbp fragment in D95 and B3, while it selected a 7.8 Kbp fragment in D79 indicating that it was the end *Eco*RI fragment. D79 also contained a *Sal*I site in the 12 Kbp *Eco*RI fragment. The 3.2 Kbp *Eco*RI fragment of A21 was used as a probe to a Southern blot of PFG separated *Not*I and *Sal*I fragments and selected NF-175, while a 5.8 Kbp probe selected D95. A *Not*I site was found in a 3.6 Kbp of A21 which was lacking in B14. This connection established the proximal end of the NB-370 fragment. A contig containing B27, B39, B8 and A77 was easily identified because of an uncharacteristically large number of *Eco*RI fragments. This contig also contained an unique *Sal*I site. Chromosome walking out from the end of D79 selected B22, which, in turn, was selected with a probe from B27. The overlapping *Eco*RI fragments are seen on Figure 30-2.

Cosmids, first selected by hybridization using the entire NB-370 fragment as a probe, containing different *Sal*I sites were identified by analyzing *Eco*RI and *Eco*RI plus *Sal*I digestion patterns from representative members of the contigs. Hybridizing small probes derived from each of these contigs back to PFG blots allowed accurate position assignment within NB-370. The relative positions of these contigs with in NB-370 could be estimated by using the *Sal*I position data. Walking out from the contigs connected the smaller contigs. Lambda phage clones 4-23 and 18-4 (Chen and Widger, 1993) were used to select cosmids at the distal end of NB-370 by hybridization. D40, E24, C57 and D84 were selected and walking from D40 both A93 and F59 were selected. Double digestion with *Eco*RI and *Sal*I or *Eco*RI and *Not*I established the order of small *Sal*I (SX-17 and SQ-43) and *Not*I (NR-23 and NS-12) genome fragments.

Conclusion

A gene encyclopedia has been constructed from overlapping cosmid clones covering about 90% of the *Synechococcus* PCC 7002 genome (Figure 30-3). A detailed, but, still partial, *Eco*RI restriction map of these clones has been built and should proven to be very useful for locating and subcloning many different genes. Linking cosmid clones have been identified for most of the *Not*I and *Sal*I marker sites of the genome, and these cosmids nail the corresponding contigs to those sites strengthening the map data. At present, there are 9 gaps in the cosmid map that sum up to a total of 290 kbp of DNA without representative cosmids. However, there are 90 cosmids in the genomic library that are not associated with contigs placed on the physical map and will be used to fill the gaps in future experiments. Gaps 1 and 3 are about 10–15 kbp in size and correspond to the DNA region containing the loci for the *rrn* operons. No cosmids have been

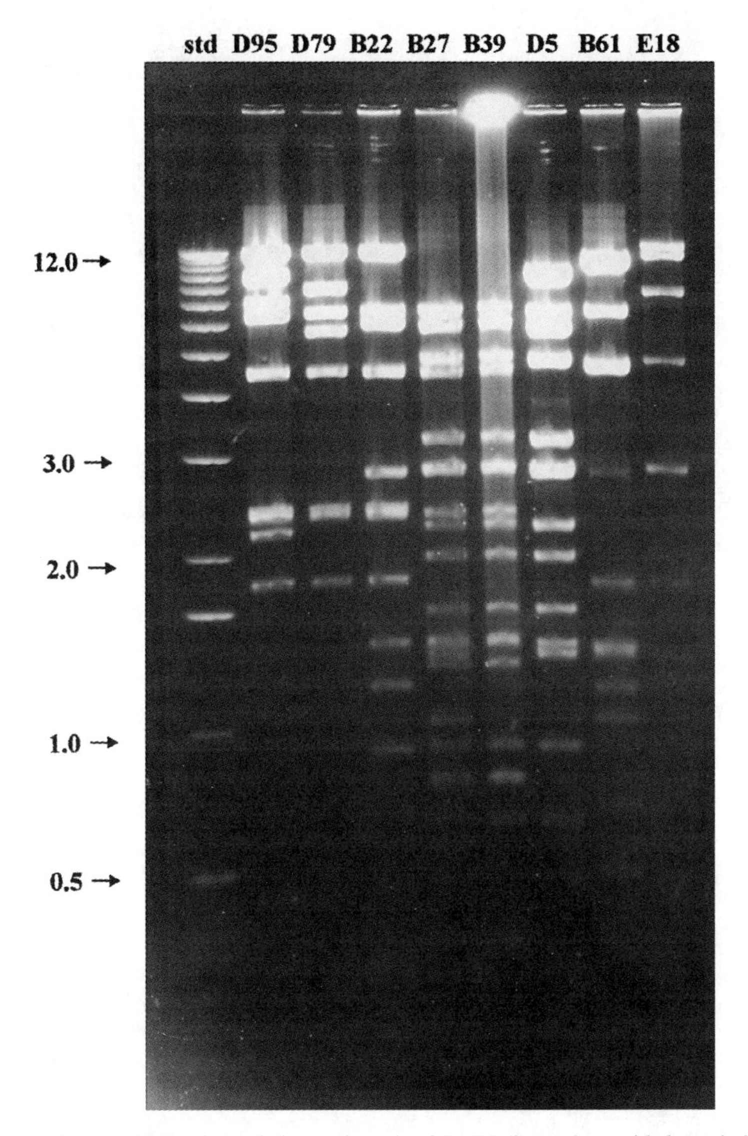

Figure 30-2. Field inversion gel electrophoresis of *Eco*RI digested cosmid clones indicating many common fragments. The field inversion gel was run by a homemade switch box consisting of a more simple circuit than reported (Lai *et al.*, 1991), a resistor was used to allow 60% of the voltage in the backward direction and switch times of 0.6 s forward and 0.2 s backward were used. Timing was controlled by an Artisan solid state Repeat Cycle Timer Model 4600. Lane std, contains 1 kbp DNA standard ladder, other lanes as labeled.

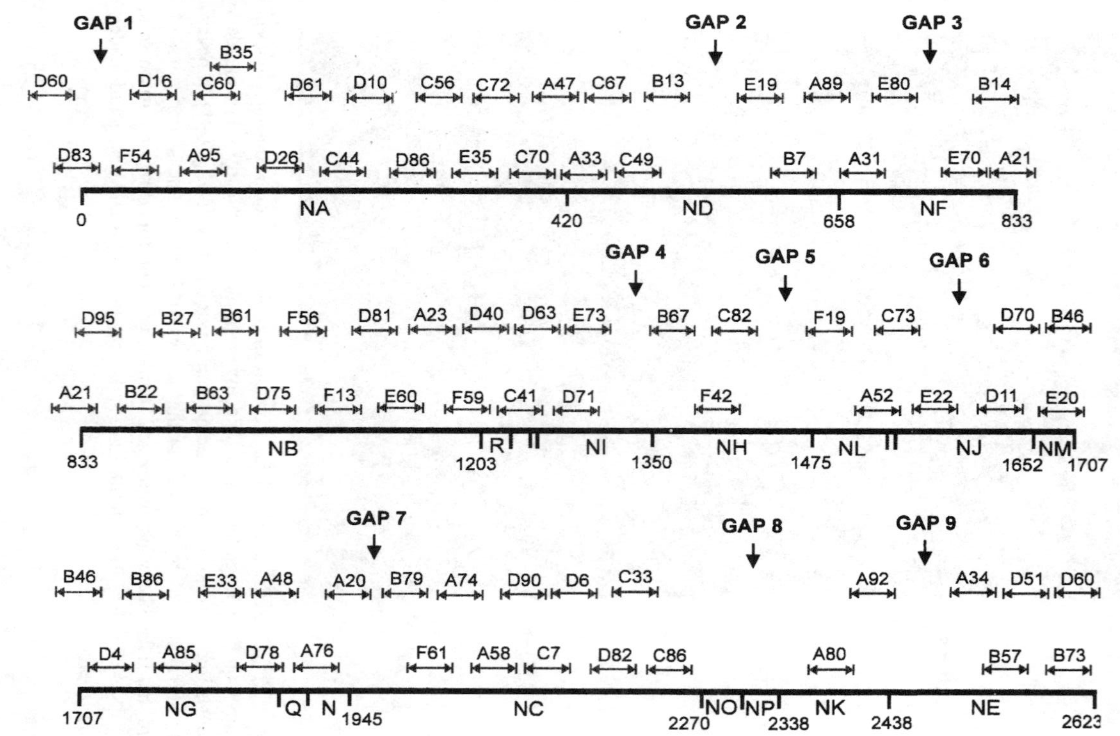

Figure 30-3. The minimal cosmid overlap map on the entire *Synechococcus* PCC 7002 genome. Several gaps are indicated. At the present time, about 90 percent of the genome is covered by contigs. The map indicates the position of the *Not*I sites and their fragment order (see section 3, 38).

identified in the library containing the *rrn* operons, so these gaps are not covered with cosmids. Instead, three *Eco*RI fragments for each one of the gaps 1 and 3, have been identified to contain the *rrn* operons and they will be used to fill the corresponding gaps. Gap 2 is approximately 30 kbp and gaps 4 and 5 in NI-NH region are about 10–15 kbp each and they should be easily filled by chromosome walking. Cosmids containing the NI-NH *Not*I site have not been identified and screening the remaining cosmids in the library for the presence of *Not*I sites should reveal clones that could fit into gap 4. In addition, there are several NI and NH selected cosmids not positioned on the map. Gaps 6 and 7 on NJ and NC are also both small less than 20 kbp each. Gaps 8 and 9 are the two largest gaps left to be covered with associated cosmids. Gap 8 is at least 50 kpb and extends over two *Not*I sites, NC/NO and NO/NP. The approach suggested for gap 4 could be employed for coverage of gap 8. Gap 9 on NE is about 50 kbp and chromosome walking with potential end fragments from the cosmids that flank this gap is again an option for covering this gap.

In conclusion, this map will be very useful in locating and ordering genes of interest as it was seen for *E. coli* (Kohara *et al.*, 1987). Most importantly, the completed cosmid map with its detailed *Eco*RI restriction fragment order will be used as the basis for random sample sequencing with the ultimate goal of a completed sequencing of the entire genome. Sampled sequencing has already been initiated in the laboratory, and for this approach *Eco*RI fragments of each cosmid from the minimal set covering the entire genome will be sequenced from both ends. This approach should identify many genes and their order on the genome. The genome sequence of the freshwater *Synechocystis* PCC 6803, a cyanobacterium moderately related to *Synechococcus* PCC 7002, is currently underway in Japan (Kaneko *et al.*, 1995). The phylogenetic comparisons of genes and genome organization between the two cyanobacteria and chloroplasts will be very useful for our understanding of unicellular cyanobacterial genome organization.

Acknowledgments

I would like to thank Drs. D. Bryant, S. Golden, S. Richter, L. Sherman and W. Vermass for hybridization probes. A special thanks to X. Tan for the generation of the cosmid library and M. Varguhese and H. Lakatos for assisting in several hybridization experiments. This research was supported by NIH GM46297 and a grant from the Texas Advanced Research Program.

References

Birkenbihl, R. P., and W. Vielmetter. 1989. Cosmid-derived map of *E. coli* strain BHB2600 in comparison to the map of strain W3110. *Nucl. Acids Res.* 17:5057–5068.

Chen, X., and W. R. Widger. 1993. The physical genome map of the cyanobacterium, *Synechococcus* PCC 7002. *J. Bacteriol.* 175:5106–5116.

Coulson, A., J. Sulston, S. Brenner, and J. Karn. 1986. Toward a physical map of the genome of the nematode *Caenorhabditis elegans. Proc. Natl. Acad. Sci. USA* 83:7821–7825.

Kohara, Y., K. Akiyama, K. Isono, 1987. The physical map of the whole *E. coli* chromosome: appllication on a new strategy for rapid analysis and sorting of a large genomic library. *Cell* 50:495–508.

Lai, E., K. Wang, N. Avdalovic, and L. Hood. 1991. Rapid restriction map constructions using a modified pWE15 cosmid vector and a robotic workstation. *Biotech.* 11:212–217.

Sankar, P., M. E. Hutton, R. A. VanBogelen, R. L. Clark, and F. C. Neidhardt. 1993. Expression analysis of cloned chromosomal segments of *Escherichia coli. J. Bacteriol.* 175:5145–5152.

Schluchter, W. M., J. Zhao, and D. A. Bryant. 1993. Isolation and characterization of the *ndh*F gene of *Synechococcus* sp. strain PCC 7002 and initial characterization of an interposon mutant. *J. Bacteriol.* 175:3343–3352.

Schluchter, W. M., and D. A. Bryant. 1992. Molecular characterization of ferredoxin-NADP+oxidoreductase in cyanobacteria: Cloning and sequence of the *pet*H gene of *Synechococcus* sp. PCC 7002 and studies on the gene product. *Biochem.* 31:3092–3102.

Sulston, J., F. Mallett, R. Staden, R. Durbin, T. Horsnell, and A. Coulson. 1988. Software for genome mapping by fingerprinting techniques. *CABIOS* 4:125–132.

Tan, X., M. Varghese, and W. R. Widger. 1994. A light-repressed transcript found in *Synechococcus* PCC 7002 is similar to a chloroplast-specific small subunit ribosomal protein and to a transcription modulator protein associated with sigma 54. *J. Biol. Chem.* 269:20905–20912.

Widger, W. R. 1991. The cloning and sequencing of *Synechococcus* sp. PCC 7002 *pet*CA operon: Implications for the cytochrome c-553 binding domain of cytochrome *f. Photosyn. Res.* 30:71–84.

PART B

Genomic Fingerprinting Methods

31

Fingerprinting Bacterial Genomes using Restriction Fragment Length Polymorphisms

David H. Demezas

Introduction

The identification of bacterial isolates to the strain level is often necessary for epidemiological and ecological studies to identify and track strains of infectious nature or monitor populations of environmental significance. Typically strains of bacteria have been differentiated on the basis of specific phenotypic traits. Techniques used to delineate strains of bacteria include biotyping, serotyping, bacteriophage typing, antibiotic resistance, or sodium dodecyl sulfate polyacrylamide gel electrophoresis (SDS-PAGE) patterns for epidemiological (see Pfaller, 1991 and references therein) and ecological studies (see Bottomley, 1992 and references therein). Though these approaches have proven to be useful, each method has inherent disadvantages (see Pfaller, 1991). In the past 10 years there has been an increase in the application of molecular biological techniques, including DNA fingerprinting, to the problems of differentiating and identifying strains of bacteria in the fields of epidemiology and microbial ecology. The advantage of DNA fingerprinting over phenotypic analysis is that the genotype, not the phenotype, is assayed; therefore it is a universally applicable approach for typing any species of bacteria.

There are two general approaches to DNA fingerprinting for typing bacterial strains. Analysis of restriction fragment length polymorphisms (RFLPs) detects DNA sequence variation by comparing the number and size of restriction fragments produced by digesting DNA with restriction endonuclease enzymes. Repetitive sequence based-polymerase chain reaction (rep-PCR) (Versalovic *et al.*, 1994), randomly amplified polymorphic DNA (RAPD) (Williams *et al.*, 1990) and the closely allied arbitrary priming-PCR (AP-PCR) (Welsh and McClelland, 1990) detect differently-sized multiple amplicons consisting of DNA sequences lying between the primers (i.e., amplification products). This chapter will focus on the application of RFLP analysis for DNA fingerprinting; the methodologies of RFLP analysis, and their application to strain typing will be discussed. The reader is referred to Chapters 33 and 34 for further discussion of the applications of PCR for DNA fingerprinting.

Principles

There are two basic approaches to RFLP analysis: 1) total genomic restriction patterns, including small and large restriction fragments and 2) hybridization patterns of genomic DNA probed with a DNA fragment to detect polymorphisms in a specific genomic region. The basic steps common to both approaches are 1) isolating genomic DNA, 2) digesting DNA with restriction enzyme(s), 3) separating DNA fragments based on size by gel electrophoresis, 4) staining gels to visualize the separated fragments, and 5) scoring for the presence or absence of restriction fragments. The salient features of each step will be discussed briefly in the following; the reader is referred to Sambrook *et al.,* (1989), Stackebrandt and Goodfellow (1991), and Ausubel *et al.,* (1992) for further details.

(1) Isolation of DNA from bacteria. Several approaches to isolating genomic DNA from bacteria have been published (Ausubel *et al.,* 1992; Stackebrandt and Goodfellow, 1991). Typically DNA is extracted from bacterial cells using modifications of Marmur's (1961) lysozyme/detergent method to lyse the cells, followed by an incubation with a nonspecific protease, and a series of phenol/chloroform/isoamyl alcohol extractions, prior to an alcohol precipitation of the nucleic acids. Various protocols have been developed to lyse the more recalcitrant Gram-positive bacteria efficiently. Pitcher *et al.* (1989) have described a method using the chaotrophic agent guanidium (iso)thiocyanate to lyse cells. A combination of an incubation with a murolytic enzyme (e.g., mutanolysin) followed by guanidium (iso)thiocyanate lysis Gram-positive bacteria effectively (Grimont and Grimont, 1991). The above approaches produce sufficiently pure DNA for further molecular biological enzymatic activities, e.g., restriction digestion, without additional purification steps, such as cesium chloride/ethidium bromide equilibrium centrifugation.

(2) Restriction digestion. A large number of restriction enzymes have been isolated from various species of bacteria (Roberts, 1988; Sambrook *et al.,* 1989) and are available commercially (e.g., New England Biolabs, Promega, or Boehringer Mannheim). Genomic DNA is digested to completion under specific conditions (e.g., salt concentration, or temperature) specified on the information sheets supplied with the restriction enzyme. Genomic DNA digested to completion yields a reproduc-

ible array of restriction fragments that is characteristic of the strain of bacteria.

(3) Agarose gel electrophoresis. Horizontal agarose gel electrophoresis is used commonly to fractionate restriction fragments based on their molecular weight (Ausubel *et al.,* 1992; Sambrook *et al.,* 1989; Stackebrandt and Goodfellow, 1991). The concentration of agarose chosen must be able to separate digested DNA in a wide range of molecular weights. Agarose gels of between 1 and 0.8% (w/v) are used routinely to separate DNA fragments of 0.7 to 20 kbp (Ausubel *et al.,* 1992; Sambrook *et al.,* 1989; Stackebrandt and Goodfellow, 1991). Although agarose is the standard medium used in for electrophoresis, polyacrylamide gels may be used for the separation of small fragments (Ausubel *et al.,* 1992; Sambrook *et al.,* 1989).

After electrophoresis the DNA is visualized by staining the gel with a dilute solution of ethidium bromide (0.5 mg/ml w/v) for a period of time. The stained gel can then be photographed using either a transmitted or reflected (incident) ultraviolet light source.

Source of Restriction Fragment Length Polymorphisms

Base substitutions within a restriction recognition site, deletions, or insertions can result in a change in the number and/or size of restriction fragments (see Figure 31-1). Base substitutions can cause a gain or loss of a restriction fragment(s). If a restriction site is lost, a new restriction fragment, equal to the sum of the two flanking restriction fragments, is created (see Figure 31-1). On the other hand, if a restriction site is gained two new restriction fragments, summing up to the size of the lost fragment, are generated. Generally, polymorphisms created by base substitutions are specific to the restriction enzyme(s) which cleave within the mutated restriction recognition site(s). Deletions and insertions affect the RFLPs of all restriction enzymes that cleave within the effected region. A deletion or insertion changes the size of the effected restriction fragment by the corresponding size of the deletion or insertion (see Figure 31-1). In a recent study, Hall (1994) determined the basis for RFLPs detected among strains of *Enterococcus faecalis.* Six of nine RFLPs observed for strains of *E. faecalis* were probably due to DNA rearrangements, two RFLPs were apparently due to point mutations and the cause of the last RFLP was undetermined.

Digestion of genomic DNA from bacteria with a restriction enzyme, that has a 6 base pair recognition site, may result in ≥1000 restriction fragments of various sizes, depending on the restriction enzyme used and the G + C content of the genome. The large number of bands makes it difficult to detect restriction fragment length polymorphisms when comparing closely-related strains. Two approaches

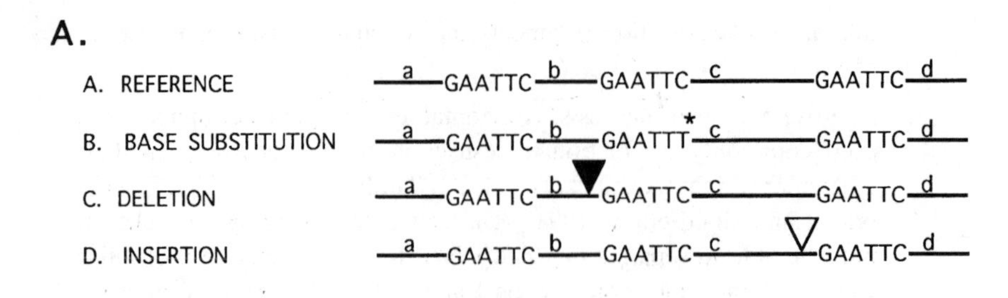

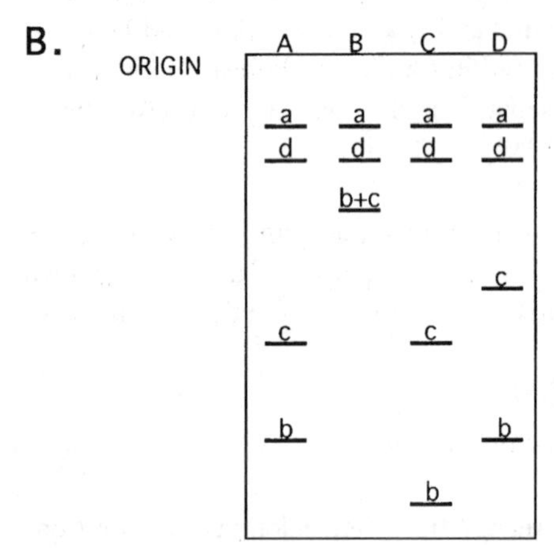

Figure 31-1. Illustration of various types of mutations (a) and their effects on digested DNA band patterns (b).
*indicates a base substitution, ▼ indicates a region with a deletion, and ▽ indicates a region with an insertion.

have been developed to simplify comparing closely-related strains: Southern blot hybridization, and pulsed field gel electrophoresis (see Chapters 24 to 26).

Southern Blotting and Hybridization

Southern (1975) published a method to detect specific DNA sequences within digested-genomic DNA separated by agarose gel electrophoresis. In the original procedure, digested DNA was electrophoresed, denatured *in situ* and transferred by capillary action from the agarose gel to nitrocellulose filters. The denatured, nitrocellulose filter-bound DNA was then hybridized to a radiolabeled DNA or RNA probe and the hybridization pattern detected by autoradiography. The resulting hybridization pattern provides a unique and reproducible fingerprint composed of relatively few bands that is much simpler to interpret than the complex banding

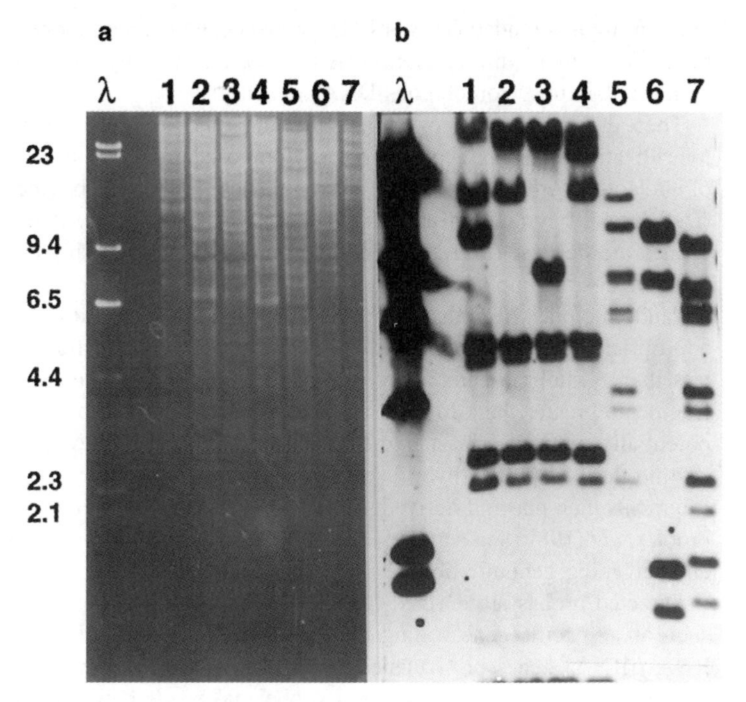

Figure 31-2. DNA fingerprint (a) and ribotype (b) patterns of *Hin*dIII-digested DNA isolated from strains of *Pseudomonas fluorescens* probed with a *P. fluorescens*-specific 23S rRNA gene (Festl *et al.,* 1986). λ, *Hin*dIII-digested phage λ DNA markers. Lanes 1 *P. fluorescens* (ATCC 948); 2, *P. fluorescens* type A (ATCC 17552); 3, *P. fluorescens* type A (ATCC 17397); 4, *P. fluorescens* type B (ATCC 17816), 5, *P. fluorescens* type B (ATCC 17467); 6, *P. fluorescens* type C (ATCC 17561); 7, *P. fluorescens* type C (ATCC 17400).

patterns of digested-genomic DNA after electrophoresis (cf. Fig. 31-2a and 2b). Southern transfer and hybridization procedures are described in detail elsewhere (Ausubel *et al.,* 1992; Sambrook *et al.,* 1989).

In recent years several modifications on the original procedure of Southern have been developed, including using nylon membranes (e.g., Hybond-N, Amersham) instead of nitrocellulose membranes. The advantage nylon offers over nitrocellulose is that nylon membranes are more durable than nitrocellulose and are less prone to tear after repeated hybridizations. Another significant advancement is the use of vacuum or electrophoretic blotting devices to transfer DNA from the gel to the nitrocellulose or nylon membrane. Usually the transfer can be completed in less than three hours compared with ≥18 hrs for capillary transfer.

(1) Types of nucleic acid probes. An essential element in Southern blot hybridization analysis is the nucleic acid probe. The probe can be

either a single-stranded DNA or RNA sequence, with a reporter moiety (e.g., ^{32}P, digoxigenin, or biotin, see below), that will hybridize with membrane-bound, homologous DNA sequences.

There are numerous possibilities for the type of probe used for hybridization analysis. Typically probes composed of 100s to thousands of nucleotides are used for RFLP analysis. These probes can be categorized broadly as random and directed probes (Table 31-1). Random probes are derived from a genomic recombinant library using either plasmid, cosmid or phage cloning vectors. Potential probes are selected as random clones from the genomic library and are screened for their appropriate specificity. Generally, plasmid clones contain fragments less than 5 kilobase pairs and cosmid and phage clones may contain fragments greater than 20 kilobase pairs. Cosmid-derived probes are potentially more discriminating than plasmid-derived probes because cosmid-derived probes hybridize with a larger number of restriction fragments than plasmid-derived probes. The reader is referred to Sambrook *et al.,* (1989) and Ausubel *et al.,* (1992) for details on constructing and screening genomic libraries (*see also* Chapter 28).

Directed probes are derived from a specific cloned DNA sequence such as a specific gene coding for an enzyme, conserved ribosomal RNA (rRNA) genes or virulence genes (Table 31-1). Knowledge of the DNA sequence of the directed probe makes their hybridization patterns readily interpretable with respect to the position of restriction sites within the probed region. Another class of probes are reiterated sequence probes (discussed in Chapter 32). There are a number of different types of reiterated sequences in the genomes of bacteria including transposons (Sherrat, 1989), insertion sequences (Galas and Chandler, 1989; *see* Chapters 4 and 20), and duplicated genes (e.g., *nif* genes; Quinto *et al.,* 1982; *see* Chapter 15). The copy numbers per genome can range from a few (*nif* genes; Quinto *et al.,* 1982; see Chapter 15) to greater than 15 (RSα reiterated sequence; Hahn and Hennecke, 1987; see Chapter 32).

(2) Preparation of labeled probes. Labeled-DNA probes are synthesized using one of the following methods: (a) end label (5′-OH terminus in DNA) with T4 polynucleotide kinase, (b) nick translation or (c) random priming (Sambrook *et al.,* 1989; Stackebrandt and Goodfellow, 1991). Traditionally DNA or RNA probes are radiolabeled with ^{32}P-nucleotides (i.e., ^{32}P-dATP or ^{32}P-dCTP) *in vitro.* Several nonradioactive labeling kits are available commercially (e.g., Genius System, Boehringer Mannheim or Photogene System, BRL). Rather than labeling the DNA probe with a radionucleotide, a chemically modified nucleotide containing a hapten (e.g., digoxigenin-dUTP, Boehringer Mannheim

Table 31-1. Examples of random, directed and reiterated sequence probes

Source of Probe	Reference
A. RANDOM PROBES	
Plasmid-cloned probes	
Mycobacterium avium complex	(McFadden *et al.*, 1987)
Pseudomonas species	(Brown *et al.*, 1990)
Cosmid-cloned probes	
Pseudomonas syringae	(Denny *et al.*, 1988)
Salmonella species	(Tompkins *et al.*, 1986)
Rhizobium leguminosarum bv. viciae	(Demezas *et al.*, 1991)
Xanthomonas campestris	(Lazo *et al.*, 1987)
Clavibacter michiganense subsp. *michiganense*	(Thompson *et al.*, 1989)
Bacillus mycoides	(Bell and Friedman, 1994)
B. DIRECTED PROBES	
Conserved protein-coding genes	
Nitrogenase *(nifHDK)* from *Rhizobium leguminosarum* bv. *trifolii*	(Schofield *et al.*, 1983)
Glutamine synthetase I (*gln*A) from *Bradyrhizobium japonicum*	(Stanley *et al.*, 1985)
Nitrite reductase *(nir)* cloned from *Pseudomonas stutzeri*	(Smith and Tiedje, 1992)
PCP-4-monooxygenase (*pcp*B) cloned from *Flavobacterium* sp.	(Karlson *et al.*, 1995)
Conserved rRNA genes	
23S rRNA from *Micrococcus luteus*	(Regensburger *et al.*, 1988)
16S rRNA	(Barry *et al.*, 1990)
16S/23S rRNA	(Grimont and Grimont, 1986)
Virulence genes	
Coagulase gene from *Staphylococcus aureus*	(Goh *et al.*, 1992)
Virulence and hypersensitive response gene *(hrp)* from *Pseudomonas solanacearum*	(Cook *et al.*, 1989)
Exotoxin A (*tox*A) from *Pseudomonas aeruginosa*	(Ogle *et al.*, 1987)
Toxin gene *(tox)* from *Corynebacterium diphtheriae*	(Groman *et al.*, 1983)
C. REITERATED SEQUENCE PROBES	
Insertion sequences	
IS6110 from *Mycobacterium tuberculois*	(van Embden *et al.*, 1993)
ISRm1 from *Rhizobium meliloti*	(Wheatcroft and Watson, 1988)
ISRm3 and ISRm5 from *R. meliloti*	(Barran *et al.*, 1994)
IS elements from *Escherichia coli*	(Nass *et al.*, 1995)
Repeated sequences	
RSα and other RS's from *Bradyrhizobium japonicum*	(Hahn and Hennecke, 1987)
Hyperreiterated sequences from *B. japonicum*	(Rodriguez-Quinones *et al.*, 1992)
Repeated sequences from *Bordetella pertussis*	(Alsheikhly and Loefdahl, 1989)

or biotin-dATP, BRL) is incorporated into the probe using standard probe labeling procedures.

There are several advantages of nonradioactive labeled DNA probes over radioactive labeled DNA probes. Nonradioactive probes do not require the licensing, expensive safety equipment (e.g., plexiglass shields or Geiger counters) or precautions required while working with radiolabeled probes. Nonradioactive probes have a shelf life of up to a year if stored at –20°C, while radiolabeled probes have a much shorter shelf life. Optimal exposure times for the detection of nonradioactive hybrids by x-ray film range from several minutes to less than an hour; for ^{32}P-labeled probes, the exposure time varies from one up to 10 days depending on the specific activity of the probe and the amount of target DNA. These advantages have to be balanced against the issue of sensitivity; though nonradioactive probes are reported to be as sensitive as radiolabeled probes (Boehringer Mannheim, The genius system user's guide for filter hybridization).

(3) Hybridization procedures. There are numerous variations on the hybridization procedure published in the literature. The differences between hybridization procedures may be summarized as follows (adapted from Sambrook *et al.,* 1989):

- Hybridization in an aqueous solution at 68°C versus in 50% formamide at 42°C.
- Continuous shaking or stationary incubation during hybridization.
- Use of one of several blocking reagents (e.g., Blotto or Denhardt's) to reduce non-specific attachment of the probe to solid surfaces.
- The stringency of hybridization and post hybridization washes.

There is no single universal hybridization procedure applicable to all probes and species of bacteria. The prime consideration are those factors that affect the melting temperature (T_m) of a hybrid formed between the probe and its target sequence. The T_m may be defined as the mid-point in the transition from duplexed DNA to denatured DNA. There are a number of critical parameters that affect T_m including ionic strength of the hybridization solution, percentage mismatched base pairs, the presence of formamide, probe length and the percent G + C in both the probe and target DNA (Stahl and Amann, 1991). Specific parameters of the hybridization conditions (e.g., hybridization temperature and ionic strength of the hybridization solution) can be adjusted to achieve the level of specificity desired. A thorough discussion of the parameters that control the hybridization reactions are beyond the scope of this chapter; there are several excellent reviews on this subject (Sambrook *et al.,* 1989; Stahl and Amann, 1991; Wahl *et al.,* 1987).

(4) Detection of hybridization signal. The DNA fragments that hybridize with a radiolabeled DNA probe are detected using autoradiography (Ausubel *et al.,* 1992; Sambrook *et al.,* 1989). The genomic restriction fragments that hybridize with a digoxigenin (or biotin)-labeled probe are detected with an anti-digoxigenin (or biotin)-alkaline phosphatase conjugate and either a chemiluminescent substrate such as Lumi-Phos (Lumigen, Inc., Detroit MI), which can be detected by using x-ray film, or a chromogenic substrate such as NBT (nitroblue tetrazolium salt solution) and X-phosphate (5-bromo-4-chloro-3-indolyl phosphate), which deposits a purple color on the membrane.

Pulsed Field Gel Electrophoresis

Pulsed field electrophoresis (PFGE) is a technique capable of resolving DNA molecules from less than 25 kilobase pairs to more than 1,000 kilobase pairs (Anand, 1986; Birren *et al.,* 1989; Lai *et al.,* 1989; Schwartz and Cantor, 1984). Bacterial genomes cleaved with restriction enzymes, that have rare recognition sequences, yield a reasonable number (10-30) of large molecular weight (20-1500 kilobase pairs) restriction fragments that can be resolved by pulsed field gel electrophoresis. The resulting band patterns are much simpler to analyze than the complex band patterns of genomic DNA digested with standard restriction enzymes (e.g., *Eco*RI). The reader is referred to Chapters 24–26 which provide a detailed discussion of pulsed field electrophoresis.

Application of RFLP Analysis for Genomic Fingerprinting

In the past 5 years there has been greater than 200 publications describing the use of RFLP analysis for typing bacterial isolates. Genomic fingerprinting by RFLP analysis will continue to be one of the major typing techniques in epidemiologic and ecologic studies.

Restriction Patterns of Genomic DNA

Restriction patterns (also referred to as BRENDA; Owen, 1989) is the simplest method to type isolates of bacteria. Genomic DNA is extracted, digested with a restriction enzyme(s), electrophoresed on an agarose gel and the band patterns compared; no DNA probes or specialized electrophoresis apparatuses are required. Restriction patterns have been used to type strains from a variety of different species of bacteria, e.g., *Spiroplasma citri* (Bove and Saillard, 1979), *Rhizobium* (Mielenz *et al.,* 1979), *Xanthomonas campestris* (Lazo *et al.,* 1987), *Campylobacter* (Bryner *et al.,* 1988), *Staphylococcus aureus* (Jordens and Hall, 1988),

Enterococcus (Hall *et al.*, 1992), *Bacillus subtilus* (Istock *et al.*, 1992), *Pasteurella haemolytica* (Murphy *et al.*, 1993), *Staphlococcus schleiferi* (Grattard *et al.*, 1993) and other bacteria (see Owen, 1989). Restriction patterns are more discriminating than other conventional typing methods based on phenotypic traits such as serotyping (Kakoyiannis *et al.*, 1984; Ogle *et al.*, 1987), antibiotic resistance profiles (Jordens and Hall, 1988) and plasmid profiles (Jordens and Hall, 1988). The difficulties with restriction patterns for typing isolates are i) the complexity of the patterns makes comparisons between isolates electrophoresed on separate gels difficult and ii) the presence or absence of plasmids may alter the banding pattern (Jordens and Hall, 1988).

Southern Hybridization Analysis

Genomic fingerprinting by Southern blot hybridization using nucleic acid probes has become a standard method to type strains of bacteria and rRNA genes have become popular nucleic acid probes for epidemiological and ecological studies (Grimont and Grimont, 1991; Schmidt, 1994; see Chap. 21). Typing strains with rRNA genes as nucleic acid probes is referred to as ribotyping. Ribosomal RNA genes are present in multiple copies, ranging from a few (mycobacteria) to 11 copies (*Bacillus* spp.); therefore several restriction fragments containing rRNA genes will hybridize with rRNA probes. Furthermore, rRNA genes are conserved among different species of bacteria (Woese, 1987). The fact that the rRNA genes are conserved makes it possible to use end-labeled 16S and 23S rRNA or labeled-cloned rRNA genes from any bacterium as probes; though the hybridization patterns obtained using different probes are not necessarily identical (see Grimont and Grimont, 1991). Grimont and Grimont (1986) used end-labeled 16S and 23S rRNA from *Escherichia coli* to detect rRNA gene restriction patterns in 41 species of phylogenetically distant bacteria and concluded that rRNA gene restriction patterns were potentially useful for distinguishing between species and strains of a species. Subsequently, ribotyping has been used to type strains of many bacterial species, e.g., *Haemophilus influenzae* (Irino *et al.*, 1988), *Listeria monocytogenes* (Baloga and Harlander, 1991), *Neisseria meningitidis* (Woods *et al.*, 1992), *Vibrio cholera* (Popovic *et al.*, 1993), *Pasteurella haemolytica* (Murphy *et al.*, 1993), *Staphylococcus schleiferi* (Grattard *et al.*, 1993) and others (see Owen, 1989; Schmidt, 1994). Various other nucleic acid probes, isolated from specific bacterial species, have been used for typing strains of bacteria by Southern hybridization (see Table 31-1 for examples). The limitations of these probes is that they generally will not hybridize with other genomic species of bacteria (Grimont and Grimont, 1991).

Pulsed Field Gel Electrophoresis

Pulsed field gel electrophoresis is an alternative approach to produce DNA fingerprints composed of relatively few bands and does not require nucleic acid probes. Pulsed field gel electrophoresis has been used to type isolates of various species of bacteria, e.g., *Pseudomonas aeruginosa* (Struelens *et al.*, 1993b), *Streptococcus pneumoniae* (Lefevre *et al.*, 1993), *Staphylococcus aureus* (Struelens *et al.*, 1993a), *Escherichia coli* (Bohm and Karch, 1992), *Enterococcus faecium* (Miranda *et al.*, 1991), *Pseudomonas syringae* and *Xanthomonas campestris* (Cooksey and Graham, 1989). Pulsed field gel electrophoresis may be more discriminating than conventional Southern hybridization analysis with randomly chosen cosmid clones (Cooksey and Graham, 1989) or rRNA genes (Prevost *et al.*, 1992). Pulsed field gel electrophoresis detects global variation in chromosome structure due to insertions or deletions and local variation in nucleotide sequences within restriction sites. In contrast Southern hybridization only detects variation within restriction sites and particular regions of the chromosome (i.e., probed regions). The limitations of pulsed field gel electrophoresis are that it requires specialized electrophoresis apparatuses and considerable experience to achieve reproducible results.

Acknowledgments

I wish to thank M. N. Vijayakumar for his critical reading of this manuscript.

References

Alsheikhly, A. R., and S. Loefdahl. 1989. Identification of a DNA fragment in the genome of *Bordetella pertussis* carrying repeated DNA sequences also present in other *Bordetella* species. *Microb. Pathog.* 6:193–201.

Anand, R. 1986. Pulsed field gel electrophoresis: a technique for fractionating large DNA molecules. *Trend Genet.* 19:278–283.

Ausubel, F. M., R. Brent, R. E. Kingston, D. D. Moore, J. G. Seidman, J. A. Smith, and K. Struhl. 1992. *Current protocols in molecular biology.* Greene Publishing & Wiley-Interscience, New York.

Baloga, A. O., and S. K. Harlander. 1991. Comparison of methods for discrimination between strains of *Listeria monocytogenes* from epidemiological surveys. *Appl. Environ. Microbiol.* 57:2324–2331.

Barran, L. R., E.S.P. Bromfield, S. Laberge, and R. Wheatcroft. 1994. Insertion sequence (IS) hybridization supports classification of *Rhizobium meliloti* by phase typing. *Mol. Ecol.* 3:267–270.

Barry, T., R. Powell, and F. Gannon. 1990. A general method to generate DNA probes for microorganisms. *Bio/technology* 8:233–236.

Bell, J. A., and S. B. Friedman. 1994. Genetic structure and diversity within local populations of *Bacillus mycoides. Evolution* 48:1698–1714.

Birren, B. W., E. Lai, L. Hood, and M. I. Simon. 1989. Pulsed field gel electrophoresis techniques for separating 1- to 50-kilobase DNA fragments. *Anal. Biochem.* 177:282–286.

Bohm, H., and H. Karch. 1992. DNA fingerprinting of *Escherichia coli* O157:H7 strains by pulsed-field gel electrophoresis. *J. Clin. Microbiol.* 30:2169–2172.

Bottomley, P. J. 1992. Ecology of *Bradyrhizobium* and *Rhizobium*. In Biological Nitrogen Fixation, G. Stacey, R. Burris and H. J. Evans, eds. pp. 293–348. Chapman and Hall, New York.

Bove, J. M., and C. Saillard. 1979. Cell biology of spiroplasmas. In *The mycoplasmas,* R. F. Whitcomb and J. G. Tully, eds. pp. 85–153. Academic Press, New York.

Brown, G., Z. Khan, and R. Lifshitz. 1990. Plant growth promoting rhizobacteria: strain identification by restriction fragment length polymorphisms. *Can. J. Microbiol.* 36:242–248.

Bryner, J. H., I. V. Wesley, and L. A. Pollet. 1988. Restriction enzyme analysis of *Campylobacter* genomic DNA. *Ann. N.Y. Acad. Sci.* 529:279–282.

Cook, D., E. Barlow, and L. Sequeira. 1989. Genetic diversity of *Pseudomonas solanacearum*: detection of restriction fragment length polymorphisms with DNA probes that specify virulence and the hypersensitive response. *Mol. Plant-Microbe Interact.* 2:113–121.

Cooksey, D. A., and J. H. Graham. 1989. Genomic fingerprinting of two pathovars of phytopathogenic bacteria by rare-cutting restriction enzymes and field inversion gel electrophoresis. *Phytopathology* 79:745–750.

Demezas, D. H., T. B. Reardon, J. M. Watson, and A. H. Gibson. 1991. Genetic diversity among *Rhizobium leguminosarum* bv. trifolii strains revealed by allozyme and restriction fragment length polymorphism analyses. *Appl. Environ. Microbiol.* 57:3489–3495.

Denny, T. P., M. N. Gilmour, and R. K. Selander. 1988. Genetic diversity and relationships of two pathovars of *Pseudomonas syringae. J. Gen. Microbiol.* 134:1949–1960.

Festl, H., W. Ludwig, and K.-H. Schleifer. 1986. DNA hybridization probe for the *Pseudomonas fluorescens* group. *Appl. Environ. Microbiol.* 52:1190–1194.

Galas, D. J., and M. Chandler. 1989. Bacterial insertion sequences. In *Mobile DNA,* D. E. Berg and M. M. Howe, eds. pp. 109–162. American Society for Microbiology, Washington, D.C.

Goh, S. H., S. K. Byrne, J. L. Zhang, and A. W. Chow. 1992. Molecular typing of *Staphylococcus aureus* on the basis of coagulase gene polymorphisms. *J. Clin. Microbiol.* 30:1642–1645.

Grattard, F., J. Etienne, B. Pozzetto, F. Tardy, O. G. Gaudin, and J. Fleurette. 1993. Characterization of unrelated strains of *Staphylococcus schleiferi* by using ribosomal DNA fingerprinting, DNA restriction patterns and plasmid profiles. *J. Clin. Microbiol.* 31:812–818.

Grimont, F., and P.A.D. Grimont. 1991. DNA fingerprinting. In *Nucleic acid techniques in bacterial systematics,* E. Stackebrandt and M. Goodfellow, eds. pp. 249–279. John Wiley & Sons, New York.

Grimont, F., and P.A.D. Grimont. 1986. Ribosomal ribonucleic acid gene restriction patterns as potential taxonomic tools. *Ann. Inst. Pasteur/Microbiol.* 137B:165–175.

Groman, N., N. Cianciotto, M. Bjorn, and M. Rabin. 1983. Detection and expression of DNA homologous to the *tox* gene in nontoxinogenic isolates of *Corynebacterium diphtheriae. Infect. Immun.* 42:48–56.

Hahn, M., and H. Hennecke. 1987. Mapping of a *Bradyrhizobium japonicum* DNA region carrying genes for symbiosis and an asymmetric accumulation of reiterated sequences. *Appl. Environ. Microbiol.* 53:2247–2252.

Hall, L.M.C. 1994. Are point mutations or DNA rearrangements responsible for the restriction fragment length polymorphisms that are used to type bacteria? *Microbiology* 140:197–204.

Hall, L.M.C., B. Duke, M. Guiney, and R. Williams. 1992. Typing of *Enterococcus* species by DNA restriction fragment analysis. *J. Clin. Microbiol.* 30:915–919.

Irino, K., F. Grimont, I. Casin, P.A.D. Grimont, and T.B.P.F.S. Group. 1988. rRNA gene restriction patterns of *Haemophilus influenzae* biogroup aegyptius strains associated with Brazilian purpuric fever. *J. Clin. Microbiol.* 26:1535–1538.

Istock, C. A., K. E. Duncan, N. Ferguson, and X. Zhou. 1992. Sexuality in a natural population of bacteria—*Bacillus subtilis* challenges the clonal paradigm. *Mol. Ecol.* 1:95–103.

Jordens, J. Z., and L.M.C. Hall. 1988. Characteristics of methicillin-resistant *Staphylococcus aureus* isolates by restriction endonuclease digestion of chromosomal DNA. *J. Med. Microbiol.* 27:117–123.

Kakoyiannis, C. K., P. J. Winter, and R. B. Marshall. 1984. Identification of *Campylobacter coli* isolates from animals and humans by bacterial restriction endonuclease DNA analysis. *Appl. Environ. Microbiol.* 48:545–549.

Karlson, U., F. Rojo, J. D. van Elsas, and E. Moore. 1995. Genetic and serological evidence for the recognition of four pentachlorophenol-degrading bacterial strains as a species of the genus *Sphingomonas. Syst. Appl. Microbiol.* 18:539–548.

Lai, E., B. W. Birren, S. M. Clark, M. I. Simon, and L. Hood. 1989. Pulsed field gel electrophoresis. *BioTechniques* 7:34–42.

Lazo, G. R., R. Roffey, and D. W. Gabriel. 1987. Pathovars of *Xanthomonas campestris* are distinguishable by restriction fragment-length polymorphism. *Int. J. Syst. Bacteriol.* 37:214–221.

Lefevre, J. C., G. Faucon, A. M. Sicard, and A. M. Gasc. 1993. DNA fingerprinting of *Streptococcus pneumoniae* strains by pulsed-field gel electrophoresis. *J. Clin. Microbiol.* 31:2724–2728.

Marmur, J. 1961. A procedure for the isolation of deoxyribonucleic acid from microorganisms. *J. Mol. Biol.* 3:208–218.

McFadden, J. J., P. D. Butcher, J. Thompson, R. Chiodini, and J. Hermon-Taylor. 1987. The use of DNA probes identifying restriction-fragment-length polymorphisms to examine the *Mycobacterium avium* complex. *Mol. Microbiol.* 1:283–291.

Mielenz, J. R., L. E. Jackson, F. O'Gara, and K. T. Shanmugam. 1979. Fingerprinting bacterial chromosomal DNA with restriction endonuclease *Eco*RI: comparison of *Rhizobium* spp. and identification of mutants. *Can. J. Microbiol.* 25:803–807.

Miranda, A. G., K. V. Singh, and B. E. Murray. 1991. DNA fingerprinting of *Enterococcus faecium* by pulsed-field gel electrophoresis may be a useful epidemiologic tool. *J. Clin. Microbiol.* 29:2752–2757.

Murphy, G. L., L. C. Robinson, and G. E. Burrows. Restriction endonuclease analysis and ribotyping differentiate *Pasteurella haemolytica* serotype A1 isolates from cattle within a feedlot. *J. Clin. Microbiol.* 31:2303–2308.

Nass, T., M. Blot, W. M. Fitch, and W. Arber. 1995. Dynamics of IS-related genetic rearrangements in resting *Escherichia coli. Mol. Biol. Evol.* 12:198–207.

Ogle, J. W., J. M. Janda, D. E. Woods, and M. L. Vasil. 1987. Characterization and use of a DNA probe as an epidemiological marker for *Pseudomonas aeruginosa. The J. Infect. Dis.* 155:119–126.

Owen, R. J. 1989. Chromosomal DNA fingerprinting—a new method of species and strain identification applicable to microbial pathogens. *J. Med. Microbiol.* 30:89–99.

Pfaller, M. A. 1991. Typing methods for epidemiologic investigation. In *Manual of clinical microbiology,* A. Balows, J. Hausler, W. J., K. L. Hermann, H. D. Isenberg and H. J. Shadomy, eds. pp. 171–182. American Society for Microbiology, Washington, D.C.

Pitcher, D. G., N. A. Saunders, and R. J. Owen. 1989. Rapid extraction of bacterial genomic DNA with guanidium thiocyanate. *Lett. Appl. Microbiol.* 8:151–156.

Popovic, T., C. Bopp, O. Olsvik, and K. Wachsmuth. 1993. Epidemiologic application of a standardized ribotype scheme for *Vibrio cholera* O1. *J. Clin. Microbiol.* 31:2474–2482.

Prevost, G., B. Jaulhac, and Y. Piemont. 1992. DNA fingerprinting by pulsed-field gel electrophoresis is more effective than ribotyping in distinguishing among methicillin-resistant *Staphylococcus aureus* isolates. *J. Clin. Microbiol.* 30:967–973.

Quinto, C., H. de la Vega, M. Flores, L. Fernandez, C. Ballado, G. Soberon, and R. Palacios. 1982. Reiteration of nitrogen fixation gene sequences in *Rhizobium phaseoli. Nature* (London) 299:724–726.

Regensburger, A., W. Ludwig, and K.-H. Schleifer. 1988. DNA probes with different specificities from a cloned 23S rRNA gene of *Micrococcus luteus. J. Gen. Microbiol.* 134:1197–1204.

Roberts, R. J. 1988. Restriction enzymes and their isoschizomers. *Nucl. Acids Res.* 16:r271.

Rodriguez-Quinones, F., A. K. Judd, M. J. Sadowsky, R.-L. Liu, and P. B. Cregan. 1992. Hyperreiterated DNA regions are conserved among *Bradyrhizobium japonicum* serocluster 123 strains. *Appl. Environ. Microbiol.* 58:1878–1885.

Sambrook, J., E. F. Fritsch, and T. Maniatis. 1989. *Molecular cloning:A laboratory manual* Cold Spring Harbor: Cold Spring Harbor Laboratory Press.

Schmidt, T. M. 1994. Fingerprinting bacterial genomes using ribosomal RNA genes and operons. *Methods Mol. Cell. Biol.* 5:3–12.

Schofield, P. R., M. A. Djordjevic, B. G. Rolfe, J. Shine, and J. M. Watson. 1983. A molecular linkage map of nitrogenase and nodulation genes in *Rhizobium trifolii. Mol. Gen. Genet.* 192:459–465.

Schwartz, D. C., and C. R. Cantor. 1984. Separation of yeast chromosome-sized DNAs by pulsed field gradient gel electrophoresis. *Cell* 37:67–75.

Sherrat, D. 1989. Tn3 and other related transposable elements: site-specific recombination and transposition. In *Mobile DNA,* D. E. Berg and M. M. Howe, eds. pp. 163–184. American Society for Microbiology, Washington, D.C.

Smith, G. B., and J. M. Tiedje. 1992. Isolation and characterization of nitrite reductase gene and its use as a probe for denitrifying bacteria. *Appl. Environ. Microbiol.* 58:376–384.

Southern, E. M. 1975. Detection of specific sequences among DNA fragments separated by gel electrophoresis. *J. Mol. Biol.* 98:503–517.

Stackebrandt, E., and M. Goodfellow. 1991. *Nucleic acid techniques in bacterial systematics.* John Wiley & Sons, New York.

Stahl, D. A., and R. Amann. 1991. Development and application of nucleic acid probes. In *Nucleic acid techniques in bacterial systematics,* E. Stackebrandt and M. Goodfellow, eds. pp. 205–248. John Wiley & Sons, New York.

Stanley, J., G. G. Brown, and D.P.S. Verma. 1985. Slow-growing *Rhizobium japonicum* comprises two highly divergent symbiotic types. *J. Bacteriol.* 163:148–154.

Struelens, M. J., R. Bax, A. Deplano, W.G.V. Quint, and A. van Belkum. 1993a. Concordant clonal delineation of methicillin-resistant *Staphylococcus aureus* by macrorestriction analysis and polymerase chain reaction genome fingerprinting. *J. Clin. Microbiol.* 31:1964–1970.

Struelens, M. J., V. Schwam, A. Deplano, and D. Baran. 1993b. Genome macrorestriction analysis of diversity and variability of *Pseudomonas aeruginosa* strains infecting cystic fibrosis patients. *J. Clin. Microbiol.* 31:2320–2326.

Thompson, E., J. V. Leary, and W.W.C. Chun. 1989. Specific detection of *Clavibacter michiganense* subsp. *michiganense* by a homologous DNA probe. *Phytopathology* 79:311–314.

Tompkins, L. S., N. Troup, A. Labigne-Roussel, and M. L. Cohen. 1986. Cloned, random chromosomal sequences as probes to identify *Salmonella* species. *J. Infect. Dis.* 154:156–162.

van Embden, J. D. A., M. D. Cave, J. T. Crawford, J. W. Dale, K. D. Eisenach, B. Gicquel, P. Hermans, C. Martin, R. McAdam, T. M. Shinnick, and P. M. Small. 1993. Strain identification of *Mycobacterium tuberculosis* by DNA fingerprinting: recommendations for a standardized methodology. *J. Clin. Microbiol.* 31:406–409.

Versalovic, J., M. Schneider, F. J. de Bruijn, and J. R. Lupski. 1994. Genomic fingerprinting of bacteria using repetitive sequence-based polymerase chain reaction. *Methods Mol. Cell. Biol.* 5:25–40.

Wahl, G. M., S. L. Berger, and A. R. Kimmel. 1987. Molecular hybridization of immobilized nucleic acids: theoretical concepts and practical considerations. *Methods Enzymol.* 152:399–406.

Welsh, J., and M. McClelland. 1990. Fingerprinting genomes using PCR with arbitrary primers. *Nucl. Acids Res.* 18:7213–7218.

Wheatcroft, R., and R. J. Watson. 1988. A positive strain identification method for *Rhizobium meliloti. Appl. Environ. Microbiol.* 54:574–576.

Williams, J.G.K., A. R. Kubelik, K. J. Livak, J. A. Rafalski, and S. V. Tingey. 1990. DNA polymorphisms amplified by arbitrary primers are useful as genetic markers. *Nucl. Acids Res.* 18:6531–6535.

Woese, C. R. 1987. Bacterial evolution. *Microbiol. Rev.* 51:221–271.

Woods, T. C., L. O. Helsel, B. Swaminathan, W. F. Bibb, R. W. Pinner, G. B.G., S. F. Collins, S. H. Waterman, M. W. Reeves, D. J. Brenner, and C. V. Broome. 1992. Characterization of *Neisseria meningitidis* serogroup C by multilocus enzyme electrophoresis and ribosomal DNA restriction profiles (ribotyping). *J. Clin. Microbiol.* 30:132–137.

32

Use of Endogenous Repeated Sequences to Fingerprint Bacterial Genomic DNA

Michael J. Sadowsky and Hor-Gil Hur

Various genomic DNA fingerprinting methods have been used to investigate epidemiologic, taxonomic, and phylogenetic relationships among microorganisms. While initial studies used classical restriction enzyme-generated DNA fingerprints of bacteria for the epidemiological analysis of nosocomial infections (Kaper *et al.,* 1982; Kristiansen *et al.,* 1986; Kuijper *et al.,* 1987; Langenberg *et al.,* 1986; Skjoid *et al.,* 1987; Tompkins *et al.,* 1987), the technique has also found great application in the epidemiological and taxonomic analysis of yeast (Panchal *et al.,* 1987; Scherer and Stevens, 1987), mycoplasmas (Chandler *et al.,* 1982; Ruland *et al.,* 1990), fungi (Koch *et al.,* 1991), viruses (Buchman *et al.,* 1978; Christensen *et al.,* 1987), several diverse bacterial species (Langenberg *et al.,* 1986; Ramos and Harlander, 1990), and humans (Gill *et al.,* 1987). In addition to medically important organisms, DNA fingerprinting techniques have also been used to study the taxonomic relatedness of agriculturally important microorganisms. These organisms include bacterial and fungal pathogens (Lazo *et al.,* 1987) as well as plant symbionts (Brown *et al.,* 1989; Glynn *et al.,* 1985; Kaijalainen and Lindstrom, 1989; Mielenz *et al.,* 1979; Sadowsky *et al.,* 1987; and Schmidt *et al.,* 1986). While a majority of the techniques can be used for the examination of both prokaryotic and eukaryotic microorganisms, this chapter concentrates on fingerprinting bacterial DNA.

DNA fingerprinting techniques can be divided into four categories based on the methods used to visualize similarities or differences between bacterial strains. Two of the techniques require the enzymatic cleavage of DNA using restriction endonucleases, and electrophoretic separation of the resulting DNA fragments. The two techniques then differ with respect to the means of examining the resultant restriction fragments: classical DNA fingerprinting uses direct ethidium bromide staining of the resultant gels and visualizes all restriction fragments, whereas restriction fragment length polymorphism (RFLP)-based analyses use Southern transfer of DNA and nucleic acid probes, which selectively hybridize to a limited number of restriction fragments (*see* Chapter 31). The third technique, rep-PCR DNA fingerprinting (Versalovic *et al.,* 1991; de Bruijn, 1992; Versalovic *et al.,* 1994; *see* Chapter 34) uses the polymerase chain reaction and primers to amplify specific portions of the microbial genome which are subsequently visualized following electrophoresis and ethidium bromide staining. The fourth technique, referred to by different authors as randomly amplified polymorphic DNA (RAPD) (Williams *et al.,* 1990), arbitrarily primed polymerase chain reaction (AP-PCR) (Welsch and McClelland, 1990) or DNA amplification fingerprinting (DAF) (Caetano-Anollés *et al.,* 1991), is similar in some respects to rep-PCR, however, arbitrary oligonucleotide primers are used to randomly initiate amplification of multiple target sites in the genome (Caetano-Anollés *et al.,* 1994). Regardless of which technique is used, the resulting banding patterns are generally unique to one or a few strains of a particular microbe and as such, can serve as a "fingerprint" for strain identification or analysis of populations. All three techniques are strictly genotype-based methods and as such, alleviate problems associated with phenotypic analyses in that DNA expression is not required. Consequently, the techniques work well on biochemically ill-defined and physiologically and morphologically nondescript microbes. Organisms having indistinguishable banding patterns can be regarded as being identical or near-identical. It should be noted, however, that while DNA fingerprints are stable over many generations of microbial growth (Ramos and Harlander, 1990; Scherer and Stevens, 1987; Schneider and de Bruijn, 1996), they are susceptible to changes caused by curing or rearrangement of indigenous plasmids, prophages, and the potential loss or rearrangement of endogenous repeated sequences.

Endogenous repetitive DNA sequences have found wide application for the fingerprinting of prokaryotic genomes. Bacterial genomes contain a variety of repetitive DNA sequences (Table 32-1). These repetitive elements typically are comprised of duplicated genes, such as rRNA, tRNA, and members of the *rhs* gene family (Lin *et al.,* 1984; Sadosky *et al.,* 1989; *see* Chapters 21 and 23), insertion sequences and transposons (Kleckner, 1981; *see* also Chapter 4), interspersed repetitive extragenic palindromes (REP) (Lupski and Weinstock, 1992) and other palindromic unit (PU) sequences (Gilson *et al.,* 1984, Gilson *et al.,* 1987), intergenic repeat units (IRU) (Sharples *et al.,* 1990) or enterobacterial repetitive intergenic consensus (ERIC) sequences (Hulton *et al.,* 1991), bacterial interspersed mosaic elements (BIME) (Gilson *et al.,* 1991), short tandemly repeated repetitive (STRR) sequences (Mazel *et al.,* 1990), and BOX elements (Martin *et al.,* 1992). The exact function of many of these repetitive sequences is unknown, although some have been postulated to be important for genome structure and function. Most well defined interspersed repetitive sequences are 26 to 400 bp in size (see Lupski and Weinstock, 1992 for a review; *see also* Chapter 5).

Since specific bacterial strains have a variable number of copies of unique repetitive elements that are integrated into different regions of the bacterial

Table 32-1. Selected endogenous repetitive sequences in diverse bacteria

Strain	Sequence[a]	Reference
Agrobacterium tumefaciens	*At*-rep[b]	Flores, *et al.* (1987)
Bordetella pertussis	*Bp*-rep[b]	McLafferty, *et al.* (1988)
Bradyrhizobium japonicum	HRS1	Rodriguez-Quiñones, *et al.* (1992); Judd, *et al.* (1993)
Bradyrhizobium japonicum	RSRjα, RSRjβ	Kaluza, *et al.* (1985)
Calothrix sp.	STRR	Mazel, *et al.* (1990)
Coxiella burnetti	*Cb*-rep[b]	Hoover, *et al.* (1992)
Deinococcus radiodurans	SRE	Lennon, *et al.* (1991)
Escherichia coli	*Ec*-rep[b]	Higgins, *et al.* (1982)
Escherichia coli	REP (or PU)	Stern, *et al.* (1984); Gilson *et al.* (1984)
Escherichia coli	IRU	Sharples, *et al.* (1990)
Escherichia coli	ERIC	Hulton, *et al.* (1991)
Escherichia coli	BIME	Gilson, *et al.* (1991)
Halobacterium halobium	*Hh*-rep[b]	Sapienza, *et al.* (1982)
Halobacterium volcanii	*Hv*-rep[b]	Sapienza, *et al.* (1982)
Klebsiella pneumonia	ERIC	Hulton, *et al.* (1991)
Mycobacterium tuberculosis	MPTR	Hermans, *et al.* (1992)
Mycobacterium tuberculosis	*Mt*-rep[b]	Ross, *et al.* (1992)
Mycobacterium tuberculosis	*Mt*-rep[b]	van Sooligan, *et al.* (1993)
Mycoplasma pneumoniae	RepMP1, RepMP2	Wenzel, *et al.* (1988)
Mycoplasma pneumoniae	RepMP2/3, RepMP4, RepMP5	Ruland, *et al.* (1990)
Myxococcus xanthus	RMX	Fujitani, *et al.* (1991)
Neisseria gonorrhoeae	*Ng*-rep[b]	Correia, *et al.* (1986)
Neisseria meningitidis	*Nm*-rep[b]	Correia, *et al.* (1986)
Neisseria meningitidis	RS1, RS2	Hass, *et al.* (1986)
Nostoc commune	STRR	Angeloni, *et al.* (1994)
Rhizobium meliloti		Flores, *et al.* (1987)
Rhizobium phaseoli	*Rp*-rep[b]	Flores, *et al.* (1987)
Rhizobium trifolii	*RtRs*	Watson, *et al.* (1985)
Rhodomicrobium vannielii	IR DNA	Russell and Mann (1986)
Salmonella typhimurium	*St*-rep[b]	Higgins, *et al.* (1982)
Salmonella typhimurium	REP (or PU)	Stern, *et al.* (1984)
Salmonella typhimurium	IRU	Sharples, *et al.* (1990)
Salmonella typhimurium	ERIC	Hulton, *et al.* (1991)
Spiroplasma citri	*Sc*-rep[b]	Nur, *et al.* (1987)
Streptococcus pneumoniae	BOX (box A, box B, box C)	Martin, *et al.* (1992)
Vibrio cholerae	ERIC	Hulton, *et al.* (1991)
Vibrio cholerae O1	VCR	Barker, *et al.* (1994)
Xenorhabdus luminescens	ERIC (or IRU)	Meighen, *et al.* (1992)
Yersinia pseudotuberculosis	ERIC	Hulton, *et al.* (1991)

[a]BIME, bacteria interspersed mosaic element; ERIC, enteric repetitive intergenic consensus; HRS, hyperreiterated sequence; IR DNA, inverted repeated DNA; IRU, intergenic repeat unit; MPTR, major polymorphic tandem repeat; REP, repetitive extragenic palindromic; SRE, SARK repeated element; RMX, repetitive DNA sequence of *Myxococus xanthus;* STRR, short tandemly repeated repetitive.

[b]Some repeated sequences that have not been given names by authors have been named by us according to convention of Lupski and Weinstock, 1992.

genome, non-identical strains tend to have a large number of polymorphic restriction fragments containing the repeated elements (van Soolingen *et al.,* 1993). Consequently, repeat-sequence-related restriction fragment length polymorphisms (RFLPs) can be used to differentiate genetically-related, non-identical, strains that could not be distinguished from each other by other criteria (Judd *et al.,* 1992). The distribution of endogenous repeat sequences in a particular strain of bacteria can be assessed by directly hybridizing cloned and labelled repeat sequences to total genomic DNA (*see* Chapter 31) or by amplifying repeat sequences in the genome by using polymerase chain reaction techniques (*see* Chapters 33 and 34). The use of endogenous repeat sequences as hybridization probes for RFLP analyses has proven to be an extremely powerful technique for strain identification, epidemiological studies, and the taxonomic analysis of prokaryotic and eukaryotic organisms (Bostein *et al.,* 1980; Brown *et al.,* 1989; Cook *et al.,* 1989; Denny, 1988; Landry and Michelmore, 1987; Sadowsky *et al.,* 1990; F. Rodriguez Quiñones *et al.,* 1992; Scherer and Stevens, 1987; Wheatcroft and Watson, 1988; Gabriel *et al.,* 1989; Hahn and Hennecke, 1987). Bacteria having their DNA altered by the presence of repeat elements will have DNA fragments of differing sizes following digestion with restriction enzymes. The fragment size differences are a direct result of changes in the number and/or location of restriction enzyme recognition sites within a given piece of DNA (*see* Chapter 31).

Fingerprinting prokaryotic DNA with endogenous repeat sequences has received a lot of attention in epidemiological and taxonomic analyses of medically important bacteria. Perhaps one of the best studied organisms in this regard is *Mycobacterium tuberculosis.* In *M. tuberculosis,* at least five different repetitive DNA elements have been shown to differentially hybridize to genomic DNA of various strains. Probes from these elements, IS6110 (van Soolingen *et al.,* 1993), IS1081 (Collins *et al.,* 1991), IS986 (Hermans *et al.,* 1990), MPTR (Hermans *et al.,* 1992), IS987 (McAdams *et al.,* 1990) and a 36 bp small direct repeat (DR) sequence (Hermans *et al.,* 1991) range in size from 158 bp to 2.4 kb and differentially hybridize to *M. tuberculosis* strains. In addition, Ross *et al.* (1992) reported several unique cloned repeated sequences which differentially hybridized to *M. tuberculosis* strains. Repetitive DNA sequences have also been found in other mycobacteria and include IS900 (Green *et al.,* 1989), IS901 (Kunze *et al.,* 1991), and IS1096 (Crillo *et al.,* 1991). In addition, novel repetitive elements have been identified in *M. leprae* (Clark *et al.,* 1989; Grosskinsky *et al.,* 1989, Woods and Cole 1990). *M. tuberculosis* contains many copies of several IS3-like insertion elements (Hermans *et al.,* 1991, McAdam *et al.,* 1990, Thierry *et al.,* 1990, and van Soolingen *et al.,* 1991). The MPTR sequence (major polymorphic tandem repeat) appears to represent one of the major repeat sequences in the *M. tuberculosis* genome (Hermans *et al.,* 1992). It is present as a short repeat of 10 bp with a 5 bp spacer region separating adjacent repeats. It has been estimated that there are about 80 different MPTR-containing regions in the *M. tuberculosis* genome (Hermans *et al.,* 1992).

Endogenous repeat elements have been used to examine the genome structure and relatedness of *Neisseria gonorrhoeae* and *N. meningitidis* strains (Correia *et al.,* 1986; Haas and Meyer, 1986). Correia and coworkers (1986) identified two types of repeat sequences in *N. gonorrhoeae.* While the first is relatively long (1.1 kb) and is present in at least two copies in the genome, the other consists of 26 bp repeat and is present in multiple copies (at least 20) in the genomes of *N. gonorrhoeae* and *N. meningitidis.*

The genomes of other medically important bacteria have also been examined by using labelled repeated sequences as probes. A 124 bp direct repeat sequence, VCR, has been found in the genome of *Vibrio cholerae* strain 01 (Barker *et al.,* 1994). Southern hybridization analysis indicated that VCR is present in 60 to 100 copies in the *V. cholerae* genome and that related sequences are only found in *V. cholerae* serotype strains and not in the genomes of other members of the *Vibrionaceae. Coxiella burnetti,* the causative agent of Q fever also contains a repeated element that resembles an insertion sequence (Hoover *et al.,* 1992). Though not tested for hybridization homology to other *C. burnetti* strains or other members of the family *Rickettsiaceae,* the 1.4 kb repeat sequence, with 7 bp inverted repeats, might be a useful tool to investigate the epidemiology and taxonomy of this group of bacteria.

Repetitive DNA elements have also been found to reside in the genome of *Bordetella pertussis* and related strains (McLafferty *et al.,* 1988). Southern hybridization analysis indicated that one of the repeated sequences, which is 1053 bp in size with 28 bp nearly-perfect inverted repeats at the termini, is present in at least 40 copies in the genome. The structure and size of the element resembles an insertion sequence. While this sequence was not present in the genomes of the related strains *B. parapertussis* and *B. bronchiseptica,* another element located outside the 1053 bp IS-like sequence is present in the genomes of all three *Bordetella* species. Both of these sequences appear useful for taxonomic and epidemiological studies.

DNA hybridization and sequencing studies have revealed the presence of at least five endogenous repeated elements in the genomes of *Mycoplasma* strains (Colman *et al.,* 1990; Ruland *et al.,* 1990, Su *et al.,* 1988, and Wenzel *et al.,* 1988). *Mycoplasma pneumoniae* contains five repeated elements, RepMP1, RepMP2/3, RepMP4, and RepMP5 which are 300, 1.8 kb, 1.1–1.5 kb, and 1.9–2.2 kb, in size, respectively. Southern hybridization/RFLP analysis indicated that Rep1 and Rep2/3 are present in 8 to 10 copies in the *M. pneumoniae* genome. Another 400 bp repeated element, CDS1, is also present in the genome in about 8 copies, however, this element has been shown to constitute part of RepMP5 (Ruland *et al.,* 1990).

Repeated DNA sequences have also been found in the genomes of *Escherichia coli* and *Salmonella typhimurium* and have been very useful for the examination of genome structure and for the fingerprinting of bacterial DNA. The repeated palindrome (REP) or palindromic unit (PU) sequences and bacterial interspersed mosaic elements (BIMEs) have been detected by DNA-DNA hybridization studies in a large number of bacterial strains (Lupski and Weinstock, 1992; Gilson *et al.,* 1984; Gilson *et al.,* 1991, Sharples *et al.,* 1990; Stern *et al.,* 1984). The use of REP sequences to fingerprint bacterial DNA will be discussed in more detail in Chapter 34.

Many non-medically important bacteria also contain endogenous inverted repeats which are useful for taxonomic and population studies. For example, in *Rhodomicrobium vannelli,* a photosynthetic member of the *Rhodospirillacaea,* approximately 7% of the genomic DNA is comprised of inverted repeat (IR) elements (Russell and Mann, 1986). Two heterogenous classes of IR elements have been reported. The large IR class contains repeated elements ranging in size from 100 to 700 bp, while the smaller class consists of elements with sizes ranging from 7 to 27 bp.

Highly repetitive DNA elements have been reported in the cyanobacterium *Calothrix* genome (Mazel *et al.,* 1990). This organism contains about 100 copies of a tandemly amplified heptanucleotide which have been termed short tandemly repeated repetitive (STRR) sequences. Of 24 cyanobacterial strains tested, 13 had at least three different STTR sequences in their genome and the STRR sequences were not present in *E. coli* or *Bacillus subtilus.* Interestingly, STRR sequences are restricted to heterocystous cyanobacterial strains (Mazel *et al.,* 1990) and thus provide important taxonomic markers.

Deinococcus radiodurans strain SARK has also been shown to contain repetitive chromosomal elements, SRE (Sark repeated element), which are 150 to 192 bp in size (Lennon *et al.,* 1991). When hybridized to DNA from *D. radiodurans,* a SRE probe produced a smear on a southern blot, suggesting that there are many copies in the genome. A unique repetitive element has also been isolated from *Myxococcus xanthus* (Fujitani *et al.,* 1991). *M. xanthus,* which undergoes multicellular development, has at least six repetitive elements, named RMX (for repetitive DNA sequence of *M. xanthus*). The RMX elements share a similar 87 bp core sequence and are present in up to 15 copies in the genome.

The genomes of the Archaea also contain endogenous repeated DNA sequences. Southern hybridization analyses indicated that *Halobacterium halobium* and *H. volcanii* contain a large number of families (at least 50) of repeated elements (Sapienza and Doolittle, 1982). The repeated elements were not found in the genome of another archaebacterial strain, *Theroplasma acidophilum.*

The application of endogenous repeated DNA sequences to fingerprint the genomes of agriculturally important organisms has received much attention. This has been due to necessities of strain identification and for epidemiological studies as well as for taxonomic and phylogenetic analyses. Many members of the family *Rhizobiaceae,* which produce root and stem nodules on leguminous plants contain repeated DNA elements (Wheatcroft and Watson, 1988, Watson and Schofield, 1985, Rodriguez-Quiñones *et al.,* 1992, Judd *et al.,* 1993, Kaluza *et al.,* 1985, Flores *et al.,* 1987). In most instances, the repeated elements have characteristics of insertion-like sequences. For example, the fast-growing microsymbiont of alfalfa, *Rhizobium meliloti,* contains 1 to 11 copies of an insertion sequence, ISRm1 (Wheatcroft and Watson, 1987). ISRm1 was present in 80% of the *R. meliloti* strain examined and produced banding patterns on southern blots that could be used to differentiate among strains (Wheatcroft and Watson 1988). In *R. meliloti* strain GR4, another insertion sequence, ISRm3, has been reported to be located within a locus conferring nodulation competitiveness (Soto *et al.,* 1992). *R. leguminosarum* bv. *trifolii* also contains a repeated DNA sequence (Watson and Schofield, 1985). The repeated sequence constitutes a reiteration of the *nif*HDK promoter region and in some cases, a reiteration of the N-terminal end of the *nif*H gene. Repeated *nif*H sequences have also been found in the *mos* genes of *R. meliloti* (Murphy *et al.,* 1993). The repeated sequence, RtRS1, only hybridized to genomic DNA from *R. leguminosarum* bv. *trifolii* strains and not to DNA from other rhizobia. Three to six copies of RtRS1 are present in the genome and different strains have different hybridization patterns. Thus, RtRS1 can distinguish among closely related strains. Reiteration of nitrogen fixation genes has also been reported in the fast-growing microsymbiont of beans, *Rhizobium phaseoli* (Flores *et al.,* 1987; *see also* Chapter 15). Other rhizobia and strains of *Agrobacterium* are also reported to contain from 2 to 10 copies of repeated elements (Flores *et al.,* 1987). Interestingly, these authors reported that reiterated sequences are not confined to the chromosome or plasmids, but are an attribute of the total genome. Reiterated sequences have also been reported in *R. leguminosarum* bv. *viceae* (Ulrich and Puhler, 1994). *R. fredii* (Krishnan and Pueppke, 1993), and *Rhizobium* sp. strain Tal 1145 (Rice *et al.,* 1994). In the latter case, the 2.5 kb insertion sequence, ISR1dTal1145-1, hybridized to genomic DNA from several *Rhizobium* strains, but nevertheless could be used to determine a strains cross-inoculation preference (Rice *et al.,* 1994).

The slow-growing microsymbionts of legumes also contain insertion-like sequences that have proven useful for strain identification and taxonomic analyses. Repeated sequences, which are similar to insertion elements, have been found in the genome of *Bradyrhizobium japonicum,* the soybean root nodule bacteria (Kaluza *et al.,* 1985, Judd *et al.,* 1993, Rodriguez-Quiñones *et al.,* 1992). Kaluza and coworkers (1985) reported the identification of two repeated elements RSRJα and RSRJβ which are clustered around the nitrogen fixation genes in *B. japonicum* strain USDA 110. The RSRJα is 1126 bp in size and is repeated 12 times in the genome, while RSRJβ is 950 bp and is repeated at least 6 times. The RSRJα and RSRJβ elements have proven very useful to characterize field isolates of *B. japonicum* (Minamisawa *et al.*, 1992). Another insertion-like element, with terminal repeats similar to RSRJα, has been reported in *B. japonicum* strain USDA 424 (Judd and Sadowsky, 1993). This sequence, HRS1, has DNA and amino acid sequence homology to *IS*1380, an insertion sequence from *Acetobacter pasteurianus* and is reiterated 18 to 21 times in the genomes of *B. japonicum* serocluster 123 strains. Since HRSI preferentially hybridizes to serocluster 123 strains and each strain has a different hybridization pattern (Figure 32-1), the IS-like elements can be used for ecological and taxonomic purposes.

While limited space does not allow a more thorough discussion, it is important to note that in addition to the symbiotic bacteria of plants, repetitive DNA has also been identified in plant pathogenic bacteria. For example, short, interspersed, repetitive DNA sequences have been isolated from *Spiroplasma citri* (Nur *et al.*, 1987). The reiterated sequences hybridized to DNA from several *Spiroplasma* strains and could be used to differentiate among *S. citri, S. kunkelli,* and a *Spiroplasm* strain infecting Vinca plants. Other plant pathogens, such as *Xanthomonas campestris, X. citri,* and *X. phaseoli* also contain repetitive DNA elements which can be used to differentiate among strains and pathovars (Denny 1988; Cook *et al.,* 1989; Gabriel *et al.,* 1989; Lazo *et al.,* 1987).

Lastly, while a majority of repeat elements have been isolated from Gram negative bacteria, there have also been reports of endogenous repeat elements in the gram positive bacteria. For example, *Streptococcus pneumoniae* contains a highly conserved DNA sequence, collectively termed BOX, which is located within intergenic regions of the chromosome. The *S. pneumoniae* genome contains about 25 BOX elements which range in size from about 40 to 60 bp (Martin *et al.,* 1992). It has been postulated that BOX sequences serve a regulatory functions.

In summary, while various genomic DNA fingerprinting methods have been used to investigate epidemiologic, taxonomic, and phylogenetic relationships

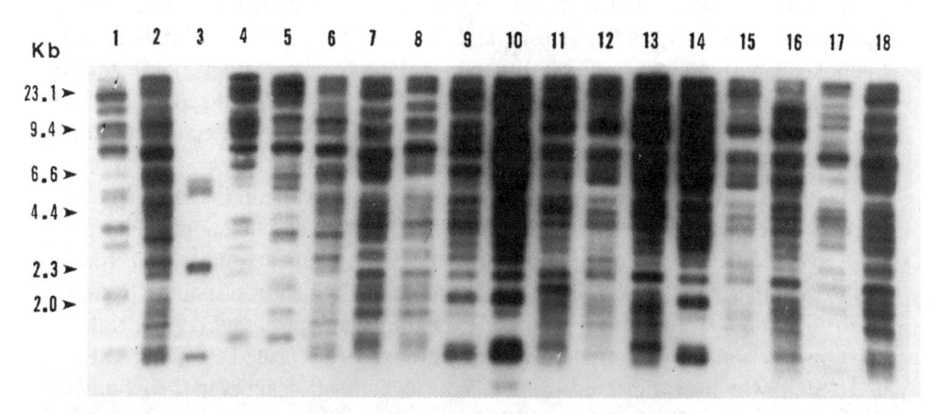

Figure 32-1. Southern hybridization of reiterated sequence HRS1 from *B. japonicum* strain USDA 438 to *Eco*R1-digested genomic DNAs from *B. japonicum* serocluster 123 strains. Lanes: 1, USDA 430; 2, USDA 424; 3, USDA 185; 4, USDA 171; 5, Becker 4-N18; 6, IN56; 7, IN78; 8, IN34; 9, WI 3058; 10, PA3; 11, USDA 432; 12, IA5; 13, IA23; 14, IA35; 15, OH6; 16, USDA 436; 17, USDA 162; and 18, USDA 123. Photograph used with permission of American Society for Microbiology.

among microorganisms, RFLP analyses coupled with the use of endogenous repeat sequence probes has proven to be an invaluable technique. Endogenous repeat sequences have been found in almost all organisms thus far examined and provide a valuable tool to identify and track medically and environmentally important microorganisms. In addition, although the function(s) of most of the endogenous repeat sequences remain unknown, they nevertheless have proven invaluable in phylogenetic studies and the analysis of genome structure and function. While most hybridization patterns obtained by using endogenous repeat sequence probes are relatively simple and easily interpreted, in some instances complex hybridization patterns are obtained by using several restriction enzyme and probe combinations. In this case, the hybridization data can be analyzed using genetic distance equations, principle coordinate, and cluster analysis to determine the degree of relatedness between strains. One PC-based computer program that can be used to perform all of the required statistical operations is NTSYS-PC (Rohlf, 1989), a commercially available, PC-based series of programs for numerical taxonomic and phylogenetic analysis.

References

Angeloni, S. V., and M. Potts. 1994. Analysis of the sequences within and flanking the cyanoglobin-encodine gene, *glbN,* of the cyanobacterium *Nostoc commune* UTEX 584. *Gene* 146:133–134.

Barker, A., C. A. Clark, and P. A. Manning. 1994. Identification of VCR, a repeated sequence associated with a locus encoding a hemagglutinin in *Vibrio cholerae* O1. *J. Bacteriol.* 176:5450–5458.

Bostein, D. R., M. S. Skolnick, and R. W. Davies. 1980. Construction of a genetic linkage map in man using restriction fragment length polymorphisms. *Am. J. Hum. Genet.* 32:314–331.

Brown, G., Z. Khan, and R. Lifshitz. 1989. Plant promoting rhizobacteria: strain identification by restriction fragment length polymorphisms. *Can. J. Microbiol.* 36:242–248.

Buchman, T. G., B. Roizman, G. Adams, and B. H. Stover. 1978. Restriction endonuclease fingerprinting of *Herpes simplex* virus DNA: a novel epidemiological tool applied to a nosocomial outbreak. *J. Infect. Dis.* 138:488–498.

Caetano-Anollés, G., and P. M. Greshoff. 1994. DNA amplification fingerprinting: a general tool with applications in breeding, identification and phylogenetic analysis of plants. In B. Schierwater, B. Striet, G. P. Wagner, and R. DeSalle, eds., pp. 17–31, *Molecular ecology and evolution: approaches and applications.* Burkhauser Verlag, Basel, Switzerland.

Caetano-Anollés, G., G. J. Bassam, and P. M. Greshoff. 1991. DNA amplification fingerprinting using very short arbitrary oligonucleotide primers. *Bio/Technol.* 9:553–557.

Chandler, D. K. S., S. Razin, E. B. Stephens, R. Harasawa, and M. F. Barile. Genomic and phenotypic analysis of *Mycoplasma pneumoniae* strains. *Infect. Immun.* 38:604–609.

Christensen, L. S., K. J. Soerensen, and J. C. Lei. 1987. Restriction fragment pattern (RFP) analysis of genomes from Danish isolates of suid herpes virus 1 (Aujezsky's disease virus). *Arch. Virol.* 97:215–224.

Cirillo, J. D., R. G. Barletta, B. R. Bloom, and Jr., W. R. Jacobs. 1991. A novel transposon trap for mycobacteria: isolation and characterization of IS*1096*. *J. Bacteriol.* 173:7772–80.

Clark-Curtiss, J. E., and M. A. Docherty. 1989. A species specific repetitive sequence in *Mycobacterium leprae*. *J. Infect. Dis.* 159:7–15.

Collins, D. M., and D. M. Stephens. 1991. Identification of insertion sequence, IS*1081*, in *Mycobacterium bovis*. *FEMS Lett.* 83:11–16.

Colman, S. D., P. C. Hu, and K. F. Bott. 1990. Prevalence of novel repeat sequences in and around the P1 operon in the genome of *Mycoplasma pneumoniae*. *Gene* 87:91–96.

Cook, D., E. Barlow, and L. Sequeira. 1989. Genetic diversity of *Pseudomonas solanacearum*, detection of restriction fragment length polymorphisms with DNA probes that specify virulence and the hypersensitive response. *Molec. Plant Microbe Interact.* 2:113–121.

Correia, F. F., S. Inouye, and M. Inouye. 1986. A 26-base-pair repetitive sequence specific for *Neisseria gonorrhoeae* and *Neisseria meningitidis* genomic DNA. *J. Bacteriol.* 167:1009–1015.

de Bruijn, F. J. 1992. Use of repetitive (repetitive extragenic element and enterobacterial repetitive intergenic consensus) sequences and the polymerase chain reaction to fingerprint the genomes of *Rhizobium meliloti* isolates and other soil bacteria. *Appl. Environ. Microbiol.* 58:2180–2187.

Denny, T. P. 1988. Differentiation of *Pseudomonas syringae* pv. tomato from *P. syringae* with a DNA hybridization probe. *Phytopath.* 78:1186–1193.

Flores, M., V. González, S. Brom, E. Martinez, D. Pinero, D. Romero, G. Davila, and R. Palacios. 1987. Reiterated DNA sequences in *Rhizobium* and *Agrobacterium* sp. *J. Bacteriol.* 169:5782–5788.

Fujitani, S., T. Komano, and S. Inouye. 1991. A unique repetitive DNA sequence in the *Myxococcus xanthus* genome. *J. Bacteriol.* 173:2125–2127.

Gabriel, D. W., M. T. Kingsley, J. E. Hunter, and T. Gottwald. 1989. Reinstatement of *Xanthomonas citri* (ex Haase) and *X. phaseoli* (ex Smith) to species and reclassification of all *X. campestris* pv. *citri* strains. *Int. J. Syst. Bacteriol.* 39:14–22.

Gill, P., J. E. Lygo, S. J. Fowler, and D. J. Werrett. 1987. An evaluation of DNA fingerprinting for forensic purposes. *Electrophoresis* 8:38–44.

Gilson, E., J.-M. Clement, D. Brutlag, and M. Hofnung. 1984. A family of dispersed repetitive extragenic palindromic DNA sequences in *E. coli*. *EMBO J.* 3:1417–1421.

Gilson, E., D. Perrin, J. M. Clement, S. Szmelcman, E. Dassa, and M. Hofnung. 1986. Palindromic units from *E. coli* as binding sites for a chromoid-associated protein. *FEBS Lett.* 206:323–328.

Gilson, E., J. M. Clement, D. Perrin, and M. Hofnung. 1987. Palindromic units, a case of highly repetitive DNA sequences in bacteria. *Trends Genet.* 3:226–230.

Gilson, E., D. Perrin, and M. Hofnung. 1990. DNA polymerase I and protein complex bind specifically to *E. coli* palindromic unit highly repetitive DNA: implications for bacterial chromosome organization. *Nucl. Acids Res.* 18:3941–3952.

Gilson, E., W. Saurin, D. Perrin, S. Bachellier, and M. Hofnung. 1991. Palindromic units are part of a new bacterial interspersed mosaic element (BIME). *Nucl. Acids Res.* 19:1375–1383.

Glynn, P., P. Higgins, A. Squartini, and F. O'Gara. 1985. Strain identification in *Rhizobium trifolii* using DNA restriction analysis, plasmid DNA profiles, and intrinsic antibiotic resistances. *FEMS Microbiol. Lett.* 30:177–182.

Green, E. P., M. L. V. Tizard, M. T. Moss, J. Thompson, D. J. Winterburne, J. J. McFadden, and J. Hermon-Taylor. 1989. Sequence and characteristics of IS*900,* an insertion element identified in human Crohn's disease isolate of *Mycobacterium paratuberculosis. Nucl. Acids Res.* 17:9603–9073.

Grosskinsky, C. M., W. R. Jacobs, Jr., J. E. Clark-Curtiss, and B. R. Bloom. 1989. Genetic relationship among *Mycobacterium leprae, Mycobacterium tuberculosis,* and candidate leprosy vaccine strains determined by DNA hybridization, identification of an M. leprae-specific repetitive sequence. *Infect. Immun.* 57:1535–1541.

Hahn, M., and H. Hennecke. 1987. Conservation of a symbiotic DNA region in soybean root nodule bacteria. *Appl. Environ. Microbiol.* 53:2253–2255.

Haas, R., and T. F. Meyer. 1986. The repertoire of silent pilus genes in *Neisseria gonorrhoeae:* evidence for gene conversion. *Cell* 44:107–115.

Hermans, P. W. M., D. van Soolingen, E. M. Bik, P. E. W. de Haas, J. W. Dale, and J. D. A. van Embden, 1991. The insertion element IS*987* from *Mycobacterium bovis* BCG is located in a hot-spot integration region for insertion elements in *Mycobacterium tuberculosis* complex strains. *Infect. Immun.* 59:2695–2705.

Hermans, P. W. M., D. van Soolingen, J. W. Dale, A. R. J. Schuitemea, R. A. McAdam, D. Cathy, and J. D. A. van Embden. 1990. Insertion element IS*986* from *Mycobacterium tuberculosis,* a useful tool for diagnosis and epidemiology of tuberculosis. *J. Clin. Microbiol.* 28:2051–2058.

Hermans, P. W. M., D. V. van Soolingen, and J. D. A. van Embden. 1992. Characterization of a major polymorphic tandem repeat in *Mycobacterium tuberculosis* and its potential use in the epidemiology of *Mycobacterium kansaii* and *Mycobacterium gordonae. J. Bacteriol.* 174:4157–4165.

Higgins, C. F., G. F.-L. Ames, W. M. Barnes, J. M. Clement, and M. Hofnung. 1982. A novel intercistronic regulatory element of prokaryotic operons. *Nature* 298:760–762.

Higgins, C. F., R. S. McLaren, and S. F. Newbury. 1988. Repetitive extragenic palindromic sequences, mRNA stability and gene expression: evolution by gene conversion?—a review. *Gene* 72:3–14.

Hoover, T. A., M. H. Vodkin, and J. C. Williams. 1992. A *Coxiella burnetti* repeated DNA element resembling a bacterial insertion sequence. *J. Bacteriol.* 174:5540–5548.

Hulton, C. S. J., C. F. Higgins, and P. M. Sharp. 1991. ERIC sequences, a novel family of repetitive elements in the genomes of *Escherichia coli, Salmonella typhimurium* and other enterobacteria. *Mol. Microbiol.* 5:825–834.

Judd, A. K., and M. J. Sadowsky. 1993. The *Bradyrhizobium japonicum* serocluster 123 hyperreiterated DNA region, HRS1, has DNA and amino acid sequence homology to IS*1380,* an insertion sequence from *Acetobacter pasteurianus. Appl. Environ. Microbiol.* 59:1656–1661.

Kaijalainen, S., and K. Lindstrom. 1989. Restriction fragment length polymorphism analysis of *Rhizobium galegae* strains. *J. Bacteriol.* 171:5561–5566.

Kaper, J. B., H. B. Bradford, Roberts, and S. Falkow. 1982. Molecular epidemiology of *Vibrio cholerae* in the U.S. Gulf coast. *J. Clin. Microbiol.* 16:129–134.

Kaluza, K., M. Hahn, and H. Hennecke. 1985. Repeated sequences similar to insertion elements clustered around the *nif* region of the *Rhizobium japonicum* genome. *J. Bacteriol.* 162:535–542.

Kleckner, N. 1981. Transposable elements in prokaryotes. *Annu. Rev. Genet.* 15:341–404.

Koch, E., K. Song, T. C. Osborn, and P. H. Williams. 1991. Relationship between pathogenicity and phylogeny based on restriction fragment length polymorphism in *Leptosphaeria maculans. Mol. Plant Microbe Interact.* 4:341–349.

Kristiansen, B. E., B. Sorensen, B. Bjorvatn, E. S. Falk, E. Fosse, K. Bryn, L. O. Froholm, P. Gaustad, and K. Bovre. 1986. An outbreak of group B meningococcal disease: tracing the causative strain of *Neisseria meningitidis* by DNA fingerprinting. *J. Clin. Microbiol.* 23:764–767.

Kuijper, E. J., J. H. Oudbier, W. N. Stuifbergen, A. Jansz, and H. C. Zanen. 1987. Application of whole-cell DNA restriction endonuclease profiles to the epidemiology of *Clostridium difficile*-induced diarrhea. *J. Clin. Microbiol.* 25:751–753.

Kunze, Z. M., S. Wall, R. Appelberg, M. T. Silva, F. Portaels, and J. J. McFadden. 1991. IS*901,* a new member of a wide spread class of atypical insertion sequences, is associated with pathogenicity in *Mycobacterium avium. Mol. Microbiol.* 5:2265–2272.

Landry, B. S., and R. W. Michelmore. 1987. Methods and applications of restriction fragment length polymorphism analysis to plants, in *Proceedings of the Conference on Tailoring Genes for Crop Improvement.* G. Bruening, J. Harada, T. Kosuge, and A. Hollaender, eds. pp. 25–44. University of California, Davis, CA.

Langenberg, W., E. A. J. Rauws, A. Widjojokusumo, G. N. J. Tytgat, and H. C. Zanen. 1986. Identification of *Campylobacter pyloridis* isolates by restriction endonuclease DNA analysis. *J. Clin. Microbiol.* 24:414–417.

Lazo, G. R., R. Roffey, and D. W. Gabriel. 1987. Pathovars of *Xanthomonas campestris* are distinguishable by restriction fragment length polymorphisms. *Int. J. Syst. Bacteriol.* 37:214–221.

Lennon, E., P. D. Gutman, H. Yao, and K. W. Minton. 1991. A highly conserved repeated chromosomal sequence in the radioresistant bacterium *Deinococcus radiodurans* SARK. *J. Bacteriol.* 173:2137–2140.

Lin, R.-J., M. Capage, and C. W. Hill. 1984. A repetitive DNA sequence, *rhs,* responsible for duplications within the *Escherichia coli* K-12 chromosome. *J. Mol. Biol.* 177:1–18.

Lupski, J. M., and G. M. Weinstock. 1992. Short, interspersed repetitive DNA sequences in prokaryotic genomes. *J. Bacteriol.* 174:4525–4529.

Martin, B., O. Humbert, M. Camara, E. Guenzi, J. Walker, T. Mitchell, P. Andrew, M. Prudhome, G. Alloing, R. Hakenbeck, D. A. Morrison, G. J. Boulnois, and J.-P. Claverys. 1992. A highly conserved repeated DNA element located in the chromosome of *Streptococcus pneumoniae*. *Nucl. Acids Res.* 20:3479–3483.

Mazel, D., J. Houmard, A. M. Castets, and N. T. de Marsac. 1990. Highly repetitive DNA sequences in cyanobacterial genomes. *J. Bacteriol.* 172:2755–2761.

McAdam, R. A., P. W. M. Hermans, D. van Soolingen, Z. F. Zainuddin, D. Catty, J.D.A. van Embden, and J. W. Dale. 1990. Characterization of a *Mycobacterium tuberculosis* insertion sequence belonging to the IS*3* family. *Mol. Microbiol.* 4:1607–1613.

McLafferty, M. A., D. R. Harcus, and E. L. Hewlett. 1988. Nucleotide sequence and characterization of a repetitive DNA element from the genome of *Bordetella pertussis* with characteristics of an insertion sequence. *J. Gen. Microbiol.* 134:2297–2306.

Meighen, E. A., and R. B. Szittner. 1992. Multiple repetitive elements and organization of the *lux* operons of luminescent terrestrial bacteria. *J. Bacteriol.* 174:5371–5381.

Mielenz, J. R., L. E. Jackson, F. O'Gara, and K. T. Shanmugam. 1979. Fingerprinting bacterial chromosomal DNA with restriction endonclease *Eco*R1: comparison of *Rhizobium* spp. and identification of mutants. *Can. J. Microbiol.* 25:803–807.

Minamisawa, K., T. Seki, S. Onodera, M. Kubota, and T. Asami. 1992. Genetic relatedness of *Bradyrhizobium japonicum* field isolates as revealed by repeated sequences and various other characteristics. *Appl. Environ. Microbiol.* 58:2832–2839.

Murphy, P. J., S. P. Trenz, W. Grzemski, F. J. de Bruijn, and J. Schell. 1993. The *Rhizobium meliloti* rhizopine *mos* locus is a mosaic structure facilitating its symbiotic regulation. *J. Bacteriol* 175:5193–5204.

Newbury, F. S., N. H. Smith, and C. F. Higgins. 1987. Differential mRNA stability controls relative gene expression within a polycistronic operon. *Cell* 51:1131–1143.

Newbury, F. S., N. H. Smith, E. C. Robinson, I. D. Hiles, and C. F. Higgins. 1987. Stabilization of translationally active mRNA by prokaryotic REP sequences. *Cell* 48:297–310.

Nur, I., D. J. LeBlanc, and J. G. Tully. 1987. Short, interspersed, and repetitive DNA sequences in *Spiroplasma* species. *Plasmid.* 17:110–116.

Panchal, C. J., L. Bast, T. Dowhanick, and G. G. Stewart. 1987. A rapid, simple, and reliable method of differentiating brewing yeast strains based on DNA restriction patterns. *J. Inst. Brew.* 93:325–327.

Ramos, M. S., and S. K. Harlander. 1990. DNA fingerprinting of lactococci and streptococci used in dairy fermentations. *Appl. Microbiol. Biotechnol.* 34:368–374.

Rice, D. J., P. Somasegaran, K. Macglashan, and B. B. Bohlool. 1994. Isolation of insertion sequence ISRLdTAL1145-1 from a *Rhizobium* sp. *(Leucaena diversifolia)* and distribution of homologous sequences identifying cross-inoculation group relationships. *Appl. Environ. Microbiol.* 60:4394–4403.

Ross, B. C., K. Raios, K. Jackson, and B. Dwyer. 1992. Molecular cloning of a highly repeated DNA element from *Mycobacterium tuberculosis* and its use as an epidemiological tool. *J. Clin. Microbiol.* 30:942–946.

Rodriguez-Quiñones, F., A. K. Judd, M. J. Sadowsky, R.-L. Liu, and P. B. Cregan. 1992.

Hyperreiterated DNA regions are conserved among *Bradyrhizobium japonicum* serocluster 123 strains. *Appl. Environ. Microbiol.* 58:1878–1885.

Rohlf, F. J. 1989. NTYSYS-pc. *Numerical taxonomy and multivariate analysis system.* Exeter Publishing, Ltd. Setauket, New York.

Ruland, K., R. Wenzel, and R. Herrmann. 1990. Analysis of three different repeated DNA elements present in the P1 operon of *Mycoplasma pneumoniae:* size, number, and distribution on the genome. *Nucl. Acids Res.* 18:6311–6317.

Russell, G. C., and N. H. Mann. 1986. Analysis of inverted repeat DNA in the genome of the *Rhodomicrobium vannielii. J. Gen. Microbiol.* 132:325–330.

Sadosky, A. B., A. Davidson, R.-J. Lin, and C. W. Hill. 1989. *rhs* gene family of *Escherichia coli* K-12. *J. Bacteriol.* 171:636–642.

Sadowsky, M. J., P. B. Cregan, and H. H. Keyser. 1990. DNA hybridization probe for use in determining restricted nodulation among *Bradyrhizobium japonicum* serocluster 123 field isolates. *Appl. Environ. Microbiol.* 56:1768–1774.

Sadowsky, M. J., R. E. Tully, P. B. Cregan, and H. H. Keyser. 1987. Genetic diversity in *Bradyrhizobium japonicum* serogroup 123 and its relation to genotype-specific nodulation of soybeans. *Appl. Environ. Microbiol.* 53:2624–2630.

Sapienza, C., and W. F. Doolittle. 1982. Unusual physical organization of the *Halobacterium* genome. *Nature* 295:384–389.

Scherer, S., and D. A. Stevens. 1987. Application of DNA typing methods to epidemiology and taxonomy of *Candida* species. *J. Clin. Microbiol.* 25:675–679.

Schmidt, E. L., M. J. Zidwick, and H. H. Abebe. 1986. *Bradyrhizobium japonicum* serocluster 123 and diversity among member isolates. *Appl. Environ. Microbiol.* 51:1212–1215.

Schneider, M., and F. J. de Bruijn. 1996. Rep-PCR mediated genomic fingerprinting of rhizobia and computer-assisted phylogenetic analysis. *World J. Microbiol. Biotechnol.* 12:163–174.

Skjold, S., P. G. Quie, L. A. Fries, M. Barnham, and P. P. Cleary. 1987. DNA fingerprinting of *Streptococcus zooepidemicus* (Lancefield group C). as an aid to epidemiological study. *J. Infect. Dis.* 155:1145–1150.

Sharples, G. J., and R. G. Llod. 1990. A novel repeated DNA sequence located in the intergenic regions of bacterial chromosomes. *Nucl. Acids Res.* 18:6503–6508.

Soto, M. J., A. Zorzano, J. Olivares, and T. Toro. 1992. Nucleotide sequence of *Rhizobium meliloti* GR4 insertion sequence ISRm3 linked to the nodulation competitiveness locus *nfe. Plant Molec. Biol.* 20:307–309.

Stern, M. J., G. F.-L. Ames, N. H. Smith, E. C. Robinson, and C. F. Higgins. 1984. Repetitive extragenic palindromic sequences, a major component of the bacterial genome. *Cell* 37:1015–1026.

Stern, M. J., E. Prossnitz, and G. F.-L. Ames. 1988. Role of the intercistronic region in post-transcriptional control of gene expression in the histidine transport operon of *Salmonella typhimurium:* involvement of REP sequences. *Mol. Microbiol.* 2:141–152.

Su, C. J., A. Chavoya, and J. B. Baseman. 1988. Regions of *Mycobacterium pneumoniae* cytadhesin P1 structural gene exit as multiple copies. *Infect. Immun.* 56:3157–3161.

Thierry, D., M. D. Cave, K. D. Eisenach, J. T. Crawford, J. H. Bates, B. Gicquel, and J. L. Guesdon. 1990. IS*6110,* an IS-like element of *Mycobacterium tuberculosis* complex. *Nucl. Acids Res.* 18:188.

Tompkins, L. S., N. J. Troup, T. Woods, W. Bibb, and R. M. McKinney. 1987. Molecular epidemiology of *Legionella* species by restriction endonuclease and alloenzyme analysis. *J. Clin. Microbiol.* 25:1875–1880.

van Soolingen, D., P. E. W. de Hass, P. W. M. Hermans, P. M. A. Groenen, and J. D. A. van Embden. 1993. Comparison of various repetitive DNA elements as genetic markers for strain differentiation and epidemiology of *Mycobacterium tuberculosis. J. Clin. Microbiol.* 31:1987–1995.

van Soolingen, D., P. W. M. Hermans, P. E. W. de Haas, D. R. Soll, and J.D.A. van Embden. 1991. Occurrence and stability of insertion sequences in *Mycobacterium tuberculosis* complex strains: evaluation of an insertion sequence-dependent DNA polymorphism as a tool in the epidemiology of tuberculosis. *J. Clin. Microbiol.* 29:2578–86.

Versalovic, J., T. Koeuth, and J. R. Lupski. 1991. Distribution of repetitive DNA sequences in eubacteria and application to fingerprinting of bacterial genomes. *Nucl. Acids Res.* 24:6823–6831.

Versalovic, J., M. Schneider, F. J. de Bruijn, and J. R. Lupski. 1994. Genomic fingerprinting of bacteria using repetitive sequence-based polymerase chain reaction. *Meth. Molec. Cell. Biol.* 5:25–40.

Watson, J. M., and P. R. Schofield. 1985. Species-specific, symbiotic plasmid-located repeated DNA sequences in *Rhizobium trifolii. Mol. Gen. Genet.* 199:279–289.

Welsh, J., and M. McClelland. 1990. Fingerprinting genomes using PCR with arbitrary primers. *Nucl. Acids Res.* 18:7213–7218.

Wenzel, R., and R. Herrmann. 1988. Repetitive DNA sequences in *Mycoplasma pneumoniae. Nucl. Acids Res.* 16:8337–8350.

Wenzel, R., and R. Herrmann. 1988. Physical mapping of the *Mycoplasma pneumoniae* genome. *Nucl. Acids Res.* 16:8323–8336.

Wheatcroft, R., and R. J. Watson. 1987. Identification and characterization of insertion sequence IS*Rm1* in *Rhizobium meliloti* JJ1c10. *Can. J. Microbiol.* 33:314–321.

Wheatcroft, R., and R. J. Watson. 1988. A positive strain identification method for *Rhizobium meliloti. Appl. Environ. Microbiol.* 54:574–576.

Williams, J. G. K., A. R. Kubelik, K. J. Livak, J. A. Rafalski, and S. V. Tingey. 1990. DNA polymorphisms amplified by arbitrary primers are useful as genetic markers. *Nucl. Acids Res.* 18:6531–6535.

Woods, S. A., and S. T. Cole. 1990. A family of dispersed repeats in *Mycobacterium leprae. Mol. Microbiol.* 4:1745–1751.

Yang, Y., and G. F.-L. Ames. 1988. DNA gyrase binds to the family of prokaryotic repetitive extragenic palindromic sequences. *Proc. Natl. Acad. Sci. (USA)* 85:8850–8854.

33

Applications of DNA and RNA Fingerprinting by the Arbitrarily Primed Polymerase Chain Reaction

Françoise Mathieu-Daudé, Karen Evans, Frank Kullmann, Rhonda Honeycutt, Thomas Vogt, John Welsh, and Michael McClelland

Introduction

The related methods Arbitrarily Primed Polymerase Chain Reaction (AP-PCR) and Random Amplified Polymorphic DNA (RAPD) were developed independently by our group (Welsh and McClelland, 1990) and by Williams *et al.* (1990). These methods were based on the observation that PCR performed at relatively low stringency (in the case of AP-PCR) or with low selectivity primers at high stringency (in the case of RAPD) yield a reproducible collection of products that depend on the template and primer sequences. Arrayed on an agarose or acrylamide gel, this collection of products can be viewed as a "fingerprint" or "bar code" for the DNA template. Because of their dependence on template sequence, AP-PCR and RAPD can be used as methods for sampling in *sequence space;* diverse applications can be imagined, many of which have been demonstrated in the literature.

AP-PCR and RAPD have been used widely in evolutionary biology, population biology, taxonomy, ecology and genetics. In evolutionary biology, AP-PCR can be used to construct reasonable phylogenetic hypotheses. In population biology, AP-PCR can be used to construct a statistical picture of species distribution, using its taxonomic capabilities for the resolution between and identification of strains. Large collections of diverse organisms or germ plasm present obvious maintenance issues; DNA fingerprinting is very useful for classification and verification of strain type. In genetics, AP-PCR has been used extensively in genetic mapping and to some extent in bulk segregant analysis. Chromosomal alterations accompanying some forms of cancer, including changes in ploidy and mutation, have been determined using AP-PCR (Peinado *et al.*, 1992; Ionov *et al.*, 1993). Sex, lineage, and other important genetic determinants can be detected using AP-PCR.

More recently, AP-PCR has been applied to the fingerprinting of RNA. The related methods, *R*NA fingerprinting by *AP-PCR* (i.e. RAP-PCR) and Differential Display (i.e. DD) are based on AP-PCR and RAPD. RAP-PCR depends on the low stringency PCR sampling of RNA sequences, using a single arbitrary primer or a pair of primers for both first and second strand cDNA synthesis; DD employs a primer of the form oligo $d(T)_{10}$-MX for first strand cDNA synthesis and an arbitrarily selected 10-mer for second strand synthesis. The anchor (MX) is intended to lock the primer to the 5′ end of the poly-A tails of most eukaryotic messages. For second strand synthesis, the short primers are usually used slightly above their T_m (i.e. 10-mers at 37°C) to discourage mis-match priming.

RAP-PCR and DD are useful methods for the discovery of differentially regulated genes. Differential gene regulation is one of the cornerstones of genetics, and identifying differentially regulated genes is an important avenue toward understanding the rules governing gene expression. Alternative methods such as differential screening of high density arrays of clones, using probes prepared in various ways, also contribute significantly to our understanding of the choreography of gene regulation.

Although this book is dedicated primarily to the biology of prokaryotes, many of the applications of arbitrarily primed PCR fingerprinting methods to other organisms can be translated into the prokaryotic experimental environment. Thus, we include a wide range of topics and invite the reader to imagine novel applications of the methods and strategies. We also refer to Chapter 34 for a detailed description of another PCR-based genomic fingerprinting method using endogenous repetitive sequences as primer sites.

DNA Fingerprinting by AP-PCR and RAPD

In AP-PCR, a single arbitrarily chosen primer, or a pair of arbitrary primers, is used in low stringency PCR to amplify genomic sequences lying between sites that best match the primers. The resulting products are then arrayed by gel electrophoresis to reveal a complex banding pattern, or "fingerprint". Because the fingerprint is generated by a primer-template, best-match scenario, its detailed structure depends on the sequence of the template. Sequence divergence between two templates gives rise to distinct fingerprints due to alterations of priming site matches or length variation between two priming sites; in some applications the copy number of the amplified sequence is an important variable. Thus, for example, variation in sequence or sequence representation that occurs during evolution, oncogenesis, life-cycle stage (i.e. ploidy) or mating (i.e. assortment) can be detected by AP-PCR. In Figure 33-1, an AP-PCR fingerprint of DNA isolated from a number of recombinant inbred mouse lines is shown. In this illustration, genomic sequence differences between the parental mouse lines (A × B) were made homozygous by repeated sibling matings and revealed by AP-PCR (see,

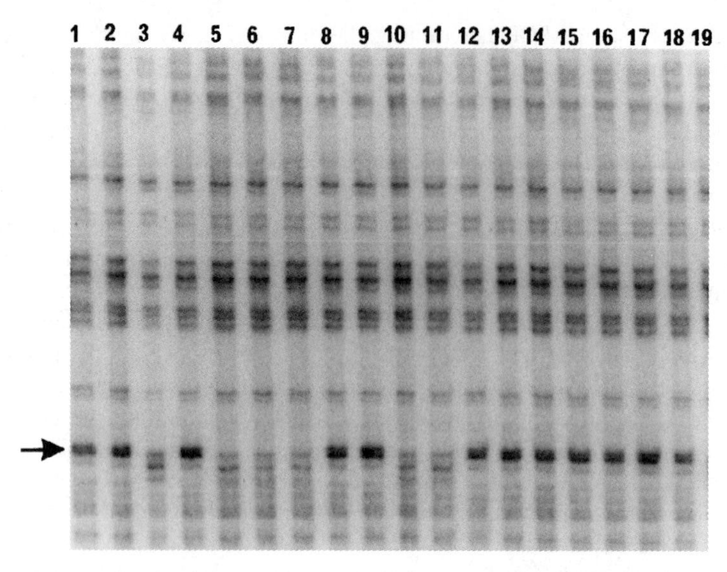

Figure 33-1. A portion of a Single Strand Conformation Polymorphism gel containing AP-PCR fingerprints of DNA from 19 recombinant inbred mice. The arrow indicates a polymorphism.

for example, Welsh *et al.,* 1991). In bacteria, differences in fingerprint patterns usually reflect evolutionary sequence divergence (e.g. Ralph *et al.,* 1993).

Variations of AP-PCR differ primarily in (1) primer choice, (2) buffers and cycling parameters and (3) fingerprint display strategies (McClelland and Welsh 1994a; Welsh *et al.,* 1995). We also mention a couple of novel strategies based on AP-PCR which deliver significantly different information, such as the procedure of Burt *et al.* (1994) for generating sequence information from arbitrarily chosen sites in the genome, and an alternative strategy for quantitative PCR, LS-PCR (Damas de Caballero *et al.,* 1994a,b).

Primer Choice

Primer design considerations include length, G+C content, sequence bias or inclusion of motifs, inclusion of convenient restriction sites or RNA polymerase promoter sequences (e.g. T7 polymerase promoter), and care that the 3′ end is not self-complimentary. Shorter primers are less costly and usually yield fingerprints with a greater signal-to-noise ratio. Longer primers are generally more promiscuous, and may work better for smaller genomes. Motif sequences can be encoded into the primer sequence in order to bias fingerprinting toward specific sequences e.g. repetitive sequences. The design of primers based specifically on endogenous repetitive sequences has been referred to as rep-PCR (*see* Chapter 34). However, almost any primer or combination of primers between 10 and

35 nucleotides in length will yield useful fingerprints under the appropriate experimental conditions. The ostensible difference between AP-PCR and RAPD generation with 10-mers is that AP-PCR is based on priming at sequences that have the best overall match, but many mismatches, while the RAPD generation with 10-mers proceeds at or above the Tm of the oligos, biasing against mismatches. However, it is generally observed that the patterns generated with 10-mers also involve mismatches. Both methods generate complex patterns having only a few to more than 100 bands, when arrayed on sequencing style polyacrylamide gels.

Buffers and Cycling Parameters

In AP-PCR, magnesium is relatively high (about 4 mM) to encourage mismatched priming. Cycling parameters (times, temperatures, and cycle numbers) are adjusted according to several criteria. Annealing temperatures must be adjusted to suit the primer length, and generally range from 35°C for 10-mers up to 50°C for longer primers. For longer primers, we use a two-tiered thermoprofile, with several low stringency steps, followed by many high stringency steps. This strategy was designed to avoid "cross-over" PCR, where one product serves as a primer initiating extension on another product, and to avoid internal AP-PCR, wherein products are not independent.

Recently, the importance of cycle number has become clearer. It appears that differences in band intensities can be erased if too many cycles are used. This arises from what we have called the "C_0t effect" (Mathieu-Daudé *et al.,* 1996). The C_0t effect occurs when, as the concentration of a product grows, its two single-stranded partners begin to anneal at a rate comparable to that at which primer extension occurs. When this happens, PCR amplification of the product slows down in those samples where the product is abundant. In samples where the product is initially rare, the C_0t effect only becomes significant at higher numbers of cycles. Thus, when quantitative comparisons are desired, it is necessary to limit the number of cycles to fewer than about 20. Occasionally, fingerprint patterns are very simple, and under these circumstances a "context effect" may be observed. A context effect can occur when a product in one of the reactions, but not the other, amplifies because another abundant product is missing due to polymorphism; this context effect product may not reflect a polymorphism, but rather an amplification efficiency artifact.

Fingerprint Display and Detection Strategies

AP-PCR fingerprints can be arrayed on either agarose or polyacrylamide gels of the sequencing or SSCP styles. A number of methods have been used to detect DNA fingerprint patterns, including radioactive labeling, ethidium fluorescence, and fluorescent-labeled primers. Silver staining has also been used (Sanguinetti *et al.,* 1994), and it is of note that silver stained bands can be cut out of the gel and successfully reamplified by PCR. Technical innovations have appeared that are relevant to the study of population structure using DNA fingerprinting. For example, Cancilla *et al.* (1992) used DNA fingerprinting by AP-PCR to distinguish between strains of *Lactococcus lactis;* in this study fluorescently labeled primers and an automated DNA sequencer were used. For strain identification, this approach exploits the high data throughput and automatic recording of data capabilities of automated DNA sequencing machines. One interesting approach to population structure study and phylogenetics is the use of AP-PCR to generate sequence information from arbitrarily chosen segments of the genome, as described below.

Novel Applications of AP-PCR

Burt *et al.* (1994) demonstrate a strategy for generating sequence information from arbitrarily selected regions of a genome. First, AP-PCR using two primers is used to generate a fingerprint; in this process, some of the resulting bands have one or the other primer sequences at their ends, while other bands have both primer sequences. These latter bands require both primers for high stringency PCR amplification, and can be identified by this property. Once identified and isolated, one or the other primer can be used to sequence the fragment. This allows the investigator to sample sequences from many places in the genome, thereby avoiding the mistake of relying entirely on a single, possibly non-representative, region of the genome in constructing a phylogenetic hypothesis.

Damas de Caballero *et al.* (1994a,b) have invented a very simple and clever use of AP-PCR. In PCR diagnostics both positive and negative controls must be included to control for errors in reactive preparation and reagent quality. Using specific primers directed toward a diagnostic sequence, they use low stringency PCR (LS-PCR) to amplify the band of interest, superimposed on an AP-PCR fingerprint. The AP-PCR fingerprint serves as a very informative control for several potential experimental errors. However, one must keep in mind the Cot effect spoken of earlier: quantitative amplification is compromised by the reannealing of products at higher concentrations.

Population Biology and Taxonomy

Natural Populations of Bacteria

The most common application of AP-PCR to prokaryotes has been in the description of natural populations of bacteria; both local and global populations of bacterial species have been characterized. The first application of AP-PCR to bacterial phylogenetics was our work with *Borrelia burgdorferi* sensu lato, the causative agent of Lyme Disease (Welsh *et al.,* 1992c). Lyme disease was known in both North America and Northern Europe, and presented different clinical manifestations in the two continents; in North America, it presented as severe arthritis, while in Northern Europe, the presentation was more complex, with both arthritis and cardiovascular complications. Our experiments revealed that *B. burgdorferi* sensu lato was actually comprised of multiple species. Our collection contained four of the currently recognized five species. One of these species, *B. burgdorferi* sensu stricto, of which the type strain B31 is a member, was the only species prevalent in North America, while three of the species, *B. b.* sensu stricto, *B. garinii* and *B. afzelli* were found in Europe, and the fourth, for which we had only one example, was found in Asia. It may be that the different clinical manifestations of the disease arise from different species of the causal agent. Table 33-1 contains a partial list of the bacterial species that have been dissected by AP-PCR, RAPD or, in some cases, other PCR fingerprinting methodologies based on the use of primers complementary to endogenous repeated elements (e.g., Louws *et al.,* 1994; *see* Chapter 34). For a listing of bacterial species characterized by rep-PCR, *see* Chapter 34 and references therein.

Nosocomial Populations

An important practical application of AP-PCR has been as a taxonomic tool for the characterization of nosocomial outbreaks of infectious agents. We performed a study of an outbreak of multi-drug resistant *S. aureus* in a local hospital (Fang *et al.,* 1993). Using AP-PCR, we were able to establish the likelihood that this intra-hospital population structure exhibited a founder effect. In those experiments, the distribution of various strains within the hospital was not consistent with a random distribution of strain types. This kind of information allows health care workers to determine whether procedures limiting within-hospital transmission of infectious disease are adequate, and to trace possible carriers. For examples of the use of rep-PCR in clinical application, *see* Chapter 34.

AP-PCR as a Preliminary Screen

For recurrent practical problems in strain identification, it is desirable to design primers for specific PCR that reveal important strain distinctions, rather than relying on the more experimentally demanding fingerprinting methods, such as AP-PCR or RAPD. One advantage of specific primer PCR over the arbitrarily primed PCR methods is that specific PCR often does not require that the organism be cultured and purified away from contaminating DNA (e.g. from infected tissue). For example, we developed a strategy for designing species specific primers that relies on the conservation of tRNA gene sequences and the more rapid genetic drift of tRNA intergenic spacer regions (Welsh and McClelland 1992; Welsh *et al.,* 1992b). However, before specific primers are designed, it is often helpful to first describe the phylogenetic relationships between the organ-

Table 33-1.

Acinetobacter	*baumanni*	Vila *et al.*, 1994
Actinobacillus	*actinomycetemcomitans pleuropneumoniae*	Slots *et al.*, 1993; Preus *et al.*, 1993; Hennessy *et al.*, 1993
Arthrobacter	*ureafaciens*	Fontana *et al.*, 1994
Borrelia	*burgdorferi*	Welsh *et al.*, 1992c
Bradyrhizobium	*japonicum*	Kay *et al.*, 1994
Brucella	*abortus*	Fekete *et al.*, 1992
	melitensis	
	suis	
	canis	
	ovis	
	noeomae	
Clostridium	*difficile*	Barbut *et al.*, 1993, 1994
Haemophilus	*somnus*	Meyers *et al.*, 1993
	ovis	
	agni	
Helicobacter	*pylori*	Talor *et al.*, 1995; Akopyanz *et al.*, 1992
Klebsiella	*pneumoniae*	Eisen *et al.*, 1995; Wong *et al.*, 1994
Legionella	*pneumophila*	van Belkum *et al.*, 1993; Gomez-Lus *et al.*, 1993; Sandery *et al.*, 1994
Leptospira	*interrogans*	Ralph *et al.*, 1993
	meyeri	
	kirschneri	
	borgpetersenii	
	noguchii	
	santarosai	
	weillii	
	biflexa	
Listeria	*monocytogenes*	Boerlin *et al.*, 1995; Black *et al.*, 1995; Gray and Kroll, 1995; Farber and Addison, 1994; MacGowan *et al.*, 1993; Lawrence *et al.*, 1993
	innocua	
	ivanovii	
	seeligeri	
	welshimeri	
	grayi	
Mycobacterium	*tuberculosis*	Linton *et al.*, 1994; Aznar *et al.*, 1995; Palittapongarnpim *et al.*, 1993a,b
Neisseria	*meningitidis*	Woods *et al.*, 1994
Proteus	*mirabilis*	Bingen *et al.*, 1993
Pseudomonas	*syringae*	Louws *et al.*, 1994
	morsprunorum	
	tomato	
Salmonella	*enteritidis*	Fadl *et al.*, 1995
	typhi	
	pullorum	
	typhimurium	

Table 33-1. (Continued)

Staphylococcus	*aureus*	van Belkum *et al.*, 1995;
	epidermidis	Struelens *et al.*, 1993; Marquet-Van Der Mee *et al.*, 1995
Stenotrophomonas	*maltophila**	Van Couwenberghe *et al.*, 1995
Vibrio	*cholerae*	Coelho *et al.*, 1995a,b
Xanthomonas	*campestris*	Louws *et al.*, 1994
	poae	
	graminis	
	translucens	
	phaseoli	
	begonia	
	pelargonii	
	citri	
	vesicatoria	
	oryzae	
Yersinia	*enterocolitica*	Rasmussen *et al.*, 1994; Odinot *et al.*, 1995a,b

Formerly known as *Xanthomonas maltophilia*

isms to be distinguished. In this way, a polymorphic region can be chosen that correctly parses the group into meaningful taxonomic sublevels. Otherwise, specific primers may indicate similarity between organisms at one locus that are actually quite different at many other loci. AP-PCR can be a very useful tool in this regard. The Lyme disease example is illustrative; once the substructure of the genus *Borrelia* was described by AP-PCR (Welsh *et al.*, 1992c), it was straightforward to design specific primers able to distinguish between the known species (Ralph *et al.*, 1993).

Some studies of populations seek to maximize the genetic diversity of a population. This is important when more detailed subsequent studies of variation between individuals is labor intensive. DNA fingerprinting can be used as an initial screen to eliminate genetically identical or very similar strains as in *Actinomycetes* isolates (Anzai *et al.*, 1993). Demeke and Adams (1994) demonstrated the use of DNA fingerprinting in germplasm analysis in *Phytolacca dodecandra* and provide a nice demonstration of how this technology can be used to rationally plan germplasm collection strategies. This study makes use of principal coordinate analysis and compares results with those obtained through other methods.

Disease Vectors

An important area in taxonomy is the identification and characterization of disease vectors. An illustration of the application of DNA fingerprinting to vector taxonomy is found in Favia *et al.* (1994), who identified polymorphisms distinguishing between different strains of the malaria vector *Anopheles gambiae*. In principle, different disease vectors represent different host environments within which pathogen subspecies structure may be highly restricted. Thus, understanding the population dynamics of a vector borne disease requires not only knowledge of the taxonomic structure of the pathogen, but also knowledge of the taxonomic structure of its vector.

Evolutionary Biology

Evolution is influenced by the size and genetic diversity populations. Ecology issues, including interaction between diverse species, are also critically important. Concise descriptions of extant population structures, and interactions between populations, have been important throughout the development of evolutionary theory, and AP-PCR is useful in this regard. It is also useful for developing phylogenetic hypotheses, as discussed below.

Before discussing phylogenetics, it may be appropriate in this section to mention the application of AP-PCR to population genetics, the rules of which pertain to many evolutionary phenomena. An important variable in the evolution of prokaryotes is whether recombination has played an important role. In some bacteria, evolutionary history appears to be consistent with clonality, with occasional recombination playing a significant role only infrequently. In other cases, recombination is thought to occur often, such that evolutionary models assuming clonality do not fit very well. For example, experiments using AP-PCR were consistent with non-clonal population structure in *Neisseria gonorrhoeae* (O'Rourke and Sprat, 1994), which appears to recombine frequently in nature.

Most evolutionary studies using AP-PCR have involved organisms other than bacteria. For example, linkage disequilibrium has been detected in *Trypanosoma brucei* (Stevens and Tibayrenc, 1995; Mathieu-Daudé *et al.,* 1995), illustrating the usefulness of DNA fingerprinting in measuring departure from panmixia. An important agronomic and species preservation issue is the genetic diversity in germplasm collections or breeding populations. Virk *et al.* (1995), for example, used RAPD analysis to determine the diversity and phylogenetic relationships within a large *Oryza sativa* germplasm collection.

Phylogenetics

Genetic distance algorithms and other procedures, such as cluster or parsimony analysis, can be used to construct hypotheses for ancestral relationships between taxonomic groups. There are several clocks in DNA sequence that can be informative at various levels of divergence (therefore, time) in evolution. For example, ribosomal sequences evolve at various rates, generally allowing genus and higher level distinctions to be made. Some proteins evolve very slowly at the amino acid level, and are highly conserved in all species. On the other hand, so called neutral DNA evolves very quickly, as it is not under selective pressure to remain unchanged. Thus, neutral DNA can be used as a clock for relatively recent evolutionary history.

Because DNA fingerprinting patterns reflect the underlying genome sequence, it has been used extensively to analyze phylogenetic structures of a wide variety of organisms. So far, however, only a few studies on bacteria explicitly attempt phylogenetic analysis. Among the studies that hypothesize specific phylogenetic structures are those of Welsh *et al.* (1992c) for *Borrelia burgdorferi* sensu lato, Daffonchio *et al.* (1994) for *Pseudomonas,* Menard and Mouton (1995) for *Porphyromonas gingivalis,* and Lee *et al.* (1994), for *Mycobacterium tuberculosis.* Analyses have been based on both genetic distance and parsimony analysis.

AP-PCR is often useful for the analysis of the *recent* evolutionary history of prokaryotes and works well in the comparison of genomes having greater than about 80% similarity, which is greater than the sequence homology between most bacterial species. Beyond 80% similarity, AP-PCR patterns can become too complex to distinguish between homologous vs. coincidental bands. Thus, for most bacterial species, AP-PCR is an excellent taxonomic tool, but not very useful for phylogenetics. However, above 80% sequence similarity, AP-PCR is an excellent phylogenetic tool. This has been explored most vigorously in non-bacterial species. Note that in many genera, speciation occurs at much less than 80% sequence similarity. There are a number of reasonable arguments against the use of AP-PCR for formal phylogenetics, including the fact that character weighting and the reversibility of character transitions are difficult to estimate, *a priori.* Empirically, however, the correspondence between phylogenetic trees constructed using these methods and those constructed using other methods tends to be quite high. In general, if one is merely building trees, and has no more general phylogenetic hypothesis, confidence is always a problem. However, as we discuss below, even a relatively sloppy approach to building phylogenetic trees can be useful as long as some independent (usually maximum likelihood) method for confidence can be evoked.

Usually, the overall goal of an experiment in evolutionary biology is not so much to build a phylogenetic tree as it is to correlate elements of an evolutionary hypothesis: phylogenetic trees subserve this more general purpose. The question might be generally phrased, "How are characteristics A and B related within the context of the phylogenetic structure of the group?" In this case, it is important to remember the kinds of error that might arise from a phylogenetic hypothesis. When a strong correlation between A and B is observed (e.g. the traits are monophyletic), the hypothesis is supported, because it is less likely that a statistically meaningless tree-building algorithm would place A and B on the same branch. If A and B do not correlate, however, one must always wonder whether the tree-building algorithm or the quality of the data was adequate. These arguments are particularly relevant when considering which tree-building algorithms are appropriate to AP-PCR data. Thus, AP-PCR data must be used with caution when constructing phylogenetic trees.

Why is DNA fingerprinting by Arbitrarily Primed PCR, RAPD, or rep-PCR (Chapter 34) technology so popular, given the obvious power of direct sequencing? Many biologists prefer DNA fingerprinting over sequencing due to its relative ease of application. This is particularly true when one considers that sequencing information, if it is to be useful in constructing a phylogenetic hypothesis for recent history, must be taken from more than one place in the genome. Thus, several loci must be sequenced in order to construct an accurate phylogenetic hypothesis. Another factor in the practical choice of methods is that, given that sequencing is somewhat more labor intensive, when faced with a large collection of unclassified strains. AP-PCR, RAPD, or rep-PCR can be used to quickly eliminate duplicates or similar strains. This initial screen might then be followed by a more extensive study by DNA sequencing.

Genetics

As we mentioned in the introduction, many of the applications of AP-PCR have been to non-prokaryotic systems. AP-PCR has been used in genetics in several ways. As a device for sampling sequence space, applications have been developed for both types of nucleic acids (i.e. DNA and RNA). AP-PCR has been particularly useful in the genetic mapping of organisms for which family size and structure can be controlled by the investigator, including the mouse, sugarcane, pine, etc. Below, we briefly discuss some of these studies in the hope that the discussion will spark innovation. Many of these studies only make sense in diploids. So, for example, the inheritance of a trait coupled with the inheritance of a genetic marker for the trait facilitates the cloning of genes that might be involved. Some aspects of these studies could, however, be transported into bacterial systems.

Mutation Analysis

The essential function of AP-PCR is to detect mutations by comparing DNA from two different sources. Most of the work in prokaryotes involving AP-PCR involves mutations that have accumulated over evolutionary time. Under some circumstances, however, mutations can accumulate very quickly, and AP-PCR might also be useful in these situations. For example, Peinado and his colleagues used AP-PCR to detect somatic mutations in hereditary non-polyposis colon cancer (Ionov *et al.,* 1993). He also demonstrated the use of AP-PCR in describing aneuploidy associated with cancer (Peinado *et al.,* 1992). Kubota *et al.* (1995) have used AP-PCR to detect mutations induced by gamma irradiation in fish.

Genetic Mapping and Genome Structure

AP-PCR can be used to generate anonymous markers that are genetically linked to specific phenotypes. Correlation between genetic markers and phenotypic characteristics is important in two general senses. First, such a marker can serve as an unambiguous prognosticator of adult characteristics without having to wait for the organism to mature. Second, DNA markers that can be linked to a trait can be used to walk toward the gene responsible for the trait. We and others (Welsh *et al.,* 1991; Birkenmeier *et al.,* 1992) have used AP-PCR to generate genetic maps of the mouse using the recombinant inbred strategy. To date, approximately 1000 markers have been placed on the genetic map of the mouse, using recombinant inbred populations. Our colleagues (Al Janabi *et al.,* 1993) developed the first extensive map for sugarcane, containing over 200 markers. This work illustrates a clear advantage of AP-PCR markers over microsatellite markers: microsatellite markers are rather expensive, each requiring sequence information followed by the synthesis of a unique set of specific primers for PCR. Most genetic mapping efforts cannot afford the expense of this approach. AP-PCR provides a cheap alternative to the microsatellite approach.

Sederoff and colleagues point out the importance of adapting genetic mapping strategies to breeding populations that have already been established for more classical types of studies. Grattapaglia and Sederoff (1994) used a two-way pseudo-testcross strategy to construct genetic linkage maps of *Eucalyptus grandis* and *Eucalyptus mophylla.* This strategy exploits 1:1 segregation of DNA fingerprinting markers that appear in only one of two highly heterozygous parents. In trees, this is a very efficient approach to genetic map construction that makes use of a previously existing pedigree structure.

Some organisms exhibit both diploid and monoploid adult phases, such as *Solanum* species. Genetic mapping in such species should be possible directly from the monoploid offspring of a single heterozygous diploid clone (Singsit and Ozias-Akins 1993).

Bulk Segregant Analysis

In genetics, the goal is often to isolate a gene involved in some process. One successful approach has been the combination of DNA fingerprinting with bulk segregant analysis. In this approach, DNA from organisms that display the trait of interest is pooled such that the only genetic marker likely to be shared by all members of the pool is one linked to the trait of interest (Williams *et al.,* 1993).

Family Structure

In genetic mapping, an important variable is the structure of the breeding population or family that is being studied. A frequently encountered problem in the study of natural populations and poorly documented domestic populations is the determination of likely family histories from the genetic characteristics of extant individuals. The general class of problems entail the identification of the most likely course of historical events from the set of all possibilities. In some cases, only an approximation to the actual family structure can be reconstructed. However, in many non-mapping studies, AP-PCR can answer important questions about relatedness. For example, Apostol *et al.* (1993, 1994) used DNA fingerprinting to distinguish between families of the mosquito *Aedes aegypti* at different oviposition sites. Using AP-PCR, they were able to resolve cohorts at different oviposition sites to estimate the number of full sibling families at a site. McCoy and Echt (1993) used DNA fingerprinting to confirm interspecific crosses in alfalfa. Interspecific crossing in plants is likely to lead to agronomically important insights. Fritz *et al.* (1994) used AP-PCR to follow interspecific crosses between models for herbivory in two species of willow *(Salix sericea* and *S. eriocephala).*

Genome Structure as it Correlates with Evolution

AP-PCR is an excellent tool for the study of genome organization. In one very clear example, hybrid speciation of wild sunflowers is accompanied by genomic reorganization, and evidence of this reorganization was found in comparative linkage maps generated with AP-PCR/RAPD markers (Rieseberg *et al.,* 1995). In general, polymorphic markers generated by AP-PCR or RAPD are inherited with expected Mendelian frequency. In the diploid organism *Pythium sylvaticum,* however, Martin (1995) observed deviation from expected frequencies consistent with meiotic instability, which was verified by karyotype analysis. One possibility suggested by Martin is that mitotic instability may allow sexual outcrossing to occur between isolates with significantly different karyotypes or ploidy (Martin, 1995). If this occurs in nature, it could profoundly effect the evolution of the organism. Valles *et al.* (1993) used fingerprinting to demonstrate genome stability in *Festuca pratensis* Huds. generated from embryogenic suspension cultures. This is important because genetic instability in some plants has been observed after regeneration from cell culture. DNA fingerprinting can be used to resolve between hypotheses for mechanisms of unreduced egg formation. In diploid alfalfa, DNA fingerprinting was used to determine whether apomeiotic 2n eggs developed through parthenogenesis (Barcaccia *et al.,* 1994).

Finally, in one of the most entertaining studies that employed DNA fingerprinting, Smith *et al.* (1992) showed that *Armillaria bulbosa,* a facultative tree-root pathogen found in Europe and eastern North American forests, is one of the oldest and largest living organisms. DNA fingerprinting was used by these authors to distinguish between individual and inbred siblings and outbred siblings. This study identified an individual estimated to occupy greater than 15 hectares, and with a mass greater than 10,000 kg. One individual was estimated to be at least 1,500 years old.

RNA Fingerprinting by AP-PCR and Differential Display

When the principles of AP-PCR were applied to cDNA, a new approach to studying differential gene expression was born. Differential Display (Liang and

Pardee 1994), and our method, RAP-PCR (Welsh *et al.,* 1992a), have become important methods for the discovery of genes whose expression correlates with other biological phenomena. Over the past year alone pieces of more than 50 genes have been cloned; genes whose expression patterns correlate with interesting biological phenomena. Below, we discuss the methods briefly, as well as several important current concerns.

RNA Fingerprinting Methods

The differential display method of Liang and Pardee (1994) employs a primer of the form oligo $d(T)_n$-MX for first strand cDNA synthesis and an arbitrarily selected 10-mer for second strand synthesis. The anchor (MX) is intended to lock the primer to the 5′ end of the poly-A tails of most eukaryotic messages. For second strand synthesis, the 10-mer primer is used near its T_m (~37°C) to discourage mis-match priming. This step following the flanking of arbitrarily chosen messages with one anchor primer and one arbitrary primer, the flanked sequences can be amplified by PCR and arrayed by electrophoresis. For fingerprinting RNA from bacteria, the anchor primer strategy is not appropriate because bacterial mRNA is usually not polyadenylated.

RAP-PCR is distinct from DD in that it uses arbitrary priming for both first and second strand cDNA synthesis (Welsh *et al.,* 1992a). First strand synthesis is initiated with an arbitrarily chosen primer at low stringency, the double stranded product is denatured, and second strand synthesis is, likewise, initiated with the same or a second arbitrary primer. Thus, RAP-PCR is appropriate for use in prokaryotes (e.g. Wong and McClelland, 1994). In eukaryotes, this approach allows the sampling of open reading frames more frequently than DD, and using two primers successively, the orientation of the product relative to message polarity can be known. This latter feature is important when performing database searches of phylogenetically distant organisms and when correct conceptual translation of a short fragment is needed.

RAP-PCR can reveal differential gene expression in a number of experimental scenarios. First, we compared gross anatomical tissues from the mouse because we knew that differentially regulated genes would be easy to find (Welsh *et al.,* 1992a). These experiments demonstrated that differentially regulated mRNAs could be detected by the method, that a few bands are concentration dependent as with AP-PCR, and that differences in banding patterns can also arise from sequence polymorphisms. The latter, differences in fingerprint patterns arising from sequence polymorphisms, do not reflect differential gene regulation. Thus, experiments where different individuals or strains are being compared must be structured so that differential gene regulation can be distinguished from polymorphisms.

Before RNA fingerprinting was developed, differential screening and subtractive hybridization were the principle methods for identifying differentially regulated genes on the basis of their mRNA abundances. Differential screening was hampered by its relative insensitivity. Subtractive hybridization was usually hampered by the difficulty of driving subtraction to completion, such that additional sorting must be done, sometimes between thousands of clones. RNA fingerprinting does not completely solve the sensitivity problem of differential screening, as discussed above. In theory, subtraction methods allow the cloning of very rare messages, but this is seldom achieved in practice. RAP-PCR and DD trade off sensitivity with technical ease. Because many very interesting biological phenomena involve some member of the 1,000 or so most abundant messages, RAP-PCR is a very powerful method.

Usually, RAP-PCR with 10-mers gives rise to products that have about 6 nucleotides of perfect match with the target mRNA sequence at the 3′ end of the primer. This was determined by an examination of known sequences that have appeared in various experiments. About a tenth of all messages will have two sites that match with the primer in the proper orientations to give a product, and when such messages are abundant, they give rise to robust fingerprint bands. Therefore, the high and moderate abundance classes dominate the fingerprint. Abundant messages are certainly interesting, but a greater variety of phenomena will be reflected in mRNA changes in the rare class. Thus, "abundance normalized sampling", that is, sampling that is independent of abundance, is an important technical goal.

The abundance normalization of sampling has been addressed in a few ways, but there is room for improvement. cDNA synthesis using reverse transcriptase can only be performed at low stringency (i.e. at <45°C, 4 mM magnesium). Under these conditions, even 10-mers (including the anchor primers, oligo $d(T)_n$-MX, used in DD) prime first-strand synthesis from many sites where the match is poor. The first improvement, therefore, would be to synthesize first strand cDNA with oligo-d(T), followed by relatively high stringency AP-PCR in *both directions* to create templates for PCR. Because these two latter steps can be done at high stringency (relative to the length of the primer) this approach may improve abundance normalization. However, at temperatures where mismatches are prohibited, 10-mer-target interactions are not very stable, while at lower temperatures, mismatches are encouraged, and it may be impossible to adjust this parameter such that only perfect matches are amplified. This is coupled with the very serious practical problem of needing to optimize conditions for each primer, which impacts negatively on throughput. Thus, some other way of solving the problem of abundance normalized sampling must be sought.

One approach we have tried is "nested arbitrary priming" (Ralph *et al.,* 1993). In this approach, a fingerprint is generated by the standard RAP-PCR protocol; then, a portion of the first fingerprint is *re-amplified* using a primer related to the initial primer, but having one or more *additional* 3′ arbitrarily chosen nucleotides. High stringency PCR amplification with such a primer results in the amplification of a subset of the original fingerprint. Because the rare class of messages has high sequence complexity, the additional nucleotide(s) is more likely to find a match in molecules from this class. We have not yet calibrated this method: it is technically demanding because it requires very high quality initial fingerprints. However, in principle, this solves the abundance normalization problem. In another approach, a fingerprint can be re-fingerprinted using an entirely different primer. Each successive fingerprinting of a previous fingerprint should result in higher selectivity, and therefore, greater abundance normalization of sampling.

Finally, the purification of bands from RAP-PCR gels has been time-consuming, but we have solved this problem. First, a band of interest is identified by RAP-PCR. Second, the fingerprint reaction that contains the product is loaded on a preparative sequencing-style gel, the gel is dried, exposed to x-ray film to localize the band, and the band is cut from the dried gel. Third, the band is eluted from the gel into 100 μl TE for several hours at 68°C, and the radioactive product is ethanol precipitated. Fourth, the precipitated material is loaded on an SSCP gel. In this step, background bands that have unrelated sequence travel to markedly different places in the gel. Fifth, the band is cut from this gel, eluted, reamplified and cloned or directly sequenced. This procedure is generally adequate to purify the product. Often, all resulting clones are of a single sequence.

Detection of Differentially Regulated Genes in Response to Developmental Cues: "In Parallel" Studies

The rationale behind every RNA fingerprinting experiment is to correlate the differential abundance of a message with some observable biological event. Genes have been found that correlate with developmental phenomena (e.g. Dalal *et al.*, 1994), neoplasia (Wong *et al.*, 1993), hormonal treatment (Ralph *et al.*, 1993), drug treatment (McClelland *et al.*, 1994b) and a wide variety of other experimental circumstances. An illustration is presented in Figure 33-2. The phenotype of this gene is very interesting even before we know anything at all about its sequence.

RNA fingerprinting can be used together with combinatorial strategies for experimental design (McClelland *et al.*, 1994b). When several experimental variables are included in the same fingerprinting experiment, genes can be identified that fit a certain pre-defined criteria. The design of the RNA fingerprinting experiment depends, therefore, on the complex relationship between phenotype and gene expression. In the end, the methods return sequences of genes that *correlate,* in mRNA abundance, with some biological phenomenon. For example, we used a combinatorial strategy to parse genes into 8 of 35 observable categories by their responses to transforming growth factor-beta (TGFβ), cycloheximide, or both. That only 8 of the 35 categories are heavily occupied implies limitations to pleiotropic mechanisms in use by TGFβ (Ralph *et al.*, 1993, McClelland *et al.*, 1994b). This strategy can certainly be employed in bacteria; RAP-PCR has been used to detect differential gene expression in *S. typhimurium* in response to oxidative stress (Wong and McClelland, 1994).

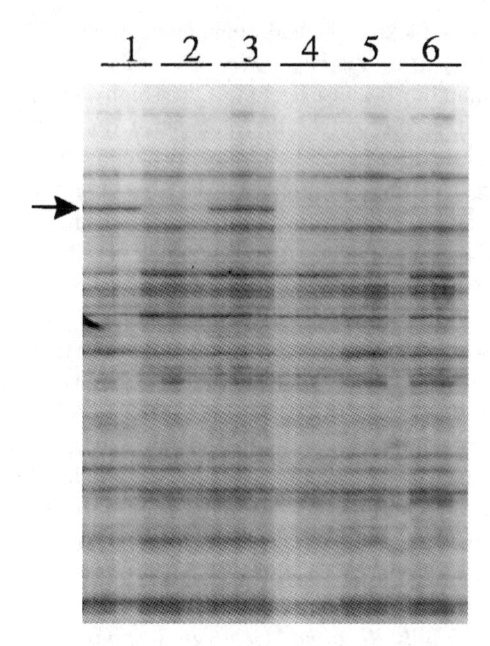

Figure 33-2. Mv1Lu cells were maintained at confluence for 16 hr. and reseeded at lower density at t=0. At 4 hr. Transforming growth factor-β was added to sample 3. At 4.5 hr., cycloheximide (Cx) was added to samples 2–6. At 5.5 hr. 1–3 were harvested, TGFβ was added to sample 4 and Cx was washed away from sample 5. At 7 hr., 4–6 were harvested. RNA was fingerprinted by RAP-PCR. The band appearing in lanes 1 and 3 informs us that Cx causes the down-regulation of the product, but that this effect is overridden by TGFβ. TGFβ prevents loss of the band, but does not promote its recovery once lost (sample 4), and the band does not recover quickly once Cx is washed away (sample 5).

References

Akopyanz, N., N. O. Bukanov, T. U. Westblom, S. Kresovich, and D. E. Berg. 1992. DNA diversity among clinical isolates of *Helicobacter pylori* detected by PCR-based RAPD fingerprinting. *Nucl. Acids Research* 20:5137–5142.

Al Janabi, S. M., R. J. Honeycutt, M. McClelland, and B. S. Sobral. 1993. A genetic linkage map of *Saccharum spontaneum* L. 'SES 208'. *Genetics* 134:1249–1260.

Anzai, Y., T. Okuda, and J. Watanabe. 1994. Application of the random amplified polymorphic DNA using the polymerase chain reaction for efficient elimination of duplicate strains in microbial screening. II. Actinomycetes. *J. of Antibiotics* 47:183–193.

Apostol, B. L., W. C. Black IV, B. R. Miller, P. Reiter, and B. J. Beaty. 1993. Estimation of the number of full sibling families at an oviposition site using RAPD-PCR markers: applications to the mosquito *Aedes aegypti. Theor. Appl. Genet.* 86:991–1000.

Apostol, B. L., W. C. Black IV, P. Reiter, and B. R. Miller. 1994. Use of randomly amplified polymorphic DNA amplified by polymerase chain reaction markers to estimate the number of *Aedes aegypti* families at oviposition sites in San Juan, Puerto Rico. *Am. J. Trop. Med. Hyg.* 51:89–97.

Aznar, J., H. Safi, J. Romero, A. Alejo, A. Gracia, and J. C. Palomares. 1995. Nosocomial transmission of tuberculosis infection in pediatrics wards. *Pediatr. Infect. Dis. J.* 14:44–48.

Barbut, F., N. Mario, M. Delmee, J. Gozian, and J.-C. Petit. 1993. Genomic fingerprinting of *Clostridium difficile* isolates by using a random amplified polymorphic DNA (RAPD) assay. *FEMS Microbiol. Ltrs.* 161–166.

Barbut, F., N. Mario, M. C. Meyohas, D. Binet, J. Frottier, and J. C. Petit. 1994. Investigation of a nosocomial outbreak of *Clostridium difficile*-associated diarrhoea among AIDS patients by random amplified polymorphic DNA (RAPD) assay. *J. of Hospital Infection* 26:181–189.

Barcaccia, G., S. Tavoletti, M. Pezzotti, M. Falcinelli, and F. Veronesi. 1994. Fingerprinting of alfalfa meiotic mutants using RAPD markers. *Euphytica* 80:19–25.

Bingen, E., C. Boissinot, P. Desjardins, H. Cave, N. Brahimi, N. Lambert-Zechovsky, E. Denamur, P. Blot, and J. Elion. 1993. Arbitrarily primed polymerase chain reaction provides rapid differentiation of *Proteus mirabilis* isolates from a pediatric hospital. *J. of Clin. Microbiol.* 31:1055–1059.

Birkenmeier, E. H., U. Schneider, and S. J. Thurston. 1992. Fingerprinting genomes by use of PCR primers that encode protein motifs or contain sequences that regulate gene expression. *Mamm. Genome* 3:537–545.

Black, S. F., D. I. Gray, D. R. Fenlon, and R. G. Kroll. 1995. Rapid RAPD analysis for distinguishing Listeria species and *Listeria monocytogenes* serotypes using a capillary air thermal cycler. *Ltrs. in Appl. Microbiol.* 20:188–190.

Boerlin, P., E. Bannerman, F. Ischer, J. Rocourt, and J. Bille. 1995. Typing *Listeria monocytogenes:* a comparison of random amplification of polymorphic DNA with 5 other methods. *Res. Microbiol.* 146:35–49.

Burt, A., D. A. Carter, T. J. White, and J. W. Taylor. 1994. DNA sequencing with arbitrary primer pairs. *Molecular Ecology* 3:523–525.

Cancilla, M. R., I. B. Powell, A. J. Hillier, and B. E. Davidson. 1992. Rapid genomic fingerprinting of *Lactococcus lactis* strains using arbitrarily primed polymerase chain reaction with ^{32}P and fluorescent labels. *Appl. and Environmental Microbiol.* 58:1772–1775.

Coelho, A., J.R.C. Andrade, A.C.P. Vicente, and C. A. Salles. 1995a,b. New variant of *Vibrio cholerae* 01 from clinical isolates in Amazonia. *J. of Clin. Microbiol.* 33:114–118.

Coelho, A., A.C.P. Vicente, M.A.S. Baptista, H. Momen, F.A.R.W. Santos, and C. A. Salles. 1995. The distinction of pathogenic *Vibrio cholerae* groups using arbitrarily primed PCR fingerprints. *Res. Microbiol.* 146:671–683.

Dalal S., J. Welsh, A. Tkachenko, D. Ralph, E. DiCicco-Bloom, M. McClelland, and K. Chanda. 1994. Rapid isolation of tissue-specific and developmentally regulated brain cDNAs using RNA Arbitrarily Primed PCR (RAP-PCR). *J. Molecular Neuroscience* 5:93–104.

Damas de Caballero, O.L.S., E. D. Neto, M. C. Koury, A. J. Romanha, and A.J.G. Simpson. 1994. Low-stringency PCR provides an internal control for negative results in PCR-based diagnosis. *PCR Methods and Applications* 3:305–307.

Daffonchio, D., G. Monaco, E. Zanardini, and C. Sorlini. 1994. AP-PCR fingerprinting of *Pseudomonas* strains isolated from different environmental matrices. *Med. Fac. Landbouww. Univ. Gent.* 59/4b:2121–2129.

Damas de Caballero, O.L.S., E. D. Neto, M. C. Koury, A. J. Romanha, and A.J.G. Simpson. 1994. Low-stringency PCR with diagnostically useful primers for identification of *Leptospira* serovars. *J. of Clin. Microbiol.* 32:1369–1372.

Demeke, T., and R. P. Adams. 1994. The use of RAPDS to determine germplasm collection strategies in the african species *Phytolacca dodecandra* (phytolaccaceae). In: *Conservation of Plant Genes II: Utilization of Ancient and modern DNA.* (R. P. Adams, J. S. Miller, E. M. Golenberg, and J. A. Adams, eds.), 131–139.

Eisen, D., E. G. Russell, M. Tymms, E. J. Roper, M. L. Grayson, and J. Turnidge. 1995. Random amplified polymorphic DNA and plasmid analyses used in investigation of an outbreak of multiresistant *Klebsiella pneumoniae. J. of Clin. Microbiol.* 33:713–717.

Fadl, A. A., A. V. Nguyen, and M. I. Khan. 1995. Analysis of *Salmonella enteritidis* isolates by arbitrarily primed PCR. *J. of Clin. Microbiol.* 33:987–989.

Farber, J. M., and C. J. Addison. 1994. RAPD typing for distinguishing species and strains in the genus *Listeria. J. of Appl. Bacteriol.* 77:242–250.

Fekete, A., J. A. Bantle, S. M. Halling, and R. W. Stich. 1992. Amplification fragment length polymorphism in *Brucella* strains by use of polymerase chain reaction with arbitrary primers. *J. of Bacteriol.* 174:7778–7783.

Fontana, J. D., S. Astolfi Fo, R. Rogelin, J. Kaiss, M.C.O. Hauly, V. C. Franco, and M. Baron. 1994. PCR protocol- and inulin catabolism-based differentiation of inulinolytic soil bacteria. *Appl. Biochem. and Biotechnol.* 45/46:269–282.

Fritz, R. S., C. M. Nichols-Orians, and S. J. Brunsfeld. 1994. Interspecific hybridization of plants and resistance to herbivores: hypotheses, genetics, and variable responses in a diverse herbivore community. *Oecologia* 97:106–117.

Gomez-Lus, P., B. S. Fields, R. F. Benson, W. T. Martin, S. P. O'Connor, and C. M. Black. 1993. Comparison of arbitrarily primed polymerase chain reaction, ribotyping, and monoclonal antibody analysis for subtyping *Legionella pneumophila* serogroup 1. *J. of Clin. Microbiol.* 31:1940–1942.

Grattapaglia, D., and R. Sederoff. 1994. Genetic linkage maps of *Eucalyptus grandis* and *Eucalyptus urophylla* using a pseudo-testcross: mapping strategy and RAPD markers. *Genetics* 137:1121–1137.

Gray, D. I., and R. G. Kroll. 1995. Polymerase chain reaction amplification of the *flaA* gene for the rapid identification of *Listeria* spp. *Ltrs. in Appl. Microbiol.* 20:65–68.

Hedrick, P. 1992. Shooting the RAPDs. *Nature* 355:679–680.

Hennessy, K. J., J. J. Iandolo, and B. W. Fenwick. 1993. Serotype identification of *Actino-*

bacillus pleuropneumoniae by arbitrarily primed polymerase chain reaction. *J. of Clin. Microbiol.* 31:1155–1159.

Ionov, Y., M. A. Peinado, S. Malkhosyan, D. Shibata, and M. Perucho. 1993. Ubiquitous somatic mutations in simple repeated sequences reveal a new mechanism for colonic carcinogenesis. *Nature* 363:558–561.

Kay, H. E., H.L.C. Coutinho, M. Fattori, G. P. Manfio, R. Goodacre, M. P. Nuti, M. Basaglia, and J. E. Beringer. 1994. The identification of *Bradyrhizobium japonicum* strains isolated from Italian soils. *Microbiology* 140:2333–2339.

Kubota, Y., A. Shimada, and A. Shima. 1995. DNA alterations detected in the progeny of paternally irradiated Japanese medaka fish *(Oryzias latipes). Proc. Natl. Acad. Sci. U.S.A.* 92:330–334.

Lawrence, L. M., J. Harvey, and A. Gilmour. 1993. Development of a random amplification of polymorphic DNA typing method for *Listeria monocytogenes. Appl. and Environm. Microbiol.* 59:3117–3119.

Lee, T. Y., T. J. Lee, and S. K. Kim. 1994. Differentiation of *Mycobacterium tuberculosis* strains by arbitrarily primed polymerase chain reaction-based DNA fingerprinting. *Yonsei. Med. J.* 35:286–294.

Liang, P., and A. Pardee. 1992. Differential display of eukaryotic messenger RNA by means of the polymerase chain reaction. *Science* 257:967–971.

Linton, C. J., H. Jalal, J. P. Leeming, and M. R. Millar. 1994. Rapid discrimination of *Mycobacterium tuberculosis* strains by random amplified polymorphic DNA analysis. *J. of Clin. Microbiol.* 32:2169–2174.

Louws, F. J., D. W. Fulbright, C. T. Stephens, and F. J. de Bruijn. 1994. Specific genomic fingerprints of phytopathogenic *Xanthomonas* and *pseudomonas* pathovars and strains generated with repetitive sequences and PCR. *Appl. and Environm. Microbiol.* 60:7.

MacGowan, A. P., K. O'Donaghue, S. Nicholls, M. McLauchlin, P. M. Bennett, and D. S. Reeves. 1993. Typing of Listeria spp. by random amplified polymorphic DNA (RAPD) analysis. *J. Med. Microbiol.* 38:322–327.

Marquet-Van Der Mee, N., S. Mallet, J. Loulergue, and A. Audurier. 1995. Typing of *Staphylococcus epidermidis* strains by random amplification of polymorphic DNA. *FEMS Microbiol. Ltrs.* 39–44.

Mathieu-Daudé F., J. Welsh, T. Vogt, and M. McClelland. 1996. DNA rehybridization during PCR: the "Cot effect" and its consequences. *Nucl. Acids Res.* 24: (in press).

Mathieu-Daudé, F., J. Welsh, M. Tibayrenc, and M. McClelland. 1995. Genetic diversity and population structure of *Trypanosoma brucei. Mol. Biochem. Parasitol.* 72:89–101.

McClelland, M., and J. Welsh. 1994a. DNA fingerprinting by arbitrarily primed PCR. *PCR Methods and Applications.* 4:S59–S65.

McClelland, M., and J. Welsh. 1994b. RNA fingerprinting by arbitrarily primed PCR. *PRC Methods and Applications* 4:S66–S81.

McClelland, M., H. Arensdorf, R. Cheng, and J. Welsh. 1994a. Arbitrarily primed PCR fingerprints resolved on SSCP gels. *Nucl. Acids Res.* 22:1770–1771.

McClelland, M., D. Ralph, R. Cheng, and J. Welsh. 1994b. Interactions among regulators of RNA abundance characterized using RNA fingerprinting by arbitrarily primed PCR. *Nucl. Acids Res.* 22:4419–4431.

McCoy, T. J., and C. S. Echt. 1993. Potential of trispecies bridge crosses and random amplified polymorphic DNA marers for introgression of *Medicago daghestanica* and *M. pironae* germplasm into alfalfa (M. sativa). *Genome* 36:594–601.

Menard, C., and C. Mouton. 1995. Clonal diversity of the taxon *Porphyromonas gingivalis* assessed by random amplified polymorphic DNA fingerprinting. *Infect. Immun.* 63:2522–2531.

Myers, L. E., S.V.P.S. Silva, J. D. Procunier, and P. B. Little. 1993. Genomic fingerprinting of *Haemophilus somnus* isolates by using a random-amplified polymorphic DNA assay. *J. of Clin. Microbiol.* 31:512–517.

Odinot, P. T., J.F.G.M. Meis, P.J.J.C. Van Den Hurk, J.A.A. Hoogkamp-Korstanje, and W.J.G. Melchers. 1995a. PCR-based characterization of *Yersinia enterocolitica:* comparison with biotyping and serotyping. *Epidemiol. Infect.* 15:269–277.

Odinot, P. T., J.F.G.M. Meis, P.J.J.C. Van Den Hurk, J.A.A. Hoogkamp-Korstanje, and W.J.G. Melchers. 1995b. PCR-based DNA fingerprinting discriminates between different biotypes of *Yersinia enterocolitica. Contrib. Microbiol. Immunol.* 13.

Palittapongarnpim, P., S. Chomyc, A. Fanning, and D. Kunimoto. 1993a. DNA fingerprinting of *Mycobacterium tuberculosis* isolates by ligation-mediated polyerase chain reaction. *Nucleic Acids Research* 21:7661–7662.

Palittapongarnpim, P., S. Chomyc, A. Fanning, and D. Kunimoto. 1993b. DNA fragment length polymorphism analysis of *Mycobacterium tuberculosis* isolates by arbitrarily primed polymerase chain reaction. *J. of Infectious Diseases* 167:975–978.

Peinado, M. A., S. Malkhosyan, A. Velazquez, and M. Perucho. 1992. Isolation and characterization of allelic losses and gains in colorectal tumors by arbitrarily primed polymerase chain reaction. *Proc. Natl. Acad. Sci.* 89:10065–10069.

Preus, H. R., V. I. Haraszthy, J. J. Zambon, and R. J. Genco. 1993. Differentiation of strains of *Actinobacillus actinomycetemcomitans* by arbitrarily primed polymerase chain reaction. *J. of Clin. Microbiol.* 31:2773–2776.

Ralph, D., M. McClelland, and J. Welsh. 1993. RNA fingerprinting using arbitrarily primed PCR identifies differentially regulated RNAs in Mink lung (Mv1Lu) cells growth arrested by transforming growth factor β1. *Proc. Natl. Acad. Sci. USA* 90:10710–10714.

Ralph, D., D. Postic, G. Baranton, C. Pretzman, and M. McClelland. 1993. Species of *Borelia* distinguished by restriction site polymorphisms in 16S rRNA genes. *FEMS Micro. Lett.* 111:239–244.

Rasmussen, H. N., J. E. Olsen, and O. F. Rasmussen. 1994. RAPD analysis of *Yersinia enterocolitica. Ltrs. in Appl. Microbiol.* 19:359–362.

Sandery, M., J. Coble, and S. McKersie-Donnolley. 1994. Random amplified polymorphic DNA (RAPD) profiling of *Legionella pneumophila. Ltrs. in Appl. Microbiol.* 19:184–187.

Sanguinetti, C. J., E. D. Neto, and A.J.G. Simpson. 1994. Rapid silver staining and recovery of PCR products separated on polyacrylamide gels. *BioTechniques* 17:3–6.

Singsit, C., and P. Ozias-Akins. 1993. Genetic variation in monoploids of diploid potatoes and detection of clone-specific random amplified polymorphic DNA markers. *Plant Cell Reports* 12:144–148.

Slots, J., Y. B. Liu, J. M. DiRienzo, and C. Chen. 1993. Evaluating two methods for fingerprinting genomes of *Actinobacillus actinomycetemcomitans. Oral Microbiol. Immunol.* 8:337–343.

Smith, M. L., J. N. Bruhn, and J. B. Anderson. 1992. The fungus *Armillaria bulbosa* is among the largest and oldest living organisms. *Nature* 356:428–431.

Stevens, J. R., and M. Tibayrenc. 1994. Detection of linkage disequilibrium in *Trypanosoma brucei* isolated from tsetse flies and characterized by RAPD analysis and isoenzymes. *Parasitology* 110:181–186.

Struelens, M. J., R. Bax, A. Deplano, W.G.V. Quint, and A. Van Belkum. 1993. Concordant clonal delineation of methicillin-resistant *Staphylococcus aureus* by macrorestriction analysis and polymerase chain reaction genome fingerprinting. *J. of Clin. Microbiol.* 31:1964–1970.

Taylor, N. S., J. G. Fox, N. S. Akopyants, D. E. Berg, N. Thompson, B. Shames, L. Yan, E. Fontham, F. Janney, F. M. Hunter, and P. Correa. 1995. Long-term colonization with single and multiple strains of *Helicobacter pylori* assessed by DNA fingerprinting. *J. of Clin. Microbiol.* 33:918–923.

Tibayrenc, M., K. Neubauer, C. Barnabe, F. Guerrini, D. Skarecky, and F. J. Ayala. 1993. Genetic characterization of six parasitic protozoa: parity between random-primer DNA typing and multilocus enzyme electrophoresis. *Proc. Natl. Acad. Sci. USA* 90:1335–1339.

Valles, M. P., Z. Y. Wang, P. Montavon, I. Potrykus, and G. Spangenberg. 1993. Analysis of genetic stability of plants regenerated from suspension cultures and protoplasts of meadow fescue (Festuca pratensis Huds.). *Plant Cell Reports* 12:101–106.

van Belkum, A., J. Kluytmans, W. van Leeuwen, R. Bax, W. Quint, E. Peters, A. Fluit, C. Vandenbroucke Grauls, A. van den Brule, H. Koeleman, *et al.* 1995. Multicenter evaluation of arbitrarily primed PCR for typing of *Staphylococcus aureus* strains. *J. Clin. Microbiol.* 33:1537–1547.

Van Belkum, A., M. Struelens, and W. Quint. 1993. Typing of *Legionella pneumophila* strains by polymerase chain reaction-mediated DNA fingerprinting. *J. of Clin. Microbiol.* 31:2198–2200.

VanCouwenberghe, C. J., S. H. Cohen, Y. J. Tang, P. H. Gumerlock, and J. Silva, Jr. 1995. Genomic fingerprinting of epidemic and endemic strains *Stenotrophomonas maltophilia* (formerly *Xanthomonas maltophilia*) by arbitrarily primed PCR. *J. of Clin. Microbiol.* 33:1289–1291.

Vila, J., A. Marcos, T. Llovet, P. Coll, and T. Jimenez de Anta. 1994. A comparative study of ribotyping and arbitrarily primed polymerase chain reaction for investigation of hospital outbreaks of *Acinetobacter baumannii* infection. *J. Med. Microbiol.* 41:244–249.

Virk, P. S., B. V. Ford-Lloyd, M. T. Jackson, and H. J. Newbury. 1995. Use of RAPD for the study of diversity within plant germplasm collections. *Heredity* 74:170–179.

Welsh, J., and M. McClelland. 1990. Fingerprinting genomes using PCR using arbitrary primers. *Nucl. Acids Res.* 18:7213–7218.

Welsh, J., and M. McClelland. 1991. Genomic fingerprints produced by PCR with consensus tRNA gene primers. *Nucl. Acids Res.* 19:861–866.

Welsh, J., and M. McClelland. 1992. PCR-amplified length polymorphisms in tRNA intergenic spacers for categorizing staphylococci. *Mol. Microbiol.* 6:1673–1680.

Welsh, J., K. Chada, S. Dalal, R. Cheng, D. Ralph, and M. McClelland. 1992a. Arbitrarily primed PCR fingerprinting of RNA. *Nucl. Acids Res.* 20:4965–4970.

Welsh, J., C. Petersen, and M. McClelland. 1992b. Polymorphisms in tRNA intergenic spacers detected by using the polymerase chain reaction can distinguish streptococcal strains and species. *J. Clin. Microbiol.* 30:1499–1504.

Welsh, J., C. Petersen, and M. McClelland. 1991. Polymorphisms generated by arbitrarily primed PCR in the mouse: Application to strain identification and genetic mapping. *Nucl. Acids Res.* 19:303–306.

Welsh, J., C. Pretzman, D. Postic, I. Saint Girons, G. Baranton, and M. McClelland. 1992. Genomic fingerprinting by arbitrarily primed PCR resolves *Borrelia burgdorferi* into three distinct phyletic groups. *Int. J. System. Bacteriol.* 42:370–377.

Welsh, J., N. Rampino, M. McClelland, and M. Perucho. 1995. Nucleic acid fingerprinting by PCR-based methods: applications to problems in aging and mutagenesis. *Mutation Research* 338:215–229.

Williams, J. G., R. S. Reiter, R. M. Young, and P. A. Scolnik. 1993. Genetic mapping of mutations using phenotypic pools and mapped RAPD markers. *Nucl. Acids Res.* 21:2697–2702.

Wong, K.K., and M. McClelland. 1994. A stress-induced gene from *Salmonella typhimurium* identified by arbitrarily primed PCR of RNA (RAP). *Proc. Natl. Acad. Sci. USA* 91:639–643.

Wong, N.A.C.S., C. J. Linton, H. Jalal, and M. R. Millar. 1994. Randomly amplified polymorphic DNA typing: a useful tool for rapid epidemiological typing of *Klebsiella pneumoniae. Epidemiol. Infect.* 113:445–454.

Woods, J. P., D. Kersulyte, R. W. Tolan, Jr., C. M. Berg, and D. E. Berg. 1994. Use of arbitrarily primed polymerase chain reaction analysis to type disease and carrier strains of *Neisseria meningitidis* isolated during a university outbreak. *J. of Inf. Diseases* 169:1384–1389.

34

Repetitive Sequence-based PCR (rep-PCR) DNA Fingerprinting of Bacterial Genomes

James Versalovic, Frans J. de Bruijn, and James R. Lupski

Introduction

Bacterial chromosomes contain multiple interspersed repetitive sequences that occupy intergenic regions at sites dispersed throughout the genome. Such blocks of noncoding, repetitive sequences can serve as multiple genetic targets for oligonucleotide probes, enabling the generation of unique DNA profiles or fingerprints for individual bacterial strains. DNA fingerprinting requires the resolution of differently sized DNA fragments derived from chromosomal or plasmid DNA by restriction endonuclease-mediated digestion and/or DNA amplification to yield a band pattern that serves as a unique identifier. These unique "bar codes" or DNA fingerprints define each bacterial chromosome without the need for measuring gene expression or enzyme function. Genotypic or molecular approaches differ with respect to the level of resolution of individual bacterial species or strains into distinct categories (Figure 34-1).

Repetitive sequence-based polymerase chain reaction (rep-PCR) was introduced by Versalovic *et al.* (1991) and yields DNA fingerprints comprised of multiple differently-sized DNA amplicons. These amplicons contain unique sequence chromosomal segments lying between repetitive sequences, the latter of which are complementary to the oligonucleotide repetitive sequence primers. In contrast to hybridization methods, whereby DNA fingerprints indicate the presence of repetitive DNA elements in differently sized chromosomal restriction fragments (*see* Chapters 31, 32), rep-PCR based DNA fingerprints reflect varying distances between oligonucleotide primer binding sites at repetitive sequence targets (Versalovic *et al.,* 1991; van Belkum, 1994; Versalovic *et al.,* 1994; Louws *et al.,* 1996). Amplicons of different sizes can be fractionated by electrophoresis and constitute the DNA fingerprint patterns specific for individual bacterial clones or strains. The rep-PCR method has been useful for DNA fingerprinting

GENUS/SPECIES	SUBSPECIES/STRAIN
Ribotyping	ARDRA
tRNA-PCR	Chromosomal RFLP
ITS-PCR	ITS Sequencing
16S rRNA sequencing	Plasmid RFLP
	Pulsed-field gel elecrophoresis (PFGE)
	RAPD
	rep-PCR

Figure 34-1. List of bacterial genotypic typing methods according to ability to distinguish genus/species or subspecies/strains.

of a large variety of prokaryotic and eukaryotic microorganisms (van Belkum, 1994; Versalovic *et al.,* 1994; de Bruijn *et al.,* 1995; Louws *et al.,* 1996). Key advantages of rep-PCR based chromosomal typing include its speed, reproducibility, convenience with respect to lack of radioisotope usage, adaptability to intact cell and native tissue preparations, and modest resource requirements of standard equipment available in molecular biology laboratories. In terms of its sensitivity, rep-PCR genomic fingerprinting has been found to be most useful in the subspecies/strain-specific level (see Figure 34-1).

rep-PCR: The Method

Various repetitive DNA elements have been described in bacterial genomes (*see* Chapter 5). Primers complementary to repetitive sequence targets have been designed and used in rep-PCR based applications (Figure 34-2) (Versalovic *et al.,* 1994; Louws *et al.,* 1996). Sizes of interspersed repetitive elements may vary from 15 bases, in the case of repeated trinucleotides (e.g. GTG), to hundreds of base pairs (Lupski and Weinstock, 1992). The Repetitive Extragenic Palindromic (REP) element (Stern *et al.,* 1984) and the *Neisseria gonorrhoeae* repetitive (*Ng*rep) element (Correia *et al.,* 1986) are 38-bp and 26-bp in length respectively. Longer repetitive elements include the 300-bp RepMP1 (Wenzel and Herrmann, 1988; Forsyth and Geary, 1996) and 400-bp SDC1 (Colman *et al.,* 1990) elements initially identified in *Mycoplasma pneumoniae.* Primers varying between 15 and 30 base pairs in length and matching the consensus sequences of conserved repeated elements have been found to be useful for PCR-based typing applications,

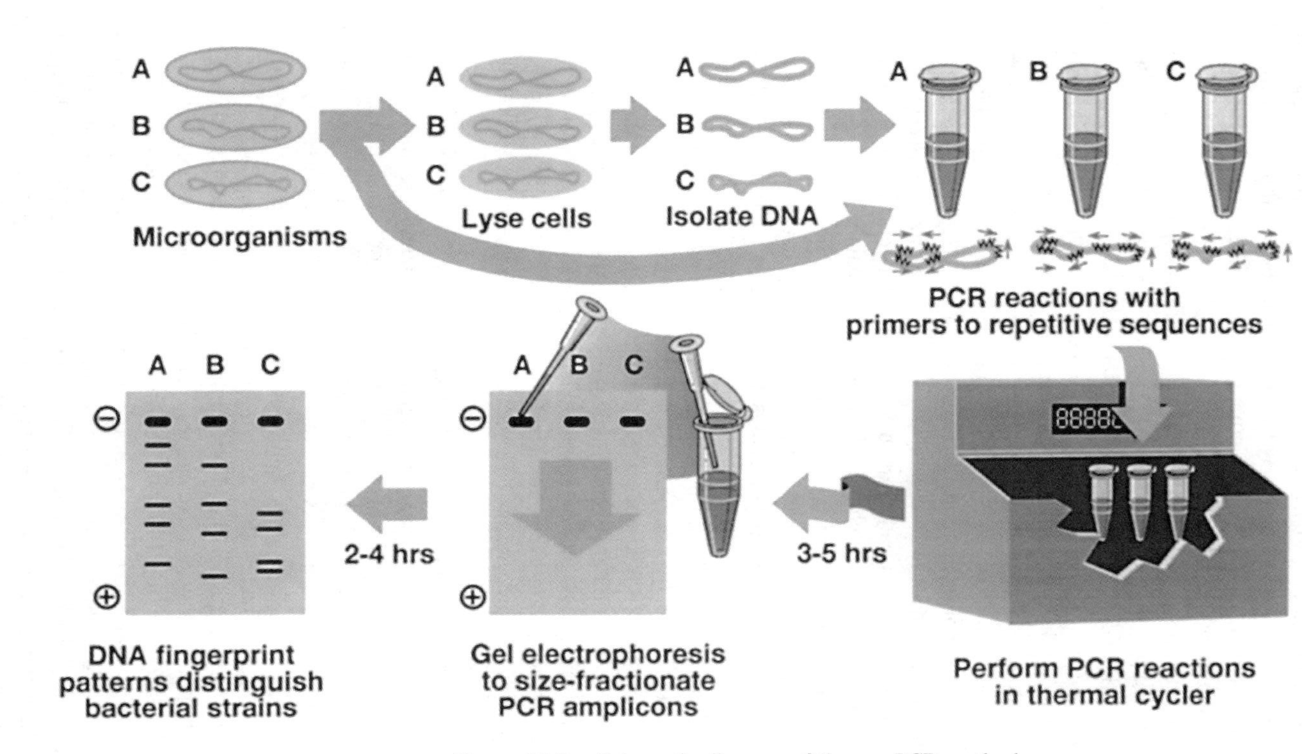

Figure 34-2. Schematic diagram of the rep-PCR method.

whereas randomized sequence based oligonucleotide primers lacked utility for DNA fingerprinting of bacteria (Versalovic and Lupski, 1995). Degenerate repetitive sequence based primers with either inosines or multiple bases at divergent positions of each repetitive element consensus sequence have been designed (Versalovic *et al.,* 1991; Versalovic *et al.,* 1994).

An alternative PCR-based DNA fingerprinting method, known as randomly amplified polymorphic DNA (RAPD) (Williams *et al.,* 1990) or arbitrary priming PCR (AP-PCR) (Welsh and McClelland, 1990) has also been described (*see* Chapter 33), employs primers of random sequence that must be tested empirically for their ability to generate DNA fingerprints. Random or arbitrary primers are often shorter in sequence (less than 20 bases in length), require less stringent annealing conditions, and therefore are occasionally not reproducible or can be subject to amplification artifacts (Ellsworth *et al.,* 1993). Nevertheless multiple studies have made use of arbitrary primers and their utility in epidemiologic studies of microbial pathogens has been amply documented (Ralph *et al.,* 1993; Welsh *et al.,* 1992; Taylor *et al.,* 1995) (*see* Chapter 33).

The key difference between PCR-based DNA fingerprinting and other PCR-based applications that yield single amplicons is that PCR-based genomic fingerprinting requires the simultaneous amplification of multiple differently sized DNA fragments. Smaller amplicons are amplified more efficiently than larger amplicons and conditions must be established to enable the simultaneous amplification of differently-sized DNA fragments ranging in size from several hundred base pairs to several kilobases. Reaction parameters that will facilitate multi-size amplicon detection are excess nucleotide triphosphates (dNTPs), excess primers, greater initial amounts of template DNA, and longer elongation times during each amplification cycle. Insufficient amounts of template genomic DNA and impure DNA preparations represent frequent causes of limited DNA fingerprint complexity.

Following amplification, PCR products must be size-fractionated to obtain DNA fingerprint patterns. Either agarose or acrylamide gel matrices may be used to separate the differently sized amplification products. For many applications, ethidium bromide-stained agarose gels provide sufficient resolution and allow detection of most rep-PCR generated amplicons (Versalovic *et al.,* 1994). Instead of intercalating dyes such as ethidium bromide, covalently attached fluorescent labels with characteristic emission spectra may be incorporated into the 5′ ends of repetitive sequence-based primers and used for detection following PCR and electrophoresis by laser scanning (Versalovic *et al.,* 1995; Del Vecchio *et al.,* 1995; Harvey *et al.,* 1995; de Bruijn *et al.,* 1995) (F. Louws, S. Rossbach, and F. J. de Bruijn, unpublished data). Fluorophore enhanced repetitive element PCR (FERP) (Versalovic *et al.,* 1995) has been applied, for example, in comparative analyses of bacterial isolates of medical importance. Since each dye possesses a characteristic emission spectrum when excited by an argon laser-generated light source, repetitive sequence based primers with different dyes can be multiplexed in the same reaction to produce complex fingerprint profiles. Alternatively, different fluorescent dyes can be used in separate PCR reactions and subsequently combined prior to electrophoresis. Gene specific and repetitive sequence derived primers with different fluorescent labels can be combined and differentially detected to maximize information yield per reaction (Versalovic *et al.,* 1995; Del Vecchio *et al.,* 1995; de Bruijn *et al.,* 1995). Simultaneous DNA fingerprinting and antimicrobial resistance gene detection (Del Vecchio *et al.,* 1995), toxin gene identification (Versalovic *et al.,* 1995), or the identification of a gene encoding catabolism of a nutritional mediator in soil microbes (de Bruijn *et al.,* 1995) have been combined in primer multiplexing reactions. Fluorescent signals are detected by photomultiplier tubes present within the DNA sequencing station and profiles are digitized and stored by a dedicated microcomputer. Acrylamide gels provide enhanced resolution of differently sized rep-PCR generated PCR amplicons.

Comparative analyses of DNA fingerprint patterns require the detection and storage of sufficiently complex patterns. The purpose of PCR-based DNA fingerprinting is to enable rapid differentiation of bacterial strains with relatively small initial amounts of sample. Once DNA fingerprint patterns are obtained, the question arises, "Are these strains clonally related or different?" Inevitably the issue becomes that of relative degrees of similarity for which more rigorous fingerprint comparison and quantitative, computer assisted methods are essential. However rapid screening of a limited series of isolates may be sufficient for epidemiologic purposes. DNA fingerprint patterns represented by ethidum bromide-stained agarose gels (Figure 34-3) may be stored on conventional photographic prints and compared by visual inspection. Qualitative visual comparisons may be sufficient to document whether bacterial isolates are identical or different from other isolates. Ethidium bromide stained images may be converted to digital information by charged coupled device (CCD) photography or scanning of photographic print data (Versalovic *et al.,* 1994; de Bruijn *et al.,* 1995; Schneider and de Bruijn, 1996). Digitized images may be converted to electrophoretograms depicting gel bands as peaks in line drawings (Versalovic *et al.,* 1995). FERP permits the automatic generation of electrophoretograms from digital fluorescent gel images (Figure 34-4). Electrophoretograms occupy significantly less disk storage space than gel images and represent the most convenient form for permanent "archival" fingerprints.

Rigorous analyses of sizeable sample sets require the systematic evaluation of relative degrees of similarity between DNA fingerprints. Multiple rep-PCR studies have included semiquantitative evaluations of similarities between bacterial isolates (Woods *et al.,* 1992; Versalovic *et al.,* 1994; van Berkum *et al.,* 1994; Leung *et al.,* 1994; Go *et al.,* 1995; Louws *et al.,* 1995; Schneider and de Bruijn, 1996). The presence or absence of bands at specific positions (specific sizes) may be scored manually (Woods *et al.,* 1992; Judd *et al.,* 1993; Reboli *et al.,* 1994; Go *et al.,* 1995; Versalovic and Lupski, 1995; Koeuth *et al.,* 1995) or

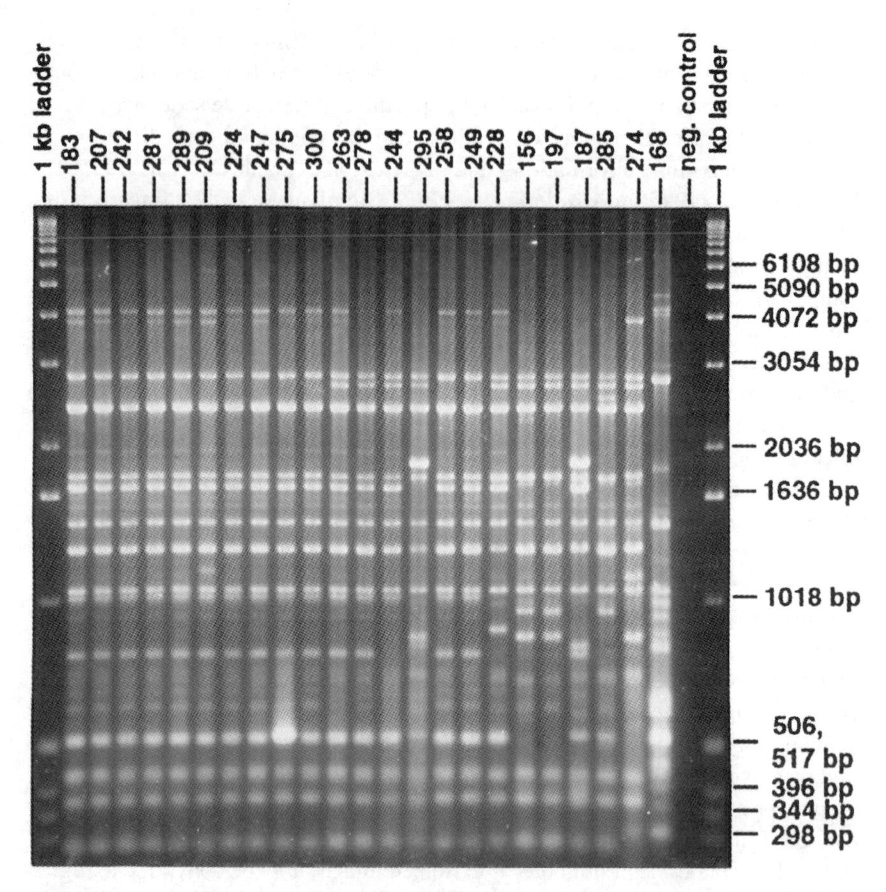

Figure 34-3. rep-PCR DNA fingerprints of *Streptococcus pneumoniae* isolates. Purified genomic DNA was prepared as described (Versalovic *et al.*, 1993) from pneumococcal isolates obtained from different human patients. The primer BOXA1R (Koeuth *et al.*, 1995) was used in the rep-PCR reactions. The negative control (neg. control) lane represents the same PCR reaction without template genomic DNA. The DNA molecular weight marker was a 1-kb ladder (Gibco-BRL). Gel was 1.5% agarose with 1x Tris-acetate-EDTA and was stained with 0.6 μg/ml ethidium bromide following electrophoresis.

▶

Figure 34-4. Fluorophore-enhanced rep-PCR electrophoretograms of *Citrobacter diversus* isolates obtained from an infected female patient. The isolate B14717 was typed by FERP with FAM (fluorophore)-labeled REP1R-I and FAM-labeled REP2-I primers on different dates (A, B, C; see insets) as described previously (Harvey *et al.*, 1995. The numbers at the top of the electrophoretogram represent laser scanning time and reflect electrophoretic mobility through the acrylamide gel matrix. These numbers represent increasing DNA fragment size from left to right. The numbers on the left represent the peak height of the fluorescent signals. Each electrophoretogram peak is analogous to a band on a gel (see Figure 34-3).

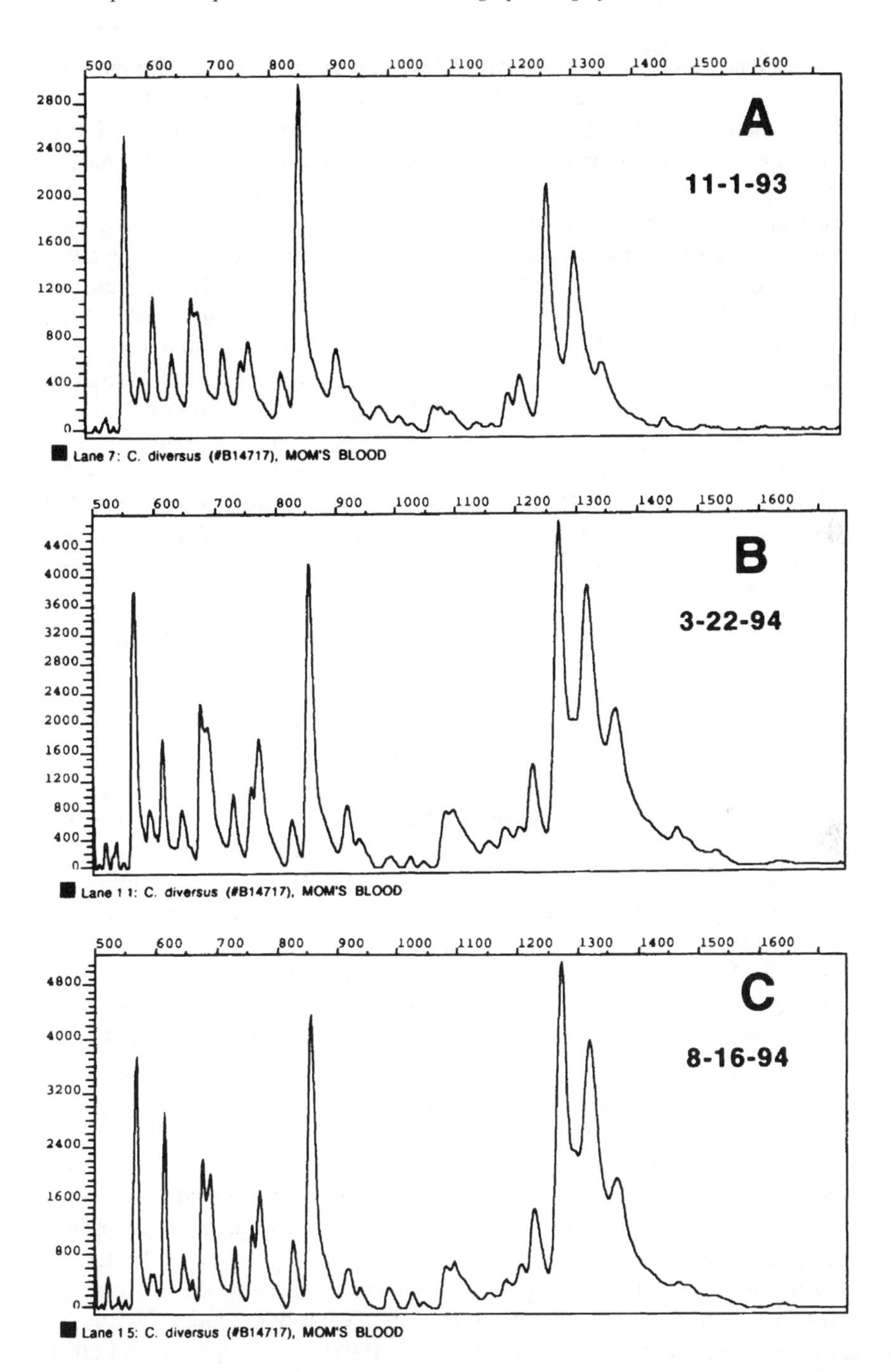
A
11-1-93
Lane 7: C. diversus (#B14717), MOM'S BLOOD
B
3-22-94
Lane 11: C. diversus (#B14717), MOM'S BLOOD
C
8-16-94
Lane 15: C. diversus (#B14717), MOM'S BLOOD

with computer-aided image analysis (Versalovic *et al.,* 1994; de Bruijn *et al.,* 1995; Schneider and de Bruijn, 1996). Manual methods require the tedious measurement of band migration distances using photographic prints. Each band position is recorded and presence or absence of a specific band is coded numerically (e.g. presence equals 1; absence equals 0). A binary code (presence or absence) matrix is produced for a given sample set and used as the basis for computer-aided pairwise sample comparisons, similarity coefficient generation, and clustering algorithms (Judd *et al.,* 1993; Go *et al.,* 1995). Alternatively, computer software systems such as AMBIS (Scanalytics, Waltham, MA, USA) or GELCOMPAR (Applied Math, Kortrijk, Belgium) will use digital images to size the bands relative to intragel standards (Versalovic *et al.,* 1994; Rossbach *et al.,* 1995; de Bruijn *et al.,* 1995; Schneider and de Bruijn, 1996). The same software systems will perform pairwise sample comparisons, similarity coefficient generation, and clustering analysis based on similarity values. FERP profiles may also be converted to "bar code" data for input into the AMBIS or GELCOMPAR software systems (de Bruijn *et al.,* 1995). The chief advantage of computer-aided analyses is the elimination of manual scoring for presence or absence of bands at each gel position.

Similarity coefficients are generated and serve as the values for graphical dendrogram construction based on cluster analysis. Different similarity coefficients can be calculated based on relative weightings of positive and negative matches (for explanations, see Sokal and Sneath, 1963; Versalovic *et al.,* 1994; Schneider and de Bruijn, 1996). For example the simple matching (SM) coefficient includes consideration of both positive and negative band matches whereas the Jaccard coefficient ignores negative matches. Most rep-PCR-based studies used the SM coefficient (van Berkum *et al.,* 1994; Leung *et al.,* 1994; Go *et al.,* 1995; Louws *et al.,* 1995; Versalovic and Lupski, 1995; Koeuth *et al.,* 1995) so as to avoid weighing the significance of positive versus negative matches. The unweighted pair group mathematical averaging (UPGMA) clustering method (Sneath and Sokal, 1973) represents the most commonly used approach for constructing graphical dendrograms (Versalovic *et al.,* 1994; Go *et al.,* 1995; Versalovic and Lupski, 1995; Koeuth *et al.,* 1995; Louws *et al.,* 1995; van Belkum *et al.,* 1996). Dendrograms depict relative degrees of similarity within a sample set in an easily visualized format. An alternative way of depicting similarity groups is by correspondence and cluster analysis (Vera Cruz *et al.,* 1995). Importantly, similarity cluster data generated by rep-PCR have been found to be consistent with data generated by other approaches including biotyping (Louws *et al.,* 1995), chromosomal restriction fragment length polymorphism (RFLP) analysis (Georghiou *et al.,* 1995; Edel *et al.,* 1995) (*see* Chapter 31), DNA:DNA hybridization (Leung *et al.,* 1994), multilocus enzyme electrophoresis (MLEE) (de Bruijn, 1992; Woods *et al.,* 1992; Versalovic *et al.,* 1993; Leung *et al.,* 1994; Strain *et al.,* 1994; Koeuth *et al.,* 1995), phage typing (van Belkum *et al.,* 1993a), plasmid profiling (Georghiou *et al.,* 1995), pulsed-field gel electrophoresis (PFGE) (Poh *et al.,* 1996) (*see* Chapters 24 to 26), and ribotyping (McManus and Jones, 1995) (*see* Chapter 21). In selected studies rep-PCR surpassed the abilities of biotyping (Reboli *et al.,* 1994), plasmid profiling (Reboli *et al.,* 1994), RAPD (Rivera *et al.,* 1995), ribotyping (Snelling *et al.,* 1996), or serotyping (Judd *et al.,* 1993) to resolve individual bacterial strains. On the other hand, pulsed-field gel electrophoresis (*see* Chapters 24 to 26), though a more cumbersome technique, may have enhanced resolving power relative to rep-PCR, as documented in at least two studies (van Belkum *et al.,* 1993b; Poh *et al.,* 1996). rep-PCR generated DNA fingerprints appear to represent intrinsic properties of a specific strain's genome because similar clustering patterns were obtained regardless of which repetitive sequence based primer set was used (Koeuth *et al.,* 1995).

rep-PCR: Applications

Rep-PCR based DNA fingerprinting has spawned multiple applications due to the ability of the rep-PCR generated DNA profiles to distinguish closely related bacterial species and strains. The majority of rep-PCR studies have demonstrated the ability to discriminate human pathogens (Versalovic and Lupski, 1996), plant pathogens, soil bacteria, and plant symbionts for the purposes of molecular epidemiology and classification. A subset of these studies have used the rep-PCR DNA fingerprinting data to establish associations between particular strains or pathovars with specific human or plant diseases. Patterns of human infection and disease transmission have also been examined by rep-PCR. Finally, species-specific molecular probes have been isolated and rapid whole cell or tissue-based approaches for strain differentiation have been described.

Molecular epidemiologic analyses of human pathogens from patients with nosocomial and community-acquired infections have been performed by rep-PCR based DNA fingerprinting. Investigation of two nosocomial outbreaks in Houston, Texas by *Enterobacter aerogenes* resulted in the implication of a predominant clone in each outbreak (Georghiou *et al.,* 1995). Though a point source was never found, each clone presumably caused one outbreak involving multiple patients in one hospital. The homogeneity found by rep-PCR for each clone was verified by alternative methods including restriction fragment length polymorphisms (RFLP) of digested chromosomal DNA, plasmid profiling, and antibiotic susceptibility patterns. Nosocomial outbreak isolates of methicillin-resistant *Staphylococcus aureus* (MRSA) (Struelens *et al.,* 1993; Del Vecchio *et al.,* 1995) and *Legionella pneumophila* (van Belkum *et al.,* 1993b; Georghiou *et al.,* 1994) were clustered by rep-PCR based DNA fingerprinting in a manner consistent with other typing methods. REP element based PCR was also found to be useful for distinguishing epidemic and sporadic multidrug resistant *Acinetobacter baumannii* isolates (Reboli *et al.,* 1994). Alternative methods such as biotyping, antibiograms, and plasmid analysis lacked utility in this study. Rep-PCR approaches may be primer dependent for specific applications. For example though REP element primers were found to be useful for strain differentiation, ERIC element primers were not helpful for distinguishing the same *Acinetobacter baumannii* isolates (Reboli *et al.,* 1994).

In addition to nosocomial outbreaks, rep-PCR based approaches have been used to examine the nature of community-acquired infections. The clonal nature of penicillin resistant *Streptococcus pneumoniae* (Versalovic *et al.,* 1993) isolates, associated with community-acquired pediatric infections, was examined by rep-PCR based DNA fingerprinting and multilocus enzyme electrophoresis (MLEE). Penicillin resistant pneumococcal isolates, though not identical, demonstrated significantly less genetic diversity than penicillin susceptible isolates. Penicillin resistant isolates from Texas were also genetically similar to highly penicillin resistant isolates from Europe. Though clonal resolution was greater in this study with MLEE, rep-PCR fingerprinting with degenerate primers complementary to the REP element (REP1R-Dt and REP2-Dt) yielded profile clusters consistent with results obtained by MLEE. Community-acquired infections caused by *Citrobacter diversus* (Woods *et al.,* 1992) were also examined by rep-PCR fingerprinting with primers complementary to the REP and ERIC elements. More rigorous quantitative analysis of rep-PCR generated DNA fingerprints provided estimations of relative degrees of similarity between local and geographically diverse isolates.

Epidemiologic analysis of transmission of microbial pathogens between individual patients and patients and animals can also be documented by rep-PCR mediated DNA fingerprint analysis. Vertical transmission of *C. diversus* isolates from mother to infant was verified by FERP (Harvey *et al.,* 1995). Bacterial isolates from the mother's blood, umbilical cord blood, and infant were found to be identical by rep-PCR fingerprinting. Possible transmission of a *Bartonella henselae* clone between a cat and the animal's HIV-positive owner was suggested by rep-PCR fingerprinting (Rodriguez-Barradas *et al.,* 1995). Another HIV-positive individual was infected with a clone of *B. henselae* distinct from the *Bartonella* strain isolated from his cat (Rodriguez-Barradas *et al.,* 1995). DNA fingerprinting by rep-PCR, in addition to fatty acid analysis and determination of growth characteristics, implied that the cat's *Bartonella* isolate belonged to a previously uncharacterized species (Clarridge *et al.,* 1995).

Investigations of disease caused either by relapse of a latent infection or reinfection with a distinct bacterial clone have also been addressed by rep-PCR based DNA fingerprinting. Recurrent pneumonia in HIV-positive patients was identified as either a relapse with a single strain or reinfection with different strains (Jordens *et al.,* 1995). Persistent infections of cystic fibrosis patients by identical clones of *Alcaligenes xylosoxidans,* as verified by rep-PCR fingerprinting, were found to be associated with exacerbations of pulmonary symptoms (Dunne, Jr. and Maisch, 1995). Recurrent infections within individual patients by *Campylobacter jejuni* have also been evaluated by rep-PCR fingerprinting (Endtz *et al.,* 1993). Individual patients were shown to have been either infected multiple times with different strains or infected persistently with the same strain. Individual patients with chronic *Helicobacter pylori* infections were found to be colonized with the same strain in different regions of the stomach (Go *et al.,* 1995).

Rep-PCR based DNA fingerprinting methods have also been used in diverse non-medical applications including DNA fingerprinting and strain differentiation of soil bacteria and plant symbionts. Oligonucleotide primers based on both the REP and ERIC elements have been used to distinguish isolates of the Gram negative soil bacterium and plant symbiont *Rhizobium meliloti* as well as other soil microbes (de Bruijn, 1992). *R. meliloti* isolates, previously clustered by multilocus enzyme electrophoresis, were found to be also similar by rep-PCR fingerprinting (de Bruijn, 1992; Rossbach *et al.,* 1995). Dendrogram analysis was used to analyze and cluster *Bradyrhizobium japonicum* serocluster 123 isolates (Judd *et al.,* 1993). Genetically and phenotypically near-identical isolates of *B. japonicum* could be distinguished and rep-PCR was postulated to be a useful method for examining competition for plant nodulation among serocluster 123 isolates.

Closely related and yet phenotypically distinct plant pathovars that cause distinct diseases can be distinguished and classified by rep-PCR. DNA:DNA hybridization and ribotyping may be more useful than rep-PCR in distinguishing bacterial genera or species but rep-PCR is clearly most useful for differentiation of closely related bacteria and surpasses host range phenotyping and cumbersome infectivity assays used to classify different pathovars and strains. DNA fingerprints generated by rep-PCR subdivided the tomato pathovar, *Xanthomonas campestris* pv. *vesicatoria,* into biologically meaningful groups and demonstrated intra-pathovar genomic variation (Louws *et al.,* 1995). *Xanthomonas oryzae* pv. *oryzae* and *X. oryzae* pv. *oryzicola* cause distinct symptoms in infected rice plants. These individual pathovars could be reproducibly distinguished by rep-PCR based DNA fingerprinting, although most previous phenotypic tests failed to differentiate them (Louws *et al.,* 1994). Both cultured whole cells and ooze from bacterial blight lesions were used in rep-PCR fingerprinting experiments of the rice pathovar *Xanthomonas oryzae* pv. *oryzae* (Vera Cruz *et al.,* 1995). Early field diagnosis of strawberry bacterial leaf-blight necessitates observation of leaf symptoms and pathogenicity tests, requiring several weeks of time and considerable expense and greenhouse space. DNA fingerprinting of the causative agent, *Xanthomonas fragariae,* by rep-PCR proved superior to ELISA in terms of specificity and was found to correspond with the results derived from more cumbersome pathogenicity tests (Opgenorth *et al.,* 1996).

Isolation of species-specific DNA probes by rep-PCR represents another application of DNA fingerprinting for the identification of bacterial pathogens. DNA fingerprinting by rep-PCR yielded species-specific DNA fragments unique to different *Campylobacter* species (Giesendorf *et al.,* 1993). Such rep-PCR-generated DNA fragments were used as species-specific probes to distinguish isolates of *Campylobacter jejuni, Campylobacter coli,* and *Campylobacter lari.* In this fashion, species-specific DNA probes may be developed without the prior need of DNA sequence information. *Helicobacter pylori,* a relative of *Campylobacter* spp., has been associated with different gastroduodenal diseases such as asymptomatic gastritis and duodenal ulcer disease. *H. pylori* isolates associated with gastritis or duodenal ulcer disease clustered into distinct groups by rep-PCR fingerprinting (Go *et al.,* 1995), suggesting that different strains were associated with specific disease entities. Individual rep-PCR amplicons were associated with each group of *H. pylori* isolates (Go *et al.,* 1995). These amplicons may contain DNA sequences specific for each group of *H. pylori* isolates and may enable DNA probe development for *H. pylori* clones associated with different gastroduodenal diseases (Go *et al.,* 1995). Rapid DNA probe development may be possible which combines DNA:DNA hybridization and PCR amplification to establish novel molecular detection and epidemiologic assays.

Widespread application of PCR-based DNA fingerprinting methods requires rapid and simplified sample preparation. Though purified DNA provides the optimal template for DNA fingerprinting, purification of genomic DNA is laborious and time consuming. Cell lysates of both human and plant pathogens cultivated in solid or liquid media have been utilized with similar success (Woods *et al.,* 1993; Louws *et al.,* 1994; Louws *et al.,* 1995; Schneider and de Bruijn, 1996; Snelling *et al.,* 1996). Several Gram-negative enterobacteria such as *Escherichia coli* and *Citrobacter diversus* may be added as cells directly to the amplification reactions and lysed during the initial DNA denaturation. Rapid "whole cell" methods may be advantageous for fingerprinting of strains that secrete nucleases (Woods *et al.,* 1993). Additional treatment steps or sonication may be required to facilitate cell lysis prior to amplification of some bacterial isolates (Woods *et al.,* 1993; Versalovic *et al.,* 1994). In addition, infected plant or animal tissues may be directly used. For example, sonicated root nodules containing the symbionts *Rhizobium meliloti* or *Rhizobium galegae* were found to yield DNA fingerprints specific for individual strains of the symbionts (Nick and Lindstrom, 1994; Schneider and de Bruijn, 1996). Moreover, infected plant tissues such as tomato leaf lesions (Louws *et al.,* 1994) or infected geranium leaves (Louws *et al.,* 1994) have been used directly in rep-PCR based DNA fingerprinting of *Pseudomonas syringae* and *Xanthomonas campestris* pathovars, respectively. Such rapid tissue based applications are useful for characterization of infections caused by single pathovars but may be complicated in cases of mixed infections. Mixtures of purified bacterial cultures have not been found to produce a simple sum of bands generated from both genomes, but instead yielded skewed representations with genomic fingerprints of single predominant isolates (Schneider and de Bruijn, 1996).

Summary and Conclusions

Interspersed repetitive DNA sequences in bacterial genomes form the basis for distinguishing bacterial strains following PCR amplification. The relative ease and speed of rep-PCR is within the reach of many laboratories performing molecular biology. A thermal cycler and agarose gel electrophoresis unit comprise the hardware required for such experiments. Rapid PCR based DNA fingerprinting methods such as RAPD or AP-PCR (*see* Chapter 33) and rep-PCR have proven useful in molecular epidemiologic investigations, bacterial strain identification, and clonal analysis of human and plant pathogens. As novel interspersed repetitive elements are discovered during the course of bacterial genome sequencing projects, applications for rep-PCR will continue to increase with respect to different prokaryotic organisms.

Acknowledgments

JV acknowledges support of the NIH Medical Scientist Training Program at Baylor College of Medicine. JRL acknowledges support for rep-PCR development from the Department of Molecular and Human Genetics and the Department of Pediatrics at Baylor College of Medicine and thanks Thearith Koeuth for technical support. FJdB acknowledges support from the NSF Center for Microbial Ecology (DIR 8809640) and the U.S. Department of Energy (DE FG0290ER20021) and wants to thank Jan Rademaker for helpful comments and Maria Schneider for technical support.

References

Clarridge, J. E., T. J. Raich, D. Pirwani, B. Simon, L. Tsai, M. C. Rodriguez-Barradas, R. Regnery, A. Zollo, D. C. Jones, and C. Rambo. 1995. Strategy to detect and identify *Bartonella* species in routine clinical laboratory yields *Bartonella henselae* from human immunodeficiency virus-positive patient and unique *Bartonella* strain from his cat. *J. Clin. Microbiol.* 33:2107–2113.

Colman, S. D., P. C. Hu, and K. F. Bott. 1990. Prevalence of novel repeat sequences in and around the P1 operon in the genome of *Mycoplasma pneumoniae. Gene* 87:91–96.

Correia, F. F., S. Inouye, and M. Inouye. 1986. A 26-base-pair repetitive sequence specific for *Neisseria gonorrhoeae* and *Neisseria meningitidis* genomic DNA. *J. Bacteriol.* 167:1009–1015.

de Bruijn, F. J. 1992. Use of repetitive (repetitive extragenic element and enterobacterial repetitive intergenic consensus) sequences and the polymerase chain reaction to fingerprint the genomes of *Rhizobium meliloti* isolates and other soil bacteria. *Appl. Environ. Microbiol.* 58:2180–2187.

de Bruijn, F. J., M. Schneider, U. Rossbach, and F. J. Louws. 1995. Automated fluorescent and conventional rep-PCR genomic fingerprinting and multiplex PCR to classify bacteria and track genes. In: *Proceedings of the Seventh International Symposium on Microbial Ecology.* Anonymous, Brazil.

Del Vecchio, V. G., J. M. Petroziello, M. J. Gress, F. K. McCleskey, G. P. Melcher, H. K. Crouch, and J. R. Lupski. 1995. Molecular genotyping of methicillin-resistant *Staphylococcus aureus* via fluorophore-enhanced repetitive-sequence PCR. *J. Clin. Microbiol.* 33:2141–2144.

Dunn, W. M., Jr., and S. Maisch. 1995. Epidemiological investigation of infections due to *Alcaligenes* species in children and patients with cystic fibrosis: Use of repetitive-element-sequence polymerase chain reaction. *Clin. Infect. Dis.* 20:836–841.

Edel, V., C. Steinberg, I. Avelange, G. Laguerre, and C. Alabouvette. 1995. Comparison of three molecular methods for the characterization of *Fusarium oxyxporum* strains. *Phytopathol.* 85:579–585.

Ellsworth, D. L., K. D. Rittenhouse, and R. L. Honeycutt. 1993. Artifactual variation in randomly amplified polymorphic DNA banding patterns. *Biotechniques* 14:214–217.

Endtz, H. P., B.A.J. Giesendorf, A. van Belkum, S.J.M. Lauwers, W. H. Jansen, and W.G.V. Quint. 1993. PCR-mediated DNA typing of *Campylobacter jejuni* isolated from patients with recurrent infections. *Res. Microbiol.* 144:703–708.

Forsyth, M. H., and S. J. Geary. 1996. The repetitive element Rep MP 1 of *Mycoplasma pneumoniae* exists as a core element within a larger, variable repetitive mosaic. *J. Bacteriol.* 178:917–921.

Georghiou, P., R. J. Hamill, C. E. Wright, J. Versalovic, T. Koeuth, D. A. Watson, and J. R. Lupski. 1995. Molecular epidemiology of infections due to *Enterobacter aerogenes:* Identification of hospital outbreak strains by molecular techniques. *Clin. Infect. Dis.* 20:84–94.

Georghiou, P. R., A. M. Doggett, M. A. Kielhofner, J. E. Stout, D. A. Watson, J. R. Lupski, and R. J. Hamill. 1994. Molecular fingerprinting of *Legionella* species by repetitive element PCR. *J. Clin. Microbiol.* 32:2989–2994.

Giesendorf, B.A.J., A. van Belkum, A. Koeken, H. Stegeman, M.H.C. Henkens, J. Van der Plas, H. Goossens, H.G.M. Niesters, and W.G.V. Quint. 1993. Development of species-specific DNA probes for *Campylobacter jejuni, Campylobacter coli,* and *Campylobacter lari* by polymerase chain reaction fingerprinting. *J. Clin. Microbiol.* 31:1541–1546.

Go, M., K. Chan, J. Versalovic, T. Koeuth, D. Y. Graham, and J. R. Lupski. 1995. Cluster analysis of *Helicobacter pylori* genomic DNA fingerprints suggests gastroduodenal disease-specific associations. *Scand. J. Gastroenterol.* 30:640–646.

Harvey, B. S., T. Koeuth, J. Versalovic, C. R. Woods, and J. R. Lupski. 1995. Vertical transmission of *Citrobacter diversus* documented by DNA fingerprinting. *Infect. Control Hosp. Epidemiol.* 16:564–569.

Jordens, J. Z., J. Paul, J. Bates, C. Beaumont, J. Kimari, and C. Gilks. 1995. Characterization of *Streptococcus pneumoniae* from human immunodeficiency virus-seropositive patients with acute and recurrent pneumonia. *J. Infect. Dis.* 172:983–987.

Judd, A. K., M. Schneider, M. J. Sadowsky, and F. J. de Bruijn. 1993. Use of repetitive sequences and the polymerase chain reaction technique to classify genetically related *Bradyrhizobium japonicum* serocluster 123 strains. *Appl. Environ. Microbiol.* 59:1702–1708.

Koeuth, T., J. Versalovic, and J. R. Lupski. 1995. Differential subsequence conservation of interspersed repetitive *Streptococcus pneumoniae* BOX elements in diverse bacteria. *Genome Research* 5:408–418.

Leung, K., S. R. Strain, F. J. de Bruijn, and P. J. Bottomley. 1994. Genotypic and phenotypic comparisons of chromosomal types within an indigenous soil population of *Rhizobium leguminosarum* bv. trifolii. *Appl. Environ. Microbiol.* 60:416–426.

Louws, F. J., D. W. Fulbright, C. T. Stephens, and F. J. de Bruijn. 1994. Specific genomic fingerprints of phytopathogenic *Xanthomonas* and *Pseudomonas* pathovars and strains generated with repetitive sequences and PCR. *Appl. Environ. Microbiol.* 60:2286–2295.

Louws, F. J., D. W. Fulbright, C. Taylor Stephens, and F. J. de Bruijn. 1995. Differentiation of genomic structure by rep-PCR fingerprinting to rapidly classify *Xanthomonas campestris* pv. *vesicatoria. Mol. Plant Pathol.* 85:528–536.

Louws, F. J., M. Schneider, and F. J. de Bruijn. 1996. Assessing genetic diversity of microbes using repetitive sequence-based PCR (rep-PCR). In Environmental Applications of Nucleic Acid Amplifications Techniques. pp. 63–64. G. Toranzos, ed. Technomic Publishing Co., Lancaster, PA.

Lupski, J. R., and G. M. Weinstock. 1992. Short, interspersed repetitive DNA sequences in prokaryotic genomes. *J. Bacteriol.* 174:4525–4529.

McManus, P. S., and A. L. Jones. 1995. Genetic fingerprinting of *Erwinia amylovora* strains isolated from tree-fruit crops and *Rubus* spp. *Phytopathology* 85:1547–1553.

Nick, G., and K. Lindstrom. 1994. Use of repetitive sequences and the polymerase chain reaction to fingerprint the genomic DNA of *Rhizobium galegae* strains and to identify the DNA obtained by sonicating the liquid cultures and root nodules. *System. Appl. Microbiol.* 17:265–273.

Opgenorth, D. C., C. D. Smart, F. J. Louws, F. J. de Bruijn, and B. C. Kirkpatrick. 1996. Identification of *Xanthomonas fragariae* field isolates by rep-PCR genomic fingerprinting. *Plant Disease* 80:868–873.

Poh, C. L., V. Ramachandran, and J. W. Tapsall. 1996. Genetic diversity of *Neisseria gonorrhoeae* IB-2 and IB-6 isolates revealed by whole-cell repetitive element sequence-based PCR. *J. Clin. Microbiol.* 34:292–295.

Ralph, D., M. McClelland, J. Welsh, G. Baranton, and P. Perolat. 1993. *Leptospira* species categorized by arbitrarily primed polymerase chain reaction (PCR) and by mapped restriction polymorphisms in PCR-amplified rRNA genes. *J. Bacteriol.* 175:973–981.

Reboli, A. C., E. D. Houston, J. S. Monteforté, C. A. Wood, and R. J. Hamill. 1994. Discrimination of epidemic and sporadic isolates of *Acinetobacter baumannii* by repetitive element PCR-mediated DNA fingerprinting. *J. Clin. Microbiol.* 32:2635–2640.

Rivera, I. G., M.A.R. Chowdhury, A. Huq, D. Jacobs, M. T. Martins, and R. R. Colwell. 1995. Enterobacterial repetitive intergenic consensus sequences and the PCR to generate fingerprints of genomic DNAs from *Vibrio cholerae* O1, O139, and non-01 strains. *Appl. Environ. Microbiol.* 61:2898–2904.

Rodriguez-Barradas, M. C., R. J. Hamill, E. D. Houston, P. R. Georghiou, J. E. Clarridge, R. L. Regnery, and J. E. Koehler. 1995. Genomic fingerprinting of *Bartonella* species by repetitive element PCR for distinguishing species and isolates. *J. Clin. Microbiol.* 33:1089–1093.

Rossbach, S., G. Rasul, M. Schneider, B. Eardley, and F. J. de Bruijn. 1995. Structural and functional conservation of the rhizopine catabolism *(moc)* locus is limited to selected *Rhizobium meliloti* strains and unrelated to their geographical origin. *Mol. Plant Micr. Interact.* 8:549–559.

Schneider, M., and F. J. de Bruijn. 1996. Rep-PCR mediated genomic fingerprinting of rhizobia and computer assisted phylogenetic pattern analysis. *World J. Microbiol. Biotechnol.* 12:163–174.

Sneath, P.H.A., and R. R. Sokal. 1973. *Numerical Taxonomy.* Freeman, San Francisco.

Snelling, A. M., P. Gerner-Smidt, P. M. Hawkey, J. Heritage, P. Parnell, C. Porter, A. R. Bodenham, and T. Inglis. 1996. Validation of use of whole-cell repetitive extragenic palindromic sequence-based PCR (REP-PCR) for typing strains belonging to the *Acinetobacter calcoaceticus-Acinetobacter baumannii* complex and application of the method to the investigation of a hospital outbreak. *J. Clin. Microbiol.* 34:1193–1202.

Sokal, R. R., and P.H.A. Sneath. 1963. *Principles of Numerical Taxonomy.* WH Freeman and Company, San Francisco.

Stern, M. J., G.F.L. Ames, N. H. Smith, E. C. Robinson, and C. F. Higgins. 1984. Repetitive extragenic palindromic sequences: A major component of the bacterial genome. *Cell* 37:1015–1026.

Strain, S. R., K. Leung, T. S. Whittam, F. J. de Bruijn, and P. J. Bottomley. 1994. Genetic structure of *Rhizobium leguminosarum* biovar trifolii and viciae populations found in two Oregon soils under different plant communities. *Appl. Environ. Microbiol.* 60:2772–2778.

Struelens, M. J., R. Bax, A. Deplano, W.G.V. Quint, and A. van Belkum. 1993. Concordant clonal delineation of methicillin-resistant *Staphylococcus aureus* by macrorestriction analysis and polymerase chain reaction genome fingerprinting. *J. Clin Microbiol.* 31:1964–1970.

Taylor, N. S., J. G. Fox, N. S. Akopyants, D. E. Berg, N. Thompson, B. Shames, L. Yan, E. Fontham, F. Janney, F. M. Hunter, and P. Correa. 1995. Long-term colonization with single and multiple strains of *Helicobacter pylori* assessed by DNA fingerprinting. *J. Clin. Microbiol.* 33:918–923.

van Belkum, A., R. Bax, P. Peerbooms, W. H. Goessens, N. van Leeuwen, and W.G.L. Quint. 1993a. Comparison of phage typing and DNA fingerprinting by polymerase chain reaction for discrimination of methicillin-resistant *Staphylococcus aureus* strains. *J. Clin. Microbiol.* 31:798–803.

van Belkum, A., M. Struelens, and W. Quint. 1993b. Typing of *Legionella pneumophila* strains by polymerase chain reaction-mediated DNA fingerprinting. *J. Clin. Microbiol.* 31:2198–2200.

van Belkum, A. 1994. DNA fingerprinting of medically important microorganisms by use of PCR. *Clin. Microbiol. Rev.* 7:174–184.

van Belkum, A., M. Sluijter, R. de Groot, H. Verbrugh, and P. W. M. Hermans. 1996. Novel BOX repeat PCR assay for high-resolution typing of *Streptococcus pneumoniae* strains. *J. Clin. Microbiol.* 34:1176–1179.

van Berkum, P., R. B. Navarro, and A.A.T. Vargas. 1994. Classification of the uptake hydrogenase-positive (Hup+) bean rhizobia as *Rhizobium tropici. Appl. Environ. Microbiol.* 60:554–561.

Vera Cruz, C. M., L. Halda, F. Louws, D. Z. Skinner, M. L. George, R. J. Nelson, F. J. de Bruijn, C. Rice, and J. E. Leach. 1995. Repetitive sequence-based PCR of *Xanthomonas oryzae* pv. *oryzae* and *Pseudomonas* species. *Intl. Rice Res. Instit. Newsletter* 20:23–25.

Versalovic, J., T. Koeuth, and J. R. Lupski. 1991. Distribution of repetitive DNA sequences in eubacteria and application to fingerprinting of bacterial genomes. *Nucleic Acids Research* 19:6823–6831.

Versalovic, J., V. Kapur, Jr. Mason, U. Shah, T. Koeuth, J. R. Lupski, and J. M. Musser. 1993. Penicillin resistant *Streptococcus pneumoniae* strains recovered in Houston, Texas: Identification and molecular characterization of multiple clones. *J. Infect. Dis.* 167:850–856.

Versalovic, J., M. Schneider, F. J. de Bruijn, and J. R. Lupski. 1994. Genomic fingerprinting of bacteria using repetitive sequence-based polymerase chain reaction. *Meth. Mol. Cell Biol.* 5:25–40.

Versalovic, J., V. Kapur, T. Koeuth, G. H. Mazurek, T. S. Whittam, J. M. Musser, and J. R. Lupski. 1995. Automated DNA fingerprinting of pathogenic bacteria by fluorophore-enhanced repetitive sequence-based polymerase chain reaction. *Arch. Pathol. Lab. Med.* 119:23–29.

Versalovic, J., and J. R. Lupski. 1995. DNA fingerprinting of *Neisseria* strains by rep-PCR. *Meth. Mol. Cell Biol.* 5:96–104.

Versalovic, J., and J. R. Lupski. 1996. Distinguishing bacterial and fungal pathogens by repetitive sequence-based PCR (rep-PCR). *LabMedica Intl.* in press.

Welsh, J., C. Pretzman, D. Postic, I. Saint Girons, G. Baranton, and M. McClelland. 1992. Genomic fingerprinting by arbitrarily primed polymerase chain reaction resolves *Borrelia burgdorferi* into three distinct phyletic groups. *Int. J. Syst. Bacteriol.* 42:370–377.

Welsh, J., and M. McClelland. 1990. Fingerprinting genomes using PCR with arbitrary primers. *Nucl. Acids Res.*18:7213–7218.

Wenzel, R. and R. Herrmann. 1988. Repetitive DNA sequences in *Mycoplasma pneumoniae. Nucl. Acids Res.* 16:8337–8350.

Williams, J.G.K., A. R. Kubelik, K. J. Livak, J. A. Rafalski, and S. V. Tingey. 1990. DNA polymorphisms amplified by arbitrary primers are useful as genetic markers. *Nucl. Acids Res.* 18:6531–6535.

Woods, C., J. Versalovic, T. Koeuth, and J. R. Lupski. 1993. Whole cell rep-PCR allows rapid assessment of clonal relationships of bacterial isolates. *J. Clin. Microbiol.* 31:1927–1931.

Woods, C. R., J. Versalovic, T. Koeuth, and J. R. Lupski. 1992. Analysis of relationships among isolates of *Citrobacter diversus* using DNA fingerprints generated by repetitive sequence-based primers in the polymerase chain reaction. *J. Clin. Microbiol.* 30:2921–2929.

PART C

Genomic Sequencing Projects

35

Organization of the European *Bacillus subtilis* Genome Sequencing Project

Ivan Moszer, Philippe Glaser, Antoine Danchin, and Frank Kunst

Introduction

The genome of *Haemophilus influenzae* (1.8 megabases) has been entirely sequenced (Fleischmann *et al.*, 1995). Several genome sequencing projects of other model organisms have been initiated (*see also* Chapters 36, 38–44). The sequencing of the *Bacillus subtilis* genome (4.2 megabases) is an attractive choice, since genetic analysis in this organism is highly advanced. The advantages of such a systematic approach have recently been described (Glaser *et al.*, 1993; Sorokin *et al.*, 1993). It allows us to get a complete blueprint of the genetic content of an organism and it contributes key information to the understanding of molecular evolution. Gram-negative and Gram-positive bacteria diverged more than 2 billion years ago (Woese, 1987), underscoring the importance of sequencing in parallel the genomes of both *Escherichia coli* and *B. subtilis.*

The Participants

In September 1989, a consortium of five European laboratories started a joint project aimed at developing the physical map, constructing appropriate DNA libraries and launching, on a pilot scale, the systematic sequencing of the *Bacillus subtilis* genome (Kunst and Devine, 1991). At present, 14 laboratories (numbers 1–7 and 9–15, Table) are supported by the European Commission in the framework of the Biotechnology programme. The laboratory of D. Karamata (laboratory number 8, Table 35-1) participates in this project as a non European Union member using funds obtained from the Swiss government. Two biotechnology companies, Genencor (USA, The Netherlands) and Novo Nordisk (Denmark, USA) participate in the European project using their own funds (laboratory numbers 16 and 17, Table 35-1). Different DNA regions were assigned to these

Table 35-1. Contact addresses of the European laboratories participating in the B. subtilis *genome sequencing project*

Laboratory number	Principal Investigator(s)	Contact address
1	A. Danchin and G. Rapoport	Institut Pasteur, 25–28, rue du Dr Roux, 75724, Paris Cedex 15, France Fax: 33 1 45 68 89 48/38 E-mail: adanchin@pasteur.fr; rapoport@pasteur.fr
2	J. Errington	Sir William Dunn School of Pathology, University of Oxford, South Parks Road, Oxford OX1 3RE, United Kingdom Fax: 44 1865 275 556 E-mail: erring@molbiol.ox.ac.uk
3	A. Galizzi and A. Albertini	University of Pavia, Via Abbiategrasso 207, 27100 Pavia, Italy Fax: 39 482 528 496 E-mail: galizzi@ipvgen.unipv.it
4	S.D. Ehrlich	Laboratoire de Génétique Microbienne, Institut de Biotechnologie, INRA, Domaine de Vilvert, 78352 Jouy-en-Josas Cedex, France Fax: 33 1 34 65 25 21 E-mail: ehrlich@biotec.jouy.inra.fr
5	K. Devine	Department of Genetics, Trinity College, Lincoln Place Gate, Dublin 2, Republic of Ireland Fax: 353 1679 85 58 E-mail: kdevine@tcd.ie
6	S. Bron	Department of Genetics, University of Groningen, Kerklaan 30, 9751 NN Haren, The Netherlands Fax: 31 50 363 23 48 E-mail: S.Bron@biol.rug.nl
7	C. Harwood and P. Emmerson	Department of Microbiology, The Medical School, University of Newcastle, Framlington Place, Newcastle-upon-Tyne NE2 4HH, United Kingdom Fax: 44 191 222 77 36 E-mail: colin.harwood@newcastle.ac.uk
8	D. Karamata and C. Mauel	Institut de Génétique et de Biologie Microbiennes, 19 rue César Roux, 1005 Lausanne, Switzerland Fax: 41 213 20 60 78 E-mail: dkaramat@igbm.unil.ch
9	F. Denizot and J. Haiech	CNRS, BP 71, Laboratoire de Chimie Bactérienne, 31 Chemin Joseph Aiguier, 13402 Marseille Cedex 09, France Fax: 33 91 71 89 14 E-mail: denizot@lcb.cnrs-mrs.fr

(Continued)

Table 35-1. Continued

Laboratory number	Principal Investigator(s)	Contact address
10	B. Oudega	Vrije Universiteit Amsterdam, Faculty of Biology, De Boelelaan 1087, Amsterdam 1081 HV, The Netherlands Fax: 31 20 444 71 23 E-mail: oudega@bio.vu.nl
11	I. Connerton	BBSRC, Institute of Food Research, Reading Laboratory, Early Gate, Whiteknights Road, Reading RG6 2EF, United Kingdom Fax: 44 1734 267 917 E-mail: ian.connerton@bbsrc.ac.uk
12	R. Borriss	Humboldt Universität, Institut für Genetik und Mikrobiologie, Warschauer Strasse 43, D 10243 Berlin, Germany Fax: 49 30 5800 528 E-mail: Rainer=Borriss@rz.hu-berlin.de
13	B. Joris	Université de Liège, Centre d'Ingénierie des Protéines, Institut de Chimie B6, Sart Tilman, B-4000 Liège, Belgium Fax: 32 41 66 33 64 E-mail: bjoris@acgt.cip.ulg.ac.be
14	S. Seror	Université Paris 11, Bât. 409, 91405 Orsay, France Fax: 33 1 69 41 78 08 E-mail: seror@igmors.u-psud.fr
15	K. Entian	Institut für Mikrobiologie, J. W. Goethe Universität, Marie Curie Strasse 9, Frankfurt am Main 60439, Germany Fax: 49 69 7982 9527 E-mail: Entian@em.uni-frankfurt.d400.de
16	E. Ferrari	Genencor International, Inc., 180 Kimball Way, South San Francisco, CA 94080, USA Fax: 415 583 8269 E-mail: EFerrari@genencor.com
17	M. Rey	Novo Nordisk, 1445 Drew Avenue, Davis, CA 95616-4880, USA Fax: 916 758 0317 E-mail: mrey@nnbt.com
18	C. Bruschi	ICGEB, Area Science Park Padriciano 99 I-34012 Trieste, Italy Fax: 39 40 375 7343 E-mail: bruschi@icgeb.trieste.it

(Continued)

Table 35-1. Continued

Laboratory number	Principal Investigator(s)	Contact address
19	A. Düsterhöft	Qiagen GmbH, Max-Volmer-Strasse 4, D-40724 Hilden, Germany Fax: 49 2103 892 222 E-mail: a.duesterhoeft@qiagen.de
20	F. Foury and B. Purnelle	Faculté des Sciences Agronomiques, Unité de Biochimie Physiologique, Place Croix du Sud, 2-20, Louvain-la-Neuve, Belgium Fax: 32 10 47 36 72 E-mail: purnelle@fysa.ucl.ac.be
21	M. Vandenbol	Department of Microbiology, Faculty of Agronomy, 6, Avenue du Maréchal Juin, B-5030 Gembloux (Belgium) Fax: 32 81 61 15 55 E-mail: microbiol@fsagx.ac.be

laboratories (Fig. 35-1); each laboratory is in charge of both cloning and sequencing of its assigned DNA region and chooses its own strategy to achieve this goal. Four additional groups, who will sequence part of the DNA fragments provided by the other participants, have recently joined the project (laboratory numbers 18–21, Table 35-1).

Thus, the original project has grown from a core of five laboratories into a major European sequencing initiative, carried out in close cooperation with the Japanese project involving the participation of seven laboratories (Devine, 1995; Kunst *et al.*, 1995; Ogasawara *et al.*, 1995). Since detailed genetic and physical maps were available for the *B. subtilis* chromosome (section 3.4; Anagnostopoulos *et al.*, 1993), it was possible to assign precisely DNA regions flanked by well identified genetic markers or, whenever possible, by already sequenced genes to the different participating laboratories, as indicated in Figure 35-1. This international subdivision of the genome was achieved by direct negotiation between the scientists, who had obtained funds to carry out sequencing projects, taking into account individual scientific interests in specific DNA regions. The decision was also taken to concentrate all efforts on the same strain: *B. subtilis* strain 168 (*see also* Chapter 50).

Cloning and Sequencing

The first results obtained gave us insight into the requirements for the cloning and analysis of large genome segments. It now appears that the main bottleneck

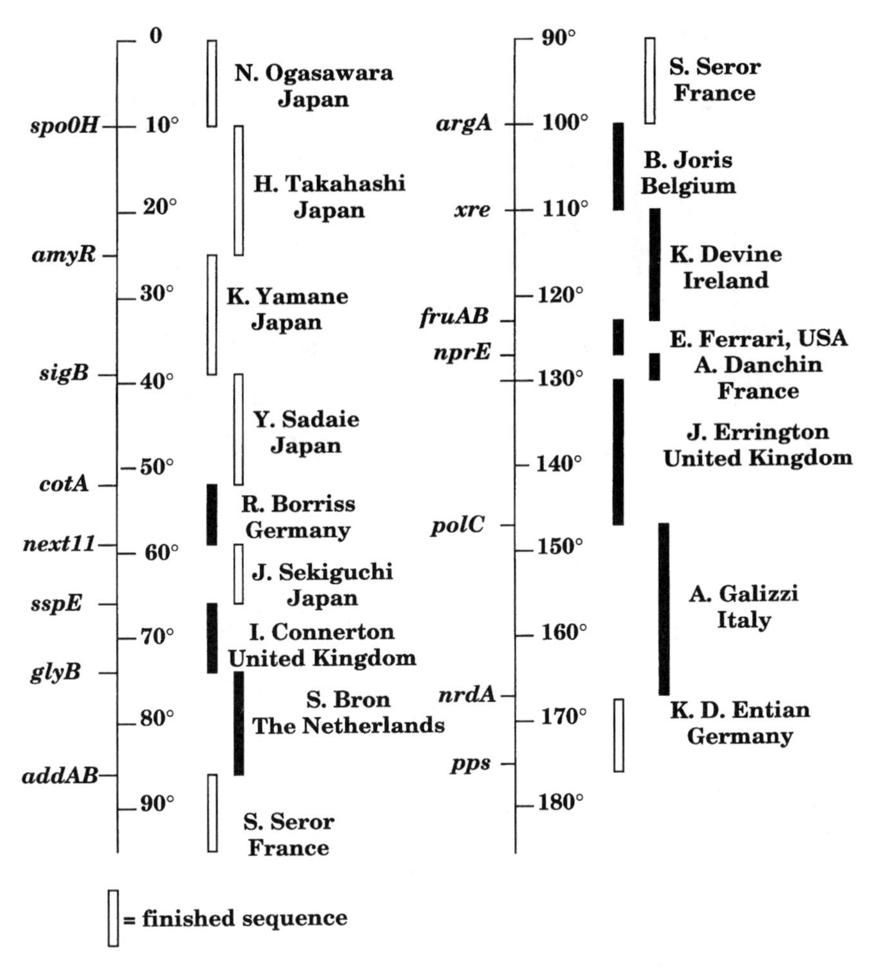

Figure 35-1. DNA regions assigned to the laboratories participating in the international *B. subtilis* genome sequencing project. The names of the principal investigators of these laboratories are indicated. The assigned DNA regions are presented as black bars or open bars flanked by the indicated genetic markers (Kunst *et al.*, 1995). The open bars indicate DNA regions which have been entirely sequenced. Part of the *odhAB-ilvA* DNA region is being sequenced by S. H. Park (Yusong, Taejon, Korea).

does not occur at the sequencing level, but rather at two other levels: the cloning and data interpretation steps.

Different strategies were successfully applied to obtain appropriate ordered sublibraries covering portions of the genome: 1) cloning into a range of plasmid, lambda and YAC vectors; 2) the use of an *E. coli* strain that lowers the copy

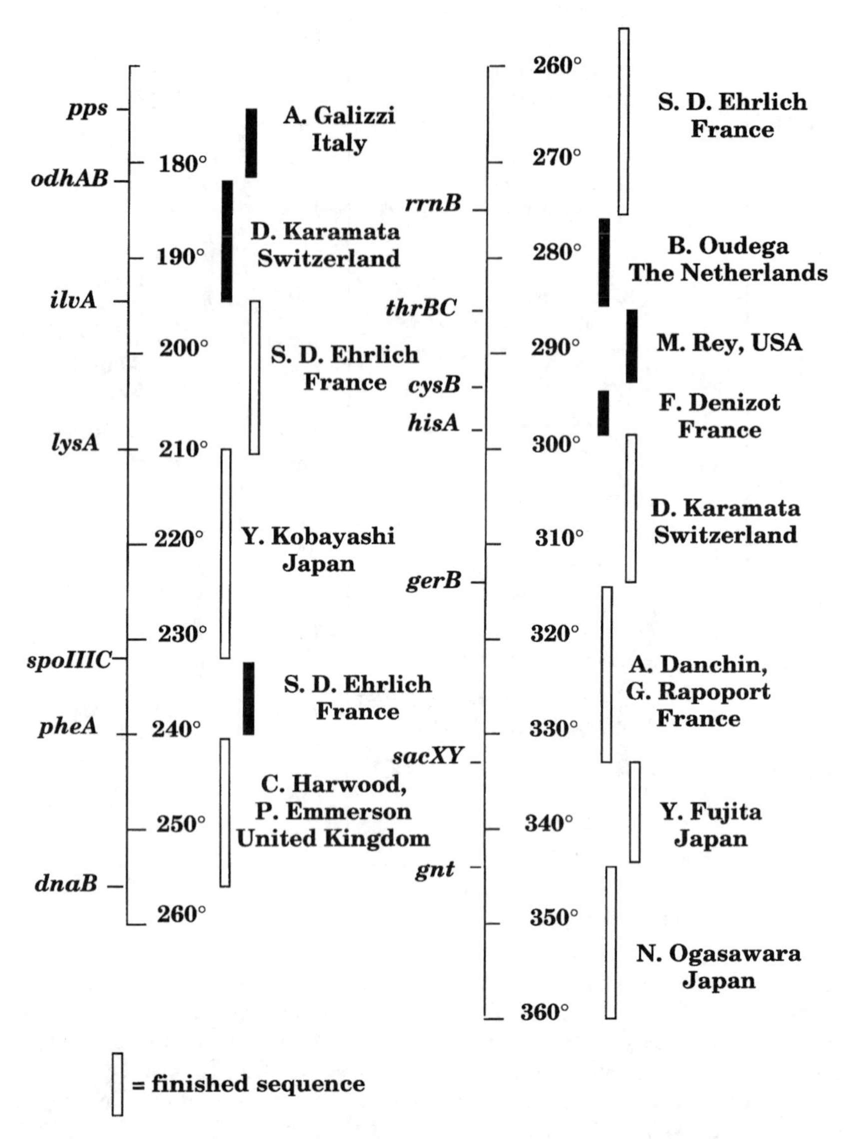

Figure 35-1. Continued

number of ColE1-based plasmids and which correspondingly reduces the potential toxic effects of cloned genes; 3) use of integrative plasmids and marker rescue in genome walking experiments; 4) *in vitro* amplification using standard, long-range (LC PCR), and inverse PCR techniques.

Since *B. subtilis* fragments cloned in *E. coli* are often unstable, attempts to construct complete libraries of overlapping lambda or cosmid clones failed.

However, *B. subtilis* DNA fragments which were not present in a lambda library we constructed, could be cloned by chromosome walking (Kunst and Devine, 1991; Glaser *et al.*, 1993). This alternative strategy is based on either plasmid rescue or marker rescue. In the first method, an *E. coli* plasmid—a pDIA5304 derivative containing a *B. subtilis* DNA fragment—was integrated through homologous recombination in the *B. subtilis* chromosome, thus conferring chloramphenicol resistance to the host. This recombinant plasmid DNA as well as adjacent *B. subtilis* DNA sequences were then excised from the chromosome using a restriction enzyme with a single site within the polylinker of the plasmid and infrequent sites in *B. subtilis* chromosomal DNA. After circularization with T4 DNA ligase, the recombinant plasmid containing adjacent DNA sequences was transformed into *E. coli* strain TP611 (Fig. 35-2). In this strain, ColE1 plasmids are maintained at a low copy number, thus avoiding high level of expression of *B. subtilis* genes. However, cloning of some *B. subtilis* DNA fragments in this host proved still unsuccessful. In these cases, a marker rescue method was applied, which consisted of integrating a selective marker into the *B. subtilis* chromosome *via* a double cross-over event. This marker, as well as adjacent DNA sequences, were subsequently excised from the chromosome and directly transformed into *B. subtilis,* using a replicative plasmid. Finally, in a limited number of cases, *B. subtilis* genes were cloned by direct complementation of specific mutations.

The YAC library constructed by S. D. Ehrlich and coworkers covered more than 98% of the *B. subtilis* chromosome (Azevedo *et al.*, 1993). However, the drawback of this method was found to be that the yield of YAC recombinant DNA after purification was usually low.

The European laboratories are each sequencing at least 20 kb of *B. subtilis* genomic DNA every year. The average error rate is presently estimated to be lower than 1 per 2000 nucleotides. It is our aim to finish the sequence of the entire *B. subtilis* genome by mid-1997.

Identified Genes

A list of putative genes, which were identified during this phase of the European project, and which encode putative proteins showing similarities with proteins present in databanks, has been published (Kunst *et al.*, 1995). Additional genes have been presented in several publications which appeared in the same issue of the journal Microbiology as the paper cited above; another issue of Microbiology dedicated in part to this project appeared in November 1996.

Data Handling and Analysis

A sequence depository has been established at the Institut Pasteur. The European laboratories deposit in this directory the sequences they produce, as well as the

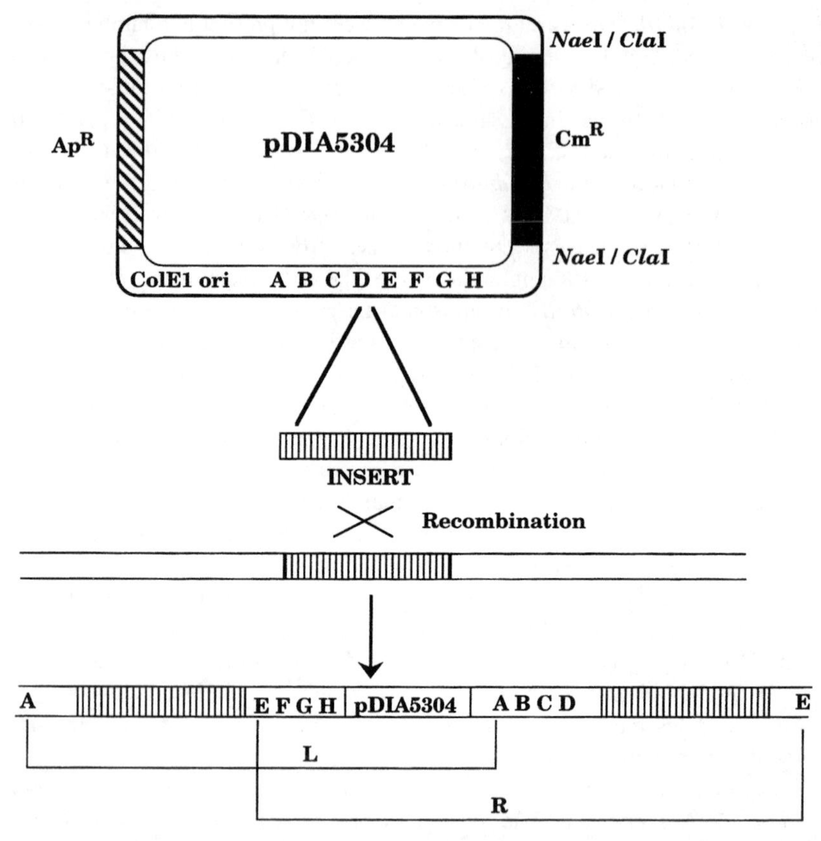

Figure 35-2. Plasmid pDIA5304 is used for genome walking experiments. It contains a ColE1 replicon and the ampicillin resistance marker (Ap^R) used for selection in *E. coli.* It has been constructed by inserting the *Cla*I fragment encompassing the pC194 chloramphenicol resistance (Cm^R) marker at the *Nae*I site of pBluescript SK(+) from the Stratagene company (Glaser *et al.*, 1993). For genome walking experiments a recombinant plasmid is constructed by inserting a *B. subtilis* DNA fragment (labeled "insert" in the Figure) at one of the sites of the polylinker (schematically indicated as ABCDEFGH). The constructed recombinant plasmid, which is unable to replicate in *B. subtilis,* is introduced in this host by transformation. The selected Cm^R recombinants contain a copy of this plasmid which is integrated in the host genome through a single cross-over event. The chromosomal DNA of such a Cm^R *B. subtilis* strain is cut with restriction enzyme A or E. The obtained fragments are then circularized with T_4 DNA ligase. Finally, the constructed recombinant plasmids L or R containing adjacent *B. subtilis* chromosomal DNA sequences are cloned in *E. coli.*

cloning and sequencing strategies used, the annotations of the DNA sequence and a list of similarities of the putative gene products identified with known proteins in public databanks.

The overall quality of the sequence is then estimated. For this purpose, a specialized database containing all known sequences of *B. subtilis* has been built (Moszer *et al.*, 1995). It gathers public sequences present in data libraries as well as confidential sequences determined in the course of the genome project. This database, named *SubtiList,* has been implemented with a relational database management system (4th Dimension), running on Macintosh and Windows-compatible computers. It is constructed around a core constituted by contigs of DNA sequence, *i.e.* overlapping sequences extracted from the EMBL databank. The redundancy is thus eliminated *(ca.* 35%), and the errors present in the original data are corrected in a concomitant step, providing a verified set of sequences and biological objects related to these sequences (genes, proteins, regulatory signals, etc.). In May 1996, *SubtiList* contained 2800 kb of non redundant *B. subtilis* DNA sequences, corresponding to 67% of the genome. The consistency of the sequences determined by the European laboratories was checked by integrating these sequences in the database and by performing various analyses (such as comparison with already published sequences, annotations of genes, etc., see below).

In the *SubtiList* database, the information is organized in such a way that all data are logically interconnected. This "linked" organization provides many advantages as opposed to the "flat file" organization of general data libraries (see also Chapter 43). In addition to removing redundancy and discrepancies found in international databanks, *SubtiList* allows easy handling and extraction of precise and correct data related to the *B. subtilis* genome (such as sequences, genes, DNA signals), as well as the establishment of relationships between a set of biological objects of a given category with the corresponding objects of another category.

Using this clean dataset, statistical and other types of analyses have been carried out. First of all, a codon usage analysis has been performed by factorial correspondence analysis, allowing the clustering of the genes of *B. subtilis* into two classes: one containing a small subset of highly expressed genes, and the other one containing the bulk of the genes (Moszer *et al.*, 1995). Another result obtained with the protein data of *SubtiList* has highlighted a strong bias against proteins exhibiting an isoelectric point around 7.5, which could correspond to the intracellular pH of *B. subtilis* (Moszer *et al.*, 1995). Finally, a DNA pattern search program has been integrated into the database software. For example, this tool allowed us to identify regulatory signals recognized by the protein FNR, involved in the regulation of gene expression in anaerobiosis (Cruz Ramos *et al.*, 1995).

In order to analyze and manage raw sequence data produced by genome sequencing projects, a task-driven cooperative system is being constructed (Médigue *et al.*, 1995). This system is designed to help the user in solving sequence analysis problems, through a dynamic choice and chaining of methods, and a graphical and interactive display of results. Several sequence analysis methods have already been integrated (Blast, GeneMark, etc.), and a wider range of programs will be added in the near future (see also Chapter 46). A 50 kb contig has already been annotated using this program (Médigue *et al.*, 1995).

A user-friendly interface has been conceived for an easy consultation of the *SubtiList* database, including graphical representation, multi-criteria searches, multi-format exports and sequence analysis capabilities. This tool has been made publicly available through the Internet network at the FTP server of the Institut Pasteur (address ftp://ftp.pasteur.fr, directory /pub/GenomeDB/SubtiList). Moreover, the data contained in *SubtiList* is accessible through the World-Wide Web at the following address: http://www.pasteur.fr/Bio/SubtiList.html.

Release of the Sequences

Sequences in the depository at the Institut Pasteur are kept confidential for a period of nine months following submission until verification has been completed satisfactorily. This process includes verification under the responsibility of two of us (A. D. and F. K.) and resequencing up to 15% of the submitted sequence by a second laboratory of the network in order to minimize errors. After this delay the sequences are released to international databanks (GenBank or EMBL).

Perspectives

This genome sequencing project is followed by a second European project aimed at the systematic analysis of *B. subtilis* genes of unknown function (contract BIO2-CT95-0278).

Acknowledgments

This work was supported by the European Commission under the Biotechnology programme (contract numbers BIO2-CT93-0272 and BIO2-CT94-2011). Biotechnology companies, including Dupont de Nemours (U.S.A., France), Genencor (U.S.A., The Netherlands), F. Hoffmann-La Roche (Switzerland), Novo Nordisk (Denmark, U.S.A.), Puratos (Belgium) and Roussel-Uclaf (France), provided additional funds. The work from our laboratories was supported by funds from the Institut Pasteur and Centre National de la Recherche Scientifique. The expert secretarial assistance of C. Dugast is gratefully acknowledged.

References

Anagnostopoulos, C., P. J. Piggot, and J. A. Hoch. 1993. The genetic map of *Bacillus subtilis*. In Bacillus subtilis *and Other Gram-Positive Bacteria, Biochemistry, Physiology, and Molecular Genetics,* pp. 425–461. A. L. Sonenshein, J. A. Hoch and R. Losick, eds. American Society for Microbiology, Washington, D.C.

Azevedo, V., E. Alvarez, E. Zumstein, G. Damiani, V. Sgaramella, S. D. Ehrlich, and P. Seror. 1993. An ordered collection of *Bacillus subtilis* DNA segments cloned in yeast artificial chromosomes. *Proc. Natl. Acad. Sci. USA* 90:6047–6051.

Cruz Ramos, H., L. Boursier, I. Moszer, F. Kunst, A. Danchin, and P. Glaser. 1995. Anaerobic transcription activation in *Bacillus subtilis:* Identification of distinct FNR-dependent and independent regulatory mechanisms. *EMBO J.* 14:5984–5994.

Devine, K. M. 1995. The *Bacillus subtilis* genome project: aims and progress. *Trends Biotechnol.* 13:210–216.

Fleischmann, R. D., *et al.* 1995. Whole genome random sequencing and assembly of *Haemophilus influenzae Rd. Science* 269:449–604.

Glaser, P., F. Kunst, M. Arnaud, M-P. Coudart, W. Gonzales, M-F. Hullo, M. Ionescu, B. Lubochinsky, L. Marcelino, I. Moszer, E. Presecan, M. Santana, E. Schneider, J. Schweizer, A. Vertès, G. Rapoport, and A. Danchin. 1993. *Bacillus subtilis* genome project: cloning and sequencing of the 97 kb region from 325° to 333°. *Mol. Microbiol.* 10:371–384.

Kunst, F., and K. Devine. 1991. The project of sequencing the entire *Bacillus subtilis* genome. *Res. Microbiol.* 142:905–912.

Kunst, F., A. Vassarotti, and A. Danchin. 1995. Organization of the European *Bacillus subtilis* genome sequencing project. *Microbiology* 141:249–255.

Médigue, C., I. Moszer, A. Viari, and A. Danchin. 1995. Analysis of a *Bacillus subtilis* genome fragment using a cooperative computer system prototype. *Gene* 165:GC37–GC51.

Médigue, C., T. Vermat, G. Bisson, A. Viari, and A. Danchin. 1995. Cooperative computer system for genome sequence analysis. *Proceedings of the Third International Conference on Intelligent Systems for Molecular Biology,* pp. 249–258.

Moszer, I., P. Glaser, and A. Danchin. 1995. *SubtiList:* a relational database for the *Bacillus subtilis* genome. *Microbiology* 141:261–268.

Ogasawara, N., Y. Fujita, Y. Kobayashi, Y. Sadaie, T. Tanaka, H. Takahashi, K. Yamane, and H. Yoshikawa. 1995. Systematic sequencing of the *Bacillus subtilis* genome: Progress report of the Japanese group. *Microbiology* 141:257–259.

Sorokin, A., E. Zumstein, V. Azevedo, S. D. Ehrlich, and P. Seror. 1993. The organization of the *Bacillus subtilis* 168 chromosome region between the *spoVA* and *serA* genetic loci, based on sequence data. *Mol. Microbiol.* 10:385–395.

Woese, C. R. (1987). Bacterial evolution. *Microbiol. Rev.* 51:221–271.

36

Sequence Features of the Genome of a Unicellular Cyanobacterium *Synechocystis* sp. strain PCC6803

Satoshi Tabata and Mituru Takanami

Introduction

Cyanobacteria, which are capable of photoheterotrophic growth, have been used as a model organism for the study of oxygenic photosynthesis, and the structure and biochemical organization of the photosynthetic apparatus have been extensively investigated and compared with those of plant plastids (Bryant, 1994). One advantage of using these microorganisms, especially the unicellular strains, is their transformable characteristic. This permits elucidation of the structure and function of genes by using various genetic engineering procedures. However, information on the genes and on their organization in the genome remains imperative. The physical and genetic maps of cyanobacterial genomes have been constructed for two strains: the heterocyst-forming *Anabaena* sp. strain PCC7120 (Bancroft, Wolk and Oren, 1989; *see also* Chapter 47) and the unicellular *Synechococcus* sp. strain PCC7002 (Chen and Widger, 1993). However, not many genes have been characterized in detail. To achieve understanding of the entire genetic system involved in the oxygenic photosynthesis, as well as the general genetic complement carried by a single organism living in a relatively simple environment, we have begun the project of sequencing the entire genome of another unicellular cyanobacterium, *Synechocystis* sp. strain PCC6803 (Kotani *et al.*, 1994; Kotani *et al.*, 1995, Kaneko *et al.*, 1995). The main reason for the selection of this strain was that the size and GC content of the genome were appropriate for the sequencing project.

Physical Map of the Genome

As the first step of the project, we constructed a physical map of the PCC6803 genome by restriction analysis, using *Asc* I (6 sites), *Mlu* I (25 sites) and *Spl* I

(31 sites), and by isolation of linking cosmid clones which covered the key restriction sites (Kotani *et al.*, 1994) (Fig. 36-1). The validity of the map was further confirmed by the assignment of ordered clones and known genes to the physical map (Kotani *et al.*, 1995). The genome size estimated from the added sizes of the restriction fragments was 3.6 mega-bases (Mb). In the course of this analysis, three extra chromosomal units were identified (Fig. 36-1). Recently, a restriction map of the same strain has been reported by Churin *et al.* (1995). The number and sizes of their *Mlu* I-restriction fragments are essentially identical to our data, but their final map is somewhat different from ours. This could be due to variations in the original strains used in the two laboratories.

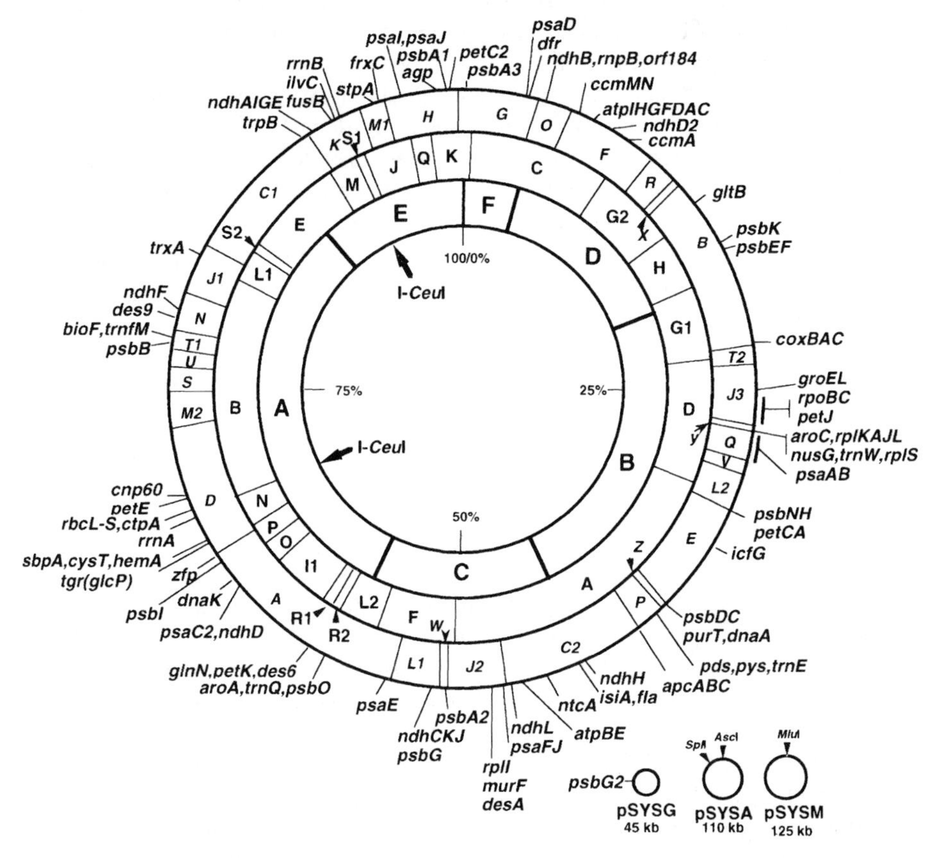

Figure 36-1. The physical map of the *Synechocystis* sp. strain PCC 6803 genome constructed by cleavage with *Asc* I (inner circle), *Mlu* I (middle circle) and *Spl* I (outer circle), and assignment of 82 reported genes and gene clusters on the map. The map position is indicated by a centesimal scale starting from the junction of *Asc* I-F and *Asc* I-E. Three plasmids identified are shown below the genome map. The details are given in Table 1 in ref. Kotani et al. (1995).

To construct the ordered clones for DNA sequencing, a cosmid library was prepared by cloning size-fractionated 40 kb *Sau*3AI fragments of the genome into the SuperCos 1 cosmid vector (Stratagene). Most of the cosmid clones carrying *Synechocystis* DNA were found to be stably maintained in *Escherichia coli* cells. Approximately 1,100 clones were sorted out by cross-hybridization, and the resulting contigs were located on the map by restriction fragment hybridization. About 90% of the genome could be covered with 127 partial overlapping clones (Kotani *et al.*, 1995). The gap regions were filled in using a λ library which was prepared by cloning fractionated 15 to 20 kb *Sau*3AI fragments of the genome into the λ DashII vector (Stratagene), and also with long PCR products which were amplified using information on the terminal sequences of contigs.

Since a total of 82 genes and gene clusters have already been isolated from this strain, most of which are involved in oxygenic photosynthesis, portions of their sequences were amplified by PCR and used as hybridization probes to blots carrying restriction fragments, ordered clones and long PCR-products. All the genes and gene clusters could be successfully assigned to particular locations on the genome, except for *psb*G2 [the gene for a subunit of NAD(P)H dehydrogenase], which was mapped to an extra-chromosomal unit of 45 kb (Fig. 36-1) (Kotani *et al.*, 1995). Since the genetic maps of some of the genes listed above, especially those for photosynthesis, have been reported for *Anabaena* sp. strain PCC7120 and *Synechocystis* sp. strain PCC7002, the interstrain gene organization was compared. No significant correlations were observed, implying that substantial rearrangement of genes occurred in the respective strains during or after establishment of the different species.

Sequencing Strategy

The most common strategy used to construct sequence contigs, consisting of shotgun sequencing followed by gap-filling via flanking sequence-primed sequencing, was adopted, but the shotgun method was modified as follows (Kaneko *et al.*, 1995). The 40 kb cosmid inserts were sonicated to two different levels and following size-fractionation, two shotgun libraries were prepared, one with inserts of approximately 700 bp (element clones) and the other with inserts of approximately 2.5 kb (bridge clones). The element clones were sequenced by a single-run, but only one of the strands was analyzed to collect the sequence information from different regions with the same number of runs. The bridge clones were single-run sequenced from both ends to assist computer assembly of the raw sequence data. By analyzing 600 to 700 element clones and 200 bridge clones, which yielded approximately 10 times raw sequence data, a single contiguous sequence could be deduced for most cosmid clones which did not carry repetitive sequences. When multiple contigs were obtained, the contigs were combined by searching bridge clones.

The DNA sequences deduced were confirmed as follows: Every 20 to 25 kb region of the genome was directly amplified from the genomic DNA by long PCR, and the occurrence of restriction sites was compared with the assembled sequences. This allowed the detection of rearrangements or deletions in the ordered clones used for sequencing within the resolution of gel electrophoresis.

Sequencing procedures were semi-automated by the combination use of ABI Catalysts 800 or Amersham Labstation and ABI DNA sequencer 373S or 377. Sequencing is currently progressing at a pace of 200–250 kb/month and as of June 30, 1995, about 40% of the entire genome has been sequenced.

Assignment of Coding Regions

The sequence features in a contiguous region of about 1 Mb encompassing map positions 64% to 92% in Fig. 36-1 were analyzed using the GCG software package. When the open reading frames (ORFs) longer than 300 bp were identified, the occurrence of ORFs was an average of 82/100 kb, so that the total number of genes in the genome longer than 300 bp was estimated to be 2,900. Most of the ORFs were found to be in the range of 300–1,500 bp (64/100 kb), but the largest ORF found so far is 12,597 bp. The ORFs as a whole occupied 86% of the 1 Mb region, indicating compact arrangement of potential coding regions in the genome. When searching for similarities, 31% of the ORFs contained sequences with some similarity (more than 30% identity of amino acid sequences) to known genes or their homologues in the GenBank/EMBL databases and an additional 18% of the ORFs showed weak similarities to reported genes.

The precise assignment of the coding regions is in progress using two different approaches: One is the prediction of likely genes using the computer program "GeneMark" developed by Borodovsky and McIninch (1993), and the other is the direct analysis of the N-terminal sequences of proteins which are resolved by high-resolution two-dimensional gel electrophoresis (2-DE) (Goer, Postel and Gunther, 1988). When total proteins were resolved by isoelectric focusing between pH 3 and 10 in the first dimension followed by sodium dodecyl sulfate polyacrylamide gel electrophoresis (15%) in the second dimension, approximately 400–500 spots could be visualized by silver staining (Fig. 36-2). Analysis of the N-terminal sequences of protein spots with relatively high intensities is in progress using a Shimadzu PPSQ-10 protein sequencer. So far, N-terminal blocks were only identified in a few proteins. It should be mentioned that several intense spots corresponded to proteins related to photosynthesis (Fig. 36-2). In Fig. 36-3, the DNA sequences corresponding to the N-terminal regions of proteins analyzed to date are shown aligned with the initiation codons. In addition to the common ATG initiation, GTG and TTG starts were detected, and in the regions about 5 to 15 bp upstream of the initiation codons, the Shine Dalgarno (SD)-like sequences were recognizable. It is therefore likely that mechanisms common to prokaryotes are controlling in translational initiation in cyanobacteria.

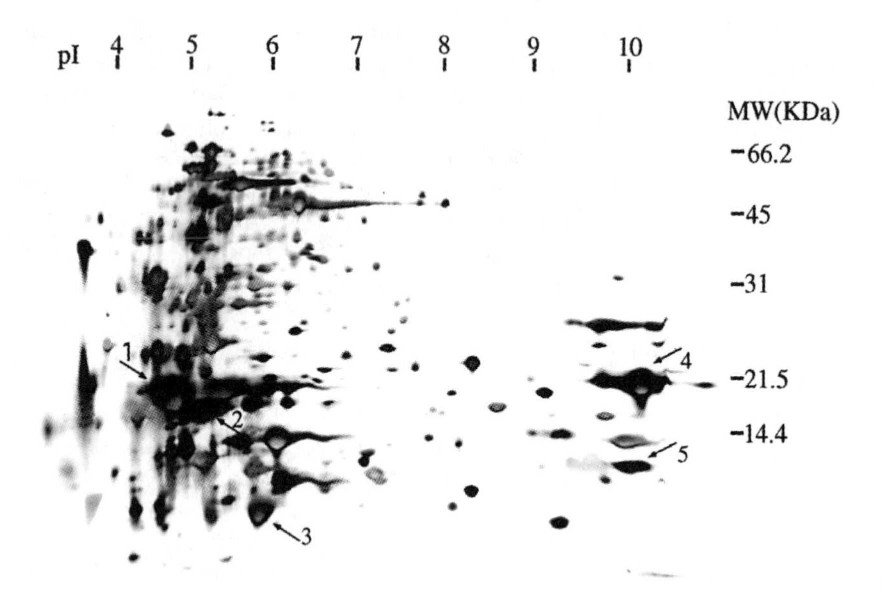

Figure 36-2. Silver-stained cellular proteins, resolved by high resolution two-dimensional gel electrophoresis. Around 400–500 spots can be visualized. Strong spots, indicated by arrows, are gene products of 1: *apc*A, 2: *apc*B, 3: *psa*C, 4: *psa*D, and 5: *psa*E.

The genome contained two *rrn* operons at map positions 68% and 93%, respectively. As the one at 93% was located outside the 1 Mb region, its sequence was determined separately. When the two *rrn* sequences were compared, each unit was 5,028 bp long and their sequences including internal spacer regions were completely identical. The rRNA genes in each operon were found to be aligned in the following order:

[16s rRNA]-[Ile-tRNA]——[23s rRNA]-[5s rRNA]

This organization is different from that reported for *S. sp.* strain PCC6301 with regard to the species of the internal tRNA, but the sequences of the rRNA genes were found to be well conserved (70% to 90% identity of DNA sequences).

Unique Sequences Identified

IS-like Elements

Within the sequenced regions, three classes of insertion sequence (IS)-like elements, tentatively named ISc100, ISc523 and ISc352, were identified. These elements were found to carry inverted repeats of 11–18 bp at both ends, and

Code No.	Genes	ORF No.	
cy009	NI	sll0822	CCAATTCTGATCAATCTCAATCACCGAAAACTTGA**ATG**GCTAAATCA A K S
cy010	*psaD*	slr0737	CCTAAAATACCTTATGAAATTCCATTTTCATCCCT**ATG**ACAGAACTC T E L
cy011	NI	slr0006	CTGTTAATCTGGAGGCAGTTTCAAGGGACAATTCA**ATG**GTTACACTA V T L
cy014	NI	sll0359	GCAATCAAGCATCATTCAACGGAAAAAATAATCCT**ATG**CCAAACGCC P N A
cy016	NI	ssl0707	ACAACAACCAGATCATTGCACGAGGAGTACCAGGT**TTG**AAAAAAGTA M K K V
cy017	NI	sll0230	AGATTGTCGTCAGCCATTACCGACGAGTAAAAGCT**ATG**AACGTGATT M N V I
cy019	NI	sll0018	AAATACCCATTAGGCAATAGAACTTAGGAGGATTT**ATG**GCTCTTGTA A L V
cy022	*psaC*	ssl0563	TTACGGAGAACCCTCGGTTAAGGAGCCGATAGTCA**ATG**TCCCATAGT S H S
cy025	*rbcS*	slr0012	TAACGTAGTCATCAGCAAGGAAAACTTTTAAATCG**ATG**AAAACTTTA M K T L
cy027	*trxA*	slr0623	ATCAGCCCATCACGAGGTTTAGAAGGATTTCCAGT**ATG**AGTGCTACC S A T
cy030	NI	ssl0352	GCTAGAATTGTCCAATCCAAAATCAGGTAGCCATT**ATG**ATTTTTCCC M I F P
cy065	NI	ssl0617	TTAACCCCCAGCGATTTTGAGAGGATAAGTAAGAC**ATG**GGATTATTT G L F
cy074	NI	slr0476	ATTAATCGAATATTTATATTAAGGAGATAAACAGC**ATG**ACTGAGGAA T E E
cy077	NI	sll0145	AGGCGAAGCAGTCGGTACGTTAGTTGGAGAAAATT**GTG**AAGTTAGCT M K L A
cy078	NI	slr0676	GGATCGACAATTTTTGAAAAATTCCTAGATTATCT**ATG**CAACAACGT M Q Q R

Figure 36-3. The DNA sequences around the translational initiation sites assigned by analysis of the N-terminal sequences of 2-DE resolved proteins. The DNA sequences are shown aligned with the initiation codons. In about a half of proteins, the N-terminal methionine has been cleaved off. The sequences corresponding to the putative SD sequence are underlined. NI: not identified gene.

outside the inverted repeats, apparent direct repeats were recognized for ISc523 and ISc352, resulting from translocation-induced duplication of the target sequence (Fig. 36-4). Most of the elements which belonged to ISc100 and ISc523 contained single common ORFs, but some of them carried segmentally split or frame-shifted forms of the conserved ORFs. The conserved ORF in ISc523 showed 42% identity to the putative transposase gene of IS1031C in *Acetobacter* (Coucheron, 1993) and 39% identity to that of Tn4811 in *Streptomyces* (Chen *et al.*, 1992), while that in ISc100 did not show any similarity to the reported sequences in the DNA databases. Their copy number in the genome was estimated to be about 20 for ISc100 and 10 for ISc523, including the rearranged forms. In contrast, the elements in the ISc352 class did not contain any unique ORF, but

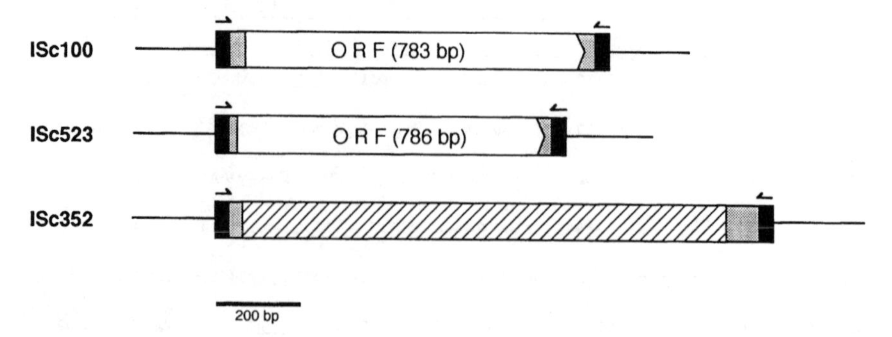

Figure 36-4. The common structures of the three classes of IS-like elements. The inverted repeats specific to IS-elements are shown by filled boxes, and ORFs by open boxes. The shaded box in ISc-352 is the region with similarity to the putative transposase gene in IS701 in *Calothrix.*

instead, each of the elements carried a region with significant similarity to the predicted transposase gene of IS701 in *Calothrix.* (Mazel *et al.*, 1991). It was also noted that one of the elements contained the insertion of ISc100 in the middle. It is therefore likely that the elements in this IS class are rearrangement products of an ancestral IS-element.

Short Repetitive Sequences

The presence of several short repetitive sequences was also noted. The most significant one is the repeat of an 150 bp segment. The consensus sequence of these repeats is as follows:

-AA****TTTT AGAGCCTGTT TGAAAAGCCC CCCTGGCCCC CCAAATTTGG
GGGGA*AACT GAGTGAAAGT CCCCCAGTAT TGCCGGAGCT TTAGTGGGGA
GATTTAGGGG GC*GAATA** *******AAA CTTT*CAAAC ACGTTCTTAG-

The approximate copy number in the genome was estimated to be 80, but the functional significance of this type of repeats is unknown.

WD-repeats

The presence of a unique repeating unit, named Trp-Asp (WD)-repeats, has recently been noted in various regulatory proteins which participate in important cellular functions of eukaryotic cells (Neer *et al.*, 1994). The WD-repeats, which usually end with Trp-Asp, have been defined as those as indicated at the top of Fig. 36-5, and the regular expression in the constant-length core region is shown above the sequence pile. On the basis of the strong conservation of these repeating units in the regulatory proteins of a wide variety of eukaryotes, but not in prokaryotes, it has been speculated that the functions associated with each were

(A)

$$\{ X_{6\text{-}94} - [GH - X_{23\text{-}41} - WD] \}^{N_{4\text{-}8}}$$

Variable length   Constant length core

(B)

```
                                                 S         (2 of 3)
                                             A   A          A              A A
                                             M   M   I      C   C CC       M M
                                          V  F   F   L D D  F   A AA       F F K
Regular                                   A  V   V   W G G  V   T TT       V VYR
expression                                S  I   I   Y N N  I   S SS       I IFN
in Core                                  GH  L··L·   F·P·P  ·L  G GG·D··L·LWD

repeat01       NECNRCHHEG PVTVLRISPS MENTPPLVLT ATTNGIAYLWS
repeat02       FHGELINVLR GHQEAITALD WSADGQYFAT ASADHTVKLWQ
repeat03       RHGEEVATLR GHEDWVRSVH FSPHHQFLVT SGQDNTARIWN
repeat04       FAGEQLTLCQ GHADWVRNAE FNCHGQILLT ASRDGTARLWD
repeat05       LEGREIGLCQ GHTSWVRNAQ FSPDGQWIVT CSADGTARLWD
repeat06       LSSQCFAVLK GHQNWVNNAL WSPDGQHIIT SSSDGTARVWS
repeat07       LHGKCLGTLR GHDHNIHGAR FSLDGQKIVT YSTDNTARLWT
repeat08       KEGTLLTILR GHQKEVYDAD FSADGRFVFT VSADQTARQWD
repeat09       ISQKDTITLT GHSHWVRNAH FNPKGDRLLT VSRDKTARLWT
repeat10       TEGECVAVLA DHQGWVREGQ FSPDGQWIVT GSADKTAQLWN
repeat11       VLGKKLTVLR GHQDAVLNVR FSPDSQYIVT ASKDGTARVWN
repeat12       NTGRELAVLR HYEKNIFAAE FSADGQFIVT ASDDNTAGIWE
repeat13       IVGREVGICR GHEGPVYFAQ FSADSRYILT ASVDNTARIWD
repeat14       FLGRPLLTLA GHQSIVYQAR FSPEGNLIAT VSADHTARLWD
repeat15       RSGKTVAVLY GHQGLVGTVD WSPDGQMLVT ASNDGTARLWD
repeat16       LSGRELLTLE GHGNWVRSAE FSPDGRWVLT SSADGTAKLWP
```

Figure 36-5. The WD-repeats identified in an ORF of 5,078 bp at map position 69%. The repeating units defined are given at the top (A), and the regular expression in the GH(Gly-His)-WD core is shown above the sequence pile (13). The repeats identified in the ORF are indicated aligned with the GH-WD core as a sequence pile (B). The consensus sequences are boxed, and the sequences corresponding to the variable length region are shown at the left. Note that the length of this region is the same for all the repeats.

fixed at a very ancient age, dating back to the eukaryotic origin (Neer *et al.*, 1993). We found that at least two clusters of the WD-repeats are present in the sequenced region of the cyanobacterial genome. The one with the most characteristic repeats was located in an ORF of 5,079 bp at position 69%, which encodes a protein of up to 1,693 amino acids (Fig. 36-5), and the other in an ORF of 3,572 bp, which can encode a protein of up to 1,191 amino acids. No information is available for the expression of these genes, but this should be possible to examine, because the clones and the vector system for gene manipulation are available. If the functional roles of the genes containing the WD-repeats could be confirmed, it may lead to the speculation that genes carrying these repeats were fixed in an organism ancestral to both prokaryotes and eukaryotes.

Conclusion

Computer analysis of a contiguous sequence of about 1 Mb from map positions 64% to 92% predicted that the genome of the unicellular cyanobacterium *Synecocystis* sp. strain PCC6803 contains approximately 2,900 ORFs longer than 300 bp. According to the assignment of the coding regions on the basis of the N-terminal amino acid sequences of 2-DE resolved proteins, it is likely that mechanisms common to prokaryotes are operating in the translational initiation of the cyanobacterial proteins. Identification of at least three classes of IS-like elements and their rearrangement products in the genome provides strong evidence that disruption and rearrangement of the genes frequently occurred during and after establishment of the species. An important finding was also that characteristic WD-repeats, which have only been detected in the regulatory proteins of eukaryotes so far, were identified within large ORFs of the cyanobacterial genome. It will be essential to prove that ORFs containing the WD-repeats actually code for functional proteins, before further speculations about the origin of these repeats can be made. The sequence determination of the entire PCC6803 genome was completed by the end of February 1996. The genome was 3,573,470 bp long, and contained a total of 3,168 potential protein coding regions which were predicted on the basis of ORF analysis, survey by the GeneMark program and similarity search. As to the structural RNA genes, 42 potential transfer RNA genes were assigned in addition to the two sets of *rrn* operons.

Acknowledgments

We thank T. Sazuka and O. Ohara for unpublished information on 2-DE resolved protein sequences; A. Tanaka, T. Kaneko and S. Sato for unpublished information on unique sequences including WD-repeats; and H. Kotani for discussion.

References

Bancroft, I., C. P. Wolk, and E. V. Oren. 1989. Physical and genetic maps of the genome of the heterocyst-forming cyanobacterium *Anabena* sp. strain 7120. *J. Bacteriol.* 171:5940–5948.

Borodovsky, M., and J. McIninch. 1993. GENEMARK: Parallel gene recognition for both DNA strands. *Comput. Chem.* 17:123–133.

Bryant, D. A., ed. 1994. *The Molecular Biology of Cyanobacteria.* Kluwer Academic Publishers, The Netherlands.

Chen, C. W., T. W. Yu, H. M. Chung, and C. F. Chou. 1992. Discovery and characterization of a new transposable element, Tn4811, in *Streptomyces lividans* 66. *J. Bacteriol.* 174:7762–7769.

Chen, X., and W. R. Widger. 1993. Physical genome map of the unicellular cyanobacterium *Synechococcus* sp. strain PCC 7002. *J. Bacteriol.* 175:5106–5116.

Churin, Y. N., I. N. Shalak, T. Borner, and S. V. Shestakov. 1995. Physical and genetic map of the chromosome of the unicellular cyanobacterium *Synechocystis* sp. strain PCC 6803. *J. Bacteriol.* 177:3337–3343.

Coucheron, D. H. 1993. A family of IS1031 elements in the genome of *Acetobacter xylinum*: Nucleotide sequences and strain distribution. *Mol. Microbiol.* 9:211–218.

Goer, R., W. Postel, and S. Gunther. 1988. The current state of two dimensional electrophoresis with immobilized pH gradients. *Electrophoresis* 9:531–546.

Kaneko, T., A. Tanaka, S. Sato, H. Kotani, T. Sazuka, N. Miyajima, M. Sugiura, and S. Tabata. 1995. Sequence analysis of the genome of the unicellular cyanobacterium *Synechocystis* sp. strain PCC 6803. I. Sequence features in the 1 MB region from map positions 64% to 92% of the genome. *DNA Research* 2:153–166.

Kotani, H., T. Kaneko, T. Matsubayashi, S. Sato, M. Sugiura, and S. Tabata. 1994. A physical map of the genome of a unicellular cyanobacterium *Synechocystis* sp. strain PCC 6803. *DNA Research* 1:303–307.

Kotani, H., A. Tanaka, T. Kaneko, S. Sato, S. Sugiura, and S. Tabata. 1995. Assignment of 82 known genes and gene clusters on the genome of the unicellular cyanobacterium *Synechocystis* sp. strain PCC 6803. *DNA Research* 2:133–142.

Mazel, D., C. Bernard, R. Schwarz, A. M. Castets, J. Houmard, and N. Tandeau de Marsac. 1991. Characterization of two insertion sequences, IS701 and IS702, from the cyanobacterium *Calothrix* species PCC 7601. *Mol. Microbiol.* 5:2165–2170.

Neer, E. J., C. J. Schmidt, R. Nambudripad, and T. F. Smith. 1994. The ancient regulatory-protein family of WD-repeat proteins. *Nature* 371:297–300.

37

A Minimal Gene Complement for Cellular Life and Reconstruction of Primitive Life Forms by Analysis of Complete Bacterial Genomes

Arcady R. Mushegian and Eugene V. Koonin

A Minimal Gene Set for Modern-Type Cellular Life Derived by Comparing the *Haemophilus influenzae* and *Mycoplasma genitalium* Genomes

The completely sequenced genome of the small parasitic bacterium *Mycoplasma genitalium* has been dubbed "the minimal gene complement" (Fraser *et al.*, 1995). Whereas this is indeed the smallest known genome of a cellular life form (*see* Chapter 40), there is no evidence it is minimal in any sense. We argue that a much closer approximation of a minimal self-sufficient genome, able to support a cell without relying on import of functionally active proteins, can be derived from the set of genes shared by the *M. genitalium* genome and the complete genome of another parasitic bacterium, *Haemophilus influenzae* (Fleischmann *et al.*, 1995).

These first two complete genomes seem to be a fortunate choice to address the problem of a minimal gene set. Since *M. genitalium* and *H. influenzae* belong to distant lineages of bacteria (Olsen *et al.*, 1994), it is plausible that only those genes that are integral for survival of any cell are conserved between the two species. Moreover, secondary simplification during adaptation towards parasitic lifestyle in both bacteria may be thought of as an additional advantage, rather than a source of confusion, for this analysis. Indeed, both *M. genitalium* and *H. influenzae* lack many metabolic pathways and regulatory networks (Koonin *et al.*, 1996a; *see* Chapters 40 to 42), while retaining the capacity to grow extracellularly, e.g. on artificial media, albeit rich ones. It should be noted that only complete genomes provide the adequate data for reconstructing minimal and hypothetical ancestral gene repertoires, as they contain a finite number of genes known to be sufficient for performing all vital tasks. At the first stage of this analysis, we refrained from attempts to deduce primitive pathways drastically different from those known in extant cells. Thus we were aiming at producing a minimal gene set for "modern-type" cellular life.

Our analysis strategy is depicted in Figure 37-1. We first compared 469 *M. genitalium* genes to 1703 *H. influenzae* genes and to general-purpose protein databases, using the combination of iterative BLAST search and motif analysis (Koonin *et al.*, 1996b), in order to delineate *orthologous* genes in two genomes. Orthologs are defined as a pair of genes from two species derived by vertical descent from a common ancestor (Fitch, 1970); there is by definition only one ortholog of a given gene in another genome. By contrast, paralogs are related genes with similar but non-identical functions evolved by gene duplications within a given lineage (Fitch, 1970). Paralogs form gene families, and there may be more than one paralog for a given gene in other genomes. Not distinguishing orthologs from paralogs can lead to erroneous conclusions concerning the status of a pathway in a given genome, and ultimately to erroneous evolutionary reconstructions (Jensen and Wu, 1996; Tatusov *et al.*, 1996).

The criteria for identifying orthologous pairs have been described (Tatusov *et al.*, 1996). Briefly, a candidate gene (or gene product) A in genome B is considered to be the ortholog of a given gene A′ in genome B′ on the basis of: 1) High sequence similarity, when sequence A′ is used to search database B, the score with sequence A is the highest, and vice versa; 2) Consistent taxonomic distribution, when sequence A′ is used to search the complete database, the score with

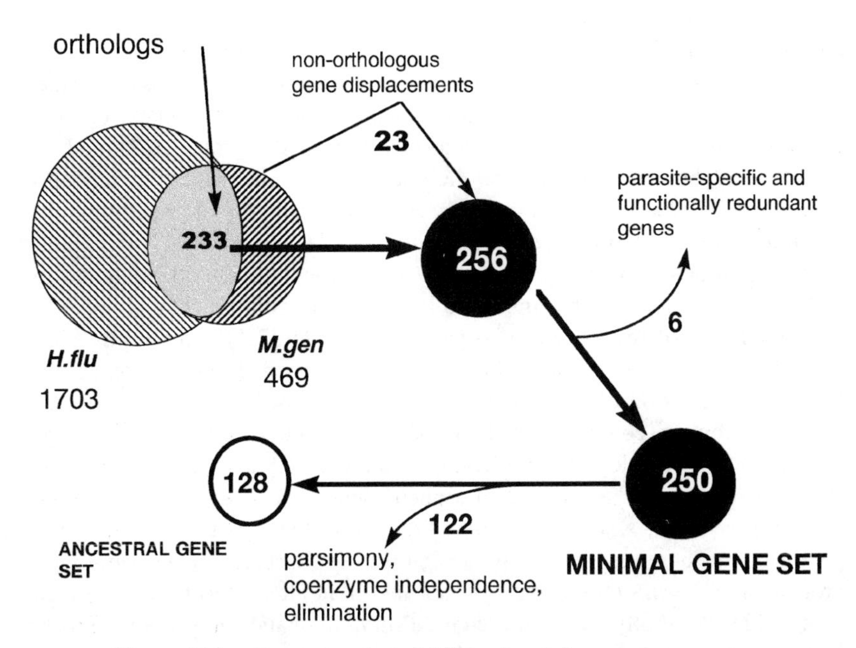

Figure 37-1. Reconstruction of Minimal and Ancestral gene sets.

A is higher than that with any homolog from a taxonomically distant species; 3) Congruent domain structure, A and A′ align along most of the lengths of both proteins, except for rare, clear-cut cases of domain rearrangement, and 4) Syntheny, A′ can be chosen among highly similar candidates if a neighbor of gene A′ in B′ is orthologous to a neighbor of gene A in B; this criterion is useful when two closely related, largely synthenic genomes are compared, whereas it was of limited use in the present study.

Using these criteria, 240 gene pairs were judged to be orthologous in *M. genitalium* and *H. influenzae.* A breakdown of these proteins into broad functional categories is shown in Table 37-1. To make functional assignments as precise as possible, the whole set of the orthologs was superimposed onto the map of general metabolic pathways, evaluated from the point of view of the known nutritional requirements of both bacteria, and checked for consistency on a case-by-case basis. As a result, about 10% of the *M. genitalium* genes were functionally re-assigned compared to the original identification (Fraser *et al.*, 1995). This may be illustrated by the case of serine (glycine) hydroxymethyltransferase (GlyA ortholog). Originally, this gene product has been categorized as the sole enzyme of amino acid metabolism in *M. genitalium* (Fraser *et al.*, 1995). In *H. influenzae* and in better-studied bacteria, like *E. coli,* serine (glycine) hydroxymethyltransferase is at the crossroads of many pathways, which lead to lipids, glyoxylate, pyruvate, aliphatic amino acids, and folate, as well as to the pools of glycine and serine themselves. However, both in *M. genitalium* and in the list of the 240 orthologous gene pairs, only one pathway that involves interconversion of glycine and serine is partially conserved, namely the cycle of folate derivatives, some of which, in turn, serve as group donors in such essential processes as biosynthesis of thymidine phosphate and formylmethionine. Thus, in the context of the *M. genitalium* metabolism, an assignment of GlyA to the group concerned with cofactor metabolism appears to be the only relevant one. Likewise, among *M. genitalium/H. influenzae* set of orthologs, the whole category of proteins involved in carbohydrate metabolism could be eliminated by unequivocal assignment of each protein either to energy conversion or to exopolysaccharide metabolism.

The set of 240 orthologs appears to lack a number of enzymes necessary for the completion of several essential pathways (Table 37-1). On the other hand, several orthologs were found in the two bacteria, for which only sequence-based functional predictions, but no biochemical data, are available. In some cases, the gaps in pathways could be filled by plausible candidates from this group of poorly studied orthologs (Table 37-1). A number of gaps remained, however. The apparent reason for this is the phenomenon we called *non-orthologous gene displacement,* whereby the same function in two species is performed by paralogous or unrelated genes. We examined the "surplus" of non-orthologous proteins in *M. genitalium* and *H. influenzae,* in an attempt to fill as many gaps as possible. Generally, we preferred *M. genitalium* candidates as they function in the context of a simpler genome, although in a few cases there were additional

Table 37-1. Minimal and Ancestral Gene Sets Deduced by Comparing M. Genitalium *and* H. Influenzae *Protein Sequences*[a]

Functional System	Minimal Gene Set for Modern-Type Cellular Life: *M. genitalium/H. influenzae* Orthologs, Supplemented with *M. genitalium* Proteins	*M. genitalium* Proteins[b]	Ancestral Gene Set: Theoretical Reduction of Minimal Set 1	*M. genitalium* Proteins[b]
Translation	Ribosomal proteins, aminoacyl-tRNA synthetases, translation factors, RNA and protein modification enzymes (pseudoU-synthetases, methyltransferases); a putative aminotransferase (MG336) to convert glutamyl-tRNA into glutaminyl-tRNA; no tRNA-nucleotidyltransferase	**95** 005, 012, 021, 026, 035, 036, 070, 081–083, 087–090, 092, 093, 104, 106, 113, 126, 136, 142, 143, 150–152, 154–169, 172–176, 178, 182, 194, 195–198, 209, 232, 234, *251*–253, 257, 258, 266, *283*, 292, 325, 334, 336, 345, 346, 361–363, 363a, 365, 367, 375, 378, 417, 418, 424–426, 433, 435, 444–446, 451, 455, 466	No protein- or rRNA-modifying enzymes, except for one pseudouridine synthase (MG209); no formylmethionine, no methionine aminopeptidase, no glutamine, no RNAase P; one Tu-like translation factor (GTPase), one peptide chain releasing factor.	**72** 005, 021, 035, 036, 070, 081–083, 087, 088, 092, 093, 113, 126, 136, 150–169, 174–176, 178, 194, 195, 197, 198, 209, 232, 234, *251*, 253, 257, 266, *283*, 292, 325, 334, 345, 346, 361–363, 367, 375, 378, 417, 418, 424, 426, 435, 444, 446, 451, 455, 466

Continued

Table 37-1. Continued

Functional System	Minimal Gene Set for Modern-Type Cellular Life: *M. genitalium/H. influenzae* Orthologs, Supplemented with *M. genitalium* Proteins	*M. genitalium* Proteins[b]	Ancestral Gene Set: Theoretical Reduction of Minimal Set 1	*M. genitalium* Proteins[b]
Replication, Recombination, Repair, and Transcription	Rudimentary repair system: no photorepair, reduced excision repair; only one DNA polymerase (polIII); 4 RNA polymerase subunits including a single sigma factor; 3 transcription factors but no termination factor Rho; no helix-turn-helix transcription regulators; ppGpp synthetase (SpoT) may enable global transcription regulation	**33** 001, 003, 004, 054, 073, 091, 094, 097, 122, 141, 177, 203, 204, 206, 244, 249, 250, 254, 259, 261, *262*, 262a, 278, 282, 308, 339–341, *351,* 358, 359, 421, 469	RNA genome; replication based on one RNA polymerase and one helicase (MG308), also involved in translation (see text).	**2** 177, 308
Nucleotide Metabolism	Synthesis of ribonucleotides from ribose, ATP, and nitrous bases; reduction of ribonucleotides to deoxyribonucleotides coupled with thioredoxin oxidation. No Ndk; two possible substitutes are MG264 (misidentified in ref. 5), and MG268. Nitrous base transport system uncertain; a candidate is the spermidine-putrescine transport system	**22** 006, 030, 042–044, 049, *052, 058,* 102, 107, 124, 127, 171, 227, 229, 231, *268,* 276, 330, 382, 434, 458	Ribonucleotide salvage system retained No deoxyribonucleotides	**16** 030, 042–044, 049, *052,* 058, 107, 171, *268,* 276, 330, 372, 382, 434, 458
Energy	Complete Embden-Meyerhof pathway, conversion of pyruvate into coenzyme A, and further into acetate; uptake and utilization of glucose, fructose, ribose, xylulose, glycerol and lactate; H^+ ATPase; (deoxy)ribose aldolase (DeoC)—a crucial link between nucleotide metabolism and energy supply	**34** *023, 033,* 038, *039,* 050, *053, 063,* 111, 112, 119, 120, 187, 215, 216, 271, 272, *273–275, 299, 300, 301,* 357, 398–405, 407, *430,* 431	Uptake and utilization of ribose and glycerol; deoC; abridged glycolytic pathway including only stages from glyceraldehyde 3-phosphate to pyruvate; no H^+ ATPase;	**10** 033, 038, *039,* 050, 187, 216, 300, 301, 407, *430*

Chaperones and Chaperone-like Proteins	All major groups of chaperones, except for HSP90[c]	**12** 019, 201, 238, 239, 305, 355, 392, 393, *408,* 448, *449,* 457	Elimination of paralogs but all families of chaperones in the Minimal set are retained[c]	**10** 019, 201, 238, 239, 305, 355, 392, 393, 448, *449*
Amino Acids Metabolism	ATP-dependent transport systems for oligopeptides and for amino acids; several proteases	**6** 077–080, 180, 391	ATP-dependent oligopeptide transport; one aminopeptidase	**5** 077–080, 391
Coenzyme Metabolism and Utilization	NAD, FAD, SAM are synthesized from exogenous monomers; ligation of lipoate onto proteins; 4 enzymes for turnover of folate derivatives. Altogether, *M. genitalium* genome codes for 8 enzymes that employ dinucleotide coenzymes (two or three utilize FAD, the others utilize NAD or NADP), two of each: lipoate-, pyridoxal-, thiamine-, and folate-dependent enzymes, and several SAM-utilizing methyltransferases	**8** 013, 047, 145, 228, 245, *270,* 383, 394	NAD synthase, SAM synthase; other cofactors not required	**2** 047, 383
Lipid Metabolism	Synthesis of lipids and phospholipids from glycerol and exogenous fatty acids; acyl carrier protein	**6** 114, 212, 287, 293, 333, *437*	In *Archaea,* fatty acids are replaced by terpenes, and genes of lipid biosynthesis are poorly characterized. Information is insufficient to derive an ancestral set	0?
Exopolysaccharides	Several glycosyltransferases involved in cell wall biosynthesis in *H. influenzae;* may be involved in capsule biosynthesis in *M. genitalium*[d]	**8** 059, 060, 066, 086, 118, 210, 224, 453	No exopolysaccharide biosynthesis?	0?

Continued

Table 37-1. Continued

Functional System	Minimal Gene Set for Modern-Type Cellular Life: *M. genitalium/H. influenzae* Orthologs, Supplemented with *M. genitalium* Proteins	*M. genitalium* Proteins[b]	Ancestral Gene Set: Theoretical Reduction of Minimal Set 1	*M. genitalium* Proteins[b]
Uptake of Inorganic Ions	Phosphate permease, Na^+ ATPase, TrkA-related Na^+ transporter, cationic/metal ATPases	**5** 065, 071, 322, 410, 411	same as in Minimal set	5
Secretion, Receptors	Signal recognition particle, preprotein translocase, a sialoglycoprotease (see text)	**5** 015, 048, 072, 170, 297	Only signal recognition particle with one GTPase instead of two	**3** 048, 072, 170
Other Conserved Proteins	Proteins with ATP-, GTP-, NAD-, FAD- and SAM-binding domains; 3 proteins with functionally uncharacterized ancient conserved regions	**16** 008, 009, 024, 056, 125, 138, 221, 247, 295, 329, 332, 335, 347, 379, 380, 387	GidA (putative methyltransferase) GidB putative dinucleotide-binding enzyme), one protein with a new ACR	**3** 009, 379, 380
Total Protein-Coding Genes		256 (234 *M. genitalium/H. influenzae* orthologs + 22 non-orthologous displacements)		128

[a]Annotated Minimal set and Ancestral set are available by anonymous ftp at ncbi.nlm.nih.gov in the directory/pub/koonin/MINSET

[b]*M. genitalium* proteins are designated as in ref. 5 (with the exception of 262a and 363a, proteins that have not been identified in ref. 5), with MG omitted for brevity. Underline indicates the apparent cases of non-orthologous gene displacement

[c]In addition to characterized families of chaperones, both Minimal set and Ancestral set include representatives of two new groups. MG449 belongs to an emerging family of putative chaperones that includes both stand-alone bacterial and eukaryotic proteins, and distinct domains of bacterial methionyl- and phenylalanyl-tRNA synthetases (E. V. K., unpublished observations). Interestingly, *H. influenzae* and *M. genitalium* each encode 2 members of this family but *H. influenzae* lacks a stand-alone version, whereas *M. genitalium* does not have a chaperone-like domain in its methionyl-tRNA synthetase. Two proteins, MG408 and MG448, are related to distinct domains of *Neisseria gonorrhae* PilB protein implicated in transcriptional regulation of pilus genes. Unexpectedly, we found that neither domain is similar to the known transcription regulators, but MG448 is homologous to peptide methionine sulphoxide reductases from bacteria and eukaryotes, whereas MG408 belongs to another highly conserved family of thioredoxin-like proteins (A. R. M. and E. V. K., unpublished observations). Thus MG408, MG448 and their homologs are likely to be involved in protein oxidative damage repair.

[d]This category includes transketolase (MG066) that is probably involved in heptose biosynthesis and has no bioenergetic role as proposed in ref. 5.

considerations favoring the choice (Koonin and Bork, 1996). In whole, 22 cases of apparent non-orthologous displacement were detected, and most of the missing links in biochemical pathways were thus tentatively provided.

From the Minimal Gene Set to the Reconstruction of a Primeval Genome

We further sought to go from the minimal gene for contemporary cellular life to one that may be inherited from a primitive, ancestral cell. Before proceeding with this, apparently major, reduction, it was necessary to evaluate the 262-gene set for consistency and robustness. Will a hypothetical lifeform carrying such a small genome be biochemically crippled? Although the minimal gene set is inevitably a simplified version of the *M. genitalium* genome, i.e. it might have been expected to be disarrayed by gene deletions, examination of individual functional systems suggests that this is not the case. Indeed, the minimal gene set encodes a virtually complete translation apparatus and reduced but apparently functionally competent systems for DNA replication, repair, and transcription (Table 37-1). Furthermore, the nutritional requirements of a hypothetical organism with the minimal gene set correspond quite well to the repertoire of the compounds that could have been synthesized abiotically in substantial amounts (Miller, 1987). Moreover, the candidate genes coding for transport systems sufficient for the uptake of all major classes of biosynthetic precursors are available. Thus, the derived minimal gene set appears to be internally consistent. The only exception is a group of coenzymes that cannot be synthesized de novo in *M. genitalium* or in the hypothetical organisms with the minimal genome. The means of their uptake in *M. genitalium* or in minimal organisms are also unclear. Yet, these coenzymes, i.e. lipoate, pyridoxal phosphate, thiamine phosphate, folate, and coenzyme A, are apparently required for the activity of several important enzymes (Table 37-1). Also, for two reactions, again involved in coenzyme metabolism, no obvious candidate genes could be found in the *M. genitalium* genome, although some of the functionally uncharacterized but conserved genes orthologs may be responsible for these reactions.

In an attempt to go from a hypothetical minimal gene complement for contemporary cellular life to what we shall provisionally call an "ancestral gene set," we applied the parsimony principle (Benner *et al.*, 1989): if a given gene is specific to the bacterial domain of life, i.e. is not found in archaea or eukaryotes, its inclusion in the minimal gene set has to be questioned. Using parsimony, we removed the PTS system of sugar transport and activation from the minimal gene set. The PTS system is not found outside of bacterial domain, and appears to be absent even from some bacteria (Postma *et al.*, 1993). A putative sugar kinase, a member of an ancient and ubiquitous protein family, capable of cytoplasmic sugar activation, is retained. Only two genes in the minimal set appeared to encode bacterial proteins specifically related to parasitic lifestyle, namely a sialoglycoprotease and a hemolysin, and were accordingly removed.

Three other suggestions coming from our preliminary parsimony analysis are much more dramatic. It appears that the 8 subunits of H^+ATPase, which are universally present in bacteria but in eukaryotes are found only as a mitochondrial gene products, may not belong to the ancestral gene set. Next, the portion of glycolysis, leading from glucose to glyceraldehyde 3-phosphate, had to be eliminated, as many archaea rely on the Entrner-Duodoroff pathway, or have other significant deviations in the enzymology of these reactions (Danson, 1988). Finally, the same logic applies to 7 key enzymes of DNA replication, namely two subunits of DNA polymerase III, initiator DNA-binding ATPase DnaA, helicase DnaB, primase DnaG, NAD-dependent DNA ligase, and single-stranded DNA-binding protein Ssb.

At the next step of reconstruction, admittedly quite a speculative exercise, we attempted to reduce the coenzyme dependency of a "minimal" organism by eliminating those pathways that can be bypassed without the use of complex coenzymes. This resulted in a hypothetical lifeform that relies on only two coenzymes (other than the indispensable ATP and GTP), namely NAD(P) and SAM. NAD synthase and SAM synthase genes are retained in the minimal gene set. Such a lifeform would lack transamination of glutamate and hydroxymethylation of glycine (both reactions being pyridoxal phosphate-dependent in known organisms), synthesis of extracellular heptoses (thiamin phosphate-dependent), and pyruvate decarboxylation (requiring four-subunit enzyme complex and 5 coenzymes). Interestingly, the latter reaction is bypassed in *Giardia,* an amitochondrial eukaryote that also lacks the H^+ATPase (Muller, 1988). Thus, it seems not unlikely that limiting the energy metabolism to the abridged glycolysis and substrate phosphorylation, in accord with parsimony principle, could render a hypothetical simple lifeform independent of many coenzymes that might not be readily available in a prebiotic environment. Several other, apparently not dramatic, modifications may reduce coenzyme dependency even further.

A special case is represented by two folate-dependent reactions still retained by a "minimal" lifeform. One of those, formylation of initiator methionine, can be excluded by the parsimony principle, as cytoplasmic Met-tRNAs in eukaryotes do not require this modification to serve in initiation. The other is biosynthesis of thymidylate from deoxyuridilate; apparently, this is a crucial step in any DNA-based organism. It is not known why this particular reaction of methyl transfer has a dedicated methyl donor, 5,10-methylene tetrahydrofolate. As a logical consequence of our previous reconstruction, we were tempted to eliminate this step. Three possibilities can be readily envisaged, which place the "minimal" organism in different evolutionary contexts. First, a ubiquitous and versatile methyl donor, SAM, might be employed for deoxyurydilate modification. Discovery of a SAM-dependent thymidylate synthase in a living organism is an intriguing possibility, which does not sound unrealistic given the functional diversity of SAM-dependent methyltransferases and their abundance even in our minimal gene set. Second, deoxyuridine might have been a legitimate component of DNA genome in a minimal lifeform; such DNA is currently known only in some bacteriophages where it is considered a secondary modification. The third possibility is a minimal lifeform that had an RNA genome; this is compatible with the exclusion of DNA replication enzymes from the ancestral gene set based on the parsimony principle (see above), and therefore seems to be the logically most consistent one in our reconstruction of the ancestral gene set. It is an RNA world model different from those most frequently discussed in that the RNA genome is postulated to encode a number of proteins, in particular a developed translation machinery.

An obvious way of further shrinking the "ancestral gene set" is by reducing families of paralogous genes to a limited number of "founders". Doing so in some of the most obvious cases, we arrived at the set of 128 genes (Table 37-1). At this stage, we choose to stop in our trip backwards in time.

Concluding Remarks

We describe two theoretical constructs—a Minimal gene complement for modern-type cellular life and an Ancestral gene set that includes a dramatic simplification of major cellular systems. It has to be emphasized that both gene sets are "Platonic ideas" in the sense that even if, as we hope, our logic in the construction of these sets has been robust, a present-day organism whose genome corresponds to the minimal gene complement is an unlikely discovery. Each organism probably needs certain number of genes coding for specific proteins that allow it to adapt to its unique niche. This is well exemplified by *M. genitalium* which may dedicate up to 20% of its genes to adhesion to the host cells (Koonin *et al.*, 1996a). Nevertheless, the minimal gene complement may be the common core for all species, and it is equally unlikely that cellular organisms are discovered that lack many genes from this set.

At the time of this writing, the first complete eukaryotic genome, that of yeast, is already available, and the first archaeal genome sequence has been determined as well. One could argue that it would be prudent to postpone the reconstruction of the minimal and ancestral gene sets until these genomes are carefully analyzed. We believe, however, that even the data on two distantly related bacterial genomes provide a solid basis for these reconstructions. This assumption will be put to test very soon. . .

References

Benner, S. A., A. D. Ellington, and A. Tauer. 1989. Modern metabolism as a palimpsest of the RNA world. *Proc. Natl. Acad. Sci. USA* 86:7054–7058.

Danson, M. J. 1988. Archaebacteria: the comparative enzymology of their central metabolic pathways. *Adv. Microb. Physiol.* 29:165–231.

Fitsch, W. M. 1970. Distinguishing homologous from analogous proteins. *Syst. Zool.* 19:99–106.

Fleischmann, R. D., M. D. Adams, O. White, R. A. Clayton, E. F. Kirkness, A. R. Kerlavage, C. J. Bult, J.-F. Tomb, B. A. Dougherty, J. M. Merrick, *et al.* 1995. Whole-genome random sequencing and assembly of *Haemophilus influenzae Rd. Science* 269:496–512.

Fraser, C. M., J. D. Gocayne, O. White, M. D. Adams, R. A. Clayton, R. D. Fleischmann, C. J. Bult, A. R. Kerlavage, G. Sutton, J. M. Kelley, *et al.* 1995. The minimal gene complement of *Mycoplasma genitalium. Science* 270:397–403.

Jensen, R. A., and W. Gu. 1996. Evolutionary recruitment of biochemically specialized subdivisions of family I within the protein superfamily of aminotransferases. *J. Bacteriol.* 178:2161–2171.

Koonin, E. V., A. R. Mushegian, and K. E. Rudd. 1996a. Sequencing and analysis of bacterial genomes. *Curr. Biol.* 6:404–416.

Koonin, E. V., R. L. Tatusov, and K. E. Rudd. 1996b. Protein sequence comparison at a genome scale. *Meth. Enzymol.* 266:295–322.

Koonin, E. V., and P. Bork. 1996. Ancient duplication of DNA polymerase inferred from analysis of complete bacterial genomes. *Trends Biochem. Sci.* 21:128–129.

Miller, S. L. 1987. Which organic compounds could have occurred on the prebiotic earth? *Cold Spring Harbor Symp. Quantitative Biol.* vol. LII, p. 17–27.

Muller, M. 1988. Energy metabolism of protozoa without mitochondria. *Ann. Rev. Microbiol.* 42:465–488.

Olsen, G. J., C. R. Woese, and R. Overbeek. 1994. The winds of (evolutionary) change: breathing new life into microbiology. *J. Bacteriol.* 176:1–6.

Postma, P. W., J. W. Lengeler, and G. R. Jacobson. 1993. Phosphoenolpyruvate: carbohydrate phosphotransferase systems of bacteria. *Microbiol. Rev.* 57:543–594.

Tatusov, R. L., A. R. Mushegian, P. Bork, N. P. Brown, W. S. Hayes, M. Borodovsky, K. E. Rudd, and E. V. Koonin. 1996. Metabolism and evolution of *Haemophilus influenzae* deduced from a whole-genome comparison with *Escherichia coli. Curr. Biol.* 6:279–291.

38

Resources for the *Escherichia coli* Genome Project

George M. Weinstock

The focused attack to determine the complete DNA sequence of the *Escherichia coli* genome was the first large scale bacterial DNA sequencing project to be undertaken. The *E. coli* genome, considerably larger (4.7 Mb) than the other bacterial genomes sequenced to date (generally less than 2 Mb), was also the first genome project undertaken by a single group, one of the first centers founded by the Human Genome Project, and was started before technology had matured to its current level. Even though *E. coli* is not the first bacterial genome to be sequenced, this project and other work on the *E. coli* genome have been seminal for microbial genomics by stimulating development of genome analysis tools and approaches.

In fact, work on the *E. coli* genome, at both the genetic and physical levels, has been going on for years. The linkage map of *E. coli* has developed over more than three decades (Berlyn *et al.*, 1996; Taylor and Trotter, 1967) and is only surpassed in detail by the actual DNA sequence. The physical map of the *E. coli* genome was one of the first to be established using pulsed field gel electrophoresis (Heath *et al.*, 1992; Perkins *et al.*, 1992; Perkins *et al.*, 1993; Smith *et al.*, 1987; *see also* Chapter 60). Libraries of cosmid (Birkenbihl and Vielmetter, 1989; Knott *et al..*, 1989; Tabata *et al.*, 1989) and phage lambda clones (Daniels and Blattner, 1987; Kohara *et al.*, 1987) were mapped and ordered around the *E. coli* chromosome, aiding in technology development and also providing more detailed restriction maps of the whole *E. coli* genome. One outgrowth of this was the first commercially available hybridization membranes (Takara Biochemicals) containing a genome-spanning miniset of clones from the Kohara library (Noda *et al.*, 1991). Numerous other features have been mapped on the whole genome scale including insertion sequences (Birkenbihl and Vielmetter, 1989; Deonier, 1996), tRNA genes (Komine *et al.*, 1990), and repeated sequence elements (Bachellier *et al.*, 1996; Dimri *et al.*, 1992). The resources that exist and continue to evolve for *E. coli* comprise a package of physical and DNA sequence information, clone libraries, and other biological resources that represent a comprehensive genome project. The aim of this article is to briefly review a number of these resources, which are summarized in Table 38-1.

The *E. coli* DNA sequencing project initially has deposited blocks of sequence information into GenBank, with substantial annotation (Blattner *et al.*, 1993; Burland *et al.*, 1993; Burland *et al.*, 1995; Daniels *et al.*, 1992; Plunkett *et al.*, 1993; Sofia *et al.*, 1994). A collaborative *E. coli* project in Japan has also made significant contributions to GenBank (Fujita *et al.*, 1994; Yura *et al.*, 1992). It is anticipated that, like other completed whole genome projects, the entire sequence will be available from GenBank when finished. In addition to GenBank, there is a World Wide Web site, the *E. coli* Genome Center, http://www.genetics.wisc.edu/Welcome.html, for the *E. coli* genome project that provides access to the sequences, allowing them to be downloaded by ftp. This site also provides information about the sequencing project and has links to other relevant sites.

Substantial DNA sequence information has also been generated over the years and in parallel with this project by the *E. coli* research community. Several databases have been developed and maintained to pool and collate these many short sequence contributions (Kroger *et al.*, 1991; Kroger *et al.*, 1993; Kroger *et al.*, 1992; Medigue *et al.*, 1990; Medigue *et al.*, 1990; Medigue *et al.*, 1991). These databases include EcoSeq (Rudd, 1992; Rudd, 1993; Rudd *et al.*, 1992; Rudd *et al.*, 1991), Colibri (Medigue *et al.*, 1993), and ECDC (Wahl and Kroger, 1995; Wahl *et al.*, 1994).

The EcoSeq sequence compilation is part of a larger project that includes a other resources (EcoMap, EcoGene, EcoProt) and tools (MapSearch, PrintMap, GeneScape, ChromoScope). These, as well as additional information, can be obtained by ftp from ftp://ncbi.nlm.nih.gov/repository/Eco. EcoMap is an integrated physical map of the *E. coli* genome, derived initially by aligning the EcoSeq sequences with the high resolution restriction map of Kohara (Kohara *et al.*, 1987) using the MapSearch program, and then refined. PrintMap is an accompanying map drawing utility program. EcoMap was included in GeneScape, a relational database using FoxBase+ for Macintosh, for viewing and editing restriction maps (Bouffard *et al.*, 1992). EcoGene is a database of information about the *E. coli* genes and their products in EcoSeq. EcoMap and EcoGene were part of the basis for the latest *E. coli* linkage map (Berlyn *et al.*, 1996), providing map coordinates, for example. ChromoScope is a network application that allows graphic viewing and text retrieval from EcoSeq datasets as well as other databases (GenBank, SWISSPROT, Medline) accessed through the NCBI Entrez application. EcoProt is a database of the predicted protein products from the genes in EcoSeq and is discussed below.

Another compilation of *E. coli* sequences is Colibri, a relational database using the 4th Dimension database application for Macintosh (Medigue *et al.*, 1993). Like the EcoSeq dataset, Colibri contains a number of other types of information in addition to the DNA sequences, such as protein sequences and map information.

Table 38-1. Some resources for the Escherichia coli *genome project.*

Resource	Description and URL	Reference
CGSC: *E. coli* Genetic Stock Center	http://cgsc.biology.yale.edu/top.html	(Berlyn, 1996)
Colibri	ftp://ftp.pasteur.fr/pub/GenomeDB/colibri	(Medigue *et al.*, 1993)
E. coli Genome Center	http://www.genetics.wisc.edu/Welcome.html	
E. coli Index	http://sun.1.bham.ac.uk/bcm4ght6/res.html	
E. COLI's Home Page	http://bio.taiu.edu/class/ksor003/home.html	
ECDC:*E. coli* Database Collection	http://susi.bio.uni-giessen.de/usr/local/www/html/ecdc.html	(Wahl and Kroger, 1995; Wahl *et al.*, 1994)
ECO2DBASE	ftp://ncbi.nlm.nih.gov/repository/ECO2DBASE	(Vanbogelen *et al.*, 1996)
EcoCyc	http://www.ai.sri.com/ecocyc/ecocyc.html	(Karp and Paley, 1994; Karp and Riley, 1993)
ECOR: set of reference strains		(Ochman and Selander, 1984)
EcoSeq	ftp://ncbi.nlm.nih.gov/repository/Eco	(Rudd, 1992; Rudd, 1993; Rudd *et al.*, 1992; Rudd *et al.*, 1991)
METALGEN	ftp://ftp.pasteur.fr/pub/GenomeDB/metalgen	(Rouxel *et al.*, 1993)
Ordered libraries		(Birkenbihl and Vielmetter, 1989; Daniels and Blattner, 1987; Knott *et al.*, 1989; Kohara *et al.*, 1987; Tabata *et al.*, 1989)
Ordered library on hybridization membrane		(Noda *et al.*, 1991)
PUMA	http://www.mcs.anl.gov/home/compbio/PUMA/Production/puma.html	
Transposon insertion set		(Singer *et al.*, 1989)

Colibri also contains other biological information about genes and proteins such as position, phenotype, and literature references, and provisions for various types of searches. Colibri can be downloaded from ftp://ftp.pasteur.fr/pub/GenomeDB/colibri.

ECDC is the *E. coli* database collection which also is a source of DNA and protein sequences and map data (Wahl and Kroger, 1995; Wahl *et al.*, 1994). This database can be used online at http://susi.bio.uni-giessen.de/usr/local/www/html/ecdc.html or downloaded by ftp from this site. The database, lilke Colibri, contains DNA and protein sequences as well as biological information related to map position and other features such as regulators. Methods for searching the database are available.

In addition to these, several other databases have been described (Kunisawa *et al.*, 1990; Perrière and Gautier, 1993; Shin *et al.*, 1992) that provide comprehensive information about the *E. coli* genome. Additional Internet sites are devoted to *E. coli* biology. The *E. coli* Index (http://sun1.bham.ac.uk/bcm4ght6/res.html) contains protocols and a summary of pathogenic properties of *E. coli* plus links to databases. The *E. coli* Home Page (http://bio.taiu.edu/class/ksor003/home.html) contains links to a variety of information about the biology of *E. coli,* online publications, and pictures.

Several types of information about the proteins of *E. coli* are available. A gene-protein database, ECO2DBASE, is based on proteins identified by two-dimensional polyacrylamide gel electrophoresis (Vanbogelen *et al.*, 1996). This database is available by ftp from ftp://ncbi.nlm.nih.gov/repository/ECO2DBASE. It correlates polypeptides with genes in the chromosome and also contains information on the expression of the polypeptides under a variety of growth conditions.

The proteins predicted from the EcoSeq sequences are entered in the SWISS-PROT database with the cross-reference designation ECOGENE and can be retrieved by searching with this keyword. These protein sequences are also available as the EcoProt dataset as part of the EcoSeq project. In addition to known proteins, the EcoProt sequences includes additional polypeptides predicted as part of a project for computer analysis of the *E. coli* genome (Borodovsky *et al.*, 1994; Borodovsky *et al.*, 1995; Borodovsky *et al.*, 1994). Results of analyses of *E. coli* proteins for function, motifs, and similarity have been presented (Riley, 1993) and are also available at the EcoSeq site (Koonin *et al.*, 1995).

With the availability of whole genome sequences there has been considerable interest in assembling descriptions of the complete metabolic pathway organization of organisms for both reference and modeling purposes. Several such databases are available for *E. coli.* METALGEN is a 4th Dimension relational database for the Macintosh available from ftp://ftp.pasteur.fr/pub/GenomeDB/metalgen (Rouxel *et al.*, 1993). It allows browsing of metabolic pathways that are linked to relevant genes and regulation as well as some modeling of the effects of growth conditions and mutations on phenotype. EcoCyc (Karp and Paley, 1994; Karp and Riley, 1993), located on the Internet at http://www.ai.sri.com/ecocyc/

ecocyc.html, draws on *E. coli* genes and pathways to provide users with genomic and biochemical information. Finally, PUMA (http://www.mcs.anl.gov/home/compbio/PUMA/Production/puma.html) offers a comprehensive set of information about enzymes and biochemical reactions for many different organisms.

While the type of information described above is in principle available for any of the organisms whose genomic DNA sequence is known, *E. coli* in addition offers enormous resources for genetic analysis, key to truly capitalizing on the genomic sequence. As already mentioned, there are several ordered clone sets (Birkenbihl and Vielmetter, 1989; Daniels and Blattner, 1987; Knott *et al.*, 1989; Kohara *et al.*, 1987; Tabata *et al.*, 1989), one of which has been made commercially available as hybridization filters (Noda *et al.*, 1991). A collection of transposon insertions, located at approximately one minute intervals, has also been assembled (Singer *et al.*, 1989). There is also a standard set of reference strains, the ECOR collection (Ochman and Selander, 1984), which is used to determine differences between natural isolates of *E. coli.* (*See also* Chapters 17 and 23). The clearinghouse for information about strains, mutants, phenotypes, and other genetic aspects is the *E. coli* Genetic Stock Center (Berlyn, 1996), located online at http://cgsc.biology.yale.edu/top.html.

Finally, it is the *in vivo* characterization that makes the genomic scale analysis of *E. coli* so attractive. As mentioned above, the two-dimensional polyacrylamide gel database contains information on the regulation of every polypeptide that can be detected by this method. More genetically oriented methods, using ordered clone sets, have also been employed to identify genes whose transcription responds to a variety of environmental conditions (Chuang and Blattner, 1993; Chuang *et al.*, 1993). An *in vivo* whole genome approach has also been described to identify sequences that serve as binding sites for proteins (Wang and Church, 1992). Finally, methods have been developed for use of transposon mutagenesis to scan for genes that are required under different conditions (Hensel *et al.*, 1995; Smith *et al.*, 1995). While the latter methods have yet to be employed in *E. coli,* they are easily extended to this organism because of its extensive genetic capability. One can therefore expect that *E. coli* will continue in its role as a leading model organism, as well as for intrinsic interest in the bacterium itself.

References

Bachellier, S., E. Gilson, M. Hofnung, and C. W. Hill. 1996. Repeated sequences. In Escherichia coli *and* Salmonella typhimurium. *Cell and molecular biology,* F. C. Neidhardt, R. Curtiss, III, J. L. Ingraham, E. C. C. Lin, K. B. Low, B. Magasanik, W. S. Reznikoff, M. Riley, M. Schaechter, and H. E. Umbarger, eds. pp. 2012. ASM Press, Washington, DC.

Berlyn, M. K. B. 1996. Accessing the *E. coli* genetic stock center database. In Escherichia coli *and* Salmonella typhimurium. *Cell and molecular biology,* F. C. Neidhardt, R. Curtiss, III, J. L. Ingraham, E. C. C. Lin, K. B. Low, B. Magasanik, W. S. Reznikoff,

M. Riley, M. Schaechter and H. E. Umbarger, eds., pp. 2489–2495. ASM Press, Washington, DC.

Berlyn, M.K.B., K. B. Low, and K. E. Rudd. 1996. Linkage map of *Escherichia coli* K-12, Edition 9. In Escherichia coli *and* Salmonella typhimurium. *Cell and molecular biology,* pp. 1715–1902. F. C. Neidhardt, R. Curtiss, III, J. L. Ingraham, E.C.C. Lin, K.B. Low, B. Magasanik, W. S. Reznikoff, M. Riley, M. Schaechter and H. E. Umbarger, eds. ASM Press, Washington, DC.

Birkenbihl, R. P., and W. Vielmetter. 1989. Complete maps of IS1, IS2, IS3, IS4, IS5, IS30 and IS150 locations in Escherichia coli K-12. *Mol. Gen. Genet.* 220:147–53.

Birkenbihl, R. P., and W. Vielmetter. 1989. Cosmid-derived map of *E. coli* strain BHB2600 in comparison to the map of strain W3110. *Nucl. Acids Res.* 17:5057–69.

Blattner, F. R., V. Burland, G. Plunkett, H. J. Sofia, and D. L. Daniels. 1993. Analysis of the *Escherichia coli* genome. IV. DNA sequence of the region from 89.2 to 92.8 minutes. *Nucl. Acids Res.* 21:5408–17.

Borodovsky, M., E. V. Koonin, and K. E. Rudd. 1994. New genes in old sequence: a strategy for finding genes in the bacterial genome. *Trends. Biochem. Sci.* 19:309–313.

Borodovsky, M., J. D. McIninch, E. V. Koonin, K. E. Rudd, C. Medigue, and A. Danchin. 1995. Detection of new genes in a bacterial genome using Markov models for three gene classes. *Nucl. Acids Res.* 23:3554–3562.

Borodovsky, M., K. E. Rudd, and E. V. Koonin. 1994. Intrinsic and extrinsic approaches for detecting genes in a bacterial genome. *Nucl. Acids Res.* 22:4756–4767.

Bouffard, G., J. Ostell, and K. E. Rudd. 1992. GeneScape: a relational database of *Escherichia coli* genomic map data for Macintosh computers. *CABIOS* 8:563–567.

Burland, V., G. D. Plunkett, D. L. Daniels, and F. R. Blattner. 1993. DNA sequence and analysis of 136 kilobases of the *Escherichia coli* genome: organizational symmetry around the origin of replication. *Genomics* 16:551–61.

Burland, V., G. R. Plunkett, H. J. Sofia, D. L. Daniels, and F. R. Blattner. 1995. Analysis of the *Escherichia coli* genome VI: DNA sequence of the region from 92.8 through 100 minutes. *Nucleic Acids Research* 23:2105–19.

Chuang, S. E., and F. R. Blattner. 1993. Characterization of twenty-six new heat shock genes of *Escherichia coli. J. Bacteriol.* 175:5242–5252.

Chuang, S. E., D. L. Daniels, and F. R. Blattner. 1993. Global regulation of gene expression in *Escherichia coli. J. Bacteriol.* 175:2026–2036.

Daniels, D. L., and F. R. Blattner. 1987. Mapping using gene encyclopaedias. *Nature* 325:831–2.

Daniels, D. L., G. D. Plunkett, V. Burland, and F. R. Blattner. 1992. Analysis of the *Escherichia coli* genome: DNA sequence of the region from 84.5 to 86.5 minutes. *Science* 257:771–8.

Deonier, R. C. 1996. Native insertion sequence elements: Locations, distributions, and sequence relationships. In Escherichia coli *and* Salmonella typhimurium. *Cell and molecular biology,* pp. 2000–2011. F. C. Neidhardt, R. Curtiss, III, J. L. Ingraham, E.C.C. Lin, K. B. Low, B. Magasanik, W. S. Reznikoff, M. Riley, M. Schaechter and H. E. Umbarger, eds. ASM Press, Washington, DC.

Dimri, G. P., K. E. Rudd, M. K. Morgan, H. Bayat, and G. F. Ames. 1992. Physical mapping of repetitive extragenic palindromic sequences in *Escherichia coli* and phylogenetic distribution among *Escherichia coli* strains and other enteric bacteria. *J. Bacteriol.* 174:4583–93.

Fujita, N., H. Mori, T. Yura, and A. Ishihama. 1994. Systematic sequencing of the *Escherichia coli* genome: analysis of the 2.4–4.1 min (110,917–193,643 bp) region. *Nucl. Acids Res.* 22:1637–1639.

Heath, J. D., J. Perkins, B. Sharma, and G. M. Weinstock. 1992. *Not*I genomic cleavage map of *Escherichia coli* K-12 strain MG1655. *J. Bacteriol.* 174:558–567.

Hensel, M., J. E. Shea, C. Gleeson, M. D. Jones, E. Dalton, and D. W. Holden. 1995. Simultaneous identification of bacterial virulence genes by negative selection. *Science* 269:400–3.

Karp, P., and S. Paley. 1994. Representation of metabolic knowledge: pathways. In *Second International Conference on Intelligent Systems and Molecular Biology,* R. Altman, D. Brutlag, P. Karp, R. Lathrop and D. Searls, eds. pp. 203–211. AAAI Press, Washington, DC.

Karp, P., and M. Riley. 1993. Representation of metabolic knowledge. In *First International Conference on Intelligent Systems and Molecular Biology,* pp. 207–215. I. Hunter, D. Searls and J. Sharlik, eds. AAAI Press, Washington, DC.

Knott, V., D. J. Blake, and G. G. Brownlee. 1989. Completion of the detailed restriction map of the *E. coli* genome by the isolation of overlapping cosmid clones. *Nucl. Acids Res.* 17:5901–12.

Kohara, Y., K. Akiyama, and K. Isono. 1987. The physical map of the whole *E. coli* chromosome: application of a new strategy for rapid analysis and sorting of a large genomic library. *Cell* 50:495–508.

Komine, Y., T. Adachi, H. Inokuchi, and H. Ozeki. 1990. Genomic organization and physical mapping of the transfer RNA genes in *Escherichia coli* K-12. *J. Mol. Biol.* 212:579–98.

Koonin, E. V., R. L. Tatusov, and K. E. Rudd. 1995. Sequence similarity analysis of *Escherichia coli* proteins: functional and evolutionary implications. *Proc. Natl. Acad. Sci. USA* 92:11921–11925.

Kroger, M., R. Wahl, and P. Rice. 1991. Compilation of DNA sequences of *Escherichia coli* (update 1991). *Nucl. Acids Res.* 19:2023–43.

Kroger, M., R. Wahl, and P. Rice. 1993. Compilation of DNA sequences of *Escherichia coli* (update 1993). *Nucl. Acids Res.* 21:2973–3000.

Kroger, M., R. Wahl, G. Schachtel, and P. Rice. 1992. Compilation of DNA sequences of *Escherichia coli* (update 1992). *Nucl. Acids Res.* 20:2119–44.

Kunisawa, T., M. Nakamura, H. Watanabe, J. Otsuka, A. Tsugita, L. S. Yeh, D. G. George, and W. C. Barker. 1990. *Escherichia coli* K-12 genomic database. *Protein Sequences Data Anal.* 3:157–162.

Medigue, C., J. P. Bouche, A. Henaut, and A. Danchin. 1990. Mapping of sequenced genes (700 kbp) in the restriction map of the *Escherichia coli* chromosome. *Mol. Microbiol.* 4:169–87.

Medigue, C., A. Henaut, and A. Danchin. 1990. *Escherichia coli* molecular genetic map (1000 kbp): update I. *Mol. Microbiol.* 4:1443–54.

Medigue, C., A. Viari, A. Henaut, and A. Danchin. 1993. Colibri: a functional data base for the *Escherichia coli* genome. *Microbiol. Rev.* 57:623–54.

Medigue, C., A. Viari, A. Henaut, and A. Danchin. 1991. *Escherichia coli* molecular genetic map (1500 kbp): update II. *Mol. Microbiol.* 5:2629–40.

Noda, A., J. B. Courtright, P. F. Denor, G. Webb, Y. Kohara, and A. Ishihama. 1991. Rapid identificiation of specific genes in *E. coli* by hybridization to membranes containing the ordered set of phage clones. *BioTechniques* 10:474–477.

Ochman, H., and R. K. Selander. 1984. Standard reference strains of *Escherichia coli* from natural populations. *J. Bacteriol.* 157:690–693.

Perkins, J., J. D. Heath, B. Sharma, and G. M. Weinstock. 1992. S*fi*I genomic cleavage map of *Escherichia coli* K-12 strain MG1655. *Nucl. Acids Res.* 20:1129–1137.

Perkins, J., J. D. Heath, B. Sharma, and G. M. Weinstock. 1993. *Xba*I genomic cleavage map of *Escherichia coli* K-12 strain MG1655 and comparative analysis of other strains. *J. Mol. Biol.* 232:419–445.

Perrière, G., and C. Gautier. 1993. ColiGene: Object-oriented representation for the study of *E. coli* gene expressivity by sequence analysis. *Biochimie* 75:415–422.

Plunkett, G. D., V. Burland, D. L. Daniels, and F. R. Blattner. 1993. Analysis of the *Escherichia coli* genome. III. DNA sequence of the region from 87.2 to 89.2 minutes. *Nucl. Acids Res.* 21:3391–8.

Riley, M. 1993. Functions of the gene products of *Escherichia coli*. *Microbiol. Rev.* 57:862–952.

Rouxel, T., A. Danchin, and A. Henaut. 1993. METALGEN.DB: metabolism linked to the genome of Escherichia coli, a graphics-oriented database. *Comput. Appl. Biosci.* 9:315–24.

Rudd, K. E. 1992. Alignment of *E. coli* DNA sequences to a revised, integrated genomic restriction map. In *A short course in bacterial genetics: A laboratory manual and handbook for* Escherichia coli *and related bacteria,* J. Miller, ed. pp. 2.3–2.43. Cold Spring Harbor Press, Cold Spring Harbor, NY.

Rudd, K. E. 1993. Maps, genes, sequences, and computers: an *Escherichia coli* case study. *ASM News* 59:335–341.

Rudd, K. E., G. Bouffard, and W. Miller. 1992. Computer analysis of *E. coli* restriction maps. In *Genome Analysis: Strategies for physical mapping,* K. E. Davies and S. M. Tilghman, eds. pp. 1–38. Cold Spring Harbor Press, Cold Spring Harbor, NY.

Rudd, K. E., W. Miller, C. Werner, J. Ostell, C. Tolstoshev, and S. G. Satterfield. 1991. Mapping sequenced *E. coli* genes by computer: software, strategies and examples. *Nucl. Acids Res.* 19:637–647.

Shin, D. G., C. Lee, J. Zhang, K. E. Rudd, and C. M. Berg. 1992. Redesigning, implementing and integrating *Escherichia coli* genome software tools with an object-oriented database system. *CABIOS* 8:227–238.

Singer, M., T. A. Baker, G. Schnitzler, S. M. Deischel, M. Goel, W. Dove, K. J. Jaacks,

A. D. Grossman, J. W. Erickson, and C. A. Gross. 1989. A collection of strains containing genetically linked alternating antibiotic resistance elements for genetic mapping of *Escherichia coli. Microbiol. Rev.* 53:1–24.

Smith, C. L., J. G. Econome, A. Schutt, S. Klco, and C. R. Cantor. 1987. A physical map of the *Escherichia coli* K12 genome. *Science* 236:1448–53.

Smith, V., D. Botstein and P. O. Brown. 1995. Genetic footprinting: a genomic strategy for determining a gene's function given its sequence. *Proc. Natl. Acad. Sci. of USA* 92:6479–83.

Sofia, H. J., V. Burland, D. L. Daniels, G. R. Plunkett, and F. R. Blattner. 1994. Analysis of the *Escherichia coli* genome. V. DNA sequence of the region from 76.0 to 81.5 minutes. *Nucl. Acids Res.* 22:2576–86.

Tabata, S., A. Higashitani, M. Takanami, K. Akiyama, Y. Kohara, Y. Nishimura, A. Nishimura, S. Yasuda, and Y. Hirota. 1989. Construction of an ordered cosmid collection of the *Escherichia coli* K-12 W3110 chromosome. *J. Bacteriol.* 171:1214–8.

Taylor, A. L., and C. D. Trotter. 1967. Revised linkage map of *Escherichia coli. Bacteriol. Rev.* 31:332–53.

Vanbogelen, R. A., K. Z. Abshire, A. Pertsemlidis, R. L. Clark, and F. C. Neidhardt. 1996. Gene-protein database of *Escherichia coli* K-12, Edition 6. In Escherichia coli *and* Salmonella typhimurium. *Cell and molecular biology,* F. C. Neidhardt, R. Curtiss, III, J. L. Ingraham, E.C.C. Lin, K. B. Low, B. Magasanik, W. S. Reznikoff, M. Riley, M. Schaechter and H. E. Umbarger, eds. pp. 2067–2117. ASM Press, Washington, DC.

Wahl, R., and M. Kroger. 1995. ECDC—a totally integrated and interactively usable genetic map of *Escherichia coli* K-12. *Microbiol. Res.* 150:7–61.

Wahl, R., P. Rice, C. M. Rice, and M. Kroger. 1994. ECD—a totally integrated database of *Escherichia coli* K-12. *Nucl. Acids Res.* 22:3450–3455.

Wang, M. X., and G. M. Church. 1992. A whole genome approach to in vivo DNA-protein interactions in *E. coli. Nature* 360:606–610.

Yura, T., H. Mori, H. Nagai, T. Nagata, A. Ishihama, N. Fujita, K. Isono, K. Mizobuchi, and A. Nakata. 1992. Systematic sequencing of the *Escherichia coli* genome: analysis of the 0–2.4 min region. *Nucl. Acids Res.* 20:3305–3308.

39

The Mycobacterial Database MycDB and the Mycobacterial Genome Sequencing Project

Staffan Bergh and Stewart Cole

Introduction

In recent years there has been a considerable upturn in interest in the genetics and microbiology of the mycobacteria (Young and Cole, 1993), and the reasons for this are two-fold. Firstly, after several decades of relative tranquility, mycobacterial diseases are assuming greater importance in terms of public health problems in the industrialized world and secondly, powerful new tools for working with mycobacteria have been developed that allow previously intractable biological and medical problems to be studied (Jacobs *et al.*, 1991).

About one third of the global population is infected with *Mycobacterium tuberculosis* and this highly contagious pathogen causes 26% of avoidable deaths and 8 million new cases of tuberculosis each year (Sudre *et al.*, 1992). The AIDS pandemic, together with a worsening socio-economic situation, has seen the resurgence of the disease in many industrialized countries, and in the United States, this has been accompanied by the appearance of multiple-drug resistant strains of *M. tuberculosis* (Bloom and Murray, 1992; Snider and Roper, 1992). In the later stages of AIDS many patients die from opportunistic infections with environmental mycobacteria such as *M. avium.* Leprosy, with about 5.5 million registered cases (Noordeen *et al.*, 1992), is not as big a problem as tuberculosis numerically, but its etiologic agent, *M. leprae,* represents a greater challenge. This obligately intracellular parasite has never been cultivated *in vitro,* and has an extremely long doubling time of around 13 days in mice.

The history of mycobacterial molecular biology can be said to have begun when Robert Koch showed for the first time that an infectious disease, tuberculosis, was caused by a microorganism (Koch, 1882) and attempted immunotherapy using an antigen preparation, ‘old tuberculin’ derived from *M. tuberculosis.* Since then, there have been many reports of preparations of mycobacterial antigens, and their cognate antibodies, and their use as immunoprophylactic or sero-diagnostic reagents. Likewise, the biology, microbiology and biochemistry of mycobacteria have also been the subject of detailed study.

In spite of this, it is only in recent years with the development of phage and plasmid vectors, transformation and recombination systems (Aldovini *et al.*, 1993; Jacobs *et al.*, 1991; Snapper *et al.*, 1990), that the tools of mycobacteriology have started to make large scale investigations fruitful. During this period there has been great technological development in general in molecular biology making the project of sequencing the entire genome of an organism conceivable. At the Institut Pasteur, ordered clone banks of *M. tuberculosis* and *M. leprae* have been constructed (*see* Chapter 68), to use them as a means of furthering mycobacteriology and as a source of DNA for sequencing their genomes (Cole and Smith, 1994; Eiglmeier *et al.*, 1993; Honoré *et al.*, 1993).

The bulk of the sequences so far has been generated by the mycobacterial genome sequencing consortium, consisting of the multiplex sequencing group led by Dr. D. R. Smith at Genome Therapeutics, Inc. (previously Collaborative Research Inc.), Waltham, Mass., and the Institut Pasteur group. As sequences emerge they are submitted concomitantly to MycDB (Bergh and Cole, 1994) and GenBank/EMBL and generally appear in an initial form which lacks annotation. To date, the sequences of 47 cosmids bearing DNA from the *M. leprae* chromosome and three from *M. tuberculosis* have been incorporated into MycDB. About 65% of the *M. leprae* genome has been sequenced, 1.7 Mbp total and the longest contiguous sequence is about 95 kbp.

MycDB

To present and visualize the data known about mycobacteria a database manager specifically developed for large scale molecular biology was chosen. The ACEDB software was written initially for use within the *Caenorhabditis elegans* genome project (Durbin and Thierry-Mieg, 1991) and has since been adapted for a number of other large projects, such as the one pursued by the *Arabidopsis thaliana* molecular biology community (Cherry and Cartinhour, 1993; Cherry *et al.*, 1992), several other plant projects and parts of the human genome project (Dunham *et al.*, 1993).

ACEDB is an 'object-oriented' database manager with facilities for displaying and analyzing most of the types of objects encountered in molecular biology, such as genes, genetic and physical maps, clones, sequences, and strains, as well as more mundane supporting material such as an abstract facility, and details about authors and colleagues. It is a mouse-driven program existing in Unix/X and Macintosh versions with many display functions. One of the great advantages of ACEDB over conventional database software is its adaptability for, as the project grows and generates new kinds of objects, the software can be made to evolve in parallel by means of subtle program changes. For instance, in the case of MycDB, we have introduced two new types of objects, Antigen and Antibody, to accommodate the requirements of the mycobacterial research community where considerable efforts have been devoted to immunological research.

The database schema (called 'the models' in ACEDB) is written in simple text-format, and can easily be changed to suit the users of the database. The model for the Antigen class is shown in Fig. 39-1. Cross-references between objects are marked by the 'XREF' keyword ('Antibody') and entries that should have a unique value are marked by 'UNIQUE' ('Code').

The general type of display is a Tree structure, a text format window with information on many different aspects of the object in question as shown by the information for the Antigen MI__GroES in Fig. 39-2. Displayed facts can be simple values such as the apparent size of protein antigens on SDS-polyacrylamide gels or the text of a Remark, or they can be a pointer to another object of the same or another class, such as an Antibody against this specific Antigen. The pointers, displayed in bold type, can be followed by double-clicking on a given pointer.

Some classes have special displays, that show the data in a more graphic manner. One major display type is Map (Fig. 39-3) that is used for physical and genetic maps. Objects that have been mapped can be set to display as default in

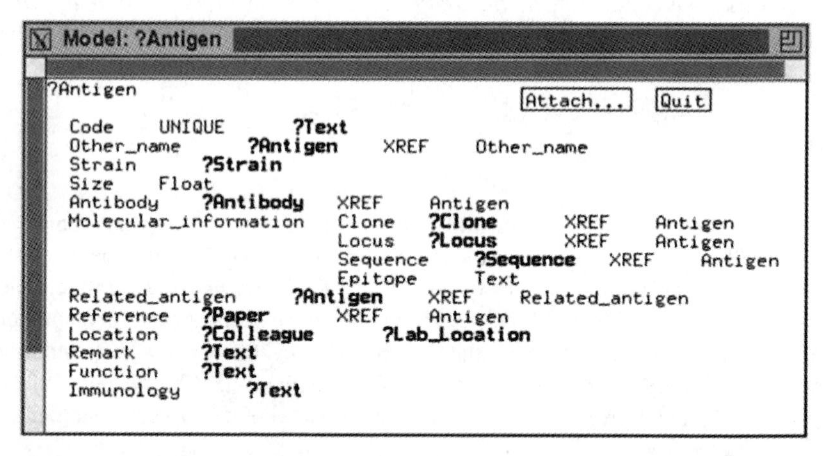

Figure 39-1. The Tree structure/model for the Antigen class, governing what can be entered about an antigen. The model is organized as a tree of 'tags', for instance Molecular__information which has the subtrees Clone, Locus, Gene, Sequence and Epitope. Note that the information can consist of other entries of the same class, permitting nesting of objects (Related__antigen). The XREF keyword directs cross-referencing between objects, determining under which tag in the cross-referenced object the pointer to this object is going to be entered. This cross-referencing is automatic. The keyword UNIQUE directs that only one instance of data is allowed under this tag, otherwise there is no restriction on the number of data fields.

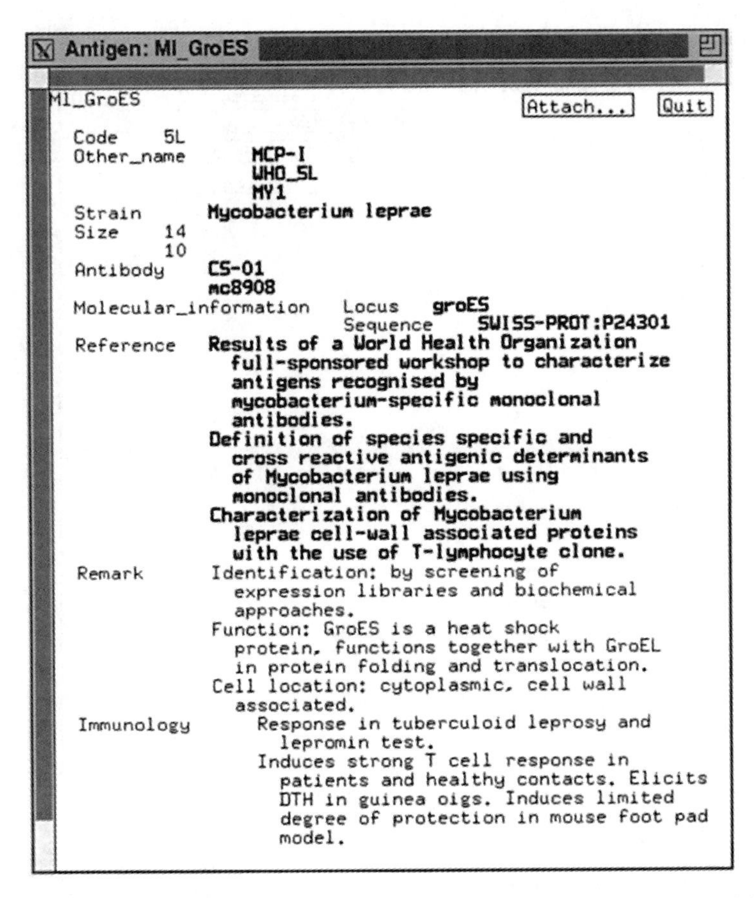

Figure 39-2. Display of the Tree for the Antigen MI__GroES. Note that not all fields are filled in, a situation typical for the sparse datasets encountered in molecular biology. Entries in bold are objects of this or another class and information about them can be viewed simply by double-clicking on the bold text itself. Double-clicking an object of Sequence type will bring up a FeatureMap display (see Figure 39.4).

a map window; if no mapping data exists, the Tree display is used. Another major display type is the FeatureMap for sequences (Fig. 39-4). In both of these displays, cross-references to other objects are shown next to the graphic representation of the object and can be double-clicked to bring up more information. The FeatureMap also includes software for analyzing the DNA sequence for open reading frames, restriction sites and homologies.

ACEDB also has powerful tools for queries: set operators, a native query language and a facility to produce relational database-type tables.

ACEDB can export data in human-readable form, in so called .ace format. In addition, sequences can be exported as fasta formatted files, and groups of

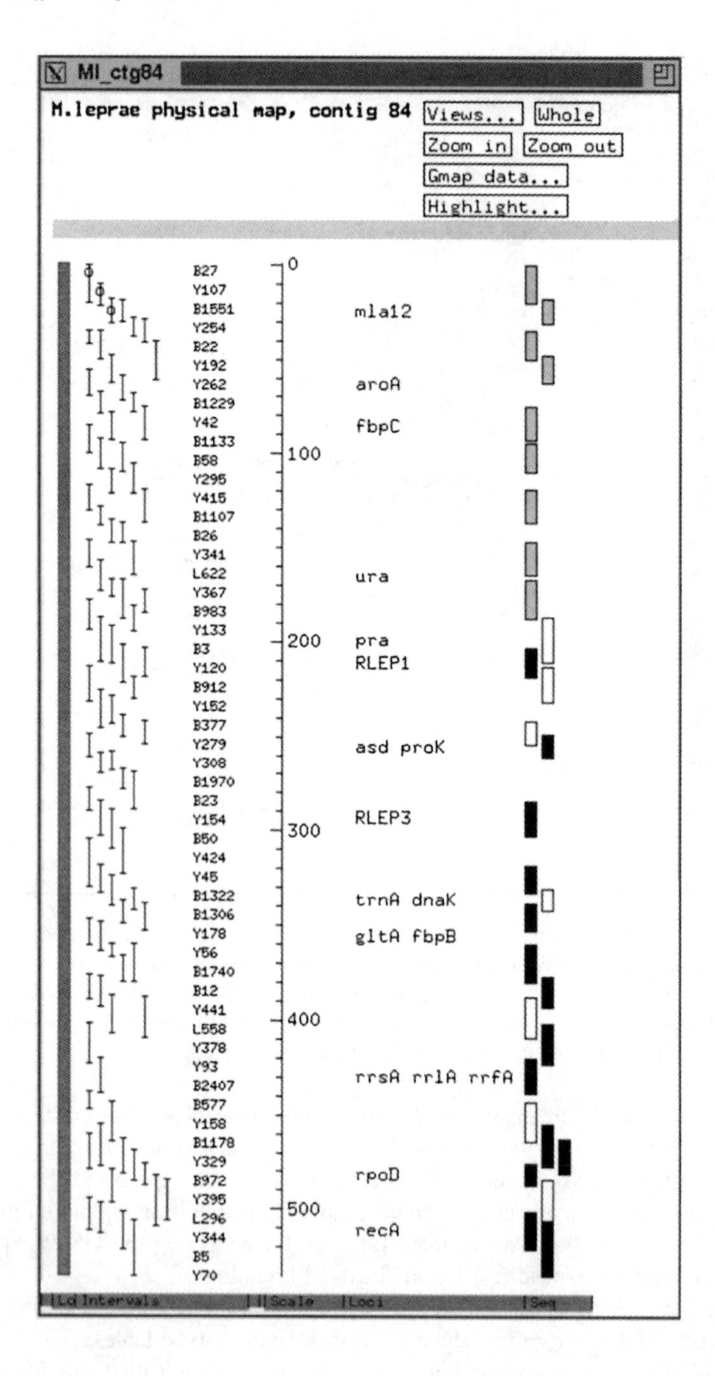
Ml_ctg84
M.leprae physical map, contig 84
Views...
Whole
Zoom in
Zoom out
Gmap data...
Highlight...
B27
Y107
B1551
Y254
B22
Y192
Y262
B1229
Y42
B1133
B58
Y295
Y415
B1107
B26
Y341
L622
Y367
B983
Y133
B3
Y120
B912
Y152
B377
Y279
Y308
B1970
B23
Y154
B50
Y424
Y45
B1322
B1306
Y178
Y56
B1740
B12
Y441
L558
Y378
Y93
B2407
B577
Y158
B1178
Y329
B972
Y395
L296
Y344
B5
Y70
0
100
200
300
400
500
mla12
aroA
fbpC
ura
pra
RLEP1
asd proK
RLEP3
trnA dnaK
gltA fbpB
rrsA rrlA rrfA
rpoD
recA
Intervals
Scale
Loci
Seq

sequences as fasta libraries. The displayed information can also be printed on a PostScript printer. Loading data can be done inside the database program by updating records directly or by reading in a text file with properly formatted records. There is an update mechanism for official updates.

Data loaded

Data loaded into the database come in part from internal sources, in part from public databases. The major classes of data are listed below.

Sequences

Sequences have been taken from the public databases GenBank/EMBL and have in some cases been scanned in from the literature. Except for most of the 16S rRNA sequences (of which there are more than 150), the sequences have been checked, connected into larger units where possible, and redundancies have been eliminated. In version 4–12 of MycDB, the total lengths of sequences from the chromosomes of *M. leprae, M. tuberculosis/M. bovis* and *M. smegmatis* are about 1.7 Mbp, 367 kbp and 64 kbp, respectively. Also represented are the results of homology searches using blastx (Altschul *et al.*, 1990; Gish and States, 1993) and the ACEDB software can display the alignments.

Genes and Loci

There are about 350 loci defined in the database, mostly coming from sequence records in EMBL and other databases as well as from the results of hybridization mapping experiments. We have named many genes that were previously unnamed or had unwieldy or unambiguous names. Thus many of the sequenced antigen genes are named as wag13, for WHO antigen 13, the gene for the antigen with IMMYC code 13. In general, genes have been named after their counterparts in *E. coli* or *Bacillus subtilis.*

◀

Figure 39-3. Map of contig 84 from the *M. leprae* clone bank. The Map display is the default way of showing clones that have been placed on the physical map by for instance fingerprinting methods. From left to right: locator (this view is zoomed out to show the entire map, thus the cursor covers the entire region); clones, with names to the right; a scale (in 'band' units; this is dependent on the map construction method); loci mapped by hybridization; and clones sequenced or being sequenced (black: finished and submitted to the database, grey: finished but sequence not yet in MycDB; white: not yet sequenced). The view can also be customized to show the result of queries on the database.

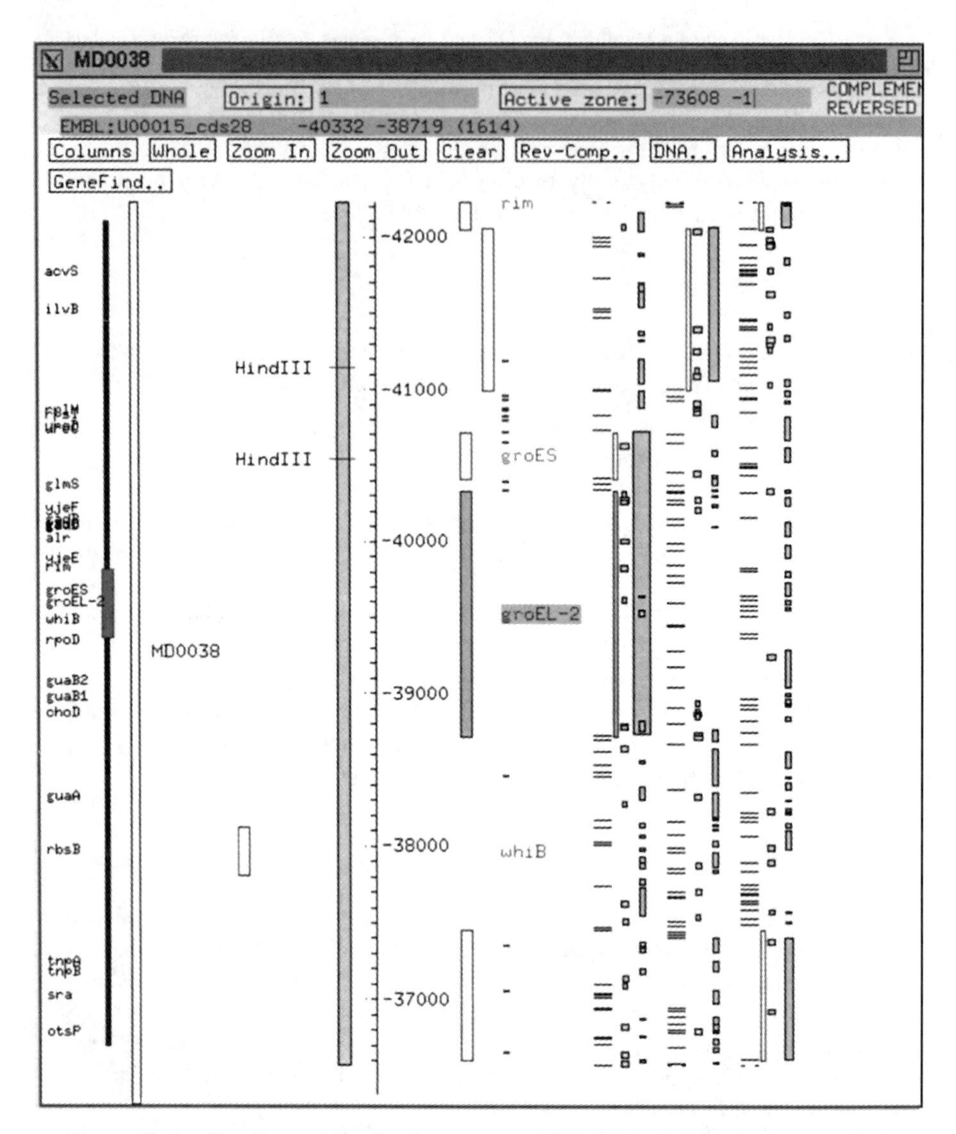

Figure 39-4. The FeatureMap for the sequence MD0038, including the sequenced *M. leprae* cosmid clones B1620 and B229. The view is zoomed in to show greater detail around the gene groEL-2, one of the genes for the heat shock protein GroEL. The different columns of the display are, from left to right: an overview of the sequence with locus names (placed in accordance to their actual positions in the sequence and thus may overlap if loci are close together) and a cursor (the box) showing the currently displayed region. This cursor can be used to move around with; coding sequences on the other strand, reading upwards; restriction map; a bar representing the currently displayed sequence; a scale; coding sequences (with locus names) and other features (such as conflicts between overlapping sequences) on this strand. The coding sequence for *groEL-2* is highlighted; stop codons, ATG codons and prediction of coding probability (repeated once for each frame). The display can be customized to show many other things, such as DNA sequence, translations, homologies etc.

Clones and Maps

Clones and physical maps presented are the ordered cosmid banks constructed as a part of the *M. leprae/M. tuberculosis* genome mapping projects, as well as maps constructed from the sequences (Fig. 39.3). Cosmids representing *M. leprae* have been assembled into 4 contigs (Chapter 68) (Eiglmeier *et al.*, 1993).

Antigens

The IMMYC list of antigens (Young *et al.*, 1992) has served as the basis for the coverage of Antigens. Additions have been made of antigens found in the literature and in the sequence databases. The code scheme proposed in (Young *et al.*, 1992) has been adopted as a standard for naming antigens, and, as mentioned above, for naming loci for antigens of unknown function.

Antibodies

Details of the antibodies in MycDB have mainly been taken from the CDC/ WHO Antibody bank and augmented with cross-references to Papers and Antigens. Some also come from the papers published from the WHO workshops (Engers et al., 1986; Khanolkar-Young *et al.*, 1992).

Papers, Authors, Colleagues

A search of MedLine results in about 6,700 references treating the molecular biology of mycobacteria (more than 26,000 about mycobacteria in general and over 74,000 dealing with tuberculosis and/or leprosy). A subset, mainly from the years since 1990 and/or including sequence data have been included. Most of the papers also have attached abstracts. A small set of addresses of people publishing in the field of mycobacteriology is also included.

Future plans

It is our intention to continue to maintain and enhance MycDB. Specific plans include: The physical maps of *M. leprae* and *M. tuberculosis* in their finished forms; continued entry and annotation of sequences; analysis of sequences, for coding potential, repetitive sequences, etc.; strain lists; expanded literature coverage; increased cross-referencing of objects. To avoid duplication of effort we also intend to highlight areas of the mycobacterial chromosomes which are currently being, or will shortly be, sequenced. We are also open to suggestions for improvements to the database schema, and to indications of new sources of data.

Regular updates to MycDB and the ACEDB software are available for searching and anonymous transfer over the Internet. Send an email message to mycdb@-

biochem.kth.se for further instructions on downloading and installation, or point your WorldWideWeb browser at http://www.biochem.kth.se/MycDB.html.

Acknowledgments

This work was initiated during a postdoctoral year for SB in the laboratory of Stewart Cole and Institut Pasteur. Many people kindly contributed their knowledge and time, among them Karin Eiglmeier (*M. leprae* physical map), Wolfgang Philipp and Sylvie Poulet (*M. tuberculosis* physical map) at the Institut Pasteur, Doug Smith at Collaborative Research, Inc., Boston (*M. leprae* and *M. tuberculosis* cosmid sequences), Jelle Thole and Douglas Young, St. Mary's Hospital, London (IMMYC Antigen compilation) and Rosalind VanLandingham and Tom Shinnick at CDC, Atlanta (CDC/WHO Antibodies). Richard Durbin, MRC, Cambridge and Jean Thierry-Mieg, CNRS, Montpellier wrote the ACEDB software and have graciously shared it and their knowledge. This work received financial support from the UNDP/World Bank/WHO Special Programme for Research and Training in Tropical Diseases and additional funds from the Association Française Raoul Follereau, the Swedish Institute and Institut Pasteur.

References

Aldovini, A., R. N. Husson, and R. A. Young. 1993. The *ura*A locus and homologous recombination in *Mycobacterium bovis* BCG. *J. Bacteriol.* 175:7282–7289.

Altschul, S. F., W. Gish, W. Miller, E. W. Myers, and D. J. Lipman. 1990. Basic local alignment search tool. *J. Mol. Biol.* 215:403–410.

Bergh, S., and S. T. Cole. 1994. MycDB: an integrated mycobacterial database. *Mol. Microbiol.* 12:517–534.

Bloom, B. R., and C. J. L. Murray. 1992. Tuberculosis: commentary on a reemergent killer. *Science* 257:1055–1064.

Cherry, J. M., and S. W. Cartinhour. 1994. ACEDB: a tool for biological information. In *Automated DNA sequencing and analysis,* M. D. Adams, C. Fields and J. C. Venter, eds. pp. 347–356. Academic Press, London.

Cherry, J. M., S. W. Cartinhour, and H. M. Goodman. 1992. AAtDB, an *Arabidopsis thaliana* database. *Plant Molecular Biology Reporter* 10:308–309, 409–410.

Cole, S. T., and D. R. Smith. 1994. Towards mapping and sequencing the genome of *Mycobacterium tuberculosis.* In *Experimental tuberculosis,* B. R. Bloom, ed. American Society for Microbiology, Washington, DC, in press.

Dunham, I., R. Durbin, J. Thierry-Mieg, and D. R. Bentley. 1994. Physical mapping projects and ACEDB. In *Guide to human genome computing,* M. J. Bishop, ed. pp. 111–158. Academic Press, London.

Durbin, R., and J. Thierry-Mieg. 1991. A *C. elegans* Database (Documentation, code and data available from anonymous FTP servers at lirmm.lirmm.fr, ftp.sanger.ac.uk and ncbi.nlm.nih.gov.)

Eiglmeier, K., N. Honoré, S. A. Woods, B. Caudron, and S. T. Cole. 1993. Use of an ordered cosmid library to deduce the genomic organization of *Mycobacterium leprae*. *Mol. Microbiol.* 7:197–206.

Engers, H. D., V. Houba, J. Bennedsen, T. M. Buchanan, S. D. Chaparas, G. Kadival, O. Closs, J. R. David, J. D. A. Embden, T. Godal, S. A. Mustafa, Y. Ivanyi, D. B. Young, S. H. E. Kaufmann, A. G. Khomenko, A. H. J. Kolk, M. Kubin, J. A. Louis, P. Minden, T. M. Shinnick, L. Trnka, and R. A. Young. 1986. Results of a World Health Organization full-sponsored workshop to characterize antigens recognised by mycobacterium-specific monoclonal antibodies. *Infect. Immun.* 51:718–720.

Gish, W., and D. J. States. 1993. Identification of protein coding regions by database similarity search. *Nat. Genet.* 3:266–272.

Honoré, N., S. Bergh, S. Chanteau, F. Doucet-Populaire, K. Eiglmeier, T. Garnier, C. Georges, P. Launois, T. Limpaiboon, S. Newton, K. Niang, P. del Portillo, G. R. Ramesh, P. Reddi, P. R. Ridel, N. Sittisombut, S. Wu-Hunter, and S. T. Cole. 1993. Nucleotide sequence of the first cosmid from the *Mycobacterium leprae* genome project: structure and function of the Rif-Str regions. *Mol. Microbiol.* 7:207–214.

Jacobs, W. R., Jr., G. V. Kalpana, J. D. Cirillo, L. Pascopella, S. B. Snapper, R. A. Udani, W. Jones, R. G. Barletta, and B. R. Bloom. 1991. Genetic systems for mycobacteria. *Meth. Enzymol.* 204:537–555.

Khanolkar-Young, S., A.H.J. Kolk, A. B. Andersen, J. Bennedsen, P. J. Brennan, B. Rivoire, S. Kuijper, K.P.W.J. McAdam, C. Abe, H. V. Batra, S. D. Chaparas, G. Damiani, M. Singh, and H. D. Engers. 1992. Results of the third immunology of leprosy/immunology of tuberculosis antimycobacterial monoclonal antibody workshop. *Infect. Immun.* 60:3925–3927.

Koch, R. 1882. *Berl. Klin. Wochenschr.* xix:221.

Noordeen, S. K., L. Lopez Bravo, and T. K. Sundaresan. 1992. Estimated number of leprosy cases in the world. *Bull. World Health Organ.* 70:7–10.

Snapper, S. B., R. E. Melton, S. Mustafa, T. Kieser, and W. R. Jacobs, Jr. 1990. Isolation and characterization of efficient plasmid transformation mutants of *Mycobacterium smegmatis*. *Mol. Microbiol.* 4:1911–1919.

Snider, D. E., Jr, and W. L. Roper. 1992. The new tuberculosis. *N. Engl. J. Med.* 326:703–705.

Sudre, P., G. ten Dam, and A. Kochi. 1992. Tuberculosis: a global overview of the situation today. *Bull. World Health Organ.* 70:149–159.

Young, D. B., and S. T. Cole. 1993. Leprosy, tuberculosis, and the new genetics. *J. Bacteriol.* 175:1–6.

Young, D. B., S. H. Kaufmann, P. W. Hermans, and J. E. Thole. 1992. Mycobacterial protein antigens: a compilation. *Mol. Microbiol.* 6:133–145.

40

Mycoplasma genitalium

Kenneth F. Bott and Claire M. Fraser

Introduction

Mycoplasmas are members of the class Mollicutes which are characterized as small wall-less bacteria; to date, over 100 species have been identified. These diverse organisms are famous for their ability to parasitize a wide range of hosts, including humans, animals, plants, insects and most notoriously, tissue culture. Several excellent reviews (Razin, 1985; Taylor-Robinson, 1995; Razin and Barile, 1985), including two superb monographs (Maniloff *et al.*, 1992; Razin and Tully, 1995) have been devoted to most aspects of the molecular biology, physiology, pathogenicity, ecology and diversity of this group.

Aside from their varied significance as pathogens, mycoplasmas are of interest because they are believed to represent a minimalist life form, having yielded to a selective pressure to reduce genome size because of an increased mutation frequency and elimination of unnecessary genes (Maniloff, 1992; *see also* Chapter 37). The genome size of mycoplasmas range from 1700 kilobases (kb) to less than 600 kb. *Mycoplasma genitalium* at 580 kb, currently has the smallest genome of any free-living organism (Colman *et al.*, 1990; Krawiec and Riley, 1990; *see* Chapter 1). Based on an extensive comparison of the primary structure of 5S and 16S rRNA genes it, like other mycoplasmas, is believed to have evolved from ancestors common to higher gram positive organisms of the *Lactobacillus/ Clostridium* branch through the loss of genetic material (Maniloff, 1992). Its low DNA G+C content (32%), has prompted some researchers to suggest that a directional selection pressure toward A+T richness has acted on a common ancestor (Maniloff, 1992; Razin, 1985). *M. genitalium* like most Mycoplasma species has evolved a specialized genetic code which uses a reduced number of isoaccepting tRNA species, by exhibiting a bias for A or T in the third position of codons, and use UGA codons to encode *tryptophan* instead of translational stops (Muto *et al.*, 1992). The practical effect of that adaptation is that expression of mycoplasma genes in conventional cloning hosts, like *E. coli,* leads to early truncation of the mycoplasma peptides. Although some researchers have used *E. coli* strains which harbor plasmid DNA encoding UGA suppressor tRNAs to prevent the premature termination (Proft and Herrmann, 1994; Renbaum *et al.* 1990), these systems are not highly efficient. Usually, successful expression of *Mycoplasma* genes in *E. coli* requires that the DNA trp codons be "repaired" in vitro.

The molecular characterization of mycoplasmas has been slow. One major hindrance to the study of these unique organisms has been the lack of a classic genetic system. Outside of the description that mycoplasmas may be manipulated in the laboratory to induce DNA uptake and undergo transformation (Hedreyda *et al.*, 1993; King and Dybvig, 1994), the process has not yet been exploited for general use as a "genetic system". Most species of mycoplasma are very fastidious, and require complex media for growth in culture. Thus, the techniques utilized by classical geneticists to define auxotrophic mutants are also not applicable. Furthermore, the pleomorphic shape, small size and tendency of mycoplasmas to aggregate makes the identification of individual clonal isolates almost impossible.

The Smallest Free Living Pathogen

At this writing *M. genitalium* is the smallest known free living pathogen (Lucier *et al.* 1994; Maniloff, 1992; Krawiec and Riley, 1990; *see* Chapters 1 and 37). Although its preferred environmental niche is a parasitic association with ciliated epithelial cells of a primate genital or respiratory tract, (Jensen *et al.*, 1994; Taylor-Robinson, 1995; Tully *et al.*, 1981) it can be cultivated in the laboratory if its highly fastidious nutritional requirements are met (Tully *et al.*, 1995). From its earliest identification as a distinct species it has been known that *M. genitalium* shares a number of important biological and serological properties with *M. pneumoniae* (Tully *et al.*, 1981; Hu *et al.*, 1987; Tully *et al.*, 1983) a significant pathogen and causative agent of atypical pneumonia. Thus the two species have a similar differentiated "flask" shape and tip structure; common attachment components (Tully *et al.*, 1983), possess a sequence homology between their adhesin genes (Yogev and Razin, 1986; Inamine *et al.*, 1989), exhibit common epitopes among their adhesin and membrane proteins and contain similar membrane glycolipids (Lind *et al.*, 1984; Hu *et al.*, 1987; Taylor-Robinson *et al.*, 1983).

They also have similarly organized genomes, (Peterson *et al.*, 1995) despite the fact that *M. pneumoniae* is 200 kb larger. In nature *M. genitalium* attaches to neuraminidase sensitive receptors via surface exposed adhesins, (Collier et al. 1990) whose major component is referred to as MgPa (Hu *et al.*, 1987). Until very recently its significance as a pathogen was questioned since there is not a reliable animal model for infection, and a paucity of experimental evidence linking it to human genital infections.

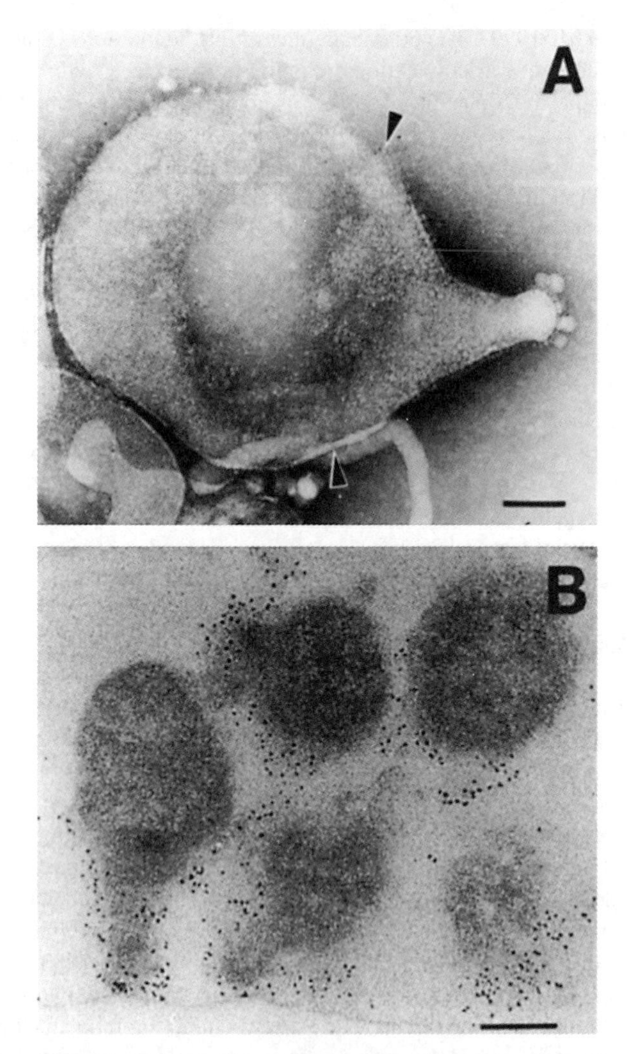

Figure 40-1. Electron micrographs of *M. genitalium* from (Hu et al. 1987) with permission of authors and Am. Soc. for Microbiol. (A) Negative staining of an intact mycoplasma cell with ammonium molybdate. Note the terminus is covered with a nap extending peripherally to the area marked by arrowheads. (B) Thin section of *M. genitalium* organisms after incubation with MgPa adhesin specific monoclonal antibody (Mg-209) and indirect staining with ferritin-conjugated rabbit antibody to mouse immunoglobin G. The lower three organisms were attached to a cover slip surface represented by the thin horizontal line. Ferritin grains are clustered around the tip terminus corresponding to the area covered with the nap as shown in A. Bars, 0.1 μm.

In the 14 years since its initial isolation, only twelve other *M. genitalium* isolates have been reported, and five of those have been described in the last year (Jensen, 1994; Tully *et al.*, 1995). Its primary localization within the human host is still under speculation. Since the initial isolation (Tully *et al.*, 1981), *M. genitalium* has been detected in arthritic joints (Horner *et al.*, 1993) and in the blood of one AIDS patient (Montagnier *et al.*, 1990a) using the polymerase chain reaction (PCR). It has been isolated from the respiratory tract and the arthritic joint of an immunodeficient patient in association with *M. pneumoniae* (Baseman *et al.*, 1988; Taylor-Robinson *et al.*, 1994) and alone from urethral swabs of patients with non-gonococcal urethritis (NGU) (Jensen *et al.*, 1994; Taylor-Robinson, 1995). Although there is convincing evidence that *M. genitalium* infection causes urogenital disease in non-human primates (Tully *et al.*, 1986), the demonstration of *M. genitalium* as a direct cause of disease in humans has been difficult. One of the major obstacles to this demonstration, besides the difficulties encountered in culturing the organism, has been the fact that *M. genitalium* is serologically cross-reactive with *M. pneumoniae* (Tully *et al.*, 1995; Lind *et al.*, 1984). Most of the early diagnostic tools for detection of mycoplasma infection relied on the use of serological tests. Lind *et al.* (1984) showed that these tests were not reliable when used to distinguish between *M. genitalium* and *M. pneumoniae,* because these two organisms share several serologically cross reactive antigens (Lind *et al.*, 1984), including the related immunodominant major adhesin protein of both species. Several research groups then turned to PCR and DNA probe techniques to detect the presence of *M. genitalium* in clinical studies of patients with NGU (Razin, 1994). Researchers using PCR initially targeted sequences from the *mpg*A operon, (Inamine *et al.*, 1989; Razin, 1994). In one study, amplification of *M. genitalium* DNA from clinical specimens indicated that this organism was present in urethral swabs of about 17% of men with NGU (Jensen *et al.*, 1993) and a second study detected *M. genitalium* in 23% of urethral samples from men with acute NGU (Horner *et al.*, 1993). No strains were isolated by culture from the PCR positive specimens (Jensen *et al.*, 1993). Researchers have more recently begun using primers designed to the 16S rRNA gene sequence in PCR detection (Jensen and Borre, 1994). Here the rationale for such a change should be questioned. In the case of *M. genitalium* a ribosomal RNA target DNA is limited by the presence of only a single ribosomal operon, while at least seven copies of the genes encoding some adhesin epitopes are present on the chromosome (Peterson, 1992; Bailey, 1995; Peterson *et al.*, 1995b; see also below). The most useful would be to select a primer directed towards an epitope not shared with the P1 adhesin of *M. pneumoniae,* and since the DNA sequence of both is known (Inamine *et al.*, 1989; Dallo *et al.*, 1989), such a project should not be difficult.

One possible explanation for the inability to culture *M. genitalium,* and the persistent and re-occurring infections seen in NGU patients (Horner *et al.*, 1993), is that this organism may obtain an intracellular localization while inside the host. Jensen et al. have suggested a possible intracellular location for *M. genitalium* when this organism is cultured with Vero Monkey Kidney cells (Jensen *et al.*, 1994). This correlates with reports which also suggested a possible intracellular location for *M. genitalium* when cultured with human lung fibroblasts (Mernaugh *et al.*, 1993). Neither of these two studies were definitive, as they were only morphological and not able to discount the possibility that the organisms which appeared to be intracellular were instead localized to invaginations of the host cell membrane. Several other mycoplasma species have been suggested to have intracellular locations (Lo *et al.*, 1993; Lo *et al.*, 1989; Taylor-Robinson *et al.*, 1991). Recently *M. genitalium* has been raised to an unexpected level of notoriety because it is not only a bonafide pathogen having possible significance as a secondary pathogen of AIDS patients (Montagnier *et al.*, 1990b), but a separate and distinct pathogenicity that distinguishes it from its closest cousin, *M. pneumoniae.* However, data relating to its preferred interaction with humans and its exact mechanism of pathogenicity are lacking. Not only is *M. genitalium* so small and pliable that it passes through normal bacteriological filters, but it also has the smallest genome size of all known free living organisms (*see also* Chapter 1). This elevates it to significance as a minimal genome for metabolism and pathogenicity. Certainly, it possesses the smallest genomic content for a species that has survived the pressures of evolution (*see also* Chapter 37).

The Smallest Free Living Prokaryote

A random sequencing analysis of the *M. genitalium* genome was undertaken by Peterson et al. to overcome the lack of a genetic system in this organism (Peterson *et al.*, 1993; Peterson *et al.*, 1991). The rationale was that an organism with such a small genome might be expected to contain a significant proportion of "essential" or evolutionary conserved genes and a minimum of "spacer" DNA. The sequences identified could be used not only as sequence tagged sites, or as markers on the existing physical map (Colman *et al.*, 1990), but might also define genes which are retained by this "minimalist-life-form" through evolution (*see also* Chapter 37). Initial observations were striking. Of more than 500 clones analyzed, a group of 291 unique contigs were characterized (Peterson *et al.*, 1993; Peterson *et al.*, 1991). Greater than 90% of random clones contained long open reading frames (ORFs). More than 40% of the random sequences could be shown to significantly match existing entries in the database. Only 11 clones were encountered which neither were homologous to RNA species nor contained ORFs of significant length.

The genomic organization of *M. genitalium* was improved to another level by the characterization of a complete set of cosmid clones and phage clones (Lucier *et al.*, 1994). Furthermore, by use of those clones as hybridization probes for the entire collection of random libraries previously analyzed (Peterson *et al.*, 1993) a preliminary genetic map was constructed (Peterson *et al.*, 1995a; *see also* Chapter 69). This marked the first "genetic" characterization of an organism where no genetics techniques could be employed. The detailed map was elucidated by first analyzing segments of DNA sequence, then relating them to a physical map derived from restriction enzyme analysis of genomic DNA.

A turning point in our understanding of small genomes was achieved in mid-1995 with the publication of the first complete nucleotide sequence from *Haemophilus influenzae* (Fleischmann *et al.*, 1995). Using the same strategy for whole genome shotgun sequencing and assembly as that described for *H. influenzae,* Fraser et al. (1995) reported the complete nucleotide sequence of *M. genitalium* in October 1995. Sequence analysis of the *M. genitalium* genome reveals a genome size of 580,073 bp (Fraser *et al.*, 1995), a value in excellent agreement with that determined by physical mapping (Colman *et al.*, 1990). The *M. genitalium* genome contains close to 470 predicted coding sequences as compared to 1,727 identified in *H. influenzae* (Fleischmann *et al.*, 1995; *see also* Chapter 37). The predicted coding sequences in *M. genitalium* display an average size of 1040 bp and cover 88% of the genome, a value similar to that found in *H. influenzae* (Fleischmann *et al.*, 1995). Minimal spacer regions were seen between ORFs and in many cases the potential ORFs overlapped each other by a few nucleotides (Bailey and Bott, 1994; Peterson *et al.*, 1993; Fraser *et al.*, 1995). These data indicate that the reduction in genome size that has occurred with the evolution of *M. genitalium* has not resulted from an increase in gene density or a decrease in gene size.

A survey of the genes in *M. genitalium* makes possible a description of the minimal set of genes required for its survival as a distinct species through evolution; and by direct comparison with *H. influenzae,* the minimal gene set required for a contemporary survival (*see* Chapter 37). Certainly, a precise determination of those numbers will require direct experimentation. Many of the potential ORFs were defined on the basis of their homology with proteins in a non-reductant bacterial protein database (NRBP) (Fraser *et al.*, 1995) or proteins from other species in GenPept, Swiss-Prot, and the Protein Information Resource (PIR). Using this approach, a total of 374 ORFs were putatively identified with 96 showing no matches to protein sequences from any other organism. One-half of all predicted coding regions in *M. genitalium* for which a putative identification could be assigned displayed the greatest degree of similarity to proteins from either a gram-positive organism or a *Mycoplasma* species. These data support the evolutionary relationship between *M. genitalium* and Gram-positive eubacteria that has been deduced from small subunit rRNA sequences (Rogers *et al.*, 1985; Weisburg *et al.*, 1989).

A limited complement of genes involved in DNA maintenance and repair plus transcriptional and translational machinery components were found, but there were no homologs to genes involved in cell wall synthesis, no cytochromes, no amino acid, purine or pyrimidine biosynthetic pathways; no TCA cycle components, and limited metabolic pathway intermediates (Fraser *et al.*, 1995). This suggests that *M. genitalium* is well adapted to scavenge the essential building blocks that it lacks the genetic capabilities to create *de novo*. The limited metabolic capacity of *M. genitalium* is in marked contrast to the complexity of the catabolic pathways in *H. influenzae* (Fleischmann *et al.*, 1995; *see also* Chapter 37), reflecting the 4-fold greater number of genes involved in energy metabolism in *H. influenzae* as compared with *M. genitalium.* Very few classic prokaryote promoter sequences were identified, however, *Mycoplasma*-specific promoter sequences have yet to be described.

One of the unexpected findings from analysis of the entire genome of *M. genitalium* was that 20% of the putative coding sequences had no database matches to any sequences in public archives including the entire *H. influenzae* genome. It is possible that these sequences may represent novel genes in *M. genitalium* and closely related species. It will be of interest to determine how many of these genes are essential for maintaining an independent species identity and what their biological role is in the cell.

Given the limited number of genes in the *M. genitalium* genome, it is perhaps somewhat surprising to find multiple "cassette elements" of repeated DNA from the major adhesin operon (see Chapter 69) scattered throughout the genome. The repetitive DNA shares significant sequence identity with two genes (orf2 and orf3) of the *mpgA* cytadherence operon (Peterson *et al.*, 1995b). The availability of the complete genomic sequence from *M. genitalium* has allowed for a comprehensive mapping of the MgPa repeats. Nine repetitive elements which range in size from approximately 400 to 2100 bp and are composites of five separate regions of the *mpgA* operon are scattered on the chromosome (Peterson *et al.*, 1995b and Chapter 69). The percent of sequence identity between the repeat elements and *mpgA* operon ranges from 78%–90%. Most elements contain a middle A+T rich region of approximately 300 nt, which has no significant homology to other *M. genitalium* sequences. With the exception of the A+T rich spacer, each segment of the cassette maintains open reading frames as if to preserve epitope cassettes. Collectively, the G+C content of the *mpgA* adhesin operon plus the repeated cassettes (exclusive of A+T spacers) is 42%; while the overall average of the entire genome is 32% G+C; strongly suggesting that the intact adhesin component was acquired by horizontal transfer from another species, before being duplicated or fragmented within the genome by recombination.

The sequences contained in the *mpgA* operon and the nine repeats scattered throughout the chromosome represent 4.5% of total genomic sequence. At first glance this might appear to contradict the expectation for a minimal genome. However, recent evidence for recombination between the repetitive elements and the *mpgA* operon has been reported (Peterson *et al.*, 1995b). Such recombination may allow *M. genitalium* to evade the host immune response through the creation of antigenic variation within the population (*see also* Chapter 14). Since *M. genitalium* survives in nature by obtaining essential nutrients from its mammalian host, an efficient mechanism to evade the immune response appears to be a necessary part of this minimal genome.

In conclusion, we are now able to analyze on an entirely new plane a living organism whose entire genomic sequence is known (Fraser *et al.*, 1995). It is a minimal organism with a single ribosomal operon, limited codon usage and unknown nutritional requirements; potentially the ideal organism for study. At this stage the species is difficult to grow as clonal progeny from single cells, especially on solid media. Genetic exchange mechanisms, mutant selection of gene alterations, transposon mutagenesis and extracellular gene expression are still under experimental development. Characterization of promoter regions, identification of regulatory sequences and genes directing species specificity, all have yet to be defined. Still many experiments that were impossible when 1995 began can now be performed on the computer, without even possessing a culture of the organism or ever concocting that extremely exacting growth medium. What directions will *M. genitalium* research take now that its entire sequence has been elucidated?

Acknowledgments

We gratefully acknowledge the technical assistance provided by Christine Doyle and Craig Swainey; and special thanks to Ping-chuang Hu for the initial incentive, encouragement and expertise relating to Mycoplasma biology. This research was supported by NIH grant AI33161.

References

Bailey, C. C. 1995. Study of highly conserved genes and species specific elements from *Mycoplasma genitalium.* [Microbiology]. Univ. No. Carolina at Chapel Hill. pp. 1–108. Ph.D.

Bailey, C. C., and K. F. Bott. 1994. An unusual gene containing a *dnaJ* N-terminal box flanks the putative origin of replication of *Mycoplasma genitalium. J. Bacteriol.* 176:5814–5819.

Baseman, J. B., S. F. Dallo, T. G. Tully, and D. L. Rose. 1988. Isolation and characterization of *Mycoplasma genitalium* strains from the human respiratory tract. *J. Clin. Microbiol.* 26:2266–2269.

Collier, A. M., J. L. Carson, P.-C. Hu, S. S. Hu, C. H. Huang, and M. F. Barile. 1990. Attachment of *Mycoplasma genitalium* to the ciliated epithelium of human fallopian tubes. *Zbl. Bakt.* S20:730–732.

Colman, S. D., P.-C. Hu, W. Litaker, and K. F. Bott. 1990. A physical map of the *Mycoplasma genitalium* genome. *Mol. Microbiol.* 4:683–687.

Dallo, S. F., A. Chavoya, C.-J. Su, and J. Baseman. 1989. DNA and protein sequence homologies between the adhesins of *Mycoplasma genitalium* and *Mycoplasma pneumoniae. Infect. Immun.* 57:1059–1065.

Fleischmann, R. D., M. D. Adams, O. White, R. A. Clayton, E. F. Kirkness, A. R. Kerlavage, C. J. Bult, J. F. Tomb, B. A. Dougherty, J. M. Merrick, K. McKenney, G. Sutton, W. Fitzhugh, C. Fields, J. D. Gocayne, J. Scott, R. Shirley, L. I. Liu, A. Glodek, J. M. Kelley, J. F. Weidman, C. A. Phillips, T. Spriggs, E. Hedblom, M. D. Cotton, T. R. Utterback, M. C. Hanna, D. T. Nguyen, D. M. Saudek, R. C. Brandon, L. D. Fine, J. L. Fritchman, J. L. Fuhrmann, N. S. M. Geoghagen, C. L. Gnehm, L. A. McDonald, K. V. Small, C. M. Fraser, H. O. Smith, and J. C. Venter. 1995. Whole-genome random sequencing and assembly of *Haemophilus influenzae* Rd. *Science* 269:496–512.

Fraser, C. M., J. D. Gocayne, O. White, M. D. Adams, R. A. Clayton, R. D. Fleischmann, C. J. Bult, A. R. Kerlavage, G. Sutton, J. M. Kelley, J. L. Fritchman, J. F. Weidman, K. V. Small, M. Sandusky, J. Fuhrman, T. R. Utterback, D. M. Saudek, C. A. Phillips, J. N. Merrick, J. Tomb, B. A. Dougherty, K. F. Bott, P. Hu, T. S. Lucier, S. N. Peterson, H. O. Smith, C. A. Hutchinson III, and J. C. Venter. 1995. The *Mycoplasma genitalium* genome sequence reveals a minimal gene complement. *Science* 270:397–403.

Hedreyda, C., K. K. Lee, and D. C. Krause. 1993. Transformation of *Mycoplasma pneumoniae* with *Tn4001* by electroporation. *Plasmid* 30:170–175.

Horner, P. J., C. B. Gilroy, B. J. Thomas, R. O. M. Naidoo, and D. Taylor-Robinson. 1993. Association of *Mycoplasma genitalium* with acute non-gonococcal urethritis. *Lancet* 342:582–585.

Hu, P.-C., U. Schaper, A. M. Collier, W. A. Clyde, Jr., M. Horikawa, Y.-S. Huang, and M. F. Barile. 1987. A *Mycoplasma genitalium* protein resembling the *Mycoplasma pneumoniae* attachment protein. *Infect. Immun.* 55:1126–1131.

Inamine, J. M., S. Loechel, A. M. Collier, R. M. Barile, and P.-C. Hu. 1989. Nucleotide sequence of the MgPa (*mmp*) operon of *Mycoplasma genitalium* and comparison to the P1 (*mpp*) operon of *Mycoplasma pneumoniae. Gene* 82:259–267.

Jensen, J. S. 1994. Sequence diversity in the *MgPa* adhesin gene of *Mycoplasma genitalium* strains. *IOM Lett.* 3:429–430 (abstract).

Jensen, J. S., J. Blom, and K. Lind. 1994. Intracellular location of *Mycoplasma genitalium* in cultured Vero cells as demonstrated by electron microscopy. *Int. J. Exp. Pathol.* 75:91–98.

Jensen, J. S., and M. B. Borre. 1994. Sequence determination of the *Mycoplasma genitalium* 16S ribosomal RNA gene and development of a species specific PCR. *IOM Lett.* 3:332–333. (abstract)

Jensen, J. S., H. T. Hansen, and K. Lind. 1994. Isolation of *Mycoplasma genitalium* strains from the male urethra. *IOM Lett.* 3:143–144. (abstract)

Jensen, J. S., R. Orsum, B. Dohn, S. Uldum, A.-M. Worm, and K. Lind. 1993. *Mycoplasma genitalium:* a cause of male urethritis? *Genitourin. Med.* 69:265–269.

King, K. W., and K. Dybvig. 1994. Mycoplasmal cloning vectors derived from plasmid pKMK1. *Plasmid* 31:49–59.

Krawiec, S., and M. Riley. 1990. Organization of the bacterial chromosome. *Microbiol. Rev.* 54:502–539.

Lind, K., B. O. Lindhardt, H. J. Schutten, J. Blom, and C. Christiansen. 1984. Serological cross-reactions between *Mycoplasma genitalium* and *Mycoplasma pneumoniae. J. Clin. Microbiol.* 20:1036–1043.

Lo, S.-C., M. S. Dawson, D. M. Wong, P. B. Newton III, M. A. Sonoda, M. F. Eagler, R. Y.-H. Wang, J. W.-K. Shih, H. J. Alter, and D. J. Wear. 1989. Identification of *Mycoplasma incognitus* infection in patients with AIDs: an immunohistochemical in situ hybridization and ultrastructural study. *Am. J. Trop. Med. Hyg.* 41:601–616.

Lo, S.-C., M. M. Hayes, H. Kotani, P. F. Pierce, D. J. Wear, P. B. Newton III, J. G. Tully, and J.W.-K. Shih. 1993. Adhesion onto and invasion into mammalian cells by *Mycoplasma penetrans:* A newly isolated mycoplasma from patients with AIDs. *Mod. Pathol.* 6:276–280.

Lucier, T. S., P.-Q. Hu, S. N. Peterson, X.-Y. Song, L. Miller, K. Heitzman, K. F. Bott, C. A. Hutchison, III, and P.-C. Hu. 1994. Construction of an orderd genomic library of *Mycoplasma genitalium. Gene* 150:27–34.

Manilof, et al. 1992. Mycoplasmas: Molecular Biology and Pathogenesis. Maniloff, J., R. N. McElhaney, L. R. Finch, and J. B. Baseman. eds. 1st edn. 1-609. Washington, DC: American Society for Microbiology. 1-55581-050-0.

Manilof, J. 1992. Phylogeny of Mycoplasmas. In: Maniloff, J., McElhaney, R. N., Finch, L. R. and Baseman, J. B. (eds.) *Mycoplasmas: molecular biology and pathogenesis,* pp. 549–559. ASM Washington, DC.

Mernaugh, G. R., S. F. Dallo, S. C. Holt, and J. B. Baseman. 1993. Properties of adhering and nonadhering populations of *Mycoplasma genitalium. Clin. Infect. Dis.* 17:S69–78.

Montagnier, L. A., D. Berneman, and D. Guetard. 1990a. Infectivity inhibition of HIV prototype strains by antibodies directed against a peptide sequence of mycoplasmas. *C. R. Acad. Sci. III* 311:425–430.

Montagnier, L. A., A. Blanchard, D. Guetard, D. Berneman, M. Lemaitre, A.-M. DiRienzo, S. Chamaret, Y. Henin, E. Bahraoui, C. Daugriet, C. Axler, M. Kirstetter, R. Roue, G. Pialoux, and D. Dupont. 1990b. A possible role of mycoplasmas as co-factors in AIDs. In: Girard, M. and Valette, L. (Eds.) *Retroviruses of human AIDs and related animal diseases: Proceedings of the colloque des cent gardes,* pp. 9–17. Lyons, France: Foundation M. Merieux.

Muto, A., Y. Andachi, F. Yamao, R. Tanaka, and S. Osawa. 1992. Transcription and translation. In: J. Maniloff, R. N. McElhaney, L. R. Finch, and J. B. Baseman. (eds.) *Mycoplasmas: molecular biology and pathogenesis,* pp. 331–347. ASM, Washington, DC.

Peterson, S. N. 1992. Characterization and analysis of the *Mycoplasma genitalium* genome. [Genetics]. Univ. No. Carolina at Chapel Hill. pp. 1–145. Ph.D.

Peterson, S. N., C. C. Bailey, E. A. Smith, K. F. Bott, and C. A. Hutchison, III. The MgPa repetitive DNA sequence of *Mycoplasma genitalium. IOM Lett.* 3:604 (abstract).

Peterson, S. N., P.-C. Hu, K. F. Bott, and C. A. Hutchison, III (1993). A survey of the

Mycoplasma genitalium genome by using random sequencing. *J. Bacteriol.* 175:7918–7930.

Peterson, S. N., T. Lucier, K. Heitzman, E. A. Smith, K. F. Bott, P. C. Hu, and C. A. Hutchison. 1995a. Genetic map of the *Mycoplasma genitalium* chromosome. *J. Bacteriol.* 177:3199–3204.

Peterson, S. N., C. C. Bailey, J. S. Jensen, M. B. Borre, E. S. King, K. F. Bott, and C. A. Hutchison, III. 1995b. Characterization of repetitive DNA in the *Mycoplasma genitalium* genome: Possible role in the generation of antigenic variation. *Proc. Natl. Acad. Sci. USA* 92:11829–11833.

Peterson, S. N., N. Schramm, P.-C. Hu, K. F. Bott, and C. A. Hutchison, III. 1991. A random sequencing approach for placing markers on the physical map of *Mycoplasma genitalium. Nucl. Acids Res.* 19:6027–6031.

Proft, T., and R. Herrmann. 1994. Identification and characterization of hitherto unknown *Mycoplasma pneumoniae* proteins. *Mol. Microbiol.* 13:337–348.

Razin, S. 1985. Molecular Biology and Genetics of Mycoplasmas (Mollicutes). *Microbiol. Rev.* 49:419–455.

Razin, S. 1994. DNA probes and PCR in diagnosis of mycoplasma infections. *Mol. Cell. Probes* 8:497–511.

Razin, S., and M. E. Barile. 1985. Mycoplasmas. Razin, S. and Barile, M. F. (eds.) IV Pathogenicity. 1-508. Academic Press, London.

Razin, S., and J. G. Tully, (eds.) 1995. Molecular and Diagnostic Procedures in Mycoplasmology, v.1 pp 1–583. Academic Press: San Diego.

Renbaum, P., D. Abrahamove, A. Fainsod, G. G. Wilson, S. Rottem, and A. Razin. 1990. Cloning, characterization, and expresion in *Escherichia coli* of the gene coding for the CpG DNA methylase from *Spiroplasma* sp. strain MQ1 (M SssI). *Nucl. Acids Res.* 18:1145–1152.

Rogers, M. J., J. Simmons, R. T. Walker, W. G. Weisburg, C. R. Woese, R. S. Tanner, I. M. Tobinson, D. A. Stahl, G. Olsen, R. H. Leach, and J. Maniloff. 1985. Construction of the *Mycoplasma* evolutionary tree from 5S rRNA sequence data. *Proc. Natl. Acad. Sci. USA* 82:1160–1164.

Taylor-Robinson, D. 1995. The history and role of *Mycoplasma genitalium* in sexually transmitted diseases. *Genitourin. Med.* 71:1–8.

Taylor-Robinson, D., H. A. Davies, P. Sarathchandra, and P. M. Furr. 1991. Intracellular location of mycoplasmas in cultured cells demonstrated by immunocytochemistry and electron microscopy. *Int. J. Exp. Pathol.* 72:705–714.

Taylor-Robinson, D., P. M. Furr, and J. G. Tully. 1983. Serological cross-reactions between *Mycoplasma genitalium* and *M. pneumoniae. Lancet* 1:527.

Taylor-Robinson, D., C. B. Gilroy, S. Horowitz, and J. Horowitz. 1994. *Mycoplasma genitalium* in the joints of two patients with arthritis. *Lancet* 13:1066–1069.

Tully, J. G., D. L. Rose, J. B. Baseman, S. F. Dallo, A. L. Lassell, and C. P. Davis. 1995. *Mycoplasma pneumoniae* and *Mycoplasma genitalium* mixture in synovial fluid isolate. *J. Clin. Microbiology* 33:1851–1855.

Tully, J. G., D. Taylor-Robinson, R. M. Cole, and D. L. Rose. 1981. A newly discovered mycoplasma in the human urogenital tract. *Lancet* i:1288–1291.

Tully, T. G., D. Taylor-Robinson, D. L. Rose, R. M. Cole, and J. M. Bove. 1983. *Mycoplasma genitalium,* a new species from the human urogenital tract. *Int. J. Syst. Bacteriol.* 33:387–396.

Tully, J. G., D. Taylor-Robinson, D. L. Rose, P. M. Furr, C. E. Graham, and M. F. Barile. 1986. Urogenital challenge of primate species with *Mycoplasma genitalium* and characteristics of infection induced in chimpanzees. *J. Infect. Dis.* 153:1046–1054.

Weisburg, W. G., J. G. Tully, D. L. Rose, J. P. Petzel, H. Oyaizu, D. Yang, L. Mandelco, J. Sechrest, T. G. Lawrence, J. Vanetten, J. Maniloff, and C. R. Woese. 1989. A phylogenetic analysis of the Mycoplasmas—Basis for their classification. *J. Bacteriol.* 171:6455–6467.

Yogev, D., and S. Razin. 1986. Common deoxyribonucleic acid sequences in *Mycoplasma genitalium* and *Mycoplasma pneumoniae* genomes. *Int. J. Syst. Bacteriol.* 36:426–430.

41

Other *Mycoplasma* sp.

Kenneth F. Bott

Mycoplasmas are members of a genus belonging to the class *Mollicutes* (soft skin). Classic properties which distinguish the genus are their rigid sterol-containing membrane and complete lack of cell wall components (Razin, 1985). Currently they are recognized as having the smallest and simplest genomes of all free living-species (Krawiec and Riley, 1990; *see* Chapters 1 and 37). Although capable of being cultivated as free-living organisms in the laboratory when their extremely fastidious nutritional requirements are met, virtually every species exists in nature in a parasitic or saprophytic relationship with some specific higher eukaryotic plant, animal or insect (Razin and Jacobs, 1992). Species were originally defined and differentiated serologically from differences in surface exposed antigens and although host specific, the parasitic associations were thought to be exclusively extracellular. With the advent of molecular hybridization and PCR as diagnostic or detection techniques there is increasing evidence that some species might be capable of intracellular associations as well (Whitcomb *et al.*, 1995; Krause and Taylor-Robinson, 1992; Mernaugh *et al.*, 1993; Taylor-Robinson *et al.*, 1991). Hence, significance of the pathogenic potential of Mycoplasmas, especially in mammalian systems, is being reevaluated. Since individual cells are pleomorphic, often filamentous, and difficult to cultivate by conventional clonal techniques on solid media, classic genetic analysis has been very slow to develop. Although genetic transformations, including transposon insertions have been described in this group (Dybvig and Alderete, 1988; Hedreyda *et al.*, 1993; King and Dybvig, 1994; Whitley and Finch, 1989; Cao *et al.*, 1994; Kapke *et al.*, 1994), the fastidiousness of species precludes the isolation of conventional auxotrophs. There has also not been a systematic use of those gene manipulation procedures to derive large collections of mutant loci "markers," or to establish genetic map distances by recombination analyses among mutants.

Since at least 24 separate species of *Mycoplasma* have been characterized by pulsed field gel electrophoresis (PFGE) and preliminary physical maps constructed to relate restriction fragment sizes and genome organization (*see* f.e. Chapter 69), more representatives of this genus than any other are now available for comparison (Herrmann, 1992; Fonstein and Haselkorn, 1995). The disappointing observation is that very few common genetic loci have been localized in those species. Collectively, this represents a group of species whose genomes range in size from 580 to 1800 kb and from 25 to 34% in G+C content; with members of the same species often showing heterogeneity (Herrmann, 1992; Krawiec and Riley, 1990). Ribosomal RNA sequence analysis and gene organization comparison among these species clearly establishes their evolutionary relatedness to ancestors common to Gram positive eubacterial species (Maniloff, 1992; Weisburg *et al.*, 1989). With the exception of two human associated species, *M. pneumoniae* and *M. genitalium* (Peterson *et al.*, 1995; Wenzel *et al.*, 1992; *see also* Chapter 40) relatively little direct comparison of genomic organization has been reported. Although those species are difficult to distinguish serologically despite a diversity in G+C content, and nearly 200 kb of additional genomic coding capacity in *M. pneumoniae*, the chromosomal organization patterns appear to be quite similar. By increasing the use of PFGE (*see* Chapters 24–26), DNA sequence analysis, and cross species comparisons of selected translation products, more markers are now being identified. Still there is a paucity of information relating these markers to physical chromosome structure. The best example of this is the recent report by Bork *et al.* (Bork *et al.*, 1995) (see also Chapter 42) characterizing 214 kb of *M. capricolum* DNA sequence from 372 non-overlapping contigs and showing good homology with more than 200 proteins from the data bases, yet the contigs were not related to specific positions on the physical map (Whitley *et al.*, 1991) or to each other. As more strains from individual species are examined, considerable genomic heterogeneity is seen (Ladefoged and Christiansen, 1992; Pyle *et al.*, 1990; Whitley *et al.*, 1990; Herrmann, 1992) (Table 41-1). This will certainly make it possible to obtain a more detailed analysis of genomic organization, rearrangements and species specific sequences. It will also be useful in interpreting phylogeny and evolution of *Mycoplasmas* as a group.

Table 41-1 summarizes pertinent references and preliminary characterization of ten separate *Mycoplasma* species in which pulse field gel analysis has been related to one or more genetic loci. Species are arranged by increasing genome size, beginning with *M. genitalium* the smallest described to date. The table also illustrates a heterogeneity of sizes between individual members of the same species, and the relative paucity of information about comparative genomics. As this manuscript is being readied for publication, reports indicate that R. Herrmann and colleagues have completed the determination of a genomic sequence from *M. pneumoniae* and are now in the process of editing its annotation in preparation for public release. Such a contribution will make *Mycoplasma* the first genus with two completed genomic sequences and provide a very strong foundation for the new discipline of Comparative Genomics.

Table 41-1.

Species	Pulse field gel analysis*	Genome Size (kb)	Comments	Reference
M. genitalium G37	XhoI (8), SmaI (8), ApaI (3)	580	map known, cosmid collection, complete genomic sequence	(Colman *et al.*, 1990; Lucier *et al.*, 1994; Peterson *et al.*, 1995; Fraser *et al.*, 1995)
M. pneumoniae	ApaI (13), SfiI (2); XhoI (25)	775; 840	*deoC, rrn, hmw3*, P1; cosmids	(Krause and Mawn, 1990; Wenzel and Herrmann, 1989; Wenzel *et al.*, 1992)
M. mobile	ApaI (2), MluI (3), BamHI (6), NruI (7)	780	no genes localized	(Bautsch, 1988)
M. hominis PG21	ApaI, SmaI, BamHI, XhoI, SalI unlike 4 others, see next entry	704	*rpoA, rpoC, rrn* (2), *tuf, gyrB, hup, ftsY*	(Ladefoged and Christiansen, 1992; Ladefoged and Christiansen, 1994)
M. hominis 4 strains	ApaI, SmaI, BamHI, XhoI, SalI # 4195, 132, 93, 7488 show heterogeneity	735–825	*rpoA, rpoC, rrn* (2), *tuf, gyrB, hup, ftsY*	(Ladefoged and Christiansen, 1992; Ladefoged and Christiansen, 1994)
M. flocculare	Asp (8), Apa(9), SalI(7)	890	5S separated from 16+23S	(Huang and Stemke, 1991)
M. fermentans (incognitus)	BglI (10), MluI (7), BamHI (12)	980	*rrn* (2) 5S RNAs separated from 16+23S	(Huang *et al.*, 1995)
M. gallisepticum	EagI (7), CeuI (2), SmaI (8)	998 to 1054	MGP1, *tuf, recA, gyr, rrn*(2?)	(Gorton *et al.*, 1995; Tigges and Minion, 1994; Scamrov and Beabealashvilli, 1991)
M. hyopneumoniae	ApaI (9), Sno (9)	1070	5S separated from 16+23S	(Huang and Stemke, 1991)
M. capricolum (kid)	ApaI- X; KpnI x; BglI	1070	see also Chapt 42	(Whitley *et al.*, 1991)
M. capricolum (ATCC 27343)	BamHI, KpnI, SalI, XhoI, ApaI, MluI, BglI	1155		(Miyata *et al.*, 1991)
M. mycoides PG50	BglI, SalI, SmaI, XhoI	1040	Placement of several markers varies; (see next entry)	(Pyle *et al.*, 1990)
M. mycoides Y, V5, Glysd, KH_3J PG-1	BglI, SalI, SmaI, XhoI	1200–1280	Different sizes but comparable placement of markers; *rrn(2) rpoC*, 17 tRNAs, *atp*; *rpn*	(Pyle *et al.*, 1990)
M. mycoides GC1176-2	BglI, SalI, SmaI, XhoI	1380	Mapped multiple insertions of Tn*916*	(Whitley and Finch, 1989)

*#s in ( ) following an entry indicates the # of restriction enzyme fragments for that enzyme. See also (Herrmann, 1992)

References

Bautsch, W. 1988. Rapid physical mapping of the *Mycoplasma mobile* genome by two-dimensional field inversion gel electrophoresis techniques. *Nucl. Acids Res.* 16:11461–11467.

Bork, P., C. Ouzounis, G. Casari, R. Schneider, C. Sander, M. Dolan, W. Gilbert, and P. M. Gillevet. 1995. Exploring the *Mycoplasma capricolum* genome: A minimal cell reveals its physiology. *Mol. Microbiol.* 16:955–967.

Cao, J., P. A. Kapke, and F. C. Minion. 1994. Transformation of *Mycoplasma gallisepticum* with Tn916, Tn4001, and integrative plasmid vectors. *J. Bacteriol.* 176:4459–4462.

Colman, S. D., P.-C. Hu, W. Litaker, and K. F. Bott. 1990. A physical map of the *Mycoplasma genitalium* genome. *Mol. Microbiol.* 4:683–687.

Dybvig, K., and J. Alderete. 1988. Transformation of *Mycoplasma pulmonis* and *Mycoplasma hyorhinis:* Transposition of Tn916 and formation of cointegrate structures. *Plasmid* 20:33–41.

Fonstein, M., and R. Haselkorn. 1995. Physical mapping of bacterial genomes. *J. Bacteriol.* 177:3361–3369.

Fraser, C. M., J. D. Gocayne, O. White, M. D. Adams, R. A. Clayton, R. D. Fleischmann, C. J. Bult, A. R. Kerlavage, G. Sutton, J. M. Kelley, J. L. Fritchman, J. F. Weidman, K. V. Small, M. Sandusky, J. Fuhrman, T. R. Utterback, D. M. Saudek, C. A. Phillips, J. N. Merrick, J. Tomb, B. A. Dougherty, K. F. Bott, P. Hu, T. S. Lucier, S. N. Peterson, H. O. Smith, C. A. Hutchison III, and J. C. Venter. 1995. The *Mycoplasma genitalium* genome sequence reveals a minimal gene complement. *Science* 270:397–403.

Gorton, T. S., M. S. Goh, and S. J. Geary. 1995. Physical mapping of the *Mycoplasma gallisepticum* S6 genome with localization of selected genes. *J. Bacteriol.* 177:259–263.

Hedreyda, C., K. K. Lee, and D. C. Krause. 1993. Transformation of *Mycoplasma pneumoniae* with Tn4001 by electroporation. *Plasmid* 30:170–175.

Herrmann, R. 1992. Genome structure and organization. In *Mycoplasmas: Molecular Biology and Pathogenesis* . J. Maniloff, R. N. McElhaney, L. R. Finch, and J. B. Baseman, eds. pp. 157–168. *Am. Soc. Microbiol.* Washington, DC.

Huang, Y., J. A. Robertson, and G. W. Stemke. 1995. An unusual rRNA gene organization in *Mycoplasma fermentans* (incognitus strain). *Can. J. Microbiol.* 41:424–427.

Huang, Y., and G. W. Stemke. 1991. Construction of the physical maps of *Mycoplasma hyopneumoniae* and *Mycoplasma flocculare* and the location of rRNA genes on these maps. *Can. J. Microbiol.* 38:659–663.

Kapke, P. A., K. L. Knudtson, and F. C. Minion. 1994. Transformation of Mollicutes with single-stranded Tn4001 DNA. *Plasmid* 32:85–88.

King, K. W., and K. Dybvig. 1994. Transformation of *Mycoplasma capricolum* and examination of DNA restriction modification in *M. capricolum* and *Mycoplasma mycoides* subsp. mycoides. *Plasmid* 31:308–311.

Krause, D., and D. Taylor-Robinson. 1992. Mycoplasmas which infect humans. In *Myco-*

plasmas: Molecular Biology and Pathogenesis. J. Maniloff, R. N. McElhaney, L. R. Finch, and J. B. Baseman, eds. pp. 417–444. Am. Soc. Microbiol. Washington, DC.

Krause, D. C., and C. B. Mawn. 1990. Physical analysis and mapping of the *Mycoplasma pneumoniae* chromosome. *J. Bacteriol.* 172:4790–4797.

Krawiec, S., and M. Riley. 1990. Organization of the bacterial chromosome. *Microbiol. Rev.* 54:502–539.

Ladefoged, S. A., and G. Christiansen. 1992. Physical and genetic mapping of the genomes of five *Mycoplasma hominis* strains by pulsed-field gel electrophoresis. *J. Bacteriol.* 174:2199–2207.

Ladefoged, S. A., and G. Christiansen. 1994. Sequencing analysis reveals a unique gene organization in the *gyrB* region of *Mycoplasma hominis. J. Bacteriol.* 176:5835–5842.

Lucier, T. S., P.-Q. Hu, S. N. Peterson, X.-Y. Song, L. Miller, K. Heitzman, K. F. Bott, C. A. Hutchison, and P.-C. Hu. 1994. Construction of an ordered genomic library of *Mycoplasma genitalium. Gene* 150:27–34.

Maniloff, J. 1992. Phylogeny of Mycoplasmas. In *Mycoplasmas: molecular biology and pathogenesis.* J. Maniloff, R. N. McElhaney, L. R. Finch, and J. B. Baseman, eds. pp. 549–559. ASM Washington DC.

Mernaugh, G. R., S. F. Dallo, S. C. Holt, and J. B. Baseman. 1993. Properties of adhering and nonadhering populations of *Mycoplasma genitalium. Clin. Infect. Dis.* 17:S69–78.

Miyata, M., L. Wang, and T. Fukumura. 1991. Physical mapping of the *Mycoplasma capricolum* genome. *FEMS Microbiol. Lett.* 79:329–334.

Peterson, S. N., T. Lucier, K. Heitzman, E. A. Smith, K. F. Bott, P. C. Hu, and C. A. Hutchison. 1995. Genetic map of the *Mycoplasma genitalium* chromosome. *J. Bacteriol.* 177:3199–3204.

Pyle, L. E., T. Taylor, and L. R. Finch. 1990. Genomic maps of some strains within the *Mycoplasma mycoides* cluster. *J. Bacteriol.* 172:7265–7268.

Razin, S. 1985. Molecular Biology and Genetics of Mycoplasmas (Mollicutes). *Microbiol. Rev.* 49:419–455.

Razin, S., and E. Jacobs. 1992. Mycoplasma adhesion. *J. Gen. Microbiol.* 138:407–422.

Scamrov, A., and R. Beabealashvilli. 1991. *Mycoplasma gallisepticum* strain-S6 genome contains three regions hybridizing with 16-S-rRNA and two regions hybridizing with 23-S-rRNA and 5-S-rRNA. *FEBS Lett.* 291:71–74.

Taylor-Robinson, D., H. A. Davies, P. Sarathchandra, and P. M. Furr. 1991. Intracellular location of mycoplasmas in cultured cells demonstrated by immunocytochemistry and electron microscopy. *Int. J. Exp. Pathol.* 72:705–714.

Tigges, E., and F. C. Minion. 1994. Physical map of *Mycoplasma gallisepticum. J. Bacteriol.* 176:4157–4159.

Weisburg, W. G., J. G. Tully, D. L. Rose, J. P. Petzel, H. Oyaizu, D. Yang, L. Mandelco, J. Sechrest, T. G. Lawrence, J. Vanetten, J. Maniloff, and C. R. Woese. 1989. A phylogenetic analysis of the Mycoplasmas—Basis for their classification. *J. Bacteriol.* 171:6455–6467.

Wenzel, R., and R. Herrmann. 1989. Cloning of the complete *Mycoplasma pneumoniae* genome. *Nucl. Acids Res.* 17:7029–7043.

Wenzel, R., Pirkl, E., and R. Herrmann. 1992. Construction of an *Eco*RI restriction map of *Mycoplasma pneumoniae* and localization of selected genes. *J. Bacteriol.* 174:7289–7296.

Whitcomb, R. F., J. G. Tully, J. M. Bove, J. M. Bradbury, G. Christiansen, I. Kahane, B. C. Kirkpatrick, F. Laigret, R. H. Leach, H. C. Neimark, J. D. Pollack, S. Razin, B. B. Sears, and D. Taylor-Robinson. 1995. Revised minimum standards for description of new species of the class Mollicutes (division Tenericutes). *Int. J. Syst. Bacteriol.* 45:605–612.

Whitley, J. C., A. D. Bergemann, L. E. Pyle, B. G. Cocks, R. Youil, and L. R. Finch. 1990. Genomic Maps of Mycoplasmas and Tn916 Insertion. *Recent Advances in Mycoplasmo Zbl. Suppl.* 20:47–55.

Whitley, J. C., A. Muto, and L. R. Finch. 1991. A physical map for *Mycoplasma capricolum* Cal Kid with loci for all known transfer RNA species. *Nucl. Acids Res.* 19:399–400.

Whitley, J. C., and L. R. Finch. 1989. Location of sites of transposon-Tn916 insertion in the *Mycoplasma mycoides* genome. *J. Bacteriol.* 171:6870–6872.

42

Mycoplasma capricolum Genome Project

P. M. Gillevet, A. Ally, M. Dolan, E. Hsu, M. S. Purzycki, R. Overbeek, E.E. Selkov, S. Smith, C. Wang, and W. Gilbert

Background

The Mycoplasmas are very small, wall-less bacteria phylogenetically related to gram-positive Eubacteria such as *Bacillus subtilis* (*see also* Chapters 40 and 41). *Mycoplasma capricolum* is an example of one of the smallest of free-living organisms (Ryan and Morowitz, 1969) with a genome estimated to be between 724 kb (Poddar and Maniloff, 1989) and 1.1 megabases. As *M. capricolum* is a parasitic organism with a truncated metabolism and can be grown in a defined medium, much of its truncated physiology has been biochemically defined (Maniloff, 1992). The acquisition of the entire genome sequence of the organism will corroborate these classic biochemical studies and allow the complete elucidation and eventual modeling of its truncated metabolism. Furthermore, the comparative analysis of this metabolic network with larger metabolic networks from organisms such as *Haemophilus influenzae* would open the door to the unprecedented opportunity to begin to analyze the minimal set of fundamental genes involved in the process we call "life." (*see* Chapter 37).

The *Mycoplasma capricolum* genome project originated at Harvard University where we developed a novel DNA sequencing strategy termed Multiplex Genomic Walking. The first two years of the Harvard project were spent on sequencing technology development, as described below. A production line to sequence the genome of *Mycoplasma capricolum* was implemented during the third year of the project and resulted in the generation of over a million raw bases of data that assembled into contigs covering some 250,000 linear bases. The project was successful in developing the technology to directly sequence genomes approaching a million bases in size and defining the standard operating procedures, informatics support, appropriate process control and quality assurance to run an integrated production facility. Lastly, the project defined two technical limitations in the walking process which affected the overall throughput of the project, specifically the oligo failure rate and the autoradiographic signal strength.

Multiplex Genomic Walking

The Multiplex Genomic Walking technique reveals the DNA sequence of the organism directly, essentially by hybridization of a Southern blot (Ohara *et al.*, 1989). To produce these sequencing blots, the genomic DNA of the entire organism is completely digested with restriction enzymes, treated with the chemical sequencing reactions, electrophoresed through a sequencing gel, and transferred to a charged nylon membrane. Each genomic restriction digest is represented on the membrane by a set of sequence lanes. When such a membrane is probed with a labeled oligonucleotide, the resulting autoradiograph displays sequencing patterns in those lanes in which the oligonucleotide has hybridized near a restriction cut. Sequence can be read out, in both directions, from the position of the oligonucleotide, on one strand of the DNA.

Figure 42-1 is a schematic that summarizes the Multiplex Genomic Walking strategy. An oligo probe is selected based on a known starting sequence and hybridized to a membrane bearing the chemically sequenced DNA from genomic restriction enzyme digests. Hybridization of this probe to the membrane reveals a single sequence ladder in each restriction enzyme digest where the probe hybridizes near the end of the restriction fragment. Sequence ladders are read in the 5′ to 3′ direction away from the probe along one strand of the DNA when the probe hybridizes near the 5′ end of a restriction fragment. Conversely, when the probe hybridizes near the 3′-end of a fragment, sequence ladders are read "backwards" in the 3′ to 5′ direction. Thus a single probing produces several "Reads" (*see* Figure 42-1) or sequence fragments which assembled into two clusters, one cluster reading in the 3′ to 5′ direction, the other cluster reading in the 5′ to 3′ direction away from the probe.

A simple majority consensus is generated from these reads and the next oligo probe is chosen from the complementary strand for the subsequent hybridization. Figure 1 illustrates a walk proceeding from the 3′ end of the contig. In this case, the second hybridization will yield a sequence cluster reading in the 3′ to 5′ direction away from the probe that provides coverage over the first step on the opposite strand. Thus as the walk proceeds on a contig, the probing of alternate strands results in multiple coverage on both strands of DNA.

Production Cycle

The production process for Multiplex Genomic Walking depicted in Figure 42-2 involves a repetitive cycle of:

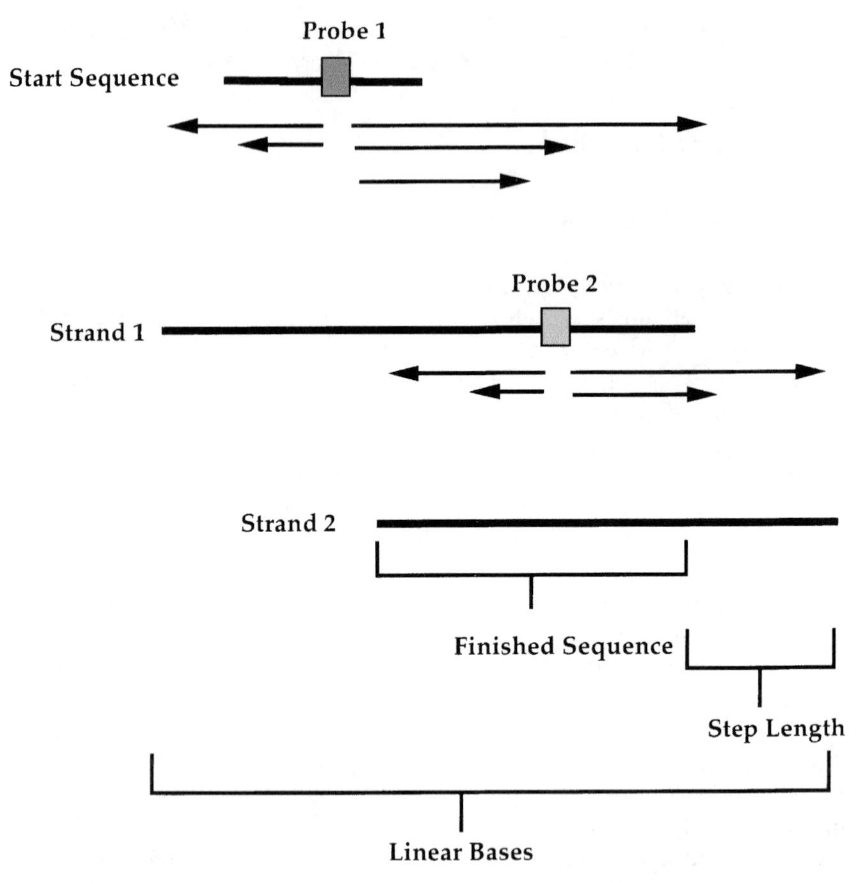

Figure 42-1. Multiplex Genomic Walking Strategy: An oligoprobe is selected based on a known sequence and hybridized to a genomic sequencing membrane. Sequence ladders read in the 5′ to 3′ direction away from the probe along one strand of the DNA when the probe hybridizes next to the 5′ restriction enzyme cut at position. Conversely, when the probe hybridizes near the 3′ restriction enzyme cut at position B, sequence ladders read "backwards" in the 3′ to 5′ direction from the position of the probe. A simple majority consensus is generated from these sequence reads and the next oligo probe is chosen from the complementary strand for the subsequent hybridization. This second hybridization yields sequence reads which extend the consensus in the 5′ to 3′ direction. We define Step Length as the difference in the lengths of Consensus 1 and Consensus 2 sequences. We define Finished sequence as those nucleotide positions in the contig where the majority consensus of both strands agree. It should be noted that the ends of each growing contig are represented by the consensus of only one strand (the region from the distal probe binding site to the end of the contig) and that these positions do not meet the criteria of Finished. Linear sequence is defined as the sum of the lengths of these individual single strand consensus sequences and the lengths of Finished sequences for all contigs.

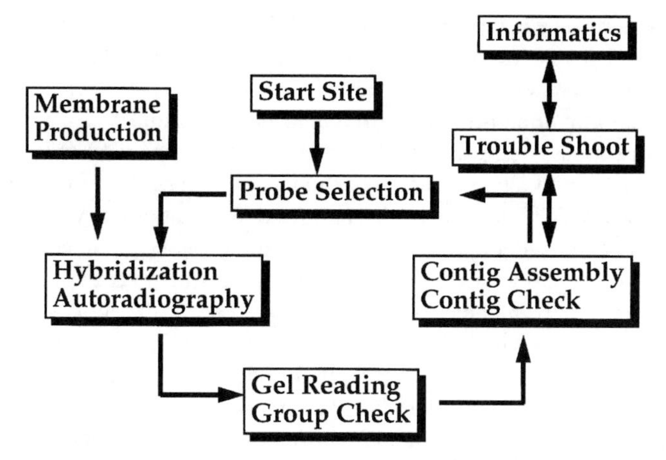

Figure 42-2. The Production Cycle: The repetitive process involved in Multiplex Genomic Walking is shown. It involves the synthesis of an oligonucleotide, the hybridization of that probe onto a membrane, washing and exposure of the membrane, stripping the membrane, reading the sequence, synthesizing a new oligonucleotide, and reprobing the membrane.

- hybridizing an oligo probe to a sequencing membrane,
- manual reading of the autorads and checking the raw data,
- assembling the contig and checking the assembly, and
- picking and synthesizing of new probes to continue the process.

Figure also illustrates the initial production of starting sequences, the membrane generation, the trouble shooting, and quality control functions of the process. The generation of each of these aspects are presented below.

Overall Process Tracking and Control

This component consists of the computational software and hardware to manage the sequence data, the bar coding, and the computerized database for the laboratory. These systems track all of the oligonucleotides, the membranes, and the sequence fragments as well as quality control information (Gillevet, 1993). The Genetic Data Environment (GDE), an X Windows based Graphic User Interface (GUI) was used for the maintenance of our internal database and automated data control systems (Smith *et al.*, 1994). This system allowed the seamless integration of a core multiple sequence editor with pre-existing external sequence analysis programs and newly developed programs into a single prototypic environment. A shell tool with the same expandable menu system as GDE was developed as an interface for technicians to input all tracking information. This model proved extremely flexible and not only allowed direct access to remote procedure calls on high-performance computers, and network-based servers, but was easily modified on a daily basis as the need arose. The latter point was critical in the development of a new technology where the needs of the system cannot predicted *a priori* (see http://uranus.gmu.edu/Mycoplasma/GDE.html).

The entire process was tracked by changing the status field associated with each individual probe. Every aspect of the walking process, from probe picking to data entry had a GUI to control it and allowed information entry in a systematic and consistent manner (Figure 42-3). Tools were developed to query the internal database, to change the status of probes, to assign membranes and probes for hybridization, and to enter data. Our internal database was a set of "tagged" fields that was linked with each of the sequences, whether that sequence corresponded to an oligonucleotide or a sequence fragment. If a field was not used by the program module being run it was not lost but was tracked by each module as an inert link. The concept of a "tagged field" allowed us to freely expand the system as the need arose. We again emphasize that this proved to be a crucial issue, as we constantly refined the walking process in the wet lab, and the informatics system was modified on a day-to-day basis.

Production of Starting Sequences

Access to about 1,000 randomly arranged sequences around the *Mycoplasma* genome was needed in order to have a continuous supply of starting points to walk from and to statistically cover the entire genome. To develop that set of sequences, we cloned *M. capricolum* DNA into M13 clones and sequenced random clones with an ABI fluorescent sequencer. We sequenced some 1,505 random clones from the organism for random start points which produced 187,309 raw bases of DNA sequence.

Oligonucleotide Production

The Multiplex Genomic Walking strategy is based on oligonucleotide hybridization to sequencing membranes. Therefore, the efficiency of the method is dependent on both the rate and cost of oligonucleotide synthesis. To address these factors, we modified the cycles on the Millipore two-column "Cyclone" synthesizer to simplify and shorten the cycle time. These machines originally had an eight-minute cycle time in a two-column mode (producing about 12 oligos per 8 hour day). We eliminated the capping steps, shortened each chemical step, used TCA to speed the detritylation, and increased the gas pressure to increase the flow rate. We developed a 90-second cycle time, for the two-column mode, and thus were able to synthesize two 20-mers in 30 minutes.

The two-column Cyclone synthesizers were fitted with a RS232 serial board and connected to the Sun computer system via a serial line so that the programs and the oligonucleotide sequences were downloaded directly. The machines were

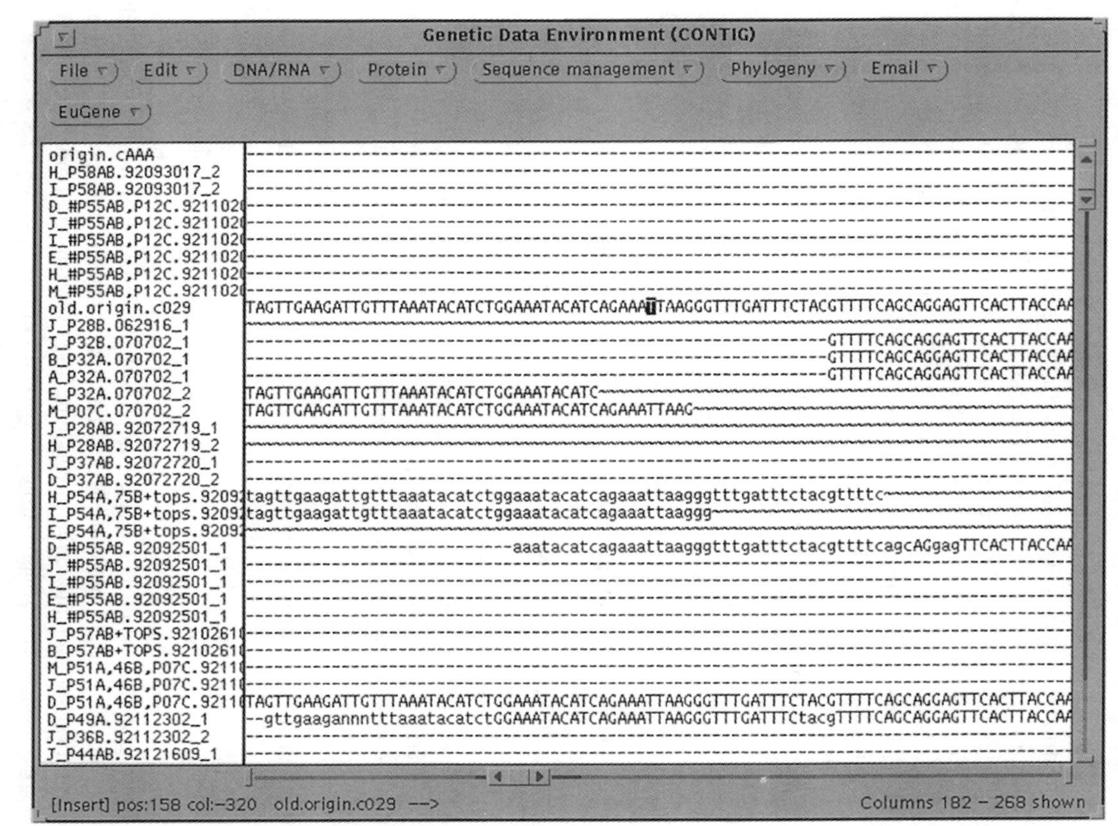

Figure 42-3. The Genetic Data Environment: GDE was used to track the entire walking process by changing the status field associated with each individual probe. A typical window is depicted showing sequence fragments going in both directions from a probe site (see http://uranus.gmu.edu/Mycoplasma/GDE.html).

controlled by a barcode reader attached to a VT100 terminal allowing automated programming from a file of oligo sequences. Using our short cycle and two modified cyclone machines, 32 to 36 oligos were synthesized per day at an estimated reagent cost of $12.70 for a 20-mer oligo. A total of 3,782 total oligonucleotides were synthesized during the whole project and 2,443 of these were used as walking probes.

Choice of Restriction Enzymes and Chemistries: Mycoplasma capricolum

(California kid) strain ATCC #27343, is a very AT-rich and thus we optimized the restriction enzymes digests to produce appropriate fragment size distributions. As current gel technologies limit read lengths to the range of 400-500 nucleotides, the optimum strategy for Multiplex Genomic Walking is to use mixtures of enzymes that cut on the average every 600 to 1,000 bases. We used nine convenient sets of enzymes for the production effort which provided the ability to "walk" across all regions of the genome.

Sequencing the AT-rich genome of *M. capricolum* (70% AT) also required modification of standard chemical sequencing methods developed for DNA with a higher GC content. Six redundant chemical modification and cleavage reactions (G, G>A, A>C, C, C+T and T) were performed for each restriction enzyme digest to ensure accurate reading of the sequence ladder. Chemical sequencing reaction conditions were optimized to (1) accommodate the higher AT content of *M. capricolum* DNA, (2) enable maximum reproducibility of the chemical sequencing ladders generated from restriction fragments of about 500 base pairs, and (3) allow simultaneous processing of multiple samples (Dolan *et al.*, 1995).

Indirect Transfer Electrophoresis

Direct blotting electrophoresis technology enables the collection of DNA fragments on a nylon membrane concurrent with their size fractionation by denaturing gel electrophoresis. We developed a modification of direct blotting electrophoresis that we call "indirect transfer electrophoresis (ITE) to produce membranes that had sequence data resolved to about 350 bases. We used instruments from Betagen, a design close to that of the original Pohl and Beck device, in which the sequencing membrane is supported by a nylon mesh. Pall Biodyne plus, a positively charged nylon, was routinely used for all the production sequencing membranes in this project.

Resolution of the sequencing ladders was enhanced by altering the gel composition to include 40% formamide in both the gel and the lower buffer chamber. These gels ran about 30% faster than conventional urea gels and we routinely resolved about 30% more sequence on the membranes with this method. A further innovation was to remove the direct contact between the membrane and the gel. ITE gels were recessed approximately two millimeters from the edge of the glass plates and this region was filled with the formamide containing buffer. As the membrane passed across the ends of the glass plates, the DNA bands left the gel, migrating undisturbed through the liquid layer, and impinged on the nylon membrane. The lack of physical contact between the gel and the membrane prevented mechanical damage to the gel by the passage of the membrane, which can tear off gel fragments. Using this technology, we routinely made sequencing membranes from which we were able to read out to 300 to 350 bases from the restriction site.

Sequence Reading and Assembly

In this component, the contigs were assembled, proofread, and new oligonucleotides were predicted and queued for synthesis. The autoradiographs were read manually and the sequence information was entered directly into a GDE interface. The set of "reads" from a single oligonucleotide were assembled within the interface and double-checked against the autoradiograph. Subsequently, these sets of "reads" were assembled onto the contigs automatically by our system.

Our sequence assembly problem was much simpler than that of a shotgun sequencing project, since the Multiplex Genomic Walking strategy is a directed walk. We were able to keep track of the growing point on each contig along with the corresponding oligo that was probing that growing point by tracking a link between each probe, each membrane, and each film in the database. Therefore, we knew from which contig each sequence cluster had been derived and consequently to which contig it had to assemble to. As described above, a hybridization generated sequence fragments that assembled into two clusters, one going 5′ to 3′ away from the probe hybridization site and the other going 3′ to 5′ away from this site. These two clusters were assembled and then compared automatically to the sequences at the ends of the contig, to determine which sequence was the complement to existing sequence and which was *de novo* data that extended the contig across the chromosome. Subsequently, a new consensus sequence was determined for each contig.

New oligos were selected using a computer routine that identifies, in the new consensus, the last restriction site (from the enzymes set used in the project) in the forward direction and then selects a 20-base long sequence for the next oligonucleotide in the region just beyond the last restriction site. The oligo selection was subject to appropriate criteria of melting temperature and secondary structure. The list of new oligonucleotides to be synthesized was sent through the computer system to the oligonucleotide synthesizers automatically and inventoried in the database.

Troubleshooting

We determined a consensus sequence, and checked each growing point, so that after one or two cycles the only portion of the sequence that was ambiguous was that portion at or near the growing point which was not yet covered on the opposite strand. The sequence behind that was confirmed on both strands and considered completed. Mismatches between the two strands, and regions covered only by one strand, were color-flagged for identification. If a mismatch between the two strands could not be resolved by re-examination of the raw data or the coverage of the other strand was not obtained by the current probing, a new oligonucleotide was selected, synthesized, and hybridized to the membranes to obtain the additional sequence information to clarify these regions.

When the growing points overlapped or crossed each other, we treated those as independent sequences. We did not form a consensus initially between the two growing points, but rather checked the sequences for accuracy. Since the two growing points crossed a given region in different directions and using different oligonucleotides, they represented independent sequences of the same region, and the agreement between those sequences represents an independent test of the accuracy of the sequencing.

Quality Control

The issue of quality control was a far greater problem that initially realized at the beginning of this project. The question of how reproducibly things were done and whether people actually knew, or recorded exactly how they did each experiment, turned out to be a much larger problem than anticipated. It was not just that records must be kept, but there must be a way of assuring that people actually did the experiments exactly the way they recorded them. These are classic issues of "General Manufacturing Process" management and without these checks one could not troubleshoot the experiment if it went wrong or be guided clearly on the development path. We developed the use of bar codes and sign-offs in SOP's "Standard Operating Protocols" (SOPs) to provide quality assurance. A large fraction of our effort went into these problems of control and assurance to verify that the machines have worked correctly or to learn why a particular procedure had stopped working.

Data Analysis

We collaborated with Peer Bork on the initial analysis of the *M. capricolum* data using the Genquiz developed at EMBL (Bork *et al.*, 1995). The assembled contigs were subjected to Blastx searches and the Blastx output was automatically parsed to check for frameshifts and artificial stop codons using the program "Frameshift". All possible open reading frames (ORFs) longer than 10 amino acids were predicted and translated. The requirement for recording ORFs within the contigs was the presence of start and stop codons; in terminal fragments only start (C-terminus) or stop codons (N-terminus) had to occur. BLAST homology searches was carried out on the resulting 1845 ORFs. DNA sequence databases were then screened for non-coding elements such as RNAs and internal repeats and all results were stored in a relational database for further statistical evaluation (see http://uranus.gmu.edu/Mycoplasma/EMBL.FunctionTable.html).

We collaborated with Ross Overbeek, Terry Gaasterland, and Natalia Maltsev to develop a browsing engine called AUTOSEQ in an attempt to incorporate heuristics into the primary identification problem. This semi-automated tool was developed to sort through the output of various search engines and putative identification of each orf is presented in a html browser (Figure 42-4a) which links to the alignments supporting the assertion (Figure 42-4b). We used the initial Genquiz identifications to set the cutoff parameters for the various tools and succeeded in automating the primary identification considerably (see http://www.mcs.anl.gov/home/gaasterl/autoseq/Reports/mcj1995/SUMMARY-mcj1995.html). The latest version of this tool called MAGPIE (see Chapter 45) has an improved user interface and an updated browsing engine (see http://www.mcs.anl.gov/home/gaasterl/magpie.html).

Evgeni Selkov and his group have worked for many years developing the database of Enzymes and Pathways which is a collection of data encoding all facts relating to enzymology and metabolism. Ross Overbeek and his group at Argonne National Laboratory have been working with Dr. Selkov, helping him to prepare the collection for distribution from within his integrated system of Phylogeny, Metabolism and Alignments (PUMA). In 1994, P.M. Gillevet proposed to examine the question of what could be learned about the metabolism of an organism from the sequence data. The first organism that was selected was *Mycoplasma capricolum* with an estimated third of the genome sequenced (see http://www.mcs.anl.gov/home/compbio/PUMA/Production/ReconstructedMetabolism/reconstruction.html).

Once one has arrived at an estimated set of metabolic pathways (see Figure 42-5), it is possible to use EMP to derive a set of enzymes that have not yet been located, but are implied by the pathways (or, more precisely, one can produce a list of implied enzymes and a list of possibly occurring enzymes). This list can be used to support a more sensitive analysis of the new sequence data. That is one takes all existing versions of these predicted enzymes and searches the new sequence data for similarities with lower thresholds (in contrast to the initial searches, which took new sequence and searched entire repositories for similarities.

Recently, the PUMA system has been expanded into a new technology that translates sequenced genome information into a consistent model of metabolic and functional organization for a bacterial cell. This technology is based on "What Is There" (WIT), which uses a sequenced genome, detected similarities, and a collection of about 2000 metabolic and functional charts from the Database on Enzymes and Metabolic Pathways (EMP) to construct a model of cellular organization represented as a set of pathway and function diagrams and assertions connecting specific regions of sequence to roles in the diagrams. This model is

A **AUTOSEQ OUTPUT FOR HIT AGAINST ELONGATION FACTOR G**

ORF	ID	LVL	FR	TOOL	WHERE	DB	SCORE	DESCRIPTION
142+660 1	EFGC_PEA	1	+3	blaize	642+1046	7+141	p(0.0)	CHLOROPLAST (EF-G)
	EFGC_SOYBN	1	+3	blaize	642+1184	94+274	p(0.0)	CHLOROPLAST PRECURSOR
	EFGM_RAT	1	+3	blaize	660+1199	46+228	p(0.0)	MITOCHONDRIAL PRECURSOR
	EFG_ANANI	1	+3	blaize	636+1193	1+186	p(0.0)	ELONGATION FACTOR G
	EFG_BACST	1	+3	blaize	636+791	1+53	p(0.0)	ELONGATION FACTOR G.
	EFG_MICLU	1	+3	blaize	654+1187	5+182	p(0.0)	ELONGATION FACTOR G
	EFG_MYCLE	1	+3	blaize	654+1232	10+204	p(0.0)	ELONGATION FACTOR G
	EFG_SPIPL	1	+3	blaize	636+1193	1+186	p(0.0)	ELONGATION FACTOR G
	EFG_THEMA	1	+3	blaize	639+1184	6+187	p(0.0)	ELONGATION FACTOR G
	EFG_THETH	1	+3	blaize	645+1280	6+216	p(0.0)	ELONGATION FACTOR G

B **LINK TO ALIGNMENTS**

```
ID   EFGC_PEA       STANDARD;      PRT;   141 AA.
          DE   ELONGATION FACTOR G, CHLOROPLAST (EF-G) (FRAGMENT).
!!!
RESULT   11     Score 786;  Match 0.0%;  Predicted No. 0.00e+00;

ID   EFGC_PEA       STANDARD;      PRT;   141 AA.
DE   ELONGATION FACTOR G, CHLOROPLAST (EF-G) (FRAGMENT).

          Matches 100;  Mismatches 20;  Partials 15;  Indels 0;  Gaps 0;

           *  .*  . *******.************** **.  **** ***.. **** ********
Db      7 RAVPLKDYRNIGIMAHIDAGKTTTTERILFYTGRNYKIGEVHEGTATMDWMEQEQERGIT 66
Qy    214 REYSLLNTRNIGIMANIDAGKTTTTERILFHTGKIHKIGETHEGASQMDWMAQEQERGIT 273

           *******.**  .  * ************.****.*******.  . *.  .*****.******
Db     67 ITSAATTTFWDKHRINIIDTPGHVDFTLEVERALRVLDGAICLFDSVAGVEPQSETVWRQ 126
Qy    274 ITSAATTAFWKNTRFNIIDTPGHVDFTVEVERSLRVLDGAVAVLDGQSGVEPQTETVWRQ 333

           *  * **** *****
Db    127 ADRYGVPRICFVNKM 141
Qy    334 ATNYRVPRIVFVNKM 348
```

Figure 42-4. Autoseq: A semi-automated tool was developed to sort through the output of various search engines. A putative identification of a orf is presented in a html browser (a) with links to the alignments supporting the assertion (b).

- **GENERAL METABOLISM**
  - **BIOENERGETICS AND METABOLISM**
    - **METABOLISM OF CARBOHYDRATES**
      - **Metabolism of polysaccharides**
      - **Metabolism of disaccharides**
      - **Metabolism of monosaccharides**
      - **Metabolism of aminosugars**
      - **Metabolism sugar alcohols**
      - **Metabolism of monocarbon compounds**
      - **Main pathways of carbohydrate metabolism**
      - **Pyruvate dehydrogenase complex**
      - **TCA**
      - **Metabolism of TCA intermediates**
    - **ATP BIOSYNTHESIS**
      - **ATP transport**
    - **METABOLISM OF AMINO ACIDS AND RELATED MOLECULES**
      - **Protein degradation**
      - **Degradation of oligopeptides**
      - **Catabolism of the amino acids**
      - **Amino Acid biosynthesis**
    - **METABOLISM OF NUCLEOTIDES AND NUCLEIC ACIDS**
      - **Degradation of the nucleic acids**
      - **Biosynthesis of nucleotides**
    - **METABOLISM OF LIPIDS**
      - **Degradation of lipids**
      - **Lipids biosynthesis**
    - **ELECTRON TRANSPORT**
      - **Oxidative phosphorylation**
    - **METABOLISM OF COENZYMES AND PROSTHETIC GROUPS**
      - **Coenzymes**
      - **Metabolism of sulfur**
      - **Phosphate metabolism**
    - **TRANSMEMBRANE TRANSPORT**
      - **Active transport**
      - **Group translocation**
      - **Other pathways of transmembrane transport**
    - **SIGNAL TRANSDUCTION**

Figure 42-5. Functional Classes of the Metabolic Network of *M. capricolum:* The putative identifications made with Genquiz and Autoseq were used to infer the set of metabolic pathways that exist in the organism. The results are accessed via an active html display that has links to the individual enzymes in each pathway along with the alignments associated with that inference (see http://www.mcs.anl.gov/home/compbio/PUMA/Production/ReconstructedMetabolism/reconstruction.html).

an attempt to reconcile the sequence data, the phylogenetic context, and the phenotypic and biochemical knowledge of the sequenced organism. An effort has been initiated to organize what is known of the metabolism for a number of the organisms for which substantial sequence has been released to the research community (see http://uranus.gmu.edu/WIT/wit.html). Several examples of the prediction from the metabolic reconstruction have been proven correct and it is hoped that further refinement of the system will enhance its robustness. Finally, as all theoretical prediction from the reconstruction must be confirmed experimentally, we are in the process of developing tools to identify critical enzymatic steps in the metabolic network that can be biochemically verified in the wetlab.

Summary

We accumulated over a million raw bases (1,039,095) of *Mycoplasma capricolum* sequence during the project at Harvard that assembled into a quarter of a million linear bases (267,686 bp). We have analyzed 372 non-overlapping contigs covering 214,528 base pairs and identified 220 open reading frames in the organism (Bork *et al.*, 1995). Only 61 frameshifts and aberrant stop codons were identified in 103,000 bases contained in the analyzed orfs indicating the error rate of our finished data is less than 10^{-3}. The identification of 220 distinct proteins revealed the minimum number of proteins encoded by the 372 contigs. At the DNA level, numerous matches with tRNA, rRNA and snRNA-like sequences were found. The current analysis of the *Mycoplasma capricolum* genome can be found on the World Wide Web at "http://uranus.gmu.edu/myc-collab.html" and the reconstruction of the metabolic network can be found at "http://www.mcs.anl.gov/cgi-bin/overbeek/Production/selkov_recon.cgi?Mycoplasma%20capricolum+evidence."

The 220 distinct proteins represent nearly half of the total number of about 500 proteins expected in *M. capricolum* (Muto, 1987). Furthermore, we identified about 35% of the known infrequently occurring restriction sites in the organism in the 215,000 bases that were analyzed indicating that the size of the *M. capricolum* genome is on the order of 765 kb and that we sequenced close to a third of the genome (Bork *et al.*, 1995).

Technical Problems with Multiplex Genomic Walking

Two related technical problems with the process as implemented at Harvard were encountered. The first was a high failure rate of hybridization, that is 40% of the probes either failed to hybridize or gave very weak, unreadable signal. The hybridization failures were due to picking oligos in inaccurate sequence at the end of the growing contig such that the oligos failed to hybridize. This problem will plague any directed sequencing approach that picks probes or primers on single stranded coverage and may be unavoidable, that is one may have to accept this inherent failure rate. Picking probes on better quality regions (multiple coverage) would help but in our technique there was a negative tradeoff between the rate of stepping forward and picking the probe from confirmed multiple covered sequence. Specifically, the step size was decreased when probe selection was from regions of multiple coverage which are further in from the end of the growing contig.

Many of the oligonucleotides that gave weak signal were synthesized from accurate sequence in retrospect and it is still undetermined why they failed to produce stronger signals. There was no canonical secondary structure involved in these probes (potential probes are checked for stability, hairpins and self complementarity before they are made) but the majority do have a high T content and it is hypothesized that these failures are due to non-canonical secondary structures (non-Watson-Crick base pairing).

The second technical problem was related to the overall signal intensity of the autorads. Signal strength was variable with many reads having only a signal/noise ratio of about three. This issue dictated that the gels be read by hand and probably contributed to the above hybridization failure rate. Apparently other factors, in addition to those criteria addressed in our probe selection programs, are components in the efficiency of hybridization and the generation of signal when oligonucleotide probes are hybridized to charged nylon membranes. The membranes could have been exposed longer to increase the signal as the membranes were only exposed overnight in the production process but this would have led to a severe disruption of the production process. We are presently looking into alternative detection methodologies to alleviate this bottleneck in the strategy.

Advantages of Technology

The project at Harvard has proven that by repeated probing of the same set of membranes one can walk around the genome of a small organism. The simple repetitive process involved in sequencing then is to synthesize an oligonucleotide, hybridize it onto a membrane, wash and expose the membrane, strip the membrane, read the sequence, synthesize a new oligonucleotide, and reprobe the membrane. The membranes were reused many times; the present set of membranes were hybridized around 70 times, with no diminution of signal strength. The repetitive process was very simple, and the ultimate rate of sequencing depended on the number of membranes that could be handled at once, and the length of the read achieved from each probing of a membrane.

A major advantage of this technology, especially as it applies to microorganisms with genomes less than a million bases, is that the organism's genome is sequenced the organism directly. This avoids ambiguities due to artifacts that could be introduced in the process of cloning and simplifies the closure problem in that there are no "unclonable" regions to analyze. A second advantage is that both DNA strands are examined directly, and thus the sequence is verified and the presence of modified bases, such as methylated C residues is readily observed. Finally, because this is a linear walking procedure, the computer assembly of the sequence is a straightforward one and does not involve the great complexities that arise with shotgun sequencing of very large organisms.

In conclusion, the technology presented in this report can be used to directly sequence small bacterial genomes or entire YACS or cosmids. Furthermore, the rationale for the Multiplex Genomic Walking will be applicable to novel sequencing methods now being developed using new fragment separation techniques and more sensitive detector systems.

Acknowledgments

We would like to thank all members of the Harvard Genome Lab, F. Barton, S.E. Brenner, R. Clark-Whitehead, N. Douglas, L. Marquez, B. Richter, J. Sartell, and J. Williams, for their efforts over the course of the *Mycoplasma capricolum* sequencing project to make this new technology a reality. We would like to thank the members of the Argonne team, T. Gaasterland and N. Maltsev, who were critical to the development of Autoseq. Finally, we would like to thanks all the members of Selkov team who have developed EMP and laid the foundation for the work on metabolic reconstruction. This work was supported by NIH Grant R01 HGD0124.

References

Bork, P., C. Ouzounis, G. Casari, C. Sander, M. Dolan, W. Gilbert, and P. M. Gillevet. 1995. Exploring the *Mycoplasma capricolum* genome: A parasite reveals its physiology. *Mol. Microbiol.* 16:955–967.

Dolan, M., A. Ally, M. S. Purzycki, W. Gilbert, and P. M. Gillevet. 1995. Large Scale Genomic Sequencing: Optimization of Genomic Chemical Sequencing Reactions. *BioTechniques* 19(2):264–273.

Gillevet, P. M. 1993. Integration of the Wet Lab and Data flow in Multiplex Genomic Walking. *Proceedings of the Second International Conference of Bioinformatics, Supercomputing and Complex Genome Analysis*, A. Hua, ed. World Scientific Publishing Co. River Edge, NJ.

Maniloff, J., R. N. McElhaney, L. R. Finch, and J. B. Baseman, eds. 1992. *Mycoplasma: Molecular Biology and Pathogenesis*, American Society for Microbiology, Washington DC.

Muto, A., F. Yamao, and S. Osawa. 1987. The genome of *Mycoplasma capricolum. Progr. in Nucl. Ac. Res.* 34:28–58.

Ohara, O., R.L. Dorit, and W. Gilbert. 1989. Direct genomic sequencing of bacterial DNA: The pyruvate kinase I gene of *Escherichia coli. Proc. Natl. Acad. Sci. USA* 86:6883–6887.

Poddar, A. K., and J. Maniloff. 1989. Determination of microbial genome sizes by two-dimensional denaturing gradient gel electrophoresis. *Nucl. Ac. Res.* 8:2889–2895.

Ryan, J. L., and H. J. Morowitz. 1969. Partial purification of native rRNA and tRNA cistrons from Mycoplasma sp. (Kid). *Proc. Natl. Acad. Sci. USA* 63(4):1282–1289.

Smith, S. W., R. Overbeek, C. R. Woese, W. Gilbert, and P. M. Gillevet. 1994. The Genetic Data Environment (GDE): An Expandable Graphic Interface for Manipulating Molecular Information. *CABIOS* 10 (6):671–675.

43

Sequence Skimming of Chromosome II of *Rhodobacter sphaeroides* 2.4.1^T

Christopher Mackenzie, Monjula Chidambaram, Madhusudan Choudhary, Kirsten S. Nereng, Samuel Kaplan, and George M. Weinstock

Rhodobacter sphaeroides is a photosynthetic member of the α-3 subgroup of Gram-negative bacteria (Woese, 1987). It is distinguished by a number of important characteristics which include at least six modes of growth. These accompany the ability of the organism to display a diverse range of metabolic activities, as well as other notable characteristics with respect to genome organization, evolution, and other processes (Table 43-1). This metabolic diversity may have evolved from a need to synthesize *de novo* a different range of compounds under each set of growth conditions.

An important characteristic of *R. sphaeroides* 2.4.1 is that it possesses two chromosomes (Suwanto and Kaplan, 1989a; b), a feature which is found in, but is not unique to, other members of this group i.e. *Brucella melitensis* (Michaux *et al.*, 1993; *see* Chapter 75) and *Agrobacterium tumefaciens* (Allardet-Servent *et al.*, 1993) and may also be considered to include the rhizobia which contain megaplasmids (Sobral *et al.*, 1991; *see* Chapter 74). The larger of the two chromosomes (3.0Mb, CI) appears to be similar to those of most other bacteria. However, we have insufficient knowledge of the smaller chromosome (0.9Mb, CII) to understand the significance of this mode of genetic organization. It is known that both chromosomes have rRNA operons (Dryden and Kaplan, 1990) and mutations causing auxotrophy have been mapped to both CI and CII (Choudhary *et al.*, 1994). In this sense CII does not appear to be specialized for a particular phase or portion of the bacterial lifestyle. A number of genes are known to be duplicated between the chromosomes, i.e., $cbbP_I$ and $cbbP_{II}$ (Hallenbeck *et al.*, 1990a; Tabita *et al.*, 1992), $cbbA_I$ and $cbbA_{II}$, $cbbG_I$ and $cbbG_{II}$ (Hallenbeck *et al.*, 1990b; Tabita, *et al.*, 1992), *hemA* and *hemT* (Neidle and Kaplan, 1993) and *rdxA* and *rdxB* (Neidle and Kaplan, 1992; Zeilstra-Ryalls and Kaplan, 1995) and in all cases to date each member of the pair is differentially regulated. Gene regulation in *R. sphaeroides* would thus seem to either contribute to its metabolic versatility or is a product of its adapation.

Table 43-1.

Reasons for studying *R. sphaeroides.*	
Growth modes	Anaerobic conditions
	—photoheterotrophy
	—photoautotrophy
	—chemoheterotrophy
	—fermentation
	Aerobic conditions
	—chemoautotrophy
	—chemoheterotrophy
Metabolism	Aminolevulinic acid synthase
	(eucaryote like)
	Rubisco (plant like)
	Calvin cycle genes
	Tetrapyrrole biosynthesis
	—heme
	—siroheme
	—bacteriochlorophyll
	—vitamin B12
	Carotenoid biosynthesis
Gas exchange	CO_2 fixation
	N_2 fixation
	N_2O, NO, NO_2, NO_3 reduction
	H_2 generation
Evolution	Related to mitochondria
	—cytochrome c_2
	—b-c_1 oxidoreductase
	—ATPase
	—cytochrome oxidase
	—enoyl-CoA hydratase
	—peripheral benzodiazopene like receptor protein
	Reaction center similar to plant photosytem I
	Yeast oxidase assembly factors
Movement	Phototactic
	—sensory transduction
	—unique flagella
Genome	2 chromosomes
	High %G+C
	Z-DNA
	Evolution of diploidy
	Chromosome replication
Bioremediation	Reduction of rare-earth oxides
Lipopolysaccharide	Non-toxic

The most direct way to learn more about CII is to determine its DNA sequence. To do this we have initially adopted a "sequence skimming" approach, where a low redundancy DNA sequence is obtained from cosmids that have been mapped around the chromosome. The sequences are used to search databases for similarities to known genes. A low resolution physical and genetic map for each cosmid is then deduced, and in this way a map of CII is eventually constructed showing the sequence similarities and possible gene functions. This is in the same spirit of abstraction as genetic maps constructed before gene cloning and sequencing became widespread. This method is also considerably less expensive both in terms of labor and materials than the approaches used to obtain high resolution sequences, yet still yields considerable information.

Methodology

The methodology developed and used for the sequence skimming approach is described in the following example. A genomic library of *R. sphaeroides* 2.4.1 was constructed in the broad host-range cosmid vector pLA2917 (Allen and Hanson, 1985) and each clone was physically mapped. A minimum set of 46 clones was selected to cover CII (Choudhary *et al.*, 1994). DNA from one of the clones (cosmid 8536) was digested with a variety of restriction enzymes to identify those that cut infrequently in the cosmid vector, while cutting the insert into fragments from a few hundred base pairs to a few kilobase pairs in size. In the case of cosmid 8536, *Bam*HI, *Eco*RI, *Bgl*II, *Not*I and *Eco*RV were used to generate a total of 25 fragments from the cosmid insert, which were then subcloned into the plasmid pBluescript. White colonies were picked into duplicate microtiter dishes and DNA micropreps (alkaline lysis minipreps scaled down to 200μl) were prepared from one of the dishes. The other dish was kept as a frozen stock. The insert in each subclone was then matched by size to a band in the original cosmid insert. This ensured that all the cosmid fragments were represented as subclones. Those containing vector fragments were discarded. Template was prepared from a subclone for each fragment and the insert ends sequenced using modified T3 and T7 primers. The sequences were tested for similarities using BLASTX (Altschul *et al.*, 1990) against the NCBI non-redundant protein database.

At this point the data was assembled into a map as follows (Fig. 43-1a). A fragment which had both ends giving matches to the database was chosen as a starting point. For example, in 8536 the T3 sequence of a 3.1kb *Bam*HI fragment gave a match to the S-adenosylmethionine synthase gene, *metK*, from *E. coli* ($P=1.8e^{-28}$) and the T7 sequence gave a match to the 5-enolpyruvylshikimate-3-phosphate synthase gene, *aroE*, from *Synechocystis* sp. ($P=4.2e^{-8}$). The BLASTX searches gave the size of the *metK* and *aroE* gene products in amino acids, and this was used to estimate the size of the *R. sphaeroides* 2.4.1 genes and their

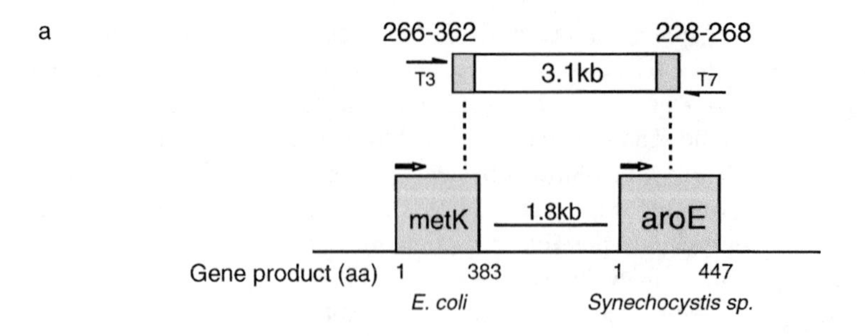

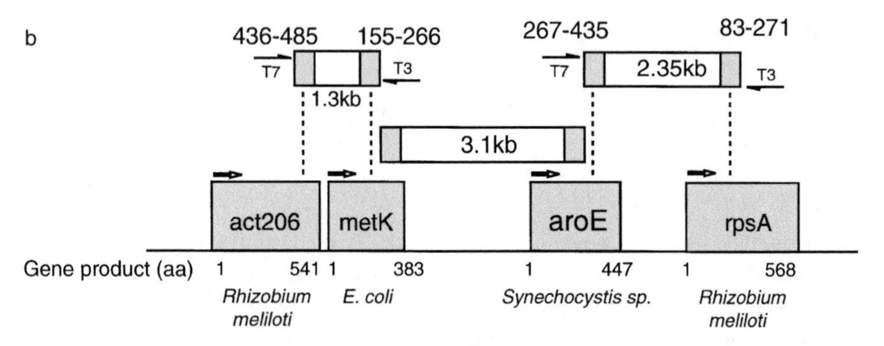

Figure 43-1. Assembly of the sequence similarity map:

(a) A 3.1kb *Bam*HI fragment was chosen as the starting point for the construction of the 8536 map. Approximately 350bp of sequence, shown as shaded boxes, was obtained from each end of the fragment using modified T3 and T7 primers of pBluescript SK (–). The T3 sequence gave a BLASTX match to amino residues 266–362 of *metK*, the T7 to residues 228–268 of *aroE.* From these data the proposed function, size, spacing, and direction of transcription of these genes in *R. sphaeroides* 2.4.1 was postulated.

(b) The T3 sequence of a 1.3kb *Bam*HI fragment also gave a match to *metK*, in this case from residues 155–266. As this amino acid sequence was continuous with that derived from the 3.1kb T3 sequence, i.e., 155-266-266-362, this suggested that this fragment lay to the left of the 3.1kb fragment. The T7 sequence match suggested that a homologue of the *R. meliloti* acid inducible gene, *act206*, lay just upstream of *metK*, with insufficient distance lying between *act206* and *metK* for an additional gene. Sequence matches also suggested that a 2.35kb *Bam*HI fragment lay to the right of the 3.1kb fragment with a homologue of *rpsA* lying downstream of *aroE.* Though *rpsA* was considered to be downstream of *aroE*, there was still sufficient space between these genes for an additional open reading frame. This possibility also exists for the region between *metK* and *aroE.*

spacing with respect to each other within CII. The BLASTX results also made a prediction, i.e., at least one other *Bam*HI fragment end should also give a database match to *metK*, and a different fragment end should match *aroE*. In both cases these new matches should be contiguous with the amino acid sequences already found by the 3.1kb fragment. These predictions were fulfilled (Fig. 43-1b). These fragments also revealed the genes upstream of *metK*, and downstream of *aroE, act206* and *rpsA* respectively, which were then added to the map. This process was repeated, first with the remaining *Bam*HI fragments, then with the fragments from the other four fragment groups. The restriction maps from the groups were then merged to give the deduced gene organization and a completed restriction map. The map was checked by assembling all the sequences with the Gelassemble program of the GCG software package (Devereux et al., 1984). This verified that the five restriction maps overlapped, as predicted at the sequence level.

The maps of two cosmids, 8801 from CI and 8536 from CII, were produced in this way (Figure 43-2). For these two cosmids, only 41 templates were prepared, and the sequences were obtained from less than 3 sequencing gels (32 lanes per gel). The sequences obtained covered 41 and 51% of the cosmid inserts respectively, (Table 43-2), with approximately 60% of the fragment ends giving database matches. This revealed 17 putative genes (Table 43-3). Two of these, *hly* (ter Huurne *et al.*, 1994) and *vacB* (Tobe *et al.*, 1992) are associated with bacterial virulence, an unexpected discovery, as *R. sphaeroides* 2.4.1 is considered a non-pathogenic organism. We also noted that cosmid 8536 contained a sequence similar to *rpsA* of *E. coli*. This gene encodes the S1 ribosomal protein, which is involved in, and essential for, binding of the ribosome to mRNA (van Duin *et al.*, 1980). Southern blot analysis suggested that this was the only copy of this gene in the *R. sphaeroides* 2.4.1 genome and that it mapped to CII (data not shown). The other 8536 derived matches suggest that CII is neither specialized for nor dedicated to any particular aspect of the organism's lifestyle, as it appears to contain a mixture of genes involved in house-keeping, aromatic amino acid biosynthesis, translation, and "virulence". Genes found on other CII cosmids (data not shown) also suggest that the genetic content of CII is typical of any other bacterial chromosome. However, much more work needs to be done before any final conclusions can be reached.

Though the use of restriction enzymes proved useful in providing physical markers along the cosmid inserts, the anticipated non-randomness of their cleavage resulted in a lack of sequence information in certain regions of the insert DNA, e.g., between *metK* and *aroE* in cosmid 8536. To overcome this problem analysis of a third cosmid (8603) followed a more random procedure. This cosmid was digested with DNaseI, the ends made flush with Klenow fragment, and the fragments sized on a gel before subcloning (Démolis *et al.*, 1995). Subclones containing cosmid insert DNA were isolated by probing with the 8603 insert, thus eliminating subclones containing only cosmid vector DNA. 80 templates

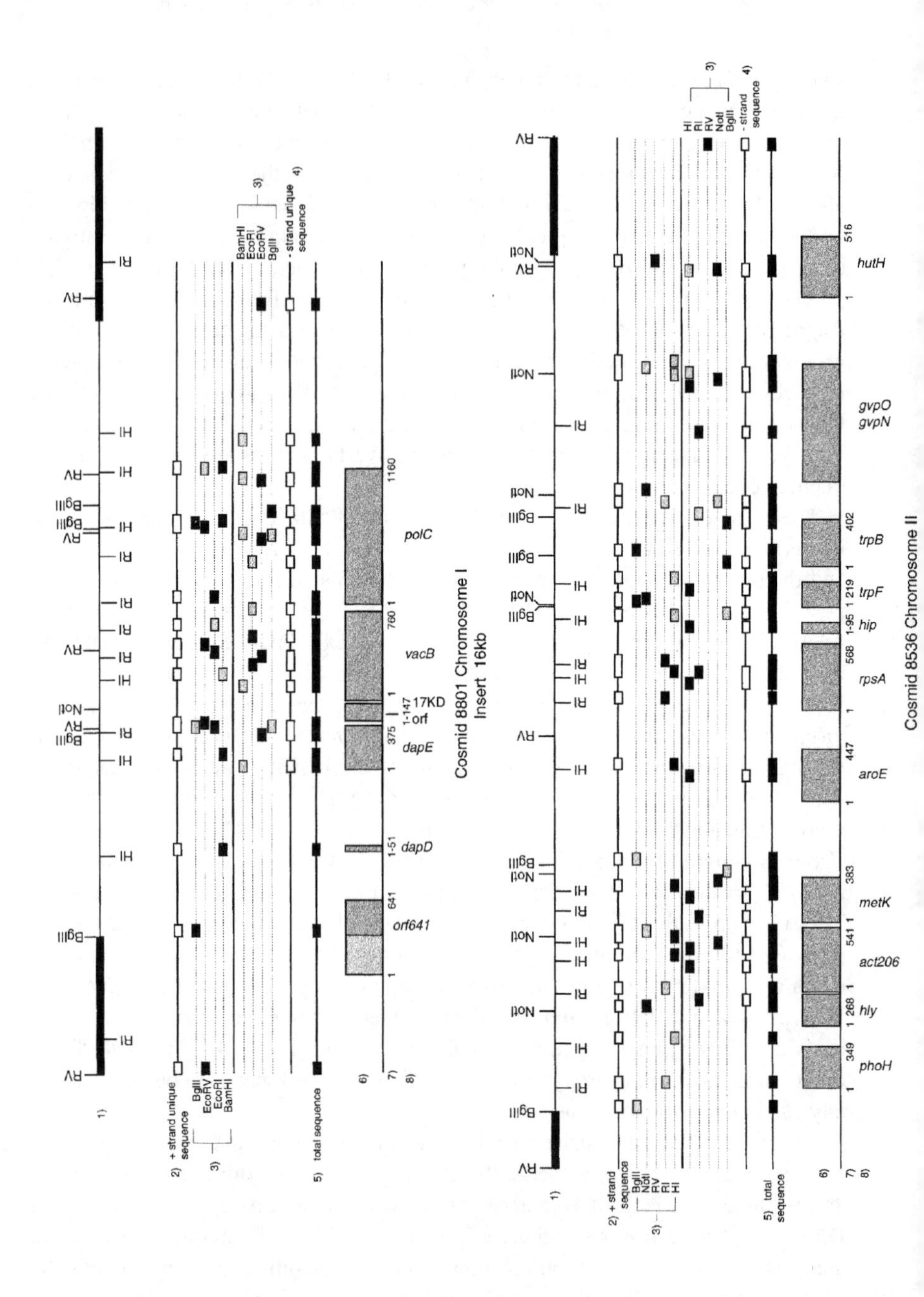
Cosmid 8801 Chromosome I
Insert 16kb
1)
2) + strand unique sequence
3)
BglII
EcoRV
EcoRI
BamHI
4) - strand unique sequence
5) total sequence
6)
7)
8)
polC
vacB
17KD orf
dapE
dapD
orf641
Cosmid 8536 Chromosome II
2) + strand sequence
NotI
RV
RI
HI
4) - strand sequence
5) total sequence
hutH
gvpO
gvpN
trpB
trpF
hip
rpsA
aroE
metK
act206
hly
phoH

◀

Figure 43-2. Proposed gene organization of cosmids 8801 and 8536

The restriction maps and gene organization of two cosmids, 8801 (chromosome I) and 8536 (chromosome II) are shown. 1) The black boxes at the ends of the restriction maps represent regions from the cosmid vector (pLA2917). The restriction enzymes shown are *Bam*HI (HI), *Eco*RI (RI), *Bgl*II, *Not*I and *Eco*RV(RV). 2) The open boxes represent the unique sequence information obtained from the cosmid plus strand. 3) The small boxes on the dotted lines represent the sequences obtained from the subcloned restriction fragment ends. The enzymes used to generate these fragments are shown at the ends of the dotted lines. The fragment sequences above or below the bold centerline were obtained from the plus or the minus strand of the cosmid respectively. Sequences giving significant BLASTX matches are shown as black boxes, those giving matches considered insignificant as gray boxes. 4) The open boxes represent the amount of unique sequence information obtained from the cosmid minus strand. 5) The black boxes represent the total unique sequence, obtained by combining the unique plus and minus cosmid sequences. 6) The large shade boxes indicate the regions of the cosmid inserts which encode putative genes. 7) The sizes of the gene products found in the database search in amino acid residues. 8) The names of the genes found in the database searches.

were prepared and their ends sequenced. This generated 48.3kb of sequence, to give a theoretical two-fold coverage of the insert. After the sequences were assembled there was a coverage of less than 1x, with approximately 50 gaps in the map. Database searches gave 6 significant matches: *micA*, (A/G-specific adenine glycosylase); a gene encoding a hypothetical ABC transporter in the *tesA* region; *tesA*, (acyl-coA thioesterase); *nifR4* (sigma-54); two orfs of unknown function and *phoH*. This last gene was also found in the neighboring cosmid 8536 and thus lies in the overlap between the two. However, without prior

Table 43-2.

	Cosmid	
	8801 (CI)	8536 (CII)
BLAST Hits / Fragment ends	20/32=62%	31/49=63%
Cosmid insert size	16kb	21kb
[1]Total sequence	9.6kb	14.7kb
[2]Insert coverage	60%	70%
[3]Unique sequence	41%	51%
[4]Proposed coding sequence	59%	69%

[1]We have 95% confidence in the first 300bp of sequence obtained from each fragment end. The total sequence is therefore 300 bp multiplied by the total number of fragment ends.

[2]This figure is the total sequence obtained expressed as a percentage of the insert size.

[3]This is the total sequence obtained when overlaps in sequence are excluded.

[4]This is the percentage of the insert for which a role has been predicted.

Table 43-3. The BLASTX results using the restriction fragment sequences.

Best match	Gene product or function	Parental organism	% of Gene Covered[1]	Best Hit Size (amino acids)[2]	% Amino Acids Identical/Positive[3]	P[4]
Chromosome I cosmid 8801						
orf641	pyruvate dehydrogenase	*Rhodobacter capsulatus*	16	87	63/77	1.6e–36
dapD	succinyl-diaminopimelate aminotransferase	*Klebsiella pneumoniae*	100	28	61/75	0.99993
dapE	N-succinyl-diaminopimelate deacylase	*Escherichia coli*	80	43	44/65	4.3e–11
17KD orf	unknown	*Bacillus subtilis*	34	57	38/57	3.8e–30
vacB	virulence-associated locus	*Shigella flexneri*	83	103	45/55	3.2e–24
polC	α-subunit of DNA polymerase III	*E. coli*	59	90	44/61	1.1e–19
Chromosome II cosmid 8536						
phoH	phosphate starvation inducible gene	*Mycobacterium leprae*	29	172	59/69	1.2e–67
hly	hemolysin	*Serpulina hyodysenteriae*	48	67	41/73	5.0e–27
act206	acid inducible gene	*Rhizobium meliloti*	86	25	72/96	2.8e–41
metK	S-adenosylmethionine synthetase	*E. coli*	48	112	57/75	5.6e–41
aroE	dehydroshikimate reductase	*Synechocystis* sp.	45	99	60/68	2.6e–32
rpsA	30S ribosomal subunit protein S1	*R. meliloti*	70	87	79/93	2.2e–63
hip	integration host factor β-subunit	*R. capsulatus*	88	55	78/85	2.2e–32
trpF	anthranilate isomerase	*Caulobacter crescentus*	100	136	35/51	5.0e–29
trpB	tryptophan synthase	*Pseudomonas aeruginosa*	61	122	78/84	3.6e–62
gvpO/N	gas vesicle operon	*Halobacterium* sp.	33	91	30/50	2.4e–08
hutH	histidine ammonia lyase	*Streptomyces griseus*	23	107	30/40	1.8e–08

[1]This percentage is an estimate. It is derived from the amount of sequence we have for the proposed homologous *R. sphaeroides* gene as a proportion of the gene found in the database search. The assumption was made that the two genes were the same size.

[2]The size of the subject region in amino acids (i.e., the sequence found in the database search) to which the highest scoring query sequence gave a match.

[3]The percentage of the query sequence amino acids found in the subject regions that were identical/similar to those of the matching subject sequence.

[4]The probability computed by the BLASTX program that the sequences found during the search were found by chance.

mapping of the clones or identification of other physical landmarks, it has not been possible to unambiguously locate these sequences in CII. Future studies are aimed at resolving this problem by binning, i.e., using cosmid insert restriction fragments as probes to map the DNaseI subclones within the cosmid prior to sequencing.

Functional Analysis

Low redundancy sequencing gives less information about genes at the base pair level than conventional sequencing strategies. However, this does not prevent us from using these sequence maps for *in vivo* functional analysis. One approach that has been used in *E. coli* aims at associating function to sequence by virtue of the regulation of gene expression (Chuang *et al.*, 1993). In this method cDNA probes are made from RNA prepared from whole cells grown under different conditions. These are hybridized to ordered clones such as our subclones for sequencing, to identify genes showing differential expression.

An alternative approach uses pools of random transposon insertions to identify genes that are essential under particular growth conditions. Insertions in required genes are lost and this can be detected by comparing pools of insertions grown with and without selection. One solution to the challenge of identifying individual insertions that are lost from a pool, and determining their position in the genome, has recently been described in *Salmonella* (Hensel *et al.*, 1995; *see also* Chapter 22). In this case each insertion was tagged with a unique oligonucleotide sequence that was used for tracking the transposons. Other methods are under development in this and other laboratories. *R. sphaeroides* 2.4.1 is well suited to this undertaking, as DNA can be introduced into this organism either by mating, transformation or electroporation, making the generation of null mutations a relatively easy task (Donohue and Kaplan, 1991).

It is thus clear that the combination of sequence skimming and functional analysis promises to provide a wealth of information about *R. sphaeroides*, 2.4.1 or any other microbe rapidly and inexpensively.

References

Allardet-Servent, A., S. Michaux-Charachon, E. Jumas-Bilak, L. Karayan, and M. Ramuz. 1993. Presence of one linear and one circular chromosome in the *Agrobacterium tumefaciens* C58 genome. *J. Bacteriol.* 175:7869–7874.

Allen, L. N., and R. S. Hanson. 1985. Construction of broad-host-range cosmid cloning vectors: identification of genes necessary for growth of *Methylobacterium organophilum* on methanol. *J. Bacteriol.* 161:955–962.

Altschul, S. F., W. Gish, W. Miller, E. W. Myers, and D. J. Lipman. 1990. Basic local alignment search tool. *J. Mol. Biol.* 215:403–410.

Choudhary, M., C. Mackenzie, K. S. Nereng, E. Sodergren, G. M. Weinstock, and S. Kaplan. 1994. Multiple chromosomes in bacteria: structure and function of chromosome II of *Rhodobacter sphaeroides* 2.4.1^{T}. *J. Bacteriol.* 176:7694–7702.

Chuang, S.-H, D. L. Daniels, and F. R. Blattner. 1993. Global regulation of gene expression in *Escherichia coli. J. Bacteriol.* 175:2026–2036.

Démolis, N., L. Mallet, F. Bussereau, and M. Jacquet. 1995. Improved strategy for large-scale sequencing using DNaseI cleavage for generating random subclones. *Biotechniques* 18:453–457.

Devereux, J., P. Haeberli, and O. Smithies. 1984. A comprehensive set of sequence analysis programs for the VAX. *Nucl. Acids Res.* 12:387–395.

Donohue, T. J., and S. Kaplan. 1991. Genetics of photosynthetic bacteria. *Meth. Enzymol.* 204:459–485.

Dryden, S. C., and S. Kaplan. 1990. Localization and structural analysis of the ribosomal RNA operons of *Rhodobacter sphaeroides. Nucl. Acids Res.* 18:7267–7277.

Hallenbeck, P. L., R. Lerchen, P. Hessler, and S. Kaplan. 1990a. Phosphoribulokinase activity and regulation of CO_2 fixation critical for photosynthetic growth of *Rhodobacter sphaeroides. J. Bacteriol.* 172:1749–1761.

Hallenbeck, P. L., R. Lerchen, P. Hessler, and S. Kaplan. 1990b. Roles of CfxA, CfxB, and external electron acceptors in regulation of ribulose 1,5- bisphosphate carboxylase/oxygenase expression in *Rhodobacter sphaeroides. J. Bacteriol.* 172:1736–1748.

Hensel, M., J. E. Shea, C. Gleeson, M. D. Jones, E. Dalton, and D. W. Holden. 1995. Simultaneous identification of bacterial virulence genes by negative selection. *Science* 269:400–403.

Michaux, S., J. Paillisson, M.-J. Carles-Nurit, G. Bourg, A. Allardet-Servent, and M. Ramuz. 1993. Presence of two independent chromosomes in the *Brucella melitensis* 16M genome. *J. Bacteriol.* 175:701–705.

Neidle, E. L., and S. Kaplan. 1992. *Rhodobacter sphaeroides rdxA*, a homolog of *Rhizobium meliloti fixG*, encodes a membrane protein which may bind cytoplasmic [4Fe-4S] clusters. *J. Bacteriol.* 174:6444–6454.

Neidle, E. L., and S. Kaplan. 1993. Expression of the *Rhodobacter sphaeroides hemA* and *hemT* genes encoding two 5-aminolevulinic acid synthase isozymes. *J. Bacteriol.* 175:2292–2303.

Sobral, B.W.S., R. J. Honeycutt, A. G. Atherly, and M. McClelland. 1991. Electrophoretic separation of the three *Rhizobium meliloti* replicons. *J. Bacteriol.* 173:5173–5180.

Suwanto, A., and S. Kaplan. 1989a. Physical and genetic mapping of the *Rhodobacter sphaeroides* 2.4.1 genome: genome size, fragment identification and gene localization. *J. Bacteriol.* 171:5840–5849.

Suwanto, A., and S. Kaplan. 1989b. Physical and genetic mapping of the *Rhodobacter sphaeroides* 2.4.1 genome: presence of two unique circular chromosomes. *J. Bacteriol.* 171:5850–5859.

Tabita, F. R., J. L. Gibson, B. Bowien, L. Dijkhuizen, and W. G. Meijer. 1992. Uniform designation for genes of the Calvin-Benson-Bassham reductive pentose phosphate pathway of bacteria. *FEMS Microbiol. Lett.* 99:107–110.

Ter Huurne, A.A.H.M., S. Muir, M. van Houten, B.A.M. van der Zeijst, W. Gaastra, and J. G. Kusters, 1994. Characterization of three putative *Serpulina hyodysenteriae* hemolysins. *Microb. Pathogen.* 16:269–282.

Tobe, T., C. Sasakawa, N. Okada, Y. Honma, and M. Yoshikawa. 1992. *vacB*, a novel chromosomal gene required for expression of virulence genes on the large plasmid of *Shigella flexneri. J. Bacteriol.* 174:6359–6367.

van Duin, J., G. P. Overbeek, and C. Backendorf. 1980. Functional recognition of phage RNA by 30-S ribosomal subunits in the absence of initiator tRNA. *European J. Biochem.* 110:593–597.

Woese, C. R. 1987. Bacterial evolution. *Microbiol. Rev.* 51:221–271.

Zeilstra-Ryalls, J. H., and S. Kaplan. 1995. Aerobic and anerobic regulation in *Rhodobacter sphaeroides* 2.4.1: the role of the *fnrL* gene. *J. Bacteriol.* 177:6422–6431.

44

Sequencing the Genome of *Sulfolobus solfataricus* P2

Christoph W. Sensen, Robert L. Charlebois, Rama K. Singh, Hans-Peter Klenk, Mark A. Ragan, and W. Ford Doolittle

The goal of our all-Canadian team is to sequence the entire 3 Mbp genome of the crenarchaeote *Sulfolobus solfataricus* P2. The work is being conducted at three sites. Robert Charlebois at the University of Ottawa is heading the team responsible for cosmid libraries, physical mapping, subcloning of cosmids into plasmids, and primary sequencing of the subclones. Mark Ragan's team at the Institute for Marine Biosciences (IMB) is involved in further primary sequencing, sequencing for polishing and finishing of cosmids, sequence assembly, and computer analysis of the sequence data. Finally, Ford Doolittle's team at Dalhousie University is characterizing select *Sulfolobus* gene products using biochemical methods. His group is also responsible for the development of shuttle vector systems between *Escherichia coli* and *Sulfolobus* as well as most of the phylogenetic analysis of the *Sulfolobus* data.

We chose to sequence the genome of *S. solfataricus* for several reasons. Thermophilic archaea are thought to be most similar to the ancestor of eukaryotes. Archaeal genes may therefore resemble eukaryotic genes more closely than do genes from bacteria, and may provide better insight into the content and organization of the early eukaryotic genome. The DNA of *S. solfataricus* P2 is easy to sequence, as its G+C content is only 35–37%. Though repeated sequences are numerous, our cosmid-by-cosmid sequencing approach dissects them from each other, resulting in few problems of sequence assembly; instead, the repeated sequences render the genome of *Sulfolobus* particularly attractive for studies on genome dynamics. Biochemically, *Sulfolobus* is also quite interesting, due in large part to its growth temperature of up to 87°C (optimum 80°C) and its preferred pH range from 2 to 5 (Grogan, 1989). Each of its proteins must function at 87°C, a feature of great interest for many industrial and biotechnological processes. Our studies so far indicate that many proteins show *in vitro* activity even above 100°C (unpublished). Adaptation to such an extreme and unstable niche will further provide ample examples of sophisticated gene regulation strategies, of interest to molecular biology. Specific physiological processes of *Sulfolobus*, such as the efficient oxidization of sulfides and of elemental sulfur, are already used by industry for several desulfurization applications (Coghlan, 1995).

The *Sulfolobus* genome project also serves as a model to build up modest but efficient genome research capabilities in Canada. This includes the sequence generation (wet-lab) and sequence analysis (computer-lab) aspects of genomics. The *Sulfolobus* team is especially focussing on distributed approaches to genome research, heavily relying on the Internet as a communication- and data transfer-tool. For the analysis of another *Sulfolobus* genome, *see* Chapter 82.

Generating the *Sulfolobus* Sequence

The *Sulfolobus* team is a small group with limited resources. All of the sequence is generated by four technicians using only two ABI 373A (Applied Biosystems Inc.) and two LiCor 4000L (LiCor Inc.) sequencers part-time. Therefore the group is specializing in efficient, least-redundant techniques for sequence generation. As the laboratories involved in the sequence generation are located in different cities, standardization of methods is vital. All sequencing reactions are currently performed using ABI's dye-terminator biochemistry or Thermosequenase biochemistry (Amersham Inc.) respectively. The exclusive use of automated sequencing techniques facilitates the group's communication over the Internet, as only trace data need to be sent from one location to another to transmit the full sequence information. Sequence assembly and polishing is performed on IMB's UNIX cluster using the Staden package (Staden, 1994). This guarantees the link between assembled sequence strings and the original trace information, essential for efficient sequence editing in a distributed genome project.

Cosmid Library Construction and Mapping

Fragments of partially *Hin*dIII-digested genomic DNA (40–45 kbp) were cloned into *E. coli* using the cosmid vector Tropist3 (De Smet *et al.*, 1993) to produce a library with an 18-fold coverage of the *S. solfataricus* P2 genome. The landmark strategy for contig mapping (Charlebois, 1993) is being used to identify overlaps between cosmids. Additionally, cosmids are being mapped to macrorestriction fragments by hybridization, in order to produce a provisional, integrated map needed to fuel the sequencing project. Cosmid clones forming a minimal set are individually subcloned into plasmid vectors for subsequent sequencing.

Subcloning of Random Fragments and Primary Sequencing

Cosmid DNA is broken into random (typically 1000–2000 bp) fragments by nebulization (i.e. spraying the dissolved DNA through an inhalation mask). After end-repair, the fragments are blunt-end ligated into *SmaI*-digested, dephosphorylated pUC18. About 50–60 sequencing reactions using the M13 universal primer are carried out per 40 kbp insert using the LiCor sequencing machines. These same templates are thereafter sequenced on the opposite strand using the M13 reverse primer. Assembly of the primary sequences using the Staden package yields 15 to 25 contigs, giving about 2.5-fold coverage of the insert.

Contig Linking and Double-stranding

The 15 to 25 contigs from the primary sequencing phase are linked into one contig via a primer-walking technique at the cosmid level, which was developed at IMB, using the ABI sequencing machines (unpublished; this protocol is accessible by non-profit researchers upon signing a non-disclosure agreement). Between 30 and 50 walking primers per cosmid are calculated (using GeneSkipper, EMBL Heidelberg, unpublished), synthesized at IMB's primer synthesis facility, and used in sequencing reactions with cosmid DNA as the template to generate a single contig. After all sub-contigs are linked, regions which were not yet sequenced on both DNA strands are identified and "double-stranded" using the same primer-walking technique on cosmids used for the contig linking phase. Remaining ambiguities are similarly resequenced. The final result is a completely polished sequence with an average coverage of about three. We estimate an error rate of approximately 1 per 5000–10000 bp. The cost per final finished base (including salaries and machine depriciation) is roughly 75 Canadian cents (US$ 0.50).

Status of the Sulfolobus *Sequencing Project*

Presently, the group is able to generate between 50 and 100 kbp of *Sulfolobus* sequence per month. As of May 1996 about 30% of the *Sulfolobus* genome has been sequenced. The largest coherent contig to date exceeds 250 kbp. Figure 44-1 shows the sequencing progress between January 1994 and May 24, 1996 (the date that this chapter was revised).

It can be anticipated, given stable funding and the increase in sequencing speed that can be projected from newly acquired resources, that the *Sulfolobus* genome will be totally sequenced no later than the middle of 1998.

Computer Analysis of the *Sulfolobus* Sequence

Because the *Sulfolobus* genomic sequence is generated cosmid by cosmid, a dedicated computational sequence analysis was possible from the very beginning of the project. We have set up an integrated computer environment suited for this purpose.

To prove that the sequence is assembled properly, pseudo-agarose gels are simulated by computer and compared to real gels. This approach allows falsely

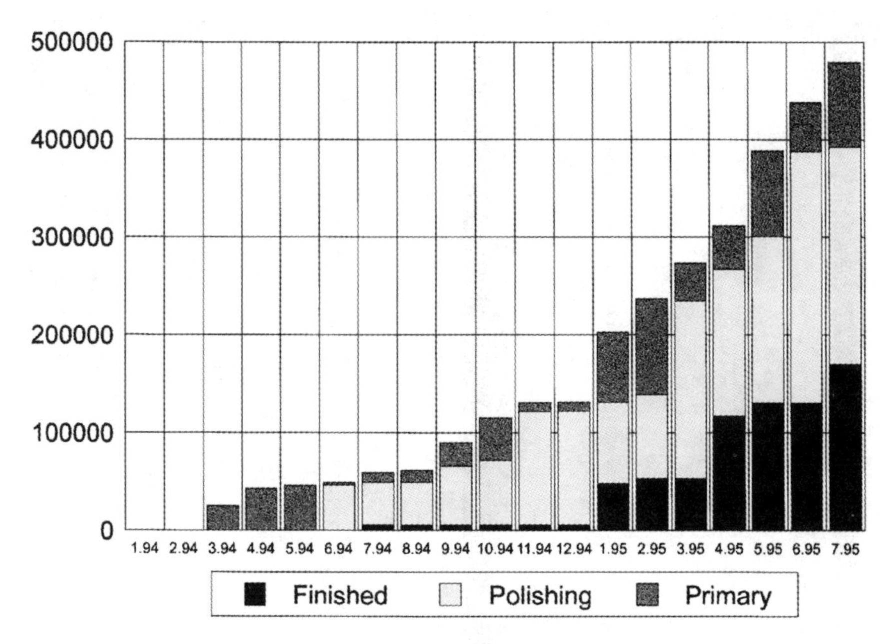

Figure 44-1. Monthly sequencing progress of the *Sulfolobus solfataricus* P2 genome project between January 01, 1994 and May 24, 1996.

The contig size to date is 897,278 bp. The contigs are grouped into three different stages:

- Primary sequencing, where random nebulized clones are sequenced;
- Polishing phase, where contigs are linked and double-stranded;
- Finished contigs, which are completely sequenced (contigs are not declared finished, if even only a single sequencing reaction is missing).

mounted repeated sequences in a contig to be detected from discrepancies between the real gel and the simulation.

Parts of the sequence analysis, especially the similarity searches against the various sequence databases, including reasoning about the search results, are already automated (Gaasterland and Sensen, 1996). This aspect is also presented in a separate chapter of this book (see Chapter 45).

Graphical display of sequence features is mostly carried out using GeneSkipper under Microsoft Windows 3.1. All open reading frames longer than a minimum size (e.g. 300 bp) are displayed in their respective frames. *Sulfolobus* can use ATG, GTG and TTG as valid start codons. The information about the open reading frame (orf) location and codon usage is complemented by information about the G+C content and pyrimidine distribution along the contig. Physical features of the DNA and the proteins are calculated and presented on the user's request. Figure 42-2 shows the display of a typical *Sulfolobus* contig and its features.

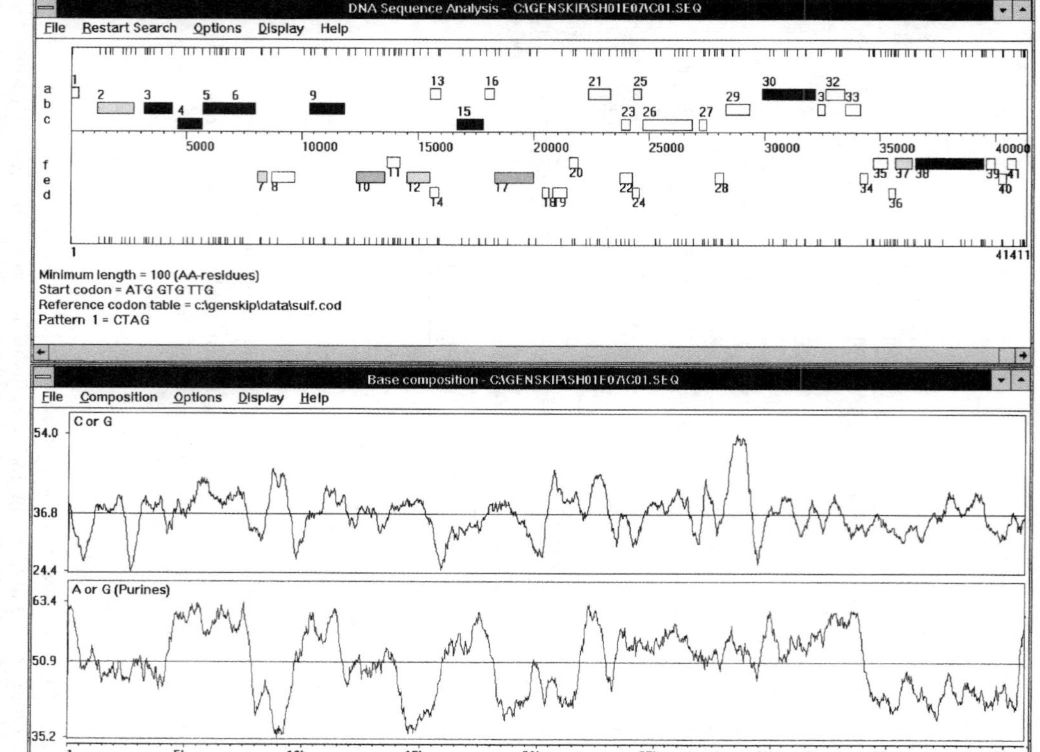

Figure 44-2. Sequence analysis of a *Sulfolobus solfataricus* contig.

The upper window shows the open reading frame (ORF) distribution (boxes 1–92) on a 100,389-bp contig. Open reading frames are located in all six possible frames. The various grey shades filling the boxes are related to the degree of match/mismatch between a codon usage table, calculated for highly expressed *Sulfolobus* genes, and that particular ORF. The bars on the upper and lower border of the inner window show the distribution of a pattern (CTAG) over the whole contig.

The upper graph in the lower window displays the G+C distribution over the entire fragment, calculated with a 500-bp sliding window. It is obvious that some open reading frames have a very different G+C content (up to 59.8%) from the average (36.4%); these high G+C frames are coding for transposable elements. In non-coding spacers between orfs, the G+C drops to between 24 and 26%.

The lower graph in the lower window displays the purine distribution (A+G) over the entire fragment, also calculated with a 500-bp sliding window. The purine distribution is biased towards the coding strand. Only 35–40% of all purines are located on the non-coding strand of the *Sulfolobus* genome.

Phylogenetic analyses of *Sulfolobus* data are mainly performed on the IMB Unix cluster. Results of the phylogenetic analyses will be published elsewhere.

Biochemical Analysis of *Sulfolobus* Genes

Selected genes of interest (e.g. having potential commercial value) are amplified from genomic DNA using the polymerase chain reaction, cloned into expression vectors and expressed in *E. coli* strains by the Doolittle laboratory. The expression studies are complemented by the analysis of isolated and purified proteins from *Sulfolobus*. These activities have to be very focused as they cannot begin to compete with the speed of DNA sequence generation: for each individual generating DNA sequence, more than a dozen biochemists might be needed to express and characterize proteins! Therefore, the *Sulfolobus* group has a major interest in collaborations with outstanding teams around the world who want to study well-defined problems using the *Sulfolobus* sequence.

We are also sequencing plasmids from *Sulfolobus* strains (Keeling *et al.*, 1996) that could be used to construct "shuttle vectors" to bring DNA back and forth between *E. coli* and *Sulfolobus*. This technology will give access to the use of a broader spectrum of *Sulfolobus* open-reading frames for commercial purposes.

Summary

Using a simple strategy and current technology, we are sequencing the genome of the archaeon *Sulfolobus solfataricus* P2. The resulting data are processed through an efficient and comprehensive semi-automated bioinformatics system, minimizing the effort required to identify the features of interest. The sequence of the *Sulfolobus* genome will provide more than a string of genes, more than an inventory of thousands of thermostable proteins, and more than a complete set of phylogenetic markers. It will provide a complete genetic description of an organism whose ancestors were poised to become eucaryotes. Comparative studies involving other archaeal genomic sequences will characterize that ancestor, and comparative studies including bacterial genomes will allow us to examine surviving features of the earliest life. Further information about the *Sulfolobus* genome project can be obtained from our World Wide Web site at: *http://www.imb.nrc.ca/imb/sulfolob/sulhom_e.html.*

Acknowledgments

For support, we thank the Canadian Genome Analysis and Technology Program, the Canadian Institute for Advanced Research, the National Research Council, and the Medical Research Council. We gratefully acknowledge our students and technicians for their outstanding skill and dedication.

This is NRC publication number 39714.

References

Charlebois, R. L. 1993. Physical mapping of genomes using the landmark strategy. In *The Second International Conference on Bioinformatics, Supercomputing and Complex Genome Analysis*, H. A. Lim, J. W. Fickett, C. R. Cantor and R. J. Robbins, eds. pp. 219–229. World Scientific, Singapore.

Coghlan, A. 1995. Bacteria go to work on old tyres. *New Scientist* 145:21.

De Smet, K.A.L., S. Jamil, N. G. Stoker. 1993. Tropist3: a cosmid vector for simplified mapping of both G-C rich and A-T rich genomic DNA. *Gene* 136:215–219.

Gaasterland, T., C. W. Sensen. 1996. MAGPIE: automated genome interpretation. *Trends in Genetics* 12:78–78.

Grogan, D. W. 1989. Phenotypic characterization of the archaebacterial genus *Sulfolobus*: comparison of five wild-type strains. *J. Bacteriol.* 171:6710–6719.

Keeling, P., H.-P. Klenk, R. K. Singh, O. Feeley, C. Schleper, W. Zillig, W. F. Doolittle, C. W. Sensen. 1996. Complete nucleotide sequence of the *Sulfolobus islandicus* multicopy plasmid pRN1. *Plasmid* 35:141–144.

Staden, R. 1994. The Staden package. In *Methods in molecular biology*, Vol 22, A. M. Griffin and H. G. Griffin, eds. pp. 9–170. Humana Press Inc., Totawa, New Jersey.

45

MAGPIE: A *M*ultipurpose *A*utomated *G*enome *P*roject *I*nvestigation *E*nvironment for Ongoing Sequencing Projects

Terry Gaasterland and Christoph W. Sensen

Motivation

A wealth of sequence analysis and interpretation tools enables interested researchers to inspect, query and, ultimately, interpret DNA sequence data. In the context of sequencing an entire genome, the task of using each tool on each bit of DNA sequence data is daunting (*see also* Chapters 19, 35 to 39, 42 to 44 and 46). Even more formidable is the task of assessing and reassessing each output from each tool in the light of what is already known. The facts that sequence information in the public databases changes regularly and that new sequence analysis tools appear frequently compound the situation further.

The MAGPIE system embodies strategies for (1) automatically monitoring changing sequence data, (2) generating requests to software tools, (3) incorporating results into a growing knowledge base about the genome, and (4) formulating reports on decisions and inferences about the genome. Figure 45-1 shows how the modules of computation in the system fit together to execute these tasks. The philosophy driving the MAGPIE system emerges from the fact that during the lifetime of a genome sequencing project, the sequence is growing and changing; the public sequence databases are also growing and changing; new and improved software tools must be used in a timely manner; and the sheer volume of interpretation tasks simply cannot be handled manually by humans.

The main engineering goal in MAGPIE has been to provide a platform for managing remote and local software requests triggered by a changing environment. The main research goal in MAGPIE has been to explore how to use deductive inference and large-scale distributed database techniques to digest the results into a human-readable and -queryable form. MAGPIE also provides the platform for ongoing research in automatic reconstruction of a minimal primary metabolism for an organism (Gaasterland and Selkov, 1995).

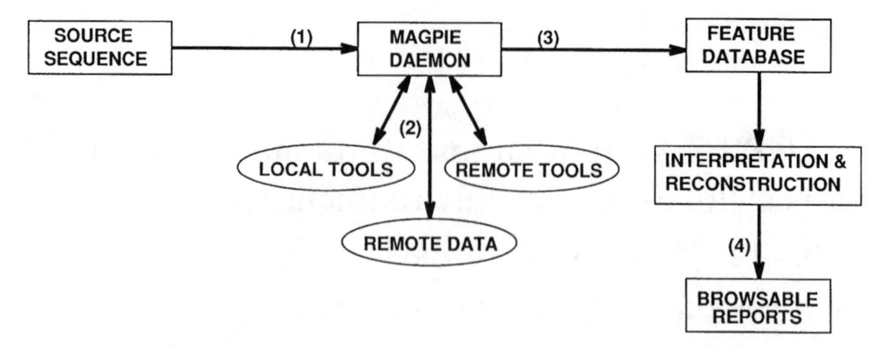

Figure 45-1. Flow of Information in MAGPIE: The MAGPIE DAEMON monitors the SEQUENCE SOURCE (*e.g.* output from the Staden sequence assembly package) for changes (1). It dispatches requests for services to LOCAL TOOLS and REMOTE TOOLS, which rely on local and REMOTE DATA (2). It then parses output into first-order logic form to create a FEATURE DATABASE about the source sequence (3). Logic-based sequence interpreters and a metabolic reconstruction engine use the encoded features to produce BROWSABLE REPORTS (4).

Sulfolobus as an Example

We shall use the *Sulfolobus solfataricus* sequencing project to illustrate the MAGPIE approach to handling volumes of increasing and changing DNA sequence data (*see also* Chapter 44). Consider a snapshot in time at which over 500,000 base pairs have been generated. Contiguous sequences (contigs) vary in length from 300 to 100,000 base pairs (bp). Contigs vary in state from "primary" (freshly read and assembled from shotgun trace data) through "being linked" and "being polished" to "finished". Some contigs will merge; others will disappear (e.g. contaminations with vector or *E. coli* sequence); others will change in quality (e.g. number of ambiguities and frameshifts) over time.

Suppose we want to send 500,000 base pairs spread through 250 contigs through an average of 10 different sequence analysis software tools per contig, and that we want to redo the request either when a contig changes or on a regular basis as sequence databases are updated. At minimum, it would be desirable to repeat requests to each software tool once a month, for a total of 30,000 output files (250 contigs * 10 tools * 12 months) over the course of a year. In reality, the number of contigs grows over the year as does the number of desirable tools. Therefore we expect a minimum of 250,000 output files per year when the entire 3.1 Mbp sequence of *Sulfolobus* is generated.

MAGPIE System Architecture

Flexibility and reconfigurability during the course of a genome project are of utmost importance. As new needs arise and as discoveries are made about how to use sequence analysis tools more effectively, users must be able to communicate this information to the system with ease. To facilitate this interactivity, configurable information must reside outside of the source code. The separation of information about tools and sequence from the system structure allows users to express choices about the following: which tools to use, when to use them, which tools are preferred under what conditions, what parameters to use, frequency of reevaluation, and preferences about groups of contigs for the system to consider (Gaasterland and Sensen, 1996). Our goal was to develop a flexible architecture that can be adapted to the changing needs of any genome project without recompiling the source code.

The MAGPIE system is implemented as a set of Perl and C executables. The reasoning and query answering components are implemented in Prolog, a logic-based declarative knowledge representation language. The package can be run under UNIX, DOS or Windows NT.

Setting up a MAGPIE project

The notion of *project* may vary from one sequencing effort to another. For example, a project may comprise an entire genome, or a project may encompass a chromosome or perhaps the genome of an organelle. For *Sulfolobus*, the notion of project corresponds to the whole genome. A MAGPIE project comprises a set of *groups* which, most generally, is a set of contigs. The notion of a group may vary from one sequence project to another; for *Sulfolobus*, a group is the set of contigs associated with a particular cosmid. MAGPIE monitors the contigs within each group in a project and takes action as they change.

Once MAGPIE has been installed on a local machine and a new account and mailqueue have been created for a project, a single command (mkproject) prompts the user for information about the project and sets up a project directory structure. Each group is created with another interactive command (mkgroup). A daemon then appears, lives in the background, monitors the state of the sequence data, and manages requests and report generation. At any stage, users can change the default configuration to reflect their local needs.

Tailoring the System

Each group in a project has an assigned state that designates its sequencing status (e.g. primary, linking, polishing, or finished). Each state has a set of tools associated with it, where a *tool* is a particular piece of software (served over the Internet or installed locally) with a particular array of input parameters. So, for example, FASTA run with two different gap penalties constitutes two different tools. Each tool, under this definition, has a unique name (e.g. FASTA100 and FASTA50) and a protocol for invoking it. Any given remote or local server may serve multiple tools. In MAGPIE, this information is stored in a set of configuration files: a GROUPCONFIG file for each group in a project plus a STATECONFIG and a TOOLCONFIG file for the project. STATECONFIG contains the set of tools called for each state. For each tool, TOOLCONFIG contains a unique ID, e-mail address or invocation command, the type and format of input, and server information. The GROUPCONFIG file lists each contig in the group together with the state of the group and the state of each contig.

A user defines new states and adds new tools by modifying the configuration files. Suppose one wanted to create a new state, say, *released*, for a group and then add a new tool to be applied in that state; the user would add the state to STATECONFIG, add the tool to TOOLCONFIG, and edit the GROUPCONFIG for the group to change the state.

Reasoning about the collection of outputs from each tool requires the tools to be calibrated against each other. In addition, different users prefer different tools for different goals. MAGPIE accommodates calibration and user preferences through a SCORECONFIG file for the project and an optional parameter in the TOOLCONFIG file. For each type of score in each tool (e.g. expectation, probability of random occurrence, cumulative score), SCORECONFIG holds the threshold scores for *levels* of confidence. Each fact gleaned from a tool is associated with a level of confidence. For example, a Blast similarity with probability $1.0e^{-40}$ might receive a confidence level of 1 (or *high*) and a Blaize similarity with probability $1.0e^{-2}$ might receive a confidence level of 3 (or *doubtful, but possible*).

Brokering Requests

Each group has a directory structure for sequences that separates *New* sequences from *Working* sequences. Whenever sequences appear in the *New* directory, they are incorporated into the Working directory, and new requests for tool services are added to a waiting queue for the project. The state for a group may be predefined or handled interactively.

The MAGPIE daemon maintains a separate request queue for each local and remote tool server. The reliability of the flow control is of critical importance to prevent flooding of community resources. Only after MAGPIE receives a response from some tool is the next request pending for that server sent out. Each incoming response is immediately parsed into first-order logic so that logical inference mechanisms can use the information to make decisions about the genome sequence (Gaasterland *et al.*, 1994; Gaasterland and Lobo, 1994). The collection of logical formulae gleaned from the responses forms a growing knowledge base about the sequence data.

Open Reading Frames

As soon as a group of contigs is in a relatively clean state, as indicated in the STATECONFIG file, MAGPIE identifies open reading frames in each contig, extracts them, translates them using organism-specific codon usage tables, and send off requests for analysis to servers that take protein sequences as input. For *Sulfolobus*, a prokaryotic approach suffices: for each stop codon in each frame, find the furthest upstream start codon that is some minimal distance away. Each such ORF is regarded as a potential coding region. Similarity searches, assessments of codon usage, secondary structure predictions, motif searches, and three-dimensional threading techniques all contribute to an attempt to associate identity with the amino acid sequence (Rayl *et al.*, 1994).

Exploring the Current State

At any given moment, the current state of the interpretation of the current sequence data is maintained as a browsable set of interconnected hypertext files. A project has a home page that presents statistics on the current state of the project (including the number of contigs in each group, the number of ORFs, the number of bases in each state) and a link to a home page for each group. A group home page presents a list of its contigs; each contig has links to its DNA sequence and, if applicable, to the translated amino acid sequence; to a detailed report of what is known so far about the sequence; to a summary report of the descriptions of the proteins involved in the strongest similarities; and optionally, to the most recent human-readable outputs from each tool. Reports have links into a hierarchical structure of supporting evidence and direct links into public sequence data records (*e.g.* though ExPasy and SRS). Contigs associated with EC numbers are linked into PUMA, a browsable world of enzymes, metabolic pathways and multiple sequence alignments (see http://www.mcs.anl.gov/home/compbio/PUMA/Production/puma.html).

Using MAGPIE for Public Data

MAGPIE is not constrained to reside inside an actual sequencing project. It can be used to inspect and interpret any body of DNA sequence data. For example, to satisfy curiosity about all known *Escherichia coli* sequences from GenBank, we put each contiguous sequence in its own group with the state *finished*. Several days later, service requests were satisfied, and we had a browsable *E. coli* space to explore. Examples of MAGPIE analysis are accessible through http://www.mcs.anl.gov/home/gaasterl/magpie.html.

Our interest in all that is known about Archaea led us to do the same for all archaeal sequences from GenBank. We organized contigs into groups by species (1.2 Mbp of sequence, divided into 50 groups including *Archaeobacterium, Archaeoglobus, Crenarchaeotal, Desulfobacter, Halococcus, Methanococcus, Thermoproteus*). The resulting set of reports created a view of archaeal proteins interconnected by similarity together with the upstream and downstream DNA sequence for each coding region. The result is a continuously updated platform for accessing integrated archaeal sequence data.

Using Logic to Interpret Sequence Data

MAGPIE handles information of varying levels of confidence from multiple tools through logical rules represented in Prolog. The rules are used to assign properties to regions of sequence and then generate human-readable reports. The rules have the form **Head—Body.** and can be read as "Head is true if Body is true" or "Head if Body".

To illustrate how information from different sources can be combined through Prolog, we shall consider a rule that combines local and global similarity evidence to discern a coding region. The following rule says "There is a coding region in a contig if there is both a global similarity and a local similarity against the same database sequence, and the confidence in the coding region is the maximum of the confidences of the similarities":

coding_region(Contig,From,To,Confidence)—
global_similarity(Contig,From1,To1,DBSequence,Confidence1),
local_similarity(Contig,From2,To2,DBSequence,Confidence2),
overlaps(From1,To1,From,To),
overlaps(From2,To2, From,To),
Confidence=maximum(Confidence1, Confidence2).

The following two rules define that Blast supplies local similarities and that Blaize supplies global similarities:

local_similarity(Contig,From,To,DBSequence,Confidence)—
similarity(Contig,From,To,*blast*,DBSequence,Confidence).

global_similarity(Contig,From,To,DBSequence,Confidence)—
similarity(Contig,From,To,*blaize*,DBSequence,Confidence).

If we wish to use both motif and similarity information in a decision about a coding region so that weak similarity information reinforces a strong motif, we could use the following:

coding_region(Contig,From,To,*high*)—
motif(Contig,From1,To1,*high*),
similarity(Contig,From2,To2,Confidence2),
Confidence2 > *shaky*.

Note that in these examples, confidence levels need not be numeric and completely ordered. A partial lattice over symbolic values also works. Annotated logic programming and qualitative query-answering techniques allow us to reason with these types of confidences (Gaasterland and Lobo, 1994).

Concluding Remarks

MAGPIE is being used with the *Sulfolobus solfataricus* project to maintain a current interpretation of the sequence data and present it through navigable hypertext files. The system enables the sequencing staff to focus their energies on regions of interest and on verification of predicted sequence properties.

The next steps in the design of MAGPIE will move the system in two directions: first, MAGPIE will run a more comprehensive body of software tools, both local and remote; and second, the MAGPIE query environment will evolve to answer questions at the level of genome organization. Currently, the system runs tools that look for DNA- and protein-sequence similarities and motifs. We plan to extend the suite of tools to include a variety of structure-prediction, structure-similarity and ligand-docking tools.

To address whole-genome analysis, we must answer the question: which tools can work together at the genome level to provide the data necessary to answer complex queries across the genome? For example, to identify operons accurately, we must be able to identify and verify potential promoter regions and relate them to sets of coding regions. Prolog as a query language allows us to pose such questions given the appropriate input from analysis tools.

Also of utmost importance is the ability to enter wet-lab confirmation or refutation of predictions. For example, suppose each of three neighboring coding regions has a small set of predicted promoters. Under the right conditions, wet-lab refutation of all but one promoter can indicate the presence of an operon. Once predicted, the operon must in turn be verified in the wet lab.

In summary, MAGPIE is a functional system that, for several sequencing projects including *Sulfolobus solfataricus*, already eases the task of using a wide suite of remote and local analysis tools across the genome to interpret sequence data. MAGPIE is dynamically reconfigurable to accommodate the preferences, conceptualizations and computing environment of a particular sequencing project. With a Prolog-based reasoning engine, the system represents a flexible, extensible, knowledge-based platform for interpreting DNA sequence at the whole-genome level.

Acknowledgments

Dr. Gaasterland was supported in part for this work by the DOE Office of Scientific Computing, contract W-31-109-Eng-38. The *Sulfolobus* project is supported by the Canadian Genome Analysis and Technology Program, the Canadian Institute for Advanced Research, the National Research Council, and the Medical Research Council.

We thank James Tee (IMB) for his programming contributions to the system; Ross Overbeek (ANL) for his wisdom and programming contributions; Mats Carlsson (SICS) for his guidance with the workings of Sicstus Prolog; and Natalia Maltsev (ANL), Mark Ragan (IMB), and Evgeni Selkov (ANL) for their comments, insights and advice.

This is NRC publication No-39715 and an ANL MCS preprint.

References

Gaasterland, T., and J. Lobo. 1994. Qualified Answers That Reflect User Needs and Preferences. In *Proceedings of the International Conference on Very Large Databases*, Santiago, Chile, 309–320.

Gaasterland, T., J. Lobo, N. Maltsev, and G. Chen. 1994. Assigning Function to CDS Through Qualified Query Answering. In *Proceedings of the Second International Conference on Intelligent Systems for Molecular Biology*, Stanford University.

Gaasterland, T., and E. Selkov. 1995. Reconstruction of Metabolic Networks Using Incomplete Information. In *Proceedings of the Third International Conference on Intelligent Systems for Molecular Biology*, Cambridge, England.

Gaasterland, T., and C. Sensen. 1996. MAGPIE: Automated Genome Interpretation. *Trends in Genetics* 12(2):76–78.

Rayl, K., T. Gaasterland, and T. Overbeek. 1994. *Automating the Determination of 3d Protein Structure.* Technical Report ANL-P417-0294, Argonne National Laboratory.

46

Integrated Genome Informatics

Antoine Danchin, Claudine Médigue, and François Rechenmann

Inventions of techniques for DNA sequencing have brought about a true revolution in our way of considering a genome as a whole: we are now able to access the complete chemical definition of a living organism (*see also* Chapter 37). This ability is, however, light years away from understanding the significance of a genome. In particular, most DNA sequence initially deal with a *local* analysis of the DNA information content, taking only into account the availability of homologous structures or functions from different organisms, in order to provide a limited view of the global properties of genomes. We shall summarize here the present approach to integrate information, at the conceptual and experimental level, emphasizing the fact that computers can be used as experimental tools, generating a new source of investigation of living organisms, namely their study *in silico* (in complement to their usual *in vivo* or *in vitro* study) (Hénaut and Danchin, 1996).

Genome studies define a scientific domain in which a steadily increasing flow of data produces very large amounts of knowledge. Informatics is clearly involved at each and every step: data acquisition, data analysis and data management. Sequences are stored and organised in data banks or data bases according to their degree of reliability and consistency. The corresponding knowledge mostly results from the application of various sequence analysis methods, which allow the investigator to single out regularities and predict functions (*see also* Chapters 38, 39 and 45). Until now, not much has been done in order to improve the use of this knowledge. We shall describe an approach which aims to provide an integrated use of informatics, in order to extract the best information content from a genome.

Data Acquisition

Sequencing short pieces of DNA (*i.e.* less than 5 kb) is still often performed manually. In general one starts from both ends of the original segment and extends the sequence using oligonucleotides, or a set of nested deletions. This can no longer be considered an easy or cost-effective approach when sequencing large segments. In this latter case it is more efficient to use a DNA library generated by shotgun cloning of ultrasonicated DNA as starting material, or use DNA fragments generated with frequently cutting restriction enzymes. These fragments can be submitted to electrophoretic separation and identified using either fluorescent hybridization probes, with on-line detection, or by autoradiography of the gel on which labelled fragments have been separated. In the case of shotgun cloning, it is important to make use of programs permitting automated alignment of fragments, generating contiguous overlaps (Staden, 1990). At this stage of data acquisition, elimination of parasitic data (e.g. vector sequences) is an important issue but not difficult to solve. In all cases a computer treatment of the raw data is necessary, and it generally is useful to use knowledge generated from previously known sequence data, to analyze and identify the likely errors in sequence (mostly insertions or deletions inducing frameshifts in coding sequences, or GC swaps). In spite of its obvious importance, this integration step is not yet necessarily implemented as an automatic approach of data acquisition, although it is clear that, manually, scientists do check for possible errors. Sequence analysis is therefore the central step in integrated genome informatics (Médigue *et al.*, 1995).

Sequence Analysis

Identification of basic "signals"

The first step in the identification of significant "signals" in sequences, is the identification of regions coding for proteins. In spite of the fact that many methods aiming at solving this problem have been published, no totally unambiguous method has been identified. In the case of model bacteria, the problem has been more simple because the codon usage, known from the first sets of sequenced genes, was found to be strongly biased. Knowing that an open reading frame (ORF, *i.e.* a region of 3n nucleotides, limited by translation termination codons, UAA, UAG or UGA) contains a high frequency of such biased codons, strongly suggests that a protein CoDing Sequence (CDS) is contained within the Open Reading Frame (ORF), in the corresponding reading frame. This method reflects only the bias derived from genes which happened to be sequenced first. Unfortunately, as found by Médigue *et al.*, there exists at least one class of genes in *Escherichia coli* which escapes such identification, a class of genes exchanged by horizontal transfer (Médigue *et al.*, 1991). Other methods should therefore be used as well. As a case in point, Fichant and Gautier proposed to base the identification of CDSs on heterogeneity in the frequency of bases present in each of the three reading frames (Fichant and Gautier, 1987). Alternatively, the GeneMark software, derived from the work of Kleffe and Borodovsky on periodical Markov chains, gives efficient CDS prediction, provided genes have been first clustered into significant classes used as training sets (Kleffe and Borodovsky, 1992).

In addition to their specific use of the code, sequences coding for proteins are also defined by the presence, upstream of the start codon, of a ribosome binding site. In 1974 Shine and Dalgarno proposed that the 5′end of bacterial mRNAs was complementary to the 3′OH terminal end of ribosomal 16S RNA (Shine and Dalgarno, 1974). This was the first example of a "consensus" sequence (AAGGAGGT), and it started the quest for sequences determining recognition processes through protein-nucleic acid interactions. In the case of translation it has been difficult however to define the domain necessary for translation initiation, because there is a clear contribution of secondary mRNA structure to initiation efficiency. Nevertheless, notwithstanding a significant role of secondary structures, the Shine-Dalgarno sequence is the only signal which has been clearly demonstrated to be relevant in translation initiation. It is, generally, closer to the consensus in strongly expressed genes than in the others. It can therefore be used for identifying CDSs in bacterial genome sequences, at least during the first round of gene identification. Less obvious signals can be identified at further steps of the analysis.

With regard to identification motifs signalling RNA genes, the situation differs for rRNA and tRNA. rRNA molecules can be easily identified because of the phylogenic constraints which operate on ribosomes. In the case of tRNAs it is more difficult, because genes could be cryptic, or expressed under very special conditions. The program of Fichant and Burks, which integrates structural data on many tRNA genes, permits scanning of the genome for putative RNA sequences (Fichant and Burks, 1991). This program is supposed to yield (in *E. coli*) less than 1, presumably false positive, positive outcome per 3×10^5 base pairs. This low value should prompt further investigation by the biologist, when a positive score, suggesting the presence of a known gene is not encountered, in order to see whether this might indicate the presence of a cryptic gene.

Predicting other signals, such as promoters, is much more difficult. In general, promoter sequences are identified as motifs corresponding to the binding site of the RNA polymerase complex involved in transcription initiation. This permitted early investigators to propose the existence of consensus regions supposed to behave as binding sites for RNA polymerase, after isolation of a collection of promoter region sequences. Bacterial promoters encompass a $\approx$ 75 nucleotides region upstream of the RNA start site. Counting from the start site, Pribnow proposed that a consensus sequence "TAT(A/G)AT", was situated around -10. Siebenlist *et al.* refined the consensus by adding a -35 region, "TTGACA", after having analyzed a collection of *E. coli* promoters (Siebenlist *et al.*, 1980). Automatic identification of promoters started with this consensus concept. However, as more bacterial promoters were collected, the very notion of consensus became more and more fuzzy, because for each position of a promoter, an exception could be found. The degeneracy is now such that a combination of -10 or -35 motifs can be found every 200 nucleotides in a random sequence. In fact the tertiary structure of DNA is certainly involved in the recognition of a promoter by RNA polymerase and it seems obvious, considering the 3D structure of the double helix axis that even if one combines -10 and -35 regions having an exact consensus on a helix having its axis pointing downwards, when the recognition asks for it pointing upwards, the corresponding sequence will not fulfill the requirements to be a promoter sequence. This means that special methods, where knowledge about the 3D structure of DNA is incorporated, must be used for predicting promoters as well as other DNA structures recognized by appropriate factors.

Sequence Alignments

As soon as protein and nucleotide sequences became available, scientists have attempted to compare sequences to each other, and to align sequences that were suspected to match, after having diverged from a common ancestor. While it was relatively easy to align similar sequences, it was often impossible to align very divergent sequences, unless some other or related sequences were also known and compared in parallel (multiple rather than pairwise alignments). This resulted in the construction of algorithms permitting alignments and multi-alignments, and providing the corresponding consensus of multiple sequences. In the case of protein alignments this requires the use of matrices meant to represent relative similarities between amino acids. Some matrices take into account the constraints on amino acid residues that derive from comparison of known 3D structures, and multiple alignments can implement knowledge from such structure. However a single matrix is used for the whole primary sequence of each polypeptide, implying that one deals with a composite view of the equivalence rules between amino acids at each position in the sequence. This cannot represent what occurs in real life because at some positions, for example, it is the basic feature which is important (meaning that K, R and H are equivalent) whereas at other positions the essential feature could be the fact that the residue is large (meaning that F, Y, R, H and W are equivalent), so that one should have an equivalence matrix for each position in the sequence. This indicates that new DNA analysis programs must still be developed in this field, despite the large number of protocols already published (see Hénaut and Danchin, 1996, for references).

Identification of Other Signals Using Complex Patterns

When many sequences are compared with each other, one often wishes to extract a significant pattern not as a consensus sequence, but, rather, as a matrix of probability reflecting the frequency of a given motif at a given place. This is reminiscent of pattern recognition problems involving learning techniques or techniques used in pattern recognition by vision. Stormo and coworkers have used an ancestor of neural networks methods, the Perceptron, for creating matrices permitting identification of the ribosome binding sites in *E. coli* messenger RNAs (Stormo *et al.*, 1982). They have been followed by many authors involved both in the study of DNA and protein sequences. Other learning techniques rest on the explicit construction of "grammars" (i.e. organized sets of combinatorial rules meant to describe the features of an object or of a phenomenon). They can be divided roughly into two classes: learning by assimilation and learning by discrimination. An example of the first type is the Calm software (Soldano and Moisy, 1985), whereas an example of the second type is the Plage software used in the identification of the relevant features of signal peptides in *E. coli* (Gascuel and Danchin, 1986). Both types are dependent on a grammar used to generate descriptors which act on the training sets. In general, grammars are essential to describe biological phenomena, and their efficiency rests heavily on the actual biological knowledge of their conceptors, providing therefore an interesting way to incorporate biological knowledge into automated procedures. For this reason one should probably always consider as the most pertinent ones, the grammars which are devoted to a specific question. An example is the work of Collado-Vides, who constructed a grammar for the identification of regulated promoters in *E. coli* (Collado-Vides, 1992).

Sequence acquisition has been growing exponentially for ten years. In parallel the power of computing has also increased in an exponential fashion, and it has been possible to foresee successful treatment of biological data that would have been impossible just a few years ago. Because DNA and protein sequences can be seen, as a first approximation, as linear texts of an alphabetical writing, they can be treated using the theories of information which are precisely to study study such texts as their favoured objects. In the following we will develop ideas showing that the alphabetic metaphor, and the corresponding treatment of information, is likely to have a very deep meaning in terms of genome. This will require progress in our reflection on the concept of information. Therefore, the sections below can only be regarded as an attempt to clarify what information is.

The Meaning of Sequence Information: Towards a Global View of Genomes

What is seen by the replication machinery: Shannon's information

One often refers to the "information" present in genomes. What does this mean? Shannon and Weaver proposed a definition that was applied to the transmission of linear sequences of integers through an electromagnetic wave channel, subject to some noise. Their concept gives a very crude view of a message, or rather of a collection of messages, explicitly considering not their meaning, but the local accuracy of their transmission. Genomes can be analyzed using this concept if one considers only the information conveyed during the process of replication, leaving aside the information involved in gene expression. This often brings interesting properties in the limelight, but one should always bear in mind the fact that the actual information content of a sequence is much richer. As case in point we will describe below other ways of considering the information content of sequences, where the global signification is taken into account more accurately (Yockey, 1992).

What is considered using Shannon's description is the probability of a message, that, in the usual description of genomes, reduces to the actualization of a succession of single letters (A, T, C, or G) at a given position of the genomic text. In the replication process, an obvious improvement would be to consider not a one letter string of the program but a string of relevant words (sequences of letters of definite length). But this assumes that one knows the words that are relevant. Analysis of the genomic text using classification methods is therefore a very important prerequisite. Among many other possibilities, several improvements in the comutation of information should be considered. For example, when investigating the information of a system made up of many components, one should take into account not only the distribution of the components in space or according to their energy states, but also the various scales at which they occur: there is some contribution in the building up of a hierarchy. And if one wishes to have information about a system, it may be important to consider the structure of the hierarchy.

Shannon's information only reflects a mathematical property of statistical analysis. However, when restricted to what assumes its mathematical form, Shannon's information can be used as a first description of DNA sequences information content, for example in the identification of coding regions in DNA. A consequence of this approach is the construction of large collection of gene sequences from a given organism. A genome is the result of a historical process which has driven a variety of genes to function collectively. Many biologists using computers have experienced the following. When one scans through a sequence library for similarity with a query sequence, one often ends up with a list of sequences that have only in common the organism to which the initial sequence belongs. This is as if there existed a *style* specific of each organism, independent of the function. Is this impression true? Several ways for investigating the overall structure of genomes have confirmed and extended this observation. This indicates that there must exist some sort of mutual interaction permitting the cell to discriminate between self and non-self.

This first level of considering the information content of a genome is rewarding. In particular it tells us that the actual selection pressure on the fine scale evolution of sequences inside genes is actually visible when one possesses a large collection of genes for analysis. Because this corresponds to a fine variation, comprising many genes at the same time, it becomes difficult to be able to propose experiments to test whether the interpretation of the observation adequately reflects reality.

Eucaryotes and Procaryotes: Algorithmic Complexity

Is it possible to go beyond this first level, and extract significant global properties of genomes? Relationships between chemical objects, such as those which are at the core of life, should be considered. In particular, the complex nature of the information carried in the DNA molecule that constitutes genes cannot be present in Shannon's information. Biology provides us with a metaphor that displaces the idea of information towards a new field, that of programming and informatics. Kolmogorov, and Chaitin and Solomonoff, have formulated the problem of the information content of a chain written with a finite alphabet as follows: a way to describe a chain of digits is to try to reduce its length so that it is accommodated into the memory of a computer using the minimum of space, without altering its meaning, or rather, without losing its information content. This is called *compressing* the chain. They proved that it is possible, given a chain, to define the shortest formal program (in terms of Turing's universal computation algorithms) that can compress an original chain, or restore it, after being compressed. The information of a chain S is then defined in this model as the minimal length of the universal program that can represent S in a compressed form (Yockey, 1992).

With this definition, a completely random chain cannot be compressed. In contrast, a chain made of repeated elements can be summarised as a very short program: the sequence of the element, and the instructions "repeat" and "end". It must be stressed here that an *apparently* random chain is sometimes compressible: this is the case of the chain comprising the digits of the decimal writing of the number π, which can be generated by a short program. In this context the information of a sequence is defined as the measure of its compressibility. Represented as sequences of letters, genes and chromosomes have an intermediate information (complexity) content: one finds local repetitions or in contrast, sequences which are impossible to predict locally. Their complexity is intermediary between randomness and repetition. The overall complexity of sequences originating from higher eucaryotes or procaryotes is very different, and it associates genomic sequences to both sides of the "uninteresting" fraction of information (repetition or randomness). The complexity of the former is more repetitive, and is usually much lower than that of the latter which looks more random.

In order to justify introduction of this new concept in genetics, we even wish to propose that this difference in algorithmic complexity is linked to the *cause* of the main morphological differences between procaryotes and differentiated organisms. The former, which occupy biotopes due to fast and varied adaptation, are forced to keep up with a small genome. This implies that several levels of meaning must often be superimposed on each other in a given sequence. The result of this constraint is that their actual algorithmic complexity is high (it looks more like a random sequence). But this constraint has an important physical consequence: superimposing signals excludes the possibility to combine recognition processes by different proteins, for it is not possible to place two different objects at the same time at the same location. In eucaryotes, the lack of limitation in the length of the DNA text permits juxtaposition of recognition signals by regulatory proteins, and allows exploration of the properties of their association following the rules of combinatorial logic. This permits differentiated expression, as a consequence of transcription control (promoters, enhancers and the like), but this may be also the reason for the multiplication of introns in higher eucaryotes, that behave as inserts without apparent signification, in the very coding regions of genes. Analysis of information in terms of algorithmic complexity permits us, therefore, to propose explanations accounting for the difference between major organisms types, while Shannon's information only gives a meaning to conservation of a DNA sequence, with a low level of errors from generation to generation, through the replication machinery.

Evolution: Bennett's Logical Depth and Critical Depth

Algorithmic complexity represents genomes as fixed, synchronous, entities. But we can go further: something more should be said about their information content. In the examples given, the concept of information did not require the notion of *value* (depending both on the object considered and on the nature of the problem that has to involve the object). A specific value of an information is its actual availability, and in the case of living organisms, this means knowledge of the paths which have been followed through evolution leading to the generation of the organisms under study. This requires introduction of the parameter of time. Using arithmetics this can be easily illustrated: although the program defining π is highly compressible, as is the program generating 01010101. . . , the information imbedded in π is richer than the one imbedded in 01010101, because calculating the exact value of the nth digit of π can be very lengthy (if n is large) so that if we had an independent means to know its value, the information thus obtained would be extremely valuable. Not only would we like to have an idea of the complexity of a sequence (this can be evaluated using Shannon's approach or the algorithmic approach), but also of the availability of the information contained in the sequence. Bennett has formalised this aspect of sequences, and named it "logical depth" (Bennett, 1988). The logical depth of a sequence S is the minimum time required for generating S using an universal Turing machine (i.e. a formal machine reading a ribbon marked with symbols, and deciding, according to its internal state, and to the nature of the symbols, to go forwards, backwards, stop, or change its internal state). This is particularly relevant in the case of DNA sequences, because all what can be inferred from the knowledge of a sequence derives from the accessibility of its actual information content. To see more repetitions in eucaryotes and more apparent randomness in procaryotes, does not account for the paths which have led to such a large overall difference in genomes.

We must stress here that this aspect of information does not completely exhaust the natural concept of information. In particular, it rests on the existence of an abstract, but perfect, Universal Turing Machine. If the building up of the machine is the result of the actualization of a program, then one should take time into account, once again, and consider only those algorithms the length of which is smaller than the lifetime of the machine as a given, unaltered, structure. A time limit, defining the algorithms permitting to construct the machine, must also be taken into account: it defines a *critical depth*, likely to be of major importance in the case of living organisms. We shall limit ourselves to the descriptions presented above, however, to consider now the specific case of living organisms.

Genomes "in silico": Information and Heredity

The arguments developed above are meant to be a substratum for reflections about the metaphor of *program* which is currently used in molecular genetics. It is known that a giant molecule, DNA, consisting of the sequential arrangement of four types of related molecular species specifies entirely the structure and dynamics of any living organism. We shall not consider here the problem of origins of such organisms but only try to investigate the nature of the relationship between the DNA sequence and the building up of an organism. As in the case of the reflection of Turing in the part of mathematics named Number Theory, one can surmise that a machine must exist in order to interpret the content of a DNA sequence. This machine is a pre-existing living cell, made of a large, but finite, number of individual components organised in a highly ordered way.

Given such a machine, a fragment of DNA, provided it contains a well-formed sequence, is necessary and sufficient to produce a specific behaviour representing part or all of the structure and dynamics of an organism. The appropriateness of the alphabetic metaphor used to describe the DNA sequence is prominent in the fact that it is possible to manipulate on paper four symbols A, T, G, C, and to organise them in such a way that, when interpreted into chemical terms (i.e. linked into the proper sequence arrangement into an artificial DNA molecule) they produce a behaviour having at least some of the properties of real natural biological entities. That the machine is, in a limited way, independent of DNA has also been indicated by the fact that it is possible in some organisms (amphibians) to replace the nucleus of a cell by a new nucleus, resulting in the construction of a new organism (i.e., both a new DNA molecule, and a new machine). However, DNA alone is absolutely unable to develop into a living organism: this is why the question of origin must be considered if one wishes to resolve this chicken and egg paradox, through the coevolution of the machine and of the program.

In order to apply the consequences of Turing's arguments it is necessary that program and data be well identified as separate entities. Biotechnology uses this fact to synthesise human proteins in bacteria. But this corresponds to small genome fragments. Virus which infect cells do the same in order to reproduce. Finally, at least as a first approximation, bacteria behave as if separating data (the cell machinery, excluding DNA) and program (the chromosome): we have found that more than a fifth of the *E. coli* chromosome is made up of DNA on which selection operates in such a way that it favours and sustains horizontal exchanges not only between bacteria of the same species, but also between distantly related organisms.

If one considers the fate of a DNA molecule in a cell it is possible to describe the various functions derived from its presence as calculable in an algorithmic way. Many routines in the process are recursive (i.e. call themselves and terminate). It is a generative process which permits generation of both the DNA molecule and the machine and the data necessary to interpret it. In this context a living organism is the algorithmic actualization of a program, the initiation of which has been obtained through a selective process, conserving only actualizations which are stable in time. And it should be noted here that this means that selection does not operate on the phenotype (data) nor on the genotype (program) but on the unfolding of the algorithm, i.e. on the organism itself, thus placing in a new light the usual paradox facing all selective theories, and standard Darwinism in particular.

We are therefore facing the puzzling observation that in order to evaluate the actual information content of a DNA molecule, one has to trace back its history, which involves an extremely slow process: the information content of a DNA sequence is so deep (in Bennett's terms) that it would require the age of life on earth to evaluate its true complexity. Bearing this in mind we should be very careful and modest not only when trying to predict the future of living organisms, but simply when analysing genomic sequences!

Managing Data to Create Biological Knowledge

Genome analysis will yield major insights into the chemical definition of the nucleic acids and proteins involved in the construction of a living organism. Further insights will come from the chemical definition of the small molecules, which constitute the building blocks of organisms, through the generation of intermediary metabolism. Finally, because life also requires in its definition the evaluation of the processes of metabolism and compartmentalization, it is important to relate intermediary metabolism to genome structure, function and evolution. This requires elaboration of systems for constructing actual metabolic pathways and when possible, dynamic modelling of metabolism, and to correlate pathways to genes and gene expression. In fact, most of the corresponding work cannot be of much use because the data on which it rests have been collected from extremely heterogenous sources, and most often obtained by *in vitro* studies.

Formal computer-based representation appears to be the only practical way to integrate and to manage the various categories of data and knowledge involved in the process of genome analysis. (*see also* Chapters 38, 39, 43 and 45)

At the present time, paper still remains the main knowledge sharing medium. This traditional medium presents however several severe drawbacks which certainly slow down the building up of scientific knowledge. It generates important time delays between knowledge submission, revision and publication on the one hand, and dispersion of the results over a multiplicity of journals and conference proceedings on the other hand. The progressive replacement of paper by electronic media certainly improves the situation by offering direct and easy access to huge volumes of scattered sources. The increasing number of World Wide Web genomic information servers clearly substantiates these advantages (*see also* Chapters 38–45).

However, the synthesis of current research, in the form of books or monographs, remains a highly time and skill demanding process: bibliographic references must be looked for, selected, read, interpreted and compared, before a text that attempts to be a consensual presentation of the state of knowledge in a specific scientific domain, may be written. One must therefore first collect, organize and actualize the existing data. In order to be effective and lasting, collecting the data should proceed through the creation of specialized databases. To manage the flow of data created by programs aiming at sequencing complete genomes, specialized databases have been developed. They make it possible not only to bypass a meticulous and time-consuming literature search, but also to organize data into self-consistent patterns through the use of appropriate procedures that aim at illustrating collective properties of genes or sequences. In addition to sequence databases, it has now also become important to create databases where the knowledge progressively acquired on intermediary metabolism is symbolized, organized and made available for consultation using to multiple criteria. Using organized data it will become possible to make *in silico* assays of plausible pathways or regulation before being in a position to make the actual test *in vivo*. Incremental and competing building up of formal knowledge bases is presented below as an efficient option, still in its infancy.

Knowledge Modeling and Knowledge Bases

Knowledge bases have long been considered only as components of the so-called "expert systems," i.e. systems which rely on large amounts of knowledge to solve problems. The existence of meaningful object-based knowledge models now allows one to develop knowledge bases for the sole purpose of knowledge representation. In this context, the first role of a knowledge base is to be an explicit and formalised model of the knowledge of a domain. Object-based models turn out to be very natural conceptual media for expressing knowledge in molecular genetics, confirming early attempts of biological knowledge representation in the MOLGEN project (Friedland and Kedes, 1985). This project led to the organization of knowledge bases into several distinct entities, termed concepts. To each concept a hierarchy of classes and sub-classes is assigned. A class gathers the generic knowledge over a set of potential instances (individual examples in a given "abstract" class). Model building of molecular genetics knowledge has thus led to the identification of four categories of knowledge that need to be represented (Rechenmann, 1995):

Descriptive Knowledge

Under descriptive knowledge we find genes, promoter or terminator as examples of elementary object classes. Their descriptions are made up of properties (for example, the level of expression of a gene) and relationships with other entities (for example, the relationship between a gene with its RBS). A description can be refined, thus introducing a new sub-class which inherits the contents of the class it specializes into. This hierarchical organization can be used by a classification procedure that looks for possible locations of an object associated with a class, in the corresponding sub classes, thus allowing a better characterization of objects. Introducing the relationship "part-of" leads to the description of composite objects, i.e. objects which can be viewed as a recursive composition of parts. The bacterial operon provides a typical example of such object classes. Its parts are the genes and the various signals which regulate its expression. At this level of description, only the structural relationships are considered.

Behavioral Knowledge

The operon displays a behavior: the expression of the genes it contains is inhibited or activated according to environmental conditions. The behavior of such objects can be described using dynamic models. However, because information on the processes involved is often lacking, qualitative models, such as systems of differential equations are more adequate candidates than quantitative models. Examples of qualitative models which have been proposed are the model of gene expression in the HIV life cycle (Koile and Overton, 1989), and the model of the behavior of the *E. coli* tryptophan operon (Karp, 1993). Representing metabolic pathways, and their connection with gene expression, are then very natural extensions. An open issue is the design of an object-based knowledge model, which would efficiently and cleanly integrate behavioral knowledge with descriptive knowledge.

Methodological Knowledge

The set of available methods for sequence analysis is steadily growing. The process of selecting and stringing together these methods, with research goals in mind, is simultaneously becoming more and more tedious. Evaluating the relevance of the results they produce is another difficulty. Paper, or even on-line documentation, is certainly necessary, but does not solve the problem. As already tested for scientific computing in engineering domains, to help the user, it is necessary to apply knowledge about the methods. Such methodological knowledge can be represented in an object-based knowledge model. (Médigue, *et al.*, 1995). It includes description of problems and their respective solving strategies, *i.e.*, given a problem, how to select the appropriate (chaining of multiple) processing method(s).

A method can be globally described as a class, the fields of which are associated with the input and output data. These classes can be organised into hierarchies, in which general methods control more specific ones. Problem-solving strategies are described as tasks. A task is associated with a problem and is, like a method, represented as a class and described by a specific set of input and output entities. To solve the task is then to solve the problem.

Let us consider the very common biological question "Where are the putative coding regions in my new DNA sequence located?". To achieve this goal, the general task *CDS-predictions* has to be carried out:

- According to its known input values and/or user choices during the solving process, this global task is first characterised (classification process). It should be specialized depending on whether a reference system permitting identification of coding regions (such as codon or tRNA frequencies tables) is available or not. This first step is finally devoted to the selection of the most adapted CDS prediction method among the set of available methods, for example the *GeneMark-analysis* tool, which uses a codon frequencies table (Borodovsky and McInnich, 1993).
- Subsequently, the selected complex task must be recursively split into sub-tasks, down to elementary tasks. This decomposition is described by an operator (sequence, selection, etc.) and the list of tasks on which this operator applies. For example, the *GeneMark-analysis* class is split into the sequence: *Matrix-selection*, then *GeneMark-computing*, then *Postscript-translation*, then *Graphic*.
- Finally, an elementary task is solved by directly executing one of the attached methods. This method is often an external procedure and its class is generally described as follows: the objects containing the data necessary for the external program to run are dumped into a file, the foreign program (such as *GeneMark*) is executed via a shell procedure, and then results are stored in the knowledge base.

Terminological Knowledge

Building a knowledge base is a representation process. The main product of this process is of course the model itself, i.e. the knowledge base. But a very important side product is the increase in the understanding of the domain by the knowledge base designer. Sharing a knowledge base thus implies to be able to explain the representation process itself and to make the representation hypotheses explicit. In order to avoid infinite recursive loops in knowledge expression, the description of the various representation choices and justifications should be expressed in natural language. In particular, a knowledge base user should be allowed to retrieve the text fragments which have been used by the designer as an aid during model building.

A first step in the direction of improving formal knowledge bases with the help of textual knowledge consists in associating hyper-text nodes to the names of the object classes and of their fields. The limitations of this connection are obvious. Only the names actually used in the knowledge base can be annotated. A user approaching the knowledge base is therefore constrained to use this vocabulary. Moreover, automatic indexing, and later retrieving, of the scientific literature which is linked to the contents of the knowledge base is very limited.

A further step thus consists in introducing a lexicon as an interface between natural language, which is used by the literature and by the users, and the formal knowledge base. The lexicon gathers the terms which are relevant to the contents of the knowledge base. A subset of these terms is the set of names of classes and their fields. Relationships between terms, such as synonyms which denote the same concept and hyponyms which denote more specific concepts, are also included, so that the user can navigate inside the lexicon before entering the knowledge base itself.

The design and implementation of an integrated cooperative environment for genome sequence analysis, relying on these four categories of knowledge to assist the biologist in his discovering process, is thus a fascinating challenge.

Conclusion

The ever increasing flow of data from genome sequencing projects led biologists to use computers essentially for their capacity to store large volumes of data and to make intensive computation. Sequence databases and sequence analysis packages are the visible part of this investment in computer science. In a very few years from now, many complete genomes of "model" living organisms will have been completely sequenced. Accordingly, the emphasis will shift from data to knowledge: how computers can help us in discovering new objects, new relationships between objects, new mechanisms? This shift has already started, as shown by the increasing complexity of sequence databases, and the use of ever more descriptive data models.

It is highly necessary however to anticipate this incoming shift. The anarchic development of data banks under the pressure of the flow of sequences has led to the present situation where a large gap has been created between the real biological needs, which could be satisfied by advanced computer technologies, and the computer tools which are actually used.

Very interesting challenges are therefore presented to the computer scientists, especially regarding knowledge representation and knowledge management.

References

Bennett, C. H. 1988. Logical Depth and Physical Complexity. *The Universal Turing Machine: a Half-Century Survey.* Oxford University Press, Oxford.

Borodovsky, M., and J. McIninch. 1993. GENMARK: parallel gene recognition for both DNA strands. *Computers and Chemistry* 17:123–133.

Collado-Vides, J. 1992. Grammatical model of the regulation of gene expression. *Proc. Nat. Acad. Sci. USA* 89:9405–9409.

Fichant, G., and C. Burks. 1991. Identifying potential tRNA genes in genomic DNA sequences. *J. Mol. Biol.* 220:659–671.

Fichant, G., and C. Gautier. 1987. Statistical methods for predicting protein coding regions in nucleic acids sequences. *Comput. Appl. Biosci.* 3:287–295.

Friedland, P., and L. K. Kedes. 1985. Discovering the secrets of DNA. *Communications of the ACM* 28:1164–1186.

Gascuel, O., and A. Danchin. 1986. Protein export in prokaryotes and eukaryotes: indications for a difference in the mechanism of exportation. *J. Mol. Evol.* 24:130–142.

Hénaut, A., and A. Danchin. 1996. Analysis and predictions from *Escherichia coli* sequences, or *E. coli in silico*. In Escherichia coli *and* Salmonella typhimurium: *Cellular and Molecular Biology*. F. C. Neidhardt, (ed.) Princeton Editorial Associates, Princeton, vol. 1, ch. 114, pp. 2047–2065.

Karp, P. D. 1993. A Qualitative Biochemistry and its Application to the Regulation of the Tryptophan Operon. Artificial Intelligence and Molecular Biology, AAAI Press, 289–324.

Kleffe, J., and M. Borodovsky. 1992. First and second moment of counts of words in random texts generated by markov chains. *Comput. Appl. Biosci.* 8:433–441.

Koile, K., and G.C. Overton. 1989. A Qualitative Model for Gene Expression. *Proc. of the 1989 Summer Computer Simulation Conference*, 415–421.

Médigue, C., I. Moszer, A. Viari, and A. Danchin. 1995. Analysis of a *Bacillus subtilis* genome fragment using a co-operative computer system prototype. Gene COMBIS, http://www.elsevier.nl/journals/genecombis/preview/. *Gene* 165:GC37–GC51.

Médigue, C., T. Rouxel, P. Vigier, A. Hénaut, and A. Danchin. 1991. Evidence for horizontal gene transfer in *Escherichia coli* speciation. *J. Mol. Biol.* 222:851–856.

Rechenmann, F. 1995. Knowledge Bases and Computational Molecular Biology. *Towards Very Large Knowledge Bases,* IOS Press, 7–12.

Shine, J., and Dalgarno, L. 1974. The 3′-terminal sequence of *Escherichia coli* 16 S ribosomal RNA: complementarity to nonsense triplets and ribosome binding sites. *Proc. Natl. Acad. Sci. USA.* 71:1342–1346.

Siebenlist, U., R. B. Simpson, and W. Gilbert. 1980. *E. coli* RNA polymerase interacts homologously with two different promoters. *Cell* 20:269–281.

Soldano, H., and J. L. Moisy. 1985. Statistico-syntactic learning techniques. *Biochimie* 67:493–498.

Staden, R. 1990. An improved sequence handling package that runs on the Apple Macintosh. *Comput. Appl. Biosci.* 6:387–393.

Stormo, G. D., T. D. Schneider, L. Gold, and A. Ehrenfeucht. 1982. Use of the 'Perceptron' algorithm to distinguish translational initiation sites in *E. coli*: Nucl. Ac. Res.,10:2997–3011.

Yockey, H. P. 1992. *Information theory and molecular biology.* Cambridge University Press, Cambridge.

SECTION 3

Physical Maps of Bacteria and Their Methods for Construction

Introduction

In this section of the book we present physical (and genetic) maps of a number of different bacteria. With the advent of pulsed field gel electrophoresis, it is now a routine matter to generate a physical map of any bacterial genome. Details of the construction of these and other maps, as well as a wealth of other information from such maps, have been reviewed by Cole and Saint Girons (1994). In this section we also summarize other information about the bacteria whose genomic maps are presented. In Table 1 we present a few of the characteristics of the genome themselves. For an additional comprehensive list of rare restriction sites, *see* Table 1 in Chapter 24. Table 2 presents miscellaneous biological information of interest to genomics. Table 3 presents genetic characteristics of these bacteria, with reference to the ability to perform genetic analysis. Clearly the ideal is a physically mapped genome that is amenable to genetic analysis. Unfortunately, this is not always the case. Finally, in Table 4 we summarize the methods of genomic fingerprinting that have been applied to these organisms. It is our hope that this material will serve in future analyses of these bacteria.

Cole, S. T., and I. Saint Girons. 1994. Bacterial Genomics. *FEMS Microbiol. Rev.* 14:139–160.

Table 1. Genome Characteristics

Organism	Strain	Genome size (kb)	GC (%)	Chrom. number	Chrom. topology	Rare cutting enzymes
Anabaena sp.	PCC 7120	7100	42.5	1	circular	AvrII, SalI, PstI, SphI, SstII, FspI, NcoI
Aquifex pyrophilus		1620	40	1	circular	CeuI, NotI, SpeI, XbaI
Bacillus subtilis	168	4188	43	1	circular	NotI, SfiI, Sse8387I, I-CeuI
Bordetella pertussis	Tohama I	3750	66	1	circular	XbaI, SpeI, SwaI, PacI, PmeI
Borrelia burgdorferi	Sh-2-82	930-955		1	linear	
Bradyrhizobium japonicum	3I1b110	8700	63	1	circular	PacI, PmeI, SwaI
Brucella melitensis	16M	3200 (2050, 1150)	59	2	circular	PacI, SpeI, XhoI, XbaI
Burkholderia cepacia	ATCC 17616	7000 (3400, 2500, 900)	67.6	3	circular	PacI, PmeI, SwaI
Burkholderia cepacia	ATCC 25416 (type strain)	8100 (3700, 3200, 1100)	67	3	circular	I-CeuI, PacI, PmeI, SpeI, SwaI
Campylobacter jejuni	TGH9011 (ATCC43431)	1700-1800	31	1	circular	SalI, SalII, SmaI, XhoI
Caulobacter crescentus	CB15	4000	67	1	circular	DraI, SpeI, AseI
Clostridium beijerinckii	NCIMB 8052	6600	27	1	circular	SfiI, RsrII, I-CeuI, SmaI, ApaI, BssHII
Clostridium perfringens	CPN50	3600	25	1	circular	ApaI, AviII, I-CeuI, KspI, MluI, SmaI, SfiI
Enterococcus faecalis	OG1RF (ATCC 47077)	2825-3093	37-40	1	circular	NotI, SfiI, AscI, SmaI
Escherichia coli	MG1655	4700	51	1	circular	NotI, XbaI, SfiI, BlnI
Haemophilus influenzae	Rd	1830	38	1	circular	SmaI, ApaI, RsrII, NotI
Halobacterium salinarium	GRB	2038	66	1	circular	AflII, AseI, DraI, HindIII, SspI
Haloferax volcanii	DS2	2920	65	1	circular	BamHI, BglII, DraI, HindIII, PstI, SspI

Table 1 (continued)

Organism	Strain	Genome size (kb)	GC (%)	Chrom. number	Chrom. topology	Rare cutting enzymes
Helicobacter pylori	UA802, UA861, NCTC11637, NCTC11639	1630-1730	35-37	1	circular	NotI, NruI
Lactococcus lactis subsp. *cremoris*	MG1363	2560	38	1	circular	NotI, ApaI, SmaI, I-CeuI
Lactococcus lactis subsp. *lactis*	IL1403	2420	38	1	circular	NotI, ApaI, SmaI, I-CeuI
Leptospira interrogans	serovar icterohaemorrhagiae, serovar pomona strain RZ11	4400-4610, 350	34-39	2	circular	NotI, AscI, SgrAI, SrfI, SseI, SfiI
Listeria monocytogenes	LO28	3150	35	1	circular	NotI, Sse83871
Methanobacterium thermoautotrophicum	Marburg (DSM 2133)	1623	50	1	circular	NheI, NotI, PmeI, I-CeuI
Mycobacterium leprae		>2800	56	1	circular	none
Mycoplasma genitalium	G37	580.067	32	1	circular	XhoI, SmaI, ApaI, AscI, I-CeuI, MluI
Myxococcus xanthus	DK1622	9454 (wild type), 9232 (DK1622)	67.5	1	circular	AseI, SpeI
Neisseria gonorrhoeae	MS11-N198 (ATCC 49759)	2330	53.3	1	circular	NheI, SpeI, BglII, PacI
Planctomyces limnophilus	DSM 3776T	5204	53	1	circular	PacI, PmeI, SwaI
Pseudomonas aeruginosa	C	6516	nd	1	circular	I-CeuI, PacI, PmeI, SpeI, SwaI
Pseudomonas aeruginosa	PAO (DSM 1707)	5933	67	1	circular	DpnI, I-CeuI, PacI, SpeI, SwaI
Rhizobium meliloti	1021	6500 (3540, 1340, 1700)	62-63	3 chromosome + megaplasmids pRme1021a (pSyma), pRme1021b (pSymb))	circular	SwaI, PacI,PmeI, I-CeuI, SpeI, AseI, DraI, XbaI

Table 1 (continued)

Organism	Strain	Genome size (kb)	GC (%)	Chrom. number	Chrom. topology	Rare cutting enzymes
Rhodobacter capsulatus	SB1003	3700		1	circular	AseI, XbaI, SpeI, DraI
Rhodobacter sphaeroides	2.4.1T	4343 (2973, 911, five plasmids)	65	2	circular	AseI, DraI, SnaBI, SpeI, SspI, XbaI, AflII
Salmonella typhi	Ty2	4780		1	circular	XbaI, BlnI, SpeI, I-CeuI
Salmonella typhimurium	LT2	4800	53	1	circular	XbaI, BlnI, SpeI, I-CeuI
Serpulina hyodysenteriae	B78T	3200	26-28	1	circular	NotI, BssHII, SalI, SmaI, EclXI
Staphylococcus aureus	NCTC 8325	2800	37	1	circular	SmaI, CspI, SgrAI, AscI
Streptococcus pneumoniae	R6	2200	39	1	circular	SmaI, ApaI, SacII
Sulfolobus tododaii	7	2760	40	1	circular	NotI, BssHII, RsrII, EagI
Synechococcus sp.	PCC 7002	2700	49.1	1	circular	NotI, SalI, SfiI, AscI
Thermus thermophilus	HB8	1741	64.7	1	circular	SspI, HpaI, NdeI, MnnI, DraI, EcoRI, EcoRV
Treponema pallidum	subsp. *pallidum* (Nichols)	1083	52.4-53.7	1	circular	NotI, SfiI, SrfI, SpeI, FscI

Presented are the genome sizes (with sizes of individual chromosomes given in parentheses) are given in kilobase pairs, mole percent G+C of the genome overall, chromosome number and topology, and a sampling of rare cutting restriction enzymes.

Table 2. Miscellaneous Biological Characteristics

Organism	Strain	Number of rRNA ops.	Comments	Repeated sequences	Prophages	Tn insertions	IS insertions
Anabaena sp.	PCC 7120	4		HIP1, unnamed, STRR1,			IS892, IS893, IS894, IS895, IS897, IS898
Aquifex pyrophilus		6		rrn			
Bacillus subtilis	168	10	Genome is being sequenced				
Bordetella pertussis	Tohama I			IS481			IS481
Borrelia burgdorferi	Sh-2-82	1-16S, 2-23S	Genome is being sequenced				
Bradyrhizobium japonicum	3I1b110	1		RSa, RSb, RSg, RSd, RSe			
Brucella melitensis	16M	3		Bru-RS1, Bru-RS2			y (IS6501 or IS711)
Burkholderia cepacia	ATCC 17616	5-6		IS401, IS402, IS403, IS404, IS405, IS406, IS407, IS408, IS411, IS415		y (TnPc1, IS402, IS403, IS404, IS405, IS411)	y
Burkholderia cepacia	ATCC 25416 (type strain)	6	2 copies of recA				
Campylobacter jejuni	TGH9011 (ATCC43431)	3		Chi, REP, ERIC			
Caulobacter crescentus	CB15	2					IS511, IS298
Clostridium beijerinckii	NCIMB 8052	14		unnamed			
Clostridium perfringens	CPN50	10			ø29, ø59		IS1136-like
Enterococcus faecalis	OG1RF (ATCC 47077)					Tn916, Tn924	IS256-like, IS6770
Escherichia coli	MG1655	7	Genome is being sequenced	y	y	clinical strains	y

Table 2 (continued)

Organism	Strain	Number of rRNA ops.	Comments	Repeated sequences	Prophages	Tn insertions	IS insertions
Haemophilus influenzae	Rd	6	Genome has been sequenced	y (uptake sequence, CAAT tandem repeats)	HP1, S2, defective Mu-like phage		
Halobacterium salinarium	GRB	1		5 sequences repeated twice			not in strain GRB, although there are many in strains NRC-1, S9
Haloferax volcanii	DS2	2		ISH51, ISH51-D			y
Helicobacter pylori	UA802, UA861, NCTC11637, NCTC11639	2	diversity in gene order and restriction map				
Lactococcus lactis subsp. *cremoris*	MG1363	6		ISS1, IS1076, IS981, IS905, rrn	øT712, TP901-1, Tuc2009, bIL285	Tn5276	ISS1, IS1076, IS981, IS905
Lactococcus lactis subsp. *lactis*	IL1403	6		IS1076, IS981, IS905, rrn	øT712, TP901-1, Tuc2009, bIL285	Tn5276	IS1076, IS981, IS905
Leptospira interrogans	serovar icterohaemorrhagiae, serovar pomona strain RZ11	2 copies of 16S and 23S, 1 copy of 5S					IS1533, IS1500
Listeria monocytogenes	LO28	6					
Methanobacterium thermoautotrophicum	Marburg (DSM 2133)	2					
Mycobacterium leprae		1		RLEP			
Mycoplasma genitalium	G37	1	Genome has been sequenced	MgPa			
Myxococcus xanthus	DK1622	4		MXARMX	Mxalpha		

Table 2 (continued)

Organism	Strain	Number of rRNA ops.	Comments	Repeated sequences	Prophages	Tn insertions	IS insertions
Neisseria gonorrhoeae	MS11-N198 (ATCC 49759)	4	Genome is being sequenced	152-bp, 26-bp, uptake seq., CR			
Planctomyces limnophilus	DSM 3776T	2					
Pseudomonas aeruginosa	C	4					
Pseudomonas aeruginosa	PAO (DSM 1707)	4		IS-PA-1, IS222	defective, encodes pyocin R2		y (IS-PA-1, IS22, IS222, IS-PA-4)
Rhizobium meliloti	1021	3		IsRm1, IsRm2, IsRm3			
Rhodobacter capsulatus	SB1003	4					
Rhodobacter sphaeroides	2.4.1T	3					
Salmonella typhi	Ty2	7					IS200
Salmonella typhimurium	LT2	7		REP, ERIC	P22, Fels1, Fels2		IS200
Serpulina hyodysenteriae	B78T	1			VSH-1		
Staphylococcus aureus	NCTC 8325	6		IS256, IS257	ø11, øL54a, ø13, ø42 (all temperate)	many	IS256, IS257/IS431, HI555
Streptococcus pneumoniae	R6	4		BOX			unnamed
Sulfolobus tododaii	7	1			y (in S. shibatae B12)		
Synechococcus sp.	PCC 7002	2					
Thermus thermophilus	HB8	2		IS1000			IS1000
Treponema pallidum subsp. *pallidum*	Nichols	2	Genome is being sequenced				

Table 3. Some Genetic Characteristics

Organism	Strain	Lib	Phage	Plasmids	Tf	El	Td	Co	Transfer from E. coli	Transfer from other	Tn endog.	Tn other
Anabaena sp.	PCC 7120		21 known (1 temperate)	a (410kb), b (190kb), g (110kb), and 5kb		y			y (conj)			Tn5
Aquifex pyrophilus												
Bacillus subtilis	168		temp: ø105, SP02, SPb, SP16, PBSX; vir: SP01, SP82G, ø29		y	y	y PBS1	y				Tn917
Bordetella pertussis	Tohama I		yes, not characterized	R plasmids in B. avium, B. bronchiseptica	y	y		y	y			Tn5
Borrelia burgdorferi	Sh-2-82			2 universally found, several others		y						
Bradyrhizobium japonicum	3I1b110					y			y	y		Tn5
Brucella melitensis	16M	IP	y			y	y		y			Tn5
Burkholderia cepacia	ATCC 17616			pTGL1, 170kb		y	y	y	y	y		Tn5
Burkholderia cepacia	ATCC 25416 (type strain)			y (100-200kb are common)		y	y	y	y	y		Tn5, Tn7
Campylobacter jejuni	TGH9011 (ATCC 43431)	IP	ø3, VFP-11, plus lytic phages	y (9.5-50kb)	y	y	ø3, VFP-11, VFP-13	y	y			miniTn3-Km
Caulobacter crescentus	CB15					y	y	y	y			Tn5, Tn7
Clostridium beijerinckii	NCIMB 8052					y			y	y		Tn1545
Clostridium perfringens	CPN50		ø29, ø59	pIP404, pIP405 in this strain, many others (2-140kb)		y		y		y		Tn916

Table 3 (continued)

Organism	Strain	Lib	Phage	Plasmids	Tf	El	Td	Co	Transfer E. coli	from other	Tn endog.	Tn other
Enterococcus faecalis	OG1RF (ATCC 47077)	IP	lytic: NPV-1, Phage 1	many: pAD1, pCF10, pAD2, pBEM10, pAM323, pAM324, pAMa1, pAMb1, pJH1	y	y		y pAMβ1	y	y	Tn916, Tn917, Tn925, Tn918, Tn5281, Tn924, Tn3871, Tn1546 (E. faecium)	y
Escherichia coli	MG1655	y	y	y	y	y	y	y	y	y	y	
Haemophilus influenzae	Rd	y	HP1C1, S2, HP3, N3		y	y				y, other Haemoph. species		Tn916
Halobacterium salinarium	GRB		many, øH	pGRB305 (305kb), pGRB90 (90kb), pGRB37 (37kb), pGRB1 (1.8kb)	y but nt on GRB							
Haloferax volcanii	DS2		HF1	pHV4 (690kb), pHV3 (442kb), pHV1 (86kb), pHV2 (6.4kb)	y			y		y		
Helicobacter pylori	UA802, UA861, NCTC 11637, NCTC 11639		HP1 temperate phage	pUA841 (2.7kb), pUA874 (5.5kb), pUA875B (15kb), pUA883 (12kb), pHPK255 (1.5kb)	y	y						
Lactococcus lactis subsp. *cremoris*	MG1363		y	y	y	y	y	y		y	y	Tn916, Tn917, Tn919
Lactococcus lactis subsp. *lactis*	IL1403		y	y	y	y	y	y		y	y	Tn916, Tn917, Tn919
Leptospira interrogans serovar icterohaemorrhagiae, serovar pomona	RZ11		LE1, LE3, LE4 lytic									

Table 3 (continued)

Organism	Strain	Lib	Phage	Plasmids	Tf	El	Td	Co	Transfer E. coli	from other	Tn endog.	Tn other
Listeria monocytogenes	LO28			y (Cd resist)	y	y		y	y	y		Tn1545, Tn916, Tn917
Methanobacterium thermoautotrophicum	Marburg (DSM 2133)		archaephage YM1	pME2001			y (YM1)					
Mycobacterium leprae		y										
Mycoplasma genitalium	G37	y										
Myxococcus xanthus	DK1622	y	lytic: Mx1, temperate: Mx4, Mx8, Mx9	n, but msDNA		y	y (Mx4, Mx8, Mx9)		y			Tn5
Neisseria gonorrhoeae	MS11-N198 (ATCC 49759)			pJD1, pMR0360, pLE2450	y	y		y pLE2450	y	y, N. meningitis, membrane vesicle mediated transfer		Tn1545-D3
Planctomyces limnophilus	DSM 3776T			yes, not named								
Pseudomonas aeruginosa	C			y (95kb)								
Pseudomonas aeruginosa	PAO (DSM 1707)		not in PAO; frequently in other strains (øCTX, many others)	not in PAO, but frequently in other strains (R and D plasmids)		y	y (many phages)	y (many plasmids)	y	y (many organisms)	y (phage D3112, Tn501, Tn2001, Tn2521)	y (Tn1, Tn5, Tn7, Tn1737)
Rhizobium meliloti	1021		DF2, CM1, phage 16-3 (temperate)	many	y	y	y (Mu, PH1M12, M1)		y			Tn5, Tn7
Rhodobacter capsulatus	SB1003	y		130-kb			GTA phage-like transduction		y			

Table 3 (continued)

Organism	Strain	Lib	Phage	Plasmids	Tf	El	Td	Co	Transfer E. coli	from other	Tn endog.	Tn other
Rhodobacter sphaeroides	2.4.1T	y	lytic: RS1, temperate: Rø-1	y (pRS2.4.1.a, pRS2.4.1.b, pRS2.4.1.c, pRS2.4.1.d, pRS2.4.1.e)	y	y		y pRS2.4.1.d, pRS2.4.1.e	y			Tn5
Salmonella typhi	Ty2						y (P22)					Tn10
Salmonella typhimurium	LT2	IP	P22, many others	pSLT	y	y	y	y	y	y (E. coli)	Tn10	many
Serpulina hyodysenteriae	B78T					y						
Staphylococcus aureus	NCTC 8325		ø11, many others (typing series)	many (pT181, pC194, pSN2, pE194, pI524, pI258, pII147, pI9789, pG01)	y	y	y	y pG01, pCRG1600		y	many (Tn551, Tn554, Tn4001, Tn552, Tn4003, IS431, IS256, Tn916)	y (Tn916, Tn917)
Streptococcus pneumoniae	R6		Dp-1, Dp-4, Ej-1	pMV158 (S. agalactiae)	y	y				y	Tn5253, IS1212, Tn1545, Tn3701	Tn916
Sulfolobus tododaii	7		n (known for other strains)	n (known for other strains of Sulfolobus)		nt (known for S. shibatae B12)					ISC1217	
Synechococcus sp.	PCC 7002	IP		possibly 7 plasmids	y							
Thermus thermophilus	HB8		øYS40 (lytic)	PTT8, PVV8	y					y		
Treponema pallidum subsp. *pallidum*	Nichols			possible plasmid								

Key:
Lib: ordered library; IP: in progress
Tf: transformation
El: electroporation
Td: transduction
Co: conjugation
Tn: transposon
nt: not tested

Table 4. Genomic fingerprinting

Organism	Strain	Ri	Ap	tRNA	Re	REA	PFG	other fingerprint
Anabaena sp.	PCC 7120							
Aquifex pyrophilus		y			y (rrn)			
Bacillus subtilis	168							
Bordetella pertussis	Tohama I		y			y		
Borrelia burgdorferi	Sh-2-82	y	y			y	y	
Bradyrhizobium japonicum	3I1b110					y		
Brucella melitensis	16M		y		y	y	y	
Burkholderia cepacia	ATCC 17616	y	y					PFGE, MLE
Burkholderia cepacia	ATCC 25416 (type strain)	y	y					PFGE, MLE
Campylobacter jejuni	TGH9011 (ATCC 43431)	y	y		y	y	y	PFGE
Caulobacter crescentus	CB15					y	y	
Clostridium beijerinckii	NCIMB 8052	y				y	y	
Clostridium perfringens	CPN50	y	y				y	
Enterococcus faecalis	OG1RF (ATCC 47077)	y				y	y	
Escherichia coli	MG1655			y	y	y	y	MLE
Haemophilus influenzae	Rd					y		
Halobacterium salinarium	GRB					y		
Haloferax volcanii	DS2					y		
Helicobacter pylori	UA802, UA861, NCTC 11637, NCTC 11639	y	y			y	y	PFGE
Lactococcus lactis subsp. *cremoris*	MG1363	y	y		y	y		PFGE
Lactococcus lactis subsp. *lactis*	IL1403	y	y		y	y		PFGE
Leptospira interrogans serovar icterohaemorrhagiae, serovar pomona	RZ11	y	y		y	y	y	PFGE
Listeria monocytogenes	LO28							
Methanobacterium thermoautotrophicum	Marburg (DSM 2133)					y	y	
Mycobacterium leprae								
Mycoplasma genitalium	G37					y		

Table 4 (continued)

Organism	Strain	Ri	Ap	tRNA	Re	REA	PFG	other fingerprint
Myxococcus xanthus	DK1622	y				y	y	msDNA
Neisseria gonorrhoeae	MS11-N198 (ATCC 49759)	y	y			y		PCR+REA, PCR+RFLP
Planctomyces limnophilus	DSM 3776T							
Pseudomonas aeruginosa	C	y	y			y	y	PFGE, MLE, esterase typing,
Pseudomonas aeruginosa	PAO (DSM 1707)	y	y			y	y	PFGE, MLE, esterase typing,
Rhizobium meliloti	1021	y			y	y		phage typing, MLE
Rhodobacter capsulatus	SB1003	y						
Rhodobacter sphaeroides	2.4.1T					y	y	PFGE
Salmonella typhi	Ty2	y						MLE, phage typing
Salmonella typhimurium	LT2		y	y	y	y		MLE, phage typing, plasmid profiles
Serpulina hyodysenteriae	B78T	y				y	y	
Staphylococcus aureus	NCTC 8325	y	y		y	y	y	phage typing, plasmid profile, antibiograms
Streptococcus pneumoniae	R6	y						autolysin PCR, serotyping, MLE, PBP patterns, PFGE
Sulfolobus tododaii	7					y		
Synechococcus sp.	PCC 7002							
Thermus thermophilus	HB8		y	y		y	y	
Treponema pallidum subsp. *pallidum*	Nichols	y		y		y		

Ri: ribotyping
Ap: AP-PCR
tRNA: REP-PCR with tRNA primers
Re: REP-PCR with other primers
REA: restriction enzyme analysis
PFG: RFLP analysis by PFGE

47

Anabaena sp. Strain PCC 7120

C. Peter Wolk

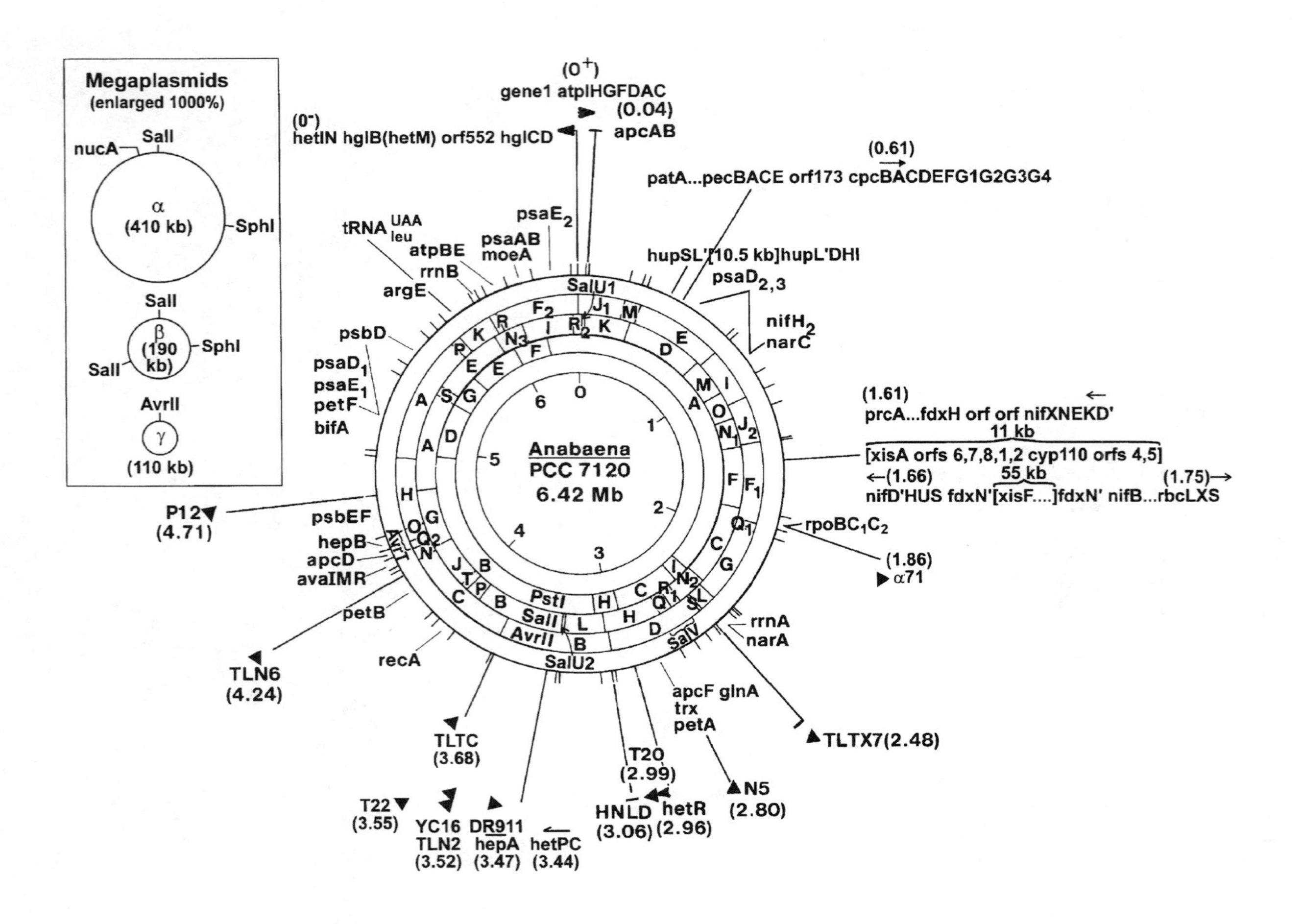

Megaplasmids
(enlarged 1000%)
nucA
SalI
α
(410 kb)
SphI
SalI
β
(190 kb)
SphI
SalI
AvrII
γ
(110 kb)
Anabaena
PCC 7120
6.42 Mb
(0+)
gene1 atpIHGFDAC
(0.04)
apcAB
(0-)
hetIN hglB(hetM) orf552 hglCD
(0.61)
patA...pecBACE orf173 cpcBACDEFG1G2G3G4
hupSL'[10.5 kb]hupL'DHI
psaD2,3
nifH2
narC
(1.61)
prcA...fdxH orf orf nifXNEKD'
11 kb
[xisA orfs 6,7,8,1,2 cyp110 orfs 4,5]
←(1.66)
55 kb
(1.75)→
nifD'HUS fdxN'[xisF....]fdxN' nifB...rbcLXS
rpoBC1C2
(1.86)
α71
rrnA
narA
TLTX7(2.48)
apcF glnA
trx
petA
N5
(2.80)
T20
(2.99)
HNLD
(3.06)
hetR
(2.96)
hetPC
(3.44)
DR911
hepA
(3.47)
YC16
TLN2
(3.52)
T22
(3.55)
TLTC
(3.68)
recA
TLN6
(4.24)
petB
avaIMR
apcD
hepB
psbEF
P12
(4.71)
psbD
psaD1
psaE1
petF
bifA
argE
rrnB
atpBE
tRNA UAA leu
psaAB
moeA
psaE2
SalU1
SalU2
SalV
PstI
SalI
AvrII
AvrT

◀

Figure 47-1. The physical map of the *Anabaena* genome was defined by Bancroft *et al.* (1989) in the form of overlapping fragments generated by the restriction endonucleases *Avr*II (*Bln*I), *Pst*I and *Sal*I, and was modified by the addition of three small *Sal*I fragments and one small *Bln*I fragment by Kuritz *et al.* (1993). As presented below, genes have been localized either to these restriction fragments (the inner circle of genetic designations) or to positions within those fragments (the outer circle of genetic designations; see the article by Kuritz and Wolk in this volume). All localizations and orientations of genes (elongated arrowheads) and of transposons (triangular arrowheads) were presented by the two references already cited, with the following exceptions: *hupL*, encoding part of an uptake hydrogenase, was localized by Matveyev and coworkers and by Golden and coworkers (see below); *hetP* by Fernández-Piñas *et al.* (1994); TLN2 and (tentatively) TLN6, loci transcriptionally activated by nitrogen deprivation, by Wolk *et al.* (1991); *bifA* (*ntcA*) by Wei *et al.* (1993); *moeA,* involved in molybdopterin biosynthesis, by K. Ramaswamy and J. W. Golden (personal communication); and *nucA* by Muro-Pastor *et al.* (1994).

Genetic contiguity has often been defined by restriction mapping and sequencing: the genel-*atpIHGFDAC* region that encodes portions of the ATP synthase, by McCarn *et al.* (1988); the region (*pecBACE orf*173 *cpcBACDEFG1G2G3G4*) that encodes a series of photosynthetic antenna pigments and linkers, by Belknap and Haselkorn (1987), Bryant *et al.* (1991), and Swanson *et al.* (1992); *patA* (Liang *et al.*, 1992) was positioned ca. 1.5 kb 5′ from this region by Y. Cai (personal communication); *hupSLDHI* and an interpolated excison (square brackets), by Carrasco *et al.* (1994; 1995a,b), Golden *et al.* (1994), Matveyev *et al.* (1994) and Matveyev and Bergman (1994); the region from *fdxH* to *rbcS,* by Böhme and Haselkorn (1988); Borthakur *et al.* (1990), Carrasco *et al.* (1994), Curtis and Haselkorn (1983), Golden *et al.* (1985, 1988), Haselkorn (1992), Haselkorn and Buikema (1992), Lammers and Haselkorn (1983), Lammers *et al.* (1986, 1990), Mazur and Chui (1982), Mevarech *et al.* (1980), Mulligan *et al.* (1988), Mulligan and Haselkorn (1989), Nierzwicki-Bauer *et al.* (1984), and Rice *et al.* (1982); *apcF glnA* by Warner *et al.* (1993); *hetCP* by I. Khudyakov and C. P. Wolk (unpublished) and Fernández-Piñas *et al.* (1994); and *hetIN hglB* (*hetM*) *orf552 hglCD* by Black and Wolk (1994) and C. C. Bauer *et al.* (unpublished; personal communication from W. J. Buikema). The roles of several of the genes identified by sequencing have been supported by the results of transpositions (see Ernst *et al.*, 1992, and Kuritz *et al.*, 1993). Numerous additional genes have been sequenced but not yet mapped.

Acknowledgments

Our mapping work on *Anabaena* sp. strain PCC 7120 has been supported by D.O.E. under grant DE-FG02-90ER20021.

References

Bancroft, I., C. P. Wolk, and E. V. Oren. 1989. Physical and genetic maps of the genome of the heterocyst-forming cyanobacterium *Anabaena* sp. strain PCC 7120. *J. Bacteriol.* 171:5940–5948.

Table 47-1. Coordinates, in kb, of the clockwise ends of restriction fragments in the map of Bancroft et al. (1989) as modified by Kuritz et al. (1993). The origin (0 kb) is arbitrarily set 9 kb clockwise from the clockwise end of fragment AvrF₂.

$SalR_2$	20	AvrS	2496	AvrO	4481
$SalU_1$	47	$SalN_2$	2534	AvrT	4497†
$AvrJ_1$	256	$SalR_1$	2604	SalG	4709
SalK	331	SalQ	2684	AvrH	4907
AvrM	346	SalV	2698*	PstB	4908
SalD	835	PstC	2951	SalA, PstD	5338
AvrE	851	AvrD	3046	SalS	5398
SalM	1084	SalH	3057	PstG	5584
AvrI	1156	PstH	3131	AvrA	5651
SalO	1253	SalL	3321	AvrP	5701
$AvrJ_2$	1421	$SalU_2$	3348	SalE	5867
$SalN_1$	1437	AvrB	3686	AvrK	5911
SalF	1826	SalB	3897	AvrR, PstE	5951
$AvrF_1$	1881	SalP	3996	$SalN_3$	6051
$AvrQ_1$	1926	SalT	4036	PstF	6231
SalC	2350	AvrC	4296	SalI	6370
PstA	2351	SalJ	4335	$AvrF_2$	6411
AvrG	2376	AvrN	4371		
AvrL, PstI	2471	$AvrQ_2$	4416		

*Position relative to $SalR_1$ and SalQ uncertain.

†Position relative to $AvrQ_2$ and AvrO uncertain.

Belknap, W. R., and R. Haselkorn. 1987. Cloning and light regulation of expression of the phycocyanin operon of the cyanobacterium *Anabaena. EMBO J.* 6:871–884.

Black, T. A., and C. P. Wolk. 1994. Analysis of a Het^- mutation in *Anabaena* sp. strain PCC 7120 implicates a secondary metabolite in the regulation of heterocyst spacing. *J. Bacteriol.* 176:2282–2292.

Böhme, H., and R. Haselkorn. 1988. Molecular cloning and nucleotide sequence analysis of the gene coding for heterocyst ferredoxin from the cyanobacterium *Anabaena* sp. strain PCC 7120. *Mol. Gen. Genet.* 214:278–285.

Borthakur, D., M. Basche, W. J. Buikema, P. B. Borthakur, and R. Haselkorn. 1990. Expression, nucleotide sequence and mutational analysis of two open reading frames in the *nif* gene region of *Anabaena* sp. strain PCC 7120. *Mol. Gen. Genet.* 221:227–234.

Bryant, D. A., V. L. Stirewalt, M. Glauser, G. Frank, W. Sidler, and H. Zuber. 1991. A small muligene family encodes the rod-core linker polypeptides of *Anabaena* sp. PCC7120 phycobilisomes. *Gene* 107:91–99.

Carrasco, C. D., K. S. Ramaswamy, T. S. Ramasubramanian, and J. W. Golden. 1994. *Anabaena-xisF* gene encodes a developmentally regulated site-specific recombinase. *Genes Develop.* 8:74–83.

Carrasco, C. D., J. A. Buettner, and J. W. Golden. 1995a. Programed DNA rearrangement of a cyanobacterial *hupL* gene in heterocysts. *Proc. Natl. Acad. Sci. USA* 92:791–795.

Carrasco, C. D., J. A. Buettner, J. S. Garcia, and J. W. Golden. 1995b. Characterization of the recombinase gene required for the *hupl* (sic) DNA rearrangement in *Anabaena* sp. strain PCC 7120. Abstr. Vth Cyanobacterial Molec. Biol. Workshop, Pacific Grove, CA, p. 26.

Curtis, S. E., and R. Haselkorn. 1983. Isolation and sequence of the gene for the large subunit of ribulose-1,5-bisphosphate carboxylase from the cyanobacterium *Anabaena* 7120. *Proc. Natl. Acad. Sci. USA* 80:1835–1839.

Ernst, A., T. Black, Y. Cai, J. M. Panoff, D. N. Tiwari, and C. P. Wolk. 1992. Synthesis of nitrogenase in mutants of the cyanobacterium *Anabaena* sp. strain PCC 7120 affected in heterocyst development or metabolism. *J. Bacteriol.* 174:6025–6032.

Fernández-Piñas, F., F. Leganés, and C. P. Wolk. 1994. A third genetic locus required for the formation of heterocysts in *Anabaena* sp. strain PCC 7120. *J. Bacteriol.* 176:5277–5283.

Golden, J. W., S. J. Robinson, and R. Haselkorn. 1985. Rearrangement of nitrogen fixation genes during heterocyst differentiation in the cyanobacterium *Anabaena. Nature (London)* 314:419–423.

Golden, J. W., C. D. Carrasco, M. E. Mulligan, G. J. Schneider, and R. Haselkorn. 1988. Deletion of a 55-kilobase-pair DNA element from the chromosome during heterocyst differentiation of *Anabaena* sp. strain PCC 7120. *J. Bacteriol.* 170:5034–5041.

Golden, J. W., C. D. Carrasco, K. S. Ramaswamy, J. A. Simon, and T. S. Ramasubramanian. 1994. Heterocyst-specific DNA rearrangements in *Anabaena* PCC 7120. Abstr. VIII Internat. Symp. Phototrophic Prokaryotes (G. Tedioli, S. Ventura, D. Zannoni, eds.), Urbino, Italy, abstr. 161A.

Haselkorn, R. 1992. Developmentally regulated gene rearrangements in prokaryotes. *Annu. Rev. Genet.* 26:111–128.

Haselkorn, R., and W. J. Buikema. 1992. Nitrogen fixation in cyanobacteria. In: G. Stacey, R. H. Burris, H. J. Evans, eds. *Biological Nitrogen Fixation* (pp. 166–190) Chapman & Hall, New York.

Kuritz, T., A. Ernst, T. A. Black, and C. P. Wolk. 1993. High-resolution mapping of genetic loci of *Anabaena* PCC 7120 required for photosynthesis and nitrogen fixation. *Mol. Microbiol.* 8:101–110.

Lammers, P. J., and R. Haselkorn. 1983. Sequence of the *nifD* gene coding for the α subunit of dinitrogenase from the cyanobacterium *Anabaena. Proc. Natl. Acad. Sci. USA* 80:4723–4727.

Lammers, P. J., J. W. Golden, and R. Haselkorn. 1986. Identification and sequence of a gene required for a developmentally regulated DNA excision in *Anabaena. Cell* 44:905–911.

Lammers, P. J., S. McLaughlin, S. Papin, C. Trujillo-Provencio, and A. J. Ryncarz. 1990. Developmental rearrangement of cyanobacterial *nif* genes: nucleotide sequence, open reading frames, and cytochrome P-450 homology of the *Anabaena* sp. strain PCC 7120 *nifD* element. *J. Bacteriol.* 172:6981–6990.

Liang, J., L. Scappino, and R. Haselkorn. 1992. The *patA* gene product, which contains a region similar to CheY of *Escherichia coli,* controls heterocyst pattern formation in the cyanobacterium *Anabaena* 7120. *Proc. Natl. Acad. Sci. USA* 89:5655–5659.

Matveyev, A. V., and B. Bergman. 1994. Organization of the uptake hydrogenase biosynthesis coding region in the cyanobacterium *Anabaena* sp. PCC 7120. Abstr. VIII Internat. Symp. Phototrophic Prokaryotes, G. Tedioli, S. Ventura, D. Zannoni, eds., Urbino, Italy, abstr. 24A.

Matveyev, A. V., E. Rutgers, E. Söderböck, and B. Bergman. 1994. A novel genome rearrangement involved in heterocyst differentiation of the cyanobacterium *Anabaena* sp. PCC 7120. *FEMS Microbiol. Lett.* 116:201–208.

Mazur, B. J., and C.-F. Chui. 1982. Sequence of the gene coding for the β-subunit of dinitrogenase from the blue-green alga *Anabaena. Proc. Natl. Acad. Sci. USA* 79:6782–6786.

McCarn, D. F., R. A. Whitaker, J. Alam, J. M. Vrba, and S. E. Curtis. 1988. Genes encoding the alpha, gamma, delta, and four F_0 subunits of ATP synthase constitute an operon in the cyanobacterium *Anabaena* sp. strain PCC 7120. *J. Bacteriol.* 170:3448–3458.

Mevarech, M., D. Rice, and R. Haselkorn. 1980. Nucleotide sequence of a cyanobacterial *nifH* gene coding for nitrogenase reductase. *Proc. Natl. Acad. Sci. USA* 77:6476–6480.

Mulligan, M. E., and R. Haselkorn. 1989. Nitrogen fixation (*nif*) genes of the cyanobacterium *Anabaena* species strain PCC 7120. The *nifB-fdxN-nifS-nifU* operon. *J. Biol. Chem.* 264:19200–19207.

Mulligan, M. E., W. J. Buikema, and R. Haselkorn. 1988. Bacterial-type ferredoxin genes in the nitrogen fixation regions of the cyanobacterium *Anabaena* sp. strain PCC 7120 and *Rhizobium meliloti. J. Bacteriol.* 170:4406–4410.

Muro-Pastor, A. M., T. Kuritz, E. Flores, A. Herrero, and C. P. Wolk. 1994. Transfer of a genetic marker from a megaplasmid of *Anabaena* sp. strain PCC 7120 to a megaplasmid of a different *Anabaena* strain. *J. Bacteriol.* 176:1093–1098.

Nierzwicki-Bauer, S. A., S. E. Curtis, and R. Haselkorn. 1984. Cotranscription of genes encoding the small and large subunits of ribulose-1,5-bisphosphate carboxylase in the cyanobacterium *Anabaena* 7120. *Proc. Natl. Acad. Sci. USA* 81:5961–5965.

Rice, D., B. J. Mazur, and R. Haselkorn. 1982. Isolation and physical mapping of nitrogen fixation genes from the cyanobacterium *Anabaena* 7120. *J. Biol. Chem.* 257:13157–13163.

Swanson, R. V., R. de Lorimier, and A. N. Glazer. 1992. Genes encoding the phycobilisome rod substructure are clustered on the *Anabaena* chromosome: characterization of the phycoerythrocyanin operon. *J. Bacteriol.* 174:2640–2647.

Warner, L. E., A. W. Stahel, and S. E. Curtis. 1993. Characterization of the region 5′ to *glnA* in *Anabaena* sp. strain PCC 7120: the *glnA* and *apcF* genes are closely linked. *Abstr. of The Cyanobacterial Workshop,* 1993, Asilomar, p. 98.

Wei, T.-F., T. S. Ramasubramanian, F. Pu, and J. W. Golden. 1993. *Anabaena* sp. strain PCC 7120 *bifA* gene encoding a sequence-specific DNA-binding protein cloned by in vivo transcriptional interference selection. *J. Bacteriol.* 175:4025–4035.

Wolk, C. P., Y. Cai, and J.-M. Panoff. 1991. Use of a transposon with luciferase as a reporter to identify environmentally responsive genes in a cyanobacterium. *Proc. Natl. Acad. Sci. USA* 88:5355–5359.

48

Aquifex pyrophilus

Rüdiger Schmitt

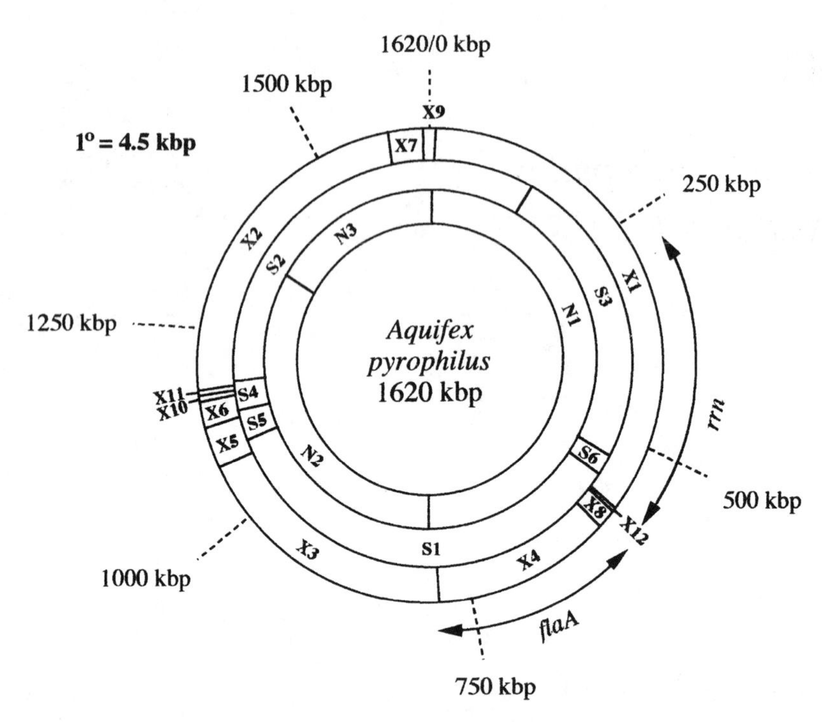

Figure 48-1. Numbering of restriction fragments in the order of decreasing sizes; N, *Not*I, S, *Spe*I, X, *Xbal*I. Map areas assigned to a single flagellin (*flaA*) and six rRNA (*rrn*) genes are indicated by arcs (Shao et al., 1994).

Aquifex pyrophilus, a marine hyperthermophilic hydrogen-oxidizing bacterium growing with an optimum temperature of 85°C, represents the earliest known divergence in the bacterial phylogenetic tree (Burggraf *et al.*, 1992; Huber *et al.*, 1992). Three rare-cutting restriction endonucleases, *Not*I, *Spe*I and *Xba*I, and contour-clamped homogeneous electric field (CHEF) electrophoresis served to determine the 1,620-kbp length of the single circular chromosome. A physical map was established (1) by cross-hybridization of electrophoretically separated and isolated restriction fragments used as probes and (2) by the construction of a linking-clone library (Smith and Condemine, 1990) that contains the 6 *Spe*I and 12 *Xba*I junctions on separate clones. A single flagellin gene (*fla*A; Behammer *et al.*, 1995) and 6 ribosomal RNA genes (rrn; Burggraf *et al.*, 1992) were assigned to distinct map areas by Southern hybridization. The *rrn* gene copy number was estimated by the fragments obtained upon digestion with I-*Ceu*I, an endonuclease recognizing a unique 19-bp sequence in the 23S rDNA (Liu *et al.*, 1993).

Small genome sizes prevalent among these thermophiles may be a consequence both of the need for fast and faithful replication at elevated temperatures and of phylogenetic antiquity, implying that small chromosomes are an archetypal feature and consistent with the hypothesis that thermophiles are representatives of a primordial form of life (Stetter, 1994). The larger genomes of more recent mesophilic species (presumably generated by gene duplication, genetic rearrangements, and horizontal exchange of genetic material) have endowed these latter organisms with a greater potential for adaptation to changing environments and for evolutionary change, as highlighted by cryptic genes, amplifiable DNA, a host of transposable elements (Leblond *et al.*, 1991; Weinstock, 1994), and efficient systems for genetic exchange, not found in the thermophiles.

Acknowledgments

This investigation was supported by the Deutsche Forschungsgemeinschaft (Schm 68/24-2).

References

Behammer, W., Z. Shao, W. Mages, R. Rachel, K. O. Stetter, and R. Schmitt. (1995) Thermostable flagella of *Aquifex pyrophilus:* analysis of a single flagellin gene and its product. *J. Bacteriol.* 177:6630–6637.

Burggraf, S., G. J. Olsen, K. O. Stetter, and C. R. Woese. (1992) A phylogenetic analysis of *Aquifex pyrophilus. Syst. App. Microbiol.* 15:352–56.

Huber, R., T. Wilharm, D. Huber, A. Trincone, S. Burggraf, H. König, R. Rachel, I. Rockinger, H. Fricke, and K. O. Stetter. (1992) *Aquifex pyrophilus* gen. nov. sp. nov., represents a novel group of marine hyperthermophilic hydrogen-oxidizing bacteria. *Syst. Appl. Microbiol.* 15:349–51.

Leblond, P., P. Demuyter, J. M. Simonet, and B. Decaris. (1991) Genetic instability and associated genome plasticity in *Streptomyces ambofaciens:* pulsed-field gel electrophoresis evidence for large DNA alterations in a limited genomic region. *J. Bacteriol.* 173:4229–33.

Liu, S., A. Hessel, and K. E. Sanderson. (1993) Genomic mapping with I-*Ceu*I, an intron-encoded endonuclease specific for genes for ribosomal RNA, in *Salmonella* spp., *Escherichia coli,* and other bacteria. *Proc. Natl. Acad. Sci. USA* 90:6874–78.

Shao, Z., W. Mages, and R. Schmitt. (1994) A physical map of the hyperthermophilic bacterium *Aquifex pyrophilus* chromosome. *J. Bacteriol.* 176:6776–80.

Smith, C. L., and G. Condemine. (1990) New approaches for physical mapping of small genomes. *J. Bacteriol.* 172:1167–72.

Stetter, K. O. (1994) The lesson of archaebacteria, in *Early Life on Earth. Nobel Symposium No. 84* (ed. S. Bengtson). Columbia University Press, New York, pp. 143–51.

Weinstock, G. M. (1994) Bacterial genomes: mapping and stability. *ASM News* 60:73–78.

49

Bacillus cereus/Bacillus thuringiensis

Anne-Brit Kolstø

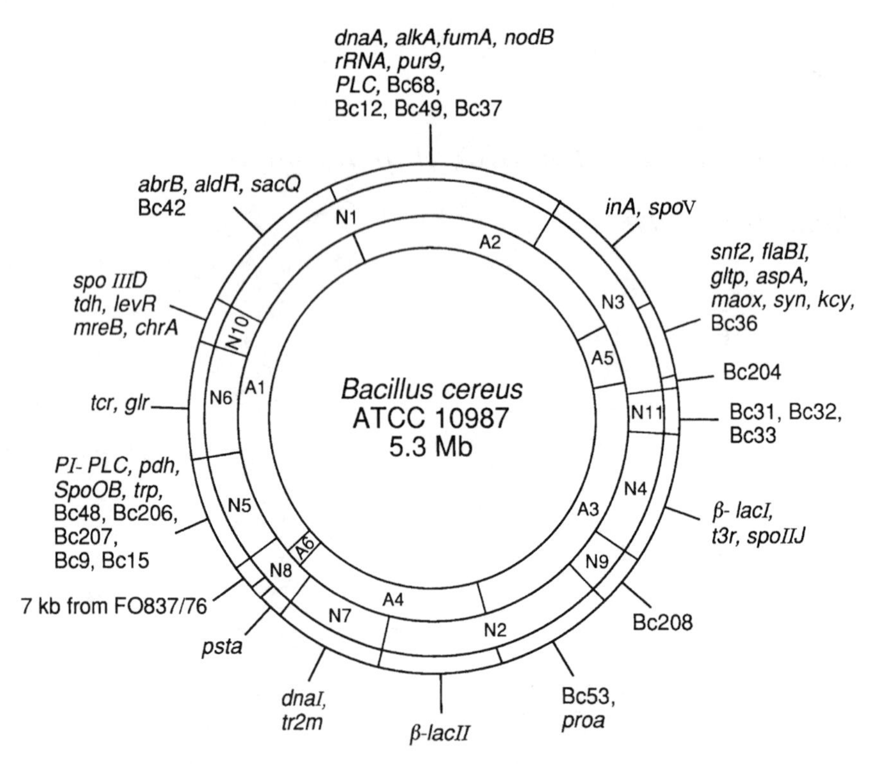

Figure 49-1. Physical map of the *B. cereus* ATCC 10987 chromosome. Anonymous DNA probes from this strain are given as numbers with BC as prefix. Gene names are given according to sequence similarities from data base searches.

Bacillus cereus/Bacillus thuringiensis (B.t.) are spore-forming soil bacteria, the only established difference being the production by B.t. of insecticidal toxins, present as intracellular crystals during sporulation (Aronsen, 1993; Priest, 1981; Carlson & Kolstø, 1993; Carlson *et al.*, 1994.) B.t. is the most widely used biopesticide in the world today. Both *B. cereus* and B.t. may produce enterotoxins as well as phospholipases and hemolysins (Drobniewski, 1993; Kramer & Gilbert, 1989; Jackson, *et al.*, 1995).

In our laboratory we have constructed physical maps of chromosomes from several *B. cereus* and B.t. strains. The size of the chromosomes appears to vary between 2.6 and 6.3 Mb (Carlson & Kolstø, 1994).

The following general approach has been used:

- Partial and complete digestion with the restriction enzymes NotI, SfiI and AscI, and combinations of these enzymes.

- Separation of the fragments by pulsed field gel electrophoresis (PFGE; *see* Chapters 24–26). Varying conditions were used to get optimal separation of all fragments (10–2000kb).
- Blotting of the DNA fragments to filters and hybridization with DNA probes.
- DNA fragments from *B. cereus* ATCC 10987 have been the main probe source. The DNA fragments have subsequently been sequenced, and genes have thus been identified.

Methods

Preparation of DNA

The method used is described in our previous publications, and is based upon the general method for PFGE described by Smith (Smith *et al.* 1988).

A brief description of the current protocol: The bacteria are grown in 10ml LB to OD_{600}=0.6. Chloramphenicol is added to final concentration 5μg/ml and the incubation continued for 1 h. The cells are harvested by centrifugation, washed and resuspended in 1.1ml buffer and mixed with 0.5ml 2.6% low melting agarose (FMC), and the solution is aliquoted into 100μl moulds. After 10 min at 4°C the agarose plugs are incubated 2-20 h at 50°C in the following solutions:

A) 2% lysozyme and 0,25 mg/ml RNase in 10 mM EDTA/25 mM Tris pH7.5

B) 1 mg/ml proteinase K and 1% sarcosyl in 0.5 M EDTA pH 8.0

The plugs are stored in proteinase K solution (solution B) at 4°C until use. Before treatment with restriction enzyme, the plugs are treated with PMSF (0.1 mM final concentration).

Digestion of DNA

Restriction enzyme (0.1–20 units) and appropriate buffer is added and one plug (or part of one plug) is incubated over night at 4°C, followed by incubation at appropriate temperature (37 or 50°C) for 10 min to 20 h. The plug is washed in TE and incubated with solution B for 2 h at 50°C before PFGE. When double digests are used, the plugs are digested successively with the enzymes.

PFGE

Beckman and BioRad apparatus are routinely used, with conditions allowing separation of small, medium or large fragments, in the size range of 10 kb to 2 Mb. The gel is stained with ethidium bromide and the fragments transferred by capillar blotting to nylon or supported nitrocellulose filters and hybridized to ^{32}P-labelled DNA fragments. A chromosomal map is built from the analysis of the data from single and double digests, after both partial and complete digestions, and after hybridization of more than 10 probes.

References

Aronsen, A. I. 1993. Insecticidal toxins. In *Bacillus subtilis* and other Gram-positive bacteria. Biochemistry, physiology, and molecular genetics. Sonenshein, A. L., Hoch, J. S. & Losick, R. eds. American Society for Microbiology, Washington, D.C., pp. 953–963.

Carlson, C. R., D. Cougant, and A. B. Kolstø (1994). Genotypic diversity among *Bacillus cereus* and *Bacillus thuringiensis* strains. *Appl. Environ. Microbiol.* 60:1719–1725.

Carlson, C. R., A. Grønstad, and A.-B. Kolstø. 1992. Physical maps of the genomes of three *Bacillus cereus* strains. *J. Bacteriol.* 174:3750–3756.

Carlson, C. R., T. Johansen, M.-M. Lecadet, and A.-B. Kolstø. 1996. Construction of a Physical Map of the *Bacillus thuringiensis* subsp. *berliner* 1715 Chromosome by pulsed field gel electrophoresis. *Microbiol.* 142:1625–1634.

Carlson, C. R., T. Johansen, and A. B. Kolsto (1996). The chromosome map of *Bacillus thuringiensis* subsp. *canadiensis* HD224 is highly similar to that of the *Bacillus cereus* type strain ATCC 14579. *FEMS Microbiol. Lett.* 141:163–167.

Carlson, C. R., and A.-B. Kostø. 1993. A complete physical map of a *Bacillus thuringiensis* chromosome. *J. Bacteriol.* 175:1053–1060.

Carlson, C. R., and A.-B. Kolstø. 1994. A small (2.4Mb) *Bacillus cereus* chromosome corresponds to a conserved region of larger (5.3 Mb) *Bacillus cereus* chromosome. *Mol. Microbiol.* 13:161–169.

Drobniewski, F. A. 1993. *Bacillus cereus* and related species. *Clin. Microbiol. Rev.* 6:324–338.

Jackson, S. G., R. B. Goodbrand, R. Ahmed, and S. Kasatiya. 1995. *Bacillus cereus* and *Bacillus thuringiensis* isolated in a gastroenteritis outbreak investigation. *Appl. Microbiol. Letters.* 21:103–105.

Kolstø, A.-B., A. Grønstad, and H. Oppegaard. 1990. Physical map of the *Bacillus cereus* chromosome. *J. Bacteriol.* 172:382–5.

Kramer, J. M., and R. J. Gilbert. 1989. *Bacillus cereus* and other *Bacillus species.* In *Food borne bacterial pathogens.* Doyle, M. P., (ed.), Marcel Dekker, New York, pp. 21–70.

Priest, F.G. 1981. DNA homology in the genus *Bacillus.* In *The aerobic endospore-forming bacteria: classification and identification.* Berkely, R.C.W. and Goodfelow, M. (eds) Academic Press, New York, pp. 33–57.

Smith, C. L., S. R. Klco, and C. R. Cantor. 1988. Pulsed field gel electrophoresis and technology of large DNA molecules pp. 41–72. *In* K. E. Davies (ed). *Genome analysis. A practical approach.* IRL Press, Oxford.

50

Bacillus subtilis 168

Mitsuhiro Itaya

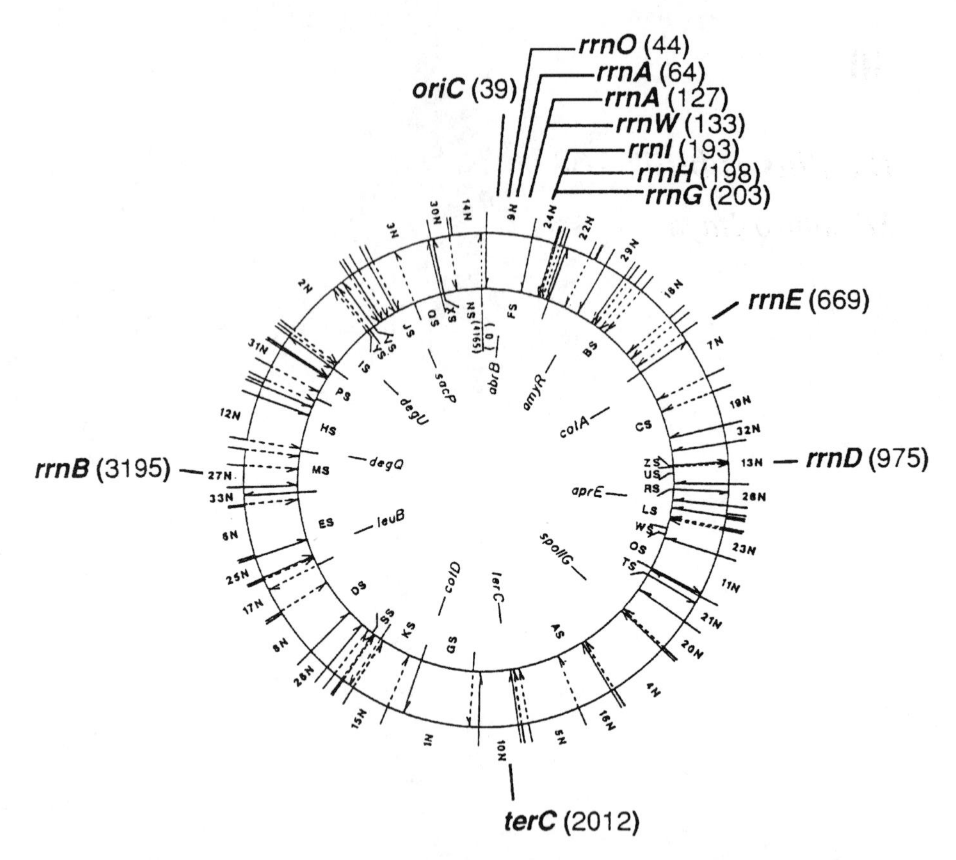

Figure 50-1. The *Not*I-*Sfi*I physical map is redrawn with minor modifications (Itaya and Tanaka, 1991). All 26 *Sfi*I fragments (AS-ZS) were aligned in the inner circle and 72 *Not*I fragments (1N-33N and smaller) in the outer circle. The *Sfi*I site between fragments FS and NS was designated as the zero point of the 4188 kb map. Locations of the region for *oriC* (Itaya, *et al.,* 1992), *terC* (Itaya and Tanaka, 1991), and ten *rrn* operons (Toda and Itaya, 1995) are indicated with distances in the parenthesis.

The highly precise physical map was constructed using a gene-directed mutagenesis method. The method allows the introduction of nucleotide sequences into a selected region of the chromosome without affecting other regions (Itaya and Tanaka, 1990). Restriction enzyme sites (*Not*I or *Sfi*I) of the chromosome thus can be eliminated or converted to different restriction enzyme sites successively. This ensured physical link of adjacent fragments that was visualized by pulsed-field gel electrophoresis. A highly precise correlation of the physical and genetic map was achieved by creation of the *Not*I and *Sfi*I sites in several gene loci of known location (Itaya and Tanaka, 1991; Itaya, 1993a). The *Not*I-*Sfi*I physical map initially constructed had a 23 kb deletion in the 18N fragment (Itaya, 1993b).

References

Itaya, M., and T. Tanaka. 1990. Gene-directed mutagenesis on the chromosome of *Bacillus subtilis* 168. *Mol. Gen. Genet.* 223:268–272.

Itaya, M., and T. Tanaka. 1991. Complete physical map of the *Bacillus subtilis* 168 chromosome constructed by a gene-directed mutagenesis method. *J. Mol. Biol.* 220:631–648.

Itaya, M., J. J. Laffan, and N. Sueoka. 1992. Physical distance between the site of type II DNA binding to the membrane and *oriC* on the *Bacillus subtilis* 168 chromosome. *J. Bacteriol.* 174:5466–5470.

Itaya, M. 1993a. Physical map of the *Bacillus subtilis* 168 chromosome, in *Bacillus subtilis* and Other Gram-positive Bacteria, A. L. Sonenshein, J. A. Hoch, and R. Losick, eds. American Society for Microbiology, Washington, D.C., pp. 463–471.

Itaya, M. 1993b. Stability and asymmetric replication of the *Bacillus subtilis* 168 chromosome. *J. Bacteriol.* 175:741–749.

Toda, T., and Itaya, M. 1995. I-*Ceu*I recognition sites in the *rrn* operons of the *Bacillus subtilis* 168 chromosome: inherent landmarks for genome analysis. *Microbiology* 141:1937–1945.

51

Bordetella pertussis Strain Tohama I

Scott Stibitz

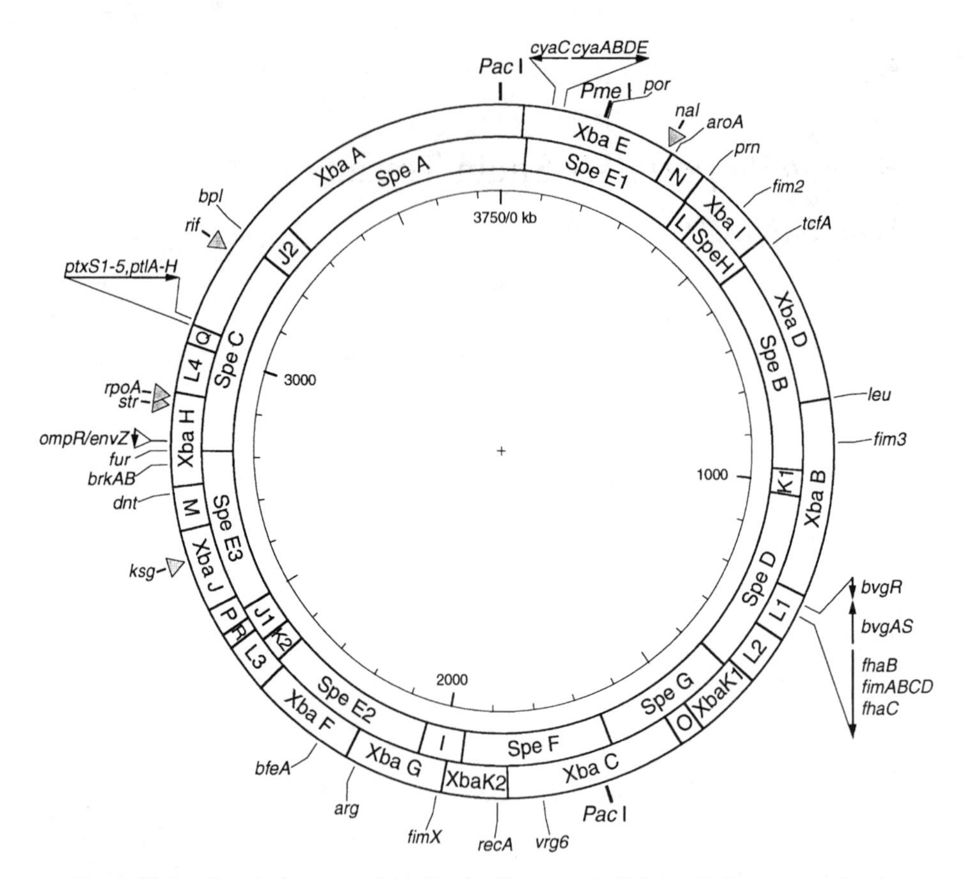

Figure 51-1. Restriction map of the *Bordetella pertussis* Tohama I chromosome for the restriction enzymes *Xba*I, *Spe*I, *Pac*I, and *Pme*I. Positions of *Pac*I and *Pme*I sites are shown by thick lines. Gene positions are denoted by thin lines. Coordinates are given in kilobase pairs, with one of the *Pac*I sites as the origin. The data used to derive this map were obtained from pulsed-field gel electrophoresis of *Xba*I and *Spe*I digests of a large number of random insertions of a plasmid suicide vector containing these restriction sites (*see also* Chapters 24–26). Insertion of the vector was mediated by homologous recombination with randomly cloned chromosomal DNA fragments (*see also* Chapter 29). Mapping of gene positions was accomplished in a similar fashion, directing insertion to specific sites using specific cloned DNA fragments. Additional data used to complete this map were obtained from the use of modified linking clones to demonstrate genetically the juxtaposition of some fragments (Stibitz and Garletts, 1992). Positions of loci defined by mutations conferring antibiotic resistance (*ksg, nal, rif, str*) as well as the *rpo*A locus were determined genetically by Hfr-mapping (Stibitz and Carbonetti, 1994) and are indicated with shaded arrowheads. The *bvgR* locus was mapped by virtue of the presence of *Xba*I and *Spe*I sites within the transposon insertions which identified it (Merkel and Stibitz, 1995). Additional genes not described by Stibitz and Garletts (1992) but subse-

quently mapped in a similar fashion are: *bfeA* (Beall and Sanden, 1995a); *bpl* (Allen and Maskell, 1996); *brkAB* (Fernandez and Weiss, 1994); *fur* (Brickman and Armstrong, 1995); *ompR/envZ* (J. Al-Meer and D. Maskell, unpublished results); *tcfA* (Finn and Stevens, 1995). The *ptl* locus has been defined genetically, and shown by DNA sequence analysis to be downstream of the ptx operon (Weiss *et al.*, 1993). The orientations of the *ptx-ptl, cya,* and *ompR/envZ* operons were determined as previously suggested (Stibitz and Garletts, 1992). In addition to the genes shown in the figure, the following homologues of *Escherichia coli* genes have been found by DNA sequence analysis of contiguous regions: *rpsM, rpsK, rpsD,* and *rplQ* are found in an operon together with *rpoA* (Carbonetti *et al.,* 1994); *groESL* is found downstream of the *brkA* gene (Fernandez and Weiss, 1995); and *recN* is found downstream of *fur* (Beall and Sanden, 1995b).

References

Allen, A., and D. Maskell. 1996. The identification, cloning and mutagenesis of a genetic locus required for lipopolysaccharide biosynthesis in *Bordetella pertussis. Mol. Microbiol.* 19:37–52.

Beall, B., and G. N. Sanden. 1995a. A *Bordetella pertussis fepA* homologue required for utilization of exogenous ferric enterobactin. *Microbiology* 141:3193–3205.

Beall, B. W., and G. N. Sanden. 1995b. Cloning and initial characterization of the *Bordetella pertussis fur* gene. *Curr. Microbiol.* 30:223–226.

Brickman, T. J., and S. K. Armstrong. 1995. *Bordetella pertussis fur* gene restores repressibility of siderophore and protein expression to deregulated *Bordetella bronchiseptica* mutants. *J. Bacteriol.* 177:268–270.

Carbonetti, N. H., T. M. Fuchs, A. A. Patamawenu, *et al.* 1994. Effect of mutations causing overexpression of RNA polymerase α subunit on regulation of virulence factors in *Bordetella pertussis. J. Bacteriol.* 176:7267–7273.

Fernandez, R. C., and A. A. Weiss. 1994. Cloning and sequencing of a *Bordetella pertussis* serum resistance locus. *Infect. Immun.* 62:4727–4738.

Fernandez, R. C., and A. A. Weiss. 1995. Cloning and sequencing of the *Bordetella pertussis cpn10/cpn60* (*groESL*) homologue. *Gene* 158:151–152.

Finn, T. M., and L. A. Stevens. 1995. Tracheal colonization factor: a *Bordetella pertussis* secreted virulence determinant. *Mol. Microbiol.* 16:625–634.

Merkel, T. J., and S. Stibitz. 1995. Identification of a locus required for the regulation of *bvg*-repressed genes in *Bordetella pertussis. J. Bacteriol.* 177:2727–2736.

Stibitz, S., and N. H. Carbonetti. 1994. Hfr mapping of mutations in *Bordetella pertussis* that define a genetic locus involved in virulence gene regulation. *J. Bacteriol.* 176:7260–7266.

Stibitz, S., and T. L. Garletts. 1992. Derivation of a physical map of the chromosome of *Bordetella pertussis* Tohama I. *J. Bacteriol.* 174:7770–7777.

Weiss, A. A., F. D. Johnson, and D. L. Burns. 1993. Molecular characterization of an operon required for pertussis toxin secretion. *Proc. Natl. Acad. Sci. USA* 90:2970–2974.

52

Borrelia burgdorferi and the Other Lyme Disease Spirochetes

Sherwood Casjens and Wai Mun Huang

Borrelia burgdorferi is a member of a cluster of closely related spirochete species, which also include *B. garinii, B. afzelii, B. japonica, B. andersonii* and several as yet unnamed types (Casjens *et al.*, 1995, Marconi *et al.*, 1995 and references therein). This group is called the "Lyme *Borrelias*" or "Lyme disease agents," even though only *B. burgdorferi, garinii* and *afzelii* are currently known to cause Lyme disease. The 945±10 kbp chromosomes of all these bacteria are linear, and genetic maps of the chromosomes from each of these species are very similar (Ojaimi *et al.*, 1994; Casjens *et al.*, 1995). The map of the chromosome of *B. burgdorferi* isolate Sh-2-82 is shown in Figure 52-1. To date, 34 gene clusters have been mapped to various restriction site intervals. These clusters contain 68 genes which are either known to be expressed in *B. burgdorferi* or have strong homology to genes whose function are known in other systems, as well as 11 complete open reading frames of unknown function.

Although the study of the molecular biology of these organisms is in its infancy (reviewed by Rosa and Schwan, 1993; Saint-Girons *et al.*, 1994), several unusual and interesting features of the Lyme agent spirochete genomes are known. (1) The chromosomes are linear, as are those of the other non-Lyme agent *Borrelias* which have been analyzed (Davidson *et al.*, 1992; Ojaimi *et al.*, 1994; Casjens and Huang, 1993; Casjens *et al.*, 1995). (2) All natural isolates carry multiple linear and circular plasmids (Hinnebusch and Tilly, 1994). (3) The termini of at least two of the linear plasmids (Hinnebusch *et al.*, 1990) and the chromosome (our unpublished results) are covalently closed, hairpin DNA structures. (4) Several of the plasmids, notably the linear 1p49 (linear, 53–60 kbp in different species) and the circular cp26 (circular, 28 kbp), are found in all Lyme disease agent isolates, and it has been argued that they should be considered to be small chromosomes, although propagation in the laboratory is possible without at least p49 (Sadziene *et al.*, 1992; Barbour, 1993; our unpublished results). (5) *Borrelia* cells may be polyploid in that at least *B. hermsii* may contain as many as sixteen

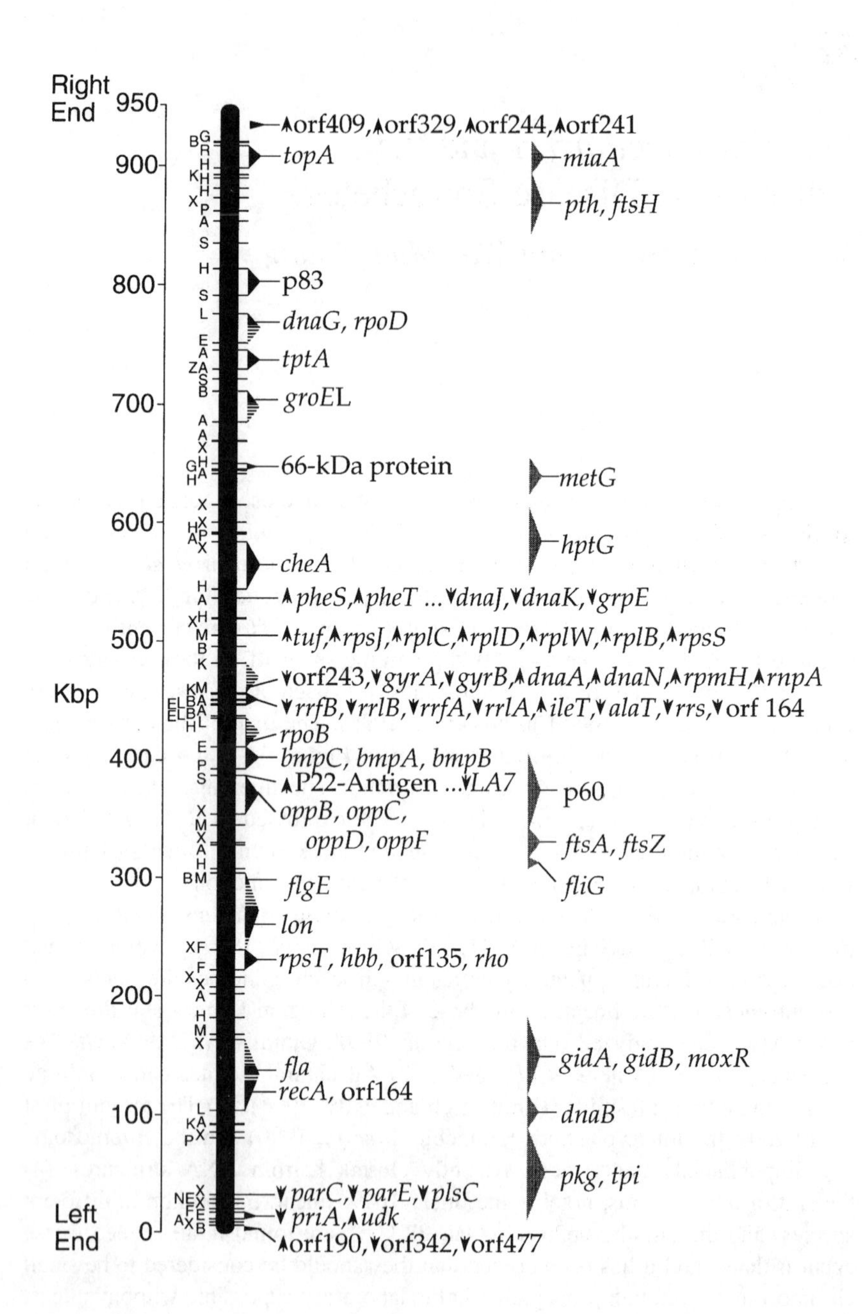

Right End
Left End
Kbp
950
900
800
700
600
500
400
300
200
100
0
orf409, orf329, orf244, orf241
topA
miaA
pth, ftsH
p83
dnaG, rpoD
tptA
groEL
66-kDa protein
metG
hptG
cheA
pheS, pheT ... dnaJ, dnaK, grpE
tuf, rpsJ, rplC, rplD, rplW, rplB, rpsS
orf243, gyrA, gyrB, dnaA, dnaN, rpmH, rnpA
rrfB, rrlB, rrfA, rrlA, ileT, alaT, rrs, orf 164
rpoB
bmpC, bmpA, bmpB
P22-Antigen ... LA7
p60
oppB, oppC,
oppD, oppF
ftsA, ftsZ
fliG
flgE
lon
rpsT, hbb, orf135, rho
gidA, gidB, moxR
fla
recA, orf164
dnaB
pkg, tpi
parC, parE, plsC
priA, udk
orf190, orf342, orf477

copies of the chromosome per cell (Kitten & Barbour, 1992; *see also* Chapter 9). (6) Genes appear to be rather highly clustered according to function on the chromosome, and gene order within operons is usually similar but not always identical to that found in other eubacteria. (7) The central location of the *dnaA* gene invites speculation that replication might originate near the center of the chromosome. The fact that 34 of 41 genes whose transcription orientation is known are transcribed away from the center supports this idea (Figure 52-1; S. our unpublished results).

◀

Figure 52-1. Physical and genetic maps of *B. burgdorferi* isolate Sh-2-82.

The following features are indicated from left to right: a scale in kbp (0 indicates the end known as the "left end"), the mapped restriction sites in strain Sh-2-82 (Casjens and Huang, 1993), and the location of all currently identified chromosomal genes. Intervals that contain the various genes are indicated as follows: black triangles, genes mapped by Casjens and Huang (1993), Casjens *et al.* (1995) and our unpublished results; stippled triangles—genes mapped by Ojaimi *et al.* (1994); triangles that include black and cross-hatched areas—the whole triangle indicates the restriction site interval in which those genes are located in strain Sh-2-82, and the black region indicates a smaller region within which the gene maps in another Lyme disease agent isolate (Casjens *et al.*, 1995; our unpublished results). Where the orientation is known, arrows indicate the direction of transcription, and the leftmost gene is nearest the lower end of the chromosome (Casjens and Huang, 1993; our unpublished results). Commas separate adjacent genes in unoriented gene clusters, and ellipses indicate <5 kbp unsequenced regions between characterized regions. Restriction enzymes are abbreviated as follows: A, *Sma*I; B, *Bss*HII; E, *Eag*I; F, *Sfi*I; G, *Sgr*AI; H, *Ehe*I; K, *Ksp*I; L, *Spl*I; M. *Mlu*I; N, *Not*I; P, *Pvu*I; R, *Rsr*II; S, *Sse*8387I; X, *Xho*I; Z, *Srf*I. The cleavage site maps for each of these enzymes is complete except for any sites that might lie within 2.5 kbp of another site cleaved by the same enzyme.

Sources for probes or sequences used in the determination of the locations of unpublished genes were the following: our unpublished results (*parA, gyrA, recA; arcA* and *orf*342 clusters); Scott Samuels and Joann Cloud (*lon*); Yuk-Ching Tse-Dinh (*topA*); Ira Schwartz (*cheA, rpoB* and *tuf* cluster); Alan Barbour and Joe Hinnebusch (*pheT, pheS*); Ron Limberger and Nyles Charon (*flgE*); Patti Rosa (*oppB* cluster); Kit Tilly (*rpsT* and *priA* clusters); Brian Stevenson (*tptA*). References and sequences for all other genes on the map can be found in Casjens and Huang (1993), Ojaimi *et al.* (1994), Casjens *et al.* (1995) and the sequence databases. The chromosomal genes are named for to their *E. coli* homologues except the p60, p83, P22-antigen, 66-kDa protein (previously called orfX) encoding genes, *bmpA/B/C* (previously called p39) and *LA7*, which encode surface or periplasmic antigens. In addition, *tptA,* appears to encode a homolog of *mglA* and *rbsA,* genes for ATPase subunits of *E. coli* ABC-type transporters, *hbb* should encode a small DNA binding protein of the Hu/Him/Hip type, and "orfZ's" indicate open reading frames of unknown function "Z" codons in length.

References

Barbour, A. 1993. Linear DNA of *Borrelia* species and antigenic variation. *Trends Microbiol.* 1:236–239.

Casjens, S., and W. M. Huang. 1993. Linear chromosome physical and genetic map of *Borrelia burgdorferi,* the Lyme disease agent. *Molec. Microbiol.* 8:967–980.

Casjens, S., H. Ley, M. DeLange, P. Rosa, and W. M. Huang. 1995. Linear chromosomes of Lyme disease agent spirochetes; genetic diversity and conservation of gene order. *J. Bacteriol.* 177:2769–2780.

Davidson, B., J. MacDougall, and I. Saint Girons. 1992. Physical map of the linear chromosome of the bacterium *Borrelia burgdorferi* 212, a causative agent of Lyme disease and localization of rRNA genes. *J. Bacteriol.* 174:3766–3774.

Hinnebusch, J., S. Bergström, and A. Barbour. 1990. Cloning and sequence analysis of linear plasmid telomeres of the bacterium *Borrelia burgdorferi. Molec. Microbiol.* 4:917–922.

Hinnebusch, J., and K. Tilly. 1994. Linear plasmids and chromosomes in bacteria. *Molec. Microbiol.* 10:917–922.

Kitten, T., and A. Barbour. 1992. The relapsing fever agent *Borrelia hermsii* has multiple copies of its chromosome and linear plasmids. *Genetics* 132:311–324.

Marconi, R., D. Liveris, and I. Schwartz. 1995. Identification of novel insertion elements, RFLP patterns, and discontinuous 23S rRNA in Lyme disease spirochetes: Phylogenetic analyses of rRNA genes and their intergenic spacers in *Borrelia japonica* sp. nov. and genomic group 21038 (*Borrelia andersonii* sp. nov.) isolates. *J. Clin. Microbiol.* 33:2427–2434.

Ojaimi, C., B. Davidson, I. Saint Girons, and I. Old. 1994. Conservation of gene arrangement and an unusual organization of rRNA genes in the linear chromosomes of Lyme disease spirochaetes *Borrelia burgdorferi, B. garinii* and *B. afzelii. Microbiology* 140:2931–2940.

Rosa, P., and T. Schwan. 1993. Molecular biology of *Borrelia burgdorferi.* In *Lyme Disease.* Coyle, P., ed. B. C. Decker, Mosby Year Book, Inc., Philadelphia, pp. 8–17.

Sadziene, A., P. Rosa, P. Thompson, D. Hogan, and A. Barbour. 1992. Antibody-resistant mutants of *Borrelia burgdorferi:* in vitro selection and characterization. *J. Exp. Med.* 176:799–809.

Saint-Girons, I., I. Old, and B. Davidson. 1994. Molecular biology of the *Borrelia,* bacteria with linear replicons. *Microbiology* 140:1803–1816.

53

Bradyrhizobium japonicum strain 3I1b110

Michael Göttfert, C. Kündig, and H. Hennecke

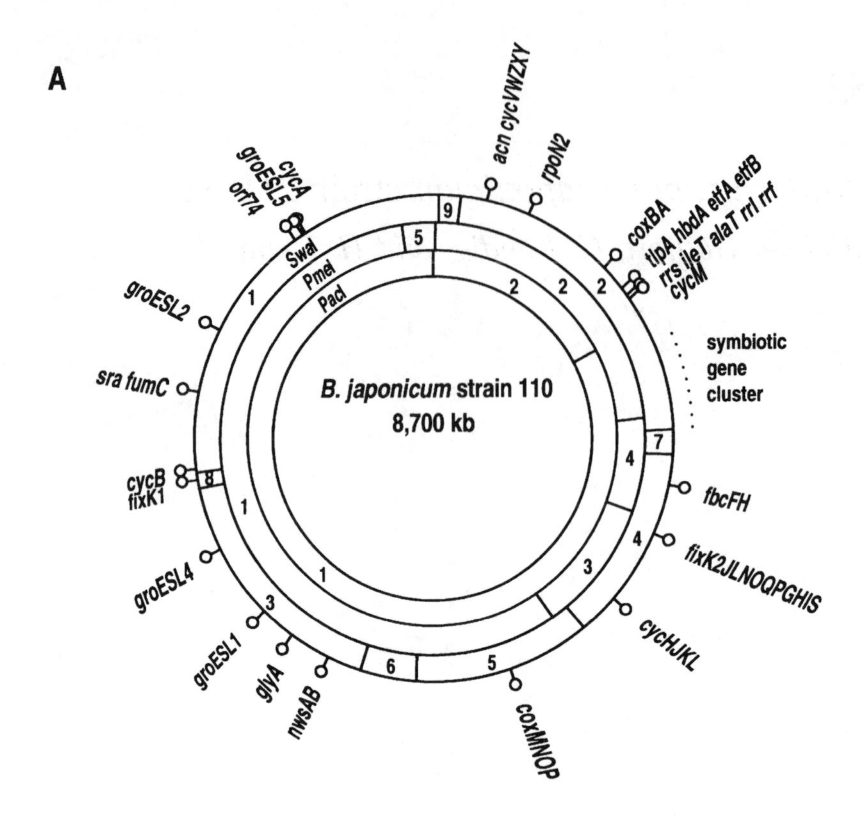
A
acn cycVWZXY
rpoN2
cycA
groESL5
orf74
coxBA
tlpA hbdA etfA etfB
rrs ileT alaT rrl rrf
cycM
SwaI
PmeI
PacI
groESL2
symbiotic
gene
cluster
sra fumC
B. japonicum strain 110
8,700 kb
cycB
fixK1
fbcFH
fixK2JLNOQPGHIS
groESL4
cycHJKL
groESL1
glyA
nwsAB
coxMNOP

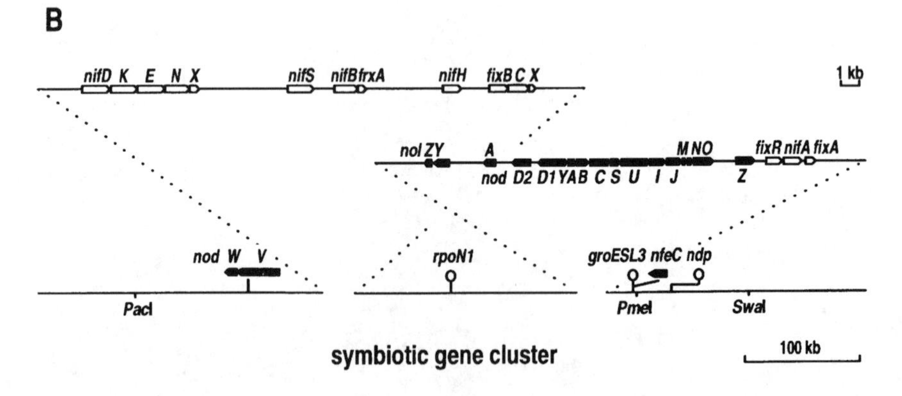
B
nifD K E N X
nifS nifB frxA
nifH fixB C X
1 kb
nol ZY
A
M N O
fixR nifA fixA
nod D2 D1 Y A B C S U I J
Z
nod W V
rpoN1
groESL3 nfeC ndp
PacI
PmeI
SwaI
symbiotic gene cluster
100 kb

◀

Figure 53-1. The genome of *Bradyrhizobium japonicum* strain 3I1b110. The map was established as reported by Kündig *et al.* (1993). Gene loci are described by Kündig *et al.* (1993) except those that follow: *acn* (Thöny-Meyer *et al.*, unpublished); *coxB* (Loferer *et al.*, unpublished); *cycA* (Bott *et al.*, 1995); *cycZY* (Thöny-Meyer *et al.*, 1995); *cycHJKL* (Ritz *et al.*, 1995); *fixGHIS* (Preisig *et al.*, unpublished); *fixK2* (Fischer, 1994); *fixNOQP* (Preisig *et al.*, 1993); *groESL1, groESL2, groESL3, groESL4, groESL5* (Fischer *et al.*, 1993); *hbdA, etfA, etfB* (Weidenhaupt *et al.*, unpublished); *ndp* (Weidenhaupt *et al.*, 1993); *nfeC* (Chun and Stacey, 1994); *nodZ* (Stacey *et al.*, 1994); *nolMNO* (Luka *et al.*, 1993); *nolYZ* (Dockendorf *et al.*, 1994). *nwsAB* (Grob *et al.* 1993); *orf74* (Weidenhaupt *et al.*, 1995); *rrs, ileT, alaT, rrl, rrf* (Kündig *et al.* 1995); *tlpA* (Loferer *et al.*, 1993). A) Correlated physical and genetic map of the *B. japonicum* chromosome. The restriction sites of *Pac*I, *Pme*I, and *Swa*I are indicated, and the fragments are numbered in order of their sizes. B) Physical and genetic map of the symbiotic gene region. Arrows show the positions, sizes, and orientations of the genes. Circles indicate the locations of genes whose orientations were not determined. Open symbols indicate genes that are supposed to be involved in symbiotic nitrogen fixation. Filled symbols indicate genes that are supposed to be involved in the nodulation process. The exact arrangement of *groESL3* and *nfeC* is unknown.

References

Bott, M., L. Thöny-Meyer, H. Loferer, S. Rossbach, R. E. Tully, D. Keister, C. A. Appleby, and H. Hennecke. 1995. *Bradyrhizobium japonicum* cytochrome c_{550} is required for nitrate respiration but not for symbiotic nitrogen fixation. *J. Bacteriol.* 177:2214–2217.

Chun, J.-J., and G. Stacey. 1994. A *Bradyrhizobium japonicum* gene essential for nodulation competitiveness is differently regulated from two promoters. *Mol. Plant-Microbe Interact.* 7:248–255.

Dockendorf, T. C., A. J. Sharma, and G. Stacey. 1994. Identification and characterization of the *nolYZ* genes of *Bradyrhizobium japonicum. Mol. Plant-Microbe Interact.* 7:173–180.

Fischer, H.-M., M. Babst, T. Kaspar, G. Acuña, F. Arigoni, and H. Hennecke. 1993. One member of a *groESL*-like chaperonin multigene family in *Bradyrhizobium japonicum* is co-regulated with symbiotic nitrogen fixation genes. *EMBO J.* 12:2901–2912.

Fischer, H.-M. 1994. Genetic regulation of nitrogen fixation in rhizobia. *Microbiol. Rev.* 58:352–386.

Grob, P., P. Michel, H. Hennecke, and M. Göttfert. 1993. A novel response-regulator is able to suppress the nodulation defect of a *Bradyrhizobium japonicum nodW* mutant. *Mol. Gen. Genet.* 241:531–541.

Kündig, C., H. Hennecke, and M. Göttfert. 1993. Correlated physical and genetic map of the *Bradyrhizobium japonicum* 110 genome. *J. Bacteriol.* 175:613–622.

Kündig, C., C. Beck, H. Hennecke, and M. Göttfert. 1995. A single rRNA gene region in *Bradyrhizobium japonicum. J. Bacteriol.*, in press.

Loferer, H., M. Bott, and H. Hennecke. 1993. *Bradyrhizobium japonicum* TlpA, a novel membrane-anchored thioredoxin-like protein involved in the biogenesis of cytochrome aa_3 and development of symbiosis. *EMBO J.* 12:3373–3383.

Luka, S., J. Sanjuan, R. W. Carlson, and G. Stacey. 1993. *nolMNO* genes of *Bradyrhizobium japonicum* are co-transcribed with *nodYABCSUIJ,* and *nolO* is involved in the synthesis of the lipo-oligosaccharide nodulation signals. *J. Biol. Chem.* 268:27053–27059.

Preisig, O., D. Anthamatten, and H. Hennecke. 1993. Genes for a microaerobically induced oxidase complex in *Bradyrhizobium japonicum* are essential for a nitrogen-fixing endosymbiosis. *Proc. Natl. Acad. Sci. USA* 90:3309–3313.

Ritz, D., L. Thöny-Meyer, and H. Hennecke. 1995. The *cycHJKL* gene cluster plays an essential role in the biogenesis of *c*-type cytochromes in *Bradyrhizobium japonicum. Mol. Gen. Genet.* 247:27–38.

Stacey, G., S. Luka, J. Sanjuan, Z. Banfalvi, A. J. Nieuwkoop, J. Y. Chun, L. S. Forsberg, and R. Carlson. 1994. *nodZ,* a unique host-specific nodulation gene, is involved in the fucosylation of the lipooligosaccharide nodulation signal of *Bradyrhizobium japonicum. J. Bacteriol.* 176:620–633.

Thöny-Meyer, L., F. Fischer, P. Künzler, D. Ritz, and H. Hennecke. 1995. *Escherichia coli* genes required for cytochrome *c* maturation. *J. Bacteriol.* 177, in press.

Weidenhaupt, M., H.-M. Fischer, G. Acuña, J. Sanjuan, and H. Hennecke. 1993. Use of a promoter-probe vector system in the cloning of a new NifA-dependent promoter (*ndp*) from *Bradyrhizobium japonicum. Gene* 129:33–40.

Weidenhaupt, M., M. Schmid-Appert, B. Thöny, H. Hennecke, and H.-M. Fischer. 1995. A new *Bradyrhizobium japonicum* gene required for free-living growth and bacteroid development is conserved in other bacteria and in plants. *Mol. Plant-Microbe Interact.* 8, in press.

54

Brucella melitensis 16M (ATCC 23456)

Sylvie Michaux-Charachon and Michel Ramuz

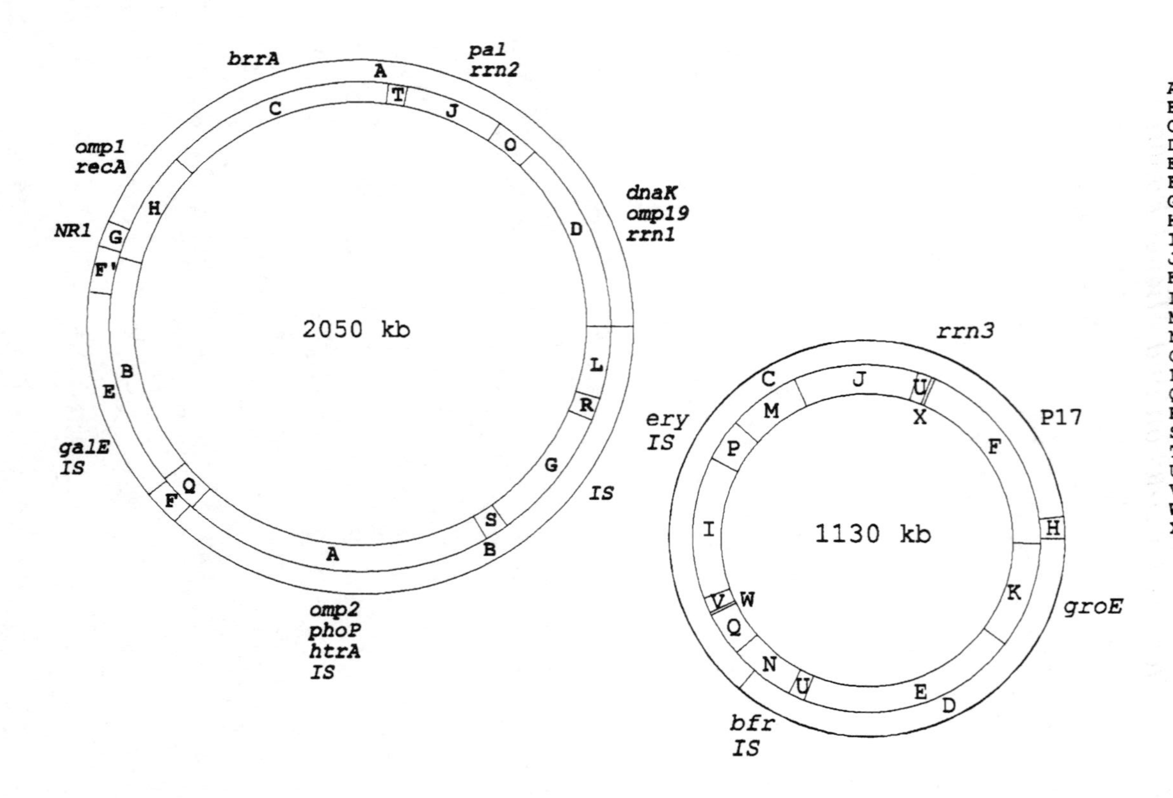

	PacI	SpeI
A	900	420
B	720	315
C	700	305
D	410	260
E	260	240
F	50	210
G	30	182
H	20	158
I		140
J		130
K		112
L		97
M		75
N		72
O		65
P		48
Q		44
R		34
S		30
T		26
U		16
V		14
W		5
X		3

◀
Figure 54-1. Restriction maps of the two chromosomes of the *Brucella melitensis* 16M genome. The circles represent the *Pac*I (outer circle) and *Spe*I maps (inner circle). The origin of the two replicons are arbitrarily fixed. The size of the two replicons is indicated in each of the maps. The sizes of *Pac*I- and *Spe*I-fragments (in kb) are indicated in the figure. Several genes were located on the maps: *rrn* (Michaux *et al.*, 1993); *omp1* and *omp2* (Ficht *et al.*, 1988); BCSP 31 (Mayfield *et al.*, 1988); *dnaK* and *groE* (Cellier *et al.*, 1992); *pal* (Tibor *et al.*, 1994), *bfr* (Denoel *et al.*, 1995), P17 (Hemmen *et al.*, 1995); *omp19* (Tibor, personal communication); *recA* (Tatum *et al.*, 1993); NR1 and *phoP* (Wren and O'Callaghan, personal communication), *htrA* (Tatum *et al.*, 1994), *ery* (Sangari *et al.*, 1994), *galE* (Petrovska, personal communication), *IS* (Ouarhani *et al.*, 1993; Halling *et al.*, 1993).

References

Cellier, M., J. Teyssier, M. Nicolas. *et al.* 1992. Cloning and characterization of the *Brucella ovis* heat shock protein DnaK functionally expressed in *Escherichia coli. J. Bacteriol.* 174:8036–8042.

Denoel, P., M. S. Zygmunt, V. Weynants, *et al.* 1995. Cloning and sequencing of the bacterioferritin gene of *Brucella melitensis* 16M strain. *FEMS Microbiol. Lett.* 361:238–242.

Ficht, T. A., S. W. Bearden, B. A. Sorva, and L. G. Adams. 1988. A 36 kilodalton *Brucella abortus* cell envelope protein is coded by repeated sequences closely related in the genomic DNA. *Infect. Immun.* 56:2036–2046.

Halling, S. M., F. M. Tatum, and B. J. Bricker. 1993. Sequence and characterization of an insertion sequence, *IS*711, from *Brucella ovis. Gene* 133:123–127.

Hemmen, F., V. Weynants, T. Scarcez, *et al.* 1995. Cloning and sequence analysis of a newly identified *Brucella abortus* gene and serological evaluation of the 17 kilodalton antigen that it encodes. *Clin. Diagn. Lab. Immunol.* 2:263–267.

Mayfield, J. E., B. J. Bricker, H. Godfrey, *et al.* 1988. The cloning, expression and nucleotide sequence of a gene coding for an immunologically important *Brucella abortus* protein. *Gene* 63:1–9.

Michaux, S., J. Paillisson, M. J. Carles-Nurit, *et al.* 1993. Presence of two independent chromosomes in the *Brucella melitensis* 16M genome. *J. Bacteriol.* 175:701–705.

Ouarhani, S., S. Michaux, J. S. Widada, *et al.* 1993. Identification and sequence analysis of *IS*6501, an insertion sequence in *Brucella* spp.: relationship between genomic structure and the number of *IS*6501 copies. *J. Gen. Microbiol.* 139:3265–3273.

Petrovska, L., personal communication.

Sangari, F. J., J. M. Garcia-Lobo, and J. Aguerro. 1994. The *Brucella abortus* vaccine strain B19 carries a deletion in the erythritol catabolic genes. *FEMS Microbiol. Lett.* 121:337–342.

Tatum, F. M., N. F. Cheville, and D. Morfitt. 1994. Cloning, characterization and construction of *htrA* and *htrA*-like mutants of *Brucella abortus* and their survival in BALB/c mice. *Microbiol. Pathogen.* 17:23–36.

Tatum, F. M., D. C. Morfitt, and S. Halling. 1993. Construction of a *Brucella abortus* RecA mutant and its survival in mice. *Microbiol. Pathogen.* 14:177–185.

Tibor, A., personal communication.

Tibor, A., V. Weynants, P. Denoel, *et al.* 1994. Molecular cloning, nucleotide sequence, and occurrence of a 16.5-kilodalton outer membrane protein of *Brucella abortus* with similarity to PAL lipoprotein. *Infect. Immun.* 62:3633–3639.

Wren, D., and O'Callaghan, unpublished results.

55

Campylobacter jejuni TGH9011 (ATCC43431)

V. L. Chan, N. W. Kim, B. Bourke, E. Hani, D. Ng, R. Lombardi, H. Bingham, Y. Hong, T. Wong, and H. Louie

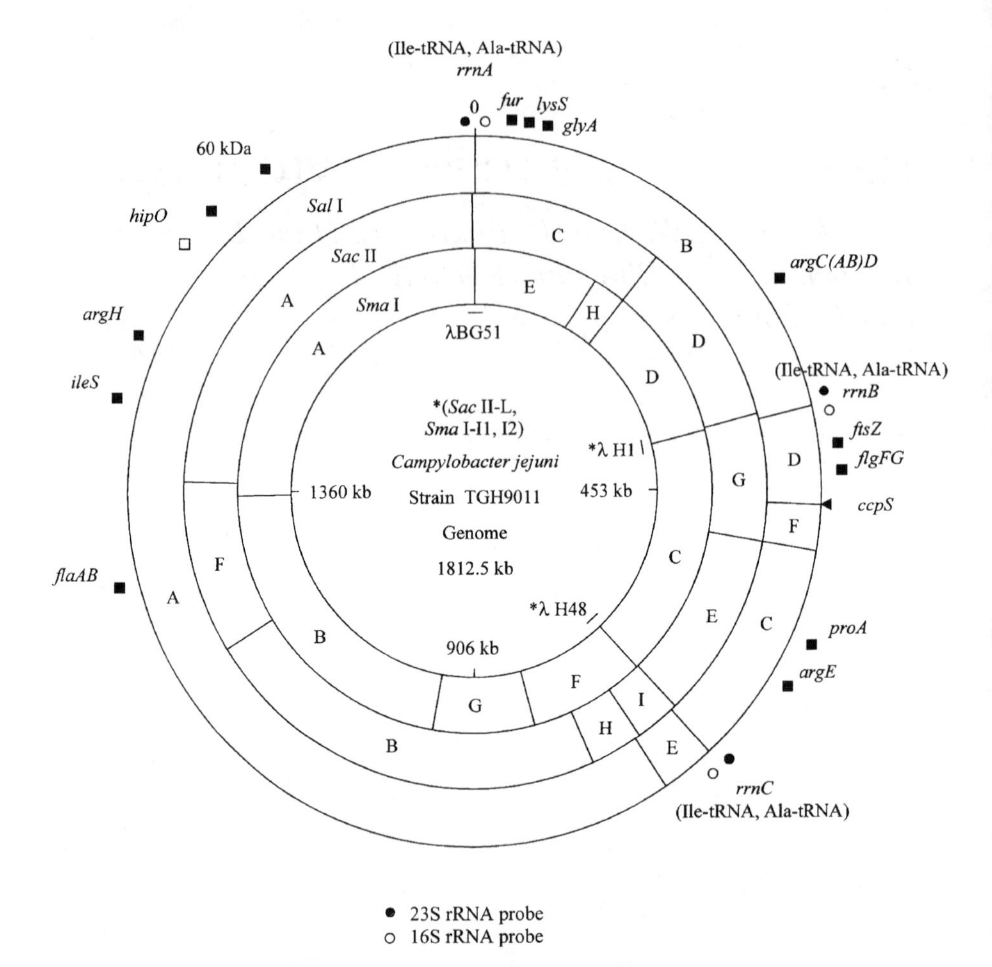

Figure 55-1. Physical map of *C. jejuni* TGH9011. The updated version of the physical map of *C. jejuni* TGH9011 with *Sal*I, *Sac*II, and *Sma*I sites is shown. To date, a total of 30 *C. jejuni* genetic markers were mapped to the specific restriction enzyme fragments and 24 of the markers were presented in Kim *et al.* (Chan, V. L., and H. Bingham, 1991). The locations of the three rRNA lambda clones (lambda BG51, H1, and H48) which correspond to the three rRNA operons in the physical map (*rrnA, rrnB,* and *rrnC,* respectively) are shown. A detailed map of *rrnA, rrnB,* and *rrnC* was presented in Kim *et al.* (Chan, V. L., and H. Bingham, 1991) and the sequence organization of the *rrnA* operon has been published (Chan, V. L., and H. Bingham, 1992). Asterisks mark the locations of the three small fragments which are contained and conserved in all three rRNA operons. Open and closed circles represent 16S and 23S rRNA probes, respectively. The *hipO* gene has been mapped to 215 or 250 kb from the *Sma*I site in the *rrnA* operon (Hani, E. K., and V. L. Chan, unpublished data) and the two possible locations are shown as the open and closed boxes.

Abbreviations and genes

argABCDEH, arginine biosynthetic genes (Chan, V. L., H. Louie, and H. Bingham (1995), Hani, E., D. Ng, and V. L. Chan, unpublished data); *ccpS,* a *Campylobacter jejuni* cytoplasmic protein gene with a *Sal*I site (Hani, E. K., and V. L. Chan 1994); *flaAB,* flagellin A and B genes (Hani, E. K., and V. L. Chan (1995), Chan, V. L., H. Bingham, and H. Louie, unpublished data); *flgFG,* basal body rod protein genes (Hong, Y., B. Bourke, H. Louie, D. Lu, and V. L. Chan 1996); *fur,* ferric uptake regulation (Hong, Y., T. Wong, B. Bourke, and V. L. Chan 1995); *ftsZ,* a cell division gene (Chan, V. L., H. Louie, and S. Rashid, unpublished data); *glyA,* Serine hydroxymethyl transferase (Khawaja, R., H. Bingham, J. L. Penner, K. Neote, and V. L. Chan 1992); *hipO,* hippuricase (Kim, N. W., H. Bingham, R. Khawaja, H. Louie, E. K. Hani, K. Neote, and V. L. Chan 1992); *ileS,* isoleucyl-tRNA synthetase (Kim, N. W., R. Gutell, and V. L. Chan 1995); *lysS,* lysyl-tRNA synthetase (Kim, N. W., R. Lombardi, H. Bingham, E. K, Hani, H. Louie, D. Ng, and V. L. Chan 1993); 60 kDa, a conserved 60 kDa gene (Chan, V. L., *et al.*, unpublished data); *proA,* gamma-glutamyl phosphate reductase (Louie, H., and V. L. Chan 1993).

Methods

The genomic map was generated using pulsed-field gel electrophoresis (PFGE; *see* Chapters 24–26) to separate a small number of large restriction fragments produced by *Sal*I, *Sac*II, and *Sma*I. The alignment of these fragments into a physical map was based largely on a cross-Southern hybridization method using radiolabeled individual PFGE-*Sma*I fragments and cloned genes to probe genomic digests created by *Sal*I, *Sac*II, and *Sma*I (Kim, N. W., R. Lombardi, H. Bingham, E. K. Hani, M. Louie, D. Ng, and V. L. Chan. (1993), Louie, H., and V. L. Chan (1996).

References

Chan, V. L., and H. Bingham. 1991. Complete sequence of the *Campylobacter jejuni glyA* gene. *Gene* 101:51–58.

Chan, V. L., and H. Bingham. 1992. Lysyl-tRNA synthetase gene of *Campylobacter jejuni. J. Bacteriol.* 174:695–701.

Chan, V. L., H. Louie, and H. Bingham. 1995. Cloning and transcription regulation of the ferric uptake regulatory gene of *Campylobacter jejuni* TGH9011. *Gene* 164:25–31.

Hani, E. K., and V. L. Chan. 1994. Cloning, characterization, and nucleotide sequence analysis of the *argH* gene from *Campylobacter jejuni* TGH9011 encoding arginosuccinate lyase. *J. Bacteriol.* 176:1865–1871.

Hani, E. K., and V. L. Chan. 1995. Expression and characterization of *Campylobacter jejuni* Benzoylglycine amidohydrolase gene (Hippuricase) in *Escherichia coli. J. Bacteriol.* 177:2396–2402.

Hong, Y., B. Bourke, H. Louie, D. Lu, and V. L. Chan. 1996. A novel *Campylobacter jejuni* cytoplasmic protein identifies a previously unrecognized *Sal*I restriction enzyme site. Submitted for publication.

Hong, Y., T. Wong, B. Bourke, and V. L. Chan. 1995. Isolation and sequence analysis of isoleucyl-tRNA synthetase gene from *Campylobacter jejuni. Microbiol.* (U.K.) 141:2561–2567.

Khawaja, R., H. Bingham, J. L. Penner, K. Neote, and V. L. Chan. 1992. Cloning and Characterization of *Campylobacter jejuni* flagella gene. *Curr. Microbiol.* 24:213–221.

Kim, N. W., H. Bingham, R. Khawaja, H. Louie, E. K. Hani, K. Neote, and V. L. Chan. 1992. Physical map of *Campylobacter jejuni* strain TGH9011 and the localization of ten genetic markers using pulsed-field gel electrophoresis. *J. Bacteriol.* 174:3494–3498.

Kim, N. W., R. Gutell, and V. L. Chan. 1995. Complete nucleotide sequences and the organization of rrnA ribosomal RNA operon in *Campylobacter jejuni* TGH9011 (ATCC 43431). *Gene* 164:101–106.

Kim, N. W., R. Lombardi, H. Bingham, E. K. Hani, H. Louie, D. Ng, and V. L. Chan. 1993. Fine mapping of three rRNA operons on the updated genomic map of *Campylobacter jejuni* TGH9011 (ATCC 43431). *J. Bacteriol.* 175:7468–7470.

Louie, H., and V. L. Chan. 1993. Cloning and characterization of the gamma-glutamyl phosphate reductase gene of *Campylobacter jejuni. Mol. Gen. Genet.* 240:29–35.

Louie, H., and V. L. Chan. 1996. Expression of the *flgFG* operon of *Campylobacter jejuni* in *Escherichia coli.* Submitted for publication.

56

Caulobacter crescentus CB15

Bert Ely

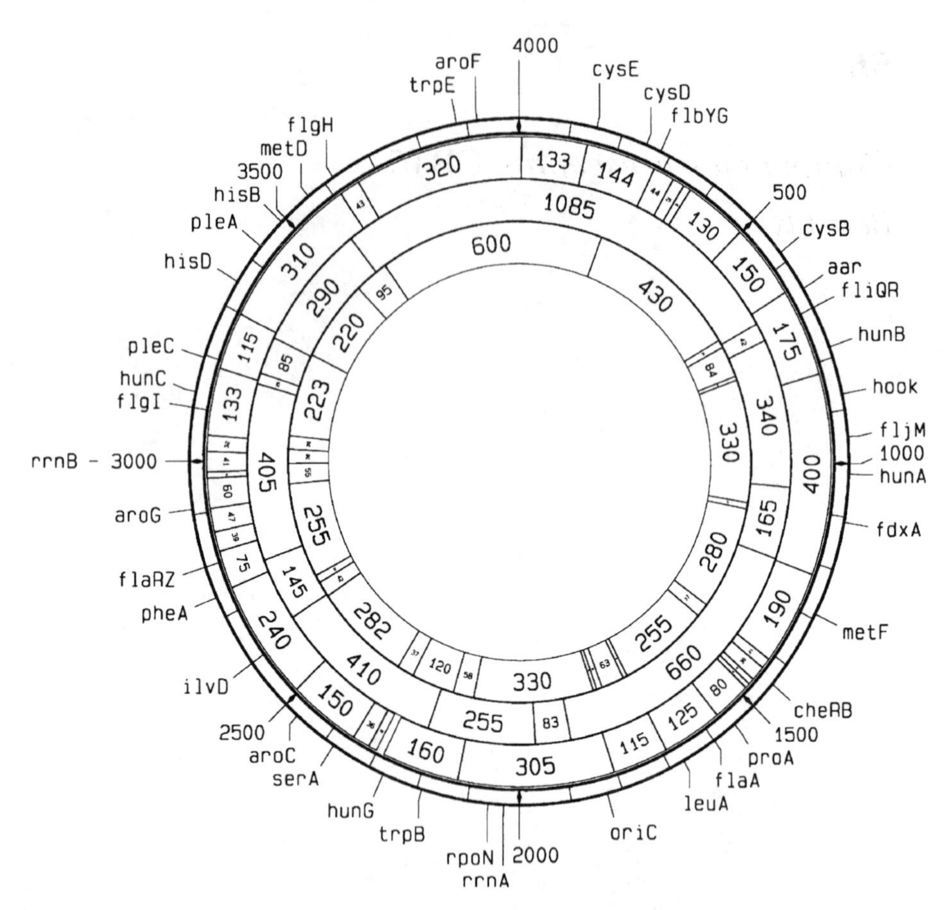

Figure 56-1. A physical and genetic map of the *Caulobacter crescentus* strain CB15 chromosome. The circles are restriction maps for *Dra*I, *Ase*I, and *Spe*I, respectively, with *Spe*I being the innermost. Pulsed field gel electrophoresis was used to determine the sizes and relative positions of restriction fragments (Ely and Ely, 1989). *C. crescentus* has a single 4000 kilobase chromosome (Ely *et al.*, 1990). Fragment sizes are in kilobases. The origin of replication (*oriC*) is located at position 1830 (Dingwall and Shapiro, 1989). Ribosomal RNA operons are located at positions 2030 and 3000 (Ely and Gerardot, 1988). Gene designations and references can be found in Ely (1993). The map positions of additional genes can be obtained from Bert Ely (ELY@BIOL.SC.EDU). From the caulobacter Homepage (www.cosm.sc.edu/caulobacter)

References

Dingwall, A., and L. Shapiro. 1989. Chromosome replication rate, origin, and bidirectionality as determined by pulsed field gel electrophorersis. *Proc. Natl. Acad. Sci. USA* 86:119–123.

Ely, B. 1993. Genomic map of *Caulobacter crescentus. Genetic Maps* 6:2.146–2.150.

Ely, B., and T. W. Ely. 1989. Use of pulsed field gel electrophoresis and transposon mutagenesis to estimate the minimal number of genes required for motility in *Caulobacter crescentus. Genetics* 123:649–654.

Ely, B., T. W. Ely, C. J. Gerardot, and A. Dingwall. 1990. Circularity of the *Caulobacter crescentus* chromosome determined by pulsed-field gel electrophoresis. *J. Bacteriol.* 172:1262–1266.

Ely, B., and C. J. Gerardot. 1988. Use of pulsed-field-gradient gel electrophoresis to construct a physical map of the *Caulobacter crescentus* genome. *Gene* 68:323–333.

57

Clostridium beijerinckii NCIMB 8052 Chromosome

Shane R. Wilkinson and Michael Young

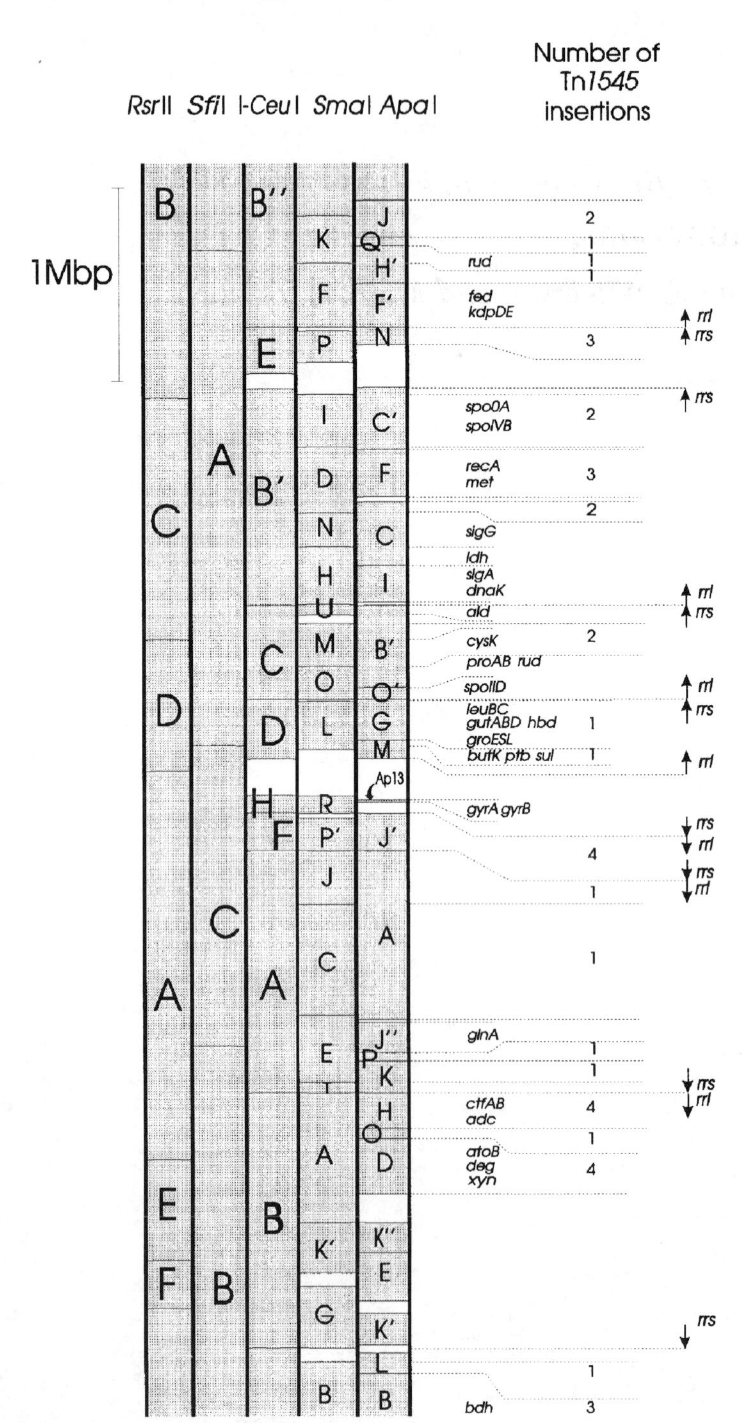
Number of Tn1545 insertions
RsrII SfiI I-CeuI SmaI ApaI
1Mbp
rud
fed
kdpDE
spo0A
spoIVB
recA
met
sigG
ldh
sigA
dnaK
ald
cysK
proAB rud
spoIID
leuBC
gutABD hbd
groESL
butK ptb sul
Ap13
gyrA gyrB
glnA
ctfAB
adc
atoB
deg
xyn
bdh
rrl
rrs

◀
Figure 57-1. Clostridium beijerinckii (formerly *Clostridium acetobutylicum*) NCIMB 8052 is a representative of the saccharolytic, solvent-forming clostridia. These organisms form four distinct taxonomic groupings within the genus *Clostridium* (Wilkinson *et al.*, 1995). *C. beijerinckii* NCIMB 8052 has a comparatively large (circa 6.6 Mbp) circular genome (*see also* Chapter 1). In Figure 57-1 it has been linearised at the approximate location of the replication terminus. The map was constructed by nucleic acid hybridization, using genes of both homologous and heterologous origin and randomly cloned DNA fragments as probes, as well as a collection of strains harbouring Tn*1545* insertions. All *Sfi*I and *Rsr*II fragment overlaps have been detected, giving an unambiguous circular map containing nine landmark restriction sites. Fragments generated by *Sma*I and *Apa*I were organised into fourteen contigs, varying in size from 625 kbp to 80 kbp, and superimposed on the map of landmark restriction sites. Many of the contigs had *Sma*I and/or *Apa*I sites within *rrn* operons at their extremities. The resulting physical map was confirmed by analysis with I-*Ceu*I, which cuts the genome into fourteen fragments at sites corresponding to the positions of *rrl* genes (*see also* Chapters 21; 24). Several gaps between adjacent *Sma*I/*Apa*I contigs that were on a previous version of the map (Wilkinson and Young, 1995) have been closed as a result of incorporation of the I-*Ceu*I data (Wilkinson and Young, 1997). The remaining gaps correspond to the locations of some of the small I-*Ceu*I, *Sma*I and *Apa*I fragments that have not yet been positioned on the map.

Most of the I-*Ceu*I fragments hybridised with both *rrs* and *rrl* probes. The exceptions were I-*Ceu*I fragment H (89 kbp), which hybridised only with the *rrs* probe, and I-*Ceu*I fragment B″, which hybridised only with the *rrl* probe (Wilkinson and Young, 1997). These data are in agreement with results obtained previously (Wilkinson and Young, 1995) indicating that the fourteen *rrn* operons of *C. beijerinckii* are transcribed confluently with direction of chromosome replication, which proceeds bidirectionally from an origin located within I-*Ceu*I fragment H to a terminus located within I-*Ceu*I fragment B″ (1258 kbp).

C. beijerinckii NCIMB 8052 has more (14) *rrn* operons than any other eubacterium studied to date (*see* Chapter 21). Nine of the *rrn* operons have been located with respect to *Sma*I and/or *Apa*I fragments on the physical map. A tenth *rrn* operon is presumably associated with an I-*Ceu*I fragment that lies between fragments B′ and E and the remaining four *rrn* operons probably lie in the interval between I-*Ceu*I fragments D and F, close to the replication origin within fragment H.

About 40 genes and a similar number of "silent" Tn*1545* insertions have been assigned approximate positions, as indicated to the right of the physical map. Many of the genes encoding enzymes concerned with volatile fatty acid and solvent production have been located on the map. They are organised in several clusters, two of which are concerned with the production of acetone (*ctfAB, adc, atoB* [= *thl*]) and butyrate (*butK, ptb*). The genes concerned with acetate production have not yet been characterised and it seems likely that several distinct alcohol dehydrogenases, one of which (*bdh*) has been located on the map, participate in butanol production. Genes encoding proteins involved in electron transfer reactions (*rud, fed*) are not clustered together. Among the housekeeping genes that have been located on the map, *gyrA* and *gyrB* are located close to the deduced position of the replication origin, as is the case in other endospore-forming bacteria (Cole and Saint-Girons, 1994). Gene symbols are the same as used previously (Wilkinson *et al.*, 1995; 1996).

References

Cole, S. T., and I. Saint Girons. 1994. Bacterial genomics. *FEMS Microbiology Reviews* 14:139–160.

Keis, S., C. F. Bennett, V. K. Ward, and D. T. Jones. 1995: Taxonomy and phylogeny of industrial solvent-producing clostridia. *International Journal of Systemic Bacteriology* 45:693–705.

Wilkinson, S. R., and M. Young. 1995. A physical map of the *Clostridium beijerinckii* (formerly *Clostridium acetobutylicum*) NCIMB 8052 chromosome. *J. Bacteriology* 177:439–448.

Wilkinson, S. R., M. Young, R. Goodacre, *et al.* 1995. Phenotypic and genotypic differences between certain strains of *Clostridium acetobutylicum. FEMS Microbiology Letters* 125:199–204.

Wilkinson, S. R., and M. Young. 1997. An I-*Ceu*I map of the *Clostridium beijerinckii* NCIMB 8052 chromosome. *Anaerobe* (submitted).

58

Clostridium perfringens CPN50

Stewart T. Cole, Sei-Ichi Katayama, Bruno Dupuy, Thierry Garnier, Brigitte Saint-Joanis, and Bruno Canard

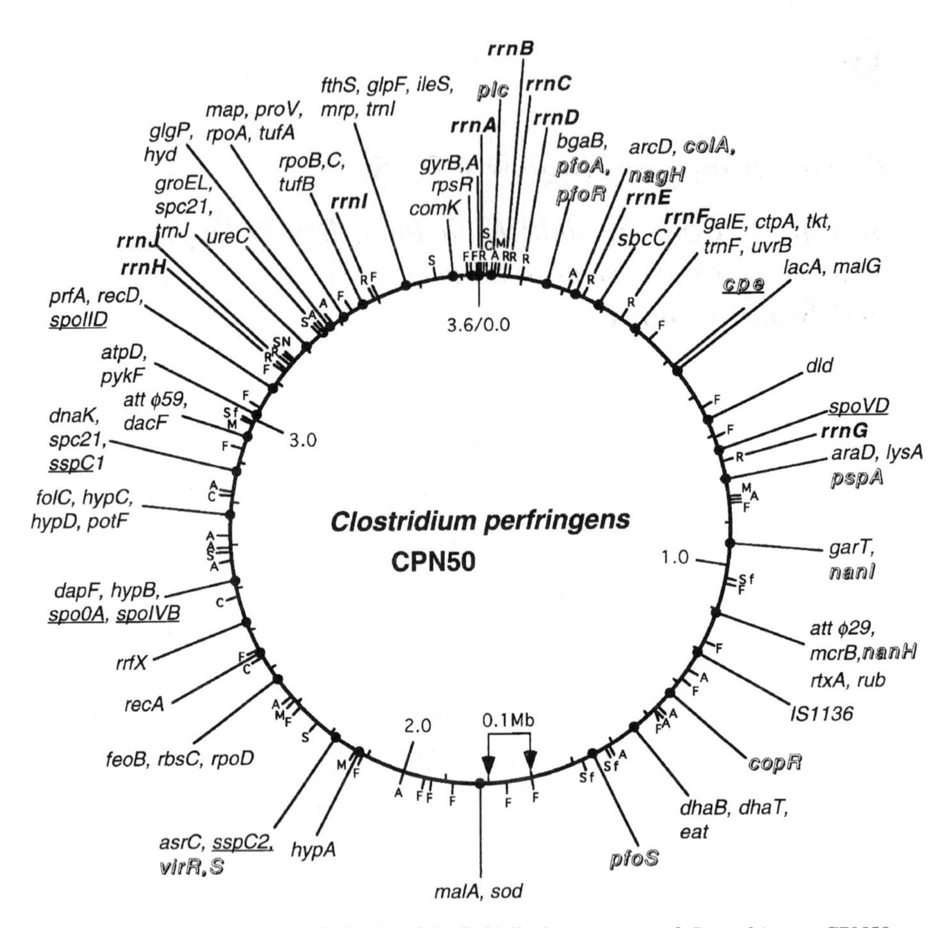

Figure 58-1. Current physical map of the 3.6 Mb chromosome of *C. perfringens* CPN50 showing the positions of the various cleavage sites, and 97 genetic markers. Restriction enzyme sites; A, *Apa*I; C, *Sac*II (*Ksp*I); F, *Fsp*I (*Avi*II); M, *Mlu*I; N, *Nru*I; R, cluster of rare sites (2*Sac*II-*Sma*I-*Nru*I-2*Sma*I-*Nru*I-I-*Ceu*I); S, *Sma*I; Sf, *Sfi*I. The *rrn* operons, harboring the R sites are shown in bold.

Genes known or likely to be associated with pathogenicity are indicated by shading, these include: *plc,* phospholipase C or α-toxin; *pfoA,* perfringolysin or Θ-toxin; *pfoR,* regulates *pfoA; colA,* collagenase or κ-toxin; *nagH,* hyaluronidase (endo-beta-N-acetylglucosaminidase) or μ-toxin; *cpe,* enteroxin produced on sporulation, gene not found in CPN50 but occasionally present at this site in some food-poisoning strains; *pspA,* putative analogue of pneumococcal surface protein A; *nanI,* large sialidase or neuraminidase; *nanH,* small sialidase or neuraminidase; *copR,* homologous to virulence gene regulator in *Pseudomonas syringae; pfoS,* putative virulence regulatory gene very similar to *pfoR;* virRS, two component regulatory system that controls expression of *plc, pfoA, colA,* plus some as yet unidentified protease, sialidase, and nuclease genes.

Genes known or likely to be expressed at different stages of sporulation are underlined: *cpe,* enterotoxin; *spoVD,* putative sporulation-specific penicillin-binding protein; *sspC2,* small acid soluble protein; *spoOA,* phosphorylation-activated transcription factor; *spoIVB,* stage IV sporulation protein B; *sspc1,* small acid soluble protein; *spoIID,* stage II sporulation protein D. The positions of genes are indicated by the circles situated in the center of the mapping intervals. When several genes map to the same site, their order is arbitrary. For details of map construction and identification and localisation of genes see the following references.

References

Canard, B., and S. T. Cole. 1989. Genome organization of the anaerobic pathogen *Clostridium perfringens. Proc. Natl. Acad. Sci. USA* 86:6676–6680.

Canard, B., B. Saint-Joanis, and S. T. Cole. 1992. Genomic diversity and organisation of virulence genes in the pathogenic anaerobe *Clostridium perfringens. Mol. Microbiol.* 6:1421–1429.

Katayama, S., B. Dupuy, T. Garnier, and S. T. Cole. 1995. Rapid expansion of the physical and genetic map of the chromosome of *Clostridium perfringens* CPN50. *J. Bacteriol.* 177:5680–5685.

59

Enterococcus faecalis

*Barbara E. Murray, Liangxia Jiang, Jianguo Xiao,
Xiang Qin, Kavindra V. Singh, Aart de Kok,
Al Claiborne, Patrice Courvalin,
and George M. Weinstock*

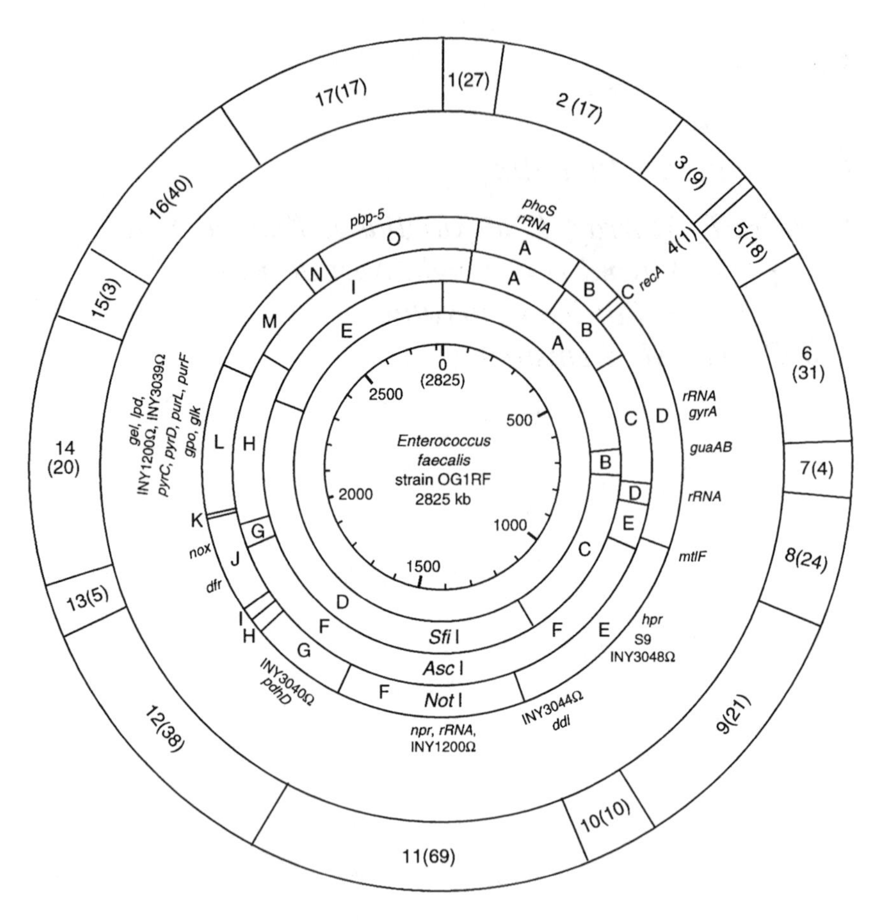

Figure 59-1. Physical map of *Enterococcus faecalis* strain OG1RF

Restriction endonuclease cleavage sites are shown for the enzymes *Sfi*I, *Not*I, and *Asc*I in the innermost concentric circles. Immediately outside of this are shown the approximate positions of those genes that have been mapped. The genes are positioned to indicate with which restriction fragments they hybridize. In addition to genes, a number of transposon insertion sites are indicated (the Ω sites). For details on the construction of this map see Murray *et al.* 1993. Generation of restriction map of *Enterococcus faecalis* strain OG1 and investigation of growth requirements and regions encoding biosynthetic function. *J. Bacteriol.* 175:5216–5223.

The outermost concentric circle shows the numbers pLAFRx and pBAC cosmid clones (in parentheses) that have been binned by hybridization to 17 different regions around the chromosome. Each region is defined by the ends of restriction fragments used in hybridizations. The insert sizes are approximately 20kb and 30kb in pLAFR and pBAC, respectively. 353 clones have been binned, representing approximately about 3.5× coverage of the chromosome. Approximately 40% of these clones have been ordered with respect to each other and a minimal set of about 140 clones is expected to be needed to cover the whole genome.

60

Escherichia coli

George M. Weinstock

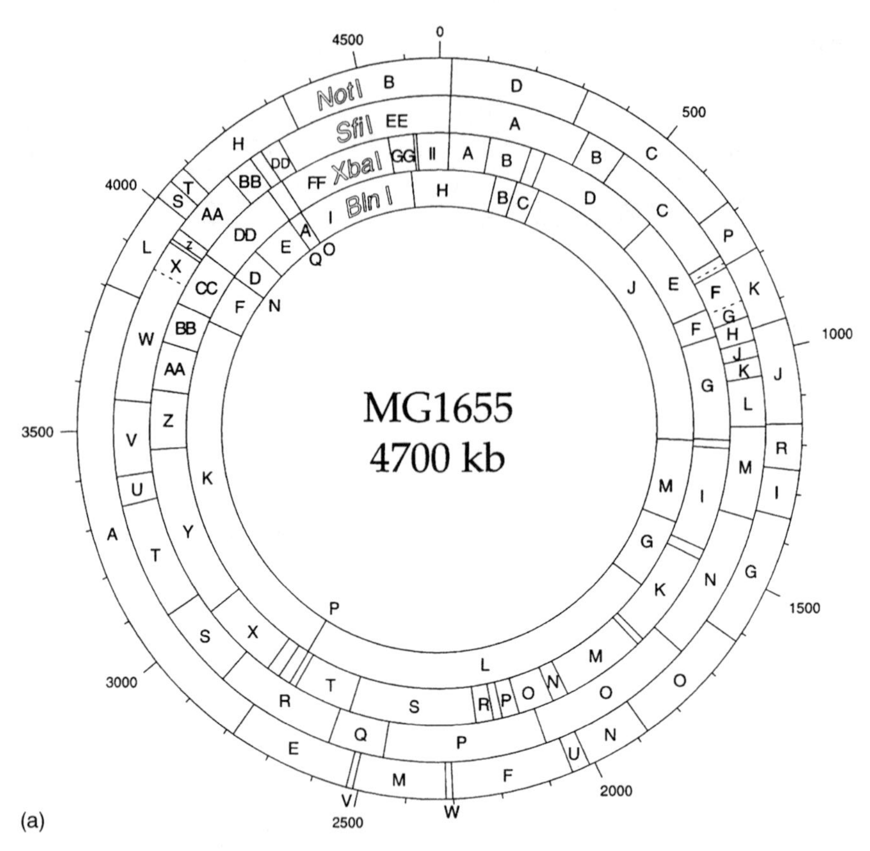

Figure 60-1. *Escherichia coli.* (a) Circular map of the chromosome of strain MG1655. The map is oriented so that 0 minutes on the genetic map and 0 kb on the physical map are at the top at the *thrABC* locus. The 3 *Sfi*A sites shown by broken lines are only partially digested due to cytosine methylation. Some fragments are not lettered due to space constraints. (b) Comparative maps of 6 *E. coli* K-12 strains. The maps have been linearized from the 0 coordinate position. The figure illustrates sites at which the strains differ from MG1655. Filled shapes represent insertions relative to MG1655, open shapes represent deletions. The size of the difference is indicated in kb by the number above the shape. The side of the shape closest to the map indicates the region where the difference is located, while the side farthest from the map indicates the size of the difference. Sites that have been deleted are indicated by a —. Extra sites are indicated by opposing arrows surrounding the site. Known correlations with genetic elements are indicated below each map. Parentheses are used to indicate when these are speculative. In the *Bln*L fragment of EMG2, two possible locations of a new site are indicated by broken lines. For additional information see J. D. Perkins, J. D. Heath, B. R. Sharma, and G. M. Weinstock, *J. Mol. Biol.* 1993. 232:419–445, *Xba*I and *Bln*I genomic cleavage maps of *Escherichia coli* K-12 strain MG1655 and comparative analysis of other strains.

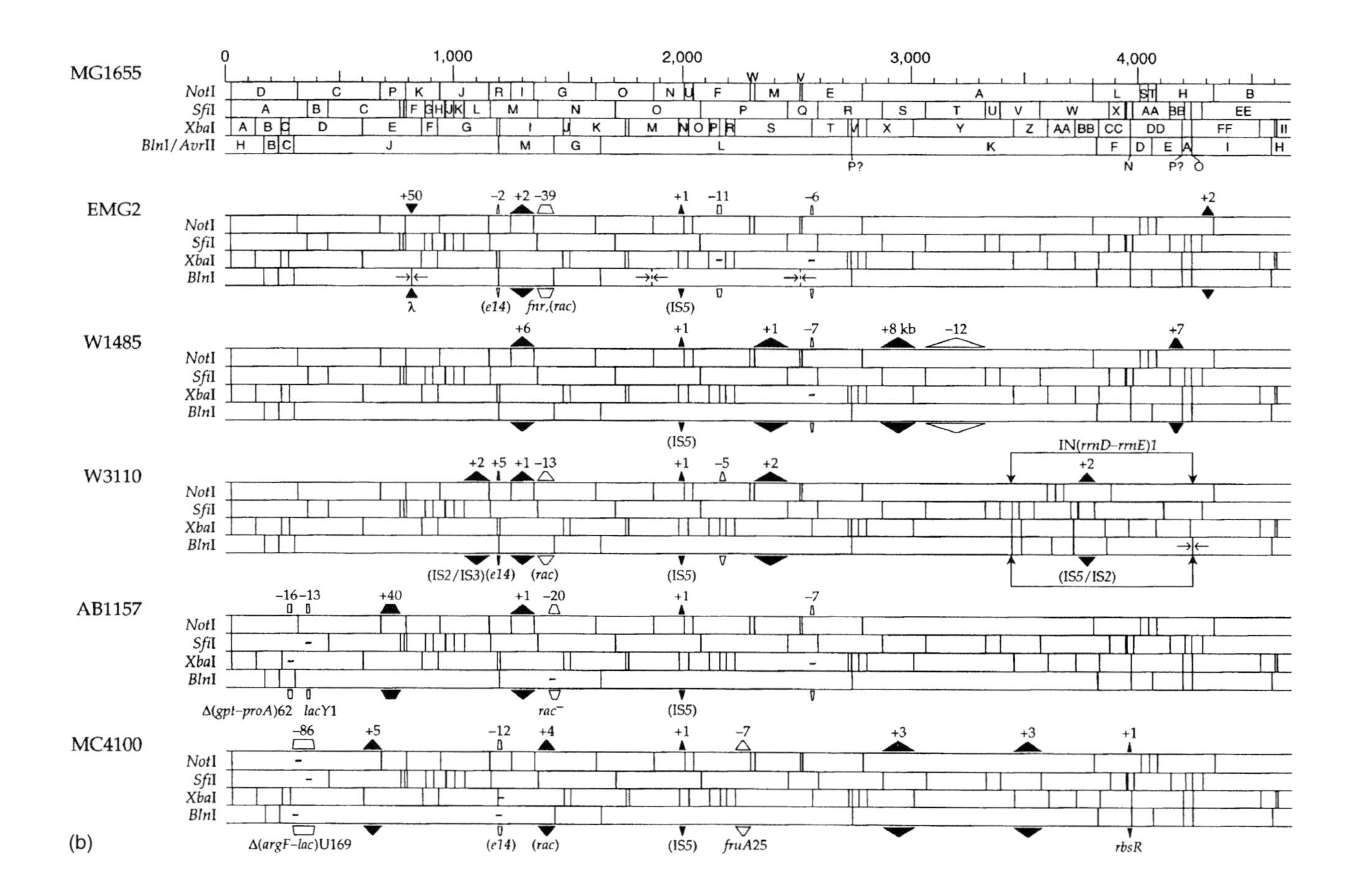
MG1655
NotI
SfiI
XbaI
BlnI/AvrII
0
1,000
2,000
3,000
4,000
EMG2
W1485
W3110
AB1157
MC4100
λ
(e14)
fnr,(rac)
(IS5)
+8 kb
IN(rrnD–rrnE)1
(IS2/IS3)(e14)
(rac)
(IS5/IS2)
Δ(gpt–proA)62
lacY1
rac⁻
Δ(argF–lac)U169
fruA25
rbsR
(b)

61

Haemophilus influenzae Rd

Sol H. Goodgal and Marilyn Mitchell

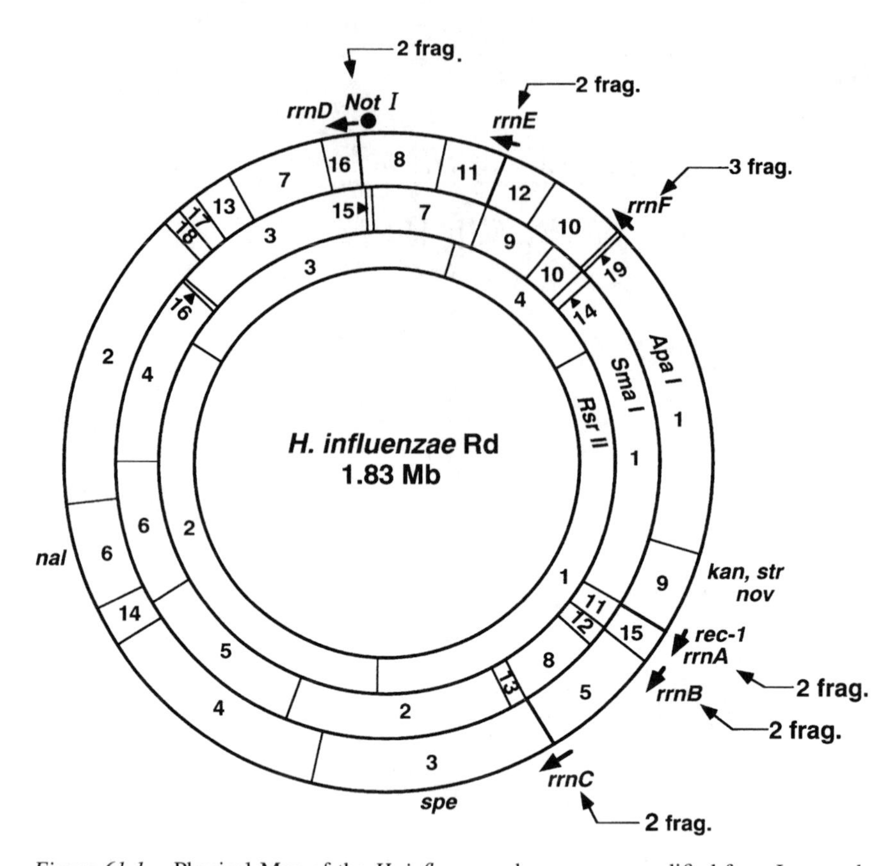

Figure 61-1. Physical Map of the *H. influenzae* chromosome modified from Lee *et al.* (1989) and Kauc *et al.* (1989). The circular map was constructed by hybridization of a labeled fragment with two or more fragments from a digest with a different restriction enzyme. The restriction enzymes ApaI, SmaI, and RsrII yielded 19, 16, and 4 discernable fragments by pulse-field electrophoresis. Additional ApaI sites have been determined from the complete DNA sequence of the *H. influenzae* chromosome (Fleischmann *et al.*, 1995) and are indicated by the numbers (2 or 3) at the intersection of larger fragments. Since the complete sequence of *H. influenzae* is available in the sequence database (Genome Sequence Database accession number L42023) only a few genetic markers are indicated, including the 6 rrn loci identified by Lee *et al.* (1989). An extensive list of all known genes is included at the end of the manuscript by Fleischmann *et al.* (1995). A physical map of *H. influenzae* type b can be found in Butler and Moxon (1990), and the map for *H. parainfluenzae* is available in Kauc and Goodgal (1989).

References

Butler, P. D., and E. R. Moxon. 1990. A physical map of the genome of *Haemophilus influenzae* type b. *J. Gen. Microbiol.* 136:2333–2342.

Fleischmann, R. D., *et. al.* 1995. Whole-genome random sequencing and assembly of *Haemophilus influenzae* Rd. *Science* 269:496–512.

Kauc, L., and S. H. Goodgal. 1989. The size and a physical map of the chromosome of *Haemophilus parainfluenzae. Gene* 83:377–380.

Kauc, L., M. Mitchell, and S. H. Goodgal. 1989. Size and physical map of the chromosome of *Haemophilus influenzae. J. Bacteriol.* 171:2474–2479.

Lee, J. J., H. O. Smith, and R. J. Redfield. 1989. Organization of the *Haemophilus influenzae* Rd genome. *J. Bacteriol.* 171:3016–3024.

62

Haloferax volcanii DS2 and *Halobacterium salinarium* GRB

Robert L. Charlebois

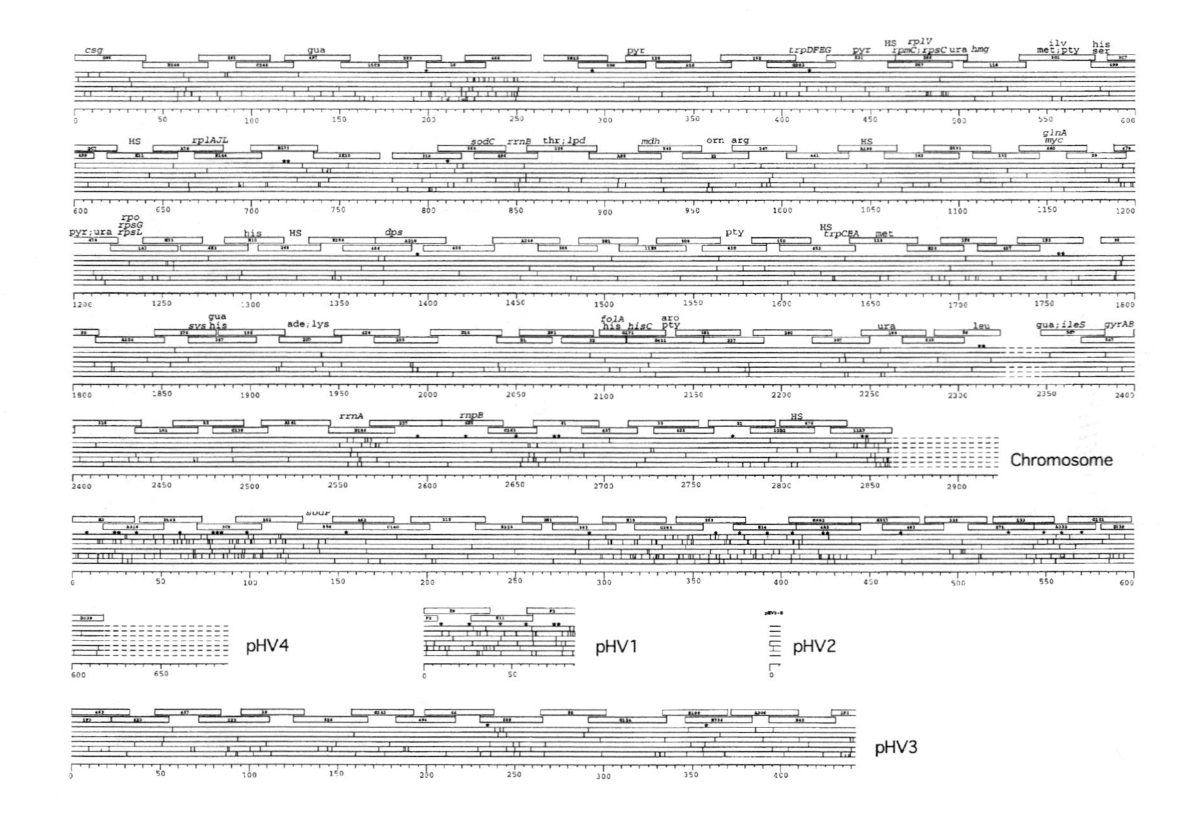

Chromosome
pHV4
pHV1
pHV2
pHV3

Contig maps for two haloarchaeal genomes have been constructed to date: *Haloferax volcanii* DS2 (Charlebois *et al.*, 1991) and *Halobacterium salinarium* GRB (St. Jean *et al.*, 1994). Additionally, chromosomal macrorestriction maps are available from *Haloferax mediterranei* ATCC 33500 (López-García *et al.*, 1992; Antón *et al.*, 1994) and from *Hb. salinarium* strains NRC-1 and S9 (Hackett *et al.*, 1994). The principal reason for producing this collection of maps has been to compare them. Haloarchaeal genomes often possess a high number and activity of insertion sequences, and there has been a common belief that the resulting genetic instability translates into genomic instability (Charlebois and Doolittle, 1989). Two chromosomal comparisons have been published, between *Hf. volcanii* and *Hf. mediterranei* (López-García *et al.*, 1995) and among three strains of *Hb. salinarium* (Hackett *et al.*, 1994), revealing marked conservation in gene order. A more distant, full-genomic comparison between *Haloferax* and *Halobacterium* has recently been completed (A. St. Jean and R. L. Charlebois (1996).

Here I present the updated maps for *Hf. volcanii* DS2 (Fig. 62-1) and for *Hb. salinarium* GRB (Fig. 62-2). They are a graphical representation of the tabular map data published elsewhere (Charlebois 1995a,b), with a few minor corrections. The strategy used to construct the physical maps is described in detail in Charlebois (1993).

◀

Figure 62-1. Physical and genetic map of the five-replicon *Haloferax volcanii* DS2 genome. It is 96% cloned as a set of overlapping cosmids, shown as boxes over the restriction map. From top to bottom, sites for *Bam*HI, *Bgl*II, *Dra*I, *Hin*dIII, *Pst*I and *Ssp*I are shown, over a scale bar in kbp. Dashed lines in the staves indicate unmapped regions, whose size we know from hybridization to *Afl*II, *Asc*I or *Dra*I macrorestriction fragments. Genes are indicated in italics, loci mapped through complementation of auxotrophs are in plain text, and putative heat shock loci are named HS. Members of the insertion sequence family ISH51 are indicated as stars. The positions of tRNA genes (Cohen *et al.*, 1992) are not shown in this version of the map since the tRNA data are being revised (C. J. Daniels, personal comm.). Data were consolidated from Charlebois *et al.* (1991), Cohen et al. (1992), Trieselmann and Charlebois (1992), López-García *et al.* (1995), and Charlebois (1995a). A more detailed map exists for pHV1 (Schalkwyk *et al.*, 1993), and pHV2 has been sequenced (Charlebois *et al.*, 1987).

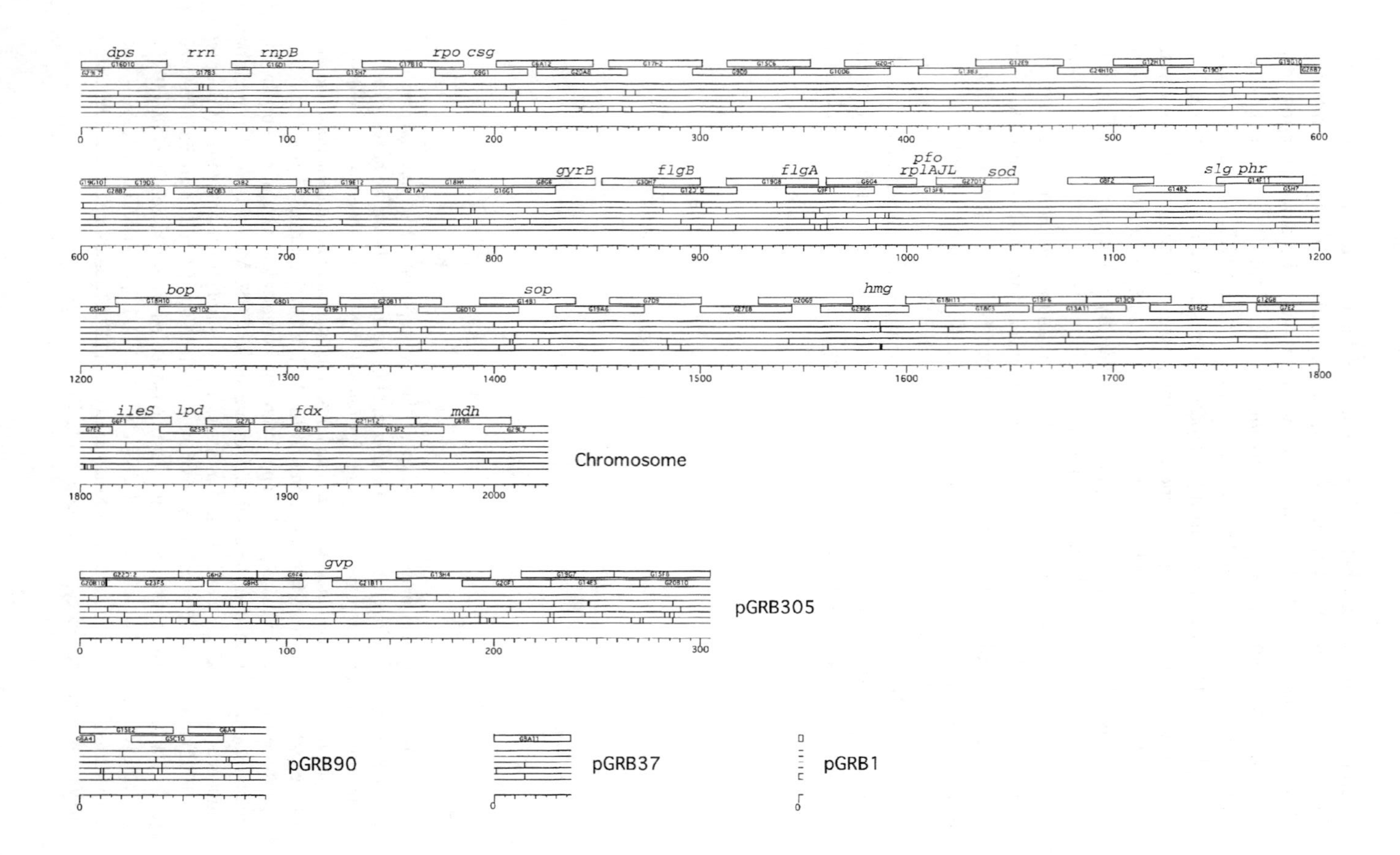

dps
rrn
rnpB
rpo csg
gyrB
flgB
flgA
pfo
rplAJL
sod
slg phr
bop
sop
hmg
ileS
lpd
fdx
mdh
Chromosome
gvp
pGRB305
pGRB90
pGRB37
pGRB1

◀
Figure 62-2. Physical and genetic map of the five-replicon *Halobacterium salinarium* GRB genome. It is 99% cloned as a set of overlapping cosmids, shown as boxes over the restriction map. From top to bottom, sites for *Afl*II, *Ase*I, *Dra*I, *Hin*dIII and *Ssp*I are shown, over a scale bar in kbp. Genes are indicated in italics. Data were consolidated from St. Jean *et al.* (1994) and Charlebois (1995b). The plasmid pGRB1 has been sequenced (Hackett *et al.*, 1990).

References

Antón, J., P. López-García, J. P. Abad, C. L. Smith, and R. Amils. 1994. Alignment of genes and *Swa*I restriction sites to the *Bam*HI genomic map of *Haloferax mediterranei. FEMS Microbiol. Lett.* 117:53–60.

Charlebois, R. L. 1993. Physical mapping of genomes using the landmark strategy, in *The Second International Conference on Bioinformatics, Supercomputing and Complex Genome Analysis,* H. A. Lim, J. W. Fickett, C. R. Cantor and R. J. Robbins, eds. World Scientific, Singapore, pp. 219–229.

Charlebois, R. L. 1995a. Physical and genetic map of the genome of *Haloferax volcanii* DS2, in *Archaea: A Laboratory Manual, Halophiles, vol.*, S. DasSarma and E. M. Fleischmann, eds. Cold Spring Harbor Laboratory Press, Cold Spring Harbor, NY, Appendix 3.

Charlebois, R. L. 1995b. Physical and genetic map of the genome of *Halobacterium* sp. GRB, in *Archaea: A Laboratory Manual, Halophiles vol.* DasSarma and E. M. Fleischmann, eds. Cold Spring Harbor Laboratory Press, Cold Spring Harbor, NY, Appendix 4.

Charlebois, R. L., and W. F. Doolittle. 1989. Transposable elements and genome structure in halobacteria, in *Mobile DNA,* M. Howe and D. Berg, eds. American Society for Microbiology, Washington, D.C., pp. 297–307.

Charlebois, R. L., W. L. Lam, S. W. Cline, and W. F. Doolittle. 1987. Characterization of pHV2 from *Halobacterium volcanii* and its use in demonstrating transformation of an archaebacterium. *Proc. Natl. Acad. Sci. USA* 84:8530–8534.

Charlebois, R. L., L. C. Schalkwyk, J. D. Hofman, and W. F. Doolittle. 1991. A detailed physical map and set of overlapping clones covering the genome of the archaebacterium *Haloferax volcanii* DS2. *J. Mol. Biol.* 222:509–524.

Cohen, A., W. L. Lam, R. L. Charlebois, W. F. Doolittle, and L. C. Schalkwyk. 1992. Localizing genes on the map of the genome of *Haloferax volcanii,* one of the Archaea. *Proc. Natl. Acad. Sci. USA* 89:1602–1606.

Hackett, N. R., M. P. Krebs, S. DasSarma, W. Goebel, U. L. RajBhandary, and H. G. Khorana. 1990. Nucleotide sequence of a high copy number plasmid from *Halobacterium* strain GRB. *Nucl. Acids Res.* 18:3408.

Hackett, N. R., Y. Bobovnikova, and N. Heyrovska. 1994. Conservation of chromosomal arrangement among three strains of the genetically unstable archaeon *Halobacterium salinarium. J. Bacteriol.* 176:7711–7718.

López-García, P., J. P. Abad, C. Smith, and R. Amils. 1992. Genomic organization of the halophilic archaeon *Haloferax mediterranei:* physical map of the chromosome. *Nucl. Acids Res.* 20:2459–2464.

López-Garciá, P., A. St. Jean, R. Amils, and R. L. Charlebois. 1995. Genomic stability in the archaea *Haloferax volcanii* and *Haloferax mediterranei. J. Bacteriol.* 177:1405–1408.

Schalkwyk, L. C., R. L. Charlebois, and W. F. Doolittle. 1993. Insertion sequences on plasmid pHV1 of *Haloferax volcanii. Can. J. Microbiol.* 39:201–206.

St. Jean, A., and R. L. Charlebois. 1996. Comparative genomic analysis of the *Haloferax volcanii* DS2 and *Halobacterium salinarium* GRB contig maps reveals extensive rearrangement. *J. Bacteriol.* 178:3860–3868.

St. Jean, A., B. A. Trieselmann, and R. L. Charlebois. 1994. Physical map and set of overlapping cosmid clones representing the genome of the archaeon *Halobacterium* sp. GRB. *Nucl. Acids Res.* 22:1476–1483.

Trieselmann, B. A., and R. L. Charlebois. 1992. Transcriptionally active regions in the genome of the archaebacterium *Haloferax volcanii. J. Bacteriol.* 174:30–34.

63

Helicobacter pylori CH19

Diane E. Taylor

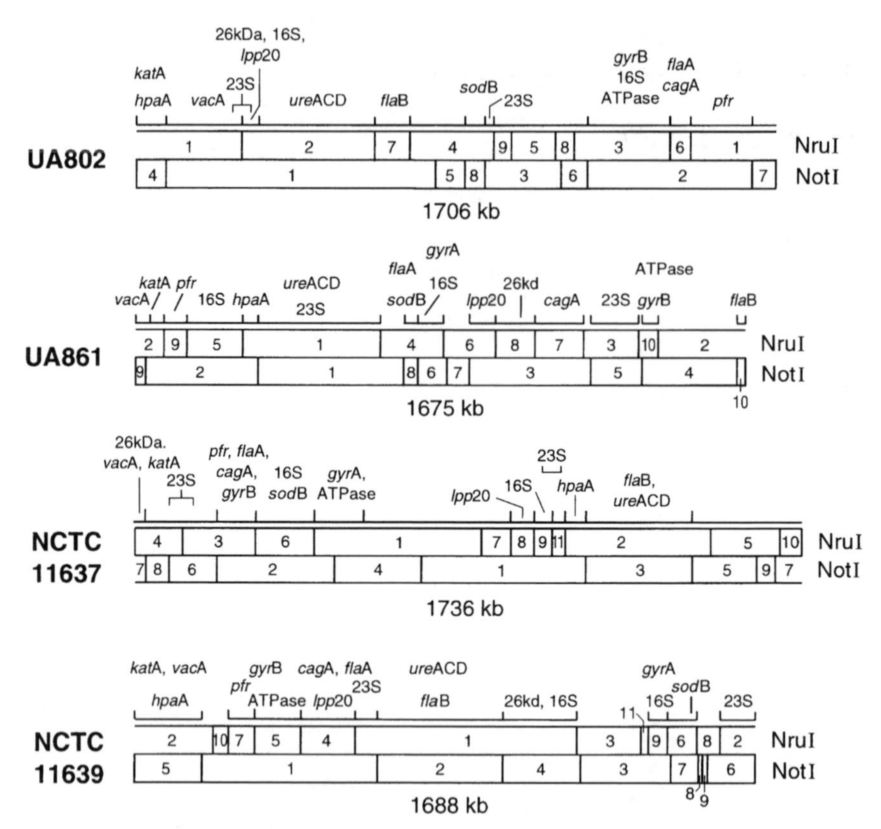

Figure 63-1. *H. pylori* genome maps of UA802, UA861, NCTC11637 and NCTC11639. *H. pylori* genomes are diverse with significant restriction fragment polymorphisms and differences in gene order, therefore, maps of four different strains are shown. Maps were constructed by pulsed-field gel electrophoresis analysis using *Not*I and *Nru*I, extraction of DNA fragments and hybridization of ^{32}P-labelled DNA fragments to Southern blots of each digest (*see also* Chapters 24–26). In addition, partial digestion fragments were used to construct the map. Position of genes was determined by hybridization of ^{32}P-labelled DNA gene probes or PCR generated probes where DNA sequences were known. The mapped genes encode as follows: *katA,* catalase; *hpaA,* adhesin; *vacA,* vacuolating cytotoxin, 16S, 23S, ribosomal RNA genes; 26 kDa, a 26 kDa protein of unknown function; lpp20, lipoprotein; *ureACD,* urease subunits; *flaA* and *flaB,* flagella probes; ATPase, copper exporting ATPase, *gyrA, gyrB,* DNA gyrase subunits; *sodB,* superoxide dismutase; *cagA,* 120 kDa protein ("cytotoxin associated protein"); *pfr,* bacterial ferritin. The *gyrA* gene could not be mapped on UA802 because neither the *gyrA* gene probe hybridized with this strain nor were *gyrA*-specific primers amplified by PCR using UA802 DNA.

References

Ge, Z., K. Hiratsuka, and D. E. Taylor. 1995. Nucleotide sequence and mutational analysis indicate that two *Helicobacter pylori* genes encode a P-type ATPase and a cation-binding protein associated with copper transport. *Molec. Microbiol.* 15(1):97–106; Jiang, Q., K. Hiratsuka, and D. E. Taylor. 1990. Variability in gene order of different *Helicobacter pylori* strains. *Molec. Microbiol.* 20(4):833–842; Kostrzynska, M., P. N. O'Toole, D. E. Taylor, and T. J. Trust. 1994. Molecular characterization of a conserved 20-kilodalton membrane-associated lipoprotein antigen of *Helicobacter pylori. J. Bacteriol.* 176 (19):5938–5948; Taylor, D. E., M. Eaton, N. Chang, S. M. Salama, 1992. Construction of a *Helicobcter pylori* genome map and demonstration of diversity at the genome level. *J. Bacteriol.* 174(21):6800–6806.

64

Lactococcus lactis subsp. *lactis* IL1403 and *Lactococcus lactis* subsp. *cremoris* MG1363

P. Le Bourgeois, M. L. Mingot-Daveran, and Paul Ritzenthaler

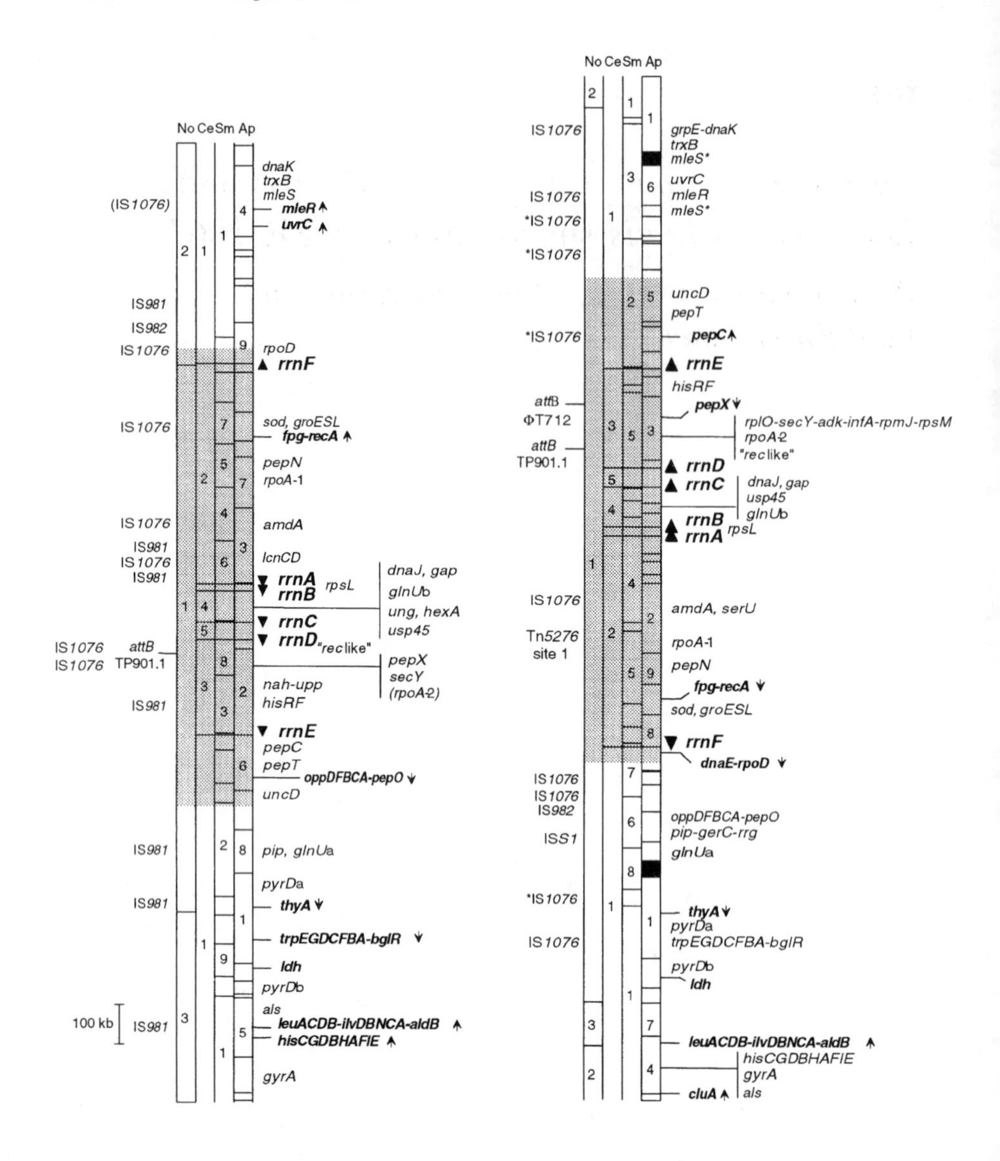
No Ce Sm Ap
dnaK
trxB
mleS
mleR
uvrC
(IS1076)
IS981
IS982
IS1076
rpoD
rrnF
sod, groESL
fpg-recA
pepN
rpoA-1
amdA
lcnCD
rrnA
rrnB
rpsL
rrnC
rrnD
"reclike"
dnaJ, gap
glnUb
ung, hexA
usp45
pepX
secY
(rpoA-2)
attB
TP901.1
nah-upp
hisRF
rrnE
pepC
pepT
oppDFBCA-pepO
uncD
pip, glnUa
pyrDa
thyA
trpEGDCFBA-bglR
ldh
pyrDb
als
leuACDB-ilvDBNCA-aldB
hisCGDBHAFIE
gyrA
100 kb
grpE-dnaK
mleS*
ΦT712
Tn5276
site 1
rplO-secY-adk-infA-rpmJ-rpsM
rpoA-2
amdA, serU
dnaE-rpoD
pip-gerC-rrg
ISS1
cluA
L. l. lactis
IL1403
L. l. cremoris
MG1363

◄

Figure 64-1. Chromosome map of *Lactococcus lactis* subsp. *lactis* strain IL1403 (Le Bourgeois *et al.*, 1992b) and *Lactococcus lactis* subsp. *cremoris* strain MG1363 (Le Bourgeois *et al.*, 1995). The physical map was constructed by random insertion of rare restriction sites (*see also* Chapters 24 and 29) using the pRL1 plasmid (Le Bourgeois *et al.*, 1992a) and indirect-end labelling. Seventy-six and eighty-four genetic markers were localized on the IL1403 and MG1363 chromosome respectively. Genes that were precisely located on the physical map by homologous recombination are indicated in bold. Arrows indicate the 5′→3′ orientation of the gene. Parentheses indicate genes which hybridized weakly with the corresponding fragment. Hatched regions correspond to the genome inversion between the two strains. Black boxes correspond to unmapped *Apa*I fragments. Asterisk indicates ambiguous location of the probe. Abbreviations: Ap, *Apa*I; Ce, I-*Ceu*I; No, *Not*I; Sm, *Sma*I.

References

P. Le Bourgeois, M. Lautier, M. Mata, and P. Ritzenthaler. 1992. New tools for the physical and genetic mapping of *Lactococcus* strains. *Gene* 111:109–114.

P. Le Bourgeois, M. Lautier, M. Mata, and P. Ritzenthaler. 1992a. Physical and genetic map of the chromosome of *Lactococcus lactis* subsp. *lactis* IL1403. *Journal of Bacteriology* 174:6752–6762.

P. Le Bourgeois, M. Lautier, L. Van den Berghe, M. Gasson, and P. Ritzenthaler. 1995. Physical and genetic map of the *Lactococcus lactis* subsp. *cremoris* MG1363 chromosome: comparison with that of *Lactococcus lactis* subsp. *lactis* IL1403 reveals a large genome inversion. *Journal of Bacteriology* 177:2840–2850.

65

Leptospira interrogans

Isabelle Saint Girons and Richard L. Zuerner

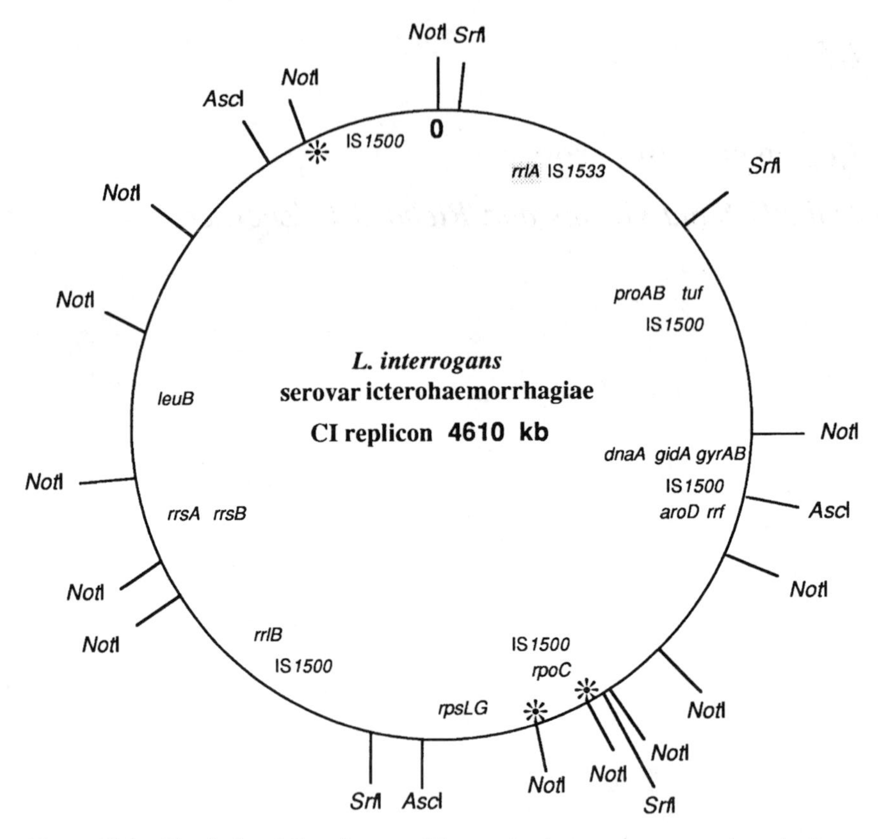

Figure 65-1. Physical and Genetic map of *Leptospira interrogans* serovar icterohaemorrhagiae strain Verdun

Physical mapping indicated the presence of two circular chromosomal replicons, one large (4610 kb) and one small (350 kb, not shown). Single and double digests and different complementary techniques were used to produce the physical map of the large chromosomal replicon for three rare-cutting endonucleases, *Not*I, *Srf*I and *Asc*I.

(1) The use of linking clones proved the contiguity of the following *Not*I fragments: I-B-H-J-H-E-F-A and G-D-F.

(2) In order to individualize DNA fragments within a *Not*I DNA band, we used partial *Not*I digestions. The results indicate that the *Not*I fragments, H, F and C were doublets. The order of the fragments was as follows: F1-D-G-I-B-H2-J-H1-E-F2-A and C1-C2

(3) *Fnu*DII methylase (5′-GGCCGC-3′) blockes *Not*I restriction sites (see the three stars on the figure) at the overlap of the endonuclease-methylase sites 5′-CGmCGGCCGC-3′ and 5′-GCGGCCGmCG-3′ (*see also* Chapter 24). The number of fragments obtained by pretreatment with *Fnu*DII methylase prior to *Not*I digestion was thus reduced from 13 to 11. This confirmed the following linkage: B-H2-J and C1-C2.

(4) The precise location of the *Asc*I and *Srf*I sites was obtained by two-dimensional pulsed field gel electrophoresis (*see* Chapter 26). These data taken together allowed to demonstrate the linkage of C2 to F1 and to deduce the linkage of C1 to A and thus the circularity of the map.

(5) For definition of the genes which have been mapped, see the last paragraph of the legend to *Leptospira interrogans* serovar pomona strain RZ11 (Figure 65-2). The *asd* gene was also located on the small (350 kbp) replicon (data not shown).

References

Baril, C., C. Richaud, J.L. Herrmann, D. Margarita, and I. Saint Girons. 1992. Scattering of the rRNA genes on the physical map of the chromosome of *Leptospira interrogans* serovar icterohaemorrhagiae. *J. Bacteriol.* 174:7566–7571.

Bourseaux-Eude, C., I. Saint Girons, and R.L. Zuerner. 1995. *IS*1500, an IS3-like element from *Leptospira interrogans. Microbiology.* 141:2165–2173.

Continued next page

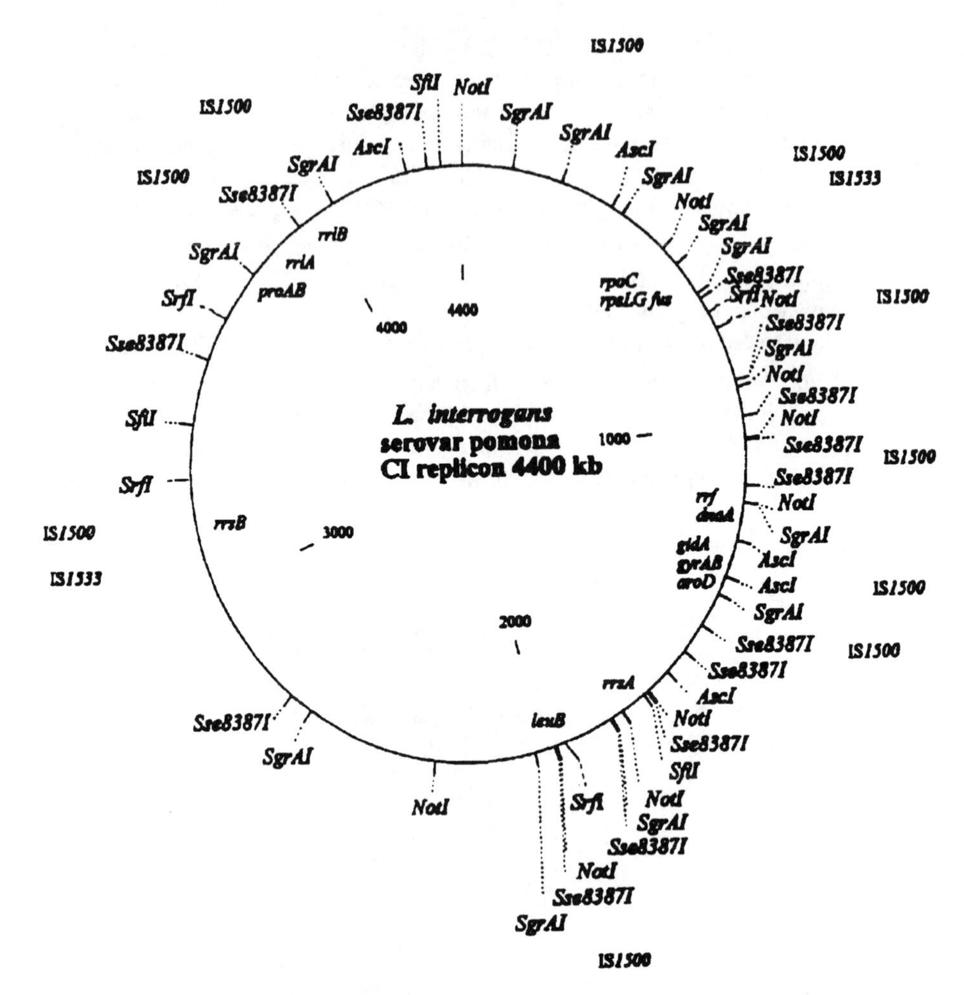

Figure 65-2. Physical and Genetic Map of *Leptospira interrogans* serovar pomona type kennewicki strain RZ11.

(A) Chromosomal replicon I (CI).

This replicon measures about 4400 Kbp in size. Markers at 1000 kbp intervals are shown on the inside of the map. The map starts with the *Not*I site located between the 2100 kbp (A) and the 550 kbp (B) fragments. All *Asc*I, *Not*I, and *Sfi*I sites are shown. Approximate location of some of the sites recognized by *SgrA*I, *Srf*I, and *Sse*8387I are shown.

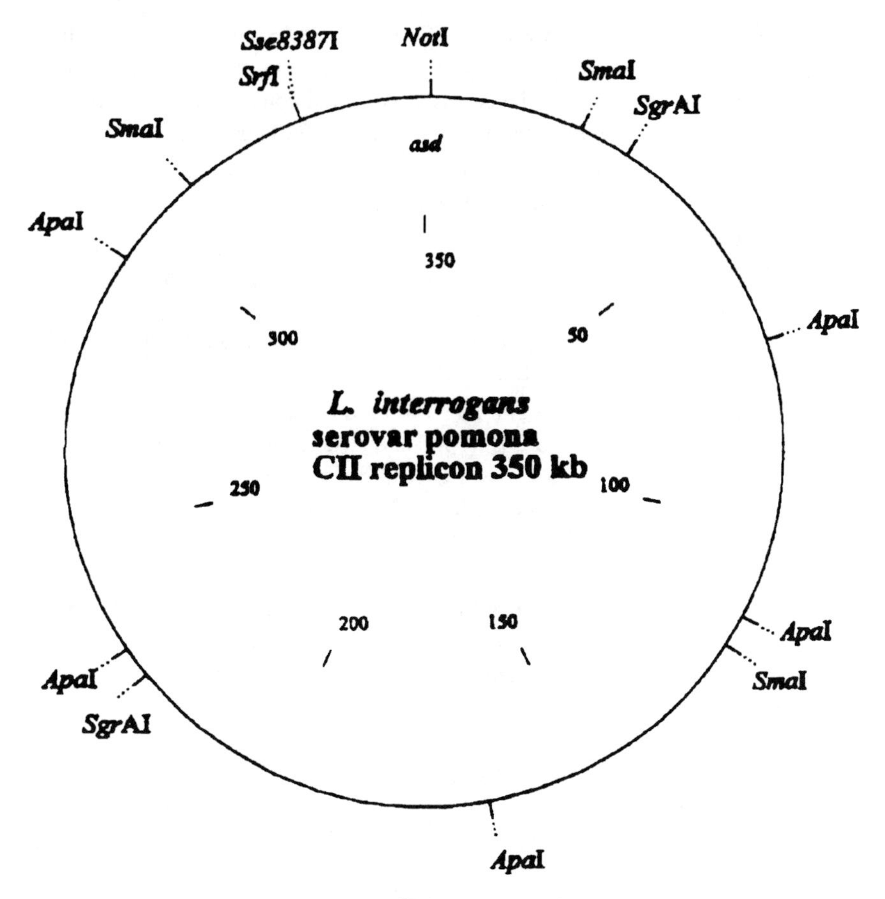

Figure 65-2. Continued

(B) Chromosomal replicon II (CII).

This replicon measures about 350 kbp in size. Markers are placed at 50 kbp intervals on the inside of the map. All *Apa*I, *Not*I, *SgrA*I, *Sma*I, *Srf*I, and *Sse*8387I sites are shown. The map is oriented with the unique *Not*I site at the top.

Different techniques were used to assemble the physical maps.

(1) Single and double restriction digestions with rare cutting enzymes allowed localization of many of the restriction sites (*see also* Chapter 24). In particular, the *Asc*I, *Sfi*I, *Srf*I, and *Sse*8387I sites were mapped in this way.

(2) Hybridization analysis using random cloned fragments dispersed around the genome was used to align fragments generated with one

restriction enzyme (primarily *Not*I) with fragments generated with other restriction enzymes. Hybridization analysis using cloned DNA fragments and PCR products was also used to confirm predicted digestion products from double restriction enzyme digests.

(3) Specific genomic restriction fragments separated by pulsed-field gel electrophoresis were used for hybridization analysis of specific regions, especially the small replicon.

(4) The generation of partial digestion products using the enzymes *Not*I and *Sse*8387I allowed fragments to be tentatively linked based on the size of the resultant fragments. The alignment of these fragments was confirmed by hybridization analysis.

(5) These techniques were combined to demonstrate the presence of two independent linkage groups (replicons). The large replicon has the *Not*I fragment order of A-B-F-I-J-H-C-K-G-E. The small replicon has single sites for *Not*I, *Srf*I, and *Sse*8387I. The order of *Apa*I restriction fragments is A-D-C-B-B′.

(6) There are two copies each of rRNA genes, *rrl* (23S ribosomal RNA) and *rrs* (16S ribosomal RNA) and one copy of the *rrf* gene (5S ribosomal RNA). Each of these genes are distributed around the map and the rRNA genes are not clustered (*see also* Chapter 21). The putative *oriC* is positioned near the *gyrAB* (DNA gyrase), *gidA*, (glucose inhibited division protein) and *dnaA* (replication protein) genes which map about 1350 kbp from the *Not*I site placed at the start of the map. Four genes, *rpoC* (RNA polymerase β′subunit) *rpsLG* (ribosomal proteins S12) and *fus* (elongation factor EF-G) encoding transcription and translation associated functions are clustered together about 650 kbp from the start of the map. Five amino acid biosynthetic genes, *leuB* (isopropylmalate dehydrogenase), *proAB* (γ-glutamyl phosphate reductase and γ-glutamyl kinase), *aroD*, 3-dehydroquinate dehydratase and *asd* (aspartate β-semialdehyde dehydrogenase) have also been mapped. Note that the *asd* gene is on the small CII replicon (Fig 2B). Approximate insertion points for the elements IS*1500* and IS*1533* are shown outside the map.

References

Zuerner, R.L. 1991. Physical map of chromosomal and plasmid DNA comprising the genome of *Leptospira interrogans*. *Nuc. Acids Res.* 19:4857–4860

Zuerner, R.L., J.L. Herrmann, and I. Saint Girons. 1993. Comparison of genetic maps for two *Leptospira interrogans* serovars provides evidence for two chromosomes and intraspecies heterogeneity. *J. Bacteriol.* 175:5445–5451.

66

Listeria monocytogenes

Pascale Cossart, Eric Michel, and Brian Sheehan

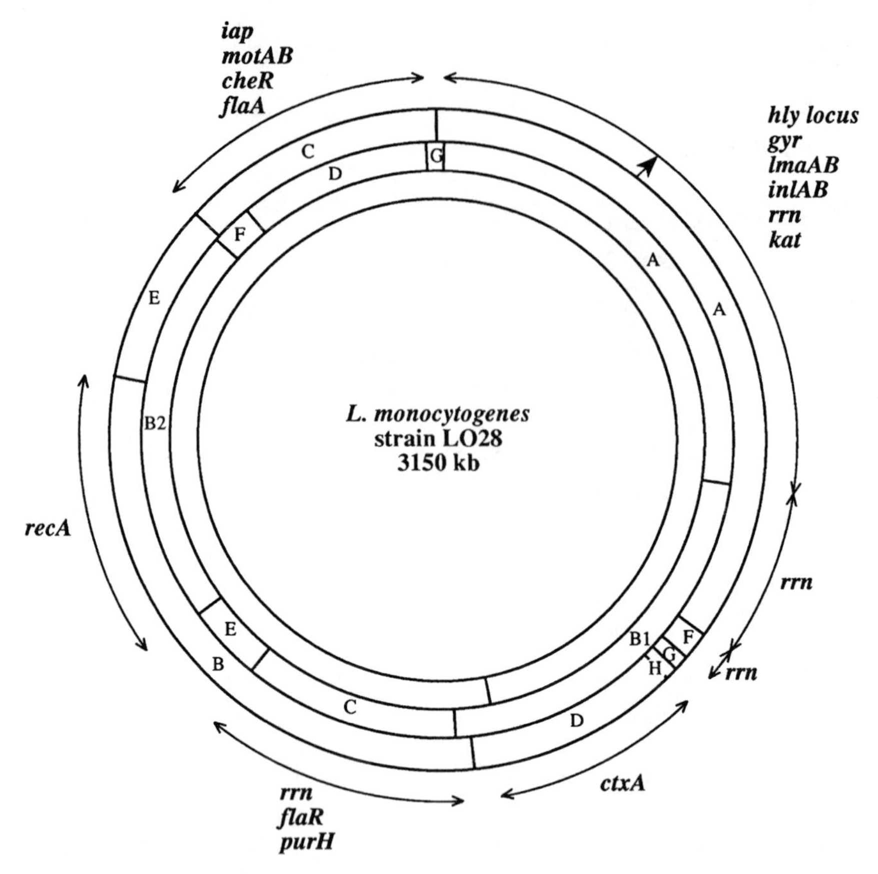

Figure 66-1. Physical map of the *Listeria monocytogenes* chromosome (strain LO28, serotype 1/2c). Most known genes are mapped (Michel E. and P. Cossart, *Journal of Bacteriology*, 1992. 174:7098–7103.

67

Methanobacterium thermoautotrophicum Marburg

Rolf Stettler and Thomas Leisinger

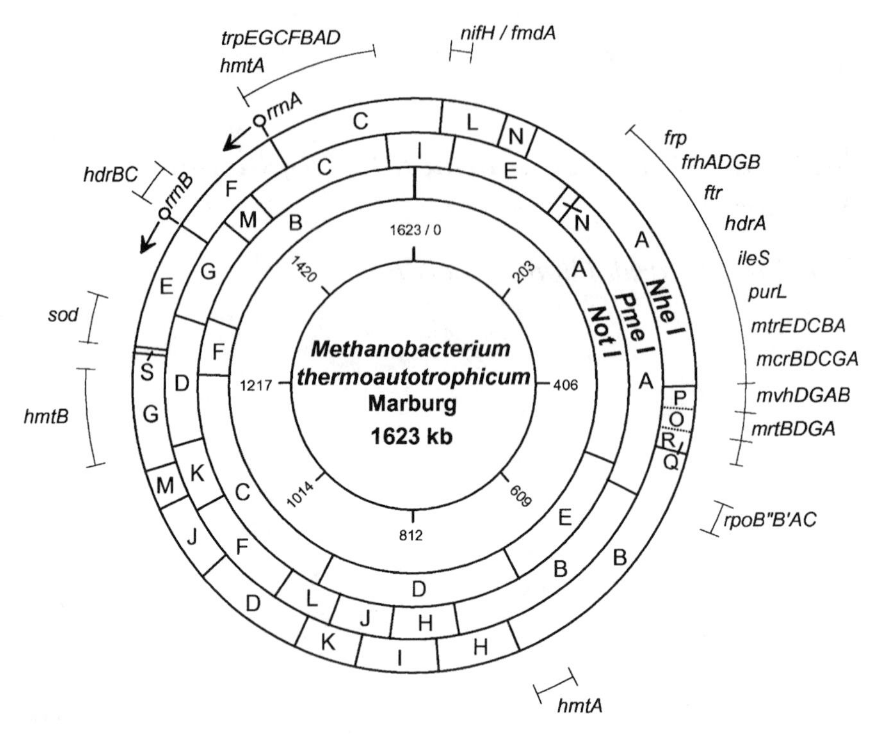

Figure 67-1. Physical and genetic map of *the M. thermoautotrophicum* Marburg chromosome, based on *Not*I, *Pme*I, and *Nhe*I restriction. (Stettler and Leisinger, 1992; Stettler *et al.*, 1995). The following methods were combined to order the fragments of the restriction enzyme digests to a physical map: (1) hybridization with isolated ^{32}P-labeled fragments and several gene probes, (2) single digests and double digests, (3) digests of isolated *Not*I fragments, and (4) partial digests. Map units are indicated in kilobases. The 0-kb position was chosen as the *Not*I site between the *Not*I fragments B and A. Approximate gene placements are indicated by bars spanning the regions in which they are located. To date, a total of 48 genes are mapped to specific restriction enzyme fragments, and the functions of the genes shown on the map are given in Stettler *et al.* (1995). The exact locations of the two *rrn* operons (*see also* Chapter 21) are designated by lollipops, and the arrows indicate the direction of their transcription. Figure reproduced by kind permission of *Archives of Microbiology.*

References

Leisinger, T., and L. Meile, 1993. Plasmids, phages and gene transfer in methanogenic bacteria in *Genetics and molecular biology of anaerobic bacteria*, (ed M. Sebald), Springer Verlag, New York, pp. 1–12.

Meile, L., P. Abendschein, and T. Leisinger, 1990. Transduction in the archaebacterium *Methanobacterium thermoautotrophicum* Marburg. *J. Bacteriol.* 172:3507–3508.

Nölling, J., D. Hahn, W. Ludwig, and W.M. De Vos, 1993. Phylogenetic analysis of thermophilic *Methanobacterium* sp: evidence for a formate-utilizing ancestor. *System. Appl. Microbiol.* 16:208–215.

Stettler, R., and T. Leisinger. 1992. Physical map of the *Methanobacterium thermoautotrophicum* Marburg chromosome. *J. Bacteriol.* 174:7227–7234.

Stettler, R., P. Pfister, and T. Leisinger, 1994. Characterization of a plasmid carried by *Methanobacterium thermoautotrophicum* ZH3, a methanogen closely related to *Methanobacterium thermoautotrophicum* Marburg. *System. Appl. Microbiol.* 17:484–491.

Stettler, R., G. Erauso, and T. Leisinger. 1995. Physical and genetic map of the *Methanobacterium wolfei* genome and its comparison with the updated genomic map of *Methanobacterium thermoautotrophicum* Marburg. *Archives Microbiol.* 163:205–210.

68

Mycobacterium leprae

*Stewart T. Cole, Staffan Bergh,
Karin Eiglmeier, Hafida Fsihi, Thierry Garnier,
and Nadine Honoré*

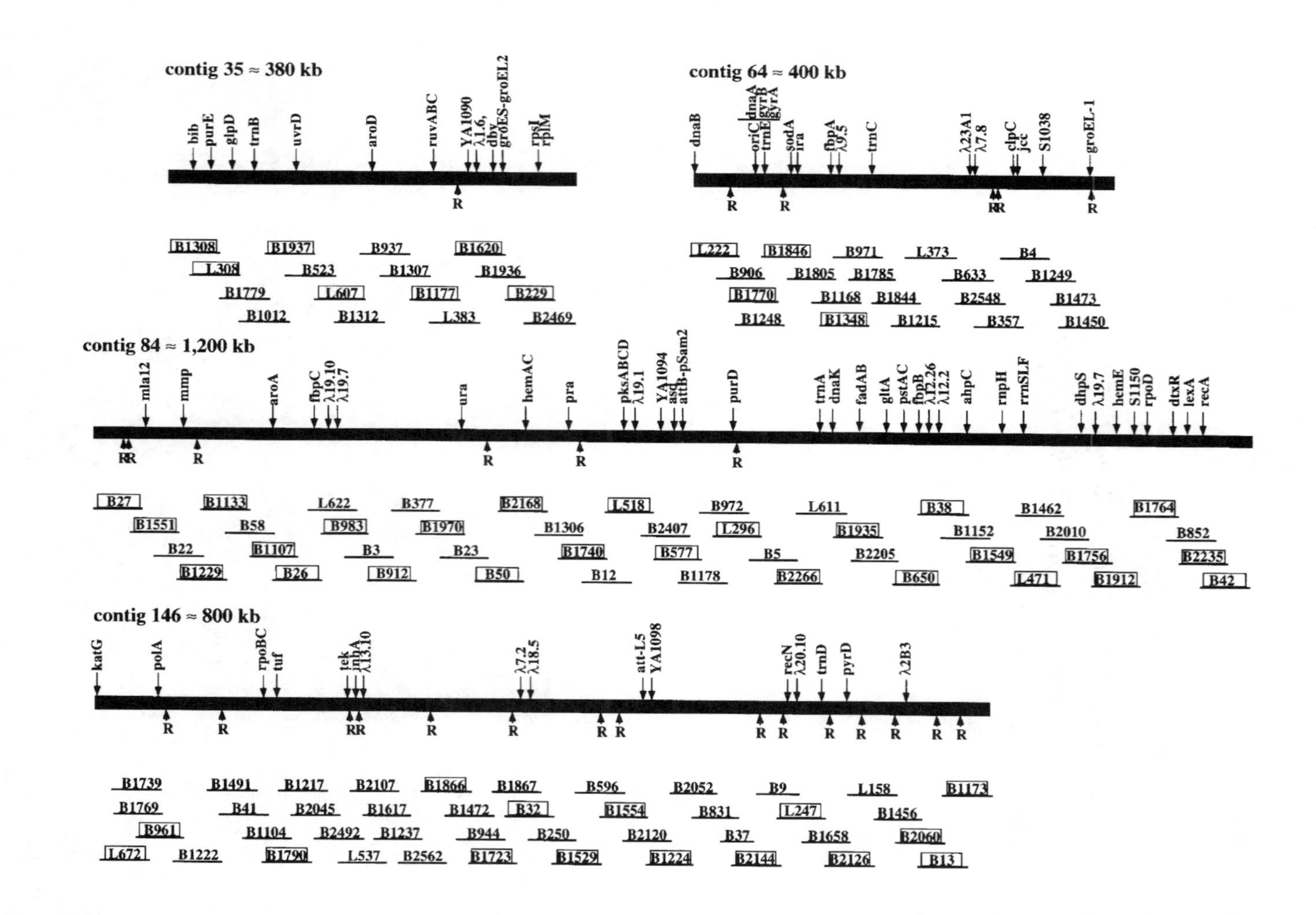
contig 35 ≈ 380 kb
bib purE glpD trnB uvrD aroD ruvABC YA1090 λ1.6, dby groES-groEL2 rpsI rplM
R
B1308 B1937 B937 B1620
L308 B523 B1307 B1936
B1779 L607 B1177 B229
B1012 B1312 L383 B2469
contig 64 ≈ 400 kb
dnaB oriC dnaA trnE gyrB gyrA sodA ira fbpA λ9.5 trnC λ23A1 λ7.8 clpC jcc S1038 groEL-1
R R RR R
L222 B1846 B971 L373 B4
B906 B1805 B1785 B633 B1249
B1770 B1168 B1844 B2548 B1473
B1248 B1348 B1215 B357 B1450
contig 84 ≈ 1,200 kb
mla12 mmp aroA fbpC λ19.10 λ19.7 ura hemAC pra pksABCD λ19.1 YA1094 asd attB-pSam2 purD trnA dnaK fadAB gltA pstAC fbpB λ12.26 λ12.2 ahpC rnpH rrnSLF dhpS λ19.7 hemE S1150 rpoD dtxR lexA recA
RR R R R R
B27 B1133 L622 B377 B2168 L518 B972 L611 B38 B1462 B1764
B1551 B58 B983 B1970 B1306 B2407 L296 B1935 B1152 B2010 B852
B22 B1107 B3 B23 B1740 B577 B5 B2205 B1549 B1756 B2235
B1229 B26 B912 B50 B12 B1178 B2266 B650 L471 B1912 B42
contig 146 ≈ 800 kb
katG polA rpoBC tuf tek inhA λ13.10 λ7.2 λ18.5 att-L5 YA1098 recN λ20.10 trnD pyrD λ2B3
R R RR R R R R R R R R R R R
B1739 B1491 B1217 B2107 B1866 B1867 B596 B2052 B9 L158 B1173
B1769 B41 B2045 B1617 B1472 B32 B1554 B831 L247 B1456
B961 B1104 B2492 B1237 B944 B250 B2120 B37 B1658 B2060
L672 B1222 B1790 L537 B2562 B1723 B1529 B1224 B2144 B2126 B13

◄

Figure 68-1. *Mycobacterium leprae*, the etiologic agent of leprosy, represents the ultimate challenge for microbiologists as this obligately intracellular parasite has an extremely long generation time (~ 13 days) and cannot be grown *in vitro* (*see also* Chapter 39). To study the genome of *M. leprae*, ordered cosmid libraries were constructed as classical genetics and pulsed field gel electrophoretic analysis proved impossible. Blunt-ended chromosomal fragments of ~40 kb were cloned into Lorist6 and recombinant cosmids ordered into contigs by a combination of fingerprinting (with *Bam*HI and *Hae*III) and hybridization mapping. Although all of the cloned genes were positioned on the contigs, four gaps remained in the map and these probably correspond to unclonable sequences.

The approximate positions of the minimal set of canonical cosmids used to construct the contig map are shown in the Figure and the locations of all genes, coding sequences for anonymous protein antigens and the *M. leprae*-specific repetitive sequence, RLEP (R) are indicated. The genome size was estimated at ~2.8 Mb based on the sum of the predicted contig sizes, however, this may well be an underestimate due to the occasionally non-random distribution of fingerprint bands. A genome sequencing project, involving initially the Institut Pasteur and Collaborative Research Inc., but later including several other academic laboratories, was initiated in 1993. At present about 60% of the genome has been sequenced and those cosmids whose sequences are known are boxed and a landmark gene identified from the sequence is shown.

For further details of clones, genes, sequences and functions consult MycDB (Chapter 39), the customised database dedicated to mycobacteria, and read the original publications:

References

Bergh, S., and S. T. Cole. 1994. MycDB—an integrated mycobacterial database. *Mol. Microbiol.* 12:517–534.

Cole, S. T. 1994. The genome of *Mycobacterium leprae. Int. J. Lep.* 62:122–125.

Eiglmeier, K., N. Honoré, S. A. Woods, B. Caudron, and S. T. Cole, 1993. Use of an ordered cosmid library to reduce the genomic organisation of *Mycobacterium leprae. Mol. Microbiol.* 7:197–206.

Fsihi, H., and S. T. Cole. 1995. The *Mycobacterium leprae* genome: systematic sequence analysis identifies key catabolic enzymes, ATP-dependent transport systems and a novel *polA* locus associated with genomic variability. *Mol. Microbiol.* 16:909–919.

Honoré, N., S. Bergh, S. Chanteau, F. Doucet-Populaire, K. Eiglmeier, T. Garnier, C. Georges, P. Launois, P. Limpaiboon, S. Newton, K. Nyang, P. del Portillo, G. K. Ramesh, T. Reddy, J. P. Riedel, N. Sittisombut, S. Wu-Hunter, and S. T. Cole. 1993. Nucleotide sequence of the first cosmid from the *Mycobacterium leprae* genome project : structure and function of the Rif-Str regions. *Mol. Microbiol.* 7:207–214.

69

Mycoplasma genitalium G-37 (ATCC 33530)

Kenneth F. Bott

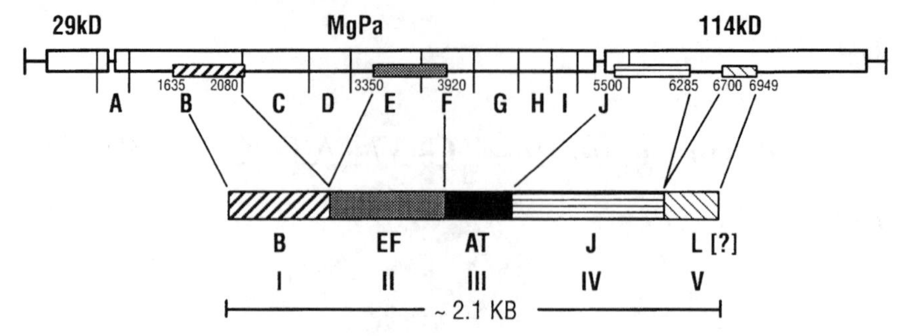

Figure 69-1. A Schematic diagram of major adhesion operon MgPa is shown on top. Numbers refer to nucleotide numbering of its original GenBank Accession # M31431. Below it, in a manner indicating how the segments are derived, is represented the cassette structure of MgPa repeats found at multiple positions on the map Regions B, EF and J were designated by Dallo and Baseman (1991). They have been more accurately defined by Peterson *et al.* (1995). In some publications those same sectors are designated I–V. An AT rich middle region of approx. 300 nt. has no other existence on the genome outside of its consistent appearance in these repeat units.

Continued next page

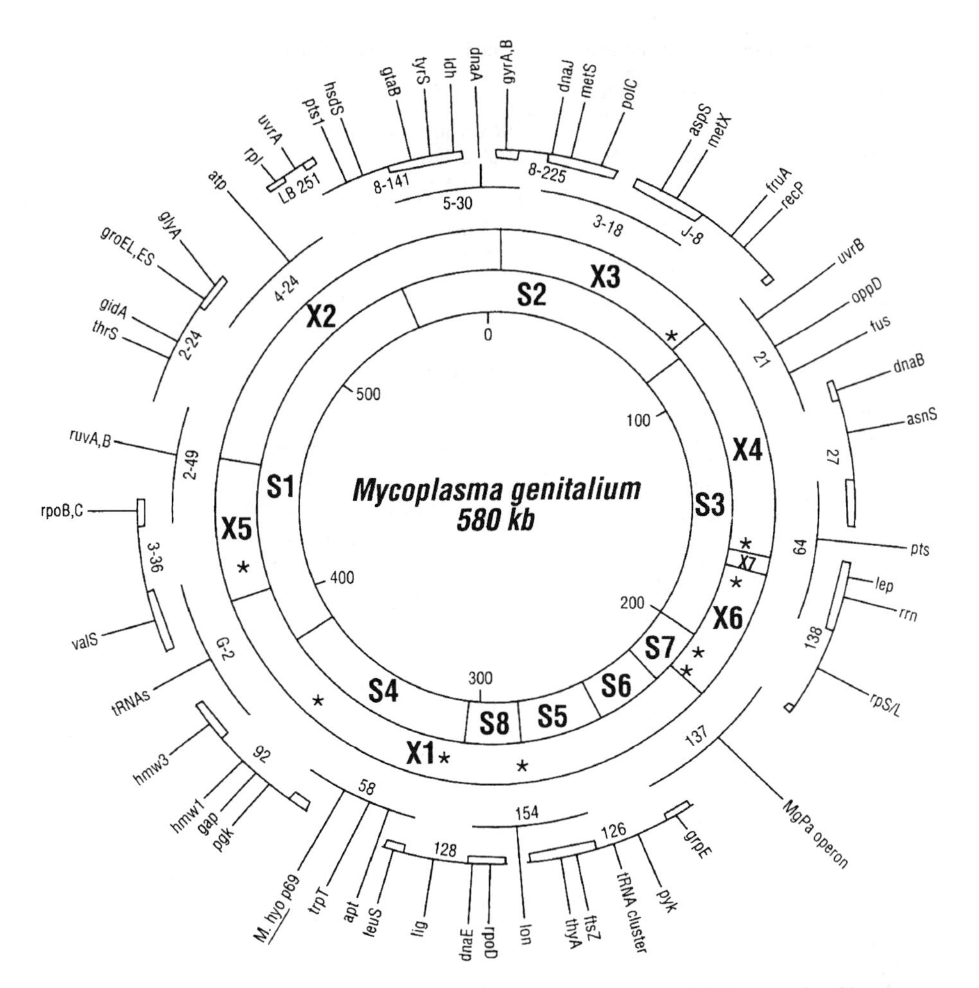

Figure 69-2. Revised Physical Map of *Mycoplasma genitalium* G-37 (*see also* Chapter 40) showing placement of Xho1 sites (X1–X7) and Sma1 sites (S1–S8) (from Colman et al., 1990). Origin (12:00) is placed between the *dna*A and *gyr* loci. Selected markers from hybridization with random clones and sequence analysis are shown relevant to the zones made by cosmid overlap (See Table). Arcs on periphery are the Lambda (LB-251) or cosmid clones from (Lucier et al., 1995). Asterisks indicate approximate positions of nine MgPa (adhesin component) repeats (Peterson *et al.*, 1995). Intact adhesin operon, MgPa is denoted. The entire genomic sequence is known (Fraser *et al.*, 1995) Genbank accession # L43967. 470 putative open reading frames were identified. All data, plus access to any genomic clone is available from The Institute for Genomic Research (TIGR) through http://www.tigr.org.

Table 69-1. Loci *of Selected* features Mapped on the genome of *Mycoplasma genitalium*

Gene	Cosmid Zone	Function
*dna*A	5–30	DNA replication initiation factor
*gyr*B A	5–30/8–225	DNA gyrase
*dna*J	8–225/3–18	Stress response protein
*met*S	8–225/3–18	Methionine tRNA synthetase
*pol*C	8–225/3–18	DNA polymerase III
*asp*S	3–18/J–8	Aspartyl tRNA synthetase
*met*X	3–18/J–8	S-Adenosylmethionine synthetase
*fru*A	J–8	PTS enzyme II fructose permease
*rec*P	J–8	Recombination
*uvr*B	21	Excision repair
*opp*D	21	Oligopeptide transport
fus	21	Elongation factor G
*dna*B	21/27	Replicative DNA-helicase
*asn*S	27	Asparaginyl tRNA synthetase
pts	64	Glucose specific permease
lep	64	Leader peptidase
rrn	64/138	Ribosomal RNA genes
rpS/L	138	Major ribosomal protein cluster
MgPa operon	137	Major adherence proteins of *M. genitalium*
grpE	137/126	Heat Shock protein
pyk	126	Pyruvate kinase
tRNA cluster	126	tRNA gene cluster
ftsZ	126/154	Cell division protein
thyA	126/154	Thymidylate synthase
lon	154	ATP-dependent protease
rpoD	154/128	Sigma factor
dnaE	154/128	DNA primase
*lig*T	128	DNA ligase
leuS	128/58	Leucine tRNA synthetase
apt	58	Adenine phosphoribosyltransferase
*trp*T	58	Tryptophan tRNA and tRNA gene cluster
M. hyo p69	58	*M. hyorhinus* 69-kDa membrane transport protein:
pgk	92	Phosphoglycerate kinase
gap	92	Glyceraldehyde 3-phosphate dehydrogenase
hmw1	92	Cytadherence-accessory protein
hmw3	92/G-2	Accessory adherence protein
tRNAs	G-2	tRNA gene cluster
*val*S	G-2/3–36	Valine tRNA synthetase
*rpo*B C	3–36/2–49	RNA polymerase subunit
ruvA,B	2–49	Holiday junction DNA helicase
*thr*S	2–24	Threonine tRNA synthetase
*gid*A	2–24	Glucose inhibited division protein
*gro*EL,ES	2–24/4–24	Chaperone/Stress response
*gly*A	2–24/4–24	Glycine hydroxymethyltransferase
atp	4–24	ATP synthetase
rpl	4–24/LB 251	Ribosomal protein cluster
*uvr*A	LB - 251	Excision repair
pts1	8–141	PEP-depend HPr protein kinase phosphoryltransferase
hsdS	8–141	Restriction-modification enzyme
gtaB	8–141/5–30	UDP-glucose pyrophosphorylase
tyrS	8–141/5–30	Tyrosine tRNA synthetase
ldh	8–141/5–30	Lactate dehydrogenase

References

Colman, S.D., P.-C. Hu, W. Litaker, and K.F. Bott. 1990. A physical map of the *Mycoplasma genitalium* genome. *Mol. Microbiol.* 4:683–687.

Dallo, S.F., and J.B. Baseman. 1991. Adhesion gene of *Mycoplasma genitalium* exists as multiple copies. *Microb. Pathog.* 10:475–480.

Fraser, C.M., J.D. Gocayne, O. White, M.D. Adams, *et al.* 1995. The *Mycoplasma genitalium* genome sequence reveals a minimal gene complement. *Science* 270:397–403.

Lucier, T.S., P.-Q. Hu, S.N. Peterson, X.-Y. Song, L. Miller, K. Heitzman, K.F. Bott, C.A., III Hutchison, and P.-C. Hu. 1994. Construction of an ordered genomic library of *Mycoplasma genitalium. Gene* 150:27–34.

Peterson, S.N., C.C. Bailey, J.S. Jensen, M.B. Borre, E.S. King, K.F. Bott, and C.A., III Hutchison, 1995. Characterization of repetitive DNA in the *Mycoplasma genitalium* genome: Possible role in the generation of antigenic variation. *Proc. Natl. Acad. Sci. USA* 92:11829–11833.

70

Myxococcus xanthus DK1622

Lawrence J. Shimkets

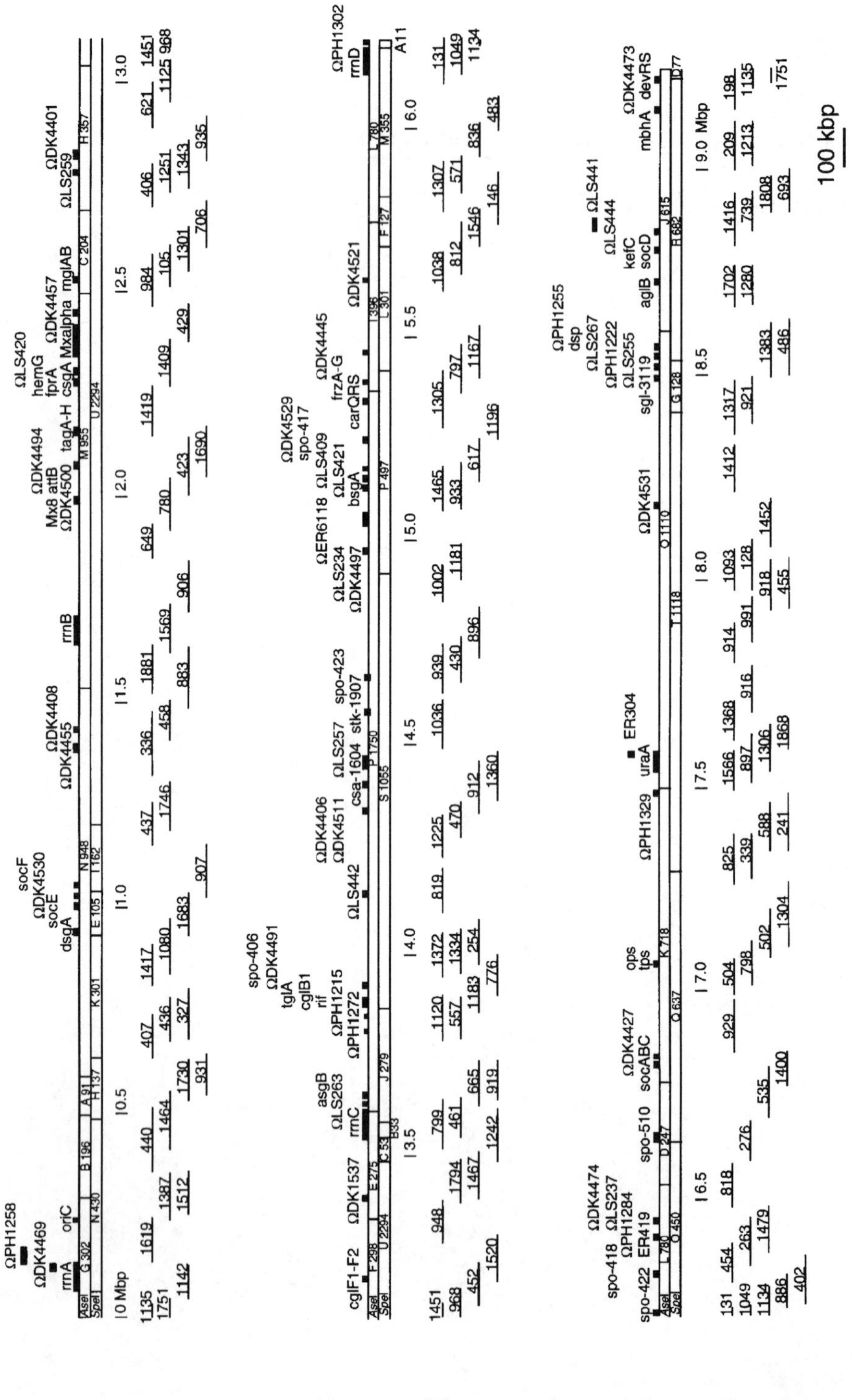

ΩPH1258 ΩDK4469 rrnA oriC dsgA socE ΩDK4530 socF ΩDK4455 ΩDK4408 rrnB Mx8 attB ΩDK4500 ΩDK4494 tagA-H csgA fprA hemG ΩLS420 Mxalpha ΩDK4457 mglAB ΩLS259 ΩDK4401

AseI SpeI G 302 N 430 B 196 A 91 H 137 K 301 E 105 N 948 I 162 M 955 U 2294 C 204 H 357

0 Mbp 0.5 1.0 1.5 2.0 2.5 3.0

1135 1619 440 407 1417 437 336 1881 649 1419 984 406 621 1451
1751 1387 1464 436 1080 1746 458 1569 780 1409 105 1251 1125 968
1142 1512 1730 327 1683 883 906 423 429 1301 1343
931 907 1690 706 935

cglF1-F2 ΩDK1537 rrnC ΩLS263 asgB ΩPH1272 ΩPH1215 rif cglB1 tglA ΩDK4491 spo-406 ΩLS442 ΩDK4511 ΩDK4406 csa-1604 ΩLS257 stk-1907 spo-423 ΩDK4497 ΩLS234 ΩER6118 bsgA ΩLS421 ΩLS409 spo-417 ΩDK4529 carQRS frzA-G ΩDK4445 ΩDK4521 rrnD ΩPH1302

AseI SpeI F 298 U 2294 E 275 C 53 B33 J 279 S 1055 P 1750 P 497 I 396 L 301 F 127 L 780 M 355 A11

3.5 4.0 4.5 5.0 5.5 6.0

1451 948 799 1120 1372 819 1225 1036 939 1002 1465 1305 1038 1307 131
968 1794 461 557 1334 470 430 1181 933 797 812 571 1049
452 1467 665 1183 254 912 896 617 1167 1546 836 1134
1520 1242 919 776 1360 1196 146 483

spo-422 spo-418 ER419 ΩPH1284 ΩLS237 ΩDK4474 spo-510 socABC ΩDK4427 tps ops ΩPH1329 uraA ER304 ΩDK4531 sgl-3119 ΩLS255 ΩPH1222 ΩLS267 dsp ΩPH1255 aglB socD kefC ΩLS444 ΩLS441 mbhA devRS ΩDK4473

AseI SpeI L 780 O 450 D 247 Q 637 K 718 T 1118 O 1110 G 128 R 682 J 615 D 77

6.5 7.0 7.5 8.0 8.5 9.0 Mbp

131 454 818 929 504 825 1566 1368 914 1093 1412 1317 1702 1416 209 198
1049 263 276 798 339 897 916 991 128 921 1280 739 1213 1135
1134 1479 535 502 588 1306 918 1452 1383 1808
886 1400 1304 241 1868 455 486 693 1751
402

100 kbp

◀

Figure 70-1. A linearized physical map of the circular *M. xanthus* DK1622 chromosome showing the locations of over 85 genetic loci. The three sets of long boxes present a composite physical map consisting of the *AseI* (upper) and *SpeI* (lower) restriction map in each set. The alphabetical name of each restriction map in each set. The alphabetical name of each restriction fragment is shown in the appropriate box along with the numerical size in kbp. The right end of the top segment is contiguous with the left end of the middle segment, and the right end of the middle segment is contiguous with the left end of the lower segment. The *AseI* sites in the far upper left and the far lower right of the figure are the identical site, shown twice for clarity. The short vertical lines below the physical map represent the distance in Mbp from the left end of the top segment. The short vertical lines below the physical map represent the locations and sizes of specific (numbered) YAC clones. To make the map more concise, only the informative, non-redundant YACs are shown. The names and locations of mapped genes and transposon insertions are shown above the long lines. A more detailed description of these genetic loci can be found in Table 70-1. The solid black boxes on the long lines represent the genetic loci named above them with the width of the box reflecting the limits of the exact location, not the size of the locus represented. Thus, the rRNA operon *rrnA* (upper lefthand corner) is located somewhere on YAC 1142. Omega (Ω) with the alphanumeric that follows it designates a genomic site identified by a particular insertion of transposon Tn5. For tracking the genealogy of a particular transposon insertion, ΩDK4414 (lower right), for example, indicates the site of a transposon insertion originally isolated in strain DK4414. The transposon itself is not actually present in the DK1622 genome which this map represents. Mxalpha is an 80-kbp prophage. Detailed *Eco*RI restriction maps exist for each of the YAC clones shown on the map and may be obtained from LJS.

Gene symbol	Mnenomic	Homologous YAC clones	Map Position (kbp)[b]	Phenotypic Trait
ade	adenine		4338 or 4617	adenine biosynthesis
aglB	adventurous gliding	1280, 1702	8723	Adventurous motility
asgA	A signal		6778-7493	histidine protein kinase/response regulator
asgB	A signal	665, 799	3662	transcriptional regulator
asgC	A signal		6778-7493	rpoD (σ^{70})
bsgA	B signal	**933**, 1465	5161	ATP-dependent protease (lon)
carQ	carotenoid synthesis	1305	5322	activator of carQRS
carR	carotenoid synthesis	1305	5322	membrane bound inhibitor of carQRS
carS	carotenoid synthesis	1305	5322	required for light activation of carB
cglB1	contact gliding	557, 1120, 1183	3844	contact-stimulated A motility
cglF1; cglF2	contact gliding	452, 786, 1468, 1520	3198	contact-stimulated A motility
csa-1604	cell surface antigen	**912, 1360**	4421	cell surface antigen
csgA	C signal	1409	2212	short chain alcohol dehydrogenase
devR	development	198, 1135, 1751	9171	required for development
devS	development	198, 1135, 1751	9171	required for development
DK4401		**935**	2848	Dev. regulated promoter
DK4406		470, 803, 912	4321	Dev. regulated promoter
DK4408		336, 458, 1341	1433	Dev. regulated promoter
DK4427		1400	6865	Dev. regulated promoter
DK4435			4245 or 5154	Dev. regulated promoter
DK4445		797, **1167**	5447	Dev. regulated promoter
DK4455		**336, 1341**	1375	Dev. regulated promoter
DK4457			2407	Dev. regulated promoter
DK4459			4162 or 4779	Dev. regulated promoter
DK4469		**1142**	110 or 126	Dev. regulated promoter
DK4473		198, 1135, 1751	9171	Dev. regulated promoter
DK4474		**263, 671, 1479**	6447	Dev. regulated promoter
DK4480			5015 or 4944	Dev. regulated promoter
DK4491		776, 1183	3914	Dev. regulated promoter
DK4492			3057 or 3304	Dev. regulated promoter
DK4494		423, **1690**	2185	Dev. regulated promoter
DK4497		**1181**	4944	Dev. regulated promoter
DK4500		**780**	1982	Dev. regulated promoter
DK4506			4180 or 4779	Dev. regulated promoter
DK4511		470, 803, 912	4321	Dev. regulated promoter
DK4514			5510 or 5595	Dev. regulated promoter

DK4521		812, 1038	5592	Dev. regulated promoter
DK4529		617, **933**, 1465	5211	Dev. regulated promoter
DK4530		**1683**	1021	Dev. regulated promoter
DK4531		**1412**	8152	Dev. regulated promoter
dnaG	DNA synthesis		6778-7493	DNA primase
dsgA	D signal	1080, 1683	895	translation initiation factor IF-3
dsp	dispersed	**486, 1383**	8538	Social motility
ER304		897, 1109, 1306, 1566, **1868**	7642	autocide-dependent development
ER419		263, **671, 1479**	6392	antibiotic TA biosynthesis
ER1310			5079	antibiotic TA biosynthesis
ER1912			5039	antibiotic TA biosynthesis
ER6118			5085	antibiotic TA biosynthesis
fprA	flavoprotein	1409	2212	pyridoxine 5'-phosphate oxidase
frzA	frizzy	797, 1305	5372	CheW-like
frzB	frizzy	797, 1305	5372	motility regulation
frzCD	frizzy	797, 1305	5372	methyl-accepting chemotaxis protein
frzE	frizzy	797, 1305	5372	phospho kinase
frzF	frizzy	797, 1305	5372	MCP methyl transferase
frzG	frizzy	797, 1305	5372	demethylation of MCP
hemG	heme biosynthesis	1409	2212	protoporphyrinogen oxidase
igl	independent gliding		3880 or 5078	motility regulation
lps-1, -2, -5	lipopolysaccharide		6080 or 6201	lipopolysaccharide biosynthesis
lps-3	lipopolysaccharide		3680 or 5279	lipopolysaccharide biosynthesis
lps-4	lipopolysaccharide		6943 or 7330	lipopolysaccharide biosynthesis
LS234		**1181**	4951	Dev. regulated promoter
LS237		263, **671, 1479**	6447	Dev. regulated promoter
LS255		**486, 1383**	8502	Dev. regulated promoter
LS257		**912, 1360**	4475	Dev. regulated promoter
LS259		406, 1343	2751	Dev. regulated promoter
LS261			3965 or 4994	Dev. regulated promoter
LS263		461, **799**	3621	Dev. regulated promoter
LS265			9570 or 9645	Dev. regulated promoter
LS267		**486, 1383**	8532	Dev. regulated promoter
LS409		617, 1465	5041	Dev. regulated promoter
LS411			5803 or 6478	Dev. regulated promoter
LS420		1409	2255	Dev. regulated promoter
LS421		**933, 1465**	5118	Dev. regulated promoter
LS441		**739, 1416**	8900	Dev. regulated promoter
LS442		819	4182	Dev. regulated promoter
LS444		1416	8873	Dev. regulated promoter
LS445			4185 or 4754	Dev. regulated promoter

mbhA	myxobacterial hemagglutinin	209, 1213	9109	lectin
mglA	mutual gliding	105, 722, 984	2503	required for A and S motility
mglB	mutual gliding	105, 722, 984	2503	stimulates mglA
Mx8 attB	Mx8 attachment	**780**	1973	Bacteriophage Mx8 integration site
Mx alpha	Mx alpha prophage	**429**	2442-2522	tandem copies of Mx alpha
ops	other protein S	504, 798	7075	Dev. regulated gene
oriC	origin	890, 1387, 1619	225	Chromosomal origin of replication
PH1215		**1120, 557**	3842	Adventurous motility
PH1222		**486, 1383**	8512	Social motility
PH1255		**486, 1383**	8542	Social motility
PH1258		**1619**	125 or 177	Social motility
PH1272		**1120, 557**	3802	Adventurous motility
PH1284		**263, 1479**	6387	Adventurous motility
PH1293			4132 or 4817	Adventurous motility
PH1302		**131, 1049, 1134**	6172	Adventurous motility
PH1329		**1566**	7476	Social motility
rif	rifampicin resistance	557, 1120, 1183	3855	high level rifampicin resistance
rpsU	ribosomal protein, small		6778-7493	30S ribosomal subunit protein S21
rrnA	rRNA	1142	75	rRNA operon
rrnB	rRNA	1569	1673	rRNA operon
rrnC	rRNA	461, 799, 1242	3542	rRNA operon
rrnD	rRNA	131, 1049	6168	rRNA operon
sgl-3119	social gliding	921	8494	Social motility
sglA1	social gliding		6778-7495	Social motility
socA	suppressor of C	535, **1400**	6354	CsgA-like short chain alcohol dehydrogenase
socB	suppressor of C	535, **1400**	6354	membrane anchor protein
socC	suppressor of C	535, **1400**	6354	negative regulator of socABC
socD	suppressor of C		8812	histidine protein kinase
socE	suppressor of C	1683	845	C signal suppressor
socF	suppressor of C	907	1045	C signal suppressor
spo-75	sporulation		9275 or 9445	essential for sporulation
spo-87	sporulation		4045 or 4914	essential for sporulation
spo-133	sporulation		1541 or 2415	essential for sporulation
spo-147	sporulation		8630 or 9170	essential for sporulation
spo-406	sporulation	**776, 1183**	3911	essential for sporulation
spo-416	sporulation		6096 or 6185	essential for sporulation
spo-417	sporulation	617	5231	essential for sporulation
spo-418	sporulation	**402**, 454, **847**	6357	essential for sporulation
spo-422	sporulation	**131, 1049, 1134**	6188	essential for sporulation
spo-423	sporulation	**430, 939**	4701	essential for sporulation
spo-510	sporulation	**276**	6654	essential for sporulation

spo-782	sporulation		3880 or 5129	essential for sporulation
spo-792	sporulation		5982 or 6295	essential for sporulation
stk-1907	sticky	**1036**	4634	cell adhesion
tagA-H	temp. sensitive aggregation	1690, 1419	2107	34ºC aggregation system
tglA		557, 1120, 1183	3861	contact-stimulated S motility
tps	true protein S	504, 798	7075	spore coat protein S
uraA	uracil	**897, 1109**, 1306, 1566	7537	orotidine-5'-phosphate decarboxylase

[a] A transposon linked to the gene of interest was mapped to a particular *Ase*I restriction fragmentby restriction enzyme analysis. Where hybridization data to the ordered YAC library is available, a more precise map location is given along with the names of the homologous YAC clones. Data is compiled from the following papers:

Chen, H.-W. A. Kuspa, I. M. Keseler, and L. J. Shimkets. 1991. Physical map of the *Myxococcus xanthus* chromosome. J. Bacteriol. 173:2109-2115.

He, Q., H.-W. Chen, A. Kuspa, Y. Cheng, D. Kaiser, and L. J. Shimkets. 1994. A physical map of the *Myxococcus xanthus* chromosome. Proc. Natl. Acad. Sci USA 91:9584-9587.

MacNeil, S. D., F. Calara, and P. L. Hartzell. 1994. New clusters of genes required for gliding motility in *Myxococcus xanthus.* Mol. Microbiol. 14:61-71.

[b] Location is relative to the *Ase*I fragment J-G junctionwhich is 1/9232 kbp; average range ± 30 kbp

[c] Bold YAC numbers refer to linkages inferred from the map structure, but not proven by hybridization.

71

Neisseria gonorrhoeae MS11-N198 (ATCC 49759)

Carol P. Gibbs

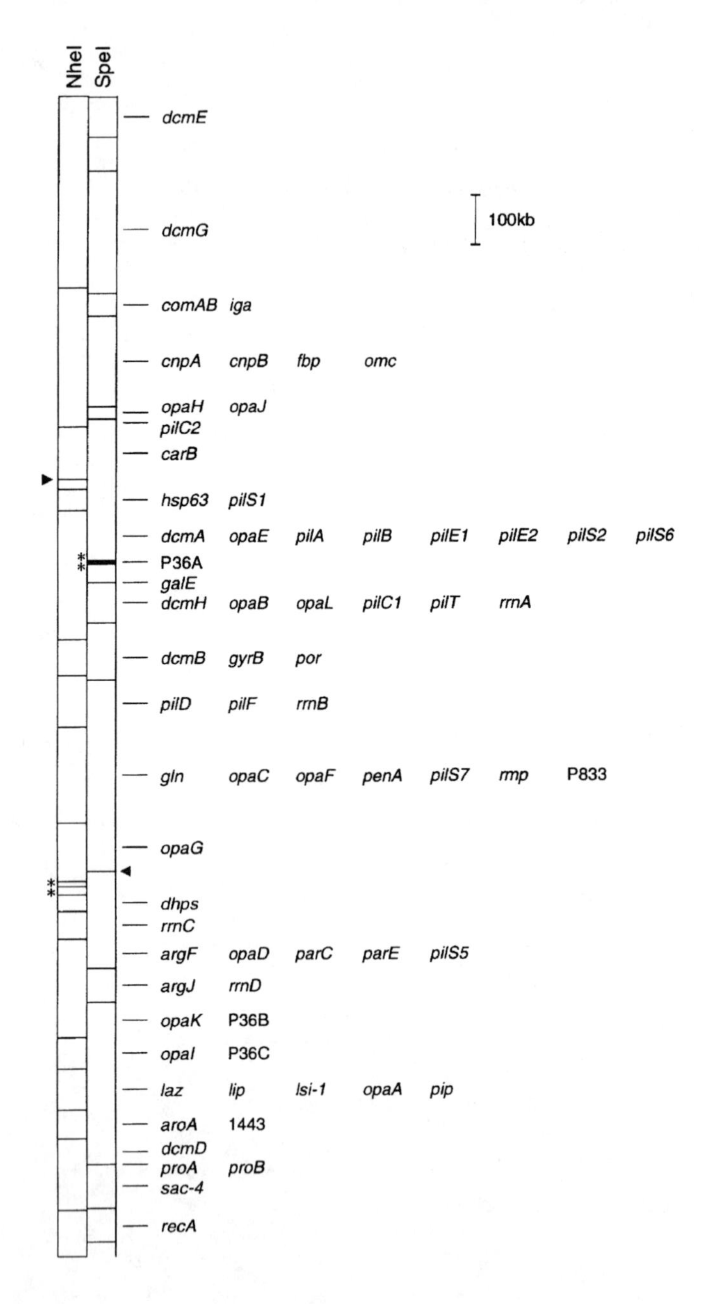
NheI
SpeI
100kb
dcmE
dcmG
comAB iga
cnpA cnpB fbp omc
opaH opaJ
pilC2
carB
hsp63 pilS1
dcmA opaE pilA pilB pilE1 pilE2 pilS2 pilS6
P36A
galE
dcmH opaB opaL pilC1 pilT rrnA
dcmB gyrB por
pilD pilF rrnB
gln opaC opaF penA pilS7 rmp P833
opaG
dhps
rrnC
argF opaD parC parE pilS5
argJ rrnD
opaK P36B
opaI P36C
laz lip lsi-1 opaA pip
aroA 1443
dcmD
proA proB
sac-4
recA

◀

Figure 71-1. Physical map of the *N. gonorrhoeae* strain MS11-N198 chromosome. The macro-restriction map was constructed using combinations of one- and two-dimensional pulsed-field gel electrophoresis of completely or partially digested chromosomal DNA (Bihlmaier *et al.*, 1991 (*see* Chapters 24–26)). The circular chromosome has been linearized at the NheI site of the smallest fragment produced by digestion with NheI and SpeI. The relative positions of the smallest fragments, marked by asterisks, are tentative. Arrowheads indicate sites that are partially resistant to cleavage. Positions of genetic loci were determined by hybridization with cloned DAN fragments; loci not included in previous maps of strain MS11-N198 are *carB* (Picard and Dillon, 1989); *dcmH* (Gunn *et al.*, 1992); *pilD, pilF, pilT* (Lauer *et al.*, 1993); and *pip* (Albertson and Koomey, 1993). Comparison with the physical map of *N. gonorrhoeae* strain FA 1090 (Dempsey and Cannon, 1994) indicates the two strains share general overall similarity, but are not identical.

References

Albertson, N.H., and M. Koomey. 1993. Molecular cloning and characterization of a proline iminopeptidase gene from *Neisseria gonorrhoeae. Mol. Microbiol.* 9:1203–1211.

Bihlmaier, A., U. Römling, T.F. Meyer, *et al.* 1991. Physical and genetic map of the *Neisseria gonorrhoeae* strain MS11-N198 chromosome. *Mol. Microbiol.* 5:2529–39.

Dempsey, J.A.F., and J.G. Cannon. 1994. Locations of genetic markers on the physical map of the chromosome of *Neisseria gonorrhoeae* FA 1090. *J. Bacteriol.* 176:2055–2060.

Gunn, J.S., A. Piekarowicz, R. Chien, *et al.* 1992. Cloning and linkage analysis of *Neisseria gonorrhoeae* DNA methyltransferases. *J. Bacteriol.* 174:5654–5660.

Lauer, P., N.H. Albertson, and M. Koomey. 1993. Conservation of gene encoding components of a type IV pilus assembly/two-step protein export pathway in *Neisseria gonorrhoeae. Mol. Microbiol.* 8:357–368.

Picard, F.J., and J.R. Dillon. 1989. Cloning and organization of seven arginine biosynthesis genes from *Neisseria gonorrhoeae. J. Bacteriol.* 171:1644–51.

72

Planctomyces limnophilus DSM 3776^{T}

Naomi Ward-Rainey, Fred A. Rainey, and Erko Stackebrandt

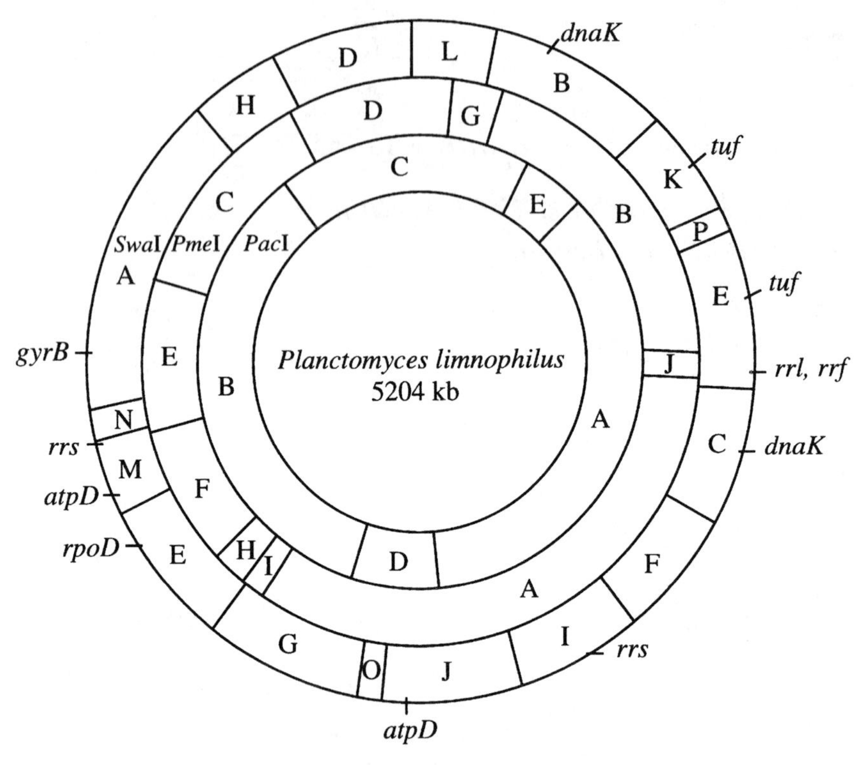

Figure 72-1. Deduced physical map of the chromosome of *Planctomyces limnophilus* DSM 3776^{T}, showing restriction sites for the enzymes *Pme*I, *Swa*I and *Pac*I. The locations of the genetic markers *rrs* (16S rRNA), *rrl* (23S rRNA), *rrf* (5S rRNA), *atpD* (membrane-bound ATP synthase, F_1 sector, β-subunit), *tuf* (protein chain elongation factor, EF-Tu), *gyrB* (DNA gyrase, subunit B), *rpoD* (RNA polymerase, σ^{70} subunit) and *dnaK* (heat shock protein) are also shown. (Ward-Rainey, N., F. A. Rainey, E.M.H. Wellington, and E. Stackebrandt. 1996. Physical map of the genome of *Planctomyces limnophilus*, a member of the phylogenetically distinct planctomycete lineage. *J. Bacteriol.* 178:1908 o 1913.)

73

Burkholderia (Pseudomonas) cepacia Strain ATCC 17616 and 25416; *Pseudomonas aeruginosa* Strain C and PAO (DSM 1707)

Ute Römling, Karen Schmidt, and Burkhard Tümmler

Burkholderia (Pseudomonas) cepacia strain ATCC 17616

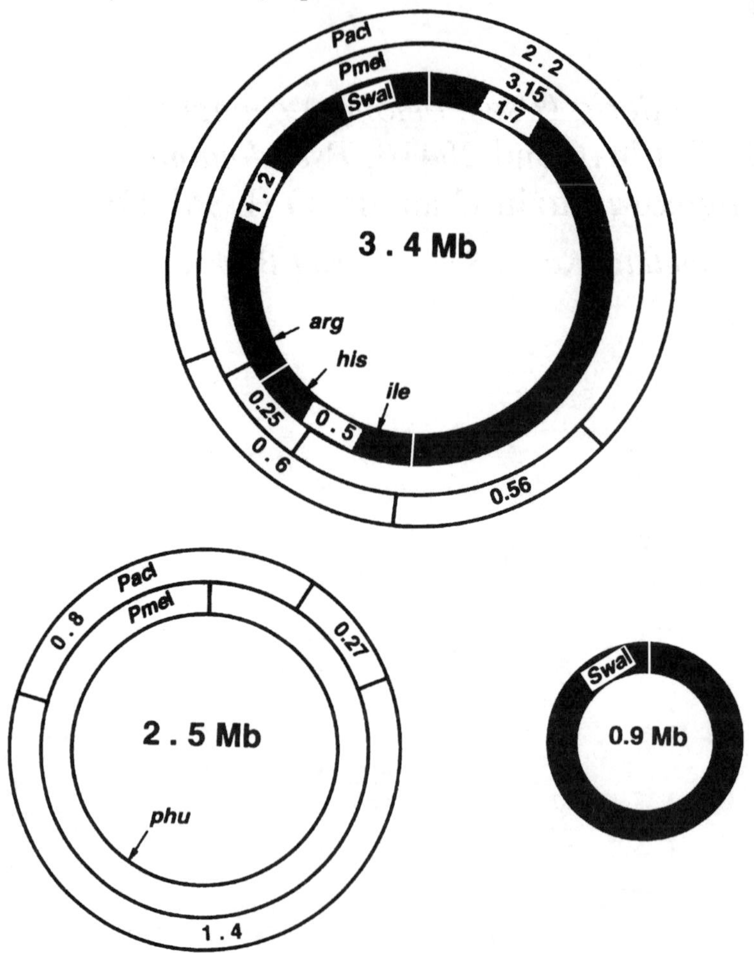

Figure 73-1. Arrangement of macrorestriction fragments constituting the three large replicons of *Burkholderia (Pseudomonas) cepacia* 17616. The concentric circles indicate the locations of *Swa*I, *Pac*I, and *Pme*I sites and fragments. The arrows show the distribution of selected genes on the 3.4- and 2.5-Mb replicons, as determined by analysis of the sites of insertion of Tn*5-751S* in mutants impaired in biosynthesis of arginine, isoleucine, or histidine or in phthalate utilization. Reprinted by permission from the authors and the American Society for Microbiology.

Reference

Cheng, H.-P., and T.G. Lessie. 1994. Multiple replicons constituting the genome of *Pseudomonas cepacia* 17616. *J. Bacteriol.* 176:4034–4042.

Burkholderia (Pseudomonas) cepacia ATCC 25416

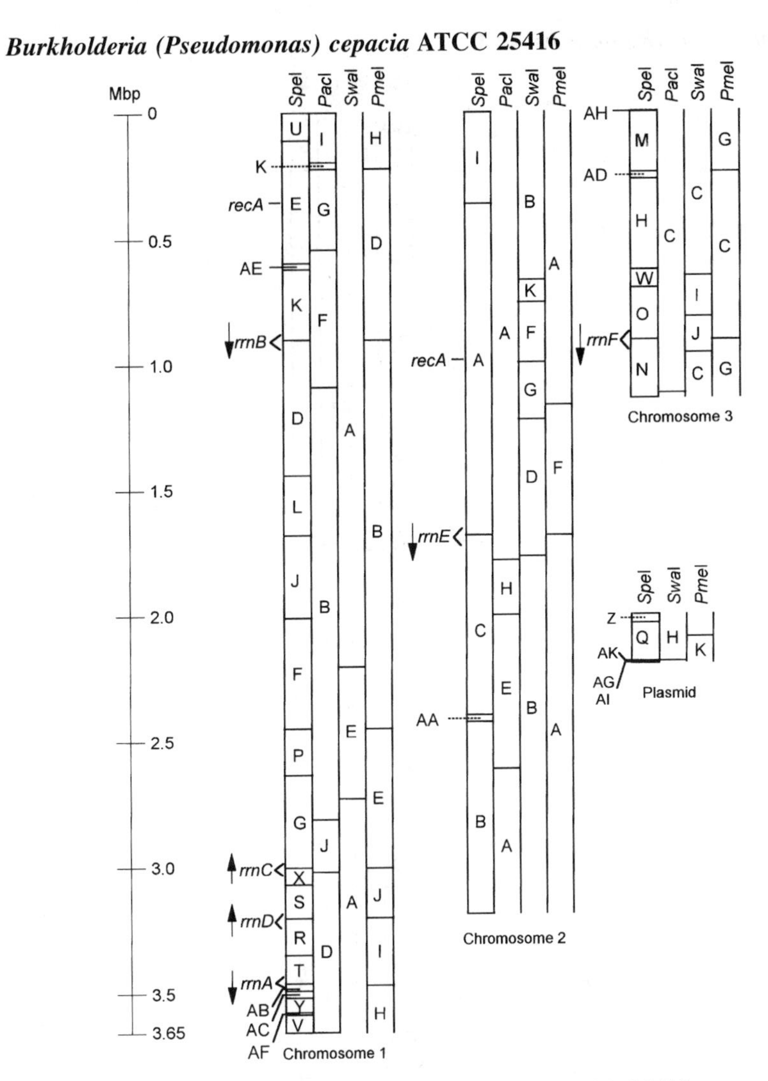

Figure 73-2. Macrorestriction map of the *B. cepacia* (ATCC 25416) genome for the enzymes *Pac*I, *Pme*I, *Spe*I, *Swa*I. Fragments are named in alphabetical order by decreasing size, the first 26 as A to Z and subsequent ones as AA to AZ. The three chromosomes and the 200 kpb plasmid are circular molecules. The arrowheads indicate the orientation of the *rrn* operons from 5′–16S rDNA to 3′–5S rDNA.

Reference

Rodley, P.D., U. Römling, and B. Tümmler. 1995. A physical genome map of the *Burkholderia cepacia* type strain. *Mol. Microbiol.* 17:57–67.

***Pseudomonas aeruginosa* strain C**

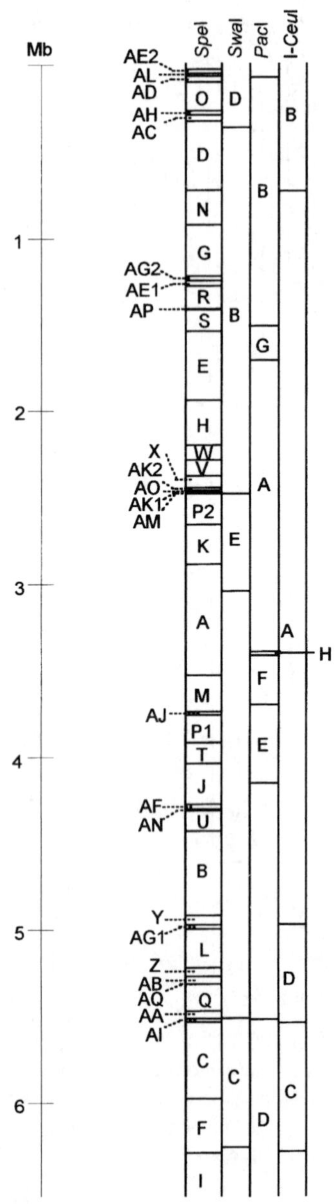

Figure 73-3. Physical map

Macrorestriction map of the *P. aeruginosa* strain C chromosome for the enzymes I-*Ceu*I, *Pac*I, *Spe*I, *Swa*I. Fragments are named in alphabetical order by decreasing size, the first 26 as A to Z and subsequent ones as AA to AZ.

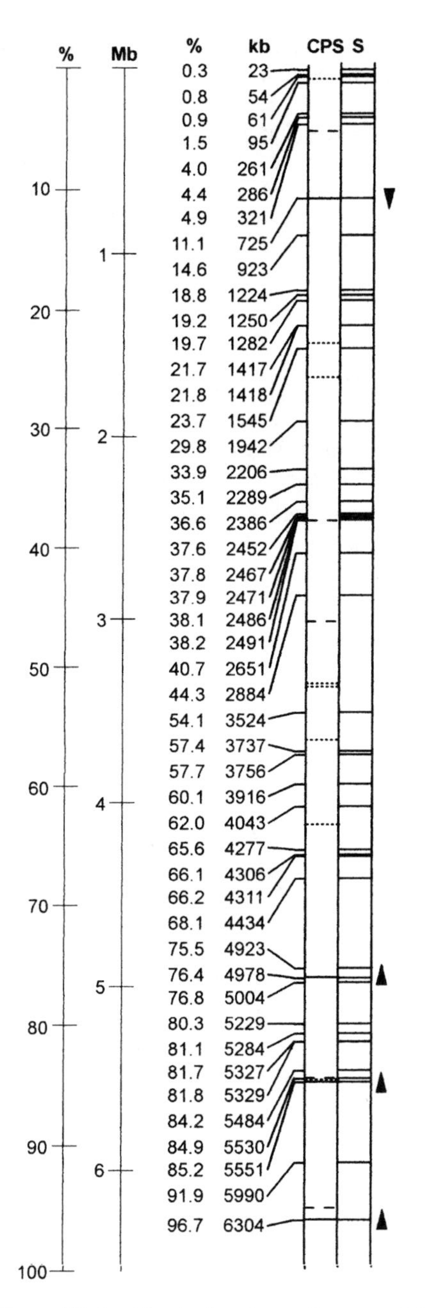

map location [%]	gene
0-0.3	*oriC*
0.3-0.8	*trpIBA*
0.9	lc1
1.5	lc2
4.0	lc3
4.9-11.1	*trpE trpGDC*
11.1	*rrnA*
11.1-14.6	*ampC* (=*blaP*)
21.8-23.7	*recA algD*-operon`(23.7)
29.8-33.9	*trpF leu-10 oru-314*
33.9-35.1	*pyrF lipA lipH*
36.6	lc4
36.6-37.6	lc5
38.2-40.7	*phoA1*
40.7-44.3	*xylS benA czr*
44.3-54.1	*pvdB fpvA hcn*
54.1-57.4	*oprF*
57.7-60.1	*fsr*
60.1-62.0	*fliA cheYZAB cheJ*
62.0-65.6	*toxA*
65.6	lc7
66.2-68.1	*phnAB*
75.5-76.4	*toxR*
76.4	*rrnB*
76.8-80.3	*pctA*
80.3-81.1	*pilA*
84.2-84.9	*hemL*
85.2	*rrnC*
91.9-96.7	*algR pstCAB phoU*
96.7	*rrnD*

Figure 73-4. Physical and genetic map of the *P. aeruginosa* strain C chromosome

The map indicates the location of I-*Ceu*I (C, straight line), *Pac*I (P, dotted line), *Swa*I (S, dashed line) and *Spe*I (S, right column) sites by absolute (kb) and relative (%) size. Genes were assigned by hybridization or probes onto macrorestriction blots (*see also* Chapters 25, 26). The *oriC-I* site is defined as the zero point of the physical chromosome. The arrowheads indicate the orientation of the *rrn* operons from 5′–16S rDNA to 3′–5S rDNA. The abbreviation lc indicates the map locations of hybridization signals with strain *P. aeruginosa* PAO—derived *Spe*I linking clones (Römling and Tümmler, 1993): 1, PAO linking fragment SpAB/SpAJ; 2, SpO/SpAB; 3, SpH/SpO; 4, SpV/SpAK; 5, SpY/SpAK; 7, SpAD/SpAH.

References

Schmidt, K., B. Tümmler, and U. Römling. 1996. Comparative genome mapping of *Pseudomonas aeruginosa* PAO with *P. aeruginosa* C, which belongs to a major clone in cystic fibrosis patients with aquatic habitats. *J. Bacteriol.* 178:85–93.

Römling, U., and B. Tümmler. 1993. Comparative mapping of the *Pseudomonas aeruginosa* PAO genome with rare-cutter linking clones or two-dimensional pulsed-field gel electrophoresis protocols. *Electrophoresis* 14:283–289.

Pseudomonas aeruginosa PAO (DSM 1707)

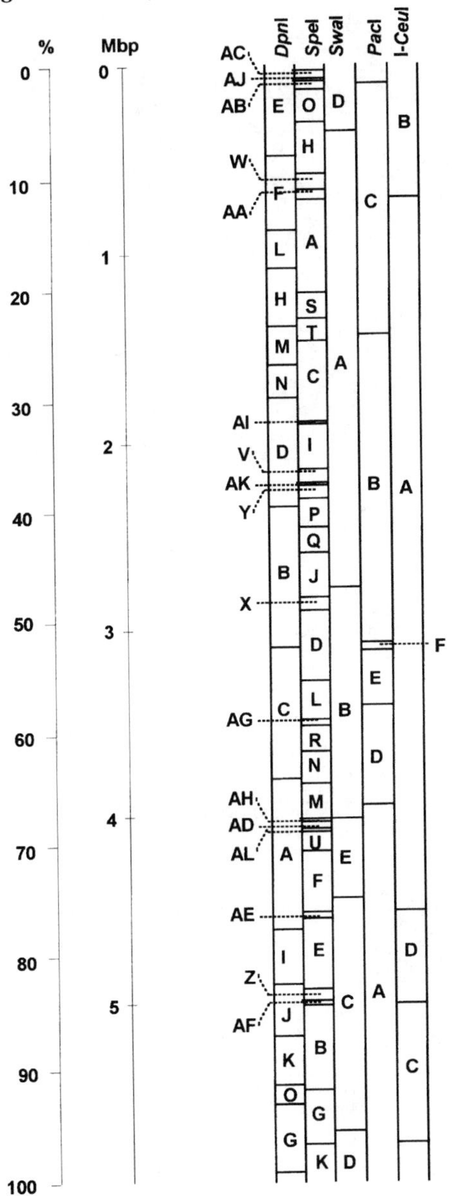

Figure 73-5. Physical map

Macrorestriction map of the *P. aeruginosa* PAO chromosome (DSM 1707) for the enzymes I-*Ceu*I, *Dpn*I, *Pac*I, *Spe*I, *Swa*I. Fragments are named in alphabetical order by decreasing size, the first 26 as A to Z and subsequent ones as AA to AZ.

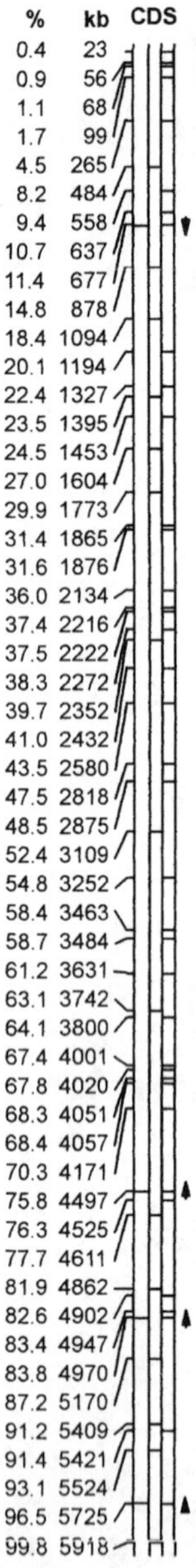

%	kb
0.4	23
0.9	56
1.1	68
1.7	99
4.5	265
8.2	484
9.4	558
10.7	637
11.4	677
14.8	878
18.4	1094
20.1	1194
22.4	1327
23.5	1395
24.5	1453
27.0	1604
29.9	1773
31.4	1865
31.6	1876
36.0	2134
37.4	2216
37.5	2222
38.3	2272
39.7	2352
41.0	2432
43.5	2580
47.5	2818
48.5	2875
52.4	3109
54.8	3252
58.4	3463
58.7	3484
61.2	3631
63.1	3742
64.1	3800
67.4	4001
67.8	4020
68.3	4051
68.4	4057
70.3	4171
75.8	4497
76.3	4525
77.7	4611
81.9	4862
82.6	4902
83.4	4947
83.8	4970
87.2	5170
91.2	5409
91.4	5421
93.1	5524
96.5	5725
99.8	5918

CDS

map location[%]	genes
0-0.4	*oriC*
0.4-0.9	*trpIBA hemF*
1.7-4.5	*atsA*
4.5-8.2	*mexA oprK pilG pilT pilU ropH yhhF*
8.2-9.4	*dnr norB norC nirQ nirS nirM nirC nirF*
4.5-9.4	*ada ccf(=denB) clrD hisV met-26 metC27 met-28 nid pilH,I,J,K proC pyrB pyrX pyr-81 ser-3 sss*
9.4-10.7	*trpE rpoD**
10.7-11.4	*trpGDC argC**
11.4	*rrnA*
11.4-14.8	*ftpA pchAB ropB ssb*
14.8-18.4	*acoA ampC ampR hemA prf1 pyo proA* t-RNAs** PIS*
18.4-20.1	*exoS* IS-PA-1
20.1-22.4	*fbp lasB**
22.4-23.5	*recA*
23.5-24.5	*algD844EGXLIJFA'* (24.5) *glpD*
22.4-24.5	*leu-8 purC6 rpoS thrC thr-59 thr-60* PIS
24.5-27.0	*vsmR vsmI*
27.0-29.9	*aro-2 cypH nosZ1 oprO oprP*
24.5-31.4	*aro-1 argF argG eda-9001 edd-1 plcN* PIS Tn*1*
31.4-31.6	*rfb*-cluster"
31.6-36.0	*leu-10 trpF oru-314 snr1, snr2, snr3, snr4 met-9011* pru-9013**
36.0-37.4	*oprI lipAH pyrF ent*
37.5-38.3	*pbpC soxR catR'*(38.3)
38.3-39.7	*pheS*
39.7-41.0	*phoA1 att*Φctx
41.0-43.5	*czr rpsB sucC sucD xylS benA ben-4* ben-3* ben-2* ben-1508* catA* catB* catC* ant-1* ant-2* ant-3** Tn*1**
43.5-47.5	*pvdB fpvA gcu-1* pvdD* pvdE* pvdA** PIS* Tn*1**
47.5-48.5	PIS* Tn*1**
48.5	major integration sites of Tn*4651* Tn*4653* (pWW0)'''
48.5-52.4	*groEL hcn opdE*
48.5-54.8	*agmR braZ* Tn*1*
54.8-58.4	*oprF exsCBAD* cysG1* cys-54* met-29* lasA1* rpoC** Tn*1**
58.4-58.7	Tn*1**
58.7-61.2	*fsr hemN nosZ2 citA anr** Tn*1**
61.2-63.1	*fliA cheYZAB cheJ*
64.1-67.4	*oprH apr** PIS*
67.4-67.8	*toxA* IS-PA-1
68.4-70.3	*phnAB fliC braC310* pvd* *PIS*
70.3-75.8	*qin oprD proS regB toxR algT*(=algU) mucA* algN*(=mucB) algM* algY* nadB* plcSR1R2* pruA* pruB* pur-70** PIS*
75.8	*rrnB*
76.3-77.7	*pctA*
77.7-81.9	*envA ftsA ftsZ murE pbpB pilA pilBCD pilRS rpoN sodA sodB*
81.9-82.6	*katA*
76.3-82.6	*clbB hisI pilE pilL pilV*
82.6-83.4	*hemL*
83.8	*rrnC*
87.2-91.2	*phaC1 phaD phaC2 aceAB dapB ponA*
83.8-91.4	*aroB aroK carA carB dapB hisIV ilvB ilvC phe-3 pilM,N,O,P pilQ pur-8001 ser-33* PIS
93.1-96.5	*ssr algR1,R2,R3 hemB phoA2 phoB phoR pstCAB phoU hemCD phoS* Polyphosphatekinase
91.4-96.5	*algC argA argH att*PS21 *hisIIA,B lysA pyrE* PIS
96.5	*rrnD*
96.5-99.8	*psaA,B,C,D*
96.5-0.4	*dnaA dnaN gidA lasI lasR oriC-II pur-136 rrpA rrpH thrB*

Figure 73-6. Physical and genetic map of the *P. aeruginosa* PAO chromosome (DSM 1707).

The map indicates the location of I-*Ceu*I (C), *Dpn*I (D) and *Spe*I (S) sites by absolute (kb) and relative (%) size. Genes that were only positioned on *Spe*I-restricted chromosomal DNA are indicated by an asterisk or a larger mapping interval. Genes harbouring a *Spe*I recognition site are designated by apostrophe. A double apostrophe marks a locus that encompasses the whole fragment. Three apostrophes indicate a locus or function that maps in the chromosomal region around the respective *Spe*I recognition site. The localization of some genes was inferred from sequence contigs that contained a tag of known map localization. The *oriC-I* site is defined as the zero point of the physical chromosome. The arrowheads indicate the orientation of the *rrn* operons from 5'–16S rDNA to 3'–5S rDNA. 272 genes are assigned of which 160 entries were already listed in the last edition of the physical map (Holloway et al., 1994). In addition the major integration sites are given for phages and transposons Tn*1*, Tn*4651* and Tn*4653* (Krishnapillai, 1993; PIS, phage integration site; Tn*1*, Tn*1* integration site). Temperate phages were isolated from clinical isolates of *P. aeruginosa* strains that plaqued on PAO. PAO lysogens of these phages were constructed and integration sites were identified by PFGE analysis of *Spe*I-restricted chromosomal DNA (Krishnapillai, 1993).

References

Holloway, B.W., U. Römling, and B. Tümmler. 1994. Genomic mapping of *Pseudomonas aeruginosa* PAO. *Microbiology* 140:2907–2929.

Krishnapillai, V. 1993. Organization of the chromosome of the bacterium *Pseudomonas aeruginosa. Current Topics in Mol. Genet.* (Life Sci. Adv.) 1:87–103.

Liao, X., I. Charlebois, C. Ouellet, M.-J. Morency, K. Dewar, J. Lightfoot, J. Foster, R. Siehnel, H. Schweizer, J.S. Lam, R.E.W. Hancock, R.C. Levesque, 1996. Physical mapping of 32 gene markers on the *Pseudomonas aeruginosa* chromosome. *Microbiology* 142:79–86.

74

Rhizobium meliloti

Rhonda J. Honeycutt and Bruno W. S. Sobral

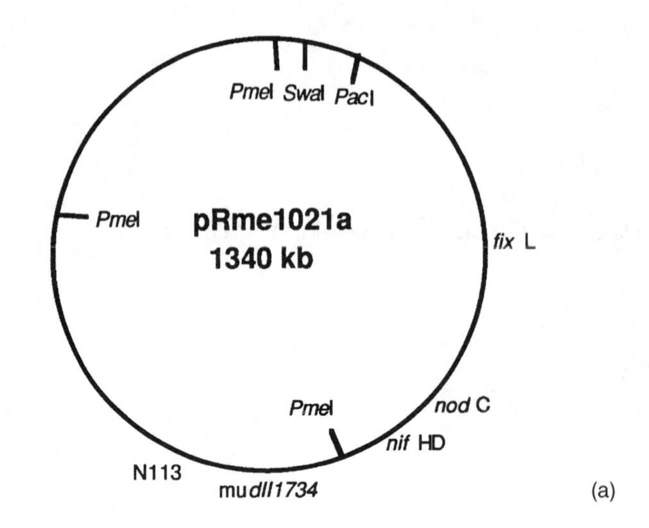
PmeI SwaI PacI
PmeI
pRme1021a
1340 kb
fix L
PmeI
nod C
nif HD
N113
mu dII1734
(a)

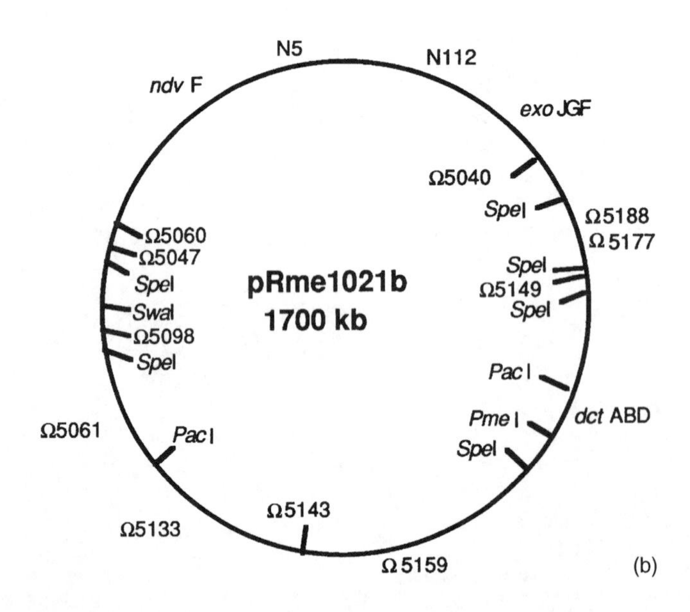
N5
N112
ndv F
exo JGF
Ω5040
SpeI
Ω5188
Ω5177
Ω5060
Ω5047
SpeI
SwaI
Ω5098
SpeI
pRme1021b
1700 kb
SpeI
Ω5149
SpeI
PacI
PmeI
dct ABD
SpeI
Ω5061
PacI
Ω5143
Ω5133
Ω5159
(b)

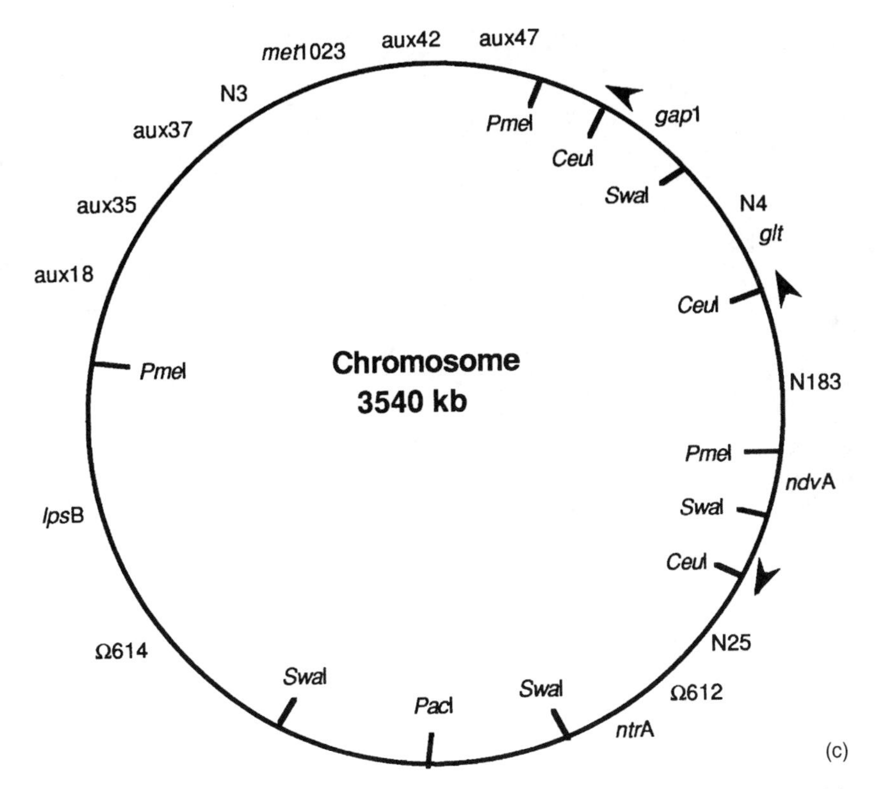

Figure 74-1. Physical maps of the replicons of *R. meliloti* 1021. (A) pRme1021a (aka pSyma) 1° approximately 3.7 kb; (B) pRme1021b (aka pSymb), 1° approximately 4.7 kb; (C) chromosome, 1° approximately 9.8 kb. Arrows indicate the orientations of the *rrn* loci. For all replicons, solid lines indicate the positions of target sites and/or transposon insertions anchored to the physical map (shown inside the circle); the positional order shown for other markers residing on the same restriction fragment is random (shown outside the circle). Honeycutt, R. J., M. McClelland, and B.W.S. Sobral. 1993. Physical map of the genome of *Rhizobium meliloti* 1021. *J. Bacteriol.* 175:6945–6952.

75

Rhodobacter capsulatus SB1003

Michael Fonstein, Elizabeth G. Koshy, Vivek Kumar, Paul Mourachov, Tatiana Nikolskaya, Michael Tsifansky, Su Zheng, and Robert Haselkorn

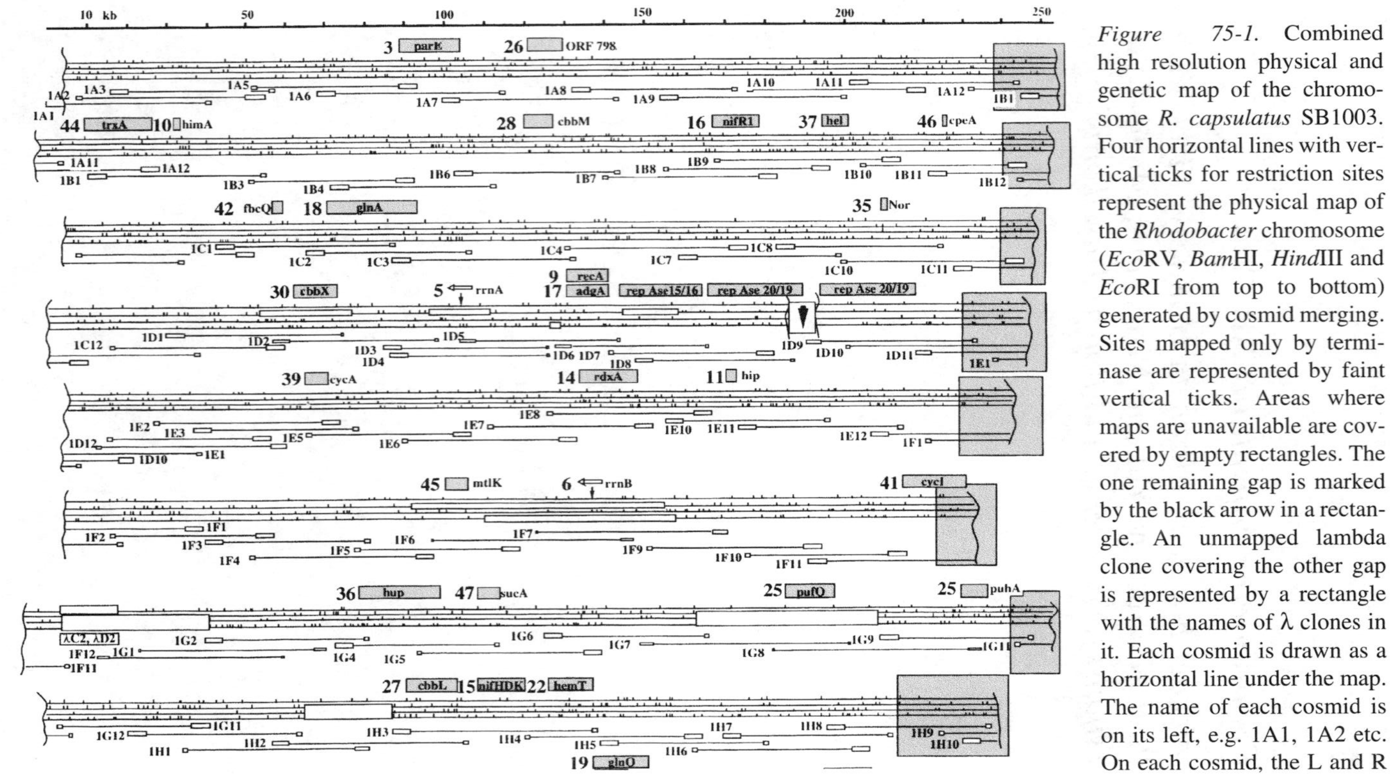

Figure 75-1. Combined high resolution physical and genetic map of the chromosome *R. capsulatus* SB1003. Four horizontal lines with vertical ticks for restriction sites represent the physical map of the *Rhodobacter* chromosome (*Eco*RV, *Bam*HI, *Hind*III and *Eco*RI from top to bottom) generated by cosmid merging. Sites mapped only by terminase are represented by faint vertical ticks. Areas where maps are unavailable are covered by empty rectangles. The one remaining gap is marked by the black arrow in a rectangle. An unmapped lambda clone covering the other gap is represented by a rectangle with the names of λ clones in it. Each cosmid is drawn as a horizontal line under the map. The name of each cosmid is on its left, e.g. 1A1, 1A2 etc. On each cosmid, the L and R

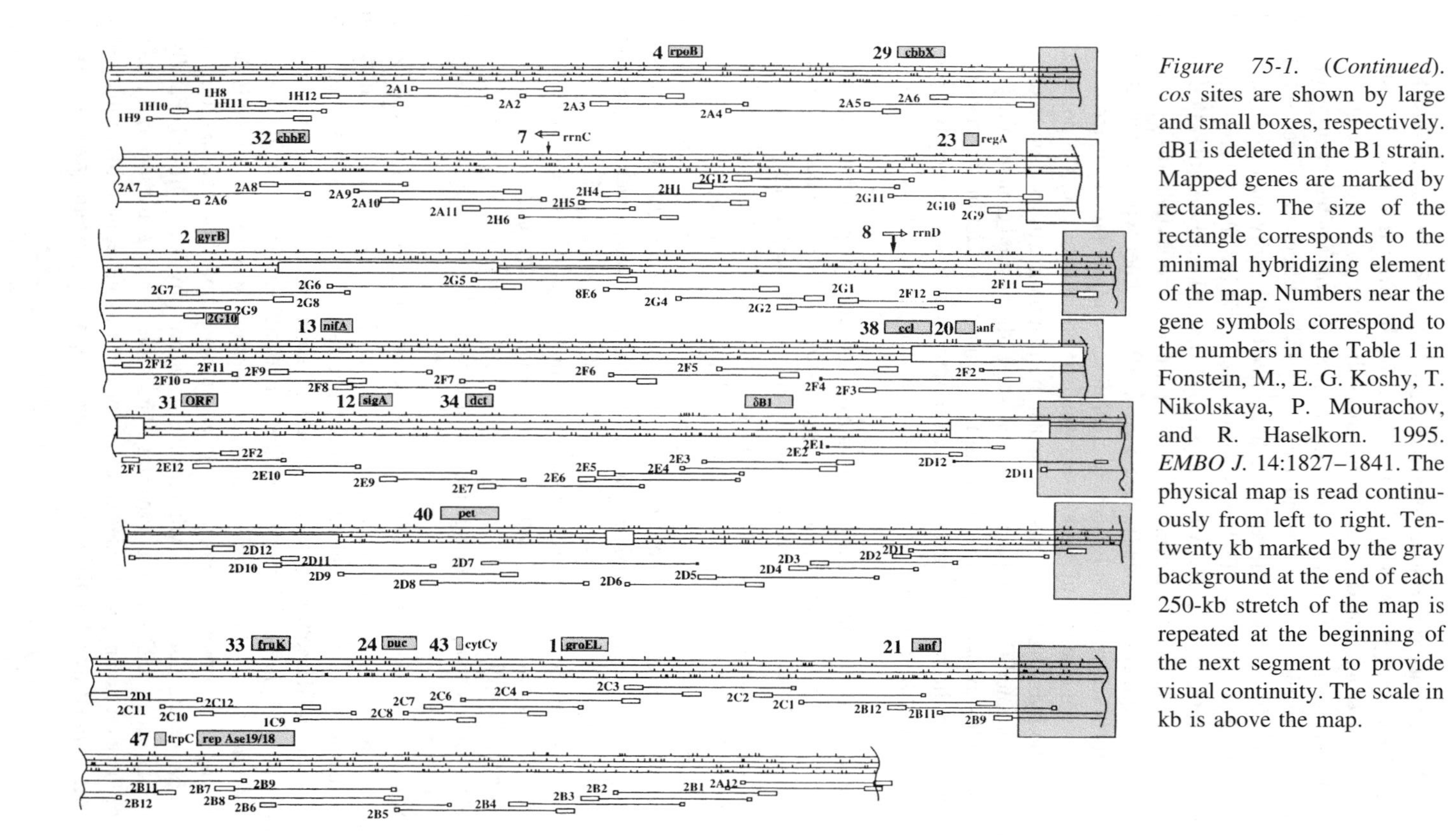

Figure 75-1. (*Continued*). *cos* sites are shown by large and small boxes, respectively. dB1 is deleted in the B1 strain. Mapped genes are marked by rectangles. The size of the rectangle corresponds to the minimal hybridizing element of the map. Numbers near the gene symbols correspond to the numbers in the Table 1 in Fonstein, M., E. G. Koshy, T. Nikolskaya, P. Mourachov, and R. Haselkorn. 1995. *EMBO J.* 14:1827–1841. The physical map is read continuously from left to right. Ten-twenty kb marked by the gray background at the end of each 250-kb stretch of the map is repeated at the beginning of the next segment to provide visual continuity. The scale in kb is above the map.

DNA fragments from a *Sau*3a partial digest of *Rhodobacter capsulatus* chromosomal DNA were size selected and cloned in the *Bam*HI site of Lorist 6 by cosmid arm cloning and clones were individually transferred to eighteen 96-well plates (*see also* Chapter 28). Using a specially constructed printing device, individual cosmid clones were replicated onto nylon filters as ordered sets (18X coverage). These sets of clones were hybridized with individual *Xba*I and *Ase*I fragments prepared from PFGE gels. Forty-one restriction sites for *Ase*I and *Xba*I were mapped onto the 3.7-Mb genome by standard analyses of PFGE blot-hybridization data and of linking cosmids, revealed as clones hybridizing with two restriction fragments at once (Fonstein *et al.*, 1992). Other types of hybridization (fragments to fragments, fragments to cosmid sets and riboprobes generated from the ends of the cosmids to cosmid sets) made it possible to vary the probe/target ratio in the hybridizations, thus distinguishing mapping artifacts connected with different repeated DNA sequences. At this stage, the cosmid clones were grouped in about 80 subcontigs, corresponding to the macro-restriction fragments from the PFG and the regions surrounding these rare sites. They formed two groups, one corresponding to the chromosome of *R. capsulatus*, the other to its 134-kb plasmid.

For detailed mapping, the *Eco*RV sites of 40 evenly spaced cosmids (based on a preliminary genetic map) were mapped using lambda terminase. Riboprobes generated by SP6- and T7-specific transcription of the ends of the inserts of these cosmids were then used as the primary probes to reveal groups of neighboring cosmids. For the next round of cosmid walking, the overhanging ends of these secondary cosmids were used to generate new probes. Finally, after about 300 riboprobe hybridizations with the first set of 864 cosmids and about the same number of *cos*-mappings, six uninterrupted cosmid contigs were mapped with high resolution.

Four of the six maps were closed using another set of arrayed cosmids. For the remaining two gaps, a lambda DASH library was screened with the last four probes (ends of two contigs). One gap was closed, and a minimal uninterrupted set of 186 cosmids mapped with high resolution covered 3.7 Mb of the chromosome of *R. capsulatus*.

Sixty-seven additional blot-hybridizations with cosmid and chromosomal digests were performed to establish the sizes of the chromosomal fragments in the areas where overlapping cosmids did not have common *Eco*RV fragments (Fonstein and Haselkorn, 1993). Several revisions of the earlier map include single cosmid shifts and inversions. One additional gap in a cosmid contig was also found, raising the possibility that the chromosome is not a contiguous circle. This gap is bordered by repeated sequences, so simple Southern blotting does not resolve the question of whether the ends are physically linked. Finally, a minimal set of 192 cosmids covering the chromosome and the large plasmid with the exact map coordinates of each cosmid was generated. More than 3000 *Eco*RI, *Bam*HI and *Hin*dIII sites have been mapped (Fonstein *et al.*, 1995).

Forty genes cloned from *R. capsulatus* and 14 probes made by PCR (based on known sequences) corresponding to nearly three hundred genes known in *Rhodobacter* (Fonstein *et al.*, 1995). were hybridized to the cosmid array and then to individual digested cosmids. Most were mapped to unique loci, while four hybridized to two locations each. The ribosomal DNA probe hybridized to four regions, explaining the presence of mystifying cosmids that link many PFGE-generated fragments. The orientation of transcription was established for the rRNA operons, revealing important details of genome structure and tentatively assigning the chromosomal *Ori* to a 200-kb segment between *rrnC* and *rrnD*. Comparison of the detailed restriction maps maps for a 2-Mb chromosomal region analyzed for two additional *Rhodobacter* strains (St. Louis and 2.3.1) revealed elements of a mosaic genome structure (Fonstein *et al.*, 1995).

References

Fonstein, M. and R. Haselkorn. 1993. Chromosomal structure of *Rhodobacter capsulatus* strain SB1003: cosmid encyclopedia and high-resolution physical and genetic map. *Proc. Natl. Acad. Sci. USA* 90:2522–6.

Fonstein, M., E. G. Koshy, T. Nikolskaya, P. Mourachov, and R. Haselkorn. 1995. Refinement of the high-resolution physical and genetic map of *Rhodobacter capsulatus* and genome surveys using blots of the cosmid encyclopedia. *EMBO Journal* 14:1827–1841.

M. Fonstein, T. Nikolskaya, and R. Haselkorn. 1995. High resolution alignment of a megabase long genome region of three strains of *Rhodobacter capsulatus. J. Bacteriol.* 177:2368–2372.

Fonstein, M., S. Zheng, and R. Haselkorn. 1992. Physical map of the genome of Rhodobacter capsulatus SB 1003. *J. Bacteriol.* 174:4070–7.

76

Rhodobacter sphaeroides 2.4.1^T

Christopher Mackenzie, Madhusudan Choudhary, Kristen S. Nereng, Erica Sodergren, Monjula Chidambaram, George M. Weinstock, and Samuel Kaplan

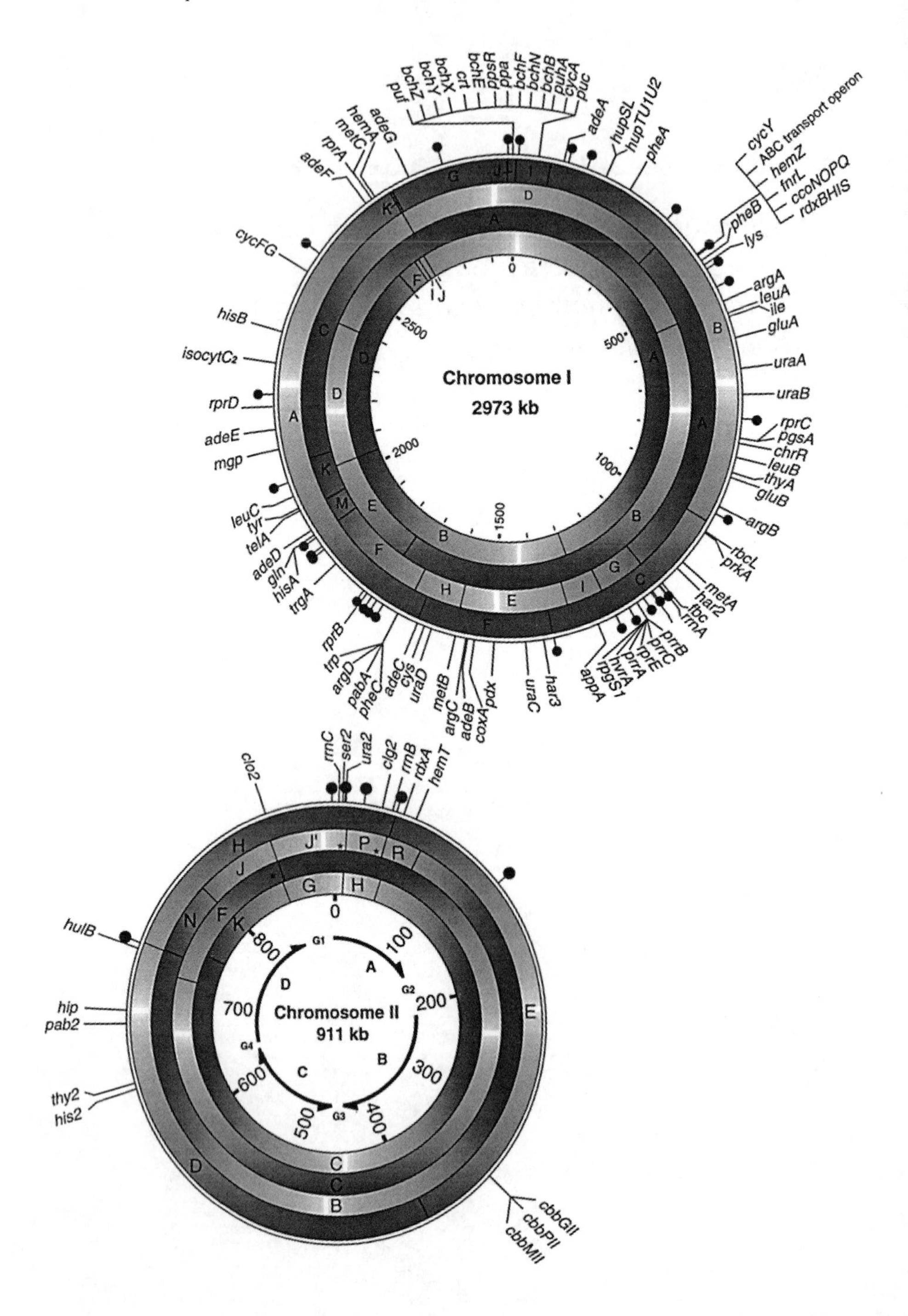
Chromosome I
2973 kb
Chromosome II
911 kb
puf
bchZ
bchY
bchX
crt
bchE
bchB
ppsR
ppa
bchF
bchN
bchB
puhA
cycA
puc
adeA
hupSL
hupTU1U2
pheA
cycY
ABC transport operon
hemZ
fnrL
ccoNOPQ
rdxBHIS
pheB
lys
argA
leuA
ile
gluA
uraA
uraB
rprC
pgsA
chrR
leuB
thyA
gluB
argB
rbcL
prkA
metA
har2
fbc
rrnA
prrB
prrC
rprE
prrA
hvrA
rpgS1
appA
har3
uraC
pdx
coxA
adeB
argC
metB
uraD
cys
adeC
pheC
pabA
argD
trp
rprB
trgA
hisA
gln
adeD
telA
tyr
leuC
mgp
adeE
rprD
isocytC2
hisB
cycFG
adeF
rprA
metC
hemA
adeG
clo2
rrnC
ser2
ura2
clg2
rrnB
rdxA
hemT
hulB
hip
pab2
thy2
his2
cbbGII
cbbPII
cbbMII

◀

Figure 76-1. Physical and genetic maps of chromosomes I and II of *R. sphaeroides* 2.4.1ᵀ. The genome consists of two circular chromosomes (I and II) of approximately 3.0, and 0.9 Mbp in size respectively (Suwanto and Kaplan, 1989a; b; see also Chapt. 43). The shaded concentric circles show (from outer to inner) the physical maps generated using the restriction enzymes *Ase*I, *Dra*I, *Sna*BI and *Spe*I. These produce fragments (in kb) with the following sizes: *Ase*I: A=1105; B=910; C=410; D=360; E=340; F=275; G=244; H=214; I=73; J=18; and K=5. *Dra*I: A=800; B=675; C=660; D=635; E=245; F=245; G=110; H=105; I=85; J=65; J′=65; K=60; M=50; N=50; P=31; and R=25. *Sna*BI: A=1225; B=1200; C=784; D=370; E=300; and F=130. *Spe*I: A=1645; B=735; C=710; D=575; F=90; G=65; H=32; K=105 I=31, and J=12. Asterisks adjacent to restriction site lines indicate the presence of more than one restriction site for that enzyme at that location. Map position zero of chromosome I is designated as the *Ase*I J-I junction (12 o'clock on the outer circle). Map position zero of chromosome II has been placed 42 kb from the end of *Ase*I fragment H within the *rrnC* region. Auxotrophic, DNA repair (*rpr*) and color (*cl*) markers were generated by Tn*5* mutagenesis then their position determined by pulsed-field gel electrophoresis (Choudhary *et al.*, 1994; Mackenzie *et al.*, 1995). Black lollipops indicate the position of Tn*5* insertions with wild type phenotypes. All other markers were sequenced, then placed by Southern hybridization. Within chromosome II the locations of four ordered cosmid clone contigs (A, B, C, D) are shown as arrowed arcs, with gaps G1–4. The arrows indicate the orientation of the contigs with respect to the chromosome II physical map. Contig C has not been oriented in this manner (Choudhary *et al.*, 1994). The ordering of chromosome I cosmid clones is currently in progress.

Acknowledgments

This work was supported by a grant from the Clayton Foundation for Research.

References

Choudhary, M., C. Mackenzie, K. S. Nereng, E. Sodergren, G. M. Weinstock, and S. Kaplan. 1994. Multiple chromosomes in bacteria: structure and function of chromosome II of *Rhodobacter sphaeroides* 2.4.1ᵀ. *Journal of Bacteriology* 176:7694–7702.

Mackenzie, C., M. Chidambaram, E. Sodergren, S. Kaplan, and G. M. Weinstock. 1995. DNA repair mutants of *Rhodobacter sphaeroides. Journal of Bacteriology* 177:3027–3035.

Suwanto, A. and S. Kaplan. 1989a. Physical and genetic mapping of the *Rhodobacter sphaeroides* 2.4.1 genome: genome size, fragment identification and gene localization. *Journal of Bacteriology* 171:5840–5849.

Suwanto, A. and S. Kaplan. 1989b. Physical and genetic mapping of the *Rhodobacter sphaeroides* 2.4.1 genome: presence of two unique circular chromosomes. *Journal of Bacteriology* 171:5850–5859.

77

Salmonella typhi

Shu-Lin Liu and Kenneth E. Sanderson

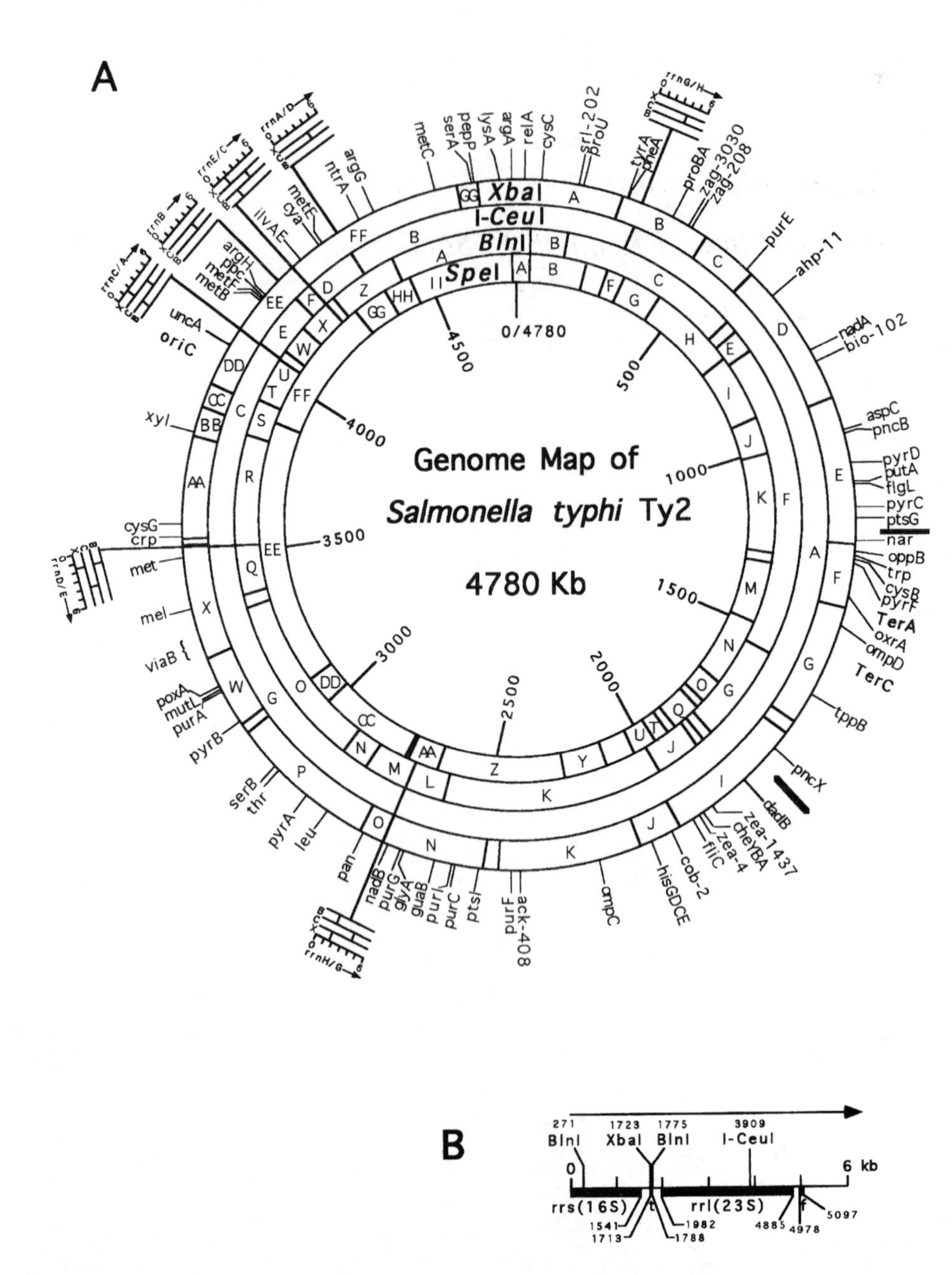

Figure 77-1. The genomic cleavage map of *S. typhi* TY2 for the endonucleases *Xba*I, I-*Ceu*I, *Bln*I, and *Spe*I (Liu and Sanderson, 1995a,b). The chromosome was determined by pulsed-field gel electrophoresis methods, using single-digestion or double-digestion of fragments and end-labelling with radioactivity when needed. (*see also* Chapter 22) The

size of the chromosome is 4780 kbs. No plasmid was detected in this strain. For three of the enzymes the fragments are given letters in alphabetical order, beginning with A at O kb; however, for I-*Ceu*I, the same letter is given as for the homologous fragment in *Salmonella typhimurium* (Liu *et al.*, 1993a) and *E. coli*, but the order of I-*Ceu*I fragments in *S. typhi* TY2 is rearranged to AGCEFDB (Liu and Sanderson, 1995a). This rearranged order is postulated to be due to homologous recombination between the *rrn* operons for rRNA (*see also* Chapters 21 and 22). Other strains of *S.typhi* have different orders of the I-*Ceu*I fragments, which indicates a high level of homologous recombination between *rrn* operons (Liu et al., 1995a). Arbitrarily, we show the large I-*Ceu*I fragment in the same location and orientation as in *S. typhimurium*, with *proA* near 500 kb and *nadB* near 2600; thus, *thr* on the fragment I-*Ceu*-G, which is normally at 0 kb on the map, is now near 3000 kb, and other genes are similarly displaced. The CCW end of the I-*Ceu*-A fragment (at *rrnG/H*) is arbitrarily placed at 300 kb (the distance of the *rrn* operon from *thr* in both *S. typhi* and *S. typhimurium*) to align the genome for convenient comparison. The enzyme I-*Ceu*I has one digestion site in the *rrl* gene for 23S-rRNA (Marshall and Lemieux, 1991) in each of the seven *rrn* operons (Liu *et al.*, 1993a). The structures of the *rrn* operons are shown in detail outside the circle; the positions of the cleavage sites in the *rrn* genes are from double digestion data. The four *rrn* genes which have *Xba*I and *Bln*I sites are inferred to have the tRNA gene for glutamyl-tRNA, and the other three to have the gene for alanyl-tRNA, by analogy with *E. coli.* The arrow indicates the direction of transcription originally determined for *E. coli.* in *rrnB* (Noller and Nomura, 1987). The relative locations of restriction sites were determined by double digestion data (Liu *et al.*, 1993a). The genes are located through mapping the *Xba*I and *Bln*I sites in Tn*10* insertions in the genes; these Tn*10* insertions, known to be in specific genes of *S. typhimurium* (Liu *et al.*, 1993a), were transduced by phage P22 from *S. typhimurium* with selection for tetracycline resistance. (*see also* Chapter 22)

References

Liu, S.-L., and K.E. Sanderson, 1995a. Rearrangements in the genome of the bacterium *Salmonella typhi. Proc. Natl. Acad. Sci. USA* 92:1018–1022.

Liu, S.-L., and K.E. Sanderson, 1995b. The genomic cleavage map of *Salmonella typhi* Ty2. *J. Bacteriol.* (in press).

Liu, S.-L., A. Hessel, and K.E. Sanderson, 1993a. The *Xba*I-*Bln* I-*Ceu*I genomic cleavage map of *Salmonella typhimurium* LT2 determined by double digestion, end-labelling, and pulsed-field gel electrophoresis. *J. Bacteriol.* 175:4104–4120.

Liu, S.-L., A. Hessel, and K.E. Sanderson, 1993b. Genomic mapping with I-*Ceu*I, an intron-encoded endonuclease, specific for genes for ribosomal RNA, in *Salmonella* spp., *Escherichia coli*, and other bacteria. *Proc. Natl. Acad. Sci. USA* 90:6874–6878.

Marshall, P., and C. Lemieux. 1991. Cleavage pattern of the homing endonuclease encoded by the fifth intron in the chloroplast subunit rRNA-encoding gene of *Chlamydomonas eugamatos. Gene* 104:1241–1245.

Noller, H.F., and M. Nomura. 1987. Ribosomes. In F.C. Neidhardt, J.L. Ingraham, K.B. Low, B. Magasanik, M. Schaechter, and H.E. Umbarger, eds., *Escherichia coli and Salmonella typhimurium: Cellular and Molecular Biology*, pp. 104–125. American Society for Microbiology, Washington, D.C.

78

Salmonella typhimurium

Shu-Lin Liu, Andrew Hessel, Michael McClelland, and Kenneth E. Sanderson

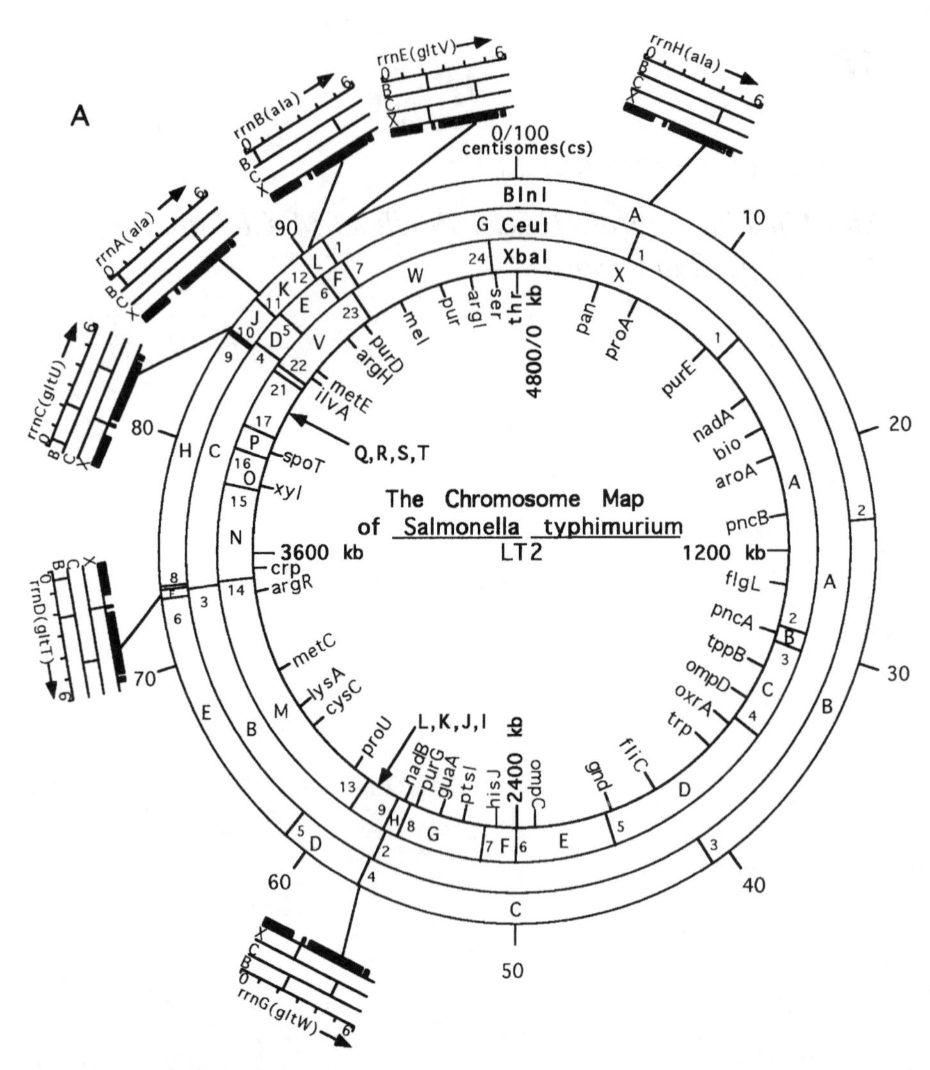

Figure 78-1. The genomic cleavage map of *S. typhimurium* LT2. This may was determined primarily by pulsed-field gel electrophoresis analysis (Liu and Sanderson, 1992; Liu *et al.*, 1993a; Wong and McClelland, 1992). The cleavage sites for the enzymes *Xba*I, I-*Ceu*I, and *Bln*I (= *Avr*II) are indicated in centisomes (CS) around the chromosome; the CS scale is shown on the outside of the circles. The I-*Ceu*I map was determined by isolation of I-*Ceu*I fragments and digestion with *Xba*I (Liu *et al.*, 1993a); all the I-*Ceu*I sites are postulated to be in 23S rRNA genes (*rrl*) (Liu et al., 1993b; Marshall and Lemieux, 1991). The I-*Ceu*I fragments are in alphabetical order, and the I-*Ceu*I sites (sC1 to sC7) are labelled. The locations of *Bln*I sites are determined by isolation of *Bln*I fragments from the gel and redigestion by *Xba*I.

The positions of genes were determined by locating additional *Xba*I and *Bln*I sites in strains with insertions of Tn*10* at known locations on the chromosome, i.e., within genes or close to genes. The phenotype associated with each gene is shown in other references (Liu *et al.*, 1993a; Sanderson *et al.*, 1995). The seven *rrn* operons are indicated in detail outside the circle at the locations determined by redigestion of I-*Ceu*I fragments by *Xba*I or by *Bln*I. The structure of the *rrn* genes and their location and orientation on the chromosome was originally determined in *E. coli* (Bachmann, 1990; Noller and Nomura, 1987; *see also* Chapter 21) and the locations were confirmed in *S. typhimurium*. The order of the genes in each *rrn* operon is indicated by the solid bar under the gene, and is *rrs* (16S-rRNA)-tRNA-*rrl* (23S-rRNA)-(5S-rRNA). The tRNA genes in the intervening regions are the same as in *E. coli*, except that *gltT* is in *rrnD* at 73 CS rather than in *rrnB* at 90 CS. The orientation of transcription indicated by the arrow on each gene is from *E. coli* (Noller and Nomura, 1987).

The eighth edition of the genetic map of *S. typhimurium*, incorporating genetic data and data from nucleotide sequences, shows the locations of over 1,000 genes on a framework derived from the data reported above (Sanderson *et al.*, 1995). A high-resolution restriction map has been constructed for part of the chromosome (Wong *et al.*, 1994). Information from these sources was used to correct a few aspects of the genomic cleavage map published earlier (Liu *et al.*, 1993a). For other details, *see* Chapter 22.

References

Bachmann, B.J. 1990. Linkage map of *Escherichia coli*, edition 8. *Microbiol. Rev.* 54:130–197.

Liu, S.-L., and K.E. Sanderson. 1992. A physical map of the *Salmonella typhimurium* LT2 genome made using *Xba*I analysis. *J. Bacteriol.* 174:1662–1672.

Liu, S.-L., A. Hessel, and K.E. Sanderson. 1993a. The *Xba*I-*Bln* I-*Ceu*I genomic cleavage map of *Salmonella typhimurium* LT2 determined by double digestion, end-labelling, and pulsed-field gel electrophoresis. *J. Bacteriol.* 175:4104–4120.

Liu, S.-L., A. Hessel, and K.E. Sanderson. 1993b. Genomic mapping with I-*Ceu*I, an intron-encoded endonuclease, specific for genes for ribosomal RNA, in *Salmonella* spp., *Escherichia coli*, and other bacteria. *Proc. Natl. Acad. Sci. USA* 90:6874–6878.

Marshall, P., and C. Lemieux. 1991. Cleavage pattern of the homing endonuclease encoded by the fifth intron in the chloroplast subunit rRNA-encoding gene of *Chlamydomonas eugamatos. Gene* 104:1241–1245.

Noller, H.F., and M. Nomura. 1987. Ribosomes. In F.C. Neidhardt, J.L. Ingraham, K.B. Low, B. Magasanik, M. Schaechter, and H.E. Umbarger, eds, Escherichia coli *and* Salmonella typhimurium: *Cellular and Molecular Biology*, pp. 104–125. American Society for Microbiology, Washington, D.C.

Sanderson, K.E., A. Hessel, and K.E. Rudd. 1995. Genetic map of *Salmonella typhimurium*, edition VIII. *Microbiol. Rev.* 59:241–303.

Wong, K.K., and M. McClelland. 1992. A *Bln*I restriction map of the *Salmonella typhimurium* LT2 genome. *J. Bacteriol.* 174:1656–1661.

Wong, K.K., R.M. Wong, and K.E. Rudd, *et al.* 1994. High resolution restriction map of a 240-kilobase region spanning 91 to 96 minutes on the *Salmonella typhimurium* chromosome. *J. Bacteriol.* 176:5729–5734.

79

Serpulina hyodysenteriae B78[T]

Richard L. Zuerner and Thaddeus B. Stanton

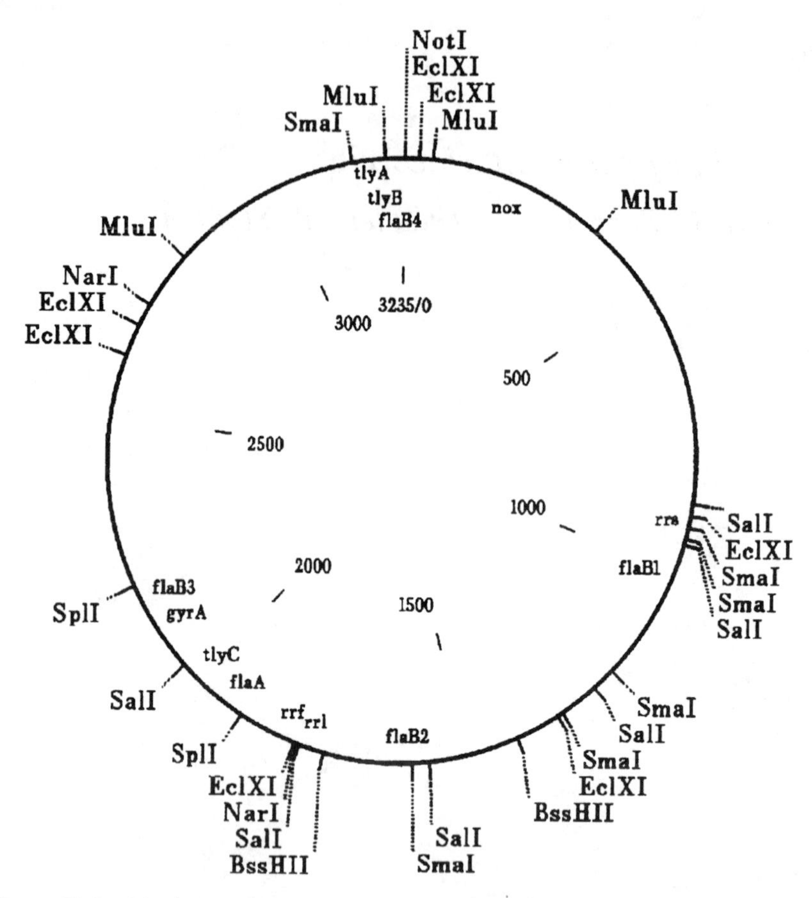

Figure 79-1. Physical and Genetic Map of *Serpulina hyodysenteriae* strain B78^{T}. The genome measures about 3235 Kb in size. Markers at 500 Kb intervals are shown on the inside of the map. The map as shown uses the unique *Not*I site as the landmark for the beginning and end of the map. All *Bss*HII, *Ecl*XI, *Not*I, *Sal*I, and *Sma*I sites are shown. Approximate locations for some restriction sites recognized by *Mlu*I, *Nar*I, and *Spl*I are shown. The putative location of the replication origin is based on the *gyrA* location and is between 2000 Kb and 2200 Kb from the *Not*I site. Ribosomal RNA genes *rrl* and *rrf* map together, but are about 860 Kb from *rrs*. Five periplasmic flagella filament protein genes have been mapped: *flaA*, which encodes a sheath protein; and *flaB*1, *flaB*2, *flaB*3, and *flaB*4, which encode core flagella proteins. Four genes thought to be important during growth in an animal host have been mapped; *nox*, which encodes NADH-oxidase; and *tlyA, tlyB*, and *tlyC* which encode genes conferring hemolytic activity to *Escherichia coli.*

80

Staphylococcus aureus

John J. Iandolo and George C. Stewart

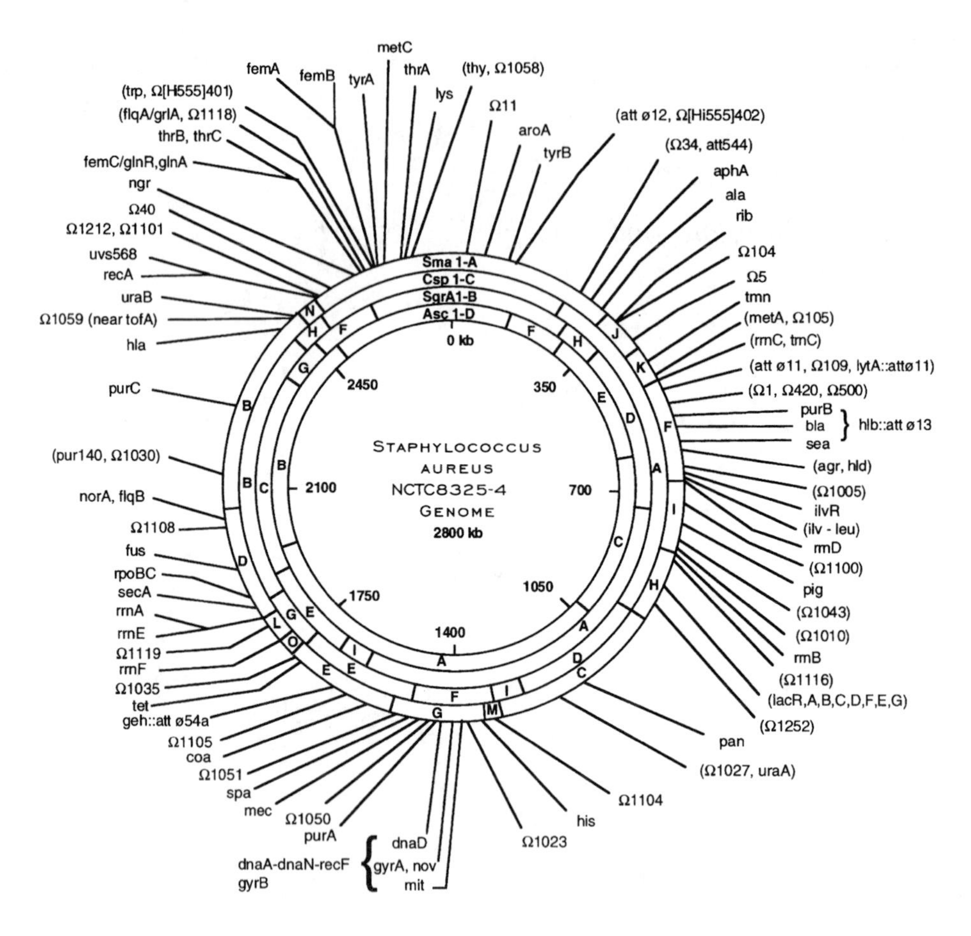

Figure 80-1. Aligned physical and genetic map of the chromosome of *Staphylococcus aureus* NCTC 8325. Information concerning individual markers is summarized in Table 1. The order of the majority of the markers shown is based on genetic analysis (see references 57–68) and DNA hybridization of *SmaI, CspI, SgrAI* and *AscI* restriction endonuclease digestions of the chromosome. The fragments were resolved by pulsed-field agarose gel electrophoresis and hybridized with appropriate probes (60). *SmaI* fragment O was identified and mapped by Wada *et al.* (95). Among the markers shown in Fig 1, *lys, thy, Ω11, Ω402, aphA, attø11, Ω420, Ω500, purB, bla, ilvR, leu, pig, uraA, his, mit, nov, dnaD, purA, fus, pur-140, tofA, ngr, thrC, flqA, Ω401, tryA,* and *metC* have not been physically mapped and their map locations are based only on multifactorial transformation analyses. The genomes of the prophages ø 12 and ø 13 contain an internal *SgrAI* restriction endonuclease recognition site. As a result, ø 12 confers an additional restriction site in *SgrAI*-B and ø 13 in *SgrAI*-D. The ø 13 site is known to be between **Ω*1*** and *agr* on *SgrAI*-D (62). In addition, each of the ribosomal RNA operons (*rrnA-F*) contains an internal *SmaI* recognition site which defines the junctions between *SmaI* L/D, H/I, K/F, and L/O). The orientation of the markers in parentheses is not known relative to the

remainder of the map. The entire chromosome is about 2,800 kilobase pairs (kb) and the individual fragments are drawn approximately to scale. Approximately 125–150 kb represents the three known prophages (ø 11, ø 12 and ø 13) that lysogenize the NCTC 8325 genome. There are several small DNA fragments that have not yet been placed on the physical map. These include *SgrAI* -G, -J, and -K representing about 300 kb, at least 2 *CspI* fragments representing about 100 kb and 2 *SmaI* fragments representing about 25 kb.

References

1. Asheshov, E. H. 1975. *J. Gen. Microbiol.* 88:132–140.
2. Bannantine, J. P. 1991. M.S. thesis. Iowa State University, Ames.
3. Barnes, I. J., A. Bondi, and K. S. Fuscaldo. 1971. *J. Bacteriol.* 105:553–555.
4. Barnes, I. J., A. Bondi, and A. G. Moat. 1969. *J. Bacteriol.* 99:169–174.
5. Behnke, D., and D. Gerlach. 1987. *Mol. Gen. Genet.* 210:528–534.
6. Berger-Bachi, B. 1983. *J. Bacteriol.* 154:479–487.
7. Berger-Bachi, B. 1995. Personal Communication.
8. Berger-Bachi, B. 1983. *J. Bacteriol.* 154:533–535.
9. Berger-Bachi, B., Barberis-Maino, A. Strassle, and F. H. Kayser. 1989. *Mol. Gen. Genet.* 219:263–269.
10. Berger-Bachi, B., and M. L. Kohler. 1983. *FEMS Microbiol. Lett.* 20:305–309.
11. Berger-Bachi, B., A. Strassle, J. G. Gustafson and F. H. Kayser. 1992. *Antimicrob. Agents and Chemother.* 36:1367–1373.
12. Betley, M. J., and J. J. Mekalanos. 1985. *Science.* 229:185–187.
13. Betley, M. J., and J. J. Mekalanos. 1988. *J. Bacteriol.* 170:34–41.
14. Betley, M. J., P. M. Schlievert, M. S. Bergdoll, G. A. Bohach, J. J. Iandolo, S. A. Khan, P. A. Pattee, and R. R. Reiser. 1990. *ASM News* 56:182. (Letter.)
15. Breidt, F., Jr., W. Hengstenburg, U. Finkeldei, and G. C. Stewart. 1987. *J. Biol. Chem.* 262:16444–16449.
16. Brown, D. R., and P. A. Pattee. 1980. *Infect. Immun.* 30:36–42.

16a. Cheung, A. L. 1995. Personal Communication.

17. Chu, M. C., B. N. Kreiswirth, P. A. Pattee, R. P. Novick, M. E. Melish, and J. F. James. 1988. *Infect. Immun.* 56:2702–2708.
18. Coleman, D. C., J. P. Arbuthnott, H. M. Pomeroy, and T. H. Birkbeck. 1986. *Microb. Pathogen.* 1:549–564.
19. Coleman, D. C., D. S. Sullivan, R. J. Russell, J. P. Arbuthnott, B. F. Carey, and H. M. Pomeroy. 1989. *J. Gen. Microbiol.* 135:1679–1697.
20. Cooney, J., M. Mulvey, J. P. Arbuthnott, and T. J. Foster. 1988. *J. Gen. Microbiol.* 134:2179–2188.
21. Dornbusch, K., H. O. Hallander, and F. Lofquist. 1969. *J. Bacteriol.* 98:351–358.
22. El Solh, N., N. Moreau, and S. D. Ehrlich. 1986. *Plasmid.* 15:104–118.

23. Fairweather, N., S. Kennedy, T. J. Foster, M. Kehoe, and G. Dougan. 1983. *Infect. Immun.* 41:1112–1117.

24. Goering, R. V., and P. A. Pattee. 1971. *J. Bacteriol.* 106:157–161.

25. Good, G. M., and P. A. Pattee. 1970. *J. Bacteriol.* 104:1401–1403.

26. Green, C. J. and B. S. Vold. 1993. *J. Bacteriol.* 175:5091–5096.

27. Gustafson, J., A. Strassle, H. Hachler, F. H. Kayser and B. Berger-Bachi. 1994. *J. Bacteriol.* 176:1460–1467.

28. Hooper, D. C. Personal Communication.

29. Iandolo, J. J. Unpublished data.

30. Janzon, L., S. Lofdahl, and S. Arvidson. 1989. *Mol. Gen. Genet.* 219:480–485.

31. Jayaswal, R. K., Y.-I. Lee, and B. J. Wilkinson. 1990. *J. Bacteriol.* 172:5783–5788.

32. Jones, J. M., S. C. Yost, and P. A. Pattee. 1987. *J. Bacteriol.* 169:2121–2131.

33. Kaida, S., T. Miyata, Y. Yoshizawa, H. Igarashi, and S. Iwanaga. 1989. *Nucl. Acids Res.* 17:8871.

34. Kehoe, M., J. Duncan, T. Foster, N. Fairweather, and G. Dougan. 1983. *Infect. Immun.* 41:1105–1111.

35. Kloos, W. E., and P. A. Pattee. 1965. *J. Gen. Microbiol.* 39:185–194.

36. Kloos, W. E., and P. A. Pattee. 1965. *J. Gen. Microbiol.* 39:195–207.

37. Kuhl, S. A., P. A. Pattee, and J. N. Baldwin. 1978. *J. Bacteriol.* 135:460–465.

38. Lee, C. Y., and J. J. Iandolo. 1986. *J. Bacteriol.* 166:385–391.

39. Lee, C. Y., and J. J. Iandolo. 1988. *J. Bacteriol.* 170:2409–2411.

40. Limpa-Amara, Y. 1978. M.S. thesis, Iowa State University, Ames.

41. Lindberg, M., J.-E. Sjostrom, and T. Johansson. 1972. *J. Bacteriol.* 109:844–847.

42. Lofdahl, S., B. Guss, M. Uhlen, L. Philipson, and M. Lindberg. 1983. *Proc. Natl. Acad. Sci. USA* 80:697–701.

43. Luchansky, J.B., and P. A. Pattee. 1984. *J. Bacteriol.* 159:894–899.

44. Mahairas, G. G., B. R. Lyon, R. A. Skurray, and P. A. Pattee. 1989. *J. Bacteriol.* 171:3968–3972.

45. Mallonee, D. H., B. A. Glatz, and P. A. Pattee. 1982. *Appl. Environ. Microbiol.* 43:397–402.

46. Martin, S. M., S. C. Shoham, M. Alsup, and M. Rogolsky. 1980. *Infect. Immun.* 27:532–541.

47. McClatchy, J. K., and E. D. Rosenblum. 1966. *J. Bacteriol.* 92:575–579.

48. McClatchy, J. K., and E. D. Rosenblum. 1966. *J. Bacteriol.* 92:580–583.

49. Morrow, T. O., and S. A. Harmon. 1979. *J. Bacteriol.* 137:374–383.

50. Ng, E. Y., M. Trucksis and D. C. Hooper. 1994. Antimicrob. Agents and Chemother. 38:1345–1355.

51. Nieuwlandt, D. T., and P. A. Pattee. 1989. *J. Bacteriol.* 171:4906–4913.

52. Novick, R. P., E. Edelman, M. D. Schwesinger, A. D. Gruss, E. C. Swanson, and P. A. Pattee. 1979. *Proc. Natl. Acad. Sci. USA.* 76:400–404.

53. Novick, R. P., S. A. Khan, E. Murphy, S. Iordanescu, I. Edelman, J. Krolewski, and M. Rush. 1981. *Cold Spring Harbor Symp. Quant. Biol.* 45:67–76.

54. O'Reilly, M., J. C. S. de Azavedo, S. Kennedy, and T. J. Foster. 1986. *Microb. Pathogen.* 1:125–138.

55. O'Toole, P. W., and T. J. Foster. 1986. *Microb. Pathogen.* 1:538–594.

56. Patel, A. H., T. J. Foster, and P. A. Pattee. 1989. *J. Gen. Microbiol.* 135:1799–1807.

57. Pattee, P. A. 1976. *J. Bacteriol.* 127:1167–1172.

58. Pattee, P. A. 1981. *J. Bacteriol.* 145:479–488.

59. Pattee, P. A. 1986. *Infect. Immun.* 54:593–596.

60. Pattee, P. A., p. 163–169, 1990. *In* K. Drlica and M. Riley (ed.), *The Bacterial Chromosome.* Amer. Soc. for Microbiol. Washington, D.C.

61. Pattee, P. A., p. 489–496, 1993. *In* A. L. Sonenshein J. Hoch and R. Losick (eds.), *Bacillus subtilis and other gram-positive bacteria.* Amer. Soc. for Microbiol. Washington, D.C.

62. Pattee, P. A. Unpublished data.

63. Pattee, P. A., and J. N. Baldwin. 1961. *J. Bacteriol.* 82:875–881.

64. Pattee, P. A., and B. A. Glatz. 1980. *Appl. Environ. Microbiol.* 39:186–193.

65. Pattee, P. A., and D. S. Neveln. 1975. *J. Bacteriol.* 124:201–211.

66. Pattee, P. A., C. J. Schroeder and M. L. Stahl. 1983. Iowa State J. Res. 58:175–180.

67. Pattee, P. A., T. Schutzbank, H. D. Kay, and M. H. Laughlin. 1974. *Ann. N. Y. Acad. Sci.* 236:175–186.

68. Pattee, P. A., N. E. Thompson, D. Haubrich, and R. P. Novick. 1977. *Plasmid.* 1:38–51.

69. Peng, H.-L., R. P. Novick, B. Kreiswirth, J. Kornblum, and P. Schlievert. 1988. *J. Bacteriol.* 170:4365–4372.

70. Phillips, S., and R. P. Novick. 1979. *Nature (London).* 278:476–478.

71. Phonimdaeng, P., M. O'Reilly, P. W. O'Toole, and T. J. Foster. 1988. *J. Gen. Microbiol.* 134:75–83.

72. Poston, S. M. 1966. *Nature (London).* 210:802–804.

73. Proctor, A. R., and W. E. Kloos. 1970. *J. Gen. Microbiol.* 64:319–327.

74. Recsei, P. A., B. Kreiswirth, M. O'Rilly, S. Schlievert, A. Gruss, and R. P. Novick. 1986. *Mol. Gen. Genet.* 202:58–61.

75. Ritz, H. L., and J. N. Baldwin. 1962. *Proc. Soc. Exp. Biol. Med.* 110:667–671.

76. Rosey, E. L., B. Oskouian and G. C. Stewart. 1991. *J. Bacteriol.* 173:5992–5998.

77. Rudin, L., J.-E. Sjostrom, M. Lindberg, and L. Philipson. 1974. *J. Bacteriol.* 118:155–164.

78. Sako, T., S. Sawaki, T. Sakurai, S. Ito, Y. Yoshizawa, and I. Kondo. 1983. *Mol. Gen. Genet.* 190:271–277.

79. Schaefler, S., W. Francois, and C. L. Ruby. 1976. *Antimicrob. Agents Chemother.* 9:600–613.

80. Schroeder, C. J., and P. A. Pattee. 1984. *J. Bacteriol.* 157:533–537.

81. Shafer, W. M., and J. J. Iandolo. 1978. *Infect. Immun.* 20:273–278.

82. Shafer, W. M., and J. J. Iandolo. 1978. *Appl. Environ. Microbiol.* 36:389–391.

83. Shortle, D. 1983. *Gene.* 22:181–189.

84. Shuttleworth, H. L., C. J. Duggleby, S. A. Jones, T. Atkinson, and N. P. Minton. 1987. *Gene.* 58:283–295.

85. Smith, C. D., and P. A. Pattee. 1967. *J. Bacteriol.* 93:1832–1838.

86. Sreedharan, S., M. Oram, B. Jensen, L. R. Peterson, and L. M. Fisher. 1990. *J. Bacteriol.* 172:7260–7262.

87. Stahl, M. L., and P. A. Pattee. 1983. *J. Bacteriol.* 154:395–405.

88. Stahl, M. L., and P. A. Pattee. 1983. *J. Bacteriol.* 154:406–412.

89. Stewart, G. C., and E. D. Rosenblum. 1980. *J. Bacteriol.* 144:1200–1202.

90. Stewart, G. C., and E. D. Rosenblum. 1980. *Antimicrob. Agents Chemother.* 18:424–432.

91. Tam, J. E. 1985. Ph.D. thesis. Iowa State University, Ames.

92. Tam, J. E., and P. A. Pattee. 1986. *J. Bacteriol.* 168:708–714.

93. Thomas, C. M., and K. G. H. Dyke. 1978. *J. Gen. Microbiol.* 106:41–47.

94. Trucksis, M., J. S. Wolfson, and D. C. Hooper. 1991. *J. Bacteriol.* 173:5854–5860.

95. Wada, A., H. Ohta, K. Kulthanen and K. Hiramatsu. 1993. *J. Bacteriol.* 175:7483–7487.

96. Wang, X., B. J. Wilkinson, and R. K. Jayaswal. 1991. *Gene.* 102:105–109.

97. Wyman, L., R. V. Goering, and R. P. Novick. 1974. *Genetics.* 76:681–702.

98. Yoshida, H., M. Bogaki, S. Nakamura, K. Ubukata, and M. Konno. 1990. *J. Bacteriol.* 172:6942–6949.

99. Yost, S. C., J. M. Jones, and P. A. Pattee. 1988. *Plasmid.* 19:13–20.

Gene symbol[a]	Phenotype	Map location[b]	Reference(s)
agr[c] *(hla)*	Accessory gene regulator of several exoproteins and toxins	F	16,45,47,48,69,74
ala[c]	L-Alanine requirement	A	46,58,64
aphA	Aminoglycoside phosphotransferase (APH3'-III)	A	22
attϕ*11*	Prophage ϕ11 integration site	F	68
attϕ*12*	Prophage ϕ12 integration site	A	40
attϕ*13*	Prophage ϕ13 integration site (in *hlb*)	F	18,19
attϕ*L54a*	Prophage ϕL54a integration site (in *geh*)	E	38,39,62
att554	Primary integration site for Tn*554*	A	53,70
bla (pen)	β-Lactamase production	F	68,72
coa	Coagulase	E	33,62,71
dna	Temperature-sensitive DNA replication	G	62,93
ermB	Impaired erythromycin resistance by Tn*551*	in Tn*551*	66
femAB[c] *(Ω2003)*	Factor essential for expression of methicillin resistance; see Ω2004	A	6-10
femC	Reduction of methicillin resistance levels	A	27
gluR	Glutamine synthetase repressor	A	27
gluA	Glutamine synthetase	A	27
flqA	Fluoroquinolone resistance	A	49
flqB	Quinolone resistance	D	50
fus	Fusidic acid resistance	D	87,88

geh[d]	Glycerol ester hydrolase (lipase)	E	38,39,62
gyrA	DNA gyrase; ciprofloxacin resistance	G	86,94
his	L-Histidine requirement (*hisEABCDG*)	G or M	35,36,46,65
hla (hly)	α-Toxin structural gene	B	23,34,54,59
hlb[d]	β-Hemolysin	F	18,19,62
hld	Delta hemolysin	F	30
hlg	Gamma hemolysin	C	20,62
ilv-leu[c]	L-Isoleucine, L-valine, L-leucine requirement (*ilvABCD-leuABCD*)	F	45,65,67,85
ilvR	Resistance to D-leucine	F	57,67
lacR	Lactose repressor	H	15,76
lacA	Galactose-6-phosphate isomerase subunit	H	15,76
lacB	Galactose-6-phosphate isomerase subunit	H	15,76
lacC	Tagatose-6-phosphate kinase	H	15,76
lacD	Tagatose-1,6-bisphosphate aldolase	H	15,76
lacE	Enzyme IIB of Lactose PTS	H	15,76
lacF	Enzyme IIA of Lactose PTS	H	15,76
lacG	Phosphobetagalactosidase	H	15,76
lip[c]	Lipoic acid requirement	A	62
lys[c]	L-Lysine requirement (*lysOABFG*)	A	3,4,65
lytA	*N*-Acetylmuramyl-L-alanine amidase (peptidoglycan hydrolase)	F	31,62,96
mec	Methicillin resistance	G	21,37,89,90
metA	L-Methionine requirement	K	44
MetC[c]	L-Methionine requirement; β-cystathionase	A	46,80
mit	Enhanced sensitivity to mitomycin C, nitrosoguanidine, and UV light	G	62,91,97
mdr	Multidrug resistance transporter	D	7

Gene symbol[a]	Phenotype	Map location[b]	Reference(s)
ngr	Apurinic endonuclease deficiency	A	92
norA	Hydrophilic quinolone resistance	D	94,98
nov	Novobiocin resistance	G	46,63,65
pan[c]	Pantothenate requirement	C	62
pbp4	Penicillin binding protein 4	D	7
pig	Golden-yellow pigment deficiency	I	57
purA	Adenine requirement	G	65
purB[c]	Adenine + guanine requirement	F	65
purC[c]	Purine requirement	B	88
purD[c]	Guanine requirement	B	62
pur-140	Purine requirement	B	43
recA	Homologous recombination deficiency	N	97
rib[c]	Riboflavin requirement	J	58,64
rif	Rifampin resistance	D	46,49
rpoB	RNA polymerase β-subunit	D	49
rrnA	rRNA operon	D/L junction	95
rrnB	rRNA operon	H/I junction	95
rrnC	rRNA operon	K/F junction	95
rrnD	rRNA operon	F/I junction	95
rrnE	rRNA operon	D/L junction	95
rrnF	rRNA operon	L/O junction	95

sak[e]	Staphylokinase production	F	5,19,78
secA	Secretion of proteins	D	29
spa	Staphylococcal protein A	G	42,56,84
sar A sar B	Exoprotein regulatory locus	D	16a
tagD	Glycerol-3-phosphate cytidyltransferase	D	7
tet	Tetracycline resistance	E	1,57,87,88
thrA[c]	L-Lysine + L-methionine + L-threonine requirement; failure to convert aspartate to aspartic β-semialdehyde	A	58,64,80
thrB[c]	L-Threonine requirement; homoserine kinase deficiency	A	46,58,65,80
thrC[c]	L-Threonine requirement; threonine synthetase deficiency	A	80
thy	Thymine requirement	A	41,46,65
tmn	Tetracycline and minocycline resistance	K	1,57,79
tofA	Temperature-sensitive osmotically remedial cell wall synthesis; D-glutamate addition defect	B	25,51
trnC	tRNA genes (27 tandem genes)	junction K/F	26,95
trp[c]	L-Tryptophan requirement (*trpABFCDE*)	A	46,65,73,75,80
tst	Toxic shock syndrome toxin 1 structural gene (see Ω401 and Ω402)	A (2 loci)	19
tyrA[c]	L-Tyrosine requirement	A	58,80
tyrB[c]	L-Tyrosine requirement	A	58,80
uraA	Uracil requirement	C	46,65
uraB[c]	Uracil requirement	B	58
uvr (uvs)	Enhanced sensitivity to UV light (recA allele)	N[f]	24,62,97
Ω1-Ω100; Ω1000-Ω1099	Silent insertions of Tn*551*	Various	43,52,58

Gene symbol[a]	Phenotype	Map location[b]	Reference(s)
Ω401 and Ω402	Insertion sites of Hi*555*[g], the heterologous insertion carrying *tst;* Ω401 maps immediately adjacent to *trp* and causes a Trp^- phenotype; Ω402 maps near *tyrB*	A	17
Ω420 (Ω42)	Insertion site of pI258	F	52,68
Ω500 (Ω50)	Insertion site of pI258	F	52,68
Ω1100-Ω1199	Silent insertions of Tn*916*	Various	32,99
Ω101-Ω120; Ω1200-Ω1299	Silent insertions of Tn*4001*	Various	44
Ω2004	Insertion site of Tn*551* that impairs Mec; may affect penicillin-binding proteins; see *femA*	A	6-10

[a]Gene designations in parentheses are old designations.
[b]Refers to specific *Sma*I restriction fragments in Fig. 1. Genes identified on a specific fragment but not shown in Fig. 1 have only been mapped by PFGE and DNA hybridization. Other genes (see Fig. 1 legend) have only been mapped by multifactorial transformation analyses.
[c]Identifies a phenotype for which insertional inactivation with Tn*551*, Tn*917*, and/or Tn*4001* is known.
[d]Controlled by negative phage conversion.
[e]Controlled by positive phage conversion.
[f]Chromosomal location of *uvr* is based on physical and genetic map data for Ω1073 and Ω1074, which exhibit greater than 50% cotransduction with *uvr* and recA.
[g]Provisional designation for an element that exhibits some of the characteristics of a transposable element, but for which definitive evidence is lacking.

81

Streptococcus pneumoniae R6

Anne-Marie Gasc and Sol H. Goodgal

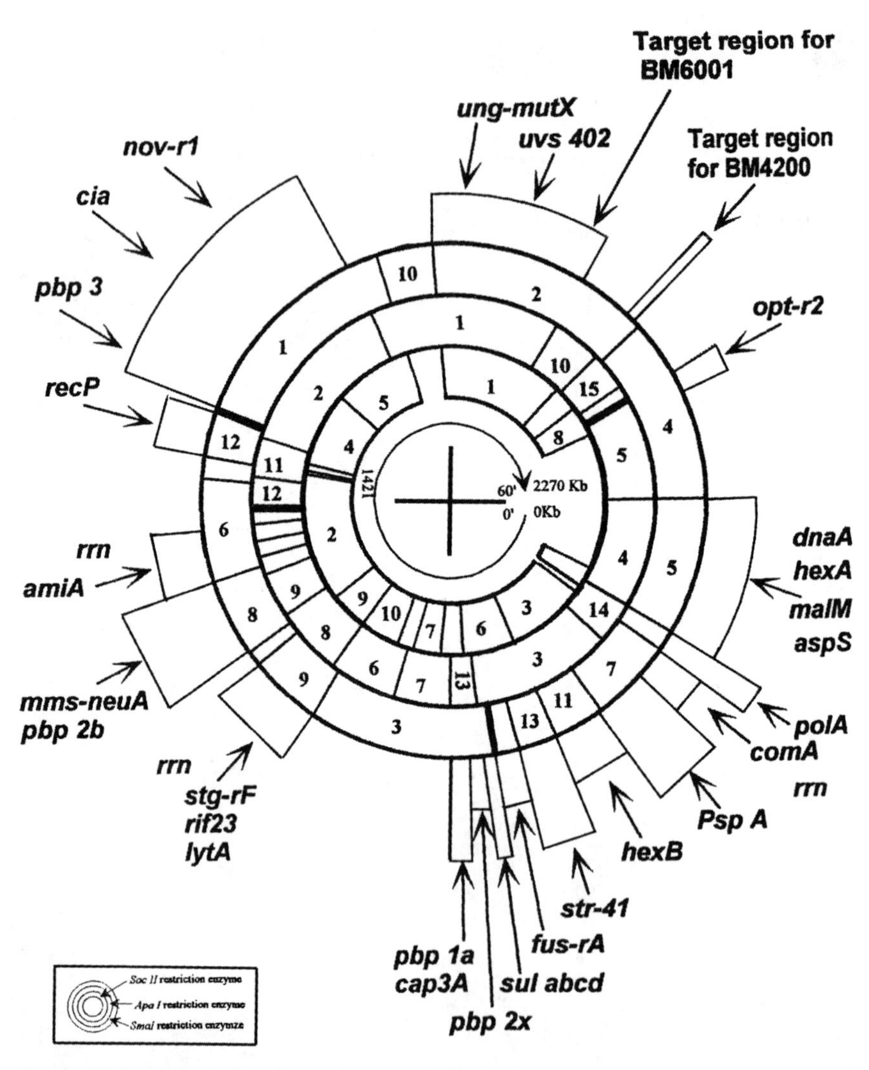

Figure 81-1. Physical and genetic map of *Streptococcus pneumoniae* R6 strain made by using *Apa*I *Sma*I and *Sac*II restriction enzyme. (Gasc, Kauc et al 1991). genes included on the map:

*hex*A reference (Martin, Prats *et al.* 1985)
*mal*M reference (Stassi, Lopez *et al.* 1981)
*dna*A reference (unpublished results)
*asp*S reference (Martin, Humbert *et al.* 1992)
polA reference (Martinez, lopez *et al.* 1986)
comAB reference (Chandler and Morrison 1987–1988)
pspA reference (McDaniel, Yother *et al.* 1987)

*hex*B reference (Prats, Martin *et al.* 1985)
str-41 reference (Tiraby and Fox 1973)
*fus-r*A reference (Tiraby and Fox 1973)
sul abcd reference (Lopez, Espinosa *et al.* 1984)
pbp 2x reference (Laible, Hakenbeck *et al.* 1989)
pbp 1a reference (Martin; Briese *et al.* 1992)
cap3A reference (Arrecubieta, Lopez *et al.* 1994)
rif23 reference (Tiraby and Fox 1973)
*stg r*F reference (Tiraby and Fox 1973)
*lyt*A reference (Garcia, Garcia *et al.* 1986)
pbp 2b reference (Dowson, Hutchison *et al.* 1989)
ami reference (Alloing, Trombe *et al.* 1990)
mms reference (Martin, Humbert *et al.* 1992)
*neu*A reference (Camara, Mitchell *et al.* 1991)
*rec*P reference (Rhee, Morrison 1988)
cia reference (Guenzi, Gasc *et al.* 1994)
*pbp*3 reference (Schuster, Dobrinski *et al.* 1990)
nov-r1 reference (Tiraby and Fox 1973)
ung reference (Mejean, Rives *et al.* 1990)
*mut*X reference (Mejean, Salles *et al.* 1994)
uvs 402 reference (Sicard and Estevenon 1990)
Target region for Ω BM6001 reference (Vijayakumar, Priebe *et al.* 1986)
Target region for Ω BM4200 reference (Guild, Hazum *et al.*)
opt-r2 reference (Tiraby and Fox 1973)

82

Sulfolobus tokodaii 7

Akihiko Yamagishi and Tairo Oshima

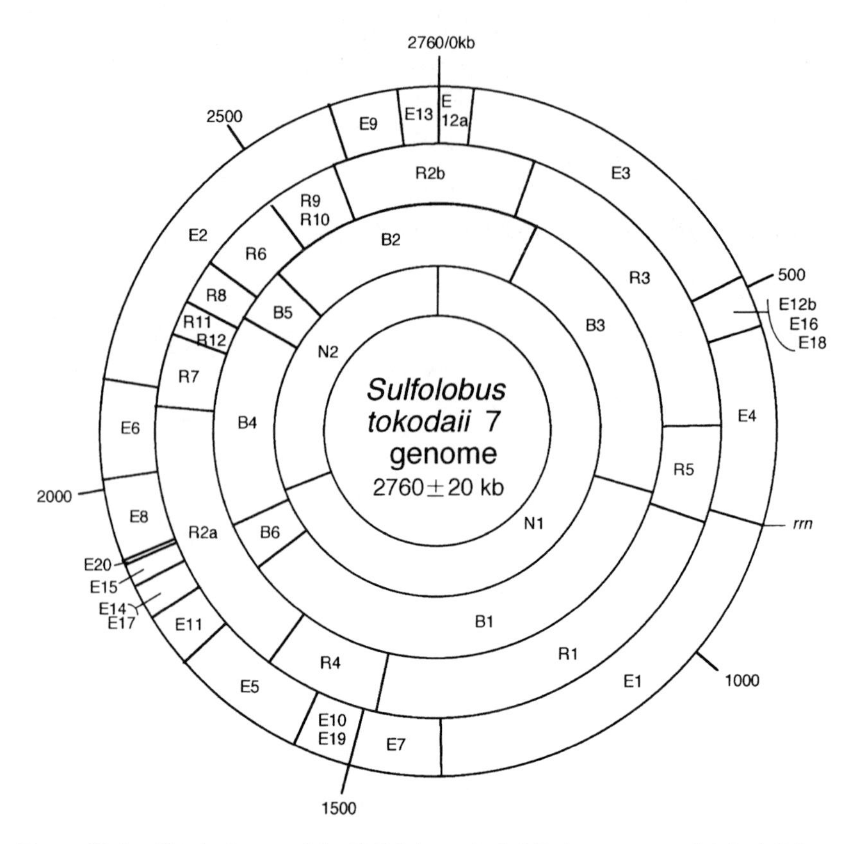

Figure 82-1. Physical map of the *Sulfolobus tokodaii* 7 chromosome. *S. tokodaii* 7 was isolated from a hot-acid spring in Japan (Inatomi *et al.*, 1983). The strain was originally identified as *S. acidocaldarius* based on the bacteriological criteria. Very recently, we have determined 16S rRNA sequence of the strain. The sequence was significantly different from those of other *Sulfolobus* species (Kurosawa *et al.* 1995; see also Chapt. 44). Therefore the strain 7 was named as *S. tokodaii*. The detailed description of the species will appear elsewhere.

The physical map of the *S. tokodaii* chromosome was constructed from the size of and hybridization data for the restriction fragments of the chromosomal DNA (Kondo *et al.* 1993). The map of the *S. tokodaii* 7 chromosome is circular (Yamagishi and Oshima 1990) and 2760 + 20 kb (Kondo *et al.* 1993). Fragment names are followings: E1–20, *Eag*I fragments; R1–12, *Rsr*II fragments; B1–6, *Bss*HII fragments, N1 and 2, *Not*I fragments. Although the order of a few fragments, i.e. R9/10, R11/12, E10/19, E12b/16/17 and E14/17, are still unsure, the position of each pair or triplet was uniquely determined.

rrn: ribosomal RNA operon. Only 16S and 23S rRNA genes are included (unpublished results).

Reader should refer to refs. (Durovic *et al.* 1994; Kurosawa *et al.* 1995) for the current understanding of the species in the genus *Sulfolobus*.

References

Durovic, P., U. Kutay, C. Schleper, and P. P. Dennis. 1994. Strain identification and 5S r RNA gene characterization of the hyperthermophilic archaebacterium *Sulfolobus acidocaldarius. J. Bacteriol.* 176(2):514–517.

Inatomi, K., M. Ohba, and T. Oshima. 1983. Chemical properties of proteinaceus cell wall from an acido-thermophile, *Sulfolobus acidocaldarius. Chemistry Lett.* 1983:1191–1194.

Kondo, S., A. Yamagishi, and T. Oshima. 1993. A physical map of the sulfur-dependent archaebacterium *Sulfolobus acidocaldarius 7 chromosome. J. Bacteriol.* 175(5):1532–1536.

Kurosawa, N., K. Ohkura, T. Horiuchi, and Y. H. Itoh. 1995. Characterization of the 16S ribosomal RNA genes and phylogenetic relationships of sulfur-dependent thermoacidophilic archaebacteria. *J. Gen. Appl. Microbiol.* 41:75–81.

Yamagishi, A., and T. Oshima. 1990. Circular chromosomal DNA in the sulfur-dependent archaebacterium *Sulfolobus acidocaldarius. Nuc. Acids Res.* 18(5):1133–1136.

83

Synechococcus PCC 7002

William R. Widger, Xiangxing Chen, and Hrissi Samartzidou

Introduction

The best established physical evidence of early life forms are stromatolites which have been dated to 3.3–3.5 × 10^9 (Awramik *et al.*, 1983; Schopf and Walter, 1987; Schopf, 1993). Although the phototrophic obligatory anaerobes belonging to the genus *Chloroflexus* can make laminated microbial mats similar to that seen in micro fossil records (Awramik *et al.* 1983; Giovannoni *et al.*, 1987), these early stromatolites are generally thought to be in large part the result of the activity of cyanobacteria. Recent micro fossil evidence (Schopf, 1993) supports this view. Molecular evidence based on ribosomal RNA sequences does not however place cyanobacteria among the earliest branchings of life but rather shows them emerging somewhat later as one of approximately twelve distinct phyla of the Domain Bacteria (Giovannoni *et al.*, 1988). Regardless of the precise time of arrival of the cyanobacteria they are of special interest to the study of early life (Lazcano and Miller, 1994) because they are the most important group of extant microorganisms in terms of relating data from the geological and molecular approaches to the study of early life.

Cyanobacteria encompass a large and diverse sub-group of Gram-negative prokaryotes (Rippka *et al.*, 1979) which maintain a distinguishing feature, the capacity for photosynthetic oxygen evolution. Comparative analysis have lead to the recognition of 22 genera which were placed into five sections with unique structural and developmental patterns (Rippka *et al.*, 1979). Detailed phylogenetic history is based on comparative studies of 16S rRNA (Giovannoni *et al.*, 1988). While unicellular and nonheterocyst forming filamentous cyanobacteria are scattered through the 16S phylogenetic tree, other morphological divisions appeared to be consistent with the 16S tree. Baeocyst formers, heterocyst forming and branched heterocyst formers cluster as distinct groups (*see also* Chapter 16).

The cyanobacterium, *Synechococcus* PCC 7002, has previously been used as a model organism for the study of phototropic behavior (Rhiel *et al.*, 1992; Schluchter and Bryant, 1992; Schluchter *et al.*, 1993). This organism grows freely in liquid culture without forming mats and is very susceptible to lysozyme digestion. It can be easily transformed (Buzby *et al.*, 1983; Essich *et al.*, 1990; Stevens and Porter, 1980) thus, homologous recombination with linearized DNA can be used to generate gene insertion or substitution mutants (Tan *et al.*, 1994) while plasmid transformation can be done with pAQ19, a small chimeric plasmid, compatible of growth in *E. coli* and PCC 7002 (Buzby *et al.*, 1983). However, plasmid recombination with genomic DNA presents problems with biphasic plasmid transformation (Porter *et al.*, 1986). *Synechococcus* PCC 7002's phylogenetic position has been established (Widger & Fox, unpublished data) by 16S rRNA sequencing where it clusters in a group near *Spirulina* PCC 6313, *Gloeocapsa* PCC 73106 and *Synechocystis* PCC 6308. It therefore is from a genetic perspective a very typical cyanobacterium in which the genomic map should be a useful contribution to studying photosynthetic organisms and early evolution.

Results

Physical Map

PCC 7002 was chosen as a target organism for genome mapping because of its small genome, 2.7×10^6 bp (Herdman *et al.*, 1979), for its easily isolated genomic DNA and its lack of small repeated sequences. A physical genome map, Figure 83-1, has been constructed with 21 *Not*I and 36 *Sal*I fragments (Table 83-1) separated using pulsed field electrophoresis techniques, (Chen and Widger, 1993). Several restriction fragments generated from *Sfi*I and *Asc*I were used to support fragment locations and assignments.

The map was assembled using linking or connecting clones for both *Not*I and *Sal*I fragments from a lambda DASH library, hybridization of probes from known genes or fragments derived from this library and hybridization of *Not*I fragments to blots of *Not*I, *Sal*I and *Not*I + *Sal*I doubly digested genomic DNA. LInking clones were screened using genomic DNA digested with *Not*I and end-labeling using α-[^{32}P]-dGTP and α-[^{32}P]-dCTP followed by digestion with *Rsa*I.

This DNA was used to probe a gridded library consisting of 1540 individual lambda recombinants. Several lambda clones were identified as having internal *Not*I sites and were used for map construction. *Sal*I linking clones were isolated by ligation of *Sal*I-*Eco*RI fragments into a pUC19 vector and using recombinants to screen the lambda library. However, linking clones isolated from the lambda library were only marginally successful in completing the physical map and were used primarily as confirmation of hybridization data. Our recent efforts have been targeted towards the completion of an overlapping cosmid contig library. *Not*I sites within cosmids are unique and have been identified by digestion with

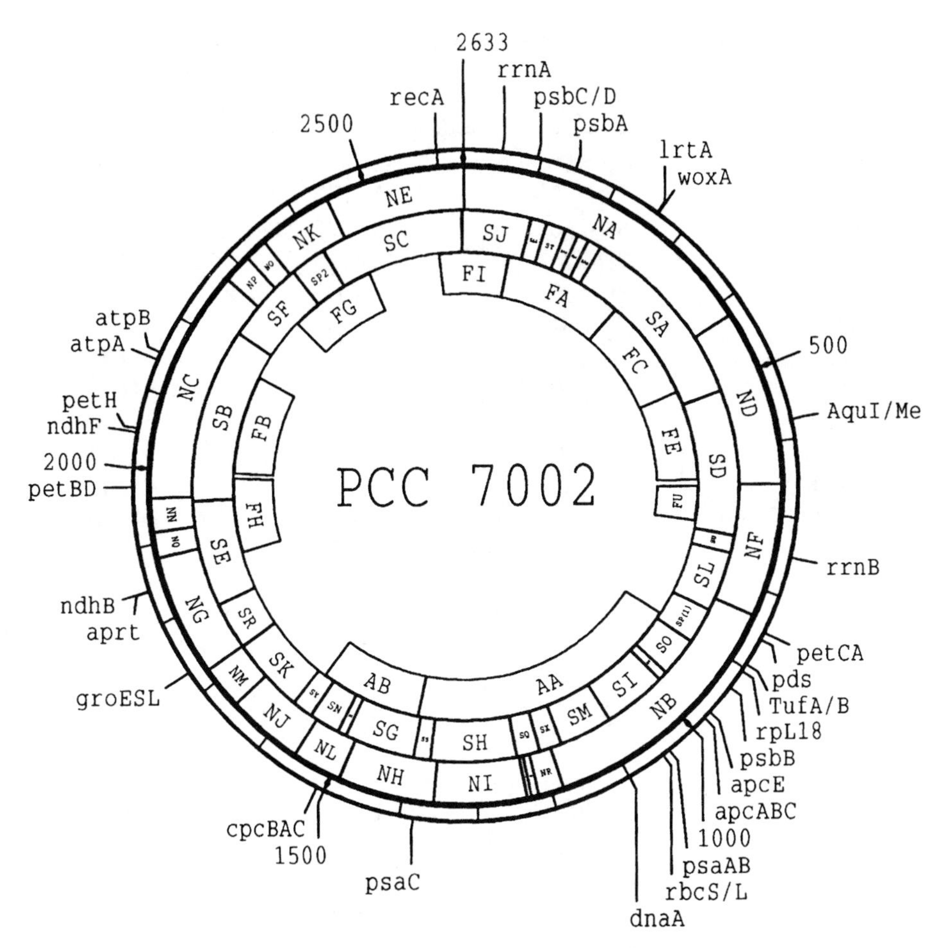

Figure 83-1. Genomic map of *Synechococcus* PCC 7002. The physical map of *Synechococcus* PCC 7002 (Chen and Widger, 1993) was assembled from 21 *Not*I (N) and 36 *Sal*I (S) restriction fragments. Two other enzymes *Sfi*I (F), 27 fragments and *Asc*I (A), 34 fragments were used to complete the circular genome and added supporting evidence to the data generated with *Not*I and *Sal*I. The summation of the restriction fragment sizes from all four enzymes were consistent and averaged to 2.7 Mbp. Base position number one of the genome was arbitrarily chosen as the first base of the largest *Not*I restriction fragment NA-420 adjoining NE-185 and proceeded clockwise around the genome.

*Eco*RI and *Not*I. Table 2 contains information about 16 of the 22 *Not*I sites in the genome. Also, 22 cosmids containing unique *Sal*I sites have been isolated.

The location of most hybridizing gene probes can easily be determined without ambiguity when hybridized to a Southern blot of *Not*I, *Sal*I and doubly digested genomic DNA. The double digested fragments allowed the extent of overlap to

Table 83-1. Restriction Fragment Sizes from Synechococcus *PCC 7002 Genomic DNA.**

*Not*I	*Sal*I	*Sfi*I	*Asc*I
NA-420	SA-290	FA-212	AA-450
NB-370	SB-270	FB-180	AB-182
NC-325	SC-230	FC-175	AC-178
ND-240	SD-225	FD-175	BD-172
NE-185	SE-165	FE-160	AE-165
NF-175	SF-150	FF-160	AF-145
NG-170	SG-125	FG-140	AG-140
NH-125	SH-120	FH-132	AH-136
NI-108	SI-117	FI-125	AI-131
NJ-105	SJ-110	FJ-123	AJ-120
NK-100	SK-98	FK-98	AK-116
NL-65S	SL-88	FL-94	AL-107
NM-55	SM-86	FM-94	AM-77
NN-43	SN-75	FN-90	AN-74
NO-30	SO-73	FO-88	AO-71
NP-25	SP(1)-55	FP-87	AP-69
NQ-23	SP(2)-55	FQ-82	AQ-67
NR-12	SQ-47	FR-79	AR-63
NS-7	SR-45	FS-79	AS-59
NT-4	SS-43	FT-64	AT-35
	ST-40	FU-62	AU-33
	SU-30	FV-55	AV-24
	SV28	FW-50	AW-23
	SW-23	FX-46	AX-22
	SX17	FY-38	AY-13
	SY-15	FZ-12	AZ-13
	SZ-12	FAA-9	AAA-11
	10 bands	FAB-7	AAB-8
	under 12 Kbp	FAC-5	AAC-6
	for 70 Kbp	FAD-4	AAD-4
			AAE-2
			AAF-1
2.63 Mbp	2.73 Mbp	2.70 Mbp	2.72 Mbp

*Nomenclature for PFG fragments are as follows N, *Not*I; S, *Sal*I; F, *Sfi*I and A, *Asc*I followed by the size order where A is the largest fragment followed by size in Kbp. NA-420 is the largest *Not*I fragment at 420 Kbp.

*Table 83-2. Cosmids Containing Not*I Sites*.

Frag	Pos	Cosmid	*E* (Kbp)	*N-E* (Kbp)	*N-E* (Kbp)	*N-E* (Kbp)
NE-NA	1	D60	18.0	14.8	3.2	
NA-ND	421	A47	10.5	8.5	2.0	
ND-NF	658	A89	6.0	3.9	2.1	
NF-NB	836	A21	3.6	2.2	1.4	
NB-NR	1204	D40	18.0	12.0	6.0	
NR-NS	1236	D62	4.0	2.4	1.6	
NS-NU	1247	C41	6.0	4.0	1.4	0.6
NU-NI	1251	C41	6.0	4.0	1.4	0.6
NI-NH	1359					
NH-NL	1556	F19	4.5	4.0	0.5	
NL-NT	1560	A52	7.3	5.0	2.3	
NT-NJ	1567	A52	7.1	5.1	2.0	
NJ-NM	1674					
NM-NG	1732					
NG-NQ	1895	D78	7.2	5.9	1.3	
NQ-NN	1929	A76	5.2	3.7	1.5	
NN-NC	1974	A20	20.0	17.1	2.9	
NC-NP	2302					
NP-NO	2328					
NO-NK	2359	D35	9.1	4.3	3.6	1.2
NK-NE	2460	A92	11.0	10.1	0.9	

*E, *Eco*RI fragment size cut with *Not*I and N-E, the resulting *Not*I-*Eco*RI fragment sizes produced by the digestion of the *Eco*RI fragment.

be determined which aided in fragment assembly by placing size limits on overlapping *Not*I and *Sal*I fragments. For example, the largest *Not*I fragment, NA-420, selected SA-290, SJ-110 and several smaller *Sal*I fragments along with the second largest double digested fragment (N-S)-200. This data showed that the overlap between NA-420 and SA-290 was approximately 200 Kbp and about 90 Kbp of SA-290 overlapped with ND-240.

*Not*I fragments around 180 Kbp (NE-185, NF-175 and NG-170) and at 100 Kbp (NI-108, NJ-105 and NK-100) were difficult to place on the map because hybridization to *Not*I filters could not distinguish one from another. *Sal*I digestion offered little help in resolution because of the size similarity in the *Sal*I (SC-230, SD-225) fragments which overlap the *Not*I fragments in question. However, double digestion with both *Not*I and *Sal*I followed by specific hybridization probes removed any degeneracy. A finer detailed physical map is currently being constructed of overlapping cosmid clones allowing a greater degree of accuracy resolving degeneracy in the physical map and in the localization of specific genes.

The updated map presented here differs slightly from the previously published version (Chen and Widger, 1993). While the order of the *Not*I fragments remain the same, several smaller *Sal*I fragments have been added or their order was changed because of the refinement imparted by the cosmid mapping. Four genes/

operons have been reassigned due to the degeneracy in the *Not*I fragments around 175 Kbp (NE-185, NF-175, NG-170). The *rrn*B operon was repositioned on the NF-175 band while the *rec*A genes was placed on NE-185. Cosmid contigs connecting NE-185 to NA-420 which contains the *rec*A gene have been identified and placed on the map by hybridization. *rrn*B is contained on an internal *Sal*I fragment found in NF-175 while the *rrn*A operon remains in its original position on NA-420 and SJ-110. Unfortunately, our cosmid library and lambda Dash libraries contains no recombinant for the rrn operons. Southern blots of *Eco*RI genomic digests probed with either 16S or 23S DNA also suggest the presence of only two rrn operons. The NG-170 fragment contains *gro*ES on an internal *Sal*I fragment and the combination of *Not*I and *Sal*I fragments hybridizing to *gro*ES is almost indistinguishable from the pattern seen for *rrn*B which is on NF-175, however cosmid contigs covering NG-170 have been constructed and *gro*ES has been placed on NG-170 and SR-45 close to the *Sal*I and *Not*I restriction sites. The *Aqu*I/Me has been changed to SD-225 / ND-240. The *atp*A/*atp*B and *ndh*F/*pet*H genes are located on the NC-325/ SB-250 overlap but their exact positions have not been established. The location of all the other genes/operons have been determined to within 20 Kbp or less. In lieu of extensive hybridization to produce a more complete map, our efforts have been concentrated to generating an overlapping cosmid contig map consisting of approximately 100 cosmids. Efforts towards the contig map are nearly 90–95% complete (unpublished data).

Extra chromosomal DNA

Seven distinct covalently closed circular DNA (cccDNA) bands were isolated from PCC 7002 by cesium chloride density gradient centrifugation. The banded DNA was distinctly separate from chromosomal DNA and contained seven separate fragments migrating on a pulsed field gel at 160, 155, 115, 55, 12, 7 and 4 Kbp. The actual sizes are unknown because the DNA was assumed to be circular when separated by electrophoresis. the cccDNAs did not comigrate at any the *Not*I fragments and bands 55 Kbp and greater were not cleaved by *Not*I. The cccDNA fragments were not easily discernible in normal PFGs run using intact or restriction digested genomic DNA. This observation suggested that the cccDNA was sub-stoichiometric to the genomic DNA. Hybridization data using the isolated cccDNA as a probe against *Not*I, *Sal*I and *Sfi*I digested genomic DNA showed strong hybridization to NA-420, SA-290 and FA-212 and nowhere else. This result was surprising and implied that the cccDNA contained sequences in common with the largest *Not*I fragment. The cccDNA probe was used to screen our gridded cosmid library and a group of nine overlapping cosmids were selected which in turn hybridized back to NA-420 only. Two possible interpretations for this are that plasmid DNA has repeated sequences found in the NA-420 region of the genome or that the cccDNA are products of a genome rearrangement event. There is no proof or supporting data for either hypothesis. The smaller cccDNA bands are most likely plasmids since chimeric plasmids have been constructed from small plasmids isolated from PCC 7002, however, the cccDNA fragment bands with apparent sizes of 55 Kbp and greater are not fully characterized.

The assembly of an overlapping cosmid library (*see also* Chapter 28) will allow greater detail in comparative genomic studies without the complete random total sequencing required to assemble a genome from a bottom up approach (Fleischmann *et al.*, 1995). The generation of tagged sequence sites in a bacterial genome can be rapid and establish gene positions and order without extensive and exhaustive sequencing.

References

Awramik, S. M., J. W. Schopf, and M. R. Walter. 1983. *Filamentous fossil bacteria 3.5* $\times$ *10^9 years old from the archean of western Australia. Precambrain Res.* 28:441–450.

Buzby, J. S., R. D. Porter, and S. E. Stevens, Jr. 1983. *Plasmid transformation in Agmenellum quadruplicatum PR-6: construction of biphasic plasmids and characterization of their transformation properties. J. Bacteriol.* 154:1446–1450.

Chen, X., and W. R. Widger. 1993. *The physical genome map of the cyanobacterium, Synechococcus PCC 7002. J. Bacteriol.* 175:5106–5116.

Essich, E., S. E. Stevens, Jr., and R. D. Porter, 1990. *Chromosomal transformation in the cyanobacterium Agmenellum quadruplicatum. J. Bacteriol.* 172:1916–1922.

Giovannoni, S. J., D. M. Ward, N. P. Revsbeeh, and R. W. Castenholz. 1987. *Obligatory phototrophic Chloroflexus: primary production in anaerobic, hot spring microbial mats. Arch. Microbiol.* 147:80–87.

Giovannoni, S. J., S. Turner, G. J. Olsen, S. Barns, D. J. Lane, and N. R. Pace. 1988. *Evolutionary relationships among cyanobacteria and green chloroplasts. J. Bacteriol.* 170:3584–3592.

Fleischmann, R. D., M. D. Adams, O. White, R. A. Clayton, E. F. Kirkness, A. R. Kerlavage, C. J. Bult, J. -F. Tomb, B. A. Dougherty, J. M. Merrick, K. McKenney, G. Sutton, W. FitzHugh, C. Fields, J. D. Gocayne, J. Scott, R. Shirley, L. -I. Liu, A. Glodek, J. M. Kelley, J. F. Weidman, C. A. Phillips, T. Spriggs, and E. Hedblom. 1995. *Whole-genome random sequencing and assembly of Haemophilus influenzae Rd. Science* 269:496–512.

Herdman, M., M. Janvier, R. Rippka, and R. Y. Stanier. 1979. Genome size of cyanobacteria. *J. Gen. Microbiol.* 111:73–85.

Lazcano, A., and S. L. Miller. 1994. *How long did it take for life to begin and evolve to cyanobacteria? J. Mol. Evol.* 39:546–554.

Porter, R. D., J. S. Buzby, A. Pilon, P. I. Fields, J. M. Dubbs, and S. E. Stevens. 1986. *Genes from the cyanobacterium Agmenellum quadruplicatum isolated by complementation: characterization and production of merodiploids. Gene* 41:249–260.

Rhiel, E., V. L. Stirewalt, G. E. Gasparich, and D. A. Bryant. 1992. *The psa genes of*

Synechococcus sp. PCC 7002 and Cyanophora paradoxa: cloning and sequence analysis. Gene 112:123–128.

Rippka, R., J. Deruelles, J. B. Waterbury, M. Herdman, and R. Y. Stanier. 1979. *Generic assignments, strain histories and properties of pure cultures of cyanobacteria. J. Gen. Microbiol.* 111:1–61.

Schluchter, W. M., J. Zhao, and D. A. Bryant. 1993. *Isolation and characterization of the ndhF gene of Synechococcus sp. strain PCC 702 and initial characterization of an interposon mutant. J. Bacteriol.* 175:3343–3352.

Schluchter, W. M., and D. A. Bryant. 1992. *Molecular characterization of ferredoxin-NADP+oxidoreductase in cyanobacteria: Cloning and sequence of the petH gene of Synechococcus sp. PCC 7002 and studies on the gene product. Biochem.* 31:3092–3102.

Schopf, J. W. 1993. *Microfossils of the early Archean apex chert: new evidence of the antiquity of life. Science* 260:640–646.

Schopf, J. W., and M. R. Walter. 1987. Early archaen *(3.3-billion to 3.5-billion-year-old) microfossils from the Warrawoona Group, Australia. Science* 237:70–73.

Stevens, S. E., Jr., and R. D. Porter. 1980. *Transformation in Agmenellum quadruplicatum. Proc. Natl. Acad. Sci. USA* 77:6052–6056.

Tan, X., M. Varghese, and W. R. Widger. 1994. *A light-repressed transcript found in Synechococcus PCC 7002 is similar to a chloroplast-specific small subunit ribosomal protein and to a transcription modulator protein associated with sigma* 54. *J. Biol. Chem.* 269:20905–20912.

84

Thermus thermophilus HB8

P. L. Bergquist and K. M. Borges

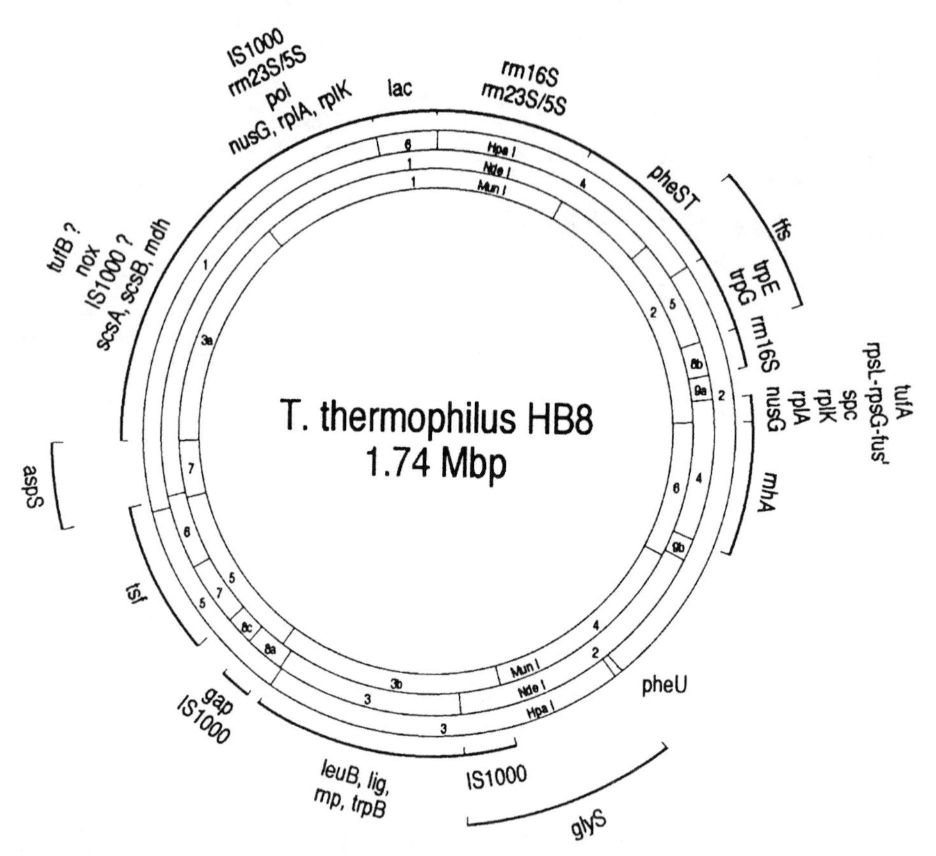

Figure 84-1. Physical map of the genome of *T. thermophilus HB8* (Borges and Bergquist, 1993; Reeves and Bergquist, unpublished). The locations of known genes were determined by hybridization with cloned NDA sequences. The relative order of *NdeI* fragments 8A and 8B are not known.

85

Treponema pallidum subsp. *pallidum* (Nichols)

Steven J. Norris, Eldon M. Walker, Jerrilyn K. Howell, Yun You, Alex R. Hoffmaster, Joe Don Heath, George M. Weinstock

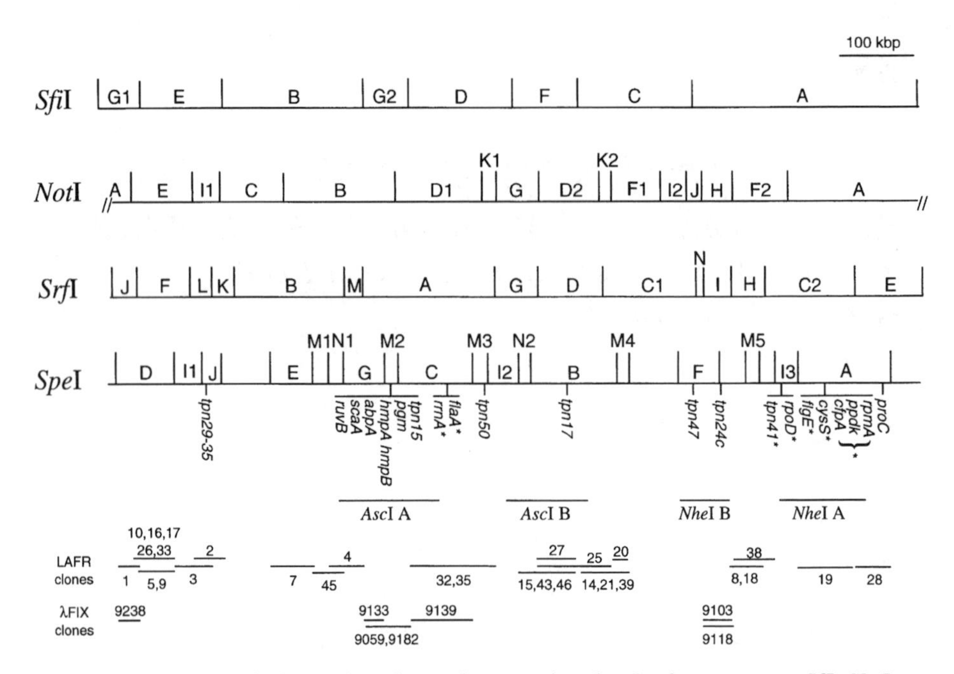

Figure 85-1. Restriction endonuclease cleavage sites for the four enzymes *Sfi*I, *Not*I, *Srf*I, and *Spe*I are shown. The gaps in the *Spe*I map are due to a number of small fragments that have not been mapped. In addition, the location of some of the larger fragments from *Asc*I and *Nhe*I digests are indicated.

Below the fragment maps are shown the approximate positions of those genes that have been analyzed. The locations indicated show which restriction fragments contain the genes, rather than a precise location. Genes that map near each other, but whose relative order has not been determined, are marked with an asterisk.

The location of clones in the cosmid vector pLAFRx and the λFIXII vector are shown at the bottom of the figure. The bars indicate the region where the clones map. The typical insert in cosmid and λFIXII clones are about 20kb and 18kb respectively. We estimate that a minimum of 450kb or about 45% of the genome is covered by the mapped clones. For details of the construction of this map, see Walker et al. 1995. Physical map of the genome of *Treponema pallidum* subsp. *pallidum* (Nichols). *J. Bacteriol.* 177:1797–1804.

Index